Laser-Solid Interactions and Transient Thermal Processing of Materials

ISSN 0272-9172

Volume 1—Laser and Electron-Beam Solid Interactions and Materials Processing,
    J.F. Gibbons, L.D. Hess, T.W. Sigmon, 1981

Volume 2—Defects in Semiconductors, J. Narayan, T.Y. Tan, 1981

Volume 3—Nuclear and Electron Resonance Spectroscopies Applied to Materials Science,
    E.N. Kaufmann, G.K. Shenoy, 1981

Volume 4—Laser and Electron-Beam Interactions with Solids, B.R. Appleton, G.K. Cellar,
    1982

Volume 5—Grain Boundaries in Semiconductors, H.J. Leamy, G.E. Pike, C.H. Seager, 1982

Volume 6—Scientific Basis for Nuclear Waste Management, S.V. Topp, 1982

Volume 7—Metastable Materials Formation by Ion Implantation, S.T. Picraux, W.J. Choyke,
    1982

Volume 8—Rapidly Solidified Amorphous and Crystalline Alloys, B.H. Kear, B.C. Giessen,
    M. Cohen, 1982

Volume 9—Materials Processing in the Reduced Gravity Environment of Space,
    G.E. Rindone, 1982

Volume 10—Thin Films and Interfaces, P.S. Ho, K.N. Tu, 1982

Volume 11—Scientific Basis for Nuclear Waste Management V, W. Lutze, 1982

Volume 12—In Situ Composites IV, F.D. Lemkey, H.E. Cline, M. McLean, 1982

Volume 13—Laser-Solid Interactions and Transient Thermal Processing of Materials,
    J. Narayan, W.L. Brown, R.A. Lemons, 1983

Volume 14—Defects in Semiconductors, S. Mahajan, J.W. Corbett, 1983

Volume 15—Scientific Basis for Nuclear Waste Management VI, D.G. Brookins, 1983

Volume 16—Nuclear Radiation Detector Materials, E.E. Haller, H.W. Kraner,
    W.A. Higinbotham, 1983

Volume 17—Laser Diagnostics and Photochemical Processing for Semiconductor Devices,
    R.M. Osgood, S.R.J. Brueck, H.R. Schlossberg, 1983

Volume 18—Interfaces and Contacts, R. Ludeke, K. Rose, 1983

Volume 19—Alloy Phase Diagrams, L.H. Bennett, T.B. Massalski, B.C. Giessen, 1983

Volume 20—Intercalated Graphite, M.S. Dresselhaus, G. Dresselhaus, J.E. Fischer,
    M.J. Moran, 1983

Volume 21—Phase Transformation in Solids, T. Tsakalakos, 1983

MATERIALS RESEARCH SOCIETY SYMPOSIA PROCEEDINGS   VOLUME 13

Laser-Solid Interactions and Transient Thermal Processing of Materials

Symposium held November 1982 in Boston, Massachusetts, U.S.A.

EDITORS:

J. Narayan
Solid State Division, Oak Ridge National Laboratory, Oak Ridge, Tennessee, U.S.A.

W.L. Brown
Bell Laboratories, Murray Hill, New Jersey, U.S.A.

and

R.A. Lemons
Los Alamos National Laboratory, Los Alamos, New Mexico, U.S.A.

NORTH-HOLLAND
NEW YORK · AMSTERDAM · OXFORD

7268-317X

PHYSICS

This work relates to Department of Navy Research Grant N00014-82-G-0112 issued by the Office of Naval Research. The United States Government has a royalty-free license throughout the world in all copyrightable materials contained herein.

It was also sponsored by the Division of Materials Sciences, U.S. Department of Energy under contract W-7405-eng-26 with Union Carbide Corporation, and by the Army Research Office under the grant DAAG29-82-M-0315. The views, opinions, and/or findings contained in this report are those of the author(s) and should not be construed as an official Department of the Army position, policy, or decision, unless so designated by other documentation.

Published by:

Elsevier Science Publishing Company, Inc.
52 Vanderbilt Avenue, New York, New York 10017

Sole distributors outside the USA and Canada:

Elsevier Science Publishers B.V.
P.O. Box 211, 1000 AE Amsterdam, The Netherlands

Library of Congress Cataloging in Publication Data

Main entry under title:

Laser-solid interactions and transient thermal processing of materials.
   (Materials Research Society symposia proceedings, ISSN 0272-9172; v. 13)
   Symposium held as part of the annual Materials Research Society meeting.
   Includes indexes.

   1. Materials—Effect of radiation on—Congresses. 2. Laser beams—Congresses. I. Narayan, J. (Jaydish) II. Brown, Walter Lyons, 1924-   . III. Lemons, R.A. IV. Materials Research Society. Meeting (1982: Boston, Mass.) V. Series.
TA418.6.L37   1983   620.1'1   83-8869
ISBN 0-444-00788-1
ISSN 0272-9172

Manufactured in the United States of America

TA418
.6
L371
1983
PHYS

Contents

Preface   xiii

Acknowledgments   xv

Symposium Photograph   xvi

Greetings from the President   xvii

PART I. **ENERGY TRANSFER AND PHASE TRANSFORMATION**

*PICOSECOND TIME-RESOLVED DETECTION OF PLASMA FORMATION
AND PHASE TRANSITIONS IN SILICON
J. M. Liu, H. Kurz, and N. Bloembergen   3

FEMTOSECOND TIME RESOLVED REFLECTIVITY OF OPTICALLY EXCITED
SILICON
R. Yen, C. V. Shank and C. Hirlimann   13

RAMAN SCATTERING WITH NANOSECOND RESOLUTION DURING
PULSED LASER HEATING OF SILICON
D. von der Linde, G. Wartmann, and A. Ozols   17

PULSED RAMAN MEASUREMENTS OF PHONON POPULATIONS: TIME
REVERSAL, CORRECTION FACTORS, AND ALL THAT
A. Compaan, H. W. Lo, A. Aydinli, and M. C. Lee   23

A DETAILED EXAMINATION OF TIME-RESOLVED PULSED RAMAN
TEMPERATURE MEASUREMENTS OF LASER ANNEALED SILICON
G. E. Jellison, Jr., D. H. Lowndes, and R. F. Wood   35

INSTABILITY OF THE ELECTRON-HOLE PLASMA AND ITS POSSIBLE
RELATION WITH MELTING
M. Combescot and J. Bok   41

TIME-RESOLVED STUDY OF SILICON DURING PULSED-LASER ANNEALING
B. C. Larson, C. W. White, T. S. Noggle, J. F. Barhorst,   43
and D. M. Mills

MEASUREMENT OF FAST MELTING AND REGROWTH VELOCITIES IN
PICOSECOND LASER HEATED SILICON
P. H. Bucksbaum and J. Bokor   51

*TIME RESOLVED MEASUREMENTS OF INTERFACE DYNAMICS DURING
PULSED LASER MELTING OBSERVED BY TRANSIENT CONDUCTANCE
M. O. Thompson and G. J. Galvin   57

TRANSIENT CONDUCTIVITY MEASUREMENTS IN PULSED ION BEAM
MELTED SILICON
R. M. Fastow, J. Gyulai, and J. W. Mayer   69

*ULTRA-HIGH SPEED SOLIDIFICATION AND CRYSTAL GROWTH IN
TRANSIENTLY MOLTEN SEMICONDUCTOR LAYERS
A. G. Cullis   75

*Invited Papers

AMORPHOUS PHASE TRAPPING AS A RESULT OF PULSED LASER
IRRADIATION OF SILICON
R. F. Wood   83

ION AND ELECTRON SPECTROSCOPY DURING PULSED LASER
IRRADIATION OF SILICON
J. P. Long, R. T. Williams, T. R. Royt, J. C. Rife, and
M. N. Kabler   89

NONLINEAR LASER MELTING OF INDIUM ANTIMONIDE AND SILICON
M. P. Hasselbeck and H. S. Kwok   97

THERMODIFFUSION - A NEW APPROACH TO PLASMA SELF-CONFINEMENT
A. Forchel, B. Laurich, and G. Mahler   105

TEMPERATURE CHARACTERIZATION OF PULSED LASER ANNEALING OF
SEMICONDUCTORS
D. L. Kwong and D. M. Kim   111

SUBSTRATE LATENT HEAT EFFECTS IN THE CALCULATIONS OF PULSED
LASER IRRADIATED THIN FILMS
J. J. Cao, K. Rose, O. Aina, W. Katz, and J. Norton   117

LASER INDUCED TEMPERATURE RISE IN SEMICONDUCTORS:
ANALYTICAL SOLUTIONS, APPLICATION TO THE TRANSIENT
A. Maruani, Y. I. Nissim, F. Bonnouvrier, and D. Paquet   123

PART II. **CRYSTALLIZATION OF AMORPHOUS SEMICONDUCTORS,
SOLID AND LIQUID PHASE PHENOMENA**

*SURVEY OF THE THERMODYNAMICS AND KINETICS OF
CRYSTALLIZATION OF Si AND Ge
D. Turnbull   131

CHARGED DANGLING BONDS AND CRYSTALLIZATION IN GROUP IV
SEMICONDUCTORS
P. J. Germain, M. A. Paesler, D. E. Sayers, and K. Zellama   135

*LASER-INDUCED SOLID PHASE CRYSTALLIZATION IN AMORPHOUS
SILICON FILMS
G. L. Olson, S. A. Kokorowski, J. A. Roth, and L. D. Hess   141

SOLID-PHASE-EPITAXIAL GROWTH OF ION IMPLANTED SILICON USING
CW LASER AND ELECTRON BEAM ANNEALING
J. Narayan, O. W. Holland, and G. L. Olson   155

*EXPLOSIVE CRYSTALLIZATION IN a-Ge AND a-Si: A REVIEW
D. Bensahel and G. Auvert   165

EXPLOSIVE RECRYSTALLIZATION OF ION IMPLANTATION AMORPHOUS
SILICON LAYERS
J. Narayan, G. R. Rozgonyi, D. Bensahel, G. Auvert,
V. T. Nguyen, and A. K. Rai   177

*Invited Papers

VISUALIZATION OF TEMPERATURE PROFILES IN a-Si CRYSTALLIZED
IN THE EXPLOSIVE MODE BY CW LASER
M. Rivier, A. Cserhati, and G. Goetz   185

SURFACE ELECTROMAGNETIC WAVES IN LASER MATERIAL INTERACTIONS
D. J. Ehrlich, S.R.J. Brueck, and J. Y. Tsao   191

LASER INDUCED PERIODIC SURFACE STRUCTURE
H. M. van Driel, J. F. Young, and J. E. Sipe   197

PICOSECOND LASER-INDUCED SURFACE TRANSFORMATIONS IN SOLIDS
P. M. Fauchet, Z. Guosheng, and A. E. Siegman   205

ALIGNED, COEXISTING LIQUID AND SOLID REGIONS IN PULSED AND
CW LASER ANNEALING OF Si
R. J. Nemanich, D. K. Biegelsen, and W. G. Hawkins   211

INTERFEROMETRIC MEASUREMENTS OF SURFACE PROPERTIES DURING
LASER ANNEALING
M. Bertolotti, A. Ferrari, C. Sibilia, and P. Jani   217

IMPURITY DIFFUSION IN SILICON ANNEALED BY SEMI-CONTINUOUS LASER
J. E. Bouree, C. Leray, and M. Rodot   221

SEARCH FOR VACANCIES IN LASER ANNEALED SILICON
H. J. Stein and P. S. Peercy   229

MICROPROBE RAMAN ANALYSIS OF PICOSECOND LASER ANNEALING OF
IMPLANTED SILICON
J. Sapriel, Y. I. Nissim, and J. L. Oudar   235

ANALYSIS OF TEMPERATURE AND STRESS PROFILES INDUCED BY A CW
LINE SCANNED ELECTRON BEAM IN <100> ORIENTED SILICON
G. G. Bentini and L. Correra   241

PART III.   **RAPID CRYSTALLIZATION, IMPURITY INCORPORATION,
AND METASTABLE SURFACES**

*ISING MODEL SIMULATIONS OF IMPURITY TRAPPING IN SILICON
G. H. Gilmer   249

*AMORPHOUS Si, CRYSTALLIZATION AND MELTING
J. M. Poate   263

IMPURITY REDISTRIBUTION IN ION IMPLANTED SI AFTER PICOSECOND
Nd LASER PULSE IRRADIATION
S. U. Campisano, P. Baeri, E. Rimini, G. Russo, and A. M. Malvezzi   273

COHERENT PRECIPITATE FORMATION IN PULSED-LASER AND THERMALLY
ANNEALED, ION-IMPLANTED Si
B. R. Appleton, J. Narayan, O. W. Holland, and S. J. Pennycook   281

*Invited Papers

*DOPANT INCORPORATION DURING RAPID SOLIDIFICATION
C. W. White, D. M. Zehner, J. Narayan, O. W. Holland,
B. R. Appleton, and S. R. Wilson   287

PULSED LASER ANNEALING OF ION IMPLANTED Ge
O. W. Holland, J. Narayan, C. W. White, and B. R. Appleton   297

SEGREGATION AND CRYSTALLIZATION PHENOMENA IN GERMANIUM
G. J. Clark, A. G. Cullis, D. C. Jacobson, J. M. Poate, and
M. O. Thompson   303

IMPURITY DISTRIBUTION PROFILES AND SURFACE DISORDER AFTER
LASER INDUCED DIFFUSION
E. Fogarassy, R. Stuck, P. Siffert, F. Broutet, and J. C.
Desoyer   311

*SURFACE STUDIES OF LASER ANNEALED SEMICONDUCTORS
D. M. Zehner and C. W. White   317

LASER-INDUCED SURFACE DEFECTS
J. M. Moison, and M. Bensoussan   329

PART IV. **TRANSIENT THERMAL PROCESSING OF SILICON FOR DEVICES**

*APPLICATIONS OF LASER ANNEALING IN IC FABRICATION
L. D. Hess, G. Eckhardt, S. A. Kokorowski, G. L. Olson, A.
Gupta, Y. M. Chi, J. B. Valdez, C. R. Ito, E. M. Nakaji, and
L. F. Lou   337

DEFECT CONTROL DURING EPITAXIAL REGROWTH BY RAPID THERMAL
ANNEALING
H. Baumgart, G. K. Celler, D. J. Lischner, McD. Robinson,
and T. T. Sheng   349

RAPID THERMAL ANNEALING OF SILICON USING AN ULTRAHIGH POWER
ARC LAMP
R. T. Hodgson, J.E.E. Baglin, A. E. Michel, S. Mader, and
J. C. Gelpey   355

FLAME ANNEALING OF ION IMPLANTED SILICON
J. Narayan and R. T. Young   361

ISOTHERMAL ANNEALING OF ION IMPLANTED SILICON WITH A
GRAPHITE RADIATION SOURCE
S. R. Wilson, R. B. Gregory, W. M. Paulson, A. H. Hamdi, and
F. D. McDaniel   369

GETTERING OF IMPURITIES BY INCOHERENT LIGHT ANNEALED POROUS
SILICON
V. E. Borisenko and A. M. Dorofeev   375

*Invited Papers

*THE CONTRIBUTION OF BEAM PROCESSING TO PRESENT AND FUTURE
INTEGRATED CIRCUIT TECHNOLOGIES
C. Hill   381

APPLICATIONS OF A CONTINUOUS WAVE INCOHERENT LIGHT SOURCE
(CWILS) TO SEMICONDUCTOR PROCESSING
H. B. Harrison, S. T. Johnson, B. Cornish, F. M. Adams, K.
T. Short, and J. S. Williams   393

EFFECT OF PULSE DURATION ON THE ANNEALING OF ION IMPLANTED
SILICON WITH A XeCl EXCIMER LASER AND SOLAR CELLS
R. T. Young, J. Narayan, W. H. Christie, G. A. van der Leeden,
D. E. Rothe, and R. L. Sandstrom   401

PULSED EXCIMER LASER (308 nm) ANNEALING OF ION IMPLANTED
SILICON AND SOLAR CELL FABRICATION
D. H. Lowndes, J. W. Cleland, W. H. Christie, R. E. Eby, G.
E. Jellison, Jr., J. Narayan, R. D. Westbrook, R. F. Wood,
J. A. Nilson, and S. C. Dass   407

SCANNING ELECTRON BEAM ANNEALING OF OXYGEN DONORS IN
CZOCHRALSKI SILICON
C. J. Pollard, J. D. Speight, and K. G. Barraclough   413

ARSENIC IMPLANT ACTIVATION AND REDISTRIBUTION IN P-TYPE
SILICON INDUCED BY PULSED ELECTRON BEAM ANNEALING
D. Barbier, M. Baghdadi, A. Laugier and A. Cachard   419

TRANSIENT ANNEALING OF ION-IMPLANTED SILICON USING A
SCANNING IR LINE SOURCE
Y. S. Liu, H. E. Cline, G. E. Possin, H. G. Parks, and
W. Katz   425

CONDUCTION IN POLYCRYSTALLINE SILICON: GENERALIZED
THERMIONIC EMISSION-DIFFUSION THEORY AND EXTENDED STATE
MOBILITY MODEL
A. N. Khondker, D. M. Kim, and R. R. Shah   431

PROCESSING OF SHALLOW (Rp < 150 Å) IMPLANTED LAYERS WITH
ELECTRON BEAMS
G. B. McMillan, J. M. Shannon, and H. Ahmed   437

SELECTIVE LASER ANNEALING FOR DEVICE PROCESSING
I. D. Calder, A. A. Naem, and H. H. Naguib   443

PULSED ELECTRON BEAM ANNEALING INDUCED DEEP LEVEL DEFECTS
IN VIRGIN SILICON
D. Barbier, M. Kechouane, A. Chantre, and A. Laugier   449

CRYSTALLIZATION OF AMORPHOUS SILICON FILMS BY PULSED ION
BEAM ANNEALING
J. Gyulai, R. Fastow, K. Kavanagh, M. O. Thompson, C. J.
Palmstrom, C. A. Hewett, and J. W. Mayer   455

*Invited Papers

PART V. **CRYSTALLIZATION OF SILICON ON INSULATORS–I**

*BEAM PROCESSING OF SILICON WITH A SCANNING CW Hg LAMP
T. J. Stultz, J. Sturm and J. F. Gibbons   463

*ZONE MELTING RECRYSTALLIZATION OF SEMICONDUCTOR FILMS
M. W. Geis, H. I. Smith, B.-Y. Tsaur, J.C.C. Fan, D. J.
Silversmith, R. W. Mountain, and R. L. Chapman   477

*DETECTION OF ELECTRONIC DEFECTS IN STRIP-HEATER
CRYSTALLIZED SILICON THIN FILMS
N. M. Johnson, M. D. Moyer, L. E. Fennell, E. W. Maby, and
H. Atwater   491

*CHARACTERIZATION AND APPLICATION OF LASER INDUCED
SEEDED-LATERAL EPITAXIAL Si LAYERS on SiO_2
M. Miyao, M. Ohkura, T. Warabisako, and T. Tokuyama   499

OXYGEN AND NITROGEN INCORPORATION DURING CW LASER
RECRYSTALLIZATION OF POLYSILICON
C. I. Drowley and T. I. Kamins   511

CW LASER ANNEALING OF ION IMPLANTED OXIDIZED SILICON LAYERS
ON SAPPHIRE
G. Alestig, G. Holmen, and S. Peterstrom   517

BEAM SHAPING FOR CW LASER RECRYSTALLIZATION OF POLYSILICON
FILMS
P. Zorabedian, C. I. Drowley, T. I. Kamins, and T. R. Cass 523

A COMPARISON OF CW LASER AND ELECTRON-BEAM
RECRYSTALLIZATION OF POLYSILICON IN MULTILAYER STRUCTURES
C. I. Drowley and C. Hu   529

PART VI. **CRYSTALLIZATION OF SILICON ON INSULATORS–II**

*LASER-INDUCED CRYSTALLIZATION OF SILICON ON BULK AMORPHOUS
SUBSTRATES: AN OVERVIEW
D. K. Biegelsen, N. M. Johnson, W. G. Hawkins, L. E.
Fennell, and M. D. Moyer   537

THE EFFECTS OF SELECTIVELY ABSORBING DIELECTRIC LAYERS AND
BEAM SHAPING ON RECRYSTALLIZATION AND FET CHARACTERISTICS
IN LASER RECRYSTALLIZED SILICON ON AMORPHOUS SUBSTRATES
G. E. Possin, H. G. Parks, S. W. Chiang, and Y. S. Liu   549

GROWTH OF SILICON-ON-INSULATOR FILMS USING A LINE-SOURCE
ELECTRON BEAM
J. A. Knapp and S. T. Picraux   557

CHARACTERISTICS OF RECRYSTALLIZED POLYSILICON ON SiO_2 PRODUCED
BY DUAL ELECTRON BEAM PROCESSING
J. R. Davis, R. A. McMahon, and H. Ahmed   563

LINEAR ZONE-MELT RECRYSTALLIZED Si FILMS USING INCOHERENT
LIGHT
A. Kamgar, G. A. Rozgonyi, and R. Knoell   569

LATERAL EPITAXIAL GROWTH OF THICK POLYSILICON FILMS ON
OXIDIZED 3-INCH WAFERS
G. K. Celler, McD. Robinson, D. J. Lischner, and T. T. Sheng   575

*CRYSTALLIZATION OF SILICON FILMS ON GLASS: A COMPARISON OF
METHODS
R. A. Lemons, M. A. Bosch, and D. Herbst   581

*ELECTRICAL CHARACTERISTICS AND DEVICE APPLICATIONS OF
ZONE-MELTING-RECRYSTALLIZED Si FILMS ON SiO_2
B.-Y. Tsaur, J.C.C. Fan, M. W. Geis, R. L. Chapman, S.R.J.
Brueck, D. J. Silversmith, and R. W. Mountain   593

THIN-FILM TRANSISTORS IN CO_2-LASER CRYSTALLIZED SILICON
FILMS ON FUSED SILICA
N. M. Johnson, H. C. Tuan, M. D. Moyer, M. J. Thompson,
D. K. Biegelsen, L. E. Fennell, and A. Chiang   605

TEMPERATURE PROFILES INDUCED BY STRIP HEATERS
M. L. Burgener and R. E. Reedy   613

PART VII. COMPOUND SEMICONDUCTORS

*TRANSIENT ANNEALING OF ION IMPLANTED GALLIUM ARSENIDE
J. S. Williams   621

NEUTRON ACTIVATION MEASUREMENTS OF As AND Ga LOSS DURING
TRANSIENT ANNEALING OF GaAs
A. Rose, J.T.A. Pollock, M. D. Scott, F. M. Adams, J. S.
Williams, and E. M. Lawson   633

RAPID ISOTHERMAL ANNEALING OF Si IMPLANTED SEMI-INSULATING
GaAs BY MEANS OF HIGH FREQUENCY INDUCTION HEATING
A. Cetronio, M. Bujatti, P. D'Eustacchio, and S. Ciceroni   641

ELECTRICAL DEFECT ANALYSIS FOLLOWING PULSED LASER
IRRADIATION OF UNIMPLANTED GaAs
D. Pribat, S. Delage, D. Dieumegard, M. Croset, P. C.
Srivastava, and J. C. Bourgoin   647

*ELECTRON BEAM PROCESSING OF SEMICONDUCTORS
H. Ahmed and R. A. McMahon   653

OPTICAL AND ELECTRICAL PROPERTIES OF Cd-Te LASER-SYNTHETIZED
L. Baufay, A. Pigeolet, R. Andrew, and L. D. Laude   665

*Invited Papers

xii

THERMAL PULSE ANNEALING OF $Hg_{1-x}Cd_xTe$
K. C. Dimiduk, W. G. Opyd, M. E. Greiner, J. F. Gibbons,
and T. W. Sigmon   671

PULSED UV LASER IRRADIATION OF ZnS FILMS ON Si and GaAs
H. S. Reehal, C. B. Thomas, J. M. Gallego, G. Hawkins, and
C. B. Edwards   677

INCOHERENT FOCUSED RADIATION FOR ACTIVATION OF IMPLANTED InP
J. P. Lorenzo, D. E. Davies, K. J. Soda, T. G. Ryan, and
P. J. McNally   683

PART VIII. **METALLIC SURFACE ALLOYS**

*LASER-GENERATED GLASSY PHASES
M. von Allmen   691

PULSED LASER MELTING OF METALLIC GLASSES
A. K. Jain, D. K. Sood, G. Battaglin, A. Carnera, G. Della
Mea, V. N. Kulkarni, P. Mazzoldi, and R. V. Nandedkar   703

PULSED PROTON BEAM ANNEALING OF Co-Si THIN FILM SYSTEMS
L. J. Chen, L. S. Hung, J. W. Mayer, and J.E.E. Baglin   709

PULSED ION BEAM INTERFACE MELTING OF PtCr AND CrTa ALLOYS
ON Si STRUCTURES
C. J. Palmstrom and R. Fastow   715

IMPURITY REDISTRIBUTION STUDIES ON LASER-FORMED SILICIDES
A. S. Wakita, T. W. Sigmon, and J. F. Gibbons   721

SUBMICRON CONDUCTOR ARRAY FABRICATED FROM A RECRYSTALLIZED
EUTECTIC THIN FILM
H. E. Cline   727

*PROCESSING/MICROSTRUCTURE RELATIONSHIPS IN SURFACE MELTING
R. J. Schaefer and R. Mehrabian   733

METASTABLE Fe(Pd) ALLOYS FORMED BY PULSED ELECTRON BEAM MELTING
D. M. Follstaedt and J. A. Knapp   745

AMORPHOUS Al(Ni) ALLOY FORMATION BY PULSED ELECTRON BEAM
QUENCHING
S. T. Picraux and D. M. Follstaedt   751

*Invited Papers

AUTHOR INDEX   757

SUBJECT INDEX   763

Preface

This book contains selected papers from the symposium on "Laser-Solid
Interactions and Transient Thermal Processing of Materials" held in Boston,
Massachusetts, November 1-4, 1982. This symposium, which was part of the annual
Materials Research Society meeting, represented a continuation of the series of
"laser annealing" symposia that started at the 1978 annual MRS meeting. The
symposium provided an international forum in which scientists from the United
States and thirteen other countries participated. The program of the conference
consisted of eight oral sessions and one poster session. The oral sessions,
which included 23 invited and 54 contributed papers, attracted audiences as
large as 350. The poster session contained 30 posters which were displayed over
the entire course of the meeting. This session was manned by the authors on the
evening of November 2, with a large number of participants taking advantage of
the opportunity to discuss new results in depth and detail. These papers have
been incorporated into different parts of the book according to their topical
content.

This symposium addressed the fundamentals of laser-solid interactions, energy
transfer, kinetics and thermodynamics of rapid phase transformations, and non-
equilibrium crystal growth, with emphasis on new transient thermal processing
methods involving incoherent light sources. Attention was focused on new
methods of preparation of materials by solid-phase crystallization and by
liquid-phase crystallization involving lateral epitaxial growth on
noncrystalline substrates. Rapid thermal annealing of elemental as well as compound semiconductors is being pursued with an ever-expanding variety of nonlaser
heat sources. At the symposium, these included scanning electron beams, quartz
halogen lamps, mercury arc lamps, flames, and graphite strip heaters, with
annealing carried out typically in times of 0.1 to 10 seconds. Reports were
presented of very high quality materials produced by this rapid, but not ultra
fast, processing. In these processing methods, the temperature gradients are
reduced and the additional time available after solid-phase-epitaxial growth is
utilized to remove residual damage.

From the fundamental aspects of laser annealing, the results of new time-
resolved experiments in pico- and subpicosecond regimes were of specific
interest. In general, time-resolved measurements of optical reflectivity and
transmission, electrical conductivity and x-ray diffraction, and TEM studies of
resulting microstructural modifications provided a consistent evidence for a
rapid (1-10 psec) conversion of electronic excitation to heat. Above a certain
threshold of beam energy density, rapid phase changes to, and subsequently from,
a molten state occur. However, time-resolved Raman measurements in the nanosecond regime continue to present a puzzling and conflicting picture as both the
measurements and their interpretation are refined. Maximum Raman temperatures
well below the melting point of silicon continue to be deduced under the conditions in which other results indicate that melting occurs.

Technologically, one of the most exciting areas discussed at this symposium
was that of lateral epitaxial growth of silicon on insulating or noncrystalline
substrates: SOI, or silicon on insulator. This crystal-growing process is
being carried out under a number of different conditions by different workers.
They involve a wide variety of heat sources from cw lasers to hot graphite
strips. These methods result in thin films of crystalline silicon with
thicknesses ranging from 0.2 to 100 micrometers and with grain sizes as large as
millimeters. Control of the lateral growth interface by manipulation of thermal
gradients has reached a high degree of sophistication as has the understanding
of the stabilization of the growth process. Very high quality devices are being

fabricated in this material which promises to have a unique role in display and VLSI (very large scale integrated) circuit devices. Progress in the fabrication of large-area solar cells with high throughput and a record high efficiency, using pulsed excimer lasers was also noteworthy.

Symposium Co-Chairmen

J. Narayan   W. L. Brown   R. A. Lemons

February 1983

Acknowledgments

We would like to thank all of the conference participants, particularly the invited speakers who provided excellent summaries of specific areas and set the tone of the meeting. They are:

H. Ahmed	M. Miyao
G. Auvert	G. L. Olson
D. K. Biegelsen	J. M. Poate
N. Bloembergen	R. J. Schaefer
A. G. Cullis	M. O. Thompson
M. W. Geis	B. Y. Tsaur
J. F. Gibbons	D. Turnbull
G. H. Gilmer	M. F. von Allmen
L. D. Hess	C. W. White
C. Hill	J. S. Williams
N. M. Johnson	D. M. Zehner
R. A. Lemons	

We are also grateful to the session chairmen who directed the sessions, guided the discussions and provided invaluable help in getting the papers refereed during the conference:

D. H. Auston	H. J. Leamy
B. G. Bagley	D. H. Lowndes
M. A. Bosch	M. A. Nicolet
J. W. Cahn	P. S. Peercy
G. K. Celler	T. O. Sedgwick
J. C. C. Fan	T. W. Sigmon
K. A. Jackson	M. Wittels
N. M. Johnson	F. W. Young, Jr.

Special thanks are also due to Ms. Susan Thomas who served as Conference Secretary. It is largely through her efforts that this book has been published. We would also like to thank Ms. S. Evans, Ms. M. Goetzee, and Ms. S. Urooman for their secretarial support.

It is our great pleasure to acknowledge with gratitude the financial support provided by the Office of Naval Research (Dr. L. R. Cooper) and the Defense Advanced Research Projects Agency (Dr. S. Roosild) through the grant number N00014-82-G-0112; the U. S. Department of Energy, Office of Basic Energy Sciences (Dr. M. C. Wittels); and the Army Research Office, Electronics Division (Drs. R. Griffith, R. Reeber) through the grant number DAAG 29-82-M-0315.

We also acknowledge additional support provided by:

Allied Corporation
Coherent, Laser Division
JEC Lasers, Inc.
Spectra Physics, Inc.
Tachisto, Inc.

Some of the conference participants:

Left Photograph: N. Bloembergen (Nobel Laureate in
  Physics, 1981), J. Narayan.

Right Photograph: S. Campisano, J. Poate, D. Turnbull,
W. Brown, B. Bagley, J. Fan, J. Narayan, R. Lemons (behind),
E. Rimini, P. Peercy.

GREETINGS FROM THE PRESIDENT OF THE UNITED STATES

THE WHITE HOUSE

Washington, DC 20500

October 26, 1982

It is my pleasure to extend greetings to the members of the Materials Research Society on the occasion of its annual meeting.

The study of materials constitutes one of the most exciting and important areas of research today. Much of today's advanced technology is made possible by developments in materials science of recent years, and tomorrow's technology will depend even more heavily on the progress you and your colleagues make in the future. From airplanes to microcircuits, materials science will play a critical role in keeping American high-technology industries strongly competitive.

In conveying my best wishes for the Society's continued success in stimulating research, I also congratulate Professor Clarence Zener as the 1982 recipient of the Society's von Hippel Award for his contributions to the field.

Ronald Reagan

Laser-Solid Interactions
and Transient Thermal
Processing of Materials

PART I
ENERGY TRANSFER
AND PHASE TRANSFORMATION

PICOSECOND TIME-RESOLVED DETECTION OF PLASMA FORMATION AND PHASE TRANSITIONS IN SILICON

J. M. LIU*, H. KURZ AND N. BLOEMBERGEN
Gordon McKay Laboratory, Division of Applied Sciences, Harvard University, Cambridge, Massachusetts 02138, USA

ABSTRACT

Picosecond time-resolved reflectivity and transmission changes of bulk silicon and silicon-on-sapphire have been measured to study the electron-hole plasma formation and phase transitions in silicon, induced by picosecond green or ultraviolet pulses. The results provide direct evidence of ultrafast energy transfer to the lattice and ultrafast phase transitions in silicon. Lattice heating up to the melting point and overheating of the melt or boiling have been observed during the picosecond pulse duration.

INTRODUCTION

The central question of how pulsed laser annealing is achieved during and following the absorption of laser energy requires a detailed treatment of the following processes: dynamics of a hot, dense electron-hole plasma; energy transfer from the electronic system to the lattice system; transient phase transitions of the lattice system; and kinetics of crystal regrowth and phase transformation.

Experimental results with nanosecond laser pulses [1-4] have already provided ample evidence to establish a thermal melting model for the physical mechanisms of pulsed laser annealing of silicon. Reflectivity changes of laser-pulse irradiated silicon were first studied on a nanosecond time scale by Auston et al. [5], and by many other investigators. Lowndes [6] has made careful measurements of reflectivity and transmission of silicon samples which support the thermal model of laser irradiation, refuting the interpretation of such measurements by Lee et al. [7]. A classical time-of-flight measurement of silicon atoms evaporated from a nanosecond laser-irradiated surface was performed by Stritzker et al. [8], which indicates a hot silicon surface, supporting our results on the emission of charged particles from picosecond laser-irradiated silicon [9].

It is clear that experiments on a picosecond time scale provide a more stringent test for the dynamics of the photoexcited electron-hole plasma and of the energy transfer to the lattice. In addition, previous reflectivity measurements with a cw monitoring laser beam revealed that the solid-liquid phase transition occurs on a subnanosecond time scale [10].

We first observed the reflected and transmitted portions of the energy of the picosecond pulses at three Nd:YAG laser wavelengths. These simple self-reflectivity and self-transmission measurements reveal changes in the optical properties which occur during a picosecond pulse. However, the observed quantities are clearly integrated averages over the spot area and the duration of the excitation pulse. Better spatial resolution and more information about the changes following the heating pulse were obtained by a pump-and-probe technique

---

*Present address: Department of Electrical and Computer Engineering, SUNY at Buffalo, Amherst, New York 14260, USA

Mat. Res. Soc. Symp. Proc. Vol. 13 (1983) ©Elsevier Science Publishing Co., Inc.

4

where a picosecond excitation pulse is followed by a tightly focused, weak probe
pulse with a variable time delay. Changes in the complex index of refraction of
silicon, induced by the heating pulse, are determined with a time resolution of
30 psec both below and above the melting point. Lattice heating, melting and
overheating of the melt or boiling have been observed during the pulse duration.
The data confirm the thermal melting model and provide more information about
the dynamics of the electron-hole plasma prior to melting and about the kinetics
of ultrafast phase transitions.

EXPERIMENT

A passively mode-locked Nd:YAG laser system was used to produce 30 psec
single pulses with Gaussian spatial and temporal profiles at the fundamental
1.064 μm wavelength. A single pulse of 20 psec duration at the second harmonic
532 wavelength can be generated by frequency-doubling of the infrared pulse at
1.064 μm in a cesium dihydrogen arsenate (CDA) crystal. An ammonium dihydrogen
phosphate (ADP) crystal can further double the green pulse to generate a 15 psec
ultraviolet pulse at the fourth harmonic 266 nm wavelength. The reflectance and
transmittance of a single pulse at each of these wavelengths were monitored by
calibrated photodiodes with appropriate line filters, yielding spatially and
temporally integrated self-reflectivity and self-transmission signals on a time
scale of the pulse duration. These pulses were incident on a silicon sample at
small angles to the normal of the sample surface. The (1/e) beam diameters on
the sample surface were measured with an *in situ* measurement technique [11].
The incident energy fluences can be varied from below 0.05 to above 2 J/cm^2.

In the pump-and-probe experiment a silicon sample was heated by a green pulse
at 532 nm or by an ultraviolet pulse at 266 nm. The changes in reflectivity
and/or transmission, induced by the heating pulses, were probed at the fundamental wavelength λ = 1.064 μm. Figure 1 shows the experimental set-up for the
pump-and-probe experiment. The infrared probe pulse was separated from the
second harmonic green pulse by a harmonic beam splitter. An ADP crystal and a
UV filter can be put in the optical path of the pump beam to generate an ultraviolet heating pulse. Two separate lenses were used to focus the pump and probe
pulses independently. The green and ultraviolet pump pulses were focused at the
sample surface to beam diameters of 240 μm and 120 μm, respectively, at the
(1/e) intensity contour. The infrared probe pulse had a diameter of only 30 μm

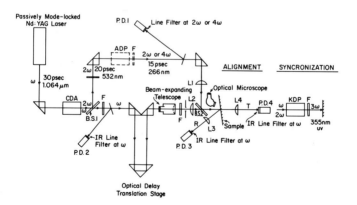

Fig. 1. Schematic diagram of the pump-and-probe experiment.

throughout the experiment, to maintain essential spatial resolution in the heating profile. A second beam splitter was used to align the focused pump and probe beams collinearly onto the sample surface.

A translation stage provided the probe pulse with a variable time delay with respect to the pump pulse. Zero time delay was determined by the generation of the sum frequency of the green pump and the IR probe in a potassium dihydrogen phosphate (KDP) crystal. The accuracy of the zero delay setting is ± 5 psec, determined by optimizing the harmonic output. The temporal resolution in this experiment is therefore 30 ± 5 psec, limited by the 30 psec duration of the infrared probe pulse.

The green and ultraviolet pulses with fluences above their respective amorphization thresholds [11] can form amorphous patterns on silicon surfaces. The unattenuated infrared pulse can anneal the amorphous structure back to crystalline. We used these phenomena to perform *in situ* alignment under an optical microscope. The alignment was found to be crucial for obtaining reliable data. It was checked from time to time during the experiment to ensure probing the *exact center* of the pump spot. Several neutral density filters were used to attenuate the probe fluence, keeping it below one thirtieth of the pump fluence during the experiment.

Lightly doped n-type and p-type bulk silicon wafers with (111) and (100) surfaces were used for reflectivity measurements. We did not observe any dependence of the results on the doping or the surface orientation. Simultaneous measurements of reflectivity and transmission changes were performed on silicon-on-sapphire (SOS) samples with (100) silicon surfaces. The thickness of the silicon film is 0.5 μm.

We used four photodiodes with appropriate line filters to monitor the pump fluence, the incident probe fluence, and the reflected and transmitted fluences of the probe pulse. Various precautions were taken against any possible artifact in our measurements. Because of the collinear configuration of the pump and probe beams, several infrared line filters were needed for each of the R and T monitoring photodiodes to completely block the pump pulse. Both the pump and probe beams were incident at 20° to the normal of the sample. Lenses collected the reflected and transmitted light. The readings from the monitoring photodiodes were registered and processed by an automatic data acquisition system. The reflectivity and transmission changes were first plotted as functions of peak energy fluence of the pump pulse at various time delays. These plots were then transformed into reflectivity and transmission changes as functions of time at various pump fluences.

RESULTS

Self-reflectivity and self-transmission

The integrated reflectivity and/or transmission changes in bulk silicon and silicon-on-sapphire for the picosecond excitation pulses themselves have been measured at three characteristic Nd:YAG laser wavelengths. Because the optical absorption depths of silicon at 1.064 μm and 532 nm are both larger than the 0.5 μm thickness of the silicon film on the SOS sample, the reflectivity and transmission of the SOS sample at these two wavelengths are determined by multiple reflections at the air/silicon and silicon/sapphire interfaces. Sapphire has negligibly small absorption coefficients at these two wavelengths, and the refractive indices of sapphire at room temperature are 1.7546 and 1.7721 at 1.064 μm and 532 nm, respectively. For pulses at very low fluences, the SOS sample has a measured reflectivity of 0.55 and a measured transmission of 0.45 at 1.064 μm, and at 532 nm a reflectivity of 0.50 and a transmission of 0.25. These values are consistent with calculations by the use of the equations of thin film optics [12]. Because of the high absorptivity, no transmitted light can be detected at 266 nm.

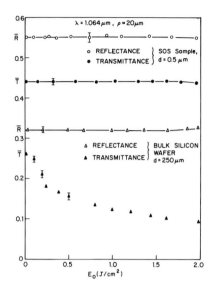

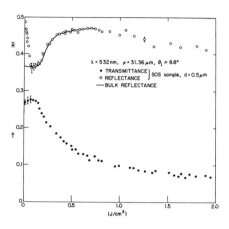

Fig. 2. Averaged self-reflectivity
and self-transmission of a 30 psec
pulse at 1.064 μm.

Fig. 3. Averaged self-reflectivity
and self-transmission of a 20 psec
pulse at 532 nm.

Figure 2 shows the results of self-reflectivity and self-transmission measurements at the fundamental 1.064 μm wavelength. Within experimental errors, no
reflectivity or transmission changes in the SOS sample were observed, up to a
fluence level of 2 J/cm². The reflectivity of a bulk silicon wafer of 250 μm
thickness also maintains a constant value. The transmission of the silicon
wafer is constant at very low fluences, but it starts to drop at about 0.1 J/cm².
It drops from the original value of about 25 percent to about 10 percent at
2 J/cm². Surface damage due to air breakdown takes place at a fluence above
2.5 J/cm². Below this critical energy fluence the crystal surface does not
undergo an optically detectable phase change. The near band gap radiation of
1.064 μm cannot deliver sufficient excess energy to melt the surface. Considerable lattice heating may take place only by free carrier absorption and
band gap shrinking.

However, by irradiation with green picosecond pulses carriers with an excess
energy of more than 1 eV are created, and the heating efficiency increases remarkably. As shown in Figure 3, the self-reflectivity of a bulk silicon sample
starts to increase when the energy fluence of the pump beam reaches 0.2 J/cm².
The initial rise of the reflectivity from the crystalline value of 0.37 to a
maximum value of 0.48 is in agreement with changes in the index of refraction
due to melting of a thin surface layer. The hot electron-hole plasma transfers
sufficient energy for melting to the lattice within the duration of the pulse
itself.

The data from an SOS sample reveal an extremely sensitive dependence of the
reflectivity and transmission on the changes in the complex index of refraction
of the silicon film. Below the critical fluence level of 0.2 J/cm², the photoexcited electron-hole plasma causes a decrease in the real part n of the index

of refraction. Temperature-induced narrowing of the band gap due to lattice heating causes an increase in the absorptivity, i.e., an increase in the imaginary part k of the index of refraction. The temperature dependence of n is negligible compared to plasma-induced changes. At 532 nm, temperature-induced changes in k dominate the contribution from free carrier absorption. Prior to melting, these changes in the complex index of refraction reduce the reflectivity from 0.5 to 0.36 ± 0.02 at about 0.1 J/cm^2. However, the transmission changes only slightly with the energy fluence, remaining rather constant around 0.25 to 0.28. Specifically, analysis of data by application of the equations of thin film optics shows that at 0.1 J/cm^2 n is reduced from the original value of 4.06 to a value of about 3.85 and k is increased from 0.035 to about 0.061, at $\lambda = 532$ nm.

As soon as the surface becomes molten, the highly absorbing liquid layer suppresses the multiple reflections completely, and the reflectivity of the SOS sample shows a behavior similar to that of the bulk sample. Correspondingly, the transmission drops exponentially to 5 percent at high fluences. Because the observed quantities are integrated averages over the area and duration of the pulse, the reflectivity and transmission data change only gradually with the fluence at fluences above the melting threshold.

Similar to the reflectivity of green pulses, the reflectivity of UV pulses at 266 nm drops at high energy fluences. Because crystalline silicon has a higher reflectivity (0.70) than liquid silicon (0.68) at 266 nm, the self-reflectivity drops for fluences above the melting threshold [11] of 0.08 J/cm^2, as expected. However, the low reflectivity level does not stay at the liquid reflectivity of 0.68. Instead, it keeps on dropping at higher fluences and goes to a level of about 0.48 ± 0.02 at fluences above 1 J/cm^2. This drop is mainly caused by overheating the molten surface and by evaporation of Si atoms, which leads to a reduction of the integrated reflectivity nearly independent of the laser wavelength.

Pump-and-probe experiments

The self-reflectivity and self-transmission data are averages over the spot area and the temporal profile of the heating pulse. More precise information of the plasma-induced transients and kinetics of phase changes should be provided by pump-and-probe experiments, where the changes of the optical properties are monitored with delayed and highly focused probe pulses on a picosecond time scale [13].

In the first two columns of Figure 4 are shown the reflectivity changes at 1.064 μm of a bulk silicon wafer as functions of the energy fluence of the pump pulse at 532 nm for various probe delays. At zero delay the observed reflectivity is an average of the change over the pump pulse duration because the probe pulse temporally overlaps the pump pulse. This explains the relative increase of the reflectivity. At a delay of 30 psec or longer, the probe pulse is gradually separated from the pump pulse. At 100 psec delay the probe pulse is temporally completely separated from the pump pulse, and an abrupt rise in the reflectivity at the threshold fluence of 0.2 J/cm^2 is observed [14,15]. This discontinuity in the carrier density must be associated with local structural changes. The reflectivity rises to 0.76 ± 0.03, characteristic of molten silicon at the probe wavelength of 1.064 μm. At higher fluences the reflectivity drops because of overheating of the molten silicon surface. This behavior is observed even at zero delay, indicating the fact that lattice melting and even overheating of the melt occur within the pulse duration of 20 psec. However, the reflectivity drop because of overheating is most obvious at a 30 psec probe delay, just after the pump pulse has terminated. The reduced reflectivity recovers to the high reflectivity level after a longer delay of a few hundred picoseconds, as the overheated liquid silicon surface cools down. The recovery time depends on the degree of overheating, which in turn is a

8

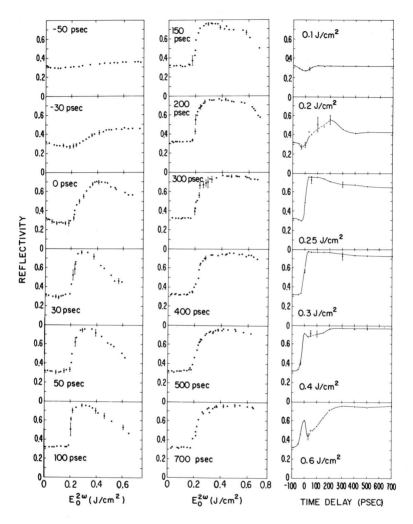

Fig. 4. Time-resolved reflectivity changes of a bulk silicon wafer monitored at 1.064 μm.

function of the pump fluence. The evidence for overheating also provides an explanation for the behavior of the self-reflectivity data at high fluence levels.

At fluences below the melting threshold we observe a small drop in reflectivity. This is an effect of the plasma-and-temperature-induced changes in the refractive index and the absorption coefficient. In the last column of Figure 4 the reflectivity changes as functions of the probe time delays at six different pump fluences are displayed. At 0.1 J/cm², below the 0.2 J/cm² melting threshold, the reflectivity decreases from 0.32 to 0.28, and this small drop recovers with a time constant of about 100 psec. At an energy fluence of 0.25 or 0.3 J/cm², above the melting threshold, the reflectivity rises from 0.32 for the crystalline silicon to 0.76 for liquid silicon during the pulse duration. At a higher fluence of 0.4 or 0.6 J/cm², the reflectivity rises in the early stage of the pump pulse, and overheating is observed during the pulse. The degree of overheating and the recovery time depend on the pump fluence. The data at the threshold fluence of 0.2 J/cm² for melting are subject to greater experimental errors because at threshold the effects of the pulse on the silicon surface are extremely sensitive to surface conditions and fluctuations in the profile of each individual laser pulse. Nevertheless, the data indicate that at the threshold the high reflectivity for the liquid phase may last for only a few hundred picoseconds. Cooling of the melt and resolidification may take place on this time scale. However, these data are not sufficient for making conclusive calculations on the upper limit of the resolidification velocity.

Figure 5 shows the reflectivity and transmission changes of a silicon-on-sapphire sample as a function of the probe time delay at two different pump fluences. The silicon film is only 0.5 μm thick, smaller than the absorption length $\alpha^{-1} = 1.25$ μm at the 532 nm pump wavelength at room temperature. A nearly uniform electron-hole plasma and heating profile is created throughout the film thickness by a green pump pulse. Because the thickness of the film is only half of the probe wavelength $\lambda = 1.064$ μm, the response of the sample to the probe pulse is extremely sensitive to changes in the optical properties of the film. The changes in the complex index of refraction in the film at 1.064 μm can be determined from the observed changes in reflectivity and transmission. Without pumping, silicon has $n = 3.56$ and $k = 3.4 \times 10^{-4}$ at the probe wavelength of 1.064 μm, and the film has a reflectivity of 0.55 and a transmission of 0.45. When the film is being heated by a pump pulse, conditions for multiple reflec-

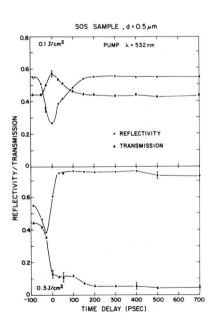

Fig. 5. Time-resolved reflectivity and transmission changes of an SOS sample at 0.1 and 0.3 J/cm², respectively.

tions change during the pulse duration as the absorption increases due to temperature-induced band gap narrowing and free carrier absorption, while the real
part of the index of refraction decreases from a plasma contribution. These
effects determine the decrease of reflectivity and the increase of transmission
in the SOS sample before melting occurs, as shown in Figure 5. At 0.1 J/cm^2,
the pump pulse induces changes in the complex index of refraction in the film to
n = 3.33 and k = 0.04 at the probe wavelength. These data are consistent with
the following events: (a) more than 80 percent of the absorbed energy is transferred to the lattice during the 20 psec pulse duration; (b) the lattice is
heated to about 900 K, which causes band shrinking and increases the lattice
absorption coefficient α_L; (c) the remaining part of the absorbed energy is
stored in the electron-hole plasma of a density of about 2-5 × 10^{20} cm^{-3} as the
band gap energy; (d) the free carrier absorption due to this electron-hole
plasma contributes about one third of the absorptivity change at λ = 1.064 μm.
The plasma absorption was found to be negligible for a probe wavelength of
532 nm; (e) Auger recombination limits the maximum electron-hole plasma density
to less than 10^{21} cm^{-3}, before melting occurs on a time scale of 10 psec.
    If we take the Auger recombination constant [16] c = 3.8 × 10^{-31} cm^6/sec, the
100 psec recovery time of the reflectivity and transmission changes gives an
averaged electron-hole plasma density of about 2 × 10^{20} cm^{-3}, in good agreement
with the value discussed above, which is obtained from independent consideration
of plasma-induced changes in the absolute magnitude of the refractive index.
    At a fluence above the melting threshold, the reflectivity drops gradually at
first because of suppression of multiple reflections. As the silicon film melts
and becomes opaque, the reflectivity exhibits a similar behavior to that of a
bulk wafer and rises to the liquid silicon reflectivity of 0.76. At the same
time the transmission drops abruptly. Experimentally, we observe a remnant
transmission of 3-5 percent, even after a long time delay. We believe this
background cannot be used to deduce the thickness of the liquid layer because
most of the transmitted light arises from scattering by evaporated material in
front of the irradiated surfaces. This background signal can be reduced by
placing a pinhole smaller than the pump spot just behind the sample, but the
signal-to-noise ratio is low.
    We also completed a series of experiments with the ultraviolet pump pulses.
The results are shown in Figure 6 for (a) a bulk sample and (b) an SOS sample.
The general features of these results confirm the results discussed above from
the green pump experiments. The ultraviolet pump experiments differ from the
green pump experiments in that the absorption depth is only 5 nm in silicon, and
the absorbed energy is thus used to heat a layer determined by the thermal diffusion length $(2Dt_p)^{1/2} \approx$ 45 nm [17]. Therefore, not much overheating is
observed at high fluences because the absorption depth does not change much on
melting and because the thermal diffusion length is much larger than the absorption depth. The high reflectivity level does not reach the full liquid reflectivity of 0.76 even at an energy fluence above the melting threshold. This
presumably could be due to a very thin molten layer of very high temperature
gradient. In this case the probe pulse may reach the liquid-solid interface,
and the reflectivity is an averaged value through the depth. However, we have
to point out the fact that the alignment and the spatial resolution in the UV
pump experiments are much more critical than in the green pump experiments.
    At low fluences no appreciable electron-hole plasma-induced effect is ob-
served because of the very thin absorption depth and because of faster heating
and the lower melting threshold of 0.08 J/cm^2.

CONCLUSIONS

    The dynamics of the electron-hole plasma and the kinetics of phase transi-
tions induced by picosecond laser pulses have been studied with picosecond time

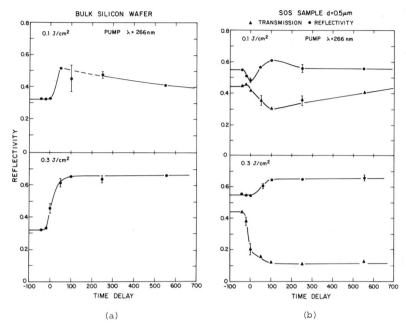

Fig. 6. Time-resolved reflectivity and transmission changes induced by UV
pulses at 266 nm.

resolution. With a green pulse at 0.1 J/cm^2, the photoinjection rate is about
2×10^{21} electron-hole pairs/cm^3 near the surface during the 20 psec pulse. The
observed electron-hole plasma prior to melting has a density smaller than the
photoinjection rate because Auger recombination limits the electron-hole plasma
density to below 10^{21} cm^{-3}. Most of the absorbed laser energy is transferred to
the lattice during the pulse duration. Lattice heating, melting, and overheating of the melt are observed during the 20 psec pulse duration. Heating
rates higher than 10^{14} °C/sec are observed upon melting. The melt front velocities may already approach the sound velocity. Cooling rates can be higher than
10^{13} °C/sec, but the resolvable resolidification velocity is only 15-20 m/sec.
Higher resolidification velocity is possible in the amorphized region. However,
our data cannot give conclusive evidence for a higher resolidification velocity
because the spatial resolution in this experiment becomes too critical at the
amorphization threshold with our small beam spot sizes. New data in the femtosecond regime are clearly desirable to further elucidate the dynamics.

ACKNOWLEDGMENT

This research was supported by the Joint Services Electronics Program under
Contract N00014-75-C-0648.

REFERENCES

1. S. D. Ferris, H. J. Leamy and J. M. Poate eds. Laser-Solid Interactions and Laser Processing (American Institute of Physics, New York 1979).

2. C. W. White and P. S. Peercy eds. Laser and Electron Beam Processing of Materials (Academic Press, New York 1980).

3. J. F. Gibbons, L. D. Hess and T. W. Sigmon eds. Laser and Electron-Beam Solid Interactions and Material Processing (North-Holland, New York 1981).

4. B. R. Appleton and G. K. Celler eds. Laser and Electron-Beam Interactions with Solids (North-Holland, New York 1982).

5. D. H. Auston, J. A. Golovchenko, A. L. Simons, C. M. Surko and T. N. C. Venkatesan, Appl. Phys. Lett. $\underline{34}$, 777 (1979).

6. D. H. Lowndes, Phys. Rev. Lett. $\underline{48}$, 267 (1982).

7. M. C. Lee, H. W. Lo, A. Aydinli and A. Compaan, Appl. Phys. Lett. $\underline{38}$, 499 (1981).

8. B. Stritzker, B. Pospieszczyk and T. A. Tagle, Phys. Rev. Lett. $\underline{47}$, 356 (1981).

9. J. M. Liu, R. Yen, H. Kurz and N. Bloembergen, Appl. Phys. Lett. $\underline{39}$, 755 (1981).

10. R. Yen, J. M. Liu, H. Kurz and N. Bloembergen, Appl. Phys. A $\underline{27}$, 153 (1982).

11. J. M. Liu, Optics Lett. $\underline{7}$, 196 (1982).

12. O. S. Heavens, Optical Properties of Thin Solid Film (Dover Publications, New York 1965), p. 46.

13. J. M. Liu, H. Kurz and N. Bloembergen, Appl. Phys. Lett. $\underline{41}$, 643 (1982).

14. J. M. Liu, H. Kurz and N. Bloembergen in: Picosecond Phenomena III, K. B. Eisenthal, R. M. Hochstrasser, W. Kaiser, A. Laubereau eds. (Springer-Verlag, Berlin Heidelberg 1982) pp. 332-335.

15. D. von der Linde and N. Fabricius in: Picosecond Phenomena III, K. B. Eisenthal, R. M. Hochstrasser, W. Kaiser, A. Laubereau eds. (Springer-Verlag, Berlin Heidelberg 1982) pp. 336-340.

16. J. Dziewior and W. Schmid, Appl. Phys. Lett. $\underline{31}$, 346 (1977).

17. N. Bloembergen in: Laser-Solid Interactions and Laser Processing, S. D. Ferris, H. J. Leamy, J. M. Poate eds. (American Institute of Physics, New York 1979) pp. 1-9.

FEMTOSECOND TIME RESOLVED REFLECTIVITY OF OPTICALLY EXCITED SILICON

R. YEN, C. V. SHANK AND C. HIRLIMANN
Bell Telephone Laboratories, Holmdel, New Jersey, USA

ABSTRACT

We report measurements of the time resolved reflectivity
of Silicon following excitation with an intense 90 femtosecond optical pulse. These measurements clearly time
resolve the process of energy transfer from the optically
excited electron-hole plasma to the crystal lattice and
subsequent melting within the first few picoseconds.

INTRODUCTION

The reflectivity of Silicon following excitation with intense laser pulses
has been the subject of extensive investigations.[8][1,2] In previous work the
dynamical processes of carrier cooling and heating of the crystal lattice have
been too rapid to be resolved in time. In the experiments reported here we
utilize recently developed femtosecond optical pulse techniques [3] to excite
a dense electron-hole plasma in crystalline Silicon and have measured the
induced reflectivity changes throughout the visible and near infrared as a
function of time and excitation energy. The results clearly reveal the formation of an electron-hole plasma with a density greater than 10^{22} cm^{-3}, the
process of energy transfer to the crystal lattice, and subsequent melting.

EXPERIMENTAL METHOD

The experimental arrangement for performing these experiments is shown in
Fig. 1. A 90 femtosecond pulse from an amplified CPM dye laser [4,5] is split
into two pulses, forming a pump and a probe pulse. The probe pulse is focused
into a cell containing D_2O to generate a white light continuum pulse. The pump
pulse is focused to about 150 microns in diameter. The continuum probe pulse
is focused onto the sample and the central 10% of the probing area is imaged
into a spectrograph with an optical multichannel analyzer vidicon array on the
output slit. The measurements are performed at a 10Hz repetition rate. A two
axis computer controlled stepper motor moves the Silicon wafer in a raster pattern so that each laser pulse sees a fresh region of the Si wafer.

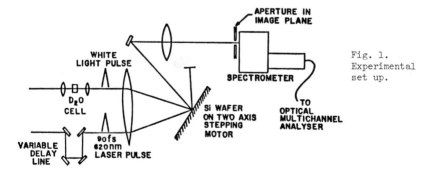

Fig. 1.
Experimental
set up.

RESULTS AND DISCUSSION

The experimental results are shown in Fig. 2. The log of the reflectivity change is plotted as a function of time for different pump intensities. The pump wavelength was 620 nm and the reflectivity was probed at three wavelengths (a) 1000 nm, (b) 678 nm and (c) 440 nm. The pulse energy threshold, E_{th}, for the formation of a clearly visible amorphous layer was 0.1J/cm^2.

We propose to interpret these reflectivity spectra in terms of a simple model. At the earliest times following excitation, the reflectivity change is expected to be dominated by the electron-hole plasma. As the pumping intensity is increased the e-h plasma becomes so dense that significant energy is transferred to lattice phonons, and the crystalline Si structure becomes unstable and the crystal melts. We further assume that melting initiates at the surface and that a melt front propagates into the bulk with a velocity v. The definition of just what melting means on this time scale is not entirely clear. For purposes of our discussion we will assume melting to have taken place when the optical properties of the material approach those of liquid Si. We realize that this model is oversimplified, but we can use it to interpret the general features of the experimentally measured reflectivity spectra.

The pump pulse is initially absorbed and creates hole-electron pairs at the Si surface in a layer whose thickness is determined by the absorption constant (α^{-1} = 3μm at 620nm). The absorption depth continually decreases as the density of pairs increases to form a dense plasma. The simple Drude expression for the refractive index of the plasma is given by,

$$N_p = N_c \left[1 - \omega_p^2/\omega^2\right]^{\frac{1}{2}} \tag{1}$$

for $\omega\tau \gg 1$ (where N_c is the crystalline Si refractive index).

At an excitation level below the melting threshold E_{th} in Fig. 2(a), the reflectivity decreases immediately following excitation, indicating that the electron-hole plasma frequency ω_p is less than the probe frequency ω. For a 1μm wavelength the plasma frequency immediately after excitation for this 0.63 E_{th} case can be determined from the magnitude of the reflectivity change shown in Fig. 2(a). From the relationship $\omega_p = (4\pi N_e e^2/m\varepsilon_c)^{\frac{1}{2}}$ we further determine the electron density to be $N_e = 5 \times 10^{21}$cm^{-3} assuming m to be the free electron mass. The slow decay of the 0.63 E_{th} data is due to carrier diffusion away from the excitation region. At this density and at higher densities we see no evidence of Auger recombination. These observations are consistent with the prediction of Yoffa that the Auger effect is screened for densities above 3×10^{21}. [7]

As the pump energy is increased it is possible for the plasma frequency to exceed the probing frequency. We see from equation (1) that the dielectric constant becomes imaginary under this condition and the reflectivity increases. Precisely this behavior is observed within a 100 femtoseconds following excitation at the higher intensities shown in Fig. 2(a). Note that at the higher probe frequencies shown in Fig. 2(b), (c) the reflectivity change starts out negative as would be expected.

Thus far we have described the reflectivity during the first few hundred femtoseconds in terms of a dense solid state plasma. At later times especially for intensities near and above E_{th} considerable structure is observed in the reflectivity spectrum seen in Fig. 2(a). The reflectivity change is observed to increase then reverse sign and then increase to a plateau value within the first picosecond and a half. We can explain this curious behavior if we consider a thin molten layer on the surface which expands in depth into the bulk with a velocity v. Using the optical properties of molten Si [6] and well known thin film optical formulas, we find that destructive interference between the wave reflected from the air melt interface and the melt solid state plasma interface acts to lower the intensity of the reflected beam. Since the Si melt is highly absorbing, the optical reflectivity becomes dominated by the optical

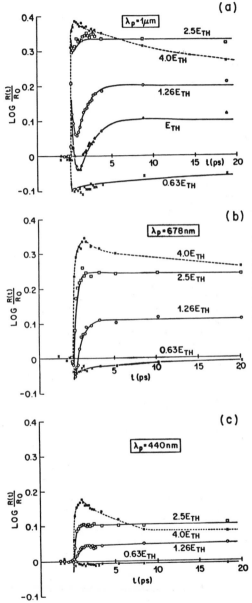

(a)

(b)

(c)

Fig. 2. Transient reflectivity data at three probe wavelengths following a 90 femtosecond excitation pulse at 620 nm. The solid lines for the 0.63 E_{th} data are calculated based on carrier diffusion into the bulk. The solid lines for E≥E_{th} are calculated based on the model described in the text. The melt front velocities for the E_{th}, 1.26 E_{th} and 2.5 E_{th} are 6.2, 9.1 and 25 (x10^5)cm/sec respectively.

properties of melted Silicon when the melt front has penetrated more than an absorption depth for the probing radiation. This gives rise to an increasing reflectivity and a final plateau reflectivity value. The peak reflectivity at 2.5 E_{th} is very close to the expected value for molten Silicon. At the lower densities near threshold the peak molten reflectivity is not observed possibly because of nonuniform melting either due to the laser beam Gaussian spatial profile and inhomogenities or because of some property of the melting process itself.

The model discussed above can be directly used to interpret the data at shorter wavelengths shown in Fig. 2(b) and (c). As we have discussed we do not see an initial rise in the reflectivity as in Fig. 2(a) consistent with the plasma frequency being less than the probing frequencies at these shorter wavelengths. We find that a single melt front velocity of 6.2×10^5 cm/sec predicts the reflectivity behavior for all observed wavelengths for an excitation of 1.0 E_{th}. This value is near the velocity of sound in crystalline Silicon. Similarly a velocity of 9×10^5 cm/sec is determined at an excitation of 1.26 E_{th}. Melting takes place exceedingly rapidly at excitation levels well above threshold. To explain the rapidly rising reflectivity at 2.5 E_{th} a velocity of 25×10^5 cm/sec is required. It is possible that bulk melting takes place at such high excitation levels. An initial energy deposition gradient due to the finite penetration depth of light gives rise to a gradient in the local melting rate as a function of depth. This can result in a very high effective melt front velocity.

At the highest excitation level of 4.0 E_{th}, the rapid decay of the reflectivity is due to catastrophic damage of the sample. At this pump level a small crater is formed.

In conclusion, we have been able to observe the reflectivity spectrum of highly excited Silicon immediately after excitation with a short 90 femtosecond optical pulse. These observations are unique in that we observe apparent melting to take place after excitation with a short pulse rather than during the excitation. We observe an unstable form of matter that exists for a time on the order of a picosecond with more than 10% of the available electrons optically excited. Finally, the dynamics of the reflectivity spectrum are consistent with a melt that is initiated at the surface and moves into a region of dense electron-hole plasma with a velocity at or above the velocity of sound.

REFERENCES

1. R. Yen, J. M. Liu, H. Kurz and N. Bloembergen, Appl. Phys. A27, 153 (1982).

2. D. H. Auston, J. A. Golovchenko, A. L. Simons, C. M. Surko, T. N. C. Venkatesan, Appl. Phys. Lett. 33, 539 (1979).

3. C. V. Shank, R. L. Fork, and R. Yen, p. 1 Picosecond Phenomena III, Springer Verlag, 1982.

4. R. L. Fork, B. I. Greene, and C. V. Shank, Appl. Phys. Lett. 38, 671 (1981).

5. R. L. Fork, C. V. Shank, and R. Yen, Appl. Phys. Lett. 41, 273 (1982).

6. V. M. Glagov, S. N. Chizhevskaya and N. N. Glagoleva, Liquid Semiconductors (Plenum Press, New York, 1969).

7. Ellen J. Yoffa, Phys. Rev. B21, 2415 (1980).

8. Article by J. M. Liu, H. Kurz and N. Bloembergen, in this proceeding.

RAMAN SCATTERING WITH NANOSECOND RESOLUTION DURING PULSED LASER
HEATING OF SILICON

D. VON DER LINDE, G. WARTMANN, AND A. OZOLS*
Universität Essen-GHS, Fachbereich Physik, 4300 Essen 1, W. Germany

ABSTRACT

We present time-resolved measurements of spontaneous anti-
Stokes and Stokes Raman scattering during pulsed laser heating
of crystalline silicon. The time-evolution of the lattice temperature is determined from the measured anti-Stokes/Stokes
intensity ratio. In a separate calibration experiment we measure the temperature dependence of the anti-Stokes/Stokes ratio
of an oven-heated silicon crystal from 300 K up to 900 K. The
phase transition occuring during laser heating is detected by
monitoring the changes of the optical reflectivity during laser
irradiation. Our data suggest that the phase transition occurs
at a lattice temperature of ~600 K.

INTRODUCTION

Spontaneous Stokes and anti-Stokes Raman scattering has recently been employed to measure the lattice temperature rise of crystalline silicon after
intense excitation by nanosecond laser pulses[1]. Raman scattering has also been
used in laser annealing experiments to study recrystallization of amorphous
layers of ion-implanted silicon[2,3]. In both experiments observations were made
which seem to be in disagreement with the thermal melting model[4] of laser
annealing: (i) the lattice temperature inferred from the anti-Stokes/Stokes
ratio was found to be well below the melting point of silicon[1,3]; (ii) Raman
signals were also detected during the laser-induced high reflectivity phase[1,2]
which is generally considered to be evidence of a surface layer of laser-molten
silicon (expected to be Raman silent).
    Raman scattering is undoubtedly a useful tool for obtaining in situ infor-
mation on phonon temperatures and changes of the lattice structure occuring
during laser heating of semiconductors. However, if quantitative temperature
information is to be inferred from anti-Stokes/Stokes Raman data correction
factors[5] involving scattering cross sections and absorption coefficients have
to be carefully considered. In addition, the temporal and spatial resolution of
the experiment as well as the reproducibility of the laser pulses play a crucial
role[6].
    We have performed Raman experiments in which the detailed temporal evolution
of the phonon temperature during the laser heating pulse is measured with a time
resolution of 1 ns. Great care was excercised to selectively collect Raman light
only from a small section of the total illuminated sample area in which the
energy density of laser exposure was uniform and accurately known. Particular
attention was also payed to the design of our laser system to ensure excellent
reproducibility of the output pulses and a very stable radial energy distribution of the laser beam.

---

*Permanent address: Institute of Physics, Latvian SSR Academy of Science, Riga,
USSR.

Mat. Res. Soc. Symp. Proc. Vol. 13 (1983) © Elsevier Science Publishing Co., Inc.

EXPERIMENTAL

We use laser heating pulses at 532 nm having a duration of 10 ns. These pulses are generated by frequency-doubling the output of a _passively_ Q-switched Nd-YAG laser which is operated at a repetition rate of 20 Hz for rapid signal accumulation. Passive Q-switching is essential for obtaining reliable single frequency operation. Mode-locking effects are avoided by using a saturable absorber dye with a recovery time of several nanoseconds. The laser pulses are free of beat notes, and the amplitude and shape of the frequency-doubled (2ω) pulses are reproducible within better than 5 percent.

Two different types of Raman experiments are performed: (i) Raman scattered light generated by the 532 nm heating pulse is measured directly; (ii) a weak third harmonic (3ω) probe pulse at 355 nm is used for Raman scattering. The UV probe pulse is coincident with the heating pulse, being generated by sum frequency mixing of the fundamental and the second harmonic beam. The UV probe pulse offers the advantage of probing only a very thin surface layer of the order of $1/2\alpha_{3\omega} \simeq 5$ nm, which is much less than the penetration depth of the heating pulse, $1/\alpha_{2\omega} > 100$ nm. The very strong resonant enhancement of the Raman scattering cross section near 3.5 eV compensates the loss of Raman signal resulting from the much smaller UV penetration depth.

The second harmonic heating beam and the collinear UV probe beam are incident at an angle of 45 degrees with respect to the [001] surface normal of the sample which is a 2" single crystalline silicon wafer. The Gaussian profile of the 532 nm beam has a full width at half maximum (FWHM) of 0.5 mm at the sample surface. The Raman light is collected in a backward scattering geometry by a high quality objective lense which produces a strongly magnified image of the sample surface in the plane of the entrance slit of a double spectrometer. Using an aperture of 1 mm diameter in front of the slit Raman light is detected only from a 160 μm circular section in the center of the Gaussian distribution. The energy density of the laser radiation varies by less than 10 percent over that section.

The incident laser light is polarized parallel to the [010] crystal axis. TO-phonon scattering is forbidden in this geometry, and Raman light due to LO-phonon scattering is polarized parallel to the [100] direction. The spectrometer bandwidth is adjusted to 0.5 nm such that the entire LO-phonon band is measured. Raman light is detected with a fast photomultiplier having a rise and fall time of 1 ns.

For each laser pulse the changes of the reflectivity of the sample are monitored simultaneously with the Raman signals. Two methods are used: (i) the second harmonic excitation beam reflected from a circular section of 50 μm diameter in the center of the laser-heated area is measured (self-reflectivity); (ii) a cw HeNe probe laser is focussed to a 10 μm spot in the center of the exposed area, and the reflected probe beam is detected. The two reflectivity measurements are simultaneously displayed on an oscilloscope (1 ns time resolution) with the cw HeNe probe signal being delayed by 30 ns. The self-reflectivity signal provides a precise measure of the onset of the phase transition relative to the laser pulse. The cw probe signal, on the other hand, shows the duration of the high reflectivity phase.

The onset and duration of the high reflectivity phase are very sensitive functions of the exposure energy. We use the reflectivity measurements primarily as a control of the constancy and reproducibility of the experimental conditions during the accumulation of Raman signals. The laser pulses are free of fluctuations in the sense that the observed time of the onset and of the duration of the high reflectivity phase is constant to within better than 1 ns.

Figure 1 depicts a series of oscilloscope traces showing the changes of the reflectivity for various laser exposure energies.

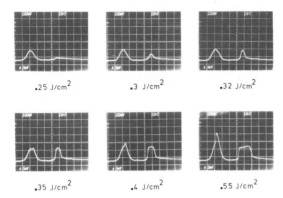

.25 J/cm^2 .3 J/cm^2 .32 J/cm^2

.35 J/cm^2 .4 J/cm^2 .55 J/cm^2

Fig. 1. Self-reflec-
tivity superimposed
on probe reflectivity
(30 ns delay). The
lower dotted line
corresponds to R=35%
for the HeNe probe.
Zero is at the very
bottom.

SIGNAL PROCESSING

Both for visible and UV Raman scattering the signals are quite weak. Typically we observe one single photon event in ten laser pulses up to about five per
pulse. For obtaining the instantaneous phonon occupation number (or phonon temperature) the precise shape of the Stokes signal, $S(t)$, and the anti-Stokes signal, $A(t)$, must be recovered from the basic single photon information by a
suitable signal processing procedure. For this purpose we use a computer-controlled 500 MHz oscilloscope equipped with a Si-multidiode target tube. The
photomultiplier output waveform obtained for each laser pulse is stored and
digitized, and the digital information is transferred to the computer for processing. Typically several thousand individual waveforms must be integrated to
recover the pulse shapes $A(t)$ and $S(t)$ with a signal-to-noise ratio of better
than one hundred.

The fidelity of the pulse shape reconstruction was carefully checked. By
tuning the spectrometer to the frequency of the heating pulse (2ω) and of the UV
probe pulse (3ω) the shapes of the second and third harmonic pulses are reconstructed from single photon photomultiplier pulses. The results are compared
with the pulse shapes measured directly with a photodetector having 0.5 ns time
resolution. Excellent agreement is obtained provided that the amplitude and
shape of the laser pulses are constant during signal accumulation. The required
stability could only be achieved with a passively Q-switched laser which was
operating in a single transverse and longitudinal mode.

RESULTS

Figure 2 depicts examples of the measured pulse shapes of Stokes and anti-
Stokes scattering (dashed and dotted curves, respectively). The solid lines
represent the incident light pulses.

Two different situations are compared in Fig. 2: a) visible Raman scattering
directly from the second harmonic heating pulse with an incident energy density
of 0.5 J/cm^2; b) a 2ω heating pulse of 0.48 J/cm^2 and Raman scattering from a UV
probe pulse of 0.02 J/cm^2. All pulse shapes are normalized to unity for ready
comparison.

The arrows in Fig. 2 mark the onset of the steep reflectivity jump observed
in the simultaneous reflectivity measurement. The total duration of the high

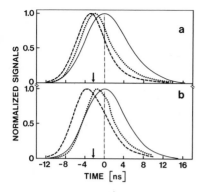

Fig. 2. Signal shapes of the incident light pulse (solid curves), Stokes (dashed curves), and anti-Stokes (dotted curves): a) Raman scattering from the 532 nm heating pulse; b) Raman scattering from 355 nm probe pulse. The ratio of anti-Stokes maximum to the Stokes maximum is 0.21 and 0.20, in a) and b), respectively.

reflectivity phase was measured to be 10 ns in both cases.

The following features of the signals in Fig. 2 should be noticed:
(1) The anti-Stokes signals are retarded with respect to the Stokes signals, indicating a distinct increase of the A/S ratio with time, and thus a rise of the phonon temperature.
(2) The Stokes signals are leading the incident light pulses. At the maximum of the incident visible and UV light pulse (t=0) the Stokes signals are down to about 65%.
(3) During the high reflectivity phase Raman scattering is strongly quenched, but still clearly observable.

It should be pointed out that the time delay between the reflectivity jump in the center and at the periphery of the 160 μm observation area was measured to be about 1 ns. The observed decrease of the Raman signal after the onset of the high reflectivity phase is much slower than one would expect if the high reflectivity phase were strictly Raman silent.

From the measured anti-Stokes and Stokes, $A(t)$ and $S(t)$, the phonon occupation number N can be obtained. If thermal equilibrium of the phonon modes is assumed, the phonon temperature can be calculated using the relation

$$S/A = C(1 + 1/N) = C \exp(h\omega_p/kT)$$

Here ω_p is the phonon frequency, and C is a correction factor given by

$$C = (t_A/t_S) \ (\omega_A/\omega_S)^3 \ (\sigma_A/\sigma_S) \ (\alpha_i + \alpha_S)/(\alpha_i + \alpha_A); \ i = 2\omega, \ 3\omega$$

The subscripts A and S refer to the anti-Stokes and Stokes frequency, respectively. The throughput ratio of the spectrometer and the spectral response of the photomultiplier tube is accounted for by (t_A/t_S).

The calculation of the frequency factor $(\omega_A/\omega_S)^3$ is trivial. However, considerable difficulties arise because the Raman scattering cross section ratio, (σ_A/σ_S), and the ratio of the absorption coefficients, $(\alpha_i + \alpha_S)/(\alpha_i + \alpha_A)$, are temperature dependent.

To overcome this problem a temperature calibration was performed. The anti-Stokes/Stokes ratio was measured for temperatures up to 800 K on a silicon crystal mounted in a furnace. Calibrated thermocouples were attached to the front and the back of the crystal to make sure that the crystal temperature is uniform. The sample is heated in an argon atmosphere to avoid oxidation.

In Fig. 3 we plot the anti-Stokes/Stokes ratio versus the inverse temperature. The solid straight line represents $\exp(-h\omega_p/kT)$. The squares are measured with the second harmonic. The energy density of laser exposure was well below

0.1 J/cm^2 such that laser heating is negligible. The correction factor C for the second harmonic is found to be independent of temperature, within our present experimental accuracy. However, from the calibration experiment we obtain C≈1.7 which is significantly greater than the value we have used in our previous work [7]. A re-examination of the various contributions to C indicated that the erroneous value was probably due to an incorrect spectral response factor of the photomultiplier.

The full circles in Fig. 3 represent the calibration obtained with the third harmonic. The strong variation with temperature of the UV correction factor is quite evident from our data. In fact, C increases from 0.44 at room temperature to 0.9 at 800 K. The room temperature value is in fair agreement with our previous value C≈0.4[7] which was estimated from literature data.

From the measured anti-Stokes and Stokes waveforms the ratios A(t)/S(t) are readily obtained. Examples of A/S-ratios are plotted as a function of time in Fig. 4a and 4b for visible and UV Raman scattering, respectively. Curves 1,2, and 3 in Fig. 4 represent, respectively, the following situations:
1) Low laser exposure energy resulting in only a minor increase of A/S.
2) Laser energy close to the threshold of the reflectivity jump.
3) Laser energy well above threshold; the arrows mark the position of the reflectivity discontinuity.

DISCUSSION

Using the temperature calibration (Fig.3) and the measured ratios A(t)/S(t) we can now discuss the temporal evolution of the phonon temperature.

Curves 1,2, and 3 in Fig.4a start from a common value of A/S≈0.14 which corresponds to T=300K. In the measurement represented by curve 3 the reflectivity jump occurs at -3ns for A/S≈0.26 which is equivalent to T≈390 K; A/S continues

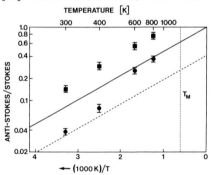

Fig. 3. Temperature calibration of A/S ratios. T_M: Melting temperature of silicon. Squares: 532 nm; Circles: 355 nm. A temperature-independent UV correction C (dashed characteristic) leads to an over-estimate of the lattice temperature.

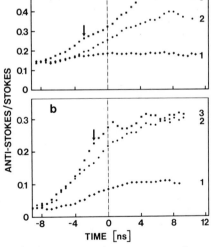

Fig. 4. A/S-ratios versus time: a) Raman scattering from heating pulse; 1: 0.14 J/cm^2, 2: 0.35 J/cm^2, 3: 0.65 J/cm^2, reflectivity jump at -3 ns; b) Raman scattering from UV probe pulse; 1: 0.02 J/cm^2 (355 nm), 2: 0.34 J/cm^2 (355 nm and 532 nm), 3: 0.48 J/cm^2 (532 nm) and 0.02 J/cm^2 (355 nm), reflectivity jump at -2 ns.

to rise, reaching a maximum of 0.5. No discontinuity of reflectivity occurs for curve 2, because the energy is slightly below threshold. Nevertheless, a value of A/S≈0.4 corresponding to T≈480K is reached 8 ns after the pulse maximum. The maximum temperature is thus substantially greater than that at the reflectivity jump.

From the UV data of Fig. 4b we find A/S values of 0.03 to 0.04 at the beginning of the pulse, in agreement with the UV room temperature value of Fig. 3. At the reflectivity jump ~2ns before the pulse maximum A/S equals 0.23, corresponding to T≈580 K (curve 3). In agreement with curve 2 in Fig. 4a a maximum value of A/S ≈0.3 much higher than that of the reflectivity jump is reached in the UV sub-threshold experiment represented by curve 2 in Fig. 4b. The observed maximum temperature is approximately 700 K.

Let as now compare temperatures inferred from the two types of experiments: (i) at the reflectivity discontinuity we have T≈390 K and T≈580 K for visible and UV, respectively; (ii) the maximum temperatures in the near threshold experiments (curve 2 in Fig. 4a and 4b) are 480 K and 700 K, although the laser energy is the same. Thus it appears that in the UV experiments higher temperatures are measured.

The disagreement can be explained by realizing that visible Raman scattering without a probe pulse provides a spatial average of the surface temperature profile; the actual surface temperature is higher. The UV pulses, on the other hand, probe only a ~5 nm thick surface layer of the crystal in which the temperature is uniform.

For an exponential temperature profile with a range given by $1/\alpha_{2\omega}$ the surface temperature would be approximately 1.5 times the measured spatial average. With this correction of spatial averaging the surface temperature from UV and visible Raman experiments are in good agreement.

CONCLUSIONS

We have obtained the temporal evolution of the phonon temperature during pulsed laser heating of crystalline silicon from Raman measurements in the visible and the UV. We also detect the phase transition usually associated with surface melting. Our data suggest that the lattice temperature at the transition in only ~600 K, in agreement with previous Raman work[1,2,3].

REFERENCES

1. H. W. Lo and A. Compaan, Phys. Rev. Lett. 44, 1604 (1980)

2. H. W. Lo and A. Compaan, Appl. Phys. Lett. 38, 179 (1980)

3. A. Compaan, H. W. Lo, A. Aydinli, and M. C. Lee, in Laser and Electron-Beam Solid Interactions and Materials Processing, ed. by J. F. Gibbons, L. D. Hess, and T. W. Sigmon (North-Holland, N.Y., 1981)

4. R. F. Wood and G. E. Giles, Phys. Rev. B3, 2923 (1981)

5. A. Compaan, H. W. Lo, M. C. Lee, and A. Aydinli, Phys. Rev. B26, 1079 (1982)

6. R. F. Wood, D. H. Lowndes, G. E. Jellison, and F. A. Modine, Appl. Phys. Lett. 41, 287 (1982)

7. D. von der Linde and G. Wartmann, Appl. Phys. Lett. 41, 700 (1982)

PULSED RAMAN MEASUREMENTS OF PHONON POPULATIONS: TIME REVERSAL, CORRECTION
FACTORS, AND ALL THAT

A. COMPAAN[*],
Max-Planck-Institut für Festkörperforschung, Heisenbergstrasse 1,
7000 Stuttgart 80, Federal Republic of Germany,

H.W. LO, A. AYDINLI[**] AND M.C. LEE,
Department of Physics, Kansas State University, Manhattan, Kansas 66506, USA

ABSTRACT

Transient optic phonon populations are measured in
crystalline Si as a function of 532 nm laser energy density.
The use of a continuously tunable pulsed dye laser as the
Raman probe allows us to obtain, under exact experimental
conditions, all correction factors necessary to extract the
phonon population without the necessity of relying on room
temperature or oven-heated conditions. We find the shift
of the 520 cm^{-1} Raman-line to be consistent with the observed
Stokes/anti-Stokes ratios indicating a maximum optic phonon
temperature of 450 ± 100°C. A discussion is also given of
the errors in several recent criticisms of the Raman results.

In this paper we describe new experiments which exploit the time reversal
invariance of the Raman effect [1] in order to obtain directly all of the temperature-dependent corrections which are needed, in addition to the readily
measured instrumental throughput corrections. Thus we are able to infer unambiguously the optic phonon population in silicon under laser annealing conditions. We demonstrate that these corrections may be obtained without requiring
the assumption of thermal equilibrium conditions in the semiconductor. We
thus avoid some of the difficulties which confuse the interpretation of other
time-resolved measurements performed under laser annealing conditions. In
addition, in this paper we describe measurements of the shift of the Stokes
Raman line and show that it is consistent with the 450°C temperature rise inferred from the Stokes/anti-Stokes ratio. Finally we critically evaluate some
of the recent attempts made to reinterpret the Raman measurements.

TIME REVERSAL INVARIANCE AND THE RAMAN RATIO

The Raman technique has generated much interest because of the fact that the
Stokes/anti-Stokes ratio is directly related to a phonon population, and because the Raman probe is non-destructive, is easily used on a nanosecond or picosecond time scale, and for some wavelengths samples less than 100 Å of the
crystal surface [2]. It is thus an attractive probe for examining phonon populations (lattice temperatures) during pulsed laser annealing (PLA). However,
a propos of laser annealing conditions, there exist almost no data at all on
the necessary corrections to the Raman ratio. In this paper we describe how we
have used the time reversal invariance of the Raman cross section to obtain,
via an additional Raman measurement, the required corrections under exactly the
conditions for which the temperature or phonon population is sought. The
technique involves a set of three Raman intensity measurements - only one more
than needed for the temperature measurement itself. The method requires the

[*]On sabbatical leave from Kansas State University
[**]Present address: Dept. of Physics, Haceteppe University, Ankara, Turkey

Mat. Res. Soc. Symp. Proc. Vol. 13 (1983) ©Elsevier Science Publishing Co., Inc.

use of a tunable dye laser and will be illustrated for pulsed-laser-excited crystalline silicon.

The Stokes Raman photon generation rate, R_S, at frequency ν_S, in the back scattered direction from the thickness element between z and z+dz below the surface is given by

$$dR_S(z) = \nu_L \nu_S^3 R_L \, e^{-\alpha_L z} \, \sigma(\nu_L,\nu_S) \, [n(z)+1] \, dz \qquad (1)$$

where R_L is the incident laser photon flux, α_L is the absorption coefficient at the laser frequency ν_L, $\sigma(\nu_L,\nu_S)$ is the Stokes Raman cross section, and $n(z)$ is the phonon occupation factor. The total signal leaving the sample is

$$R_S = \nu_L \nu_S^3 R_L \int_o^d \{\sigma_S(\nu_L,\nu_S,z) \, [n(z)+1] \, e^{-(\alpha_L+\alpha_S)z} \, dz \qquad (2)$$

where α_S is the absorption coefficient of the scattered Stokes light and d is the sample thickness. It is often possible experimentally to obtain a Raman probe depth $(\alpha_L+\alpha_S)^{-1}$ which is much smaller than the characteristic length for variation of the phonon population or Raman cross section. Under these circumstances and for $d \gg (\alpha_L+\alpha_S)^{-1}$,

$$R_S = \frac{\nu_L \nu_S^3 R_L \, \sigma_S(\nu_L,\nu_S) \, (n+1)}{\alpha_L+\alpha_S} . \qquad (3)$$

A similar expression obtains for the anti-Stokes photon generation rate. Hence the ratio of Stokes to anti-Stokes scattering is

$$\frac{R_S}{R_{AS}} = \frac{\nu_S^3 \, \sigma_S(\nu_L,\nu_S) \, (\alpha_L+\alpha_{AS})}{\nu_{AS}^3 \, \sigma_{AS}(\nu_L,\nu_{AS}) \, (\alpha_L+\alpha_S)} \, \frac{(n+1)}{n}. \qquad (4)$$

Extraction of a phonon population, n, thus requires only knowledge of the frequency-dependent σ and α factors in addition to the trivial ν^3 terms and a calibration of the throughput of the spectrometer system. The σ and α terms, however, are readily obtained by performing one additional Raman measurement [1] at a new laser frequency $\nu_{L'}$, chosen to be the same as the original anti-Stokes frequency ν_{AS}. The ratio of the two Stokes rates is then formed

$$\frac{R_S}{R_{S'}} = \frac{\nu_L \nu_S^3 R_L \, \sigma_S(\nu_L,\nu_S) \, (\alpha_L+\alpha_{S'})}{\nu_{L'} \nu_{S'}^3 R_{L'} \sigma_S(\nu_{L'},\nu_{S'}) \, (\alpha_L+\alpha_S)} \qquad (5)$$

where, since $\nu_{L'} = \nu_{AS}$ and $\nu_{S'} = \nu_L$, necessarily $\alpha_{L'} + \alpha_{S'} = \alpha_{AS} + \alpha_L$.

We now make use of the fact that the new Stokes cross section $\sigma_S(\nu_{L'},\nu_{S'})$ is identical to the original anti-Stokes cross section $\sigma_{AS}(\nu_L,\nu_{AS})$ by time reversal invariance. (The time reversal symmetry of the Raman cross section has been discussed by Loudon [3] and experimentally verified under strong resonance conditions by Compaan et al. [4]). Consequently, if the incident photon flux is held constant, the ratio of Eq. (5) gives exactly the product of correction factors required in Eq. (4) with no knowledge of lattice temperature required nor even that thermal equilibrium or steady state conditions be present. Thus, with the use of a dye laser, the Raman measurement can be made self calibrating and the Raman ratio can then be related unambiguously to the phonon population, n, via the factor $(1+1/n)$ (see Eq. (4)).

If the phonon system is in thermal equilibrium, then this factor is given very simply by $1+1/n = \exp(h\nu_o/kT_L)$, where ν_o is the phonon frequency and T_L is a lattice temperature. We shall find it convenient to describe our results for the phonon population in terms of such a temperature although we have no direct evidence that this concept is valid during pulsed laser annealing. This point is discussed further in the conclusions section.

THE MEASUREMENT OF CORRECTION FACTORS AND TEMPERATURE

For the experiments we have used a double beam system with a doubled Nd:YAG laser (λ = 532 nm) for excitation with a 600 μm spot size at the 0.4 nm thick crystalline Si wafer and a N_2-laser-pumped dye laser, electronically delayed by 30 nsec and operating at 405 nm to produce the Raman probe pulse [3]. The 532 nm and 405 nm pulse durations were 10 ns and 7 ns full width at half maximum respectively. The 405 nm beam was focussed to a 200 μm diameter spot at the center of the 532 nm spot to minimize lateral temperature gradients. The 405 nm wavelength was chosen to give a very short Raman probe depth $(a_L+a_S)^{-1} \simeq 700$ Å in room temperature Si (At higher temperatures the absorption increases still further). Further details of the laser timing and data acquisition system are described in Ref. 3.

In order to obtain the correction factors for the Stokes/anti-Stokes ratio at 405 nm ($\bar{\nu}_L$ = 24691 cm^{-1}) the laser frequency was shifted up by one phonon frequency ($\Delta\bar{\nu}$ = 520 cm^{-1}) to $\bar{\nu}_{L'}$ = 25211 cm^{-1} (396.6 nm). The ratio of the two Stokes peaks, corrected for the ν^3 factors and instrumental throughput, and normalized to the same incident laser photon flux, is shown as the quantity F in Fig. 1a, where

$$ F = \frac{(\alpha_{L'}+\alpha_{S'})}{(\alpha_{L}+\alpha_{S})} \frac{\sigma_S(\nu_L,\nu_S)}{\sigma_{S'}(\nu_{L'},\nu_{S'})}. \tag{6} $$

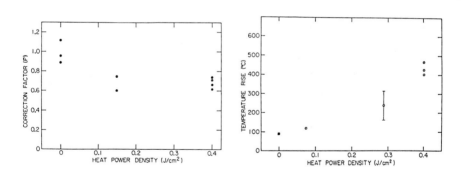

Fig. 1 a) Correction factor, F, at 405 nm and b) temperature rise measured by Stokes/anti-Stokes ratios as a function of heat pulse power at 532 nm. Probe delay 30 nsec.

As a function of increasing 532 nm excitation power, this factor shows a slight decrease which may be anticipated on the basis of the known spectral dependences of α and σ at room temperature and the temperature shift of the band gap. Empirically the slopes of dσ/dν and dα/dν are positive in the region near 405nm (3.06 eV) [12,13] since the frequency is approaching direct, allowed transitions (E_0' and E_1 gaps) near 365 nm (3.4 eV). Inspection of Equations (3) and (4) then shows that although the increasing α tends to suppress the anti-Stokes signal, the rising σ tends to enhance the anti-Stokes signal. The corrections thus act in opposite directions and one expects F to be close to unity. However, as resonance is approached in most semiconductors, one typically [5] finds the slope dσ/dν to be larger than dα/dν because σ involves the product of two electron-radiation field operators (one for virtual absorption, one for emission) whereas α involves only one [3,4,5].

In the case of Si under intense excitation, the temperature rise from the 532 nm pulse is expected to heat the sample. Since the temperature shift dE_1/dT is negative, the direct gap will approach the Raman probe frequency and the product of correction factors F may be expected to decrease. Other effects due to high photo-generated carrier densities will no doubt also be present and this emphasizes the importance of an empirical measurement of these corrections. The usefulness of the time reversal symmetry argument is that it allows one to measure this correction factor directly and simply, and thus permits the extraction of an unambiguous phonon occupation factor from the Stokes/anti-Stokes ratio.

The temperature as inferred from the Stokes/anti-Stokes ratio with a laser wavelength λ = 405 nm is shown in Fig. 1b over the same range of pump powers as the correction factor shown in Fig. 1a. In fact all conditions have been maintained constant for the two measurements. The only adjustments applied to the ratio, in addition to the factor F, are the $ν^3$ terms and corrections for instrumental throughput which were measured with a calibrated tungsten lamp. Note that the highest lattice temperature occurs at 0.4 J/cm^{-2} of 532 nm power where the duration of the high reflectivity phase is 20-30 nsec. When higher laser powers are used the duration of the high reflectivity phase exceeds the 30 ± 5 nsec probe pulse delay and no Raman signal is observed. Fig. 2 shows the decrease in the Stokes Raman count as the pump power is raised. The rapid fall of the signal at 0.4 J/cm^2 and above is additional confirmation of our earlier observation that with frequency doubled YAG laser excitation the high reflectivity phase is Raman silent [2].

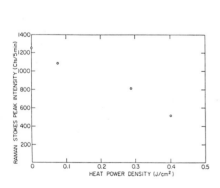

Fig. 2. Integrated Stokes Raman signal vs. 532 nm power.

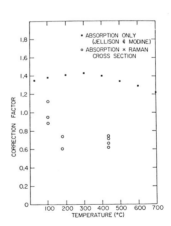

Fig. 3. Correction factors vs. temperature.

For 532 nm, an energy density of $\sim$0.4 J/cm^2 is sufficient to bring the silicon surface to the melting temperature if all the laser-deposited energy appears in the lattice and negligible diffusion occurs within roughly the pulse duration. The results of these Raman measurements thus appear to indicate that the silicon system behaves in an abnormal way under these conditions. Either the energy has diffused much further than expected, or the energy remains trapped in the electronic system for a time much longer than expected on the basis of normal electron-lattice relaxation rates, or the optic phonon population which is probed by the Raman scattering receives much less than its share of the total energy which the entire phonon system receives from the photoexcited carriers.

THE ISSUE OF CORRECTION FACTORS

As the preceeding discussion shows, the Raman Stokes/anti-Stokes ratio has been obtained in such a way that all necessary correction factors are explicitly and independently measured. Wood et al. [6,7,8] have recently criticized our earlier measurements [9,10,11] in which we used room temperature correction factors for reducing all of the data. From the beginning we pointed out the potential problems thus introduced; however, we argued on the basis of the known frequency dependence of α and σ that the temperature-dependent effects would tend to cancel (vide supra). Wood et al. discussed only the temperature-dependent effects of α, extrapolating to the melting point a temperature dependence based on measured values at 10 K and 300 K [6,7]! They then obtain "correct" temperatures in place of our "incorrect" values. However the recent ellipsometric measurements of Jellison and Modine [12] show that this extrapolation to high temperatures is badly in error. In addition our measurements of the product of both α and σ corrections (see Fig. 1) show as expected that the σ correction decreases even faster with increasing temperature than α increases. In Fig. 3 we compare the α correction obtained from the data of ref. 12 with the product of corrections, F, plotted as a function of lattice temperature rather than as a function of laser power (as seen in Fig. 1a). (Note, here this correction was already applied to the data of Fig. 1b to allow the data of Fig. 1a to be plotted vs. temperature). The observed behavior of the product of corrections, viz, a decrease with increasing temperature, can be qualitatively understood from the fact that the temperature dependence of both α and σ are controlled by the temperature dependence of the direct gap in Si (see preceeding section). Our probe wavelength of 405 nm was deliberately chosen to minimize these corrections.

At other wavelengths, however, the temperature dependence may be very strong. For example, von der Linde and Wartman [14] have recently reported nanosecond time-resolved Raman measurements obtained with probe wavelengths of 532 nm and 355 nm. The temperatures they infer using room temperature corrections are strikingly different for the two wavelengths. However, from the temperature dependence of the E_1' and E_1 band gap in Si and from the observed frequency dependence of α and σ, we would predict that at 532 nm (as in our measurements at 405 nm) the effects of α and σ would tend to cancel. But at 355 nm, $\alpha(\nu)$ is flat [12] and therefore insensitive to the temperature shift of the gap, while $\sigma(\nu)$ is rapidly falling [13]. Therefore the product of corrections, F, should be rapidly falling with increasing temperature. We have recently measured the temperature dependence of the product of corrections, F, in oven-heated Si using the 351 nm Ar laser line. Our results show that F decreases from 2.3 ± 0.2 at room temperature and reaches a constant 1.05 ± 0.10 between 300 and 500°C. Thus the authors of ref. 14 seriously overestimate the temperature by using a correction factor F = 2.44 for all of their data. A correction factor of 1.05, as indicated by our data, brings down their maximum inferred temperature from 3000 K to 690 K. Thus the two measurements of von der Linde and Wartman at

532 nm and 355 nm now show good agreement with each other and also excellent agreement with our measurements at 405 nm [1,2,9].

We emphasize, however, that even if one knows experimentally both the α and σ corrections as a function of temperature under thermal equilibrium, there is no guarantee that these would be the proper corrections to apply to Raman data obtained under pulsed laser annealing conditions. On the other hand, the time-reversal data do apply because the experimental conditions are identical. Our experience with the time-reversal technique shows that one can obtain the product of corrections to ±10% and consequently a total error in the inferred temperature of ±100 K when T $\sim$ 700 K, again with no need to assume thermalization between carriers and phonons, or within the phonon system itself.

Most of our work has emphasized Raman ratio measurements on bulk crystalline Si wafers where our results give $T_L(max) \simeq 450 \pm 100°C$ [1,2,9]. However, we also reported a temperature measurement on ion-implantation-amorphized Si [11]. Here the inferred temperature appears to be slightly higher ($T_L \simeq 600 \pm 200$ C) although the errors in this measurement were larger. (This is a consequence of limits on the quantity of material, since each new pulse must see virgin implanted material). Wood et al. [7,8] chose to "correct" this more poorly known measurement. They have made two serious errors. First they introduced only the temperature dependence of α, ignoring the compensating dependence of σ. Second they applied instrumental throughput corrections from our first reported measurement [9] to this measurement on amorphous Si [11] which was done with a new detector system. Consequently the arguments of these papers are incorrect. In addition, refs. 7 & 8 ignored completely the more accurate data obtained on crystalline Si where the discrepancy with the melting model calculations is even larger.

THE ISSUE OF FLUCTUATIONS, POOR QUALITY SPOTS AND RIPPLES

One of the first attempts to rationalize the Raman results with the normal melting model of PLA was the suggestion that possibly poor transverse spot quality might give rise to coexisting regimes of molten and cool solid regions. Therefore at the 1980 MRS symposium we arranged with J. Narayan for a rigorous study of our laser annealed samples by electron microscopy, optical interference (Nomarski) microscopy, SIMS and Rutherford backscattering. The initial work on our Raman samples was followed up by studies on a variety of In, B, and As-implanted or diffused samples which we annealed with our 485 nm dye laser at a variety of pulse energies from below threshold up to $\sim$1.2 J/cm. The detailed conclusions of this work were first shown to us at the 1981 MRS symposium [15] although results had been reported earlier [16,17].

The most detailed report of this study [18] shows quite clearly that the central region of the annealed spots of the samples used for the Raman measurements is very uniformly annealed with no evidence of hot or cold spots. Nevertheless a central thrust of references 15-18 was that our dye laser system had large pulse-to-pulse fluctuations. We quote: "We believe this discrepancy is due to pulse-to-pulse variations in energy density for at least about 50% of the laser spots" (original emphasis) [16,18]. In addition, the authors of ref. 18 claim that the central annealed region has a size which varies from 20 to only 60 μm in contrast with our reported 150 x 180 μm half power points.

This report stimulated us to reexamine several samples used for the reported Raman measurements [10,11]. (The original sample portion sent to J. Narayan is no longer available for our inspection; however, we have examined the remaining fraction of that sample and several others used between June and November, 1980).

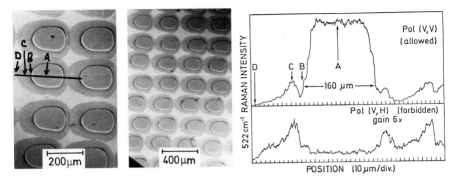

Fig. 4. Nomarski views of spot pattern from As-implanted sample used on 11-11-80.

Fig. 5. Micro-Raman spatial scan of a typical spot. Spectrometer set at 522 cm^{-1} shift from laser line at 21838 cm^{-1}.

Fig. 4 shows Nomarski photos of such a sample from a location chosen at random. The central annealed region is seen to measure typically 175 μm x 120 μm. To determine the nature of the various regions we have scanned a 10 μm spot of a 457.9 nm Ar^{+} laser line across one of these spots while detecting the 522 cm^{-1} Raman line. The results are shown in Fig. 5 for two polarization conditions. The input laser beam was incident on a (001) face with polarization along [110] and scattered polarization was analyzed in both allowed [110] and forbidden [1$\bar{1}$0] polarizations. The results clearly show that throughout the central region epitaxial regrowth has occurred since the top 0.2 μm probed by the laser has the substrate orientation. (Note that refs. 16 and 18 demonstrated that the 200 keV As implantation had amorphized the top 0.18 μm of these samples.) The strong dip in the Raman scattering for the allowed polarization occurs exactly at the thin ring which surrounds the inner region. This ring gives strong elastic scattering of light and apparently arises from growth which nucleates in a thin layer of heavy damage just below the 0.18 μm amorphous layer. The outer annular region is apparently polycrystalline since the Raman scattering is essentially unpolarized. Here the growth has apparently nucleated within the amorphous layer. Thus we find that the phase transition which induced recrystallization penetrated to a depth of at least 0.18 μm over an area of 190 μm x 140 μm in excellent agreement with our original estimate of beam size.

How then do we understand the results reported in refs. 15 - 18? Several of the samples used in that study were irradiated (at their request) at much lower energy densities. We suggest that their conclusions may have been drawn from samples irradiated very near the threshold for recrystallization in which case the central anneal spots would be small and furthermore minor power fluctuations would have created large changes in spot size, particularly if the beam profile was very smooth and very flat. Furthermore Fig. 13 of refs. 16 and 18 apparently either has an error in the scale of a factor of two, since the indicated spot spacing is half our normal repeat spacing or the figure was made from one of the samples specially prepared using a half-step scan mode. This figure apparently also contributed to the misjudgement of the central anneal spot size.

The photos of Fig. 4 also clearly show that the pulse-to-pulse reproducibility is excellent. The spot size variation is actually considerably better than the ±5% power fluctuations we originally claimed for the dye laser. Again it should be emphasized that the data of Figs. 4 and 5 were obtained at random on the original samples.

It has recently been suggested that the possible existence of surface ripples might explain the observed low Raman temperatures [19]. This is a variant of the poor spot homogeneity argument with coexisting molten and cool solid regions. Although these surface ripples are commonly observed with PLA, we find no optical evidence for such ripples in our spots except near occasional surface defects or dust particles. Narayan et al. [15-18] likewise in cross sectional TEM show very uniform anneal across individual spots. We have recently examined the diffraction of 351 nm light from these original spots and again find no evidence of any diffraction, including spacings characteristic of the two incident laser wavelengths 485 and 405 nm.

SHIFT OF THE RAMAN LINE

The multispectral detection capability of the OMA-2 vidicon lends itself naturally to a study of the laser-induced shift of the crystalline Raman peak. This provides additional evidence on the conditions present in silicon following the 532 nm excitation. Figure 6 shows both Stokes and anti-Stokes peaks for very low laser power and for 0.4 J/cm^2. Probe delay was 30 nsec and all other parameters were the same as for Fig. 1b. A distinct shift of the peak is clearly evident and a slight increase in width. If one assumes a shift equal to that of oven-heated Si [20] the observed shift of 13 cm^{-1} is consistent with a crystal temperature of 470°C. There are several reasons to be cautious about inferring a temperature in this way including the fact that the laser-excited region is clamped by the surrounding unexcited material. Furthermore there may be phonon frequency changes due to the high electronic excitation such as those calculated recently by Biswas and Ambegaokar [21] or observed in p-type Si by Cerdeira et al. [22]. Nevertheless, the shift of the Raman line is at least consistent with the lattice temperature indicated from the Raman ratio.

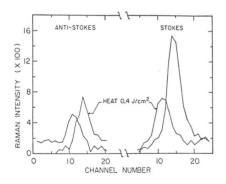

Fig. 6. Stokes and anti-Stokes spectra for 532 nm heat pulse power of 0 and 0.4 J/cm^2. Probe at 405 nm. Integration time 15 minutes for anti-Stokes and 5 minutes for Stokes.

CONCLUSIONS

A central emphasis in this paper is that our Raman measurement of transient phonon populations is a technique which can be self calibrated. It can be performed in such a way that all correction factors are explicitly measured. In addition it readily gives nanosecond resolution and probes less than 700 Å of the surface region. Furthermore it requires no prior assumption of the existence of thermal equilibrium.

We have chosen to express the optic phonon population in terms of a lattice "temperature" although the Raman data themselves provide no evidence in themselves that a well defined lattice temperature exists. The Raman measurement is sensitive only to optic branch phonons of wave vector k = $4\pi n/\lambda$= 1.5 x 10^6 cm^{-1} where n = 5.0 is the real part of the index of refraction at λ = 405 nm. The optic phonon populations further from the zone center have not been measured nor have populations of the acoustic branches. With this caveat in mind, it must be emphasized that the measurement of a very small phonon population (relative to that at melting) at only one point on the optic branch is quite sufficient to invalidate the assumptions built into the normal thermal melting model of laser annealing, viz., that there is complete thermalization of electron and phonon systems.

How then is one to understand this optic phonon population characteristic of only $\sim$400°C? It is a population measured just as the crystal leaves the high reflectivity phase and measured with a system which produces all of the usual features observed in PLA - not only epitaxial regrowth of an amorphous layer but also precipitate and extended defect dissolution in crystalline Si and the cell structure formation due presumably to interfacial instability [18].

Certainly all of the post anneal structural evidence gives conclusive proof that a phase transformation has occurred. One can not begin to explain the observed phenomena without some rapidly moving interface between two phases. It is only the nature of the second phase and the identification of it as the normal molten phase of Si which is called into serious question by the Raman results. We have suggested that Van Vechten's proposal of a dense plasma driven phase change is a possible alternative starting point [23]. Recent picosecond measurements [24] appear to indicate that a very dense plasma is present in laser-excited Si but whether its density is large enough to cause a phase transformation is not clear. It is also not clear that such a plasma will persist long enough to be important in the nanosecond regime. Finally, we believe it is now crucial to obtain information on the behavior of the rest of the phonon system to see whether these low phonon populations also characterize, e.g., the acoustic branches, and to determine whether there is any softening of these modes which might lead to a phase transition [21].

ADDITIONAL NOTE

In the following paper, Jellison, et al. [25] claim to show that statistical fluctuations of some arbitrary amounts can account for the low temperatures observed in our Raman measurements[2]. We have always recognized the presence of, e.g., laser power fluctuations, but remain convinced that they cannot substantially alter our primary conclusion that the lattice temperature is far lower than expected from the thermal model. (Not having access to any calculated results of temperature profiles vs. time for 8 ns, 485 nm pulses or for 10 or 20 nsec 532 nm pulses we have not stated our results in terms of a quantitative temperature discrepancy.) Unfortunately the analysis of Jellison, et al. is flawed on several points, including some of the errors described earlier, such as the use of improper instrumental throughput factors for the particular Raman data they try to fit.

However, there are two additional problems with the analysis of ref. 25. First, we commend them for including effects related to the spectral dependence of the Raman cross section, σ, which were neglected in their earlier analyses [6,7,8]. It should be emphasized, however, that the σ values they now use are not measured but rather semiempirically calculated via derivatives of ellipsometric data. This method is certainly a reasonable first approximation toward fitting the dispersion of the Raman σ; however, as Renucci, et al. [13] found, there are significant discrepancies between the observed σ and a semi-empirical fit using the derivative of the susceptibility. These differences may arise, e.g., from frequency or wavevector dependence of the deformation potential electron-phonon interaction. Thus the tabulated values of σ in Table 1, ref. 25, should be regarded with considerable caution. Furthermore, the correction to the Raman ratio due to this factor is not related directly to σ itself but essentially to its frequency derivative so that errors in σ are likely to be magnified in the correction thus calculated. A far better procedure is to measure this and the other corrections empirically which we have done [2]. Furthermore where the thermal model (HEATING5) predicts temperatures higher than $\sim700C$, attempted fits to the Raman data are dependent on extrapolated values of all the optical constants (see Tables II and III of ref. 25). The appropriate optical constants of Si have not been measured above this temperature.

The second and probably largest problem with the analysis of ref. 25 is that feeding the single parameter of energy density (e.g. 0.8 J/cm^2) to the code HEATING5 does not result in the proper duration for the high reflectivity phase (predicted melt duration). We have relied on reflectivity duration as a primary experimental monitor of power stability, beam uniformity, etc. As described in ref. 2, the reflectivity, continously monitored during the experiment, showed a flat-topped portion centered at 80 nsec duration and the Raman data were obtained at the point where the reflectivity had recovered essentially to that characteristic of crystalline Si. That is, even allowing for some beam inhomogeneity, parts of the sample surface remained in the high reflectivity state until the time of measurement by the probe pulse, and no parts of the probed area left the high reflectivity state more than thirty nanoseconds before the midpoint of the probe pulse. We believe it is disingenuous not to use these observed values as the starting point for analysis of fluctuations. As the analysis stands, the duration of the high reflectivity is only ~25 nsec and thus it naturally finds low temperatures at 120 nsec.

Finally we must point out that the data shown earlier in this paper, which use time reversal symmetry arguments, avoid any ambiguities associated with the σ or α corrections. In addition the close agreement shown by the Raman data of von der Linde, et al. (preceeding paper), which were obtained with 1 nsec resolution during the excitation pulse, further demonstrates that large systematic corrections due to fluctuations are not present in our data.

ACKNOWLEDGEMENTS

The support of the US Office of Naval Research (contract no. N00014-80-C-0419) for the K.S.U. work is gratefully acknowledged. One of the authors (A.C.) is indebted to the Alexander von Humboldt Foundation for support and to P. Zwicknagel and J. Trodahl for assistance with the Nomarski and micro-Raman work, respectively, at the M.P.I.

REFERENCES

1. A. Compaan, H.W. Lo, M.C. Lee and A. Aydinli, Phys. Rev. B26, 1079 (1982).

2. A. Compaan, A. Aydinli, M.C. Lee and H.W. Lo, in Laser and Electron-Beam Interactions with Solids, B.R. Appleton and G.K. Celler, eds. (Elsevier, New York, 1982), p. 43.

3. R. Loudon, Proc. R. Soc. London A 275, 218 (1963), W. Hayes and R. Loudon, Scattering of Light by Crystals (Wiley, New York, 1978), p. 31.

4. A. Compaan, A.Z. Genack, H.Z. Cummins and M. Washington in Light Scattering in Solids, edited by M. Balkanski, R.C.C. Leite and S.P.S. Porto (Flammarion, Paris, 1975), p. 39.

5. J.F. Scott, R.C.C. Leite and T.C. Damen, Phys. Rev. **188**, 1285 (1969); W. Richter, in Solid State Physics (Vol. 78 of Springer Tracts in Modern Physics, G. Höhler, ed.) Springer-Verlag, Berlin (1976); P.F. Williams and S.P.S. Porto, Phys. Rev. **B8**, 1782 (1973).

6. R.F. Wood, M. Rasolt and G.E. Jellison, Jr., Op.cit. ref. 2, p. 61.

7. R.F. Wood, D.H. Lowndes, and G.E. Giles, Op.cit. ref. 2, p. 67.

8. R.F. Wood, D.H. Lowndes, G.E. Jellison and F.A. Modine, Appl. Phys. Lett. **41**, 287 (1982).

9. H. W. Lo and A. Compaan, Phys. Rev. Lett. **44**, 1604 (1980).

10. H. W. Lo and A. Compaan, Appl. Phys. Lett. **38**, 179 (1981).

11. A. Compaan, H.W. Lo, A. Aydinli and M.C. Lee, in Laser and Electron-Beam Solid Interactions and Materials Processing, Gibbons, Hess and Sigmon, eds. (Elsevier, New York, 1981), p. 151.

12. G.E. Jellison and F.A. Modine, Appl. Phys. Lett. **41**, 180 (1982).

13. J.B. Renucci, R.N. Tyte and M. Cardona, Phys. Rev. **B11**, 3885 (1975).

14. D. von der Linde and G. Wartman, Appl. Phys. Lett. (to be published, Oct. 1982); D. von der Linde, G. Wartman, and A. Ozols (preceeding paper).

15. J. Narayan, Op.cit. ref. 2, p. 141.

16. J. Narayan and J. Fletcher, in Defects in Semiconductors, Narayan and Tan, eds. (North Holland, New York, 1981), p. 431.

17. J. Narayan, in Microscopy of Semiconducting Materials, 1981, Cullis and Joy, eds. (Institute of Physics Conf. Ser. 60), p. 101.

18. J. Narayan, J. Fletcher, W.W. White and H. Christie, J. Appl. Phys. **52**, 7121 (1981).

19. R.J. Nemanich, D.K. Biegelson and W.G. Hawkins (private communication).

20. M. Balkanski, R.F. Wallis and E. Haro, Physical Review (to be published).

21. R. Biswas and V. Ambegaokar, Phys. Rev. **B26**, 1980 (1982).

22. F. Cerdeira, T. Fjeldly and M. Cardona, Phys. Rev. **B8**, 4734 (1973).

23. J.A. Van Vechten, Op. cit. ref. 2, p. 49.

24. J.M. Liu, H. Kurz and N. Bloembergen, Appl. Phys. Lett. **41**, 643 (1982); D. von der Linde and N. Fabricius, Appl. Phys. Lett. (Nov. 15, 1982).

25. G.E. Jellison, Jr., D.H. Lowndes, and R.F. Wood (following paper, this conference).

A DETAILED EXAMINATION OF TIME-RESOLVED PULSED RAMAN
TEMEPERATURE MEASUREMENTS OF LASER ANNEALED SILICON

G. E. JELLISON, JR., D. H. LOWNDES, AND R. F. WOOD
Solid State Division, Oak Ridge National Laboratory, Oak Ridge, Tennessee 37830

ABSTRACT

Raman temperature measurements during pulsed laser
annealing of Si by Compaan and co-workers are critically
examined. It has been shown previously that the Stokes to
anti-Stokes ratio depends critically upon the optical properties of silicon as a function of temperature. These
dependences, coupled with the large spatial and temporal
temperature gradients normally found immediately after the
high reflectivity phase, result in large variations in the
calculated temperature depending upon the probe laser pulse
width and the pulse-to-pulse and spatial variations in the
annealing pulse energy density.

INTRODUCTION

Pulsed laser annealing is proving to be a useful technique for many semiconductor device processing applications. From a more fundamental standpoint,
the nonequilibrium thermodynamic processes which occur during the laser annealing process are also currently of great interest [1]. Two competing models
have been presented to explain the phenomena associated with pulsed laser
annealing: 1) the melting model [2], wherein the laser annealing process is
described in classical thermodynamic terms, and the plasma annealing model [3],
wherein the annealing behavior is believed to take place via the electron-hole
plasma created by the laser beam. The melting model successfully describes
detailed results of many experiments which imply that the surface region of the
sample melts; these include time-resolved measurements of optical reflectivity
and transmission, electrical conductivity and x-ray diffraction studies as well
as post-annealing dopant profile measurements [1,4]. Compaan and co-workers
[5], however, have published results of pulsed Raman experiments which are in
apparent conflict with the predictions of the thermal melting model.
In a Raman scattering experiment, one can measure the light emitted after
either the emission (Stokes process) or the absorption (anti-Stokes process) of
a phonon. The phonon population, determined by Bose-Einstein statistics, is
given by $n_0(\omega_0,T) = [\exp(\hbar\omega_0/kT-1]^{-1}$, where ω_0 is the phonon frequency ($\hbar\omega_0 =$
0.064 eV for the $\Gamma_{25'}$ optical phonon in Si) and T is the lattice temperature.
Therefore, the ratio of the intensities of the Stokes and anti-Stokes components ($R(S/AS)$) can be extremely T-dependent.
The determination of T from $R(S/AS)$, however, is not a simple matter. Wood
and co-workers [6] have pointed out that 1) the T-dependence of the optical
properties of Si, 2) the large T gradients that are present during laser
annealing, and 3) the statistical nature of the experiments of Ref. 5 make a
simplified treatment of the experimental Raman data difficult to justify. In a
more recent work, Jellison et al. [7] have shown in detail the effects of the
T-dependent optical properties of silicon on $R(S/AS)$. This paper presents a

*Research sponsored by the Division of Materials Science, U. S. Department of
Energy under contract W-7405-eng-26 with Union Carbide Corporation.

correspondingly detailed examination of the statistical nature of Raman measurements and examines the effects of pulse-to-pulse and spatial variations of annealing energy density, finite probe pulse width, and pulse-to-pulse probe energy variations, in the presence of high spatial and temporal T gradients.

The analysis presented here is generally valid for any wavelength and material. However, numerical examples concentrate on λ = 405 nm because 1) this is the probe laser wavelength used by Compaan and co-workers [5], thereby allowing us to make some direct comparisons and 2) this wavelength also corresponds to a region of resonant Raman scattering, so that effects due to changes in the optical properties with temperature and wavelength are large.

THE RAMAN INTENSITY AT CONSTANT TEMPERATURE

The intensity of the Stokes component in a Raman experiment performed in the backscattering geometry on a sample at temperature T having a high optical absorption coefficient α is given by

$$ I_S = Kt_S\omega_L\omega_S{}^3 \frac{\sigma_S(T)}{\alpha_L(T)+\alpha_S(T)} \cdot \frac{(1-R_S(T))}{n_L(T)n_S(T)} \cdot (n_0(\omega_0,T)+1) , \qquad (1) $$

where the subscripts L and S refer to the laser and Stokes frequencies, respectively. ω is the frequency, t the spectrometer throughput factor, σ the Raman matrix element squared (sometimes referred to as the cross section), R the normal-incidence reflectance, and n the index of refraction. K is a numerical constant which includes the intensity of the probe light. A similar expression can be determined for the anti-Stokes (AS) component by replacing S → AS and n_0+1 → n_0. Note that Eq. 1 includes an extra $(1-R_S(T)))/n_S2(T)$ term from that normally given, which converts the intensity of the scattered light from inside the crystal to outside the crystal.

As has been noted in Eq. 1, σ, α, R, and n are all denoted explicitly as functions of T. These optical functions have been determined at several temperatures by Jellison and Modine [8]; we reproduce their data for a λ_L = 405 nm probe pulse as well as the S component (λ_S = 413.6 nm) and the AS component (λ_{AS} = 397.6 nm) in Table I. In applying these data to a laser annealing experiment, extrapolations must be made to the melting point of Silicon (T_m = 1683 K); these are discussed in Refs. 7 and 8. Note that our object here is not to determine T from the work of Compaan and co-workers, but rather to emphasize the importance of including T-dependent optical properties and of recognizing the statistical nature of the pulsed Raman experiments.

THE RAMAN INTENSITY FOR A GENERAL LASER ANNEALING EXPERIMENT

A realistic treatment of Raman T measurements during pulsed laser annealing requires inclusion of several additional factors. Because signals in these experiments are weak, and signal averaging is needed, one must account for
   1) pulse-to-pulse variations in the annealing pulse and probe pulse energy density and in the probe pulse delay time.
In addition, since the signal collection optics perform a spatial and time integration, one must also take into account
   2) Spatial (x-y) variations in both the annealing and probe pulses, and
   3) the temporal width of the probe pulse.
The second point is particularly difficult to handle because spatial variations over distances as small as 1 μm can make significant differences.

TABLE I
The optical functions of Si appropriate for Raman scattering experiments performed at 405 nm (= 3.062 eV) taken from Ref. 8. The S, L, and AS columns
refer to the Stokes component (413.7 nm or 2.998 eV), the laser frequency
(405 nm) and the anti-Stokes component (396.7 nm or 3.126 eV). The quantities
tabulated are the absorption coefficient α, the Raman matrix element σ, the
index of refraction n, and the normal incidence reflectivity R.

| T | $\alpha(\times 10^5$ cm$^{-1})$ | | | σ(arb. units) | | n | | R | |
K	S	L	AS	S	AS	S	AS	S	AS
10	0.42	0.52	0.68	0.81	1.39	5.16	5.59	.457	.486
300	0.70	0.90	1.23	1.42	2.44	5.28	5.75	.466	.497
333	0.80	1.05	1.47	1.68	3.03	5.27	5.76	.464	.497
363	0.86	1.15	1.66	2.07	3.98	5.26	5.74	.463	.498
468	1.16	1.58	2.34	3.26	6.10	5.37	5.90	.473	.510
571	1.55	2.14	3.22	4.65	8.91	5.50	5.99	.484	.520
676	2.04	2.84	4.10	5.27	9.93	5.62	5.99	.493	.525
773	2.69	3.56	4.91	5.31	8.06	5.65	5.90	.498	.529
874	3.39	4.34	5.64	5.08	6.16	5.66	5.75	.504	.528
972	4.09	5.06	6.26	4.83	5.03	5.63	5.59	.507	.526
Error	± 2-5%			± 10%		± .03		± .002	

In order to parameterize these effects, Eq. 1 becomes

$$dI_S^i(x,y,t) = K'\omega_L\omega_S^3 t_S \ \frac{I_L^i(x,y,t-t_0^i)\sigma_S(T^i)}{\alpha_L(T^i)+\alpha_S(T^i)} \cdot \frac{(1-R(T^i))}{n_L(T^i)n_S(T^i)} \ (n_0(\omega_0,T^i)+1)dxdydt \ . \tag{2}$$

The intensity has been expressed as a differential intensity from a unit area
at position x,y, and at a time t after the initialization of the i-th annealing
pulse. The sample T is replaced by a temperature profile $T^i{\to}T^i(x,y,t)$ and the
intensity of the i-th probe pulse is given by $I_L^i(x,y,t-t_0^i)$, where t_0^i
is the delay time (varying from pulse to pulse). Note that we have made the
simplifying assumption that T is z-independent. For experiments that use a
probe laser wavelength for which $\alpha >10^5$ cm^{-1}, T variations occurring deeper
than ~100 nm will not be observed. Melting model calculations of T - depth
profiles indicate that this is indeed the case for any time after the end of
the high reflectivity phase. Since these are just the conditions of the
experiments that we wish to simulate, Eq. 2 is justified.
To simulate the necessary signal averaging for pulse-and-probe experiments,

$$I_S = \sum_{i=1}^{M} \int dI_S^i(x,y,t) \ , \tag{3}$$

where the integration is over all x, y and t, and M is the total number of
pulses, which can typically be as large as 10^4 [5]. The determination of the
AS intensity is made by the substitution of AS for S and $n_0(\omega_0,T)$ for
$(n_0(\omega_0 T)+1)$ into Eqs. 2 and 3. Therefore, R(S/AS) can be expressed as

$$R(S/AS) =\left\{\sum_{i=1}^{M_S} \int dI_S^i(x,y,t)\right\}/\left\{\sum_{i=1}^{M_{AS}} \int dI_{AS}^i(x,y,t)\right\} \ . \tag{4}$$

Equation 4 is to be compared to that used by Compaan and co-workers [5]:

$$R(S/AS) = \left(\frac{\omega_S}{\omega_{AS}}\right)^3 \left(\frac{t_S}{t_{AS}}\right) \left(\frac{\sigma_S}{\sigma_{AS}}\right) \left(\frac{\alpha_L + \alpha_{AS}}{\alpha_L + \alpha_S}\right) \exp\left(\frac{\hbar\omega_0}{kT}\right) = F \exp(\hbar\omega_0/kT) \quad . \tag{5}$$

The expression given in Eq. 5 neglects:
1) The n and R corrections, which can be as large as 18%,
2) The T-dependence of all optical parameters, and
3) Pulse-to-pulse and spatial variations in the probe and annealing pulses.
In addition, Compaan and co-workers [5] set $F = 1$ which is reasonable at room temperature but not at higher temperatures.

In order to quantitatively examine the effects of their approximations a numerical simulation was performed using Eq. 4. T-time profiles were obtained using melting model calculations described elsewhere [9] and shown in Fig. 1. We assume incident annealing energy densities of 0.6, 0.8, and 1.0 J/cm^2, a triangular annealing pulse of width 20 ns and temperature-dependent optical properties appropriate for a frequency-douled Nd:YAG annealing laser; these conditions correspond closely to the most recent experiments of Compaan and co-workers [5].

Since it is impossible to know in detail the intensity of both the annealing and probe laser beams as a function of x, y, and t as well as their pulse-to-pulse variations, we will assume that these deviations from the average values can be simulated by integrating over annealing and probe beam intensity and delay time probability distribution functions which are Gaussian and statistically independent. Therefore, the intensity of the Stokes component is

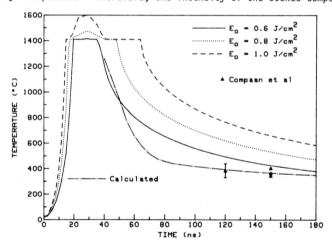

Fig. 1. Temperatures calculated using HEATING5 vs time for laser annealing with energy densities of 0.6, 0.8, and 1.0 J/cm^2 pulses from a frequency-doubled Nd:YAG laser, pulse width 20 nsec. Also shown are the uncorrected temperatures T_{u_2} (dash-dotted line) calculated assuming $E_0 = 0.8$ J/cm^2, $\Delta E_0 = 0.08$ J/cm^2, $\Delta\tau = 3.5$ nsec, and $F = 1$ (see text).

$$I_S(t_0,E_0) = K't_S\omega_L\omega_S^3 \int_0^\infty P(t_1-t_0) \int_0^\infty P(t-t_1) \int_0^\infty P(E-E_0) \int_0^\infty \frac{P(E_p-E_p^0)E_p\sigma_S(T)}{\alpha_L(T)+\alpha_S(T)}$$

$$\cdot\ (n_0(\omega_0,T)+1)\ \cdot\ \frac{1-R_S(T)}{n_L(T)n_S(T)}\ \ dE_p dEdtdt_1\ . \tag{6}$$

The quantities E_p^0, E_0, and t_0 are the averages of probe pulse energy density, the anneal pulse energy density, and the probe pulse delay, respectively. The functions P represent Gaussian probability distribution functions. The temperature is given by $T = T(t,E)$; that is, the spatial and pulse-to-pulse variables are now replaced by a single energy variable. The intensity of the AS component can be obtained by replacing S by AS and n_0+1 by n_0 in Eq. 6 and R(S/AS) can then be calculated. From this realistic R(S/AS) an uncorrected temperature T_u can be calculated from Eq. 5, assuming, with Compaan and co-workers, F=1. T_u will be a function of 4 quantities in this approximation:
1) the delay time t_0,
2) the average energy density E_0,
3) the probe pulse time Gaussian width $\Delta\tau$,
and 4) the annealing pulse energy Gaussian width ΔE_0.
Because of the assumption of statistical independence, variations of the probe energy density have no effect, and the delay time uncertainty and the probe pulse time width are included in one parameter, $\Delta\tau$.
Figure 1 also shows the uncorrected temperature T_u as a function of time after the initialization of the annealing pulse, for $\Delta E_0 = 0.08$ J/cm^2 and $\Delta\tau = 3.5$ ns. As can be seen, T_u starts near the melting point and then decreases rapidly to <500°C. However, it should be pointed out that the very high temperatures may not be observed, since the S and AS intensities are small at high T (see Table II). The cause of this decreased intensity at elevated temperature is the T-dependence of the optical properties: As T is increased, α

TABLE II
The intensity of the S and AS components as a function of time after a 0.8 J/cm^2 Nd:YAG frequency-doubled laser pulse. The probe pulse width was chosen to be $\Delta\tau = 3.5$ ns and the energy broadening $\Delta E_0 = 0.08$ J/cm^2.

Time ns	Actual Temp. (°C)	Uncorrected Temp. T_u (°C)	Intensities (arb. units)	
			Stokes	anti-Stokes
40	1410	1236	0.70	0.42
50	1294	950	5.62	3.00
60	1030	678	10.18	4.57
70	905	516	12.30	4.70
80	818	456	13.36	4.72
90	753	431	14.26	4.87
100	700	416	15.23	5.08
110	657	404	16.26	5.32
120	620	393	17.34	5.56
130	586	382	18.39	5.80
140	568	373	19.33	6.00
150	533	366	20.14	6.17

increases and σ decreases, the latter by as much as a factor of 5. As can be seen from Table I, σ peaks at 400-500°C; therefore, any averaging over a wide T range will bias the measurement toward T ~ 400-500°C. This latter effect is emphasized in Table III, where we tabulate T_u vs ΔE_o and $\Delta\tau$ for t_o = 50 ns. As can be seen, for t_o = 50 ns T_u decreases substantially as either ΔE_o or $\Delta\tau$ is increased. The extreme sensitivity of T_u for t_o = 50 ns is due to the large T-time gradients that exist at the time of the end of surface melting (see Fig. 1), and to the tendency of the optical properties to bias the measurements toward lower T.

In conclusion, we have shown that pulse-to-pulse variations during pulsed Raman temperature measurements can lead to large errors in the inferred temperature. These variations, coupled with the temperature-dependent optical properties of silicon at λ = 405 nm, very strongly imply that Compaan and co-workers [5] have severely underestimated the lattice temperature immediately after the high reflectivity phase.

TABLE III
Uncorrected temperatures (T_u;°C) obtained at 50 ns after a 0.8 J/cm^2 frequency-doubled Nd:YAG laser pulse, as a function of energy broadening ΔE_o and probe pulse width $\Delta\tau$.

ΔE_o (J/cm^2)	0	$\Delta\tau$ 3.5	7.0	10.5
t_o=50 ns (T=1294°C)				
0.00	1355	1106	899	803
0.04	1218	1090	915	794
0.08	1009	950	845	758
0.12	879	830	766	716
0.16	782	727	691	666

REFERENCES

1. Laser and Electron Beam Interactions with Solids, B. R. Appleton and G. K. Celler eds. (North-Holland, New York 1982).

2. See for example, R. F. Wood and G. E. Giles, Phys. Rev. B 23, 2923 (1981); R. F. Wood, D. H. Lowndes, and G. E. Giles, Ref. 1, pp. 67-72.

3. See for example, J. A. Van Vechten, J. Phys. C 41, 4 (1980) and references therein.

4. See for example, several papers this volume.

5. H. W. Lo and A. Compaan, Phys. Rev. Lett. 44, 1604 (1980); H. W. Lo and A. Compaan, Appl. Phys. Lett. 38, 179 (1981); A. Compaan, A. Aydinli, M. C. Lee, and H. W. Lo, Ref. 1, pp. 43-48.

6. R. F. Wood, M. Rasolt, and G. E. Jellison, Jr., Ref. 1, pp. 61-66; R. F. Wood, D. H. Lowndes, G. E. Jellison, Jr., and F. A. Modine, Appl. Phys. Lett. 41, 287 (1982).

7. G. E. Jellison, Jr., D. H. Lowndes, and R. F. Wood, manuscript in preparation.

8. G. E. Jellison, Jr. and F. A. Modine, J. Appl. Phys. 53, 3745 (1982); G. E. Jellison, Jr. and F. A. Modine, Appl. Phys. Lett. 41, 180 (1982); G. E. Jellison, Jr. and F. A. Modine, submitted to Phys. Rev. B.

9. R. F. Wood, manuscript in preparation.

INSTABILITY OF THE ELECTRON-HOLE PLASMA AND ITS POSSIBLE RELATION WITH MELTING

MONIQUE COMBESCOT, JULIEN BOK
Groupe de physique des solides de l'Ecole Normale Supérieure,
24 rue Lhomond, 75005 Paris, France

ABSTRACT

We show that the softening of the T.A. phonon mode produces an instability in the electron-hole plasma of a silicon-like semiconductor and suggest that this instability can be the mechanism for melting of this type of material. This allows us to estimate the characteristics of a laser pulse which are necessary to modify the usual melting, i.e. which would lead to a substantial decrease of the melting temperature.

The interest of laser-annealing in the manufacturing of semiconductor devices of submicronic size has produced a large amount of work on that subject. A few years ago Van Vechten [1] has suggested that, as the energy is transferred from the laser beam to the lattice via the creation of electron-hole (e-h) pairs, those pairs may modify, if their density is large enough, the lattice stability, and lead to a melting at a temperature T lower than $T_m \sim 1700°K$. This idea is in fact directly related to previous work by Heine and Van Vechten [2], who explained the decrease of the band gap E_g of silicon-like semiconductors, by the softening of the T.A. phonon mode in the presence of a large density of e-h pairs.

We came back [3] to this original idea, now well accepted for the variation $E_g(T)$ at low temperature and considered first the simple heating, in a furnace for example, of Si or Ge, without an extra energy source such as the absorbed photons in laser annealing.

In this simplest case, we follow Heine and Van Vechten, and calculate the total free energy F of a silicon sample at temperature T, having N_a atoms and N e-h pairs in a volume V. F is composed of an atomic part $F_a(N_a, V, T)$, a plasma part $F_{eh}(N, V, T)$ and a phonon part $F_{ph}(N_a, N, V, T)$ which depends on both the atom and the e-h pair density due to the change in the phonon frequency with n = N/V. Considering only the T.A. phonons, as done by Heine and Van Vechten, and for which the frequency dependence can be written, in a first approximation as

$$\omega(n) = \omega_0 \left(1 - \alpha \frac{n}{n_0}\right) \qquad (1)$$

where $n_0 = 5\times10^{22} cm^{-3}$ is the density of Si atoms and α is a constant, the phonon free energy reads

$$F_{ph} = \sum_i kT \, Ln(1 - e^{\frac{-\hbar\omega i}{kT}}) \simeq kT \sum_i Ln \frac{\hbar\omega i}{kT}$$

$$\sim N_a \, kT \, Ln(1 - \alpha\frac{n}{n_0}) + \text{terms independent of n .} \qquad (2)$$

At constant T and V, the e-h pair number N adjusts for an isolated system, such that the e-h pair chemical potential $\mu = \frac{\partial F}{\partial N}\Big|_{N_{A,V,T}}$ is zero.

Mat. Res. Soc. Symp. Proc. Vol. 13 (1983) © Elsevier Science Publishing Co., Inc.

$$\mu(n,T) = \frac{\partial F}{\partial N} = E_{G_0} - \frac{\alpha kT}{1 - \frac{\alpha n}{n_0}} + \mu_e + \mu_h \tag{3}$$

E_{G_0} is the band gap at T = 0 and μ_e and μ_h are the free electron and hole chemical potential ($\mu_e \sim kT \ln n$ at low density and $\mu_e \sim n^{2/3}$ at large n).

Considering only the low T case, where $\alpha \frac{n}{n_0} \ll 1$, Heine and Van Vechten have concluded that eq.(3) that due to the softening of the phonons by the e-h pairs, one finds an apparent band gap which decreases with T as

$$E_G(T) = E_{G_0} - \alpha kT \tag{4}$$

and that consequently, in the non degenerate limit, the intrinsic carrier density at equilibrium is

$$n = \sqrt{N_e N_h} \; e^{-E_G(T)/2kT} \tag{5}$$

N_e and N_h being two constants depending on the band structure.

But looking at the same equation (3), one notes that something pathological appears when n goes to n_0/α. In fact one can draw the set of curves F(n,T) for various T, and one sees that at small T, F has a minimum for a density n given by equation (5), but above a certain temperature T*, this minimum disappears and no stable density for an intrinsic e-h gas can be found. The value of T* depends on α and the band structure parameters which are not known around T_M. Using their low T values, one finds T* of order 2000°K, i.e. of the order of T_m. A more accurate estimation of T* would need a precise knowledge of the variation of all branches of phonons with n. But the best way to test if this instability in the e-h plasma is indeed the reason for melting should come from an experiment: the measure of the e-h plasma density close to T_M. One predicts that, at T*, the density n* is of order $10^{21} cm^{-3}$, i.e. two orders of magnitude larger than the usually quoted value of $2 \; 10^{19} cm^{-3}$ which comes from an extrapolation of eq. (5). As the change in density is rather large, and almost independent on α and the masses, one calls for an experimentalist to test if the melting of silicon or germanium is due to that instability, or if it is of a more classical kind, like the creation of vacancies in the lattice.

If one now comes to the question of whether or not the e-h pairs created by the laser are playing a role in laser annealing, one first expects that the laser should produce a density of pairs at least the order of the one already present in the sample at thermal equilibrium, which is of order $10^{21} cm^{-3}$. One can show using a hydrodynamical study [4] of the e-h plasma expansion that such densities are not reached with typical 1 joule 10 nsec pulses, but might be reached with picosecond pulses; the essential feature which limits the density created by a pulse being the very fast Auger recombination, which is of the order of the pulse duration for picosecond pulses only. On the other hand, it is true that if the laser pulse is so short that an e-h density of order 10^{21} to $10^{22} cm^{-3}$ can be piled up in the sample before recombining, then the sample will lose its stability and becomes fluid even at T = 0.

REFERENCES

1. J. A. Van Vechten, J. Phys. (Paris) Colloquy 41, C4-15 (1980).

2. V. Heine, J. A. Van Vechten, Phys. Rev. B 13, 1622 (1976).

3. M. Combescot, J. Bok, Phys. Rev. Lett. 48, 1413 (1982).

4. M. Combescot, Phys. Lett. 85 A, 308 (1981).

TIME-RESOLVED STUDY OF SILICON DURING PULSED-LASER ANNEALING*

B. C. LARSON, C. W. WHITE, T. S. NOGGLE, J. F. BARHORST,
Solid State Division, Oak Ridge National Laboratory, Oak Ridge, TN 37830
and

D. M. MILLS
Cornell High Energy Synchrotron Source and School of Applied and Engineering
Physics, Cornell University, Ithaca, NY 14850

ABSTRACT

Near surface temperatures and temperature gradients have
been studied in silicon during pulsed laser annealing. The
investigation was carried out using nanosecond resolution x-
ray diffraction measurements made at the Cornell High Energy
Synchrotron Source. Thermal-induced-strain analyses of
these real-time, extended Bragg scattering measurements have
shown that the lattice temperature reached the melting point
during 15 ns, 1.1-1.5 J/cm^2 ruby laser pulses and that the
temperature of the liquid-solid interface remained at that
temperature throughout the high reflectivity phase, after
which time the surface temperature subsided rapidly. The
temperature gradients below the liquid-solid interface were
found to be in the range of 10^7°C/cm.

INTRODUCTION

Real-time investigations [1] of pulsed laser annealing in silicon have been
carried out using optical reflectivity, optical transmission, Raman scattering,
electrical conductivity, and x-ray diffraction. These studies have provided
rather detailed information on the time scale of the annealing process under
varying laser conditions, and significant progress has been made in
understanding the transient structural and temperature aspects of the process as
well. Electrical conductivity [2] and nanosecond resolution x-ray diffraction
[3] measurements have provided depth information on the liquid phase of the
annealing process and the x-ray diffraction measurements have provided time
resolved temperature profiles in silicon following laser annealing as well. In
this paper we report time resolved x-ray diffraction measurements of the temperature and temperature gradients in silicon during the annealing process.

THEORY

The theoretical framework for Bragg scattering of x-rays from crystals with
one-dimensional near-surface strains is well understood [4] in the context of
dynamical diffracton theory. The x-ray reflectivity for such a system is given
by the following differential equation

*Research sponsored by the Division of Materials Sciences, U.S. Department of
Energy under contract W-7405-eng-26 with Union Carbide Corporation. The CHESS
portion of this research was supported by the National Science Foundation.

Mat. Res. Soc. Symp. Proc. Vol. 13 (1983) Published by Elsevier Science Publishing Co., Inc.

$$i \frac{dX}{dA} = X^2(1+ik) - 2X(y+ig) + (1+ik) \tag{1}$$

where $X(A) = X_1 + iX_2$ is the normalized scattering amplitude as a function of the reduced spatial (or depth) coordinate, A. For thermal induced strain the effect of thermal motion (not considered in ref. 3) on the scattering factor must be considered and is introduced through the Debye-Waller factor, e^{-M}. In this case, the x-ray scattering factor $f = f'+if''$ is replaced by fe^{-M} and A is then written as

$$A = \frac{r_e\, f(\psi)e^{-M}\lambda t}{V_c\, \sin(\Theta_B)} \tag{2}$$

As discussed in more detail elsewhere [4] r_e is the classical electron radius, t is the depth in the crystal, V_c is the unit cell volume, ψ is the x-ray scattering angle, and Θ_B is the Bragg angle for the crystal. λ is the x-ray wavelength and g and k are slowly varying parameters given by

$$g = \frac{-f''(0)}{f'(\psi)e^{-M}} \quad \text{and} \quad k = \frac{f''(0)}{f'(0)} \tag{3}$$

The parameter M in the Debye-Waller factor can be written as [5]

$$M = \frac{6h^2\, T}{mk_B\Theta_D^2} \, [\phi(\theta_D/T) + \theta_D/4T] \, \frac{\sin^2(\Theta_B)}{\lambda^2} \tag{4}$$

where T is the crystal temperature, θ_D is the Debye temperature of the sample, ϕ is the Debye function [5], m is the mass of the atoms, h is Planck's constant, and k_B is Boltsman's constant. The reduced coordinate, y, specifying the angular deviation of the crystal from the Bragg angle is given by

$$y(A) = \frac{\pi V_c\, \sin(2\Theta_B)(\Delta\Theta+\varepsilon(A)\, \tan(\Theta_B))}{\lambda^2\, r_e f'(\psi)e^{-M}} - \frac{f'(0)}{f'(\psi)e^{-M}} \tag{5}$$

where $\Delta\Theta$ is the angular deviation of the sample from the Bragg angle and $\varepsilon(A)$ is the thermal strain in the crystal as a function of A. The reflectivity $R(\Delta\Theta)$ (i.e., the ratio of the number of x-rays scattered to the number of incident x-rays) is given in terms of the scattering amplitude at the crystal surface (A=0) by

$$R(\Delta\Theta) = \left| X(0) \right|^2 \tag{6}$$

$R(\Delta\Theta)$ is calculated numerically using Eq. 1 and the strain profile $\varepsilon(A)$. Since $\varepsilon(A)$ is related to the lattice temperature through the thermal expansion coefficient, $\alpha(T)$, the calculations can be fit to the measured x-ray scattering with the temperature profile as the fitting parameter. The one-dimensionality of the strain (for thin layers) causes an enhancement [6], η, of the strain normal to the crystal surface. Therefore,

$$\varepsilon(A) = \int_{T_0}^{T} \alpha(T)\eta\, dT \tag{7}$$

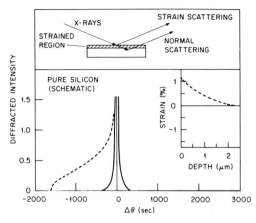

Fig. 1. Schematic of scattering geometry and a schematic
picture of extended Bragg scattering resulting from a strain
distribution.

where η varies only slightly with temperature and is ~1.70 and ~1.39 for the
$\langle 100 \rangle$ and $\langle 111 \rangle$ orientations of silicon, respectively. T_0 is the ambient temperature of the crystal.

EXPERIMENT

The experiment was carried out using the Al beam line of the Cornell High
Energy Synchrotron Source (CHESS) in essentially the same manner used in a previous experiment [3], making use of the pulsed nature of the synchrotron. Ruby
laser pulses of 15 ns (FWHM) duration and energy densities varying from 0.5 -
1.5 J/cm² were used. Both the incident laser energy and the amount of laser
energy reflected from the sample were measured so that the amount of energy
deposited as well as the incident energy density could be determined. The duration of the high reflectivity phase was measured using a He-Ne laser in order to
provide a direct link to other real-time measurements. Time resolution was
achieved by synchronizing the firing of the laser such that the laser pulse
reached the silicon surface the desired time before a probing x-ray pulse and
then recording the number of x-rays scattered from the x-ray pulse. The jitter
in the laser firing resulted in time resolution of ± 5 ns. Figure 1 is a
schematic representation of the experiment, where x-ray scattering measurements
are indicated near a Bragg reflection and the effect of near-surface thermal
strain is depicted as causing an extension of the Bragg profile on the negative
$\Delta\Theta$ side of the Bragg peak. Figure 2 shows the shape of the ruby laser pulse and
the time convention used in the experiment (i.e., t=0 at the peak of the laser
pulse).

RESULTS

Figure 3 shows x-ray scattering profiles measured 5 ns and 35 ns after 1.15
J/cm² ruby laser pulses. The scattering intensity resulting from the laser
pulses ranges out to ~450 sec. for both measurement times and in both cases a
plateau-like region is present between -300 and -200 sec. The plateau in the 35
ns measurement has a higher intensity by nearly a factor two. No scattering was
found on the positive $\Delta\Theta$ side of the Bragg reflection. The solid and dashed
lines in Fig. 3 represent x-ray scattering calculations made using Eqs. 1,6,7
in conjunction with the temperature profiles in Fig. 4, which were obtained by
iteratively fitting the calculations to the measured x-ray data.

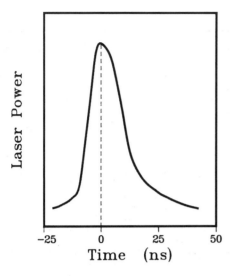

Fig. 2. Shape of ruby laser pulse showing the time measurement convention.

Similar results are shown in Fig. 5 for 1.5 J/cm² laser pulses and the temperature profiles used in generating the curves in Fig. 5 are shown in Fig. 6. The x-ray scattering data in Fig. 5 are rather similar to those in Fig. 3 for 1.15 J/cm², and consequently the respective temperature profiles are similar as well. However, as indicated in Table I, the situations for the two cases are really quite different in that the 1.5 J/cm² measurements were made at times when substantial melted layers should be present, while the times for the 1.15 J/cm² measurements correspond to times approximately at the start of melting and at the end of resolidification when only thin melted layers would be expected. In Table I, E_{in} is the total energy in the incident laser pulses and E_{abs} is the energy that would have been absorbed (or deposited) in the crystal at the measurement time, t. E_{abs} was determined using the shape of the laser pulse in

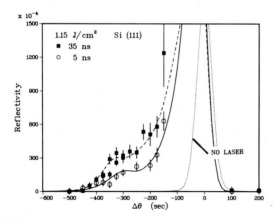

Fig. 3. Time-resolved x-ray scattering near the (111) reflection of ⟨111⟩ oriented silicon after 1.15 J/cm² laser pulses.

TABLE I
Overview of Laser and Energy Conditions.

E_{in} (J/cm^2)	t (ns)	E_{abs}	E_{Latt}	d_{melt} (µm)	∇T (10^7 °C/cm)	τ_{HRP} (ns)
0.51±0.05	10	0.26	0.26	--		--
1.15±0.1	5	0.33	0.33	∿0	2.1	45±10
	35	0.45	0.44	∿0	1.04	
1.50±0.1	20	0.56	0.29	∿0.4	1.8	100±10
	55	0.58	0.45	∿0.2	0.85	
	155	0.58	0.57	∿ 0		

Fig. 2 together with the results of the primary beam reflectivity measurements which showed that ≈60% of 1.15 and 1.5 J/cm^2 pulses was reflected while only ≈37% of the 0.51 J/cm^2 pulses was reflected from the silicon surface. The reflectivities of crystalline and liquid silicon were taken to be 0.35 and 0.7, respectively. E_{Latt} refers to the thermal energy under the temperature profiles in Figs. 4, 6, and d_{melt} was derived by assuming the absorbed energy not found in the lattice was present in the form of a liquid silicon layer, equilibrated to 1410°C. The temperature gradients were obtained from the slopes of the temperature profiles at the liquid-solid interface (i.e., Depth = 0) in Figs. 4, 6. Although error limits are not indicated, the uncertainties in determining E_{abs} and E_{Latt} are estimated to be at least ± 0.05 J/cm^2. Therefore, the degree of energy balance and the calculated melt depths should be regarded only as indicators of the consistency of the thermal analysis for these conditions and not as accurate determinations of the melt depths. τ_{HRP} is the melt lifetime.

The 155 ns temperature profile in Fig. 6 (indicating surface temperatures well below the melting point) was obtained by fitting the 155 ns scattering data for 1.5 J/cm^2 laser pulses in Fig. 7. 155 ns represents a time well after the HRP disappears for these laser pulses; therefore, the less than 1410°C temperatures and relatively low temperature gradients are consistent with surface cooling and heat flow into the bulk. This case is analogous to the 100 ns temperature measurements reported in ref. 3; the inclusion of Debye-Waller factors and improved laser pulse uniformity results in lower temperature gradients for

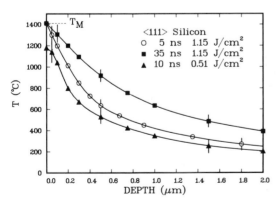

Fig. 4. Lattice temperature profiles corresponding to the x-ray scattering in Figs. 3,8.

48

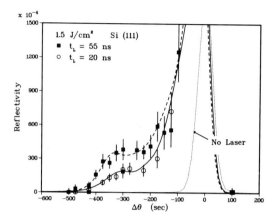

Fig. 5. Time-resolved x-ray scattering near the (111) reflection of <111> oriented silicon after 1.5 J/cm² laser pulses.

the present case, however. The energy balance indicated in Table I for 155 ns is consistent with the disappearance of the liquid layer as well, in contrast to the energy imbalance in the 20 and 55 ns measurements.

Measurements made for 0.51 J/cm² laser pulses (below the melt threshold) are shown in Fig. 8. These data show x-ray scattering limited to ∿350 sec. and they are accompanied by a calculated scattering curve obtained using the temperature profile shown in Fig. 4. Rather high temperatures are indicated; however, the surface temperature apparently remained below the melting point. This is consistent with the lower reflected fraction for the 0.51 J/cm² pulses referred to above, as the switch to higher reflectivity would not occur without melting.

The ∿10⁷°C/cm temperature gradients listed in Table I for the 1.15 J/cm² pulses after 35 ns and 1.5 J/cm² pulses after 55 ns correspond to ∿7-8 m/s regrowth velocities when analysed in terms of the conduction of the latent heat of crystallization away from the liquid-solid interface using the relation $V = K_S(Grad\ T)/L\rho$. In this equation K_S is the thermal conductivity of the solid, Grad T is the temperature gradient at the interface, L is the latent heat of crystallization, and ρ is the density. Although this expression gives an upper

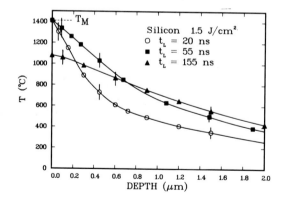

Fig. 6. Lattice temperature profiles corresponding to the x-ray scattering in Figs. 5,7.

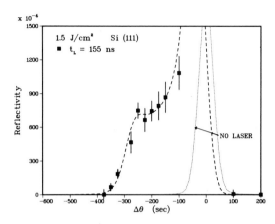

Fig. 7. Time-resolved x-ray scattering near the (111) reflection of ⟨111⟩ oriented silicon after 1.5 J/cm² laser pulses.

limit, velocities in this range are expected [7] from thermal model calculations. However, detailed calculations for the specific conditions used should
be made before making detailed comparisons. The temperature gradients for
measurements made during the melt-in phase are not so simply related to the melt
front velocity.

The uncertainties in the temperature profiles obtained by fitting the scattering measurements do not permit a definitive answer to the question of
overheating or undercooling of the silicon lattice during the melt-in and
regrowth cycles. While very large undercooling and overheating can clearly be
ruled out for these conditions, values ∿50-75°C would fall within the uncertainties of the temperature profiles. More precise, higher resolution x-ray measurements should be able to address this question more quantitatively, however.

In summary, we find that the results of these measurements fit directly into
the thermal melting model of laser annealing. The liquid-solid interface temperatures, the temperature gradients at the interface, the surface cooling

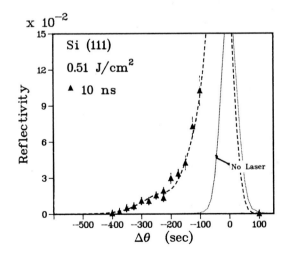

Fig. 8. Time-resolved x-ray scattering near the (111) reflection of ⟨111⟩ oriented silicon after 0.51 J/cm² laser pulses.

and heat flow into the bulk after the HRP, and the thermal energy retained in the temperature profiles all point to the thermal model interpretation. On the other hand, the suggestion of only ~400°C temperature rises during laser annealing inferred from Raman measurements [8] seems to be inconsistent with the results presented here. While the reason for the inconsistency is not known at present, we believe the overall compatibility of the x-ray results with the thermal hypothesis is compelling evidence for thermal melting during pulsed laser annealing.

CONCLUSIONS

Time-resolved x-ray diffraction measurements made on pulsed laser annealed silicon have shown lattice temperatures to rise to the melting point during 1.15-1.5 J/cm^2 laser pulses and have shown that the liquid-solid interface temperature is maintained at (or near) the melting point during the HRP. Interface temperature gradients were found to be ~10^7°C/cm during the HRP. Surface temperatures were found to decrease rapidly after the disappearance of the HRP, and for laser pulses below the HRP threshhold, surface temperatures were found to be less than the melting point of silicon. Undercooling or overheating, if present, were $\lesssim$ 100°C for the conditions studied in this investigation. It is concluded that high temperatures and thermal melting occur during pulsed laser annealing of silicon.

ACKNOWLEDGMENTS

We would like to thank F. W. Young, Jr., for useful discussions, R. F. Wood for making unpublished calculations available, and the CHESS staff for their help and cooperation during the experiment.

REFERENCES

1. Laser and Electron Beam Interactions with Solids, ed. by B. R. Appleton and G. K. Celler, North Holland, New York, 1982.
2. G. J. Galvin, M. O. Thompson, J. W. Mayer, R. B. Hammond, N. Paulter and P. S. Peercy, Phys. Rev. Lett. 48, 33 (1982).
3. B. C. Larson, C. W. White, T. S. Noggle and D. Mills, Phys. Rev. Lett. 48, 337 (1982).
4. B. C. Larson and J. F. Barhorst, J. Appl. Phys. 51, 3181 (1980).
5. Int. Tables for X-Ray Crystallography, Vol. II, The Kynoch Press, Birmingham, England, 1959, p. 241.
6. A. Fukuhara and Y. Takano, Acta Crystallogr., Sect. A33, 137 (1977).
7. D. Lowndes, G. E. Jellison, Jr., and R. F. Wood, Phys. Rev. (in press).
8. A. Compaan, A. Aydinli, M. C. Lee and H. W. Lo, Ref. 1, p. 43.

MEASUREMENT OF FAST MELTING AND REGROWTH VELOCITIES IN
PICOSECOND LASER HEATED SILICON

P. H. BUCKSBAUM AND J. BOKOR
Bell Telephone Laboratories, Holmdel, New Jersey, USA

ABSTRACT

Fast regrowth of amorphous silicon from liquid silicon
films has been directly observed in a time resolved picosecond
laser melting experiment. Liquid films up to 100 nm thick were
formed on crystalline substrates with 15 picosecond 248 nm
pulses from a KrF* excimer laser. The film thickness as a
function of time was probed directly by observing attenuation of
1.64 μm 15 psec light pulses transmitted through the melt.
Melting and regrowth velocities were compared to a heat
diffusion model, and evidence for melt undercooling was
observed.The resolidified silicon was amorphous at all values of
incident laser intensity.

INTRODUCTION

Time resolved studies of pulsed laser irradiated silicon
were first carrier out on nanosecond time scales by Auston et.
al.[1], who observed reflectivity changes, presumably associated
with melting of the silicon surface. More recently, many
investigators have studied the problem of transient heating of
silicon through reflectivity and transmission measurements, as
well as other techniques such as Raman scattering and electrical
conductivity [2]. Controversy has arisen over the time required
for equilibration of the laser-excited electrons and the lattice
[3]. After much detailed work, a consensus is now emerging which
supports a standard thermal model of silicon melting on time
scales much shorter than 20 picoseconds [4].
The fast lattice heating which occurs when silicon is
irradiated with picosecond pulses, permits a study of the
thermodynamics of phase transitions under extreme conditions of
very high thermal gradients and heat flows. These processes have
been investigated on nanosecond time scales by Cullis et al [5],
where a direct relationship was established between the velocity
of the liquid-solid interface as the liquid cools, and the solid
phase which is formed. It appears that if crystalline silicon is
melted by a pulsed laser, amorphous silicon is formed from the
melt when the regrowth velocity exceeds 10-20 m/sec. For lower
velocities, epitaxial silicon regrowth occurs. The regrowth
velocities fit the predictions of a simple thermal model, and
the amorphous regrowth is thought to be a first order phase
transition from the liquid, which may accompany an undercooling
of the melt. The purpose of the current experiment is to
begin to test these ideas in the picosecond regime, by directly
investigating the dynamics of the melting and refreezing process
which follows laser heating by picosecond pulses.
High regrowth velocities result from depositing the

Mat. Res. Soc. Symp. Proc. Vol. 13 (1983) ©Elsevier Science Publishing Co., Inc.

laser energy in as short a depth and short a time as possible.
This experiment employs 15 psec laser pulses at 248 nm, where
the absorption depth is 5.5 nm in silicon crystal, and 14 nm
in the liquid. The melt is probed in transmission and reflection
simultaneously with a 1.64 μm 15 picosecond pulse produced by
transient stimulated Raman scattering of a portion of the
ultraviolet laser output. The silicon is quite transparent to
1.64 μm light; the metallic liquid, however, attenuates the
light with a skin depth related to its electrical conductivity.
Using this technique, we have been able to directly measure melt
thicknesses with 30 psec resolution.

EXPERIMENT

        The source laser for both the melting and the probe
pulse is a KrF* amplified excimer laser system [6]. The
amplifier produces 248 nm 15-20 picosecond pulses with up to 20
mJ of energy per pulse at a repetition rate of 10 Hz. The KrF*
beam is divided using a dielectric beam splitter (figure 1).
Half of the beam is used to melt the silicon surface; the other
portion goes through a variable delay, and is then
down-converted to 1.6 μm to be used as a probe of the melted
film.

Figure 1. Experimental
setup. 1,2,3: Ge photo-
diodes; 4: uv laser; 5:
optical delay line; 6:
thallium heat pipe; 7:
laser beam homogenizer;
8: 3 inch crystalline
silicon wafer.

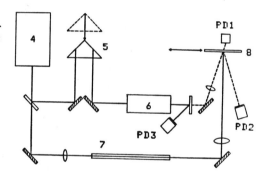

        A melting beam spot of approximately uniform intensity
is made with a simple beam homogenizer consisting of a 64 cm
long pyrex capillary with a 500 μm bore. The laser is focused
into the bore, and the output is imaged onto the <100> or <111>
surface of a polished crystalline silicon substrate, 75 mm
diameter by 0.5 mm thick, which is mechanically translated
during the measurement to insure that a new silicon surface is
exposed on every laser pulse. Substrate preparation includes
cleaning with solvents and stripping the oxide in hydrofluoric
acid when the material is first received from the supplier. No
special effort is made to keep oxide from regrowing, and the
experiments are performed in the air. The method used to
down-convert a portion of the 248 nm light to 1.64 μm for use as
a picosecond probe in this experiment is transient stimulated
electronic Raman scattering in thallium vapor. Ten mJ of 248 nm
light produces 0.1 mJ of output at 1.64 μm, which can easily be

separated from the pump light and other Raman scattering and
fluorescence in the vapor by a series of Corning glass filters,
polished silicon filters, and apertures. The probe pulse is
separated in time from the pump pulse by means of an optical
delay line placed before the Raman cell, with approximately
three nsec of total delay. After filtering, the 1.6 μm light is
focused to a 100 μm diameter spot on the surface of a silicon
wafer in the middle of the melting spot. The light is monitored
using three germanium photodiodes, PD1, 2, and 3. PD1 is placed
behind the silicon to detect transmission. A 100 μm pinhole is
used to screen other sources of scattered light. PD2 detects
light specularly reflected from the silicon surface. PD3
monitors the ir laser energy per pulse. In addition to these
three channels of detection, the energy in the melting spot is
monitored on every pulse. The normalized signal energies
(PD1/PD3) and (PD2/PD3) are signal averaged and recorded.
Reflection and transmission data have been taken as a function
of time delay between the melting pulse and the probe, for
various laser intensities, on <111> and <100> surfaces (figure
2).

Figure 2.
Transmission (+) and
reflection (o) at
1.64 μm as a
function of time for
incident uv laser
fluence of 43 mJ/cm^2.

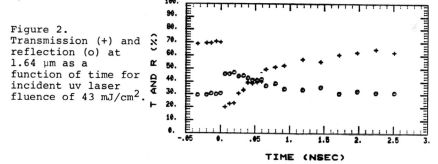

The attenuation constant α in metallic liquid silicon is
simply related to k(ω), the imaginary part of the complex
dielectric constant n(ω):

$$\alpha = 2\omega k(\omega)/c$$

In turn, n(ω) depends on the carrier density N and relaxation γ,
expressed in the complex dielectric function:

$$\varepsilon(\omega) = n(\omega)^2 = 1 - \frac{4\pi N e^2}{m\omega(\omega+i\gamma)}$$

where e and m are the electron charge and mass, respectively. γ
may be derived from the DC conductivity

$$\sigma = \frac{Ne^2}{m\gamma}$$

if N is assumed equal to four carriers per silicon atom. The
value of k(ω) thus derived is in excellent agreement with
ellipsometry measurements of n(ω) in the wavelength range
0.4-1.0 m [7]. At the probe wavelength in this experiment,
calculation yields k(ω)=8.4.
Speckle in the pump beam produces fluctuations in the
liquid depth which must be interpreted by averaging over

54

transmission in different parts of the melt spot. The thickness
of the amorphous regrowth layer was determined by Rutherford
backscattering by J.Poate and D.Jacobson. This was compared to
the minimum transmission through the melt, solving Maxwell's
equations for transmission through a metal film between two
dielectrics to find the transmission as a function of depth. The
inhomogeneity was verified by cross sectional TEM on some
samples, where amorphous depths up to 100 nm in some places were
seen. The resulting smeared transmission function, applied to
the transmission data, was used to calculate the melt thickness
as a function of time (figure 3).

Figure 3. Melt
thickness vs. time
for three values of
uv incident laser
fluence: +,87 mJ/cm^2;
o, 43 mJ/cm^2;□ , 25
mJ/cm^2.

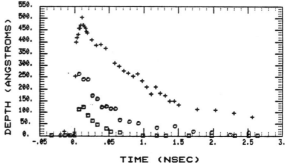

        The slope of the melting curve determines the melt front
velocity. The deepest melts reveal that melt-in occurs in about
100 psec with velocities of up to 400 m/sec or more. Melt out
velocities are also quite fast, well into the regime of
amorphous regrowth, which is expected for velocities above
15-20m/sec. Data were compiled at a number of different
intensities incident on the silicon, and the maximum melt depth
was a linear function of energy, whose slope was independent of
the orientation of the underlying crystal (figure 4).

Figure 4. Melt depth
vs. uv laser
fluence; o, <100>
crystal orienta-
tion;□,<111>
orientation.

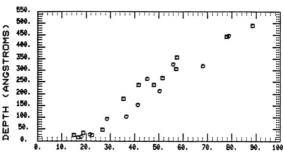

COMPARISON TO THEORY

We have compared our data to a one dimensional heat diffusion calculation of M.Thompson [8] that has been successful on nanosecond time scales. The model assumes measured values for the thermal conductivity and heat capacity of the crystal and liquid, and the latent heat of melting, and uses theoretical values for the melting temperature and latent heat of the amorphous solid (see table 1). No attempt is made to incorporate new phenomena which may become important for short pulses, such as finite electronic relaxation times of high carrier densities. Cooling through the surface by evaporation is ignored. The calculated regrowth velocity peaks at early times and for values of laser intensity close to the melting threshold, then begins falling, reaching a value of 40 m/sec after about 500 psec, roughly independent of melt depth. The data show the same general behavior, however, we find that the initial regrowth velocity is not strongly dependent on melt depth; in fact, the data show the highest initial regrowth velocity, about 60 m/sec, for the deepest melt. After about 200 picoseconds, the regrowth velocity slows to values consistently lower than the calculation predicts. One possible interpretation is that at this point, the regrowth has not kept up with heat flow, leading to undercooling of the melt. At the end of the melting process, the curves show a long tail, which may indicate that inhomogeneities in the melt beam due to speckle cause different parts of the probed area to resolidify at slightly different times, leading to smearing of the end of the melting pulse as the liquid breaks up into puddles during the last few picoseconds.

Another discrepancy between the theory and the data occurs in the melt depth vs. intensity, where the data show a deeper melt for the same laser intensity. Using the accepted values for the latent heat and crystal heat capacity we find 6.9 kJ/cm^3 must be absorbed to melt the solid starting at room temperature. The ratio obtained from figure 4, assuming a liquid surface reflection of 67% (calculated from density and conductivity data), is 4.9 kJ/cm^3. The discrepancy may be due to a depression of the ultraviolet reflectivity for the highly overheated liquid (see below).

Finally, the melt-in phase of the liquid may be compared to the thermal model. The calculation show melt-in velocities of nearly 1000 m/sec, whereas the data show slower speeds, around 400 m/sec. The discrepancy may be partially due to the finite time resolution of the laser probe. This is particularly interesting when one considers the extreme temperatures of the melt in the early stages. All of the laser energy is deposited in the first 15 psec in the first 20 nm of substrate. Assuming thermalization is very fast, and taking a constant liquid heat capacity, this produces an overheated liquid in excess of 10^4K. In the first 30 psec, the black body radiation, which peaks in the near ultraviolet, should emit several nanojoules of light. In fact, we have observed a faint white flash coming from the melt spot, and this is under further investigation.

CONCLUSION

In conclusion we have directly observed the evolution of melted silicon on bulk crystal substrates which are heated by 248 nm 15 psec laser pulses, and compared the melt-in and regrowth velocities to a simple heat diffusion model. Regrowth velocities, although faster than those observed previously, were slower than predicted. Finally, we find evidence for extreme overheating of the liquid in the early stages of melt-in.

We gratefully acknowledge enthusiastic discussion and assistance from J.M.Poate, M.Gibson, W.Brown, H.Craighead, T.Wood, C.V.Shank, R.Yen, D.Jacobson, and S.Chu of Bell Laboratories in Holmdel and Murray Hill, New Jersey, and M. Thompson at Cornell University.

REFERENCES

[1] D.H.Auston, C.M.Surko, T.N.C.Venkatesan, R.E.Slusher, and G.A.Goluvchenko, Appl. Phys. Lett. $\underline{33}$, 437 (1978).

[2] An excellent compilation of current work in this field is Laser and Electron-Beam Interactions with Solids, ed. B.R.Appleton and G.K.Celler, Materials Research Society Symposia Proceedings, vol.4, New York: North Holland, (1982).

[3] H.W.Lo and A.Campaan, Phys. Rev. Lett. $\underline{44}$, 1604, (1980); J.A. Van Vechten, Laser and Electron Beam Interactions with Solids, Op. Cit., pp.49-60.

[4] J.M.Liu, H.Kurz, and N.Bloembergen, Appl. Phys. Lett. $\underline{41}$, 643, (1982); R.Yen, J.M.Liu, H.Kurz, and N.Bloembergen, in Laser and Electron Beam Interactions with Solids, Op. Cit., pp.37-42,29-35,3-11; Also, see the very recent work of R.Yen, C.V.Shank, in this volume.

[5] A.G.Cullis, H.C.Webber, N.G.Chew, J.M.Poate, and P.Baeri, Phys. Rev. Lett. $\underline{49}$, 219 (1982). A.G.Cullis, H.C.Webber, and N.G.Chew, Laser and Electron Beam Interactions with Solids, Op. Cit., pp.131-140.

[6] P.H.Bucksbaum, J.Bokor, R.H.Storz, and J.C.White, Optics Letters, $\underline{7}$, 399 (1982).

[7] K.M.Sharev, B.A.Baum, and P.V.Gel'd, Sov. Phys. Solid State $\underline{16}$, 2111 (1975).

[8] M.O.Thompson, in this volume.

TIME RESOLVED MEASUREMENTS OF INTERFACE DYNAMICS DURING PULSED LASER MELTING
OBSERVED BY TRANSIENT CONDUCTANCE

MICHAEL O. THOMPSON AND G. J. GALVIN
Department of Material Science, Cornell University, Ithaca, NY 14853

ABSTRACT

The transient conductance technique has been used in a
detailed study of the liquid-solid interface dynamics during
pulsed laser melting of Si and silicon-on-sapphire. Average
melt and regrowth velocities, as well as the maximum melt
depth, can be obtained with the technique. The measurements
are found to agree well with a computer simulation based on
a thermal model of the melt and subsequent solidification.
The melt-in velocity has been observed to exceed 200 m/sec.
Under 2.5 ns UV irradiation, the critical velocity for
amorphization of <100> Si has been measured at 15 m/sec.

INTRODUCTION

The ability of laser irradiation to anneal damage in ion-implanted surface
layers, and recently to form new phases, has been intensely studied during the
past 5 years [1,2]. Although many of the phenomena are now well characterized,
the understanding of the fundamental mechanisms continues to cause considerable
controversy even today. During the past year, we have developed a probe based
on the electrical conductivity of molten Si [3-7]. With this new probe, it is
now possible to obtain time-resolved quantitative measurements of melt depths
and regrowth velocities. We have applied the technique to the study of melt and
regrowth in bulk Si and silicon on sapphire (SOS) using various pulse widths
with both visible and UV irradiation. The melt dynamics can be compared with
predictions based on a thermal model of the process. By studying the dynamics
during 2.5 ns irradiation in the UV, we have obtained important information on
the depths and velocities associated with the formation of the amorphous phase
directly from the melt [8,9]. Finally, these experimental observations, coupled
with a reasonable match to computer predictions, leave essentially no doubt that
melting is the primary mechanism involved in pulsed irradiation above 2 ns pulse
widths. Indeed, other current work now suggests that this upper time limit on
plasma effects is probably 3 orders of magnitude lower [10].

TECHNIQUE

A number of the differences in physical properties between crystalline and
molten Si are shown in table I. A major bonding change occurs during melting,
transforming Si from a semiconducting covalent solid to a metallic liquid.
Many of the property differences are simply a manifestation of this bonding
change. The release of the outer shell electrons to the conduction state gives
rise to large increases in the surface reflectivity, the absorption coefficient,
and the electrical conductivity. Two of these property changes have previously
been used to observe and confirm melting during laser annealing. By monitoring
the surface reflectivity, Auston et.al. [14] were able to observe the onset of
melt and measure the total duration of deeper melts in Si and GaAs during pulsed
annealing. Likewise, Lowndes et.al. [15] exploited the absorption coefficient
change using transmission measurements to obtain similar results. Although both
of these techniques provide time-resolved data, neither is able to yield

Mat. Res. Soc. Symp. Proc. Vol. 13 (1983) ©Elsevier Science Publishing Co., Inc.

58

TABLE I
Comparison of physical properties between crystalline Si and the melt. The
optical properties are for ruby radiation at 694 nm.

Phase	Bonding	Density[11]	Reflectivity[12]	Absorption[12] Coefficient	Resistivity[13] at 1685 K
Crystal	Covalent	2.33 gm/cm^3	35%	~10^4 cm^{-1}	2400 μΩ-cm
Liquid	Metallic	2.53 gm/cm^3	72%	~10^5 cm^{-1}	80 μΩ-cm

quantitative information on the volume of molten material present. However, by
monitoring the rise in electrical conductance caused by the different
resistivities of molten and crystalline Si, the melt volume can be obtained.
The technique is simple enough, requiring only a time-resolved measurement of
the electrical resistance of a thin silicon bar. Due to the large conductivity
jump, the resistance is determined primarily by the thickness of the liquid
layer on the surface, and thus is directly related to the melt depth by the
known molten resistivity. From such measurements of the molten volume, melt-in
velocities, peak melt depths and regrowth velocities can easily be obtained.

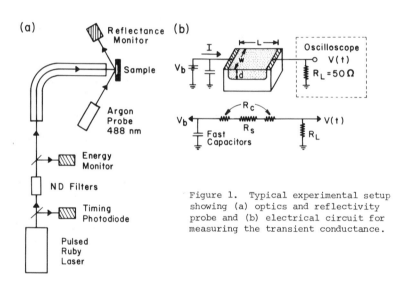

Figure 1. Typical experimental setup
showing (a) optics and reflectivity
probe and (b) electrical circuit for
measuring the transient conductance.

EXPERIMENTAL

A typical experimental setup is shown in fig. 1. Ruby lasers have been used
to provide irradiation at 694 nm with pulse widths from 2.5 to 30 ns, and, with
frequency doubling, similar pulse widths in the near UV at 347 nm. Photodiodes
monitor the pulse energy and provide consistent timing for the transient
measurements. Since this transient conductance technique probes the melt over
the entire length of a Si resistor, a quartz homogenizer [16] was essential to
provide uniform illumination and melting for these experiments. Additionally,
a CW argon probe laser at 488 nm monitored the surface reflectance during

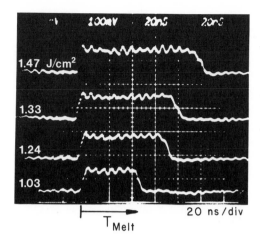

Figure 2. Transient surface reflectance from Si irradiated with a 25 ns FWHM ruby laser at 694 nm. The onset and termination of melt are clearly visible for the various energies. The reflectance probe is at 488 nm.

irradiation. A typical series of these reflectance transients on Si under 25 ns ruby irradiation is shown in fig. 2. The reflectance provides accurate markers of initiation and termination of the melt which can be correlated with the transient resistance.

Figure 1b details the sample geometry and the electrical circuit. Samples of bulk Si were prepared by cutting thin bars (L/W ~ 20) from wafers of polished Si. Additional sample resistors were photolithographically patterned (L/W from 20 to 120) on 0.5 µm silicon on sapphire (SOS). The photoconductivity of bulk Si posed a serious problem since the photo-currents induced by the laser could easily mask the conductance of the thin molten layer. However, gold doping to solid solubility at 1550 K ($\sim 10^{17}$ cm^{-3}) sufficiently reduced the photoconductive lifetime to allow observation of the resolidification. The large density of defects in SOS reduced the intrinsic photoconductive lifetime to such low values that the Au doping was unnecessary. Indium or evaporated aluminum provided low resistance contacts to the sample ends. Intentionally, the illuminated area included a portion of both contacts to assure complete and uniform melt across the resistor. With a thin uniform surface melt layer, the resistance of the sample is given approximately by $R_s = \rho(L/W)d^{-1}$, where ρ is the molten resistivity (80 µΩ-cm), L/W is the resistor aspect ratio and d is the melt thickness. This dynamic resistance was measured by a simple voltage divider as diagrammed in fig. 1b. One end of the sample was held at a fixed potential (V_b), with high frequency capacitors mounted in parallel with the power supply to provide the currents drawn during the short transient. This bias voltage is divided between the sample (R_s), any contact resistances (R_c) and a load resistor (R_L). In general, the 50 ohm termination on an oscilloscope was used as the load resistor which allowed the load voltage V(t), to be observed without ringing or reflections. These voltage traces were digitized for subsequent computer analysis. The observed voltage is related to the sample and contact resistances by $R_s + R_c = R_L \{[V_b/V(t)] - 1\}$. Due to the nonlinearity of this equation, the error associated with this measurement technique increases rapidly as the sample resistance falls below the load resistance. A large sample resistance is also desirable to minimize errors caused by the unknown contact resistances. Consequently, to accurately monitor deep melts, it is necessary to use large L/W ratios to maintain a large resistance.

RESULTS AND DISCUSSION

A series of voltage transients for bulk Si under 25 ns ruby irradiation is shown in fig. 3a. Below the melt threshold energy density near 0.8 J/cm^2, the transient is due to photocurrents only, depicted by the cross-hatched area. At high energy densities the molten conductance is obvious in the extended tail, though the photopeak unfortunately remains obscuring the initial 115 ns of melt. The peak height does, however, allow corrections due to contact resistance (R_c = 2-3 Ω typically) to be included. Adding this small correction, the voltage transients can be converted to an apparent melt depth via

$$d(t) = \rho \, \frac{L}{W} \, \{R_L [V_b/V(t) -1] - R_c\}^{-1}. \tag{1}$$

Since the non-linearity of this voltage to depth conversion would magnify small errors, we do not attempt to subtract the photopeak to obtain depth information prior to 115 ns. Beyond the 115 ns though, the remaining photopeak is adequately small and imposes little error on the converted depth. Along with the photo-current, there is an additional contribution (error) to the apparent depth due to normal electronic conduction in the hot solid behind the melt interface. Fortunately, due to the steep gradients, the exponential dependence of conductivity on temperature and the factor of 30 decrease in conductivity compared to the melt, this thermal contribution is usually less than 5% of the maximum melt depth. The long tail in the depth transient following the termination of the melt arises from this contribution however.

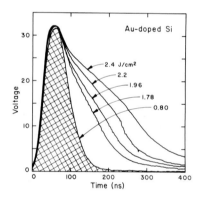 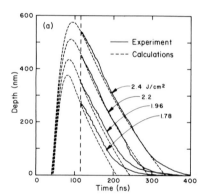

Figure 3. (a) Voltage transients for Au-doped Si under 25 ns ruby irradiation. The crosshatched curve shows the photoresponse just below the melt threshold. (b) Solid lines show the converted melt depth as a function of time. The vertical line at 110 ns is the duration of the photoconductivity. Dashed lines show computer simulations at the same energies.

The past years have also seen tremendous gain in the ability to accurately computer simulate laser melting. Thermal conductivity and specific heat are well known as functions of temperature [11,17]. Peercy and Wampler [18] measured by direct calorimetry the energy coupling of the incident laser into a sample, confirming the applicability of normal low intensity reflectivities (roughly 35% solid and 72% liquid) in this high intensity regime. These values can be included in a very simple thermal model assuming local ideal thermalization of the incident laser energy, as Baeri [19] and Wood and Giles [20] have previously described. Recently, Jellison and Modine [21] measured the temperature dependence of the absorption coefficient in bulk Si at low intensities, providing experimental values for the only remaining free parameter in the numerical simulations. Including these parameters, the threshold of 0.8 J/cm^2 for bulk Si melt under 30 ns ruby irradiation is correctly predicted. The dashed curves in fig. 3b are computer simulations for the experimental energies. The agreement is quite good during the regrowth, differing significantly only for the lowest energies. It should be stressed that such agreement is reached using equilibrium thermal and low intensity optical constants. Considering the agreement, there seems little need to invoke drastic modification of these parameters for the time scales involved.

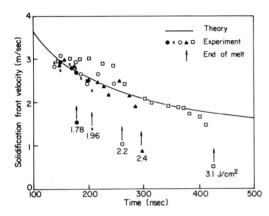

Figure 4. Comparison of a simple heat flow estimate (solid curve) of the regrowth velocity as a function of time and the numerically differentiated regrowth velocity as obtained from fig. 3b. Reprinted from Galvin et.al. [5].

The thermally dominated nature of the melt regrowth can be further demonstrated by obtaining an approximate expression for the regrowth velocity. Such an expression is useful not only for insight into the mechanism, but also as a practical means of estimating regrowth velocities. For thermal diffusion lengths ($2\sqrt{Dt}$, where D is the thermal diffusivity) long compared to the laser absorption depth, the regrowth is governed by heat diffusion only, requiring the latent heat (ΔH = 4206 J/cm^3) released at the interface to be balanced by thermal conduction into the bulk. The temperature gradient ($\partial T/\partial z$) is approximated by a linear drop from the melting point (T_m = 1685 K) to room temperature over the thermal diffusion length, with D and K (thermal conductivity) taken near the melting point ($\sim$ 0.1 cm^2/sec and 0.22 W/cm-K respectively). Figure 4 shows the comparison between this simple estimate,

$$v(t) = \frac{K}{\Delta H}\frac{\partial T}{\partial z} \sim \frac{K}{\Delta H}\frac{(T_m - 298)}{2\sqrt{Dt}} = \frac{36.3}{\sqrt{t\,(ns)}} \text{ (m/sec)}, \qquad (2)$$

and the velocity obtained by numerically differentiating the data from fig. 3b. Again, within the limits of the noise, the data agree quite well.

The bulk Si transient conductance measurements very clearly demonstrate the thermal nature of laser annealing during the regrowth. However, the initial stages of melting, which are equally interesting, are hopelessly obscured by the photocurrents. The same transient current measurements made on 0.5 μm SOS samples are able to provide information in this intriguing region. The photoconductive lifetime of SOS is only about 200 ps, which allows the entire melt signal to be observed. An oscilloscope trace showing the surface reflectance R(t) and the transient voltage V(t) is shown in fig. 5a ($\sim$110 nm

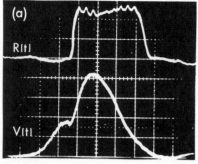
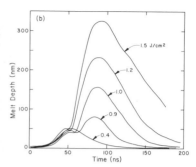

Figure 5. (a) Oscilloscope trace showing the reflectance (upper trace) and the transient voltage (lower trace) for 0.5 μm SOS under 25 ns ruby irradiation. The first peak in the transient voltage is the photocurrent response. (b) A series of the voltage transients converted to show the time-resolved melt depth.

melt). The first small peak in the voltage transient is due to the photo-induced carriers, with the melt conductivity clearly visible rising from this photopeak, allowing now measurement of the melt velocity and peak depth as well as the regrowth velocity. The onset and termination of the melt as determined by the high reflectivity phase correlates extremely well with the transient conductance. In fig. 5b, a number of SOS transients have been converted to melt depths. Again, the first peak is only an apparent melt depth due to photocurrents. The melt depth can be verified on SOS by completely melting the Si layer and observing the saturation resistance. The thickness calculated from the transient conductance by this technique agrees with other measurements (stylus, RBS) within about 10%. The melt-in velocity was measured and scales linearly with energy from 5 m/sec near threshold (0.65 J/cm^2) to above 13 m/sec at melt-through. Above threshold the peak melt depth increased linearly with energy until the entire Si film was molten at 1.5 J/cm^2. average regrowth velocity remained relatively constant, ranging from 3.3 m/sec to about 2.8 m/sec near melt-through. Unfortunately, under ruby irradiation the threshold varied among samples, ranging from 0.5 J/cm^2 to 1.0 J/cm^2 for 25 ns irradiation, apparently depending on thickness, preparation, and supplier. Much of the variation is caused by the defect structure which strongly modifies the absorption coefficient in the Si film near the interface. Though the computer simulations properly predict the melt characteristics, we have been unable to accurately match the energy due to these variations.

It is interesting also to probe the reflectance of the rear surface of the SOS during irradiation at an energy great enough to completely melt the Si layer. Figure 6 is an oscilloscope trace taken with the argon reflectivity probe laser focused on both the front and interface surfaces. Melt-through was obtained with ruby irradiation at 25 ns FWHM and 1.6 J/cm^2. The upper trace is the surface reflectance showing a melt of approximately 270 ns duration. This surface reflectance lacks the characteristic sharp falling edge due to surface damage of the Si film at the energy density used. The lower trace shows the Si-sapphire interface reflectance with the melt present for nearly 65 ns (T_L). The increase of carriers in the Si causes the initial jump in the interface reflectance and marks the beginning of the laser pulse. The time delay between the onset of melt reflectance at the surface and interface gives an estimate for the melt velocity. This value, 500nm/40ns = 13 m/sec, is in good agreement with velocities obtained from transient conductance. A rough estimate of the regrowth velocities can also be obtained, yielding 3.3 m/sec (~180 ns). Though

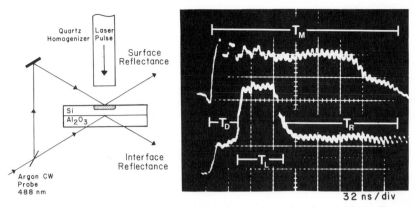

Figure 6. Oscilloscope trace of surface (upper trace) and interface (lower trace) reflectance on SOS during irradiation at 1.6 J/cm^2, sufficient to completely melt the 0.51 μm Si layer.

this value is about 15% high, it is still reasonable considering the uncertainty in determining the melt termination.

The ability to observe melts within the laser pulse on SOS allowed the technique to probe the extremely high velocities associated with short pulse irradiation in the UV at 347 nm. In particular, the critical velocity for the formation of the amorphous phase from the melt has now been directly observed. Figure 7 shows a series of transients for 2.5 ns UV irradiation on SOS, converted to melt depth as a function of time. Curve 1 shows the photoconductive response just below melt threshold. Despite the short photoconductive lifetime of the SOS, the photocurrents become important again under these conditions since they scale with intensity and thus inversely with the pulse width. Consequently, the initial 4 ns of melt data (roughly the laser pulse) is not shown. The time resolution of the transient conductance technique is well demonstrated in this figure, limited mainly by laser produced electrical noise and oscilloscope bandwidth. The maximum melt depth is plotted for various pulse widths in fig. 8. This depth is observed to rise linearly with power

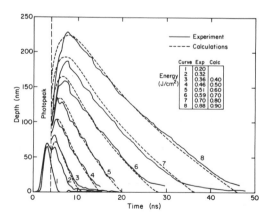

Figure 7. Melt depth curves (solid lines) for SOS irradiated with 2.5 ns UV irradiation. The vertical line at 4 ns is the duration of the photoconductivity. The dashed lines are computer simulations for similar melt depths.

above threshold, as predicted by the simple thermal model. Further, the absorption depth (few tens of nanometers) of the UV radiation deposits the incident energy in the surface region only; hence the energy coupling problem associated with irradiation at the ruby fundamental is avoided. Consequently, it is possible to numerically simulate the melt, using the simple thermal model with reflectivities of 54% solid and 72% liquid. These simulations are shown in fig. 7 by the dashed curves, with the energies chosen to provide a reasonable match to the experimental melt depth. Although there is some difficulty with the energy calibration, the match of melt depth curves with the computer simulations is excellent. Below 0.3 J/cm^2, computer simulations are not attempted as the melt interface breaks down and amorphous rather than crystalline Si forms.

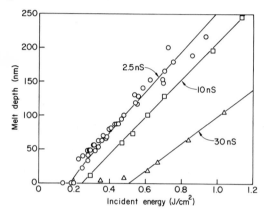

Figure 8. Maximum melt depth determined from the maxima in the transient conductance trace for various UV irradiation pulse widths.

It is clearly visible from fig. 7 that the regrowth velocity drops dramatically with increasing energy. In fig. 9, the average regrowth velocity is plotted against the melt depth for the 2.5 ns irradiation. The use of melt depth rather than incident energy as the abscissa provides a more accurate measure of deposited energy, and thus is less sensitive to changing reflectivities and calibrations. The data are plotted with open circles for energies above the critical amorphization energy, and solid circles for those involving the regrowth of amorphous material. It has been postulated that the amorphous phase nucleates above a critical velocity which corresponds to a melt undercooling below the equilibrium melting point of the amorphous phase [22]. This critical velocity for 2.5 ns UV irradiation on (100) Si is observed to be 15 m/sec, with an estimated error of ±10%. It is also important to correlate this phenomena with TEM measurements [23] to obtain further information on the kinetics of the amorphous solidification. Although the amorphous layer shows a maximum thickness of 14 nm, the conductivity shows considerably more Si (40-50 nm) to be molten. This behavior strongly suggests that formation of the amorphous phase is nucleation limited. The observed undulating nature of the amorphous-crystalline interface may suggest that nucleation occurs by interface breakup, and spreads laterally as the amorphous phase is nucleated. The small islands of amorphous material observed near the lower threshold may be the initial stage of this breakdown. Alternately, it has been proposed that these phenomena may be explained by considering the time required to undercool the liquid to the amorphous melting temperature. In this case, the undulations are caused by minor variations in the undercooling time.

Considerable study has been made of the regrowth phase of laser melting, leaving the equally fundamental subject [24] of the rapid melt-in velocity

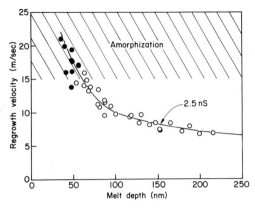

Figure 9. Regrowth velocity as a function of peak melt depth for the 2.5 ns UV irradiation. The solid circles represent energies forming amorphous material, with open circles for crystalline regrowth. The hatched area above 15 m/sec is the velocity range for formation of the amorphous phase from the melt.

largely neglected. The use of SOS permits direct observation of this advancing melt-solid interface as well. The highest melt-in velocities were observed with 2.5 ns fundamental ruby radiation. Along with the high powers available in the fundamental, the longer absorption length at 694 nm creates shallow temperature gradients near the surface, effectively further increasing the power available from the laser to the advancing interface. Figure 10 shows an oscilloscope trace for 1.07 J/cm^2 incident on 0.51 μm SOS, an energy density sufficient to completely melt the Si layer. The peak of the photoconductive response occurs near the onset of melt, since the molten layer absorbs the energy and prevents the carrier-production in the crystal. The long, flat region corresponds to the conductance of a completely molten Si layer. At the time scale shown, no appreciable regrowth can be seen. The rise time from the photopeak indicates an average melt velocity in excess of 200 m/sec. This velocity can be compared to estimates based on the thermal model. In the simple model, the temperature gradients available for heat transport in the liquid (to the interface) are limited by vaporization at the surface, effectively clamping the surface near the boiling point of Si (3540 K). Balancing this gradient with the latent heat requirement at the interface, the surface temperature limit implies a saturation melt velocity given by $K(\partial T/\partial z)/\Delta H$, where K is the thermal conductivity in the

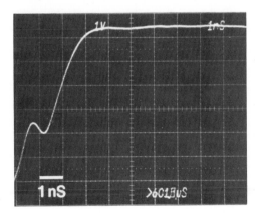

Figure 10. Transient conductance trace of 0.5 μm SOS during irradiation with a 2.5 ns pulse at 694 nm. The flat region corresponds to the conductance of a completely molten layer 0.5 μm in depth. The onset of melt occurs near the peak of the photoconductive response. The average melt-in velocity, 200 m/sec, is given by the rise time, 2.5 ns.

66

melt. The measurement of an average 200 m/sec requires a minimum thermal conductivity of 110 W/m-K, compared with literature values ranging from 70 to 210 W/m-K [17].

CONCLUSIONS

The transient conductance provides a very simple yet quantitative technique for time-resolved measurements of the volume of molten Si during pulsed laser melting. The detailed dynamics of the melt-solid interface motion have been observed for pulsed irradiation from 2.5 to 30 ns pulse widths. The use of SOS provides a dramatic reduction in the photoconductive response allowing melt-in as well as regrowth to be observed. Computer simulations based on a simple thermal model have been shown to match well with the experimental data in bulk Si and for UV irradiation in SOS. Average melt-in velocities above 200 m/sec have been observed for 2.5 ns ruby irradiation. The critical velocity for amorphization of Si with 2.5 ns UV irradiation has been observed at 15 m/sec. As the technique is refined, it will provide important data for further studies of impurity segregation and phase transitions.

ACKNOWLEDGEMENTS

We wish to acknowledge the many collaborators in this work: J. W. Mayer at Cornell; P. Peercy at Sandia; R. Hammond at Los Alamos; A. Cullis, H. Webber, and N. Chew at RSRE; and J. Poate and D. Jacobson at Bell Labs. We also acknowledge K. Carey at Hewlett Packard and O. Marsh at Hughes for supplying SOS used in these experiments. One of us (MOT) gratefully acknowledges support from a National Science Foundation graduate fellowship. The work at Cornell has been funded in part by the National Science Foundation (L. Toth).

REFERENCES

1. See the past proceedings of this conference.

2. Laser Annealing of Semiconductors, J. M. Poate, J. W. Mayer, Eds. (Academic Press, New York, 1982).

3. G. J. Galvin, M. O. Thompson, J. W. Mayer, R. B. Hammond, N. Paulter and P. S. Peercy, Phys. Rev. Lett. 48, 33 (1982).

4. M. O. Thompson, G. J. Galvin, J. W. Mayer, R. B. Hammond, N. Paulter and P. S. Peercy, in Laser and Electron-Beam Interactions with Solids, B. R. Appleton, G. K. Celler, Eds. (North Holland, New York, 1982), pp. 207-214.

5. G. J. Galvin, Michael O. Thompson, J. W. Mayer, P. S. Peercy, R. B. Hammond and N. Paulter, to be published Phys. Rev. B. (Jan. 1983).

6. Michael O. Thompson, G. J. Galvin, J. W. Mayer, P. S. Peercy and R. B. Hammond, accepted Appl. Phys. Lett.

7. Michael O. Thompson, J. W. Mayer, A. G. Cullis, H. C. Webber, N. G. Chew, J. M. Poate and D. C. Jacobson, submitted Phys. Rev. Lett.

8. P. L. Liu, R. Yen, N. Bloembergen and R. T. Hodgson, Appl. Phys. Lett. 34, 864 (1979).

9. A. G. Cullis, H. C. Webber, N. G. Chew, J. M. Poate and P. Baeri, Phys. Rev. Lett. 49, 219 (1982).

10. C. V. Shank, R. T. Yen and C. Hirlimann in these proceedings.

11. CRC Handbook of Chemistry and Physics, (CRC Press, Boca Raton, 1980).

12. P. Baeri in Laser Annealing of Semiconductors, J. M. Poate, J. W. Mayer, Eds. (Academic Press, New York, 1982), chapter 4.

13. Liquid Semiconductors, V. M. Glazov, S. N. Chizhevskaya, and N. N. Glagoleva, Eds. (Plenum Press, New York, 1969).

14. D. H. Auston, C. M. Surko, T. N. C. Venkatesan, R. E. Slusher and J. A. Golovchenko, Appl. Phys. Lett. 33, 437 (1978).

15. Douglas H. Lowndes, G. E. Jellison and R. F. Wood, in Laser and Electron-Beam Interactions with Solids, B. R. Appleton, G. K. Celler, Eds. (North Holland, New York, 1982), pp. 73-78.

16. A. G. Cullis, H. C. Webber and P. Bailey, J. Phys. E. 12, 688 (1979).

17. Thermal Conductivity of the Elements: A Comprehensive Review, C. Y. Ho, R. W. Powell, P. E. Liley, J. Phys. Chem. Ref. Data 3 Supp. 1, I-589 (1972).

18. P. S. Peercy and W. R. Wampler, Appl. Phys. Lett. 40, 768 (1982).

19. P. Baeri, S. U. Campisano, G. Foti and E. Rimini, J. Appl. Phys. 50, 788 (1979).

20. R. F. Wood and G. E. Giles, Phys. Rev. B 23, 2923 (1981).

21. G. E. Jellison and F. A. Modine, Appl. Phys. Lett. 41, 180 (1982).

22. F. Spaepen and D. Turnbull, in Laser Annealing of Semiconductors, J. M. Poate, J. W. Mayer, Eds. (Academic Press, New York, 1982), chapter 2.

23. A. G. Cullis, H. C. Webber and N. G. Chew, in Laser and Electron-Beam Interactions with Solids, B. R. Appleton, G. K. Celler, Eds. (North Holland, New York, 1982), pp. 131-140.

24. B. Chalmers, Principles of Solidification, (Krieger Publishing Company, Huntington, New York, 1964), pp. 84-86.

TRANSIENT CONDUCTIVITY MEASUREMENTS IN PULSED ION BEAM MELTED SILICON

R.M.FASTOW, J.GYULAI*, J.W.MAYER   DEPARTMENT OF MATERIALS SCIENCE AND
ENGINEERING, Cornell University, Ithaca, N.Y. 14853

ABSTRACT

A pulsed proton beam, ~200 ns in duration, has been used to melt and regrow single crystal silicon. The protons had an energy of 300 kev, yielding a measured energy density of 0.8-2.0 J/cm^2. The method of transient conductivity has been used to determine the melt depths, melt durations, and regrowth velocities. The measured values for 2.0 J/cm^2 were, respectively, 1.7 μm, 2 μsec, and 1.4 m/sec.

Computer generated melt curves were compared to experiment with good agreement. The energy required to initiate melt was determined, and a linear dependence of melt depth with energy has been observed.

INTRODUCTION

Recently it has been shown that transient conductivity measurements could be used to measure melt depth and regrowth velocities in laser melted silicon [4,8]. The idea underlying this technique was that the electrical conductivity of liquid silicon is 30 times greater (80 μohm-cm) than that of solid silicon at the melt temperature. By measuring the change in conductivity, one could thus deduce the melt depth. Typical laser pulse durations studied so far have been from 2-30 ns, with melt depths up to 4,000 A^o, and regrowth velocities of 2.7 m/sec. It has also been shown, by comparison with computer simulations, that on this time scale melting is a purely thermal phenomena. In this paper the method of transient conductivity has been applied to measure the melt depth and regrowth velocity of pulsed ion beam melted silicon. Again, our data agrees with a purely thermal melt model. However, the time scales and melt depths are all an order of magnitude larger than those in the laser annealed case. This is consistent with the deep penetration depth of protons at 300 kev (3 μm), and the longer pulse duration (200 ns vs. 30 ns).

The purpose of this experiment was not only to confirm the thermal melting model, but also to investigate the potential of using an ion beam for annealing and regrowth. The excellent coupling of energetic ions to any material makes pulsed ion beam annealing (PIBA) ideal for highly reflective surfaces. For example, there is no change in stopping power between solid and liquid silicon. Metallic substrates are also obvious candidates for ion

*permanent address:Central Res. Inst. of Physics H-1525 Budapest

Mat. Res. Soc. Symp. Proc. Vol. 13 (1983) ©Elsevier Science Publishing Co., Inc.

beam annealing, both because of the direct coupling, and because of quench rates of 10^8 deg(C)/sec. Recent work has also been done [7] in using the pulsed ion beam to anneal metal contacts on silicon. It has been observed that by carefully choosing the energy, one can melt only the interface between a thick metal (up to 2,000 A°) and the substrate [2]. Experiments have also been carried out which measure the diffusion of As and Ge in ion annealed Si and SOS [3,6]. Because of the excellent regrowth characteristics, with the implanted species going substitutional, it would be very useful to know the melt duration and regrowth velocity.

EXPERIMENTAL

A Marx bank generator was used to apply high voltages across a diode, which accelerated protons. The Marx bank consisted of ten 0.16 uF capacitors, with an upper series voltage of 500 kev. The accelerating diode could be operated between 200-450 kev, and incident energy densities of up to 3 J/cm^2 were obtainable. Three different ion species have been used: hydrogen, boron, and barium. Consequently the penetration range could be varied from 3 to .15 um. The diode was run in a vacuum of 5×10^{-5} torr. The incident beam energy density was set both by the Marx bank voltage, and by varying the distance from the sample to the diode. A beam divergence of approximately 2 degrees has been measured.

In the transient conductivity experiment, bulk <100> Si was used for sample preparation. To shorten the carrier lifetime in the silicon, gold was evaporated on the wafer and annealed at 1,300 °C. Indium contacts 2mm wide x 1 cm long were then soldered onto the Si. The contact resistance, as estimated from the saturation level in the conductivity measurements, was 1.5 ohms. To achieve high length/width ratios, a thin slit made from two glass cover slips was placed between the contacts. The length of the slit was 10 mm, and the width was varied from .165 mm to 1.5 mm. The data presented below was taken with the .165 mm slit, yielding a L/W ratio of 61. There were difficulties in using smaller slits because of charging effects which distorted the conductivity transients. To minimize electrical noise and sample charging, the silicon and the cover slips were enclosed in an aluminum box. A small hole was drilled in the front to allow the beam to enter.

During firing, the Marx bank voltage was monitored with an oscilliscope. The total beam energy density was determined by measuring the temperature rise of a 2.3 mil thick nickel film. The film, measuring 5mm x 5mm was placed outside the aluminum box, at a distance of 5mm from the sample. Thin chromel-alumel wires, which were spot welded to the back, supported the film. Temperature rise was measured with an X-Y recorder, and the cooling time constant was 30 sec. The accuracy of this method is estimated at ± 5 percent, primarily due to the difficulties in measuring the thickness of the film. Small beam inhomogenieties, curvature of the film, and the added mass of the chromel-alumel wires also added to the uncertainty. The ion beam dose was calculated from the energy and voltage data to be 5×10^{13} H/cm^2. The current density, during the pulse, had an average value of

30 amps/cm^2.

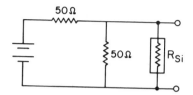

50 Ω

50 Ω   R$_{Si}$

Figure 1. Equivalent circuit used for transient conductivity measurements. Bias voltage was 45 V.

The equivalent circuit used in measuring the change in conductivity is shown above. The voltage was measured by a differential amplifier across the sample. Ceramic capacitors were placed across the battery for quick response. The 50 ohm resistor parallel to the sample was required for impedance matching the cables, and to prevent charging of the sample. At a 1 μm melt depth with l/w=60, the resistance of the silicon will drop to 48 ohms. Consequently, a large portion of the voltage trace will be linear.

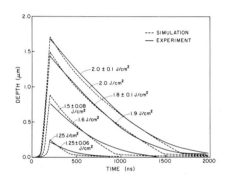

Figure 2. Melt depth vs. time, for various incident energies.

RESULTS

Figure 2 shows plots of melt depth vs. time. In each experimental plot, the photopeak was subtracted from the voltage signal. It was found that the photopeak was constant over the power range of 0.7 to 0.95 J/cm^2. The photo signal lasted 600 ns, which was equal to the entire length of the diode discharge. The computer simulation, used for comparison, was a straightfoward one dimensional thermal conduction calculation. The temperature dependance of the thermal diffusivity was included, and the ion beam was simulated as a 200 ns monoenergetic constant power pulse. The appropriate energy loss cross sections for H into Si were included [1]. There is excellent agreement between simulation and experiment for the highest energy densities. At lower energy densities the agreement is only fair. This deviation is probably an effect of the beam shape, the spread in energy of the hydrogen atoms, and the presence of carbon impurities in the beam.

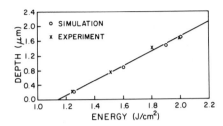

Figure 3.Maximum Si melt
depth vs. incident energy.

    The graph above shows maximum melt depth vs. incident energy density,
both experimentally and theoretically. Extrapolating this curve to low
powers yields a melt threshold of 1.1 J/cm^2.

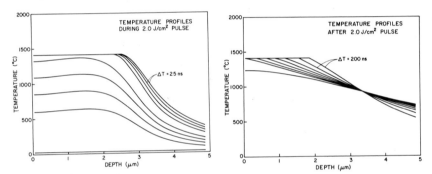

Figure 4. Computer simulation of substrate temperature profiles
(a) During the pulse. (b) After the pulse.

The linear relation between energy density and maximum melt depth can be
explained by examining the temperature vs. depth profiles. If one considers
that the first part of the beam heats the silicon to melting temperature, and
the last part supplies the latent heat, then the amount of silicon melted
should be proportional to the incident energy density. Experimental results
and computer simulations put the proportionality factor at .5 J/cm^2/μm for
bulk silicon. This is close to the latent heat of .42J/cm^2/μm. The
difference is accounted for by thermal diffusion into the bulk.

    Computer simulations of the temperature profiles during heating and
cooling (shown above) show that, indeed, during the first 100 ns of the
pulse, the Si was brought up to near melting temperature. The second half of
the pulse provided the energy for melting. For energies above 2.6 J/cm^2,
where the melt depth is deeper than the proton range, more energy would be
needed to heat the substrate. According to computer simulations, in this

region the depth vs. energy curve will become non-linear.

In pulsed laser beam melting, the regrowth velocity is limited by the rate at which heat is removed from the molten silicon [5]. In the case of a ruby laser, where the absorption depth is 1.0 µm, very steep thermal gradients are set up. Thus the regrowth velocity can be approximated by solving the one dimensional thermal diffusion equation of a point source. Galvin, et al. have done this with the results that

$$v = \frac{1.15 \times 10^{-3}}{\sqrt{t}} \quad m/sec \qquad (1)$$

where t is in seconds

This equation adequately describes nanosecond pulse laser annealing, for times much greater than the laser pulse duration. For example, it predicts a regrowth velocity of 2.6 m/sec for 200 ns after the pulse.

In pulsed ion beam melting, the regrowth velocities are slower than those observed with pulsed laser melting. The slower velocities are associated with gradual thermal gradients, produced by the relatively long pulse lengths and deep penetration. Figure #5 shows the regrowth velocity vs. time for a pulse of energy density 2.0 J/cm^2. The maximum regrowth rate is about 1.4 m/sec, and the regrowth velocity drops to about one quarter that value in about 1600 ns.

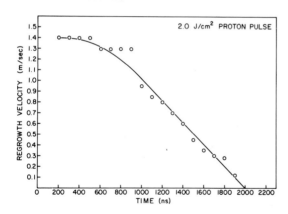

Figure 5. Regrowth velocity vs. time. Incident energy was 2.0 J/cm^2. Data was taken from figure #1.

SUMMARY

We see then, that the data obtained from transient conductivity measurements is in good agreement with computer simulations of pulsed ion beam annealing. The melt depth is proportional to incident energy density, the proportionality factor being $.5J/cm^2/\mu m$ for bulk silicon. The regrowth velocity is almost a factor of 2 slower than for laser melting. By varying the incident ions, it should be possible to increase the regrowth velocity to that obtained with laser annealing. Interesting questions remain, however, concerning how the melt is initiated. Does the melt always begin at the surface, as one dimensional computer simulations predict, are there tubes of hot liquid which form along the ion tracks, or is the interior of the silicon superheating?

ACKNOWLEDGEMENTS

We wish to acknowledge David Hammer and the Cornell University Plasma Physics Lab for the use of their pulsed ion beam. Thanks also go to Mike Thompson for the use or his computer simulations. This work was supported by the NSF (L.Toth).

REFERENCES

1. H.H. Anderson, J.F. Ziegler,
   Hydrogen Stopping Power and Ranges in All Elements. vol 3 (Pergamon Press, New York 1977).

2. J. Baglin and R. Hodgson,
   Proc. 5th Int. Conf. on Ion Beam Annealing (Sydney, 1981)

3. J.Baglin,
   Laser-Solid Interactions and Transient Processing of Materials (Elsevier Science Publishing Co., Inc., N.Y. 1983)

4. G. Galvin, M.Thompson, and J.W. Mayer,
   Physical Review Letters 48, 1,33 (1982)

5. G.Galvin, M.Thompson, and J.W. Mayer,
   Physical Review B. (To be published)

6. J.Gyulai, R.Fastow, and K.L.Kavanagh,
   Laser-Solid Interactons and Transient Processing of Materials (Elesvier Science Publishing Co., Inc., N.Y. 1983)

7. C.Palmstrom, R.Fastow, and J.W. Mayer,
   Laser-Solid Interactions and Transient Thermal Processing of Materials (Elsevier Science Publishing Co., Inc., N.Y. 1983)

8. M.Thompson, G. Galvin,and J.W. Mayer,
   Laser and Electron-Beam Interactions with Solids (Elsevier Science Publishing Co., Inc., N.Y. 1982),pp.209-214.

ULTRA-HIGH SPEED SOLIDIFICATION AND CRYSTAL GROWTH IN TRANSIENTLY MOLTEN SEMI-
CONDUCTOR LAYERS

A.G. CULLIS
Royal Signals and Radar Establishment,
Malvern, Worcs. WR14 3PS, England

ABSTRACT

The use of Q-switched laser melting techniques to investigate
new rapid solidification phenomena is described. It has been
found that Si, Ge, GaP and GaAs can give rise to orientation-
dependent, kinetically-controlled defect generation processes
during fast recrystallization from the melt. Indeed, these
materials yield amorphous phases at sufficiently high solidification rates. Ultra-fast pulsed melting permits the study of
the basic thermodynamic properties of amorphous solids. It is
shown that amorphous Si melts to give a normal, low viscosity,
undercooled liquid and that novel explosive crystal growth
processes can occur in this low temperature regime.

INTRODUCTION

Radiation pulses from Q-switched and mode-locked lasers can be used to transiently melt semiconductor surfaces for such short times that new, high speed
solidification phenomena are amenable to study. The velocity of the liquid-
solid interface during solidification can be at least as high as 20m/sec: this
is about five orders of magnitude faster than, for example, conventional crystal
growth rates. Work which employs these extreme conditions is exploring limiting
behavior during crystal growth and is demonstrating the way in which growth
breakdown can occur. The present paper will focus on the high speed solidification of elemental Si and will outline the crystal growth and amorphization
regimes. Direct comparisons will be made with the behavior of other semiconductors; namely Ge, GaP and GaAs. In addition, important features of the
amorphous-liquid phase transition in Si will be described.

EXPERIMENTAL

Samples used in the present experiments were mostly single crystals of Si, Ge,
GaP and GaAs in either (001) or (111) orientation. Some of the Si samples were
initially implanted with high doses of either As^+ or In^+ ions. Transient surface
melting was carried out with spatially uniform [1] 2.5nsec radiation pulses
obtained from a pulse-chopped, ruby laser system. The radiation wavelength was
that of either the fundamental (694nm) or second harmonic (347nm) and the sample
background temperature was either 293K or 77K. The structure of the irradiated
materials was studied in the transmission electron microscope (TEM). Samples
were thinned to electron transparency by sequential mechanical polishing and
low voltage Ar^+ ion machining. Emphasis was placed on cross-sectional studies
in order to demonstrate the stratified structure changes which occurred.

AMORPHIZATION AND DEFECT GENERATION

When resolidification of laser-melted Si occurs with interface velocities
of < 5m/sec crystalline material generally reforms with high perfection if the
substrate is defect-free. However, with higher interface velocities, orientation-

dependent thresholds are found beyond which atomic reordering cannot occur. Under the most extreme conditions resolidification gives an amorphous phase. This was first demonstrated in the work of the Harvard [2] and IBM [3] groups by use, respectively, of picosecond and nanosecond UV pulses. We have recently quantified the conditions required for this transformation [4] in studies which employed large-area, uniform, 2.5nsec UV laser pulses. For (001) Si, Figs. 1a and b show that a very thin ($\sim$100Å) amorphous surface layer is produced just above the threshold for melting ($\sim$0.15J/cm^2), the thickness of the layer initially increasing as the radiation energy density is raised. However, amorphization abruptly ceases to occur above $\sim$0.3J/cm^2, beyond which excellent single crystal Si reforms (see Fig. 1c). For (111) Si amorphous layer formation continues from just above the threshold for surface melting up to $\sim$0.55J/cm^2 (see Figs. 1d,e,f). Above the latter threshold, resolidification gives crystalline material but, unlike the case of (001) Si, the solid formed on the (111) orientation is highly defective and contains a high density of inclined microtwin lamallae and stacking faults, as shown in Fig. 1g.

One of the most important factors determining the solidification behavior is the speed with which it occurs. This is illustrated in Fig. 2 which shows the computed variation in the velocity of the 1685K (normal melting temperature) isotherm as a function of irradiation energy density with the assumption of negligible melt undercooling [4]. It is immediately clear that the energy density thresholds which were identified previously correspond to thresholds in the resolidification interface velocity. In particular, higher velocities are needed for the formation of an amorphous solid on (001) Si ($\gtrsim$18m/sec) than on (111) Si ($\gtrsim$15m/sec). While crystalline solid is formed below these thresholds,

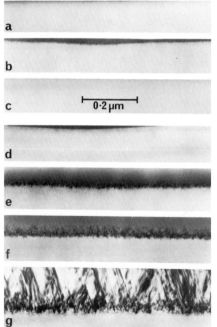

Fig. 1. Cross-sectional TEM images of Si layers irradiated with 2.5nsec UV pulses. Si (001): (a) 0.20, (b) 0.27 and (c) 0.40J/cm^2; Si (111): (d) 0.20, (e) 0.5, (f) 0.55 and (g) 0.9J/cm^2. (After Cullis et al [4].)

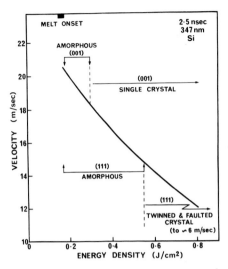

Fig. 2. Computed peak velocity of 1685K isotherm as a function of pulse energy density showing surface melting threshold and orientation-dependent amorphization and defect formation regimes. (After Cullis et al [4].)

profuse defect nucleation occurs specifically on (111) Si for resolidification interface velocities down to about 6m/sec (achieved in these experiments with 16nsec UV pulses to avoid possible surface damage with energetic shorter pulses).

It is especially important to compare the velocity predictions of Fig. 2 with the velocities that actually occur during the laser-melting process. Recently, the development of a new transient conductance technique [5] has enabled the velocity of the liquid-solid interface to be directly measured. Typical results

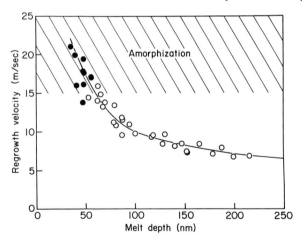

Fig. 3. Measured interface resolidification velocity as a function of maximum melt depth for (001) Si irradiated with 2.5nsec UV laser pulses. (After Thompson [6].)

for 2.5nsec laser pulses incident on (001) Si are given by the curve in Fig. 3
[6]. The threshold interface velocity for amorphization is seen to be ~15m/sec.
This is close to the computed value for (001) Si and gives confidence that the
fundamental aspects of the laser-melting model are correct. The fact that the
measured interface velocity for (001) Si is ~15% lower than its computed counter-
part may be quite significant. If the laser-induced melt became substantially
undercooled before the nucleation of amorphous Si the reduced thermal gradients
would depress the resolidification velocity in just this manner. Further
evidence for melt undercooling during high velocity solidification will be given
below.

Up to this time, most work has been directed towards elucidating the solidification behavior of elemental Si. We have recently extended our investigations
to other semiconducting materials [7]. Figure 4 shows (111) Ge samples irradiated (at 77K) with 2.5nsec UV pulses. At 0.23J/cm^2 transient melting leads to
the formation of a thin, undulating amorphous Ge layer at the crystal surface.
However, when the resolidification velocity was decreased by increasing the
irradiation energy density to 0.4J/cm^2 amorphous Ge ceased to form and, instead,
a dense array of inclined microtwins was produced. This change from an amorphization regime to a defect generation regime is precisely the same behavior as
that exhibited by (111) Si at a critical threshold velocity. In fact, calculations [7] indicate that the (111) Ge threshold may be a few meters per second
higher than that of (111) Si.

A further extension of the work to rapid quenching of compound semiconductors
has also been carried out [7]. Results for GaP and GaAs are shown in Fig. 5.
In both cases, amorphous layers could be produced on crystalline (111) substrates
at sufficiently high resolidification rates (Figs. 5a and c). For both materials,
a reduction in the solidification interface velocity yielded a regime of defect
nucleation and growth (Figs. 5b and d). It is also interesting to note that the
analogy with Si is essentially complete since the broad defect generation regime
did not exist for any of these materials (Ge, GaP and GaAs) with the (001)
crystal orientation. As noted previously [4], this orientation dependence is
likely to be kinetically controlled since simple rotational twin formation is
prevalent only on (111) crystal surfaces. Such errors in epitaxy would be quickly compounded when adjacent twin nuclei touch, due to twin boundary mismatch.

It is of considerable interest to determine the nature of laser-quenched
amorphous semiconductors. Indeed, for Si the amorphous phase has structural

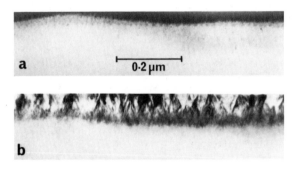

Fig. 4. Cross-sectional transmission electron images of (111) Ge laser
annealed at 77K: (a) 0.23J/cm^2 and (b) 0.4J/cm^2. (After Cullis et al [7].)

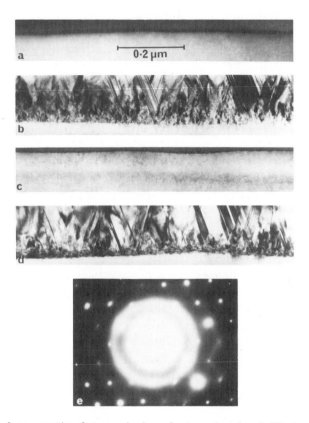

Fig. 5. Cross-sectional transmission electron images of 77K laser annealed
(111) GaP: (a) 0.27J/cm^2, (b) 0.4J/cm^2; and (111) GaAs: (c) 0.14J/cm^2,
(d) 0.4J/cm^2. (e) Transmission electron diffraction pattern from amorphous
GaP layer showing diffuse ring pattern - spots were produced by underlying
matrix crystal. (After Cullis et al [7].)

properties which are very similar to those of normal ion-implantation-produced
material [4]. Also, the laser-quenched amorphous phases of Ge, GaP and GaAs
appear to have the expected disordered structures [7]. For example, Fig. 5e
shows an electron diffraction pattern from the quenched amorphous GaP: this
pattern exhibits diffuse rings which indicate that structural order on a scale
of ≳10Å does not exist.

MELTING AND RESOLIDIFICATION OF AMORPHOUS Si

There has been a great deal of speculation about the thermodynamic properties
of amorphous Si. In particular, it has been proposed [8,9] that amorphous Si
melts by a first order phase transition to give normal (metallic) liquid Si at a
temperature T_{a-l} substantially below the crystal melting temperature T_{c-l}. The

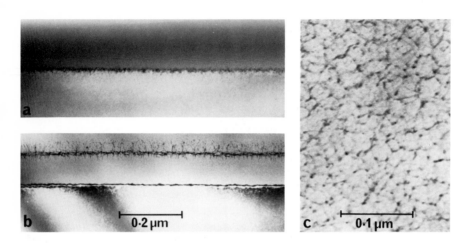

Fig. 6. Transmission electron images of In⁺ ion implanted Si: (a) cross-section of as-implanted amorphous layer; (b) cross-section after 77K laser annealing - note In segregation band in amorphous layer; (c) plan-view showing cellular In segregation structure. (After Cullis et al [13].)

value of T_{a-l} has been experimentally estimated [10,11] to lie in the range 1150K to 1450K, although the magnitude of $T_{c-l} - T_{a-l}$ is still in dispute [12]. Indeed, many features of the high speed melting and regrowth phenomena shown by amorphous Si have remained unclear. Therefore, the present experiments were designed to give a more detailed understanding of the physical processes involved.

With a low energy density 2.5nsec laser pulse it is possible to melt a pre-existing amorphous Si layer such that it resolidifies again in the amorphous state [13,14]. To determine that melting has actually taken place it is convenient to observe the behavior of an impurity previously introduced into the amorphous material. An example is shown in Fig. 6 where an amorphous Si layer formed by In⁺ ion implantation (Fig. 6a) has an initially uniform damage structure, the impurity being atomically dispersed throughout it. However, after laser melting down to the region of the crystal interface, the implanted In segregates into a narrow band within the reformed amorphous layer. A solidification sequence based on growth from both internal and external interfaces accounts for the final In distribution [13] which is shown in Fig. 6b. Of course, since In is a low solubility impurity one might expect the resolidification interface to become unstable due to the occurrence of constitutional supercooling in the transient melt [15]. In fact, this clearly occurs since, as shown in Fig. 6c, the In segregation band exhibits the characteristic lateral cellular pattern within the amorphous Si. This is thought to be the first observation of impurity segregation during the direct solidification of an amorphous solid. The associated impurity diffusion behavior (including cell formation) indicates that amorphous Si melts to give a normal, low viscosity liquid so that the process is not, for example, a glass transition [13].

The initial liquid formed by melting amorphous Si is also believed to be substantially undercooled. Evidence for this has, in part, been obtained by observations of the recrystallization behavior of amorphous layers when these are annealed with 30nsec laser pulses [16]. However, new evidence has also emerged

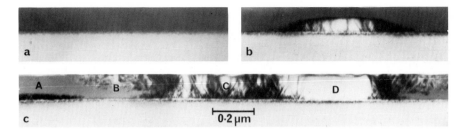

Fig. 7. Cross-sectional transmission electron images of As^+ ion implanted Si: (a) as-implanted amorphous layer; (b) and (c) laser annealed at $0.25J/cm^2$ showing explosive, local crystal growth (as at A and D in (c)). (After Cullis et al [13].)

in the ultra-high speed regime since novel explosive crystal growth phenomena have been observed in the presence of both positive [13] and negative [17] temperature gradients. An example of unstable recrystallization behavior is given in Fig. 7 where an As^+ ion implanted amorphous layer irradiated with a 2.5nsec ruby laser pulse is seen to yield explosive, local island crystal growth at the internal matrix interface (Figs. 7b and c). This process is entirely consistent with the penetration of a strongly undercooled melt to the crystal interface where nucleation and regrowth of crystal material then occurs. However, the melt temperature may be so low that the slope of the crystal growth rate vs. temperature relationship is positive (the predicted relationship [18] exhibits a maximum several hundred Kelvins below T_{c-1}). Under these conditions crystal growth can be unstable since the emission of latent heat of solidification leads to a local temperature rise which, in turn, increases the growth rate further. This crystal growth model accounts for the explosive growth events shown in Fig. 7 [7,13] and gives considerable confidence that T_{a-1} is hundreds of Kelvins below T_{c-1}.

ACKNOWLEDGEMENTS

The author would like to acknowledge close collaboration over several years with Hugh Webber, Nigel Chew and Don Hurle (RSRE) and with John Poate (Bell Labs). He also greatly appreciates stimulating interaction and discussions with Ken Jackson (Bell Labs), Pietro Baeri (Catania University) and Mike Thompson (Cornell University).

REFERENCES

1. A. G. Cullis, H. C. Webber and P. Bailey, J. Phys. E: Sci. Instrum. 12, 688 (1979).

2. P. L. Liu, R. Yen, N. Bloembergen and R.T. Hodgson, Appl. Phys. Lett. 34, 864 (1979).

3. R. Tsu, R. T. Hodgson, T. Y. Tan and J. E. Baglin, Phys. Rev. Lett. 42, 1356 (1979).

4. A. G. Cullis, H. C. Webber, N. G. Chew, J. M. Poate and P. Baeri, Phys. Rev. Lett. 49, 219 (1982).

5. G. J. Galvin, M. O. Thompson, J. W. Mayer, R. B. Hammond, N. Paulter and P. S. Peercy, Phys. Rev. Lett. 48, 33 (1982).

6. M. O. Thompson, this proceedings volume.

7. A. G. Cullis, H. C. Webber and N. G. Chew, Appl. Phys. Lett. in the press (1983).

8. F. Spaepen and D. Turnbull, in: Laser-Solid Interactions and Laser Processing - 1978, S. D. Ferris, H. J. Leamy and J. M. Poate eds. (Amer. Inst. Phys., New York, 1979) pp. 73-83.

9. B. G. Bagley and H. S. Chen, in: Laser-Solid Interactions and Laser Processing - 1978, S. D. Ferris, H. J. Leamy and J.M. Poate eds. (Amer. Inst. Phys., New York, 1979) pp. 97-101.

10. P. Baeri, G. Foti, J. M. Poate and A. G. Cullis, Phys. Rev. Lett. 45, 2036 (1980).

11. E. P. Donovan, F. Spaepen, D. Turnbull, J. M. Poate and D.C. Jacobson, Appl. Phys. Lett. in the press (1983).

12. S. A. Kokorowski, G. L. Olson, J. A. Roth and L. D. Hess, Phys. Rev. Lett. 48, 498 (1982).

13. A. G. Cullis, H. C. Webber and N. G. Chew, Appl. Phys. Lett. 40, 998 (1982).

14. A. G. Cullis, H. C. Webber and N. G. Chew, J. Crystal Growth, in the press.

15. A. G. Cullis, H. C. Webber, J. M. Poate and N. G. Chew, J. Microsc. 118, 41 (1980).

16. P. Baeri, G. Foti, J. M. Poate and A. G. Cullis, in: Laser and Electron-Beam Solid Interactions and Materials Processing, J. F. Gibbons, L. D. Hess and T. W. Sigmon eds. (North Holland, New York, 1981) pp. 39-44.

17. G. H. Gilmer and H. J. Leamy, in: Laser and Electron Beam Processing of Materials, C. W. White and P. S. Peercy eds. (Academic Press, New York, 1983) pp. 227-233.

18. K. A. Jackson, in: Proceedings of NATO Institute on Surface Modification and Alloying, J. M. Poate and G. Foti eds. (Plenum Press, New York, 1983).

AMORPHOUS PHASE TRAPPING AS A RESULT OF PULSED LASER IRRADIATION OF SILICON[*]

R. F. Wood
Solid State Division, Oak Ridge National Laboratory, Oak Ridge, TN 37830

ABSTRACT

It is concluded that large interface undercoolings of ~ 300 deg are not likely to occur during pulsed laser annealing and that the observed liquid to amorphous phase transition is not a purely thermodynamic effect. It is then shown that the formation of the amorphous phase can be understood on the basis of a kinetic rate model which makes large undercoolings of the interface unnecessary.

INTRODUCTION

Several groups [1-3] have reported that irradiation of crystalline (c) Si with pulsed lasers can result in formation of the amorphous (a) phase. The work of Cullis et al. [3] is of particular interest because the a phase was formed over large areas on both <100> and <111> substrates; also heat flow calculations indicated that solidification velocities v of ~ 20 m/sec were required to produce the effect. In discussing this phenomenon, some authors [3] have assumed, from the results of thermodynamic calculations [4], that it is necessary for the liquid-solid (ℓ-s) interface to be undercooled by $\geq$ 300 deg. Here it is shown that such large undercoolings are unlikely to be realized under the conditions of the experiments and that therefore purely thermodynamic arguments may not be sufficient to explain the observed phenomena. Instead, a kinetic rate model is used to show that "trapping" of defects and of the a phase at high v can occur. The model employs the concept of effective v-dependent activation energies for interfacial processes [5]. The idea is that when new layers of atoms are added to the solid at an increasingly rapid rate the potential barrier against localized rearrangements of atoms into the c phase increases rapidly. When the activation energy for such rearrangements in a thin layer becomes great enough, they cannot occur in the passage time of the ℓ-s interface and a disordered phase forms.

CLASSICAL CRYSTAL GROWTH THEORY

Variations of an equation for the crystal growth velocity of the form

$$v = K^{\ell s}\left\{1 - \exp[- L_c \Delta T_i / kT_m T_i]\right\} \tag{1}$$

appear frequently in the literature [6]. T_m is the melting temperature, T_i is the "temperature of the interface", $\Delta T_i \equiv T_m - T_i$ is the interface undercooling, and L_c is the latent heat of crystallization. $K^{\ell s}$ is a rate constant for transitions of atoms from the liquid to the solid. The difficulty in applying Eq. (1) arises from the lack of knowledge of $K^{\ell s}$ and the problem of determining T_i. To make progress, it is customary to introduce the expression

[*] Research sponsored by the Division of Materials Science, U.S. Department of Energy, under contract W-7405-eng-26 with Union Carbide Corporation.

Mat. Res. Soc. Symp. Proc. Vol. 13 (1983) Published by Elsevier Science Publishing Co., Inc.

for v obtained from heat flow equations. Calculations [7] of the temperature
T in the surface region during pulsed laser annealing of Si usually show that,
after the laser pulse, T in the liquid drops quickly to T_m and remains there
while latent heat conduction is determined by the gradient of T in the solid
and the thermal properties of the material. The velocity of the planar interface is then given by

$$\frac{dQ}{dt} = \frac{\Delta x L_c d}{\Delta t} = v L_c d = K_s G_{si} \quad . \tag{2}$$

The volume of material from which latent heat evolves during Δt is Δx times
unit area, d is the density, K_s is the thermal conductivity of the solid, and
G_{si} the gradient of T in the solid at the interface. The calculations also
show that G_{si} is practically constant near the interface, and hence

$$G_{si} \simeq (T_m - T_i)/\Delta X_{mi} = \Delta T_i / \Delta X_{mi} \quad , \tag{3}$$

in which ΔX_{mi} is the distance over which T drops from T_m to T_i.
    The subscript i on G_{si} in Eqs. (2) and (3) means that the gradient of T is
to be evaluated "in the solid" at the ℓ-s interface. Finite difference calculations of heat flow usually require that $T = T_m$ at the interface. However,
such a condition, and the implied specification of the position of the interface, would result in ΔT_i being zero, and hence Eq. (1) would predict no movement of the interface. To get around this difficulty, it can be recognized
that the interface boundary condition is too restrictive and that the
"interface" need not be at T_m. The G_{si} can still be taken from the heat flow
calculations (as a first approximation) and the position of the interface
adjusted to give a ΔT_i in Eq. (1) such that v in Eqs. (1) and (2) agree.
To carry out calculations with Eqs. (1) - (3), it was assumed that $K^f = 100$
m/sec in Eq. (1); this choice will be discussed briefly below. Values of ΔT_i,
and hence T_i, were used in Eq. (1) to determine v; G_{si} was then determined by
Eq. (2), and ΔX_{mi} from Eq. (3). The quantity L_c/kT_m has the value of 3.64 in
Si, d is 2.3 g/cm^3, and $K_s \simeq 0.3$ W/cm deg at T_m. G_{si} is of the order of 10^7
deg/cm for v in the range of interest here [7]. Comparison of sets of v and
G_{si} values, calculated as described above, with sets taken from melting model
calculations show that there is good agreement over a wide range of v. Heat
flow calculations carried out for 2.5 nsec pulses of a frequency doubled ruby
laser (Ref. 3) verified that energy densities of 0.25 J/cm^2 could yield
$v \simeq 20$ m/sec. The value of G_{si} determined from these calculations and the
value needed in Eq. (2) to obtain v = 20 m/sec were in very good agreement.
From Eq. (1), it could then be determined that $\Delta T_i \simeq 100$ deg for v = 20 m/sec.
Hence, it seems reasonable to conclude that, unless heat flow calculations are
grossly misleading, ΔT_i values of only ~ 100 deg give the observed amorphization of Si.
    In general, $K^{\ell s}$ in Eq. (1) may have the form

$$K^{\ell s} = A_{\ell s} \exp(-U_{\ell s}/kT) \quad . \tag{4}$$

$U^{\ell s}$ is usually assumed to be related to the activation energy for self-
diffusion in the liquid. Wood [5] estimated $U_{\ell s}$ for diffusion of Al and P in
molten Si to be ~ 1.4 eV; and self-diffusion in Si should yield a similar
value. He also found that $K^{\ell s} = 100$ m/sec gave satisfactory agreement between
theory and experiment for several aspects of dopant segregation in Si at v ~ 4
m/sec. Young [8] utilized the relationship between self-diffusion and viscosity to extract values of $U_{\ell s}$ and $K^{\ell s}$ from the viscosity data of Glasov et al.
[9]; he found that $U_{\ell s} \simeq 1.5$ eV and $K^{\ell s} = 130$ m/sec, in perhaps fortuitously

good agreement with the estimates of Wood. Figure 1 shows v calculated from Eq. (1) as a function of ΔT_i when $U_{\ell s}$ is assumed to have the values indicated. For each value of $U_{\ell s}$, $A_{\ell s}$ was determined by requiring that $K^{\ell s} = 100$ m/sec for $v \simeq 4$ m/sec and $\Delta T_i = 20$ deg. From Fig. 1, it can be seen that v increases with ΔT_i to a maximum, and then decreases as the undercooling begins to strongly effect the ℓ-s transition rate; this effect was already displayed in Ref. 6 for copper. A rigorous interpretation of Fig. 1 is not obvious, but the suggestion is that it would be possible to freeze the solid into a true liquid configuration at very large ΔT_i. However, if this were to happen presumably the latent heat would become very small, which would likely require a modification of the model leading to Eq. (1) and hence to the calculations based on it. The velocity on Fig. 1 has been referred to as the crystallization velocity because L_c was used in Eq. (1). Interfacial processes may slow down <u>crystallization</u>, but Eq. (1) does not predict a transition directly to the amorphous state.

RATE EQUATIONS

A simplified set of rate equations which can describe the incorporation of defects into the growing solid is given by

$$v = K_d^{is}C_d^i + K_c^{is}C_c^i - K_d^{si}C_d^s - K_c^{si}C_c^s \; ; \quad vC_d^s = K_d^{is}C_d^i - K_d^{si}C_d^s \qquad (5a,b)$$

$$K_c^{si}C_c^s + R_{dc}C_d^i - K_c^{is}C_c^i = 0 \; ; \quad C_d + C_c = 1 \; . \qquad (5c,d)$$

The system has been divided into liquid (ℓ) and solid (s) regions separated by an interface (i) region. Disordered material enters region i from ℓ and, if time and the various rate constants permit, is transformed into the c phase

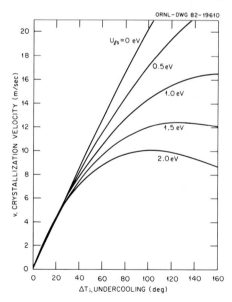

Fig. 1. Effects of $U_{\ell s}$ on crystal growth velocity.

before passing to region s. The degree of disorder in regions i and s is expressed by the concentration of defects c_d^i and c_d^s, respectively; concentrations of the c phase are given by c_c^i and c_c^s. The exact nature of the defects need not be specified at this time. A rapid build up of c_d^s is considered a precursor to the appearance of the a phase. Rate constants are denoted by K^{si}, etc., in which si indicates that material is transferred from s→i. Equations for ℓ→i transitions could be introduced but they are not independent of Eqs. (5). R_{dc} gives the rate at which defects in region i undergo processes (reorientation, recombination, rearrangement) which annihilate them. R_{dc} will be made v-dependent in a manner described below. A rate constant for the reverse type of processes has been neglected for simplicity and under the assumption that the perfect c phase is much more stable than the defect phase.

Even Eqs. (5) involve too many rate constants for the simple phenomenological approach being taken here, and one further approximation is required. The important physical content of the equations will be determined by the relative magnitudes of the K and R rate constants, and differences between K_d and K_c in any one region are of secondary importance. Hence, it will be assumed that $K_d^{is} = K_c^{is}$ and $K_d^{si} = K_c^{si}$, and the subscripts deleted. The simple result

$$c_d^s(v) = \frac{v}{v + R_{dc}(v)} \tag{6}$$

can then be obtained from Eqs. (5). Regardless of the simplicity of the model and the somewhat drastic approximations which have been made, the physical reasonableness of Eq. (6) is apparent. For small v, C_d will be small if $R_{dc}(v)$ is large, but on the otherhand, if $R_{dc}(v)$ becomes small as v becomes large, $c_d^s(v)$ can increase rapidly with v. A concentration of defects must soon be reached at which an amorphous phase begins to form.

Guided by the ideas in Ref. 5, R_{dc} will be written as

$$R_{dc}(v) = A_{dc} \exp(-U_{dc}(v)/kT) , \tag{7}$$

in which $U_{dc}(v)$ is an effective activation energy for the annihilation of defects. $U_{dc}(v)$ is dependent on v because the more quickly new layers of material are added to the solid the more rapidly the potential barrier against removal of a defect increases. $U_{dc}(v)$ can be written as

$$U_{dc}(v) = U_{dc}^o + \Delta U_{dc} f(v) . \tag{8}$$

U_{dc}^o is the activation barrier against removal of defects when v → 0. The function f(v) is required to have the limiting behavior f(v) → 0 as v → 0 and f(v) → 1 as v becomes large, so that $U_{dc}(v) → U_{dc}^o + \Delta U_{dc}$ for large v. A simple analytic approximation having the correct limiting behavior is

$$f(v) = 1 - (v_o/v)[1 - \exp(-v/v_o)] , \tag{9}$$

where v_o is to be determined by the available experimental data. A justification for Eq. (9) will be given in a later paper. It is now well established that, if v < 3-4 m/sec and if the melt front penetrates entirely any amorphous layer that may be present, a nearly perfect crystal is regrown on <100> Si. This implies that U_{dc}^o is quite small, and for simplicity it can be made zero.

An estimate of ΔU_{dc} can be obtained from the following considerations. When v is high, atoms at the interface will be covered with newly solidified material before they and their neighbors have time to achieve their crystalline configuration. In the hypothetical limit of infinite v, an atom at the a-c interface would see an environment similar to that at any a-c interface, regardless of how the material is formed. This suggests that $U_{dc}(v)$ for large v should be approximately the same as the activation energy E_{ac} for the a→c transformation of an a layer on a c substrate during furnace annealing. Several groups have studied the regrowth of a-Si layers on c-Si substrates. Here the results of Csepregi et al. [10] will be used because those authors give data for regrowth of layers on <100>, <110>, and <111> substrates. They found that v_{ac} is given by

$$v_{ac} = v_{ac}^{o} \exp(-E_{ac}/kT) \qquad (10)$$

with $E_{ac} = 2.35 \pm 0.1$ eV and v_{ac}^{o} dependent on the substrate orientation. Equation (10) is to be the limiting form of Eqs. (7)-(9) at high v, so the identifications $\Delta U_{dc} = E_{ac}$ and $A_{dc} = v_{ac}$ can be made, and $R_{dc}(v)$ written as

$$R_{dc}(v) = v_{ac}^{o} \exp\left\{-E_{ac}\left\{1-(v_{o}/v)[1-\exp(-v/v_{o})]\right\}/kT\right\} . \qquad (11)$$

Figure 2 shows the results of using Eq. (11) in Eq. (6) to calculate $C_{d}^{s}(v)$. From the data in Ref. 10, it was found that, $v_{ac}^{o}(100) = 6.5\times10^{4}$ m/sec, $v_{ac}^{o}(110) = 2.5\times10^{4}$ m/sec, and $v_{ac}^{o}(111) = 3.3\times10^{3}$ m/sec; E_{ac} was put equal to 2.4 eV. The parameter v_{o} was fixed at 23 m/sec by requiring that $C_{d}^{s} = 7\%$ when v = 20 m/sec in a <100> direction. This value of C_{d}^{s} was chosen primarily for illustrative purposes and it is doubtful that the Si lattice could contain such a high concentration of defects without transforming to a highly disordered or a phase. For ion-implanted Si there is some evidence that the onset of the a phase, as monitored by the divacancy concentration [11] and by ESR [12], occurs at values of $C_{d}^{s} < 1\%$. However, van Vechten [13] has argued that in one form of a-Si, an equivalent vacancy concentration of ~ 5% may be present. The line on Fig. 2 at this value of C_{d}^{s} would seem to be a generous upper limit to the onset of the a phase.

SUMMARY AND CONCLUSIONS

A simple phenomenological model based on kinetic rate theory and employing a "reorientation" or "recombination" rate constant with an effective velocity-dependent activation energy was introduced to describe the c→a transition. This model does not require the large undercoolings apparently necessitated by

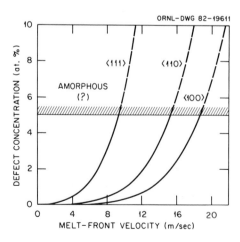

Fig. 2. Defect concentration as a function of v and substrate orientation.

88

purely thermodynamic considerations in order to account for the c→a transformation. However, it must be recognized that the thermodynamic calculations are subject to some uncertaintities and the differences between the melting points of a-Si and c-Si may not be as great as calculated in Ref. 4. Moreover, the results of Kokorowski et al. [14] and Knapp and Picraux [15] which apparently imply a very small difference between the two melting points have not yet been satisfactorily explained. For the present then, the model described here and the ideas behind it, are offered as a possible alternative to the more conventional, purely thermodynamic, approach.

REFERENCES

1. P. L. Liu, R. Yen, N. Bloembergen, and R. T. Hodgson, Appl. Phys. Lett. 34, 864 (1979).

2. R. Tsu, R. T. Hodgson, T. Y. Tan, and J. E. Baglin, Phys. Rev. Lett. 42, 1356 (1979).

3. A. G. Cullis, H. C. Weber, N. G. Chew, J. M. Poate, and P. Baeri, Phys. Rev. Lett. 49, 219 (1982).

4. B. G. Bagley and H. S. Chen, in Laser Solid Interactions and Laser Processing-1978, ed. by S. D. Ferris, H. J. Leamy, and J. M. Poate, AIP Conference Proceedings No. 50 (American Institute of Physics, New York, 1979), p. 97; F. Spaepen and D. Turnbull, ibid, p. 73.

5. R. F. Wood, Appl. Phys. Lett. 37, 302 (1980); R. F. Wood, Phys. Rev. B 25, 2786 (1982).

6. K. A. Jackson and B. Chalmers, Can. J. Phys. 34, 473 (1956).

7. R. F. Wood and G. E. Giles, Phys. Rev. B 23, 2923 (1981).

8. F. W. Young, Jr., private communication.

9. V. M. Glasov, S. N. Chizhevskaya, and N. N. Glagoleva, Liquid Semiconductors, (Plenum Press, N.Y., 1969).

10. L. Csepregi, E. F. Kennedy, J. W. Mayer, and T. W. Sigmon, J. Appl. Phys. 49, 3906 (1978).

11. H. J. Stein, F. L. Vook, D. K. Brice, J. A. Borders, and S. T. Picraux, Radiation Effects 6, 19 (1970).

12. B. L. Crowder, R. S. Title, M. H. Brodsky, and G. D. Petit, Appl. Phys. Lett. 16, 205 (1970).

13. J. A. van Vechten, Bull. Amer. Phys. Soc. 27, 365 (1982).

14. S. A. Kokorowski, G. L. Olsen, and L. D. Hess, J. Appl. Phsy. 53, 921 (1982).

15. J. A. Knapp and S. T. Picraux, Appl. Phys. Lett. 38, 873 (1981).

ION AND ELECTRON SPECTROSCOPY DURING PULSED LASER
IRRADIATION OF SILICON

J.P. LONG, R.T. WILLIAMS, T.R. ROYT, J.C. RIFE, AND M.N. KABLER*
Naval Research Laboratory, Washington, D.C. 20375

ABSTRACT

Ion mass spectrometry, charged particle yields, and
kinetic energy distributions of electrons and ions are used
to characterize silicon wafers and vacuum-cleaved silicon
surfaces under conditions related to laser annealing. We
find that alkali metals dominate the positive ion emission
from chemically-cleaned wafers, a mass-72 peak tentatively
identified as Si_2O^+ comprises the main ion emission from the
cleaved surface, and ion and electron temperatures can be
derived from the energy distribution curves, although the
Si_2O^+ emission implies more than a simple thermal
evaporation process.

INTRODUCTION

Recently, quantitative studies of laser-induced emission of charged
particles have examined the physical mechanisms involved in absorption and
thermalization or other disposition of photon energy in semiconductors [1-3].
These investigations relate particularly to pulsed laser annealing of silicon,
but also touch on related areas of laser cleaning and laser sputtering and can
be extended to other semiconductors. We have previously reported studies of
electron emission from laser-excited II-VI compounds with picosecond time
resolution [4,5].

In the work of Liu et al [1], ion yields from silicon wafers irradiated by
20 ps pulses of 532-nm light were found to be comparable in magnitude to
electron yields over a wide range of laser fluence. Another indication of
correlation of ion and electron yields from silicon wafers is their
simultaneous decrease by orders of magnitude during the course of 10^4 laser
shots on the same spot in the 1 to 3 J/cm^2 range [2], a behavior suggesting
laser clean-up effects. However, the commonly accepted view that a few tens
of laser shots at 1 to 2 J/cm^2 virtually completes cleaning, as documented by
Auger analysis [6], presents a puzzle which is one of the points addressed by
the present investigation.

We report the first determination of the emitted ion species and accurate
energy distribution curves (EDCs) of both ions and electrons from silicon
under conditions of pulsed laser annealing. Samples prepared by vacuum
cleavage were compared to wafer surfaces prepared by standard chemical
procedures. We have discovered that alkali metals dominate the charge
emission from wafers for several thousand shots under conditions described
below. The emission of a mass 72 ion, tentatively identified as Si_2O^+,

* Sachs Freeman and Assoc., Bowie, MD 20715

Mat. Res. Soc. Symp. Proc. Vol. 13 (1983) Published by Elsevier Science Publishing Co., Inc.

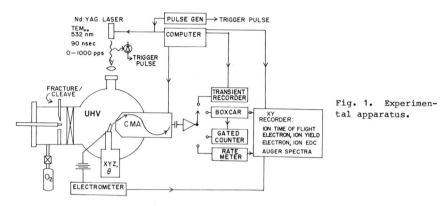

Fig. 1. Experimental apparatus.

has also been observed in all samples. Given the temperature of 1600 ± 300 K deduced from the EDCs, the presence of Si_2O^+ is remarkable as discussed below. Finally, comparison of the electron emission from 20-ps and 90-ns laser pulses has been made and an apparent dependence of charge emission on laser intensity at constant laser fluence is found.

EXPERIMENT

As illustrated in Fig. 1, the experiments were performed in an ultra-high vacuum chamber (base pressure 3 x 10^{-10} torr) equipped with a double-pass cylindrical mirror analyzer (CMA), gas admission manifold, and in situ sample cleavage apparatus. For most of the experiments reported, 90 ns pulses of 532 nm light from a frequency doubled Nd:YAG laser, operating in the TEM_{oo} mode and from 0 to 1 kHZ, were focused to a 100µ diameter as verified by scanning a 25µ pinhole mounted with the sample.

The CMA, fitted with a channeltron electron multiplier (CEM), could be configured for ion spectroscopy as described by Traum et al. [7]. The CEM output was fed to a transient recorder for ion time-of-flight (TOF) determinations of ion mass, to a gated counter for low signal levels, or to a boxcar averager for detection of signals that would otherwise saturate the counter.

Some of the silicon wafers were chemically cleaned by a peroxide/acid/distilled water process in the NRL microelectronics facility. Others were obtained from Monsanto in a sealed clean-pack, opened in a clean room facility, and transferred to the vacuum with only a few minutes exposure to ordinary laboratory atmosphere. No further cleaning was performed once the samples were in vacuum. The silicon samples for vacuum cleavage were high purity (500 ohm-cm) material cut into sticks about 2 x 2 x 10 mm^3. The cleaved surfaces contained steps but were comprised mainly of the natural [111] cleavage faces.

To address intensity dependence of charged particle emission, apart from fluence dependence, measurements were also made with 20-psec pulses of 532-nm light from a passively mode-locked Nd:YAG laser.

RESULTS

Fig. 2 shows how the electron and the ion emission decrease as a given spot on a [111] n-type wafer is subjected to repeated laser pulses. The plateau observed at the highest emission may be a consequence of space charge

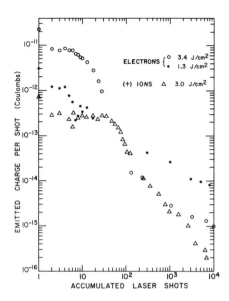

Fig. 2. Charge emission vs accumulated laser dose for a [111] n-type wafer.

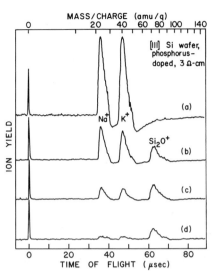

Fig. 3. Ion TOF spectra at 2.3 J/cm² for various laser doses. Spectra are averaged over shots (a) 2-11, (b) 200-300, (c) 400-500, (d) 1500-1600. Laser photodiode pulse marks t = 0.

limitations, but the important feature is a clean-up effect extending for thousands of shots. Identification of the ions was performed by time of flight (TOF) measurements such as those of Fig. 3. The parameters in the formula relating mass to TOF [7] were determined subsequent to these experiments by introducing small quantities of Na, K, and Cs salts on a silicon target. No alkali samples had previously been used in this chamber. Na and K were unobservable by Auger electron spectroscopy, but ancillary measurements of parts of the same wafers by secondary ion mass spectrometry (SIMS) demonstrated the presence of Na and K in quantities typical of the alkali "background" found in SIMS before sputter cleaning. In contrast to the wafers, the ion emission for a cleaved surface, shown in Fig. 4, exhibits only a very small K⁺ peak which was removed after several scores of shots. The ubiquitous presence of alkali ions in laser induced emission is a consequence of their low ionization potential, which is expected to result in near unity ionization at the melting temperature of Si. As the integrated number of ions emitted in all the shots of Fig. 2 represents less than 10^{-2} of a monolayer fully ionized, laser induced ion emission is a powerful method for detection of alkali contamination.

Our irradiated spots above 1 J/cm² exhibited ripples, the amplitude of which increased with increasing shot-number as observed by interference contrast microscopy. This surface modification, although it depends on accumulated laser dose, is not a likely cause for the dose dependence of charge emission in light of our discovery of alkali controlled charge emission and alkali clean-up with laser dose, a conclusion supported by the constant electron yield from cleaved surfaces. Surface modification may play a role in the third ion peak discussed below.

92

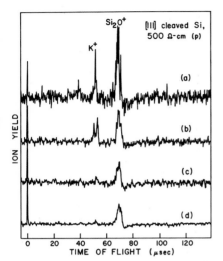

Fig. 4. Ion TOF spectra at 2.8 J/cm² from cleaved silicon averaged over shots (a) 1, (b) 2-11, (c) 12-21, (d) 32-81.

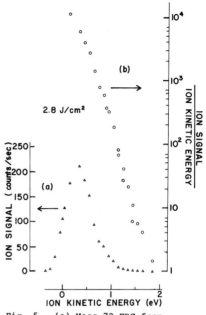

Fig. 5. (a) Mass 72 EDC from cleaved silicon at 2.8 J/cm² replotted (b) for the determination of temperature from slope of the high energy tail. Pulse pile-up is responsible for curvature.

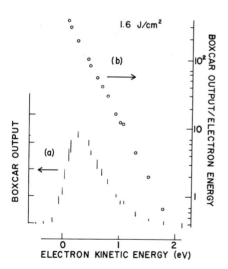

Fig. 6. (a) Electron EDC from cleaved silicon at 1.65 J/cm² replotted (b) as described in Fig. 6 and text.

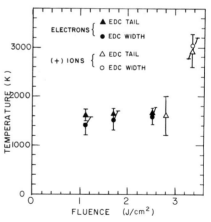

Fig. 7. Temperature determined from EDCs vs fluence. See Fig. 6 for definition of "tail."

In both Figs. 3 and 4, the third mass peak at approximately 72 behaves differently than the alkali peaks. Its erasure by repeated laser pulses is slower than for the alkalis. The peak returns on exposure to the UHV ambient. Oxygen admitted to the chamber accelerates the recovery of the peak, but not beyond its saturation value reached after about 40 minutes at 10^{-9} torr ambient. Hence a limited number of adsorption sites corresponding to a monolayer coverage or less appear to be involved. Given this oxygen association, we have tentatively assigned the mass-72 peak to Si_2O^+. This ion is a principal effluent during the sputtering of oxygen implanted Si [8].

To a sensitivity of 10^{-3} of the Si_2O^+ peak, itself measured to be less than 7×10^{-17} C/shot, we observe no mass peak for Si^+ up to 2.8 J/cm^2. Given temperatures in the range of 1600 K as deduced below, the vapor pressure of silicon, and a work function, ϕ, of about 5 eV, the absence of Si^+ (ionization potential 8.15eV) is at least consistent with the Saha-Langmuir equation. The ionization potential of Si_2O can be expected to be > 9.5 eV [9]. Its dominance in the cleaved silicon ion spectra is therefore not readily described in terms of thermal evaporation from a simple molten surface describable by the Saha-Langmuir equation.

Representative EDCs for ions and electrons are displayed in Figs. 5 and 6. For thermionic emission one expects a charge $Q(E_k) \propto E_k exp(-E_k/kT)exp(-\phi/kT)$, where E_k is the kinetic energy. In part (b) of each figure is shown a plot of $log(Q/E_k)$ vs E_k from the slope of which the effective temperature is obtained directly. The rapid rise of $exp(-\phi/kT)$ with temperature insures only the hottest material over the beam profile is measured. Fig. 7 shows temperatures obtained in this way for several incident fluences. The points below 2000 K were taken on a cleaved sample. Fig. 7 also includes temperatures measured from the width of the EDC, which is more susceptible than the slope to counter saturation, but still gives reasonable agreement. At fluences below 1 J/cm^2, the thermionic emission falls rapidly, eventually reaching the multiphoton regime. Above about 2.5 J/cm^2 the temperature evidently begins to rise more rapidly.

The total yields for the cleaved surface are plotted in Fig. 8. The electron data were obtained by calibrating the boxcar against the electrometer at high fluence, and in turn matching the counter data to the boxcar data in overlapping regions at much lower fluence. In this way, calibrated sensitivity to single electron emission was achieved. The ion points assume a conversion efficiency in the CEM equal to electrons; the dotted line above the ion data represents an upper limit determined with an electrometer. The

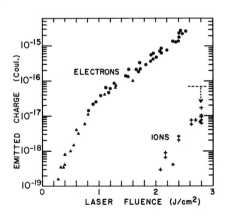

Fig. 8. Electron and ion yield vs laser fluence from cleaved silicon.

94

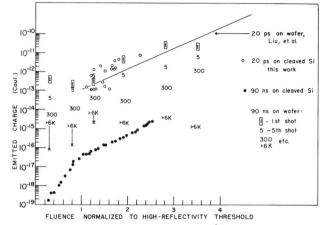

Fig. 9. Electron
yield for various
experimental
conditions.

electron yield of Fig. 8 above 1 J/cm^2 is consistent with the Richardson-
Dushman equation given our EDC temperatures, electron emission-time measured
with the transient recorder, and spot size. Without modeling for the time and
spatial dependence of temperature on the silicon surface and incorporating the
reflectivity rise above 1 J/cm^2 (a significant factor in the knee of Fig. 8),
more detailed extractions of temperature from total electron yield are
unwarranted.

Fig. 8 represents electron emission from a clean cleaved surface with a 90 ns
pulse length. Our preliminary measurements on such a surface with 20 ps,
532 nm pulses are shown in Fig. 9 along with the 20 ps data of Liu et al. [1],
taken on a wafer. Plotted in addition are our data points taken on a wafer
with 90 nsec pulses at different shot numbers (e.g. 1st, 5th, 300th, 6000th).
The electron emission has been normalized to a laser spot area of $1.4x10^{-4} cm^2$.
The horizontal axis has been normalized to the threshold fluence for high
reflectivity, i.e. 1 J/cm^2 at 90 ns and 0.2 J/cm^2 at 20 ps. The picosecond
data appear to be in agreement whether wafers or cleaved surfaces were used.
This suggests that alkalis, though probably present on all wafers which are
given only a standard chemical cleaning, are not the main factor in the
disparity of electron yield between 20 ps and 90 ns laser pulses. Instead, the
difference of three orders of magnitude in electron emission suggests a strong
intensity dependence at equivalent fluence even after differences in heat
dissipation have been approximately accounted for by scaling to the high
reflectivity threshold. However, it is worthwhile to note the overlap of the
90 ns first-shot-on-wafer emission with the 20 ps emission. Mechanisms
involving laser-heated plasmas in both solid and gas phases are being
investigated. In any case, further experiments of this sort on carefully
controlled surfaces are needed, especially as Fig. 2 suggests that a few laser
shots on an ordinary wafer surface are not sufficient cleaning where
desorption and thermal emission of charged particles are concerned.

ACKNOWLEDGEMENTS

We wish to thank J.A. Van Vechten for many stimulating discussions,
M.C. Peckerar, D. McCarthy and others on the NRL microelectronics staff
for experimental advice and assistance, R.J. Colton for SIMS analysis,
J.V. Gilfrich for x-ray fluorescence analysis, and C.L. Marquardt, S.C. Moss,
V.M. Bermudez, and V.H. Ritz for consultation on the experiments.

REFERENCES

1. J.M. Liu, R. Yen, H. Kurz, and N. Bloembergen, Appl. Phys. Lett. 39,
    755 (1981).

2. R.T. Williams, M.N. Kabler, J.P. Long, J.C. Rife, and T.R. Royt, Proc.
    1981 MRS Symposium on Laser and Electron Beam Interactions with Solids,
    edited by B.R. Appleton and G.K. Celler (North-Holland, New York, 1982),
    p. 97.

3. M. Bensoussan, J.M. Moisson, B. Stoesz, and C. Sebenne, Phys. Rev. B23,
    992 (1981).

4. R.T. Williams, J.C. Rife, T.R. Royt, and M.N. Kabler, J. Vac. Sci.
    Technol. 19, 367 (1981).

5. R.T. Williams, T.R. Royt, J.C. Rife, J.P. Long and M.N. Kabler, J. Vac.
    Sci. Technol. 21, 509 (1982).

6. D.M. Zehner, C.W. White, G.W. Ownby, Appl. Phys. Lett. 36, 56 (1980).

7. M.M. Traum and D.P. Woodruff, J. Vac. Sci. Technol. 17, 1202 (1980).

8. K. Wittmaack, Surface Science 112, 168 (1981).

9. T.L. Gilbert, W.J. Stevens, H. Schrenk, M. Yoshimine, P.S. Bagus, Phys.
    Rev. B8, 5977 (1973). See also the chemical trend in J. Phys. Chem.
    Reference Data, 6, Supplement (1977).

NONLINEAR LASER MELTING OF INDIUM ANTIMONIDE AND SILICON

MICHAEL P. HASSELBECK AND H. S. KWOK
Department of Electrical and Computer Engineering
State University of New York at Buffalo, Amherst, New York   14260   USA

ABSTRACT

Pulsed 10.6μm TEA CO_2 laser light has been used to melt
the semiconductors silicon and InSb. Measurements indicate
that generation of free carriers necessary for melting may
take place by nonlinear processes such as two-photon absorption or intraband avalanche ionization. If the semiconductor is sufficiently doped, melting may also result from
linear free carrier absorption. In all cases, it appears
that the molten depth exceeds several μm, which is much
greater than obtained with lasers of shorter wavelength.

INTRODUCTION

The physical mechanism by which a pulsed laser melts a semiconducting solid
is now well established. The transfer of laser energy to the lattice takes
place in an indirect process consisting of the simultaneous (1) generation of a
dense electron-hole plasma, (2) excitation and heating of these free carriers
by the laser field and (3) transfer of the energy from the free carriers to the
lattice. The first condition has been achieved almost exclusively by direct
single photon interband transitions, so that the laser photon of frequency ν
must satisfy $h\nu \geq E_g$, where E_g is the semiconductor bandgap energy. In a series
of experiments, we discovered that this condition on the laser frequency can be
circumvented for the CO_2 laser where the photon energy is only 0.12 eV.
The important observation is that pulsed laser-induced melting occurs at
high intensities which are quite close to the material damage threshold. At
such high intensities, nonlinear carrier generation mechanisms can be very
effective in generating the free carriers. In addition, if the semiconductor
is heavily doped to begin with, melting can occur via absorption by these
existing carriers.
We have demonstrated the above ideas for three semiconductor systems: (1)
intrinsic InSb, where the nonlinear process of two-photon absorption was used
to generate the free carriers; (2) highly doped Si, where the initial free carriers are used to melt the solid; and (3) intrinsic Si, where the highly nonlinear process of avalanche ionization took place.

LASER SYSTEM

The CO_2 laser system used in the experiments is illustrated schematically in
Fig. 1. A transversely excited, atmospheric pressure (TEA) double discharge
laser head, and a longitudinally pulsed, low pressure (∿6 torr) section provided polarized, temporally smooth, single longitudinal mode laser pulses of
∿70 ns duration and peak power of ∿8 MW. The oscillator is grating tuned and
spatially filtered for operation in the lowest order transverse mode. The
P(20) 10.6μm CO_2 laser line was used exclusively in these experiments. The
picosecond pulses were generated by optical free induction decay (OFID) [1].
This method of active pulse shaping is based on the high frequency modulation
of the laser pulse and the subsequent filtering by means of an optical resonant

Mat. Res. Soc. Symp. Proc. Vol. 13 (1983) ©Elsevier Science Publishing Co., Inc.

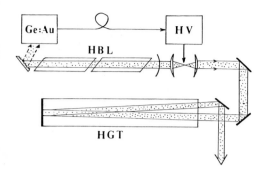

Fig. 1. The OFID laser system.
HBL: hybrid TEA laser,
HV: high voltage Marx
     generator,
HGT: hot CO_2 gas tube.

absorber. The high frequency modulation was obtained by a plasma shutter which
can be triggered with subnanosecond jitter [2]. The pressure in the gas cell
can be varied to give pulses of duration 30-250 ps. By emptying or filling the
CO_2 gas cell, smooth TEA nanosecond pulses, or picosecond OFID pulses can be
obtained to perform the experiments.

INDIUM ANTIMONIDE

   InSb is a well-known two-photon absorber at the CO_2 laser frequency [3]. It
provides an excellent test of our idea of circumventing the $h\nu \geq E_g$ condition.
Measurements were made of the self-reflectivity of intrinsic InSb ($n_i \sim 10^{16} cm^{-3}$)
with both nanosecond and picosecond pulses. The linear polarization of the
laser light allowed alignment of the crystal at Brewster angle incidence to
minimize reflected light. The reflected signal was detected with a Ge:Au crystal cooled to 77°K. Modification of the refractive index of the material by
the laser light will then manifest itself as an increased reflected signal [4].
The refractive index may be altered by changes in the electron hole pair (EHP)
concentration, large temperature change, or melting. The index of refraction
is given by

$$n = \left[\varepsilon_o (1 - N/N_c) \right]^{1/2}$$
(1)

where ε_o is the low frequency dielectric constant, N is the plasma concentration. N_c is the critical density given by

$$N_c = \varepsilon_o m^* \omega^2 / 4\pi e^2$$
(2)

in which ω is the laser frequency and m^* the reduced effective mass. For the
CO_2 laser-InSb system, N_c is $1.9 \times 10^{18} cm^{-3}$. This small value of the critical
density will enable the plasma effect to be observed before any material change
can occur.
   In the experiment, the laser pulses were focused onto the crystal with a
25.4 mm Ge lens and the crystal was rotated until minimum reflection of the CO_2
laser light was obtained. A nonzero minimum reflectivity of $\sim 4\%$ can be explained by noting that the focusing cone of the Gaussian beam does not allow
the entire spatial profile of the pulse to be at Brewster incidence. The calculated and experimentally determined incidence angles were both found to be
about 75°. The intensity incident on the crystal was varied by means of CaF_2
attenuators. Absolute calibration of the incident intensity was made by attenu-
ating the laser pulses until air breakdown at the focal point of the lens occurred on $\sim 50\%$ of the shots, corresponding to a peak intensity of $8 \times 10^9 GW/cm^2$
±20%. The damage threshold of the semiconductor was established by observing a
faint spark on the surface of the crystal. For i-InSb this corresponds to

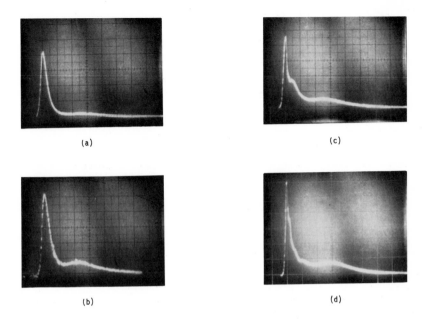

(a)

(c)

(b)

(d)

Fig. 2. Time-resolved reflected pulse from InSb. (a) 1 MW/cm^2,
(b) 15 MW/cm^2, (c) and (d) 40 MW/cm^2. 50 ns/div

$\sim$38 MW/cm^2.

Figure 2 shows the time-resolved reflected TEA laser pulse at several laser intensities. At low intensity (a), the reflected pulse resembles the incident TEA pulse. At higher intensities (b), the tail shows higher reflectivity, (c) and (d) correspond to intensities exactly at the damage threshold. Rapid modulation in the reflected pulse can be observed and can be used as a signature for catastrophic damage.

The integrated self-reflectivity of the 70 ns pulses as a function of incident intensity is plotted in Fig. 3. The onset of enhanced reflection is consistent with the two-photon absorption coefficient β of 0.23 cm/MW. The time rate of change of the carrier concentration is given by [3]

$$\frac{\partial N}{\partial t} = \frac{\beta I^2}{2\hbar\omega} - \frac{N}{\tau} - \alpha_r N (N + N_o)$$ (3)

where I is the incident laser intensity and τ and α_r allow for recombination. A dynamic Burstein shift of the bandgap energy should occur at carrier densities larger than $0.8 \times 10^{18} cm^{-3}$, thereby disabling the two-photon absorption process. However, the concurrent rise in lattice temperature may substantially modify the shape of the Fermi-Dirac carrier distribution and allow two-photon absorption to continue. Additionally, the carrier generation process may continue via thermal ionization at elevated temperatures or by avalanche ionization. It is believed that melting starts to occur at $\sim$7 MW/cm^2 by examination of the crystal surface.

The experiment was repeated using 75 ps OFID pulses. The integrated reflec-

100

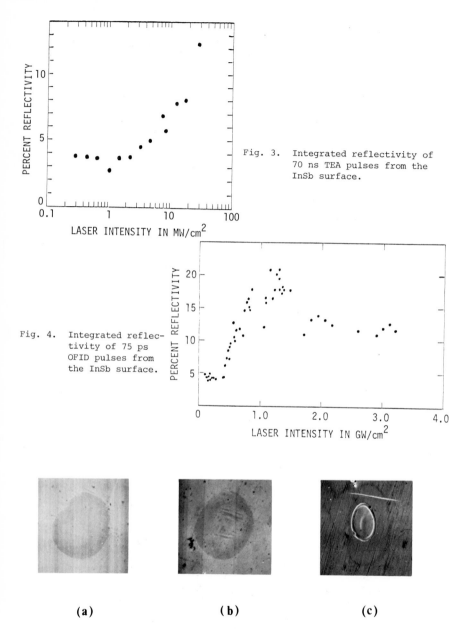

Fig. 3. Integrated reflectivity of 70 ns TEA pulses from the InSb surface.

Fig. 4. Integrated reflectivity of 75 ps OFID pulses from the InSb surface.

(a)   (b)   (c)

Fig. 5. SEM picture of the InSb surface after single shot 75 ps pulse irradiation. (a) 3 GW/cm^2, (b) 4 GW/cm^2, (c) 5 GW/cm^2. Distance across each picture is 100μm.

tivity is plotted as a function of intensity in Fig. 4. Again, the onset of enhanced reflection at 0.4 GW/cm^2 is consistent with a two-photon absorption process of equation (3). However, the reflectivity decreases from 20% at ∿1.06 GW/cm^2 to about 10% at ∿2.0 GW/cm^2. The decrease in reflectivity can be attributed to melting of the crystal. This curve should be contrasted with the one obtained in a similar experiment using 20 ps 532 nm laser pulses where the increased reflectivity was due entirely to melting [5]. This can be explained by the much larger N_c in that case.

Evidence to substantiate that melting had taken place was obtained by examining the surface after single shot irradiation at various intensities with a scanning electron microscope (Fig. 5). At intensities below ∿1.3 GW/cm^2, no permanent modification of the surface could be identified. The SEM photograph of the dark spot in Fig. 5(a) corresponds to an intensity of 3.0 GW/cm^2 indicating the surface, although still planar, has been modified. At higher intensities, surface ripples, as seen in Fig. 5(b), appeared on the surface. At still higher intensity, the surface plasma is hot enough that a faint spark is visible on the surface, with the resulting shock wave forming a crater on the molten InSb. An SEM photograph of the surface after exposed to a single shot of intensity 5.0 GW/cm^2 is shown in Fig. 5(c). The nature of the spot in Fig. 5(a) has not yet been determined, but possible explanations for its appearance include (1) changes in the relative composition of In and Sb, (2) modification of the crystalline structure, (3) surface oxidation and (4) evaporation of the surface layer. Therefore, for picosecond pulses, there appears to be a range of intensities as low as a factor of 3 below the damage threshold at which InSb can be melted without damage.

A comment is in order concerning the absence of a reflectivity decrease due to melting for nanosecond pulses. The reflected signal is simply a convolution of the incident intensity time profile and the temporal reflectivity of the surface. The inducement of a large free carrier concentration and subsequent melting of the lattice during the course of a single TEA laser pulse will not be distinguishable upon integration by the detection system. On the other hand, the picosecond self-reflectivity gives better temporal resolution as to the instantaneous nature of the semiconductor surface, enabling the observation of plasma effect and melting sequentially.

HIGHLY DOPED SILICON

Similar experiments were performed with 200μm thick Sb doped n-Si (N_D = 1.5 x $10^{18} cm^{-3}$) cut in the (111) direction. In the initial configuration, the self-reflectivity of the crystal was monitored when irradiated with nanosecond pulses

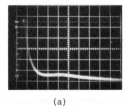 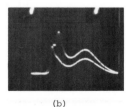

(a)   (b)

Fig. 6. Time-resolved reflected pulse from n-Si, (a) incident TEA pulse, (b) reflected pulses at ∿80 MW/cm^2. 50 ns/div

using the Brewster angle arrangement described in the preceding section.
Oscilloscope traces of the time-resolved reflected signal are shown in Fig. 6.
Fig. 6(a) depicts the shape of the single longitudinal mode laser pulse as reflected linearly from the Si surface. Figure 6(b) shows two superimposed
traces of reflected pulses just below the damage threshold with the pulse tails
much enhanced. As with InSb, the damage threshold was defined by the appearance of a faint spark on the crystal surface. It is believed that the modified
reflectivity observed is not due to a dense plasma but due to physical material
changes in the crystal.

Transmission of the 10.6μm laser light through the sample at normal incidence
was measured using nanosecond and picosecond pulses as a function of intensity
shown in Fig. 7. There exists a large region of nonlinear transmission in both
cases, as low as a factor of 8 below the damage threshold for the TEA pulses.
This measurement provides a definition of the region of melting without damage
which is an important parameter for semiconductor processing.

To confirm the existence of melting, a pump and probe experiment was done
using an unpolarized cw HeNe laser at 633 nm to monitor the silicon surface [6].
The 70 ns CO_2 laser pulses were focused normally on the sample surface using a
152 mm BaF_2 lens. The HeNe light was focused to the center of the CO_2 laser
spot at a 45° incidence using a 25.4 mm focal length lens. This light was detected using a Si photodiode having a response time of $\lesssim$3 ns. Figure 8(a) and
(b) are time-resolved oscilloscope traces of the transient HeNe reflected signal
at intensities of 64 and 78 MW/cm^2 respectively. The reflectivity rises in
$\sim$200 ns to a constant value of 70 ±5%, in agreement with molten reflectivity of
silicon observed previously [5,6]. The existence of a flat top peak in the reflected signal of duration longer than the laser pulse is also consistent with
these previous observations of molten silicon.

Upon examining the irradiated silicon surface, it was found that some morphological change has resulted from the melting process. Even at intensities 4
times lower than the damage threshold which corresponds to the formation of a
crater, a faint ring can be observed together with a smooth central region. The
structure of the recrystallized surface is still under investigation. However,
it was found that by overlapping the laser shots, the edge of the rings can be
re-smoothed. Obviously, further work needs to be done to characterize the molten silicon layer. One of the important questions to be addressed is whether
the regrown solid is single crystalline, polycrystalline or amorphous.

The transient HeNe reflectivity trace of Fig. 8(b) using nanosecond pump
pulses indicates a melt duration 10 times longer than obtained using visible

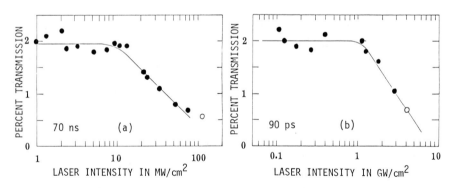

Fig. 7.   Transmission of TEA and OFID pulses through the n-Si wafer.
Open circles correspond to surface breakdown.

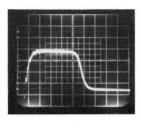 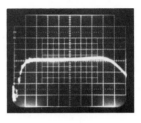

(a)   (b)

Fig. 8. Transient reflectivity of cw HeNe induced by CO_2 TEA pulses at
(a) 64 MW/cm^2, (b) 78 MW/cm^2. Horizontal scale: 0.5μs/div,
Verticle scale: 15%/div.

lasers of similar duration. The recrystallization time of the molten surface
associated with the falling edge of the pulse in Fig. 8 (∿.5μs) is also much
longer than reported for pulsed visible laser annealing. The above observations
are consistent with much deeper molten layers and slower heat diffusion compared
to lasers having $h\nu \geq E_g$.

The melting in the present case of highly doped silicon may be explained by
the large free carrier absorption cross section at 10.6μm. Using a simple
Joule heating model, the heat deposited in the solid W is given by

$$\frac{dW}{dt} = \frac{e^2 \tau_e}{m^*(1 + \omega^2 \tau_e^2)} \left| E(t) \right|^2 N(t) = N\sigma I \qquad (4)$$

where E(t) and I are the laser field and intensity, τ_e is the electron energy
transfer time, and σ the free carrier absorption cross section. For $\tau_e > 10^{-13}$s,
ωτ>>1, σ is directly proportional to λ^2. This may explain why similar observations are not made at the Nd:YAG wavelengths and shorter. Note that diffusion
has been neglected in equation (4) for simplicity.

Equation (4) leads to a temperature increase of the solid surface. To complete the picture, one has to take into account the thermal generation of free
carriers. This "thermal runaway" model may explain the apparent success of the
CO_2 laser in melting n-Si. However, the presence of avalanche ionization should
not be ruled out. The important parameters are the free carrier absorption
cross section and the initial doping concentration. The latter dependence could
be experimentally determined by repeating the above experiments for samples with
different doping concentrations. Progress is also being made in this direction.

INTRINSIC SILICON

The low initial concentration of intrinsic silicon (1.5 x 10^{10}cm^{-3}) should
not be large enough for the "thermal runaway" to occur. However, preliminary
experimental results indicate that the solid can also be melted using picosecond CO_2 laser pulses. It was found that i-Si displays a decrease in transmission at a factor of ∿10 below the damage threshold when exposed to 90 ps
pulses. With 70 nanosecond pulses, an increased transient reflectivity of HeNe

light was only observed with a concurrent visible spark on the surface. Also,
a decrease in transmission of the TEA CO_2 pulse could not be obtained without
simultaneously damaging the sample. Therefore, it is apparent that melting for
the case of i-Si is a highly nonlinear, intensity-dependent process.

Evidently, the much higher fields applied during the picosecond pulse are
generating nonequilibrium EHP concentrations by a nonlinear, intensity-dependent
process. The possibility that intraband avalanche ionization is responsible for
the EHP generation was investigated numerically using the well-known avalanche
ionization coefficients available for silicon. Initial calculations indicate
order of magnitude agreement between the intensity at which increased transmission is observed and the threshold for melting where excess electrons are
generated by an avalanche process. Further experiments, including transient
reflectivity using the HeNe probe laser and 10.6μm picosecond self-reflectivity,
are being planned. Investigations are also proceeding to study the effect of
changing the laser frequency to 9μm where the Si-O stretching mode absorbs
strongly.

CONCLUSION

In this paper, evidence has been presented to suggest that a pulsed TEA CO_2
laser may be used to melt semiconductors even without an interband transition
mechanism. The intensity range for which melting occurs without damage is comparable to that observed with pulsed visible annealing where $h\nu \gtrsim E_g$. The duration of the molten phase was also found to be considerably longer implying much
deeper molten layers. Further work needs to be done to characterize the regrown solid and the quality of the surface. The CO_2 laser may be a competitive
alternative to visible laser annealing not only because of the much higher energy efficiency, higher repetition rate and low cost, but also because of the
deeper molten layer, which may have implications for vertical integration.

This work was supported by the National Science Foundation under Grant
No. ECS8106007. We also thank Professor W. A. Anderson for the use of the SEM.

REFERENCES

1. H. S. Kwok, E. Yablonovitch and N. Bloembergen, Phys. Rev. A23, 3094 (1981).

2. H. S. Kwok and E. Yablonovitch, Appl. Phys. Lett. 30, 158 (1977) and Appl.
    Phys. Lett. 27, 583 (1975).

3. A. Gibson, C. Hatch, P. Maggs, D. Tilley and A. Walker, J. Phys. C: Sol.
    St. Phys. 9, 3259 (1976).

4. M. Hasselbeck and H. S. Kwok, to be published in Appl. Phys. Lett.

5. J. M. Liu, R. Yen, H. Kurz and N. Bloembergen, Appl. Phys. Lett. 39, 755
    (1981).

6. D. Auston, C. Surko, T. Venkatesan, R. Slusher and J. Golovchenko, Appl.
    Phys. Lett. 33, 437 (1978).

THERMODIFFUSION - A NEW APPROACH TO PLASMA SELF-CONFINEMENT

A. FORCHEL, B. LAURICH
Physikalisches Institut, Teil 4, Universität Stuttgart,

G. MAHLER
Institut für theoretische Physik, Universität Stuttgart,
Pfaffenwaldring 57, D-7000 Stuttgart 80, Germany

ABSTRACT

From a model for the thermodiffusion of high density
plasmas an electronic mechanism for plasma self-confinement is derived. Due to the increase of the confinement densities with the density of states masses
this effect should be most important in laser annealing experiments in Si but much weaker in GaAs. A fast
diffusion of the surface generated plasma is found to
enhance the impact ionization rate.

INTRODUCTION

The physics of pulsed laser annealing of ion-implanted amorphized
semiconductors is discussed mainly using two controversial models.[1,2]
For the thermal model of laser annealing one usually assumes that the
energy transferred originally from the laser beam to the electronic system of the semiconductor is passed immediately to the lattice. If the
excitation is sufficiently strong the lattice is heated above the equilibrium melting point and it recystallizes after the laser is switched
off.[1]

Van Vechten et al.[2] have proposed an alternate plasma annealing
model. They emphasize that the high excitation transfers a significant
fraction of electrons from bonding states in the valence band into
antibonding states in the conduction band. For electron-hole plasma
densities of several $10^{21} cm^{-3}$ in Si calculations predict a softening
of the transverse accoustic phonon mode and the semiconductor changes
from the solid to the liquid phase.[2,3] This leads to a plasma density
dependent melting temperature $T_m(n)$ which can be substantially lower
than the equilibrium melting temperature: For Si and $n=8 \cdot 10^{21} cm^{-3}$ the
calculations extrapolate to $T_m=0K$. Melting temperature reductions for
several hundred Kelvin are expected to arise for densities of about
$1 \cdot 10^{21} cm^{-3}$ and should be detectable experimentally.

The maximum plasma densities reached in experiments are mainly controlled by i) the fast spatial diffusion (i.e. expansion) of the electron-hole plasma and ii) the Auger recombination in the plasma.

The purpose of this paper is twofold. First we present theoretical
results for the diffusion of the plasma under the combined influence of
temperature and density gradients. This thermodiffusion approach predicts for sufficiently high excitation intensities a negative effective
thermodiffusivity D_{eff} which leads to a strong carrier confinement.[4]
Particularly in Si the confinement is found to effectively counterbalance the density reduction due to the plasma expansion.

Mat. Res. Soc. Symp. Proc. Vol. 13 (1983) ©Elsevier Science Publishing Co., Inc.

The second part addresses the influence of the plasma diffusion on the recombination probabilities. We calculate the statistical factors of Auger recombination and impact ionization as functions of the drift velocity of the carriers. Assuming a parabolic model for the band structure and a wave vector independent matrix element we find for drift velocities of the order of $4 \cdot 10^7 \text{cm/s}$ a change from Auger recombination to impact ionization..

PLASMA-CONFINEMENT BY THERMODIFFUSION

Inhomogeneous carrier generation in high excitation experiments leads to a spatially varying density and temperature. The corresponding gradients in the temperature and the chemical potential $\mu(n,T)$ of the plasma induce most generally the flow of a particle current j_m and of a heat current j_q. These currents can be related to generalized thermodynamial forces X_i by use of phenomenological Onsager relations:[4]

$$j_m = L_{mm} X_m + L_{mq} X_q \qquad (1)$$

$$j_q = L_{qm} X_m + L_{qq} X_q \qquad (2)$$

with $\quad X_m = \nabla\left(-\frac{\mu}{T}\right) \quad X_q = \nabla\left(\frac{1}{T}\right)$

By a kinetic approach the Onsager coefficients can be related to the isothermal ambipolar diffusivity, the internal energy and the chemical potential of the plasma. The relationship of the Onsager coefficients and the chemical potential leads to a dependence of the transport equations on the semiconductor band structure. We solve the Onsager relations assuming an effective one-dimensional geometry. Experimentally this assumption is verified by the particularly strong gradients of the excitation perpendicular to the excited surface and the comparatively weak lateral gradients. For simplicity we assume that the heat current vanishes. This means that the processes of plasma heating by the Auger effect counterbalance approximately the energy transfer to the lattice.

Using the dependence of the chemical potential on density and temperature eqns.(1,2) are transformed into equations for the density and the temperature gradients:

$$\frac{dn}{dx} = D_{eff}^{-1} \cdot j_m \qquad (3)$$

$$\frac{dT}{dx} = T^2 \cdot L_{qm} \cdot N^{-1} \cdot j_m \qquad (4)$$

$$D_{eff}^{-1} = \frac{-T}{\left(\frac{\partial\mu}{\partial n}\right)_T \cdot N} \cdot \left(L_{qq} + \left[T \cdot \left(\frac{\partial\mu}{\partial T}\right)_n - \mu \right] \cdot L_{qm} \right) \qquad (5)$$

$$N = L_{qq} \cdot L_{mm} - L_{qm}^2 > 0$$

Here we have written the equation for the density gradient in formal analogy to the simple Fick's law of isothermal diffusion $dn/dx = -D^{-1}j_m$. For arbitrary values of the density and temperature at the surface $n(x=0)$ and $T(x=0)$ spatial profiles of density and temperature can be calculated from the simultaneous solution of eqns.(3-5). We calculate the Onsager coefficients as functions of n and T from an ideal Fermi gas model. For the isothermal ambipolar diffusivity D and the particle current spatially constant values are used which define an effective length scale for the profiles $x \sim D/j_m$.

Fig. 1: Spatial profiles
of the plasma density in
Si under laser annealing
conditions (calculated).
Dashed line: Isothermal
diffusion. Solid line:
diffusion under the
condition $j_q=0$.

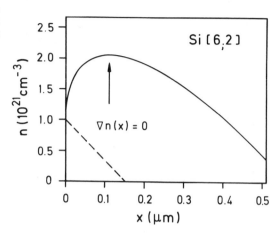

Fig.1 displays calculated spatial distributions of the plasma density
in Si for an initial density $n(x=0)=1\cdot10^{2j}cm^{-3}$, $T(x=0)=1200K$, $D=10cm^2/s$
$j_m=7\cdot10^{26}(cm^2s)^{-1}$. The dashed curve depicts a density profile calculated
for these parameters from Fick's law assuming isothermal diffusion. As
expected the density decreases rapidly with increasing distance from
the surface. The solid line shows the corresponding profile if thermodiffusion effects are included. In contrast to the isothermal case the
thermodiffusion leads to the formation of a pronounced density maximum
in the sample and to very extended density profiles. For the example
depicted in Fig. 1 the number of carriers in the high density region of
the sample is increased by about a factor of ten if thermodiffusion is
taken into account. As j_m and hence the excitation intensity is the
same in both cases Fig. 1 demonstrates that the thermodiffusion effects
give rise to a strong plasma confinement.

For the experimental study of a plasma induced reduction of the
lattice melting temperature the knowledge of materials favoring thermodiffusive confinement would be useful. Experimentally one finds that
the threshold intensities for laser annealing of Si, Ge and GaAs are
the same within a factor of two. At the densities used in pulsed laser
annealing experiments the Auger effect governs the recombination. Since
the Auger coefficients of these materials are rather similar the equivalent threshold energies correspond to very comparable carrier densities at the onset of laser annealing. Assuming that in all materials
plasma annealing were observed the calculation for Si [3] calibrates the
density at the annealing threshold to about $10^{21}cm^{-3}$.

In contrast to the independence of the densities required for laser
annealing of the particular material the calculated confinement by
thermodiffusion is strongly band structure dependent as shown in Fig.2.
Here we have plotted the maximum plasma densities $n_{max}(T)$ for GaAs, Ge,
and Si; i.e. the solutions of $dn/dx(n_o,T)=0$ from eqn.(4) are used to
represent the entire profiles. The combined bandstructure and temperature dependence of n_{max} is approximately given by:

$$n_{max} = 4.2 \cdot 10^{16} (\mu_D T)^{3/2} cm^{-3} \qquad (6)$$

where $\mu_D^{-1} = m_{de}^{-1} + m_{dh}^{-1}$

is the reduced density of states mass, in m_0; T in Kelvin. This relation is shown for the three materials in Fig. 2.

Fig. 2: Temperature dependence of the densities at maximum plasma confinement in Si, Ge and GaAs.

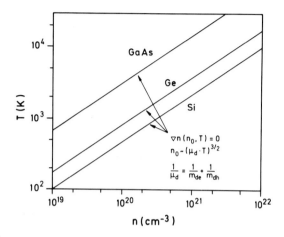

In Si the plasma temperature for $n_{max}=10^{21} cm^{-3}$ amounts to about 2000K. Therefore the comparatively small temperature difference of about 500K between the electronic and the lattice system would be sufficient to obtain a carrier density for which the melting temperature of the crystal is expected to decrease by about 300K.[3] In GaAs a confinement density of $10^{21} cm^{-3}$ is reached for temperatures of 10^4 K only. These temperatures (which often exceed the entire kinetic energy of the carriers), if achieved at all, will drop rapidly to the lattice temperatures thus preventing the occurance of noticable plasma confinement effects. So we conclude that the plasma annealing effects due to thermodiffusion should be obtained most easily for Si but would be more difficult to observe in GaAs.

AUGER RECOMBINATION AND IMPACT IONIZATION OF FAST DIFFUSING PLASMAS

Experimental values of Auger coefficients are usually obtained from moderately excited highly doped materials [5],i.e. the majority carrier density is homogeneous in the sample. In laser annealing experiments, on the other hand, electrons and holes are generated by the excitation pulse. Due to the gradients in density and temperature plasma thermodiffusion arises. The velocities of the plasma diffusion have been obtained for moderately excited GaAs [6] and Si [7] at low temperatures (T<30K). In both materials the drift velocities approach the saturated drift velocities usually measured from carriers in high electric fields.

As the distribution function represents the net momentum of the carriers one expects significantly different distributions for a plasma without and with diffusion. The drift induced changes are shown schematically in Fig. 3. In the "equilibrium" ($v_D=0$) system the carriers occupy states up to the quasi Fermi levels. In the ambipolar drifting plasma the carriers occupy states above the Fermi level in the direction of the drift. Opposite to the drift direction the population is reduced. Significant changes of the distribution are expected if the Fermi velocity and the drift velocity are comparable.

Fig. 3: Drift effects on the distributions of electrons and holes - schematic.

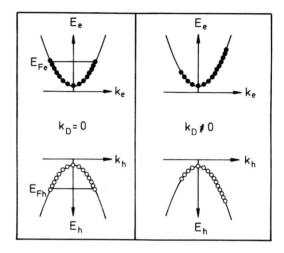

At low temperatures (T≤50K) and moderate densities (n≤10^{19}cm^{-3}) the plasma drift influences mainly the emission and transmisson spectra. For the high densities and temperatures typical for laser annealing we discuss here the influence of the plasma drift on the Auger coefficient.

Due to the plasma drift important variations of the combined Auger recombination and impact ionization rates may arise under laser annealing conditions.

The recombination probability for a phonon assisted Auger process can be written as [8]:

$$W = A \int |M|^2 S \, \delta(E_1 + E_2 - E_1 - E_2) \, d\vec{k}_1 , d\vec{k}_2 , d\vec{k}_1 d\vec{k}_2$$

A = constant, M = transition matrix element

using the δ-function to describe energy conservation. Here the indices 1 and 2 describe the initial states and 1′ and 2′ the final states in the valence and the conduction bands and

$S(n,T,v_D) = S_A - S_I$ is the statistical factor with

$S_A = f(E_1) \cdot f(E_2) \cdot f(E_1,) \cdot \{1-f(E_2,)\}$

$S_I = \{1-f(E_1)\} \cdot \{1-f(E_2)\} \cdot \{1-f(E_1,)\} \cdot f(E_2,)$

S_A = Auger recombination, S_I = impact ionization (the inverse process), $f(E) = f(E,n,T,v_D)$ "shifted Fermi" distributions

A general solution of the integral is very complex, due to e.g. the wave vector dependence of the matrix element. For the sake of simplicity we are considering in the following a parabolic band structure and k-independent matrix elements only. In this case the integration extends only over a statistical factor. Both statistical factors depend strongly on the distribution functions. In order to estimate the drift effects on the combined Auger recombination and impact ionization we calculate the net probability from both using Fermi functions which have been rigidly displaced in momentum space by the drift vector $k_D(v_D)$ from the band edges:

$$R(n,T,v_D) = B \int S(n,T,v_D) \sqrt{E_1} \sqrt{E_2} \sqrt{E_1}, \sqrt{E_2}, \delta dE_1 dE_2 dE_1' dE_2'$$

In Fig. 4 we show the ratio $R(n,T,v_D)/R(n,T,0)$ as a function of the drift velocity. The parameters of the model band structure used in the calculation are $m_e = m_h = 1$ and a band gap energy $E_g = 600 meV$. For a temperature of 1000K and densities of $1 \cdot 10^{20} cm^{-3}$, $5 \cdot 10^{20} cm^{-3}$, and $1 \cdot 10^{21} cm^{-3}$ the sum of Auger rate and impact ionization rate is found to depend strongly on the carrier drift velocity (see Fig. 4). For high densities and low velocities a slight increase of the Auger rate is obtained from our model. At high enough velocities (about $4 \cdot 10^7 cm/s$) the impact ionization exceeds the Auger recombination.

Fig. 4: Ratio of the difference of the combined Auger recombination and impact ionization probabilities for a plasma with and without diffusion as a function of the drift velocity. The parameter is the plasma density. See text.

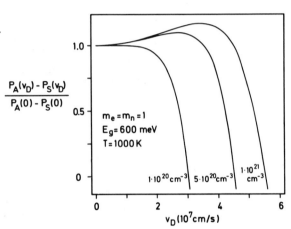

SUMMARY

We have shown that the two important mechanisms which limit the values of the plasma densities available in high excitation experiments provide probably less severe restrictions as previously believed. In Si the thermodiffusive confinement counterbalances effectively the spatial expansion. Under the conditions of pulsed laser annealing the plasma drift is expected to decrease the Auger rate.

REFERENCES

1. W.L.Brown, Proc. of the 15th Int. Conf. on the Phys. of Semicond., in J. Phys. Soc. Jpn. 49, Suppl. A 1271 (1980).
2. J.A.Van Vechten, R.Tsu, F.W.Saris, Phys. Lett. 74A, 422 (1979).
3. J.Bok, Phys. Lett 84A, 448 (1981).
4. G.Mahler, G.Maier, A.Forchel, B.Laurich, H.Sanwald, W.Schmid, Phys. Rev. Lett. 47, 1855 (1981).
5. J.Dziewior, W.Schmid, Appl. Phys. Lett. 31, 346 (1977).
6. K.M.Romanek, H.Nather, J.Fischer, E.O.Göbel, J. Lum. 25, 585 (1981).
7. A.Forchel, B.Laurich, to be published.
8. A.Haug, Sol. State Electr. 21, 1281 (1978).

TEMPERATURE CHARACTERIZATION OF PULSED LASER ANNEALING OF SEMICONDUCTORS

D.L. KWONG* AND D.M. KIM**
* Dept. of Electrical Engineering, University of Notre Dame,
Notre Dame, Indiana 46556
**Dept. of Electrical Engineering, Rice University, Houston, Texas 77251

ABSTRACT

The dynamic characteristics of lattice temperature, T attained in Si under the influence of a high power laser beam irradiation are examined analytically over a wide range of laser wavelengths, pulse intensities and durations. The strongly temperature-dependent material parameters, such as optical absorption coefficient $\alpha(T)$ and thermal diffusivity $D(T)$ are incorporated in heat diffusion equation, and their nonlinear coupling effect on the ensuing temperature in laser-processed semiconductors are examined. Specifically, the threshold pulse energy for surface melting is characterized as a function of both material and annealing laser beam parameters. This analytic description of pulsed laser heating of semiconductors should provide considerable insight into the transient heating phenomena and localized material modifications.

INTRODUCTION

The unique characteristics of laser beam processing technique in fabrication of semiconductor devices could prove to be essential for the realization of VLSI and/or VHSI[1-4]. Full utilization of its inherent advantages is contingent upon the quantitative understanding of the underlying physical mechanisms. A quantity of interest is the transient depth profile of lattice temperature, T which ultimately determines the electrical characteristics of laser-processed semiconductors. The transient depth profile of T generated in Si annealed by a high-power laser beam is mainly determined by two parameters (i) the energy deposition depth, $\alpha^{-1}(T)$ and (ii) the heat diffusion length, $(D\tau)^{1/2}$, where $\alpha(T)$ is the optical absorption coefficient, $D(T)$ is the thermal diffusivity, and τ is the laser heating time. For the case of Si, these two parameters drastically decrease with increasing T[5,6]. Hence the deposition depth of the laser energy is reduced rapidly during laser pulse, and the heat conduction into the bulk of the sample is concomitantly slowed down. In this paper, the dynamics of T attained in Si during pulsed laser annealing is analytically characterized over a wide range of laser wavelengths, pulse intensities and durations, incorporating both temperature-dependent $\alpha(T)$ and $D(T)$.

LASER HEATING OF SILICON

For typical nonosecond laser annealing, the screening effect of phonon emission does not play a significant role in the coupling of laser energy to the lattice[7,10]. As a result, the absorption depth of the laser beam mainly determines the heating profile and the temperature, T grows in time

Mat. Res. Soc. Symp. Proc. Vol. 13 (1983) ©Elsevier Science Publishing Co., Inc.

according to the following time-dependent, nonlinear heat diffusion equation:

$$\frac{\partial T}{\partial t} = \frac{\partial}{\partial x} D(T) \frac{\partial T}{\partial x} + \frac{(1-R)}{c\rho} I\alpha(T)e^{-\alpha(T)x} \qquad (1)$$

Here, since the specific heat, $c(T)$ and mass density, $\rho(T)$ are weakly dependent on T, we regard these two quantities to be constant. The boundary conditions are:

$$\frac{\partial T}{\partial x}\bigg|_{x=0} = 0, \; T(x \to \infty, t) = 0 \qquad (2)$$

For the case of Si, the data of the thermal diffusivity $D(T)$ can be approximated by[5]

$$D(T) = \frac{D_R}{1+aT}$$

with $D_R = 0.94$ cm^2 sec^{-1} and $a = 0.0072/°C$. The measured absorption coefficient $\alpha(T)$ can be approximately fitted as[6]

$$\alpha(T) = \alpha_R e^{T/T_R}$$

where

$$\alpha_R = 1340 \text{ cm}^{-1} \qquad T_R = 427°C \qquad \text{at} \qquad 693 \text{ nm}$$

$$\alpha_R = 5020 \text{ cm}^{-1} \qquad T_R = 430°C \qquad \text{at} \qquad 532 \text{ nm}$$

$$\alpha_R = 9130 \text{ cm}^{-1} \qquad T_R = 434°C \qquad \text{at} \qquad 435 \text{ nm}$$

T can be examined conveniently by partitioning it into two parts:

$$T(x,t) = T_0(x,t) + \Delta T_0(x,t) \qquad (3)$$

Here, T_0 is assumed to grow in time due to the primary laser heating with a constant optical absorption coefficient, α_0 and to diffuse in space with a constant thermal diffusivity, D_0; viz.

$$\frac{\partial T_0}{\partial t} = D_0 \frac{\partial^2 T_0}{\partial x^2} + \frac{I(1-R)}{c\rho} \alpha_0 e^{-\alpha_0 x} \qquad (4)$$

The values of D_0 and α_0 are regarded as parameters. The solution of T_0 can be obtained by using Green's function technique, and an extensive discussion of T_0 was presented previously.[8,9] The nonlinearity of the problem arising from $D(T)$ and $\alpha(T)$ can be incorporated by the secondary source function. Now we insert Eqs. (3) and (4) into Eq. (1), and quasilinearize the resulting equation by partitioning ΔT_0 further as

$$\Delta T_0(x,t) = T_1(x,t) + \Delta T_1(x,t) \qquad (5)$$

where T_1 obeys the equation

$$\frac{\partial T_1}{\partial t} = D_0 \frac{\partial^2 T_1}{\partial x^2} + S_1(T_0) \tag{6}$$

with

$$S_1 = [D(T_0) - D_0] \frac{\partial^2 T_0}{\partial x^2} + \frac{\partial D(T_0)}{\partial x} \frac{\partial T_0}{\partial x} \tag{7}$$

$$+ \frac{I(1 - R)}{c\rho} (\alpha(T_0)e^{-\alpha(T_0)x} - \alpha_0 e^{-\alpha_0 x})$$

Note that S_1 results from: (i) the local departure of net influx of heat associated with the temperature dependence of thermal diffusivity from D_0, i.e., $D(T_0) - D_0$, and (ii) the departure of local lattice heating rate $\alpha(T_0)\exp(-\alpha(T_0)x)$ from $\alpha_0\exp(-\alpha_0 x)$, namely $\alpha(T_0)\exp(-\alpha(T_0)x) - \alpha_0\exp(-\alpha_0 x)$. S_1 becomes an explicit function of material depth x, heating time t, laser power density I, and the parameters α_0, D_0 via $T_0(x,t)$. Again T_1 can be treated exactly with the use of the Green's function.

In our analysis the optimal values of effective thermal diffusivity, D_0 and optical absorption coefficient, α_0 are determined self-consistently. Since S_1 is specified in terms of T_0, which, in turn, depends on D_0 and α_0, S_1 depends on the values of D_0 and α_0. Now D_0 and α_0 are chosen in such a way that S_1 is minimal. The detailed algebra for this determination will be reported elsewhere, and the final results are presented here. When the values of D_0 and α_0 thus found are inserted back into the analytic expression of T_0, one obtains the complete, explicit expression for the dominant temperature term. The values of α_0 and D_0 given in terms of both material parameters (a, c, R, ρ, α, D) and laser beam parameters (I, t) is one of the interesting results of our analysis.

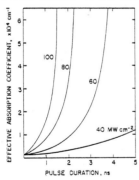

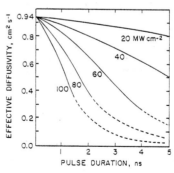

Fig. 1 Effective optical absorption coefficient, α_0 vs. heating time for different pulse intensities at 694 nm.

Fig. 2 Effective diffusivity, D_0 vs. heating time for different pulse intensities at 694 nm. The dashed line is the extrapolation of the single phase annealing.

Figures 1 and 2 show the values of α_0 and D_0 as a function of heating time for several different pulse intensities. With increasing heating time, the lattice temperature rises, and the effective mean thermal diffusivity should decrease, while the effective optical absorption coefficient increases rapidly. Furthermore, the increase in α_0 results in raising the temperature near the surface of the sample due to the enhanced heating rate therein. This, in turn, decreases D_0 and the heat diffusion into the bulk material is slowed down. This slowing down of the diffusion will further raise the surface temperature and therefore will significantly enhance the absorption coefficient. The dynamic feedback of these two processes reduces substantially the energy deposition depth of the laser beam during the irradiation and also confines the absorbed energy near the surface region of the material. Under these conditions, the surface heating efficiency is considerably enhanced, and the threshold pulse energy for the surface melting is drastically reduced. The other important feature is that the rate with which D_0 decreases and α_0 increases in time depends sensitively on the operating pulse intensity. For example, at $I = 40$ MW cm^{-2} and pulse duration 1 nsec, $D_0 = 0.88$ cm^2/sec., and $\alpha_0 = 5 \times 10^3$ cm^{-1}, while at $I = 100$ MW cm^{-2} with the same pulse duration D_0 is reduced to 0.3 cm^2/sec. and α_0 is increased to 2×10^5 cm^{-1}. Because T_0 becomes large with a small D_0 and a large α_0 one can clearly observe that the efficiency of lattice heating depends sensitively on the pulse intensity used.

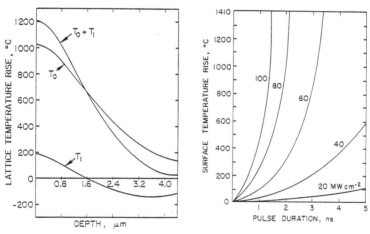

Fig. 3 Depth profile of temperature rise.

Fig. 4 Surface temperature rise vs. heating time for different pulse intensities at 694 nm.

In Fig. 3 we present the depth profile of $T_0 + T_1$, T_0 and T_1 for a fixed pulse intensity and energy. As can be clearly observed, the first order correction term, T_1 is much smaller than T_0. The maximum contribution of T_1 never exceeds 20% of T_0 for pulse intensities varying from 10 to 100 MW cm^{-2}. This rapid convergence is clearly due to the optimal determination of effective absorption coefficient and thermal diffusivity in our perturbation scheme. For practical purposes one may, therefore, approximate the laser heated lattice temperature rise by T_0 together with α_0 and D_0. The effect of the above mentioned nonlinear heating phenomena can be more clearly seen in Fig. 4, where we have plotted the surface temperature rise as a function of several different pulse intensities. At 100 MW cm^{-2}, the effective thermal diffusion time, $\tau = 4\alpha_0^2 D_0 t$ is long enough to increase the heated volume beyond the energy deposition depth, α^{-1}. Because of this rapid transfer of energy into the bulk material the rise in temperature near the surface is small. With increasing pulse intensity, and therefore with increasing heating rate, a rapid reduction in D_0 ensues, and the heat diffusion time slows down considerably. Furthermore, due to the shrinking of the energy deposition depth, the laser deposited energy is mostly confined near the surface. These two effects, therefore, drastically enhance the growth rate of the surface temperature.

The characteristics of pulsed laser heating of a Si lattice is summarized in Fig. 5, in which the threshold laser energy for the onset of surface melting is plotted as a function of pulse intensity for differentt laser wavelengths. At 694 nm wavelength for 20 MW cm^{-2} the energy needed is about 0.42 Joule cm^{-2}, while only 0.2 Joule cm^{-2} is sufficient to melt the surface at 100 MW cm^{-2}. This suggests that a pulsed ruby laser, having energy $\sim$ 1.0 Joule cm^{-2} and time duration $\sim$ 20 nsec is capable of melting the surface of Si wafer and of propagating the melt front well beyond the energy deposition depth. The thermal melting model for annealing is consistent with our results. In addition, the threshold pulse energy for surface melting depends sensitively on the operating frequency of the laser beam used. This is seen in Fig. 6, in which the threshold laser energy required to melt the silicon surface is plotted as a function of wavelength for several different intensities. The dependency of threshold pulse energy on wavelength is most pronounced at low intensity region. As the pulse intensity is increased however, the difference in threshold energy becomes small, which is consistent with experimental results.

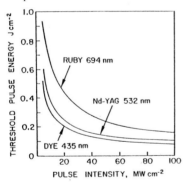

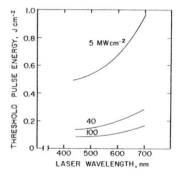

Fig. 5 Threshold pulse energy for surface melting vs. pulse intensity at several wavelengths.

Fig. 6 Threshold pulse energy for surface melting vs. wavelength for different pulse intensities.

CONCLUSION

We have presented in this paper an analytical description of the dynamic characteristics of lattice temperature rise in Si annealed by a high power laser beam. Specifically, the effect of the strong coupling between rapidly reduced thermal diffusivity and optical absorption depth during the laser beam irradiation has been discussed and the kinetics of the depth

profile of the lattice temperature and the threshold pulse energy for surface melting are characterized in terms of both the material and laser beam
parameters. This analysis provides considerable insight into the transient
heating phenomena and localized material modifications that can be achieved
by exposing a semiconductor to pulsed laser irradiation, and gives a rigorous analytical description of temperature distribution and a precise estimation of the pulse duration required for the surface of the sample to reach melt
condition at a given pulse intensity. Our calculation did not incorporate
the sudden jump in surface reflectivity value, which occursduring the laser
pulse. This may raise the values of our threshold pulse energy for the onset of surface melting approximately by a factor of 2. Our analytical
method can be applied to the real-life processes involved in semiconductor fabrication with the generalizations such as two-phase annealing,
cooling of lattice temperature after the laser beam irradiation, and   impurity redistribution. This will be reported elsewhere.

REFERENCES

1. Laser-Solid Interactions and Laser Processing -1978 (Material Research
    Society, Boston), edited by S.D. Ferries, H.J. Leamy, and J.M. Poate
    (AIP Conference Proceedings, New York, 1979), vol. 50.

2. Laser and Electron Beam Processing of Materials, edited by C.W. White
    and P.S. Peersey (Academic, New York, 1980).

3. Laser and Electron-Beam Solid Interactions and Material Processing,
    edited by J.F. Gibbons, L. Hess, and T. Sigmon (North-Holland, New York,
    1981).

4. Laser and Electron-Beam Interactions with Solids, edited by B.R.
    Appleton and G.K. Celler (North Holland, New York, 1982).

5. C.J. Glassbrenner and G.A. Slack, Phys. Rev. $\underline{134}$, 1058 (1964).

6. G.E. Jellison, Jr. and F.A. Modine, ORNL/TM 8002 (1982).

7. A. Lietola and J.F. Gibbons, Appl. Phys. Lett. $\underline{34}$, 332 (1979).

8. D.M. Kim, D.L. Kwong, R.R. Shah, and D.L. Crosthwait, Jr., J. Appl.
    Phys. $\underline{52}$, 4995 (1981).

9. D.M. Kim and D.L. Kwong, IEEE J. Quantum Electron. $\underline{QE-18}$, 224 (1982).

10. E.J. Yoffa, Phys. Rev. $\underline{B21}$, 2445 (1980).

SUBSTRATE LATENT HEAT EFFECTS IN THE CALCULATIONS FOR PULSED
LASER IRRADIATED THIN FILMS

J.J. CAO AND K. ROSE
Center for Integrated Electronics, Rensselaer Polytechnic Insti-
tute, Troy, New York 12181
O. AINA, W. KATZ, AND J. NORTON
General Electric, Corporate Research and Development, Schenectady,
New York 12309

ABSTRACT

We present a numerical model for the calculation of the temperature rise caused by pulsed laser irradiation of a thin film/sub-
strate structure. This model includes phase changes in both the
thin film and the substrate. The inclusion of phase changes
results in more complex thermal behavior and significantly affects
melt durations. This model was applied to the AuGe/GaAs system.
Morphological observation using the scanning electron microscope
and SIMS profiles provides experimental verification for the
numerical calculations.

INTRODUCTION

Thermal models of laser beam annealing allow calculations of
the space and time dependence of temperature in a material. These
temperature distributions can be used to predict crystallization
and atomic redistribution. The accuracy of the prediction depends
on the sophistication of the model.

Thermal models of all degrees of sophistication have been
developed for the laser beam annealing of homogeneous materials.
These vary from the simple one-dimensional analytical model of
Ready [1] to numerical models which include temperature dependent
parameters, melting, melt boundary motion, and evaporation [2,3].
Although analytical models have been developed for bimaterial
structures [4,5] no numerical calculations have been made for such
systems.

We have constructed a numerical model for bimaterial structures
which takes into account the latent heat of melting of both the
thin film and the substrate. This model has been applied to the
laser annealing of AuGe films on GaAs substrates for ohmic contact
formation. The temperature rise as a function of laser energy
density as calculated by an analytical model will be compared to
calculations by numerical models which take GaAs melting into consideration. The numerical model predicts a shorter melt duration
than the analytical model.

The numerical calculations were verified by SEM examination and
SIMS profiling. Calculations of the expected outdiffusions
obtained from the numerical model are shown to correlate well with
values estimated from the SIMS profiles. The experimental techniques, SEM examinations of morphology, and SIMS profiles have been
reported previously [6].

Mat. Res. Soc. Symp. Proc. Vol. 13 (1983) ©Elsevier Science Publishing Co., Inc.

NUMERICAL THERMAL CALCULATIONS

The temperature distribution caused by a laser beam incident on a sample can be obtained by solving the heat flow equation. If all thermophysical parameters are constant, this is a linear differential equation with analytical solutions. However, if these parameters are temperature dependent, if there is a phase change (such as melting) or a moving boundary condition, the equation is non-linear and cannot be solved analytically. Numerical techniques are required.

Integrating the heat flow equation over t and x we obtain an expression for the energy absorbed per unit area. To adapt it for computer calculation, the integrations are replaced by summations over sufficiently small time intervals, Δt. Thus

$$E_a(t) = \Sigma \; \rho c \; \Delta T \; \Delta x \; - \; \Sigma \; K \frac{\Delta T}{\Delta x} \; \Delta t \qquad (1)$$

where $\Delta T = \int \frac{\partial T}{\partial t} dt$. Equation (1) is a statement of the energy balance within the material. It indicates that part of the absorbed laser energy is stored within the material, giving a temperature rise, ΔT. The rest of the absorbed laser energy flows away from the surface at a rate proportional to the temperature gradient, as shown in the second term of Equation (1). From Equation (1) one can calculate the temperature rise for a layer Δx at the time intervals Δt. In our calculation $\Delta t = 0.1$ns and $\Delta x = 150$ nm is the thickness of the AuGe film. Because of the high absorption coefficient of AuGe, all the energy is absorbed within the AuGe film.

When the temperature rises to the AuGe eutectic temperature, the latent heat of melting of AuGe must be considered in the calculations. This can be done holding the temperature in a layer constant until the energy for melting has been supplied by the laser beam. Thus, the temperature will not rise until the difference between the absorbed laser energy and the heat flow from the melting layer equals the latent heat. After this, the temperature is allowed to rise.

When the temperature at the AuGe/ GaAs interface increases to the GaAs melting point (1238°C), the GaAs layer below it begins to melt. The interface between liquid and solid GaAs gradually moves from the AuGe/GaAs interface into the GaAs. At the time I-1, GaAs has been melted to a depth of X_m while a thin layer, of thickness ΔX_m is being melted at the liquid/solid interface. The thickness of GaAs, ΔX_m melted during time interval I is given by,

$$K_2 \; \Delta t \; \frac{T(I-1) \; + \; \Delta T(I) \; -1238}{X_m} \; -Q(I) = \Delta X_m \rho_2 L_2 \qquad (2)$$

T(I-1) is the temperature of the first layer during the time interval I-1, $\Delta T(I)$ is the temperature increase in the same layer at time interval I, and Q(I) is the heat flow away from the liquid/solid interface into the solid GaAs. K_2, ρ_2, and L_2 are the thermal conductivity, density and latent heat of GaAs.[2] Equation (2) is a statement of energy balance within the layer ΔX_m. The energy used to melt the layer is equal to the net flow of heat into the layer.

The energy balance for the whole material is

$$E^\circ = Q(I) + \Delta X_m \, \rho_2 L_2 + \Delta T(I) \, [c_1 \rho_1 \, \Delta_x + \tfrac{1}{2} \, c_2 \rho_2 X_m] \quad (3)$$

where E° is the laser energy absorbed within the time interval I. The temeprature increase at the AuGe/GaAs interface, $\Delta T(I)$, and the melt depth increase in the GaAs layer, ΔX_m, can be calculated for each time interval from equations (2) and (3). The GaAs continues to melt until the temperature at the surface decreases to 1238°C. The GaAs begins to solidify at this point and the solid/liquid interface moves back towards the surface.

After GaAs resolidification, the AuGe temperature starts to decrease from 1238°C to 363°C. When the surface temperature decreases to 363°C, the AuGe begins to resolidify. The latent heat of resolidification released during a time interval is conducted away into the GaAs. In this time interval the temperature of the AuGe remains constant.

RESULTS

Figures 1 and 2 show curves for the AuGe/GaAs interface temperature obtained from analytical and numerical calculations, respectively. The thermal model which gave Figure 1 was discussed previously [6]. It assumed a rectangular laser pulse, a reflectivity of 0.74, and constant thermophsical parameters for the AuGe layer. The numerical model assumed the same reflectivity and thermophysical parameters but applied triangular laser pulses of the same energy density and half width as the corresponding rectangular pulses. Wood and Giles [3] have shown that this is a good approximation to a gaussian pulse shape. (Our numerical model gave results in good agreement with Wood and Giles [3] when applied to their situations.)

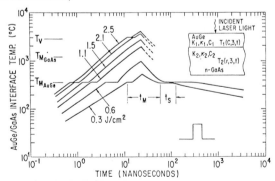

Figure 1: Temperature Rise at AuGe/GaAs Interface. Analytical Calculation

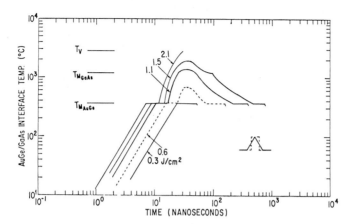

Figure 2:   Temperature Rise at AuGe/GaAs Interface.
Numerical Calculation

Comparing the analytical and numerical results we see that the
temperature is always lower for the numerical calculation. For
example, at 0.6 and 1.1 J/cm^2 the maximum interface temperature is
700 and 1400°C for the numerical model compared with 1000 and
2000°C for the analytical model. Including the latent heat associated with melting GaAs flattens the temperature peak.

Calculation of the melt duration is shown in Figure 3. This is
the sum of the time the melt stays above the melting point and the
solidification time, $(t_M + t_S)$. The melt duration calculated by
the numerical technique is a factor of 4 lower than that calculated from the analytical technique.

Atomic redistributions determined by SIMS profiling can be correlated with the numerical calculations as shown in Figure 4. The
extent of outdiffusion can be estimated from the SIMS profiles in
Fig. 3. of Ref. [6] as the distance of the intercept for a particular energy density from the intercept for the as deposited
sample. The intercepts were determined by extrapolating the
sloping parts of the profiles to the depth where the Ga signal
drops to e^{-1} of the signal intensity at the interface for the as
deposited sample.

The extent of outdiffusion can be obtained from the time-dependent interface temperature by estimating the characteristic dif-
fusion length, L_D. L_D is obtained from $L_D^2 = \Sigma_i^N L_{Di}^2$ by dividing
the melt duration into a sequence of intervals of length Δt.
L_{Di}^2 = 4 D(T) Δt where T is the interface temperature associated
with the ith time interval. $D(T) = 2.7 \times 10^{10} T^{1.7}$ is determined
by assuming that the diffusion coefficients of As and Ga in molten
AuGe will be the same as the self-diffusion coefficient of molten

AuGe [7]. Considering the difficulty of accurarely calculating
and measuring the parameters and the fact that no adjustable parameters were used, the agreement between the numerical calculations
and the experimental values in Figure 4 is quite good.

These results are consistent with SEM studies of the dependence
of surface and subsurface damage on laser energy density. Fig. 4
indicates that Ga and As outdiffuse close to the surface at 1.5 J
cm^{-2}, but the AuGe alloy should not vaporize until 2.1 J cm^{-2}.
Fig. 1 of Ref. [6] shows that damage begins to appear at the surface for 1.5 J cm^{-2} and is clearly seen for 2.1 J cm^{-2}.

CONCLUSIONS

We have shown that one can numerically calculate the evolution
of temperature with time for a laser annealed bimaterial system.
These results differ substantially from those obtained by more
approximate analytical calculations. Furthermore, these results
are in quantitative agreement with SEM images of contact morphology and SIMS measurements of outdiffusion. We conclude that
numerical calculations are necessary to accurately model melting
and phase boundary shifts in bimaterial systems.

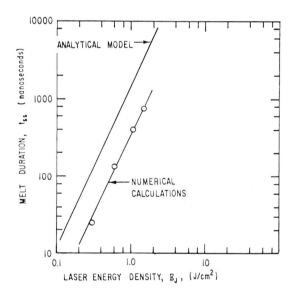

Figure 3: Laser Energy Dependence of Melt Duration for
           Analytical and Numerical Calculations.

122

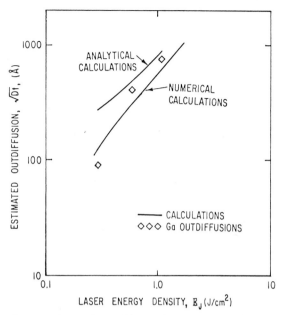

Figure 4:   Estimated Outdiffusion Lengths in the AuGe Melt

REFERENCES

[1]   J.F. Ready, J. Appl. Phys. 3b, 462 (1965).

[2]   C.M. Surko, et al., Appl. Phys. Lett. 34, 635 (1979).

[3]   R.F. Wood, G.E. Giles, Phys. Rev. B 23, 2923 (1981).

[4]   D. Maydan, Bell Sys. Tech. J., 50, 1761 (1971).

[5]   H.G. Parks, Ph.D. Thesis, Renssealer Polytechnic Institute, (1980).

[6]   O. Aina, W. Katz, G. Smith, P. Norton, K. Rose, "Laser and Electron Beam Interactions with Solids", B.R. Appleton and G.K. Cellar, eds. (North-Holland, NY 1982) p. 671.

[7]   R.B. Fair, J. Appl. Phys. 50, 6552 (1979).

LASER INDUCED TEMPERATURE RISE IN SEMICONDUCTORS : ANALYTICAL
SOLUTIONS, APPLICATION TO THE TRANSIENT

ALAIN MARUANI, Y.I. NISSIM, F. BONNOUVRIER and D. PAQUET
Centre National d'Etudes des Télécommunications*,196 rue de Paris 92220
BAGNEUX (FRANCE)

ABSTRACT

It is shown how the systematic use of the method of integral
transforms greatly simplifies the calculation of the temperature
rises in laser irradiated media. In general, this method leads
ultimately either to analytical results or to very simple numerical integrals (e.g. no poles, exponential kernels).
We focus here on the analytical results, and discuss some aspects of CW laser heating, for large surface absorption,
including radial dependance, depth dependance and transient
nonlinearities. The new results derived in this treatment
are in good agreement with experimental data from other
studies.

INTRODUCTION

The problem of the temperature rise in semiconductors is of both practical
and fundamental interest, stimulating much experimental and theoretical work
[1,4]. While the direct determination of the temperature is not an easy task,
especially in the transient regime, the calculations lean generally on the Green
function method, which in principle solves any kind of problem. This
method consists of expanding the solution in a series of plane waves. The genera-
lity of this approach can mask the simplifications that could be provided by
inspection of the symmetries of the system. On the other hand the use of the
integral transform method is the natural way to take into account those symmetries since its consists basically of expanding the solution in terms of functions naturally suited to them. A good example of application of that concept is
the case of a laser beam with a circular symmetry. Obviously that symmetry is
reflected in the temperature rise. Since the Fourier transform of the radial
part of usual functions is their Hankel transform it is clear that Hankel transforms, as was shown by Lax [1], are a very good approach of this kind of problem.
The systematic use of integral transforms for the calculation of temperature
rise of semiconductors is proposed here. It will be shown that it enables i) to
obtain previous analytical results, ii) to give an analytic expression to previous numerical results, iii) to provide new solutions. In the first section the
formalism is presented and discussed, in the second one some analytical results
are shown for the CW case and in the third one, the results for the transient
regime is introduced. Those new results are closely connected to the studies of
scanning laser beams. Other practical examples will be examined in a forthcoming
publication [5].

---

* Laboratoire associé au CNRS (LA250)

Mat. Res. Soc. Symp. Proc. Vol. 13 (1983) ©Elsevier Science Publishing Co., Inc.

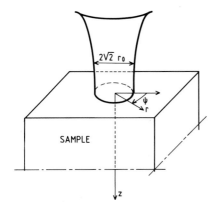

Fig. 1 : Geometry of the problem

THE FORMALISM

The starting point is the linear Fourier equation for the temperature rise T. We assume that the laser beam is described by its intensity distribution in space and time : $I = I_0 \, f(r/ro) \, \Phi(t/\tau)$ where ro and τ are some characteristic parameters of the beam and Io includes reflectivity terms. We introduce the reduced quantities $R = r/ro$, $Z = z/zo$, $\theta = t/\tau$, $W = \alpha ro$ and $\eta = K\tau/Cr_0^2$ where α is the linear absorption coefficient, C the specific heat and K the thermal conductivity. Let $D = \alpha I_0 r_0^2/K$ be a characteristic temperature of the system, then the equation reads, for the system represented in fig.1 :

$$\frac{1}{R} \frac{\partial}{\partial R}\left(\frac{R\partial T}{\partial R}\right)+ \frac{\partial^2 T}{\partial Z^2} - \frac{1}{\eta}\frac{\partial T}{\partial \theta} = - D \, f(R) \, \Phi(\theta) \, \exp\text{-}WZ \qquad (1)$$

submitted to the initial condition : $(\partial T/\partial Z)_{Z=0} = 0$ \qquad (2)

The relevance of this condition is discussed in [5]. The integral transforms are now introduced : Since we deal with semi-infinite samples, T is Laplace transformed with respect to Z : $T(R,Z,t) \rightarrow T_L(R,p,t)$; since the laser intensity has no angular dependance, T is Hankel transformed with respect to R: $T_L(R,p,t) \rightarrow T_{L,H}(\lambda,p,t)$; finally since the system is time invariant T is Fourier transformed with respect to the time : $T_{L,H}(\lambda,p,t) \rightarrow T_{L,H,F}(\lambda,p,\omega) \sim \tilde{T}(\lambda,p,\omega)$. The same transforms are performed on the source term and one is left with the algebraic equation equivalent to (1) and (2) :

$$(p^2-\lambda^2 - \frac{i\omega}{\eta}) \, \tilde{T}(\lambda,p,\omega) - pT_{H,F}(\lambda,Z=0,\omega)- \frac{\partial T_{H,F}(\lambda,Z=0,\omega)}{\partial Z} = - D \, \frac{F(\lambda) \, \phi(\omega)}{W + p} \qquad (3)$$

where $F(\lambda)$ $(\phi(\omega))$ is the Hankel (Fourier) transform of $f(R)$ $(\Phi(\theta))$. The method consists in solving (3) for T and then going back to the original, ordinary variables. In principle this method would involve the theory of complex integration of several variables. While the details of the calculation are presented in [5], this presentation will be limited to some specific cases.

THE CASE OF CW LASER HEATING - LINEAR TEMPERATURE RISE

If we assume the laser to be constant for all times, (3) is solved as :

$$T(\lambda,p) = \frac{p\,T_H(\lambda,\ Z=0) + \frac{\partial T_H}{\partial Z}(\lambda,Z=0) - \frac{DF(\lambda)}{W+\lambda}}{p^2 - \lambda^2} \tag{4}$$

the first inverse transform is performed on Z, and involves $\exp\pm\lambda Z$. The non divergence of T for infinite Z and the initial condition (2) enable us to solve for the unknown $T_H(0)$ and $(\partial T_H/\partial Z)_{Z=0}$. Then going back to R it is found that :

$$T(R,Z) = D\int_0^{\infty} Jo(\lambda R)F(\lambda)\ \frac{W\exp(-\lambda Z)-\lambda\exp(-WZ)}{W^2 - \lambda^2}d\lambda\ , \tag{5}$$

a formula already derived by Lax [1]. It was not necessary here to consider explicitly the homogeneous equation associated with (1), since initial conditions were accounted for through the Laplace transform in Z. If the laser is described by $f(r) = \exp\text{-}r^2/2r_0^2 = \exp\text{-}R^2/2$, its Hankel transform is $F(\lambda)=\exp\text{-}\lambda^2/2$ (The conventions used here are slightly different from that of ref[1] : in this reference, the dimension of λ is an inverse length and that of F a surface. Here, both λ and F (λ) are dimensionless).

Since the lowest typical value for W is 10, the surface absorption model appears very reasonable. In that limit, one can derive a good, approximate analytical expression for (5), which are presented in [5]. We will just formulate two results of practical importance :

i) Radial dependance

The integration for T(R,Z=0) is straightforward, and provides :

$$T(R,Z=0)=(\frac{\Pi}{2})^{\frac{1}{2}}\ \frac{D}{W}\ (\exp - \frac{R^2}{4})\ I_0(\frac{R^2}{4}) = T_{max}\ \eta(R) \tag{6}$$

where I_0 (x) is the zeroth order modified Bessel function.

ii) Depth dependance

The integration for T(R=0,Z) is also straightforward and provides :

$$T(R=0,Z) =(\frac{\Pi}{2})^{\frac{1}{2}}\ \frac{D}{W}\ (\exp \frac{Z^2}{2})\ \ \text{erfc}\ (\frac{Z^2}{2})^{\frac{1}{2}} \tag{7}$$

Where erfc is the complementary error function. Both expressions (6) and (7) are also the result of the integrations of expressions calculated by Nissim et al. (4), who considered that problem through a Green function method, and treated the results numerically. Expressions (6) and (7) can be approximated very accurately by elementary functions.

TRANSIENT REGIME - TRUE TEMPERATURE RISE

The technology of material laser processing involves the use of scanning laser beams. Two problems of crucial importance are to be answered i) what is the maximum reasonable scanning speed (i.e. such that no correction for the incident power has to be performed) and, for that speed, ii) what is the true (i.e. nonlinear) temperature rise. This problem can be approached from the point of view of the transient regime by heating with a sudden, but fixed laser. Instead of heating the sample with a beam with a time dependence:

$$\Phi(t) = 1 \left[\phi(\omega)=\frac{\delta(\omega)}{\sqrt{2\Pi}}\right], \text{ the time dependance } \Phi(t)=Y(t) \left[\phi(\omega)=\frac{-i}{\sqrt{2}\Pi}(PP(\frac{1}{\omega})+i\Pi\delta(\omega)\right]$$

is used, where $Y(t)$ is the Heaviside unit step function, and PP stands for principal part.

The second question is handled by the Kirchhoff transform, which gives the true temperature rise, providing that the linear regime, and the temperature dependance of the thermal conductivity are known. The integration of (3) deals with imaginary cuts in the complex plane. But in the case of highest interest : $T(R=0, Z=0 ; t)$ which concerns the maximum temperature rise, the solution is remarkably simple. Within the surface absorption model and with a thermal conductivity written as $K(T)=A/(T-B)$ [3], the true temperature is :

$$T(t) = T_1 \left[\exp(\frac{2}{\Pi})^{\frac{1}{2}} \frac{K}{299} \frac{P_{abs}}{2\Pi Kr_0} \text{ atan } \frac{1}{r_0} (\frac{2Kt}{C})^{\frac{1}{2}}\right] - T_2 \qquad (8)$$

with the values $T_0=27°C$, and $T_1= 199$, $T_2= 174$, $K=K(T_0)$, $T(t)$ is expressed in Celsius degrees.

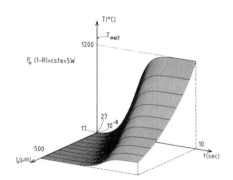

Fig.2 : Maximum true temperature induced in silicon, on all logarithmic 3-D representation.

Fig. 2 shows a 3-D plot of the actual temperature of the sample as a function of the laser beam waist and the time. The incident power is constant. This graph is very suggestive of the nonlinearities of the system ; however, the dependance on the beam waist, is best shown if one considers experimental parameters such that the stationary temperature as constant. Fig.3 shows the actual temperature as a function of time, with different beam waists and for constant final temperature $T(t=\infty)=1200°C$. It is seen that moderate changes of r_0 shift the curves in a significant way. Fig.4 represents t_2, the instant for which the temperature rise has reached 90 % of its maximum value, as a function of r_0. The curve is parabolic. Note the relatively large values of t_2 on this graph ; the insert of fig.4 concerns typical values used in laser annealing

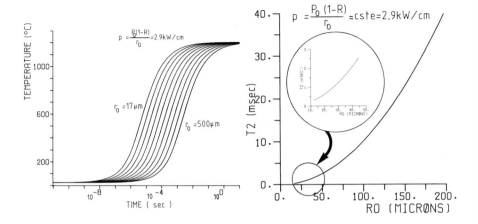

Fig.3 : Temperature rise for different beam waists vaying geometrically from 17 to 500 μm. The same final temperature is kept for all r_0.

Fig.4 : Variation of t_2 (time at which the temperature rise reaches 90% of its maximum value) as a function of beam waist.

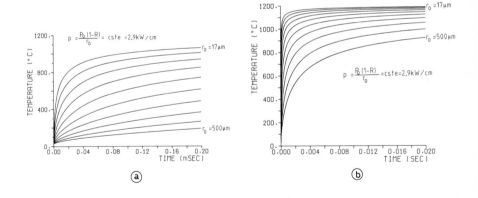

Fig.5 : Dynamics of the heating for two different time scales a) short scale : b) large scale.

128

for instance, for r_0 = 17 µm and P_{abs} ~ 5W ; t_2 is around 300 µm.
  Actually a graphical representation logarithmic in time is somewhat misleading in the sense that it favors smaller times : fig.2 and 3 would suggest that
the initial temperature rise is smooth, whereas the initial time derivative of
the temperature is infinite. Fig.5 shows the true temperature as a function of
time, in natural units, and emphasizes the abrupt initial temperature rise.
From those curves, it is possible to derive a value for the maximum reasonable
speed in scanning experiments. For instance with P_{abs} =5W and r_0=17µm, one does
not need to correct for the incident power as soon as the time exceeds 0.1msec
which corresponds to a scanning speed of about 20cm.sec^{-1}(a value determined
empirically and independantly in experimental studies of silicon annealing
[6]).

CONCLUSION

  Systematic use of integral transforms provides great simplifications in the
calculation of the temperature rise in laser heated systems. The simplification
originates from the fact that the use of basic functions adapted to the symmetry of the system is equivalent, in some respect, to a preintegration of the
general formulae obtained by the Green function method. In principle, any kind
of problem is solved through a development of the solution of eq.(3). However
problems of practical interest are immediately solved either analytically (or
in the most general case through rapidly convergent numerical integrals). The
solution for the transient, presented here is one of the possible illustrations
of the versability of the method.

REFERENCES

1. M. Lax, J. Appl. Phys. 48, 3919 (1977)

2. H.E. Cline and T.R. Anthony, J. Appl. Phys. 48, 3895 (1977)

3. Y.I. Nissim, A. Lietoila, R.B. Gold and J.F. Gibbons, J. Appl. Phys. 51,
   279 (1980)

4. S.A. Kokorowski, G.L. Olson and L.D. Hess in "Laser and Electron Beam Solid
   Interactions and Material Processing" (J.F. Gibbons, L.D. Hess and
   T.W. Sigmon eds) p.139 (1981)

5. A. Maruani, Y.I. Nissim, D. Paquet and F. Bonnouvrier to be published.

6. F. Ferrieu and G. Auvert J. of Appl. Phys. (in Press).

PART II
CRYSTALLIZATION OF AMORPHOUS SEMICONDUCTORS, SOLID AND LIQUID PHASE PHENOMENA

SURVEY OF THE THERMODYNAMICS AND KINETICS OF CRYSTALLIZATION OF Si and Ge

D. TURNBULL
Division of Applied Sciences, Harvard University, Cambridge, MA 02138, USA

ABSTRACT

The thermodynamic interrelations of the crystalline (c-sc), amorphous semiconducting (a-sc) and liquid metal (ℓm) states of Si and Ge, based on the existing thermal data, are reviewed. Then the kinetics of the interfacial processes in the growth of c-sc and a-sc into undercooled ℓm and the conditions for transition from c-sc to a-sc growth are discussed.

The main content of my oral presentation was set forth in papers which have appeared or will appear elsewhere, see especially in references 1 and 2. Here I will give only a summary of the points developed in those papers which seem most pertinent to this symposium.

We are concerned with the thermodynamic and kinetic interrelations of the crystalline semiconducting, c-sc, amorphous semiconducting, a-sc, and liquid metal, ℓm, phases of Ge and Si at virtually zero pressure. The present inform- ation indicates that a-sc is less stable than c-sc from 0°K to temperatures well in excess of, $T_{c\ell}$, the thermodynamic melting point of c-sc, but a-sc is more stable than the amorphous metallic phase at 0°K. It follows that, if crystallization were bypassed, there would be some temperature, $T_{a\ell}$, well under $T_{c\ell}$ below which transition from undercooled ℓm to a-sc would be thermodynamic- ally favored. Also it is known that the transitions a-sc $\rightarrow$ c-sc and ℓm $\rightarrow$ c-sc take place by nucleation and growth and that the melting of c-sc occurs hetero- geneously, i.e., by movement of a c-sc-ℓm interface.

Perhaps the major thermodynamic issues are whether or not the a-sc, which, as formed, is generally in a configurationally frozen state, can relax to and exist in a metastable state and the nature of the transition from such a meta- stable state to ℓm--e.g. would the transition be thermodynamically continuous or discontinuous? If a-sc has a structure, e.g. as modelled by a continuous random network (CRN), such that its crystallization must occur reconstructively, it should be, just as are amorphous SiO_2 and GeO_2, capable of reaching a fully relaxed metastable state.

It was suggested (1) that the time constant for configurational relaxation of the a-sc state is not likely to exceed that for a diffusive jump in the c-sc phase. On this basis and the assumption that the activation energy, Q, for a diffusive jump far exceeds the thermal energy the following expression was obtained for the upper limiting temperature, T_x, at which relaxation should be virtually complete:

$$T_x \overset{\sim}{<} \left[\frac{Q\dot{T}}{R} \tau(T_x) \right]^{1/2} \tag{1}$$

where $\dot{T}$ is the heating rate-assumed constant, R is the gas constant and $\tau(T_x)$ is the time constant for relaxation at T_x. At the usual heating rates T_x should be of the order of the glass temperature, T_g, of a-sc which had been estimated (2) by an analogous procedure.

When a-sc specimens are heated, rapid crystallization usually intervenes in some temperature range centering at a temperature T_{kc} somewhat below T_x. It is instructive to scale these temperatures with $T_{c\ell}$: thus $T_{rx} = T_x/T_{c\ell}$ and $T_{rkc} = T_{kc}/T_{c\ell}$. On this scale T_{rkc} is typically of the order of 0.05 less than the estimated T_{rx}.

Mat. Res. Soc. Symp. Proc. Vol. 13 (1983) ©Elsevier Science Publishing Co., Inc.

The experimentally determined enthalpies of crystallization, ΔH_{ac}, of a-sc Ge and Si are reviewed in reference 1. Excepting one reported measurement, the several determinations of ΔH_{ac} of a-Ge films prepared by a considerable variety of techniques are in good agreement and average about 0.31 of the enthalpy of crystallization $\Delta H_{c\ell}$, of the ℓm phase. This agreement and the relatively small displacements of T_{rx} from T_{rkc} suggest that the measured ΔH_{ac}'s closely approach that of the fully relaxed a-sc state. In actual magnitude, ΔH_{ac} of a-sc, amorphized by ion damage, was roughly equal to that of a-Ge (see reference 3 for latest measurements and review). However, its scaled (with $\Delta H_{c\ell}$) value, 0.235, was considerably below that of a-Ge. The lowest values of T_{rkc}, relative to T_{rx}, were exhibited by films amorphized by ion damage, probably reflecting lower impurity drag on crystal growth owing to the higher purity of these films. The calorimetrically determined (3) rates of epitaxial regrowth into ion amorphized films of both Si and Ge proved to be in good agreement with the rates obtained by Czepregi $et\ al.$ (4) by monitoring the position of the a-c interface by Rutherford back scattering.

Calculation of the thermodynamic nature of the hypothetical transition between ℓm and fully relaxed a-sc, and the temperature at which it occurs, requires, besides ΔH_{ac}, the heat capacity of a-sc and its temperature dependence and a model evaluation of the transition entropy, ΔS_{ac}, at some temperature. The method of calculation and its application to the Ge a-sc $\rightarrow$ ℓm transition were given independently by Bagley and Chen (5) and Spaepen and the author (6). Assuming that the transition was first order the calculated equilibrium temperature for Ge was $T_{a\ell} \sim 0.78\ T_{c\ell}$. A similar calculation (3) using the recently determined ΔH_{ac} measurements gave $T_{a\ell} \sim 0.84\ T_{c\ell}$ for Si.

According to these calculations the ℓm phase could transform to a-sc, by a nucleation and growth process, when undercooled to temperatures below $T_{a\ell}$. Actually the ℓm phase of Ge has been undercooled to temperatures as low as $0.8\ T_{c\ell}$ with no measurable crystal nucleation or evidence, from either flow or crystal growth behavior, of a-sc formation. These observations seem to rule out a continuous, but not discontinuous, ℓm $\rightarrow$ a-sc transition in Ge within the temperature range $(1.0-0.8)\ T_{c\ell}$.

In the steady state analysis of the movement of a planar crystallization front in a pure liquid, the crystal-liquid interfacial undercooling, $T_{c\ell} - T_i$, where T_i is the crystal-liquid interfacial temperature, is proportional to the negative of the thermal gradient, $(\text{grad } T)_i$, in the crystal at the interface. When the interfacial rearrangement rates are very high, as in growth in pure undercooled melts, the displacements of T_i from equilibrium, though definite, are quite small for gradients of the usual magnitude. However, in high energy fluence processing $(\text{grad } T)_i$ can become so large that even when the interfacial process is extremely rapid T_i can be displaced far from equilibrium (1). Thus there was the possibility that in such processing of Si and Ge crystals T_i during the regrowth phase could be depressed well below $T_{a\ell}$. In this event either a-sc or c-sc, depending on kinetics and $T_{a\ell} - T_i$, could be deposited during regrowth into the ℓm overlay. Spaepen and the author (7) have suggested that because of lesser configurational constraints the interfacial rearrangement frequency of ℓm $\rightarrow$ a-sc should be more rapid than that of ℓm $\rightarrow$ c-sc. Thus while the thermodynamic factor would always favor regrowth to c-sc the kinetic factor would favor deposition of a-sc. Therefore at sufficiently large $T_{a\ell} - T_i$ a-sc would be deposited by regrowth following its initiation at the c-sc-ℓm interface.

Actually there have been several demonstration (8,9,10,11), beginning with that of Liu $et\ al.$ (8), that, under certain conditions a-sc is deposited from ℓm overlays on c-Si following high energy laser pulses of several to 10^3 picoseconds. The reverse transition, a-sc $\rightarrow$ ℓm, could occur during rapid heating to $T_{a\ell}$ if the frequency of nucleation, homophase plus heterophase, remained so low (e.g. $< 10^4/cm^3sec$) that crystallization would not be appreciable. If the transition is first order a large superheating, over $T_{a\ell}$, should be required (2) for measurable $homophase$ nucleation of ℓm in fully relaxed a-sc but heterogeneous nucleation should occur at internal defects (e.g. voids) in the film or on its external surface, if oxide free, at only slight superheating. In view of the rapid $\ell m \rightarrow$ a-sc regrowth kinetics the growth of ℓm; once nucleated, should be extremely rapid. Indirect evidence of melting of a-sc Si to ℓm at temperatures well below $T_{c\ell}$ was reported by Baeri et $al.$ (12) and Cullis et $al.$ (13) and was reviewed by Poate (14) in this symposium. In contrast Kokoroski et $al.$ (15) (see also this symposium (16)) report that the regrowth rate of c-sc $\rightarrow$ a-sc in Si persists, within experimental error, on a singly activated Arrhenius course from low temperatures to temperatures well beyond the calculated $T_{a\ell}$ and nearing $T_{c\ell}$. However, the author pointed out (2) that, provided the temperatures were correctly evaluated, this behavior would require that the hypothetical equilibrium temperature, $T_{a\ell}$, between a-sc and c-sc must lie far above $T_{a\ell}$ with the corollary that $T_{a\ell}$ must lie well below $T_{c\ell}$. Such apparent superheating of a-sc is very difficult to explain unless it is assumed that the $nucleation$ of the ℓm phase was somehow suppressed owing to the absence of such heterogeneous nucleation sites as e.g. internal voids or oxide free external surfaces.

ACKNOWLEDGMENT

This paper was written during the author's 1982 stay at the National Bureau of Standards as a visiting Scientist in the Center for Materials Science. It is based on research supported in part by a Harvard NSF grant NSF DMR-80-12933 and N.B.S.

REFERENCES

1. D. Turnbull, J. de Physique, Colloque, in press (1983).

2. D. Turnbull, Mats. Res. Soc. Symp. Proc. (ed. by S.T. Picraux and W.J. Choyke), North Holland, Amsterdam 7, pp. 103-108 (1982).

3. (a) E.P. Donovan, F. Spaepen, D. Turnbull, J.M. Poate and D.C. Jacobson, submitted to Appl. Phys. Letters.
   (b) E.P. Donovan, Ph.D. thesis, Harvard University, Cambridge, Mass.,1982.

4. L. Czepregi, E.F. Kennedy, J.W. Mayer and T.W. Sigmon, J. Appl. Phys. 49, 3906 (1978).

5. B.G. Bagley and H.S. Chen, Am. Inst. Phys. Conf. Proc. 50, 97 (1979).

6. F. Spaepen and D. Turnbull, Am. Inst. Phys. Conf. Proc. 50, 50 (1979).

7. F. Spaepen and D. Turnbull, "Laser Annealing of Semiconductors" (ed. J.M. Poate and J.W. Mayer), pp. 15-42, Academic Press, N.Y. (1982).

8. P.L. Liu, R. Yen, N. Bloembergen and R.T. Hodgson, Appl. Phys. Lett. 34, 864 (1979).

9. R. Tsu, R.T. Hodgson, T.Y. Tan and J.E. Baglin, Phys. Rev. Lett. 42, 1356 (1979).

10. J.M. Liu, R. Yen, E.P. Donovan, N. Bloembergen and R.T. Hodgson, Appl. Phys. Lett. 38, 617 (1981).

11. A.G. Cullis, A.C. Webber, N.G. Chew, J.M. Poate and P. Baeri, Phys. Rev. Lett. 49, 219 (1982).

12. P. Baeri, G. Foti, J.M. Poate and A.G. Cullis, Phys. Rev. Lett. 45, 2036 (1980).

13. A.G. Cullis, H.C. Webber and N.G. Chew, Appl. Phys. Lett. 40, 998 (1982).

14. J.M. Poate, this volume.

15. S.A. Kokorowski, G.L. Olson and L.D. Hess, J. Appl. Phys. 53, 94 (1982).

CHARGED DANGLING BONDS AND CRYSTALLIZATION IN GROUP IV SEMICONDUCTORS

P.J. GERMAIN,* M.A. PAESLER,+ AND D.E. SAYERS
Department ofPhysics, North Carolina State University, Raleigh, NC 27650
K. ZELLAMA
Université Paris VII, Groupe de Physique des Solides de L'E.N.S.
Place JUSSIEU, 75221 Paris Cedex 05

ABSTRACT

Crystallization of amorphous Ge (or Si) has been studied as a function of temperature and the flux of ionizing radiation (or doping). The crystallization growth rate v_g takes on the form $v_g = v_0 \exp(-E/kT)$ where v_0 is an increasing function of flux (or doping). We propose the following to explain these data: A concentration of mobile dangling bonds (DBs) exists in the bulk and near the amorphous-crystalline (a-c) interface. Ionization and doping induce transitions from the uncharged state D^0 to the charged states D^+ and D^-. The process controlling crystallization resulting in the above activation energy is discussed. Only certain sites on the a-side of the a-c interface are available for crystallization, and these sites are those which have captured DBs. The charged D^+ and D^- states have a larger capture cross section than the uncharged D^0 state. Increased concentrations of charged DBs results in an enhancement of the prefactor in the above equation.

INTRODUCTION

The issue of energy flow during pulsed laser annealing experiments performed on ion-implanted Si has lain at the heart of a fundamental controversy [1,2]. One school of thought suggests that a photo-generated electron-hole plasma may carry a large fraction of the energy of a short ($\sim$10ns) laser pulse, and that the network temperature is increased by only about 300K at 10 ns after the pulse. Many other authors conclude that the energy of the laser pulse is effectively immediately transferred to the network. They report network temperature rises of $\sim$1000K. We propose that there is a third channel into which energy may flow and that consideration of this channel may be crucial to the understanding of crystallization phenomena as well as many atomic transport effects. This channel is provided by the large number of localized states in the band gap of the amorphous (a) Si prior to laser irradiation. Presuming the thermal annealing model, we submit that an appreciable fraction of the thermalized electrons and holes are trapped in gap states producing an ionization. Our model may also impact upon the plasma annealing model. The notion of band gap ionization has been used to interpret permanent photo-conduction [3]. As far as the network temperature is concerned, assuming a thermal annealing model, consideration of the ionization of band gap states results in a 0.1K effect on the network temperature.

DETERMINATION OF CRYSTALLIZATION GROWTH VELOCITIES

Ionization-enhanced crystallization in a-Ge. From electron microscopy [4] and conductivity [7] measurements, the crystallization growth velocity v_g for a-Ge has determined for a large range of temperature T. Summarized in Figure 1a these results have been plotted as log v_g versus 1000/T. The activation energy of v_g is found to be $\sim$1.5 ev, agreeing with photoemission data [6].

*on leave from Université Paris VII, address above
+MAP acknowledges the support of the General Electric Corporation.

Germain et al. [5] have observed that a flux ϕ = 0.9 μA/cm2 of 1 MeV electrons enhances v_g (see Figure 1a). In addition, these authors measured v_g at fixed T, varying the electron flux (see Figure 2a). From these results, Germain et al. [8] have written v_g in the form

$$v_g = v_0(\phi) \exp(-E/kT) \tag{1}$$

where v_0 (ϕ) does not depend on T and is an increasing function of ϕ, and E is independent of T and ϕ.

Thermal crystallization of doped and un-doped a-Si. Csepregi et al. [9] have used channeling to observe the recrystallization of ion-implanted Si. The results are plotted in Figure 1b. The variation of v_g at a constant T as a function of the concentration of implanted dopant (P) is shown in Figure 2b, where v_g increases with P concentration up to a doping level $C_D \sim 2 \times 10^{20}$ cm^{-3}.

Above this concentration, v_g is not strongly dependent on doping. For no doping and for B or As doping, the activation energies are equal to 1.5 eV within

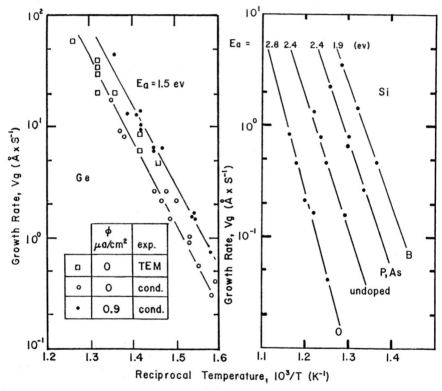

Reciprocal Temperature, $10^3/T$ (K^{-1})

Fig. 1a. Crystallization growth rate v_g for a-Ge as a function of reciprocal temperature for samples with and without electron ionization. Data are from reference 4 (squares) and reference 5 (circles).

Fig. 1b. Crystallization growth rate v_g for a-Si as a function of reciprocal temperature for several doping conditions. Data are from reference 6.

the range of the error. For doping with B, a slightly lower (1.9eV) activation
energy was determined. From the data of ref. 9, we estimate the error in the
activation energies to be $\sim$ 0.2 eV [10]. Thus within the range of the error,
allelements resultin nearly equal activation energies with O having a slightly higher and B perhaps a slightly lower value. The results of the channeling
experiments may be expressed in the form

$$v_g = v_o(C_D) \exp (-E'/kT) \qquad (2)$$

where $v_o(C_D)$ does not depend on T and is an increasing function of the doping
level C_D. Furthermore, E' is independent of T and may vary slightly with doping.
   cw laser induced crystallization. Gibbons and co-workers [11] have measured
v_g of self-implanted Si under laser radiation for 800 $\leq$ T $\leq$ 900° C. They compare their results to those of ref. 9 and find an enhancement of two orders of
magnitude. The Hughes group [12] have made measurements of the cw laser-
induced SPE regrowth rate (v_g in our notation) of As implanted a-Si using time-
resolved reflectivity. On calculating the threshold power necessary to raise
the surface temperature to the melting point, these authors find this value to
be in good agreement with experiments. The simultaneous use of doping and irradiation does not seem to significantly affect v_g. Recent Hughes work (see
following paper) has extended these investigations to include undoped Si. We
incorporate these data into our model elsewhere. [13].

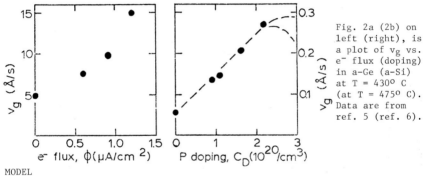

Fig. 2a (2b) on
left (right), is
a plot of v_g vs.
e^- flux (doping)
in a-Ge (a-Si)
at T = 430° C
(at T = 475° C).
Data are from
ref. 5 (ref. 6).

MODEL

   The dangling bond (DB) is the defect which is most frequently proposed in
attempts to understand the properties of group IV amorphous semiconductors. An
atom which is covalently bonded to only three neighbors has only seven electrons, D^0 in our notation, and will either: i) give up an unpaired electron to
the matrix, thus playing the role of donor, D^+ or ii) capture an electron, thus
playing the role of acceptor D^-. Because of the electrostatic repulsive
energy associated with the arrival of the eighth electron, the energy level
corresponding to the acceptor will be higher than that of the donor. A band
structure model has been proposed by Mott [14] which presents the electronic
density of states for the DB defects. In this model there are two bands of
localized states in the gap, one of donor states and one of acceptor states.
This compensation produces the pinning of the Fermi level. In order to move
the Fermi level, it is necessary to introduce a concentration of dopants which
is at least of the order of the concentration of the DBs, C_{DB}. If one dopes with
an increasing concentration of P or As (B) atoms, then for $C_D \lesssim C_{DB}$, the
Fermi level will remain pinned, but for $C_D \gtrsim C_{DB}$, the Fermi level will move
towards the conduction (valence) band and the concentration of negatively
(positively) charged DBs will increase.

The effect of radiation on an updoped a-group IV semiconductor by energetic
(1MeV) electrons [5] or photons [11] can be understood as follows. Free electron-hole pairs (in the case of ref. [7] $\sim 10^{19}$ cm^{-3}s^{-1}) are generated by the
radiation, and if we admit that an appreciable fraction of the DBs are initially in the neutral state, then the free electron-hole pairs generated will be
rapidly trapped on the DBs, changing their charge states to plus or minus. We
might describe this interaction as $2D^o \rightarrow D^+ + D^-$.

For the laser experiments [12] one must consider the effects of a laser beam
on a sample with $C_D \sim C_{DB}$. We show below that for such doping most of the DBs
near the a-c interface will be in the D^- state prior to irradiation. If such a
sample is irradiated at a low level, then some DBs will participate in the reaction $D^- \rightarrow D^o$ and the growth rate would be smaller, that is, v_g would be a
decreasing function of the flux. If however high level radiation is present
then in the limit where the effect of doping is negligible, most DBs will be in
the charged states and the growth rate should be an increasing function of the
flux. The Hughes experiments either correspond to a situation for which the
effects of doping and photo-ionization or "photo-doping" roughly cancel, or
where the range of doping was simply not large enough to result in a measure-
able effect given the size of the error. [15].

Diffusion of dangling bonds. Thomas et al. [16] prepared a-Si under UHV
conditions and showed for samples deposited at room temperature that 7×10^{19}
cm^{-3} randomly distributed and isolated spins in the bulk may be inferred. They
further showed that the spin concentration decreases with annealing temperature from which they were led to the conclusion that DBs diffuse from the bulk
toward the surface as a result of annealing [17]. The results of ref. 9 on
the crystallization of ion-implanted a-Si may be compared to the results of
Thomas et al. because of the purity of the bombarded samples. Indeed, Mayer
[18] has shown that the samples evaporated by Thomas [16] and studied during
crystallization by Zellama [19] have the same behavior as the samples of ref. 9.

The microscopic growth mechanism at the a-c interface. Theoretical models
[20], [21] of the surface energy of c-Si agree on one point: the DBs of the
surface atoms are not neutral, rather there is an alternation of doubly occu-
pied and empty states, i.e. D^- and D^+. We know of no calculations of the surface energy of the external surface of a-Si, but find it reasonable to trans-
pose the crystalline results to the a-surface. Thus we propose that the surface DBs on both sides of the interface present an alternation of doubly occu-
pied and empty configurations. We suggest that as crystallization proceeds,
charged DBs diffuse (driven, if applicable, by an electric field) towards the
a-c interface. A non fully co-ordinated charged atom at the interface would
capture the diffusing DBs resulting in a decrease of the co-ordination number
of the atom on the a-side of the interface, allowing it to jump to the c-side.
An atom that has captured a DB shall be defined as being "available" for crystallization. The exact nature of the diffusion, capture and jump are not proposed, for example several DBs might be captured at one site. Presumably a
similar mechanism involving neutral DBs exists, but the efficiency for such a
mechanism is lower. Crucial to our model are only: i) the diffusion of
charged DBs, ii) capture at the a-c interface resulting in a reduction of the
co-ordination number of the a-interfacial atom, and iii) jump of this atom to
the c-side.

Origin of the growth rate activation energy. Assuming a steady state of
crystallization, the growth rate v_g takes the form

$$v_g = \alpha d \nu' \tag{3}$$

where

$$\nu' = \nu \ \exp \left(\frac{-\Delta G}{kT} \right) \left[\left(1 - \exp \ \left(\frac{\Delta g'}{kT} \right) \right) \right] ; \tag{4}$$

d is the distance of the jump across the a-c interface; ν is an atomic frequency; ΔG is the free energy barrier; $\Delta g'$ is the difference in free energy between the a- and the c- state; and α is the fraction of atoms on the a-side available for crystallization.

This fraction α may be a function of several parameters, e.g. the orientation of the crystal [22], impurities, dangling bond concentration, etc. We treat here the effect of DBs on α holding all other relevant parameters affecting α as fixed. This is the situation which applies to the experiments we are discussing.

If there are N_S surface atoms, N_a of which are available for crystallization, then $\alpha = N_a/N_S$ and the rate of change of N_a is given by

$$\frac{dN_a}{dt} = (N_S - N_a)\nu_{DB} - N_a\nu'. \tag{5}$$

In the above, $N_S - N_a$ is the number of sites which may capture a DB and ν_{DB} is a frequency characteristic of the capture. ν_{DB} should be thermally activated. The frequency ν_{DB} which globally describes the phenomenon should be a function of : i) the concentration of DBs in the space charge of the bulk, ii) the diffusion of DBs in the bulk, and iii) the magnitude of the electric field at the interface. The first term on the right hand side of Equation (5) is related to the creation of available sites. The second term represents an annihilation due to crystallization. Assuming a steady state, $dN_a/dt = 0$, Equation (3) and (5) lead to

$$v_g = (\frac{\nu'\nu_{DB}}{\nu' + \nu_{DB}})d. \tag{6}$$

We consider three limiting cases of Equation (6):

a) For $\nu' \ll \nu_{DB}$, the expression for v_g depends only on ν', the frequency of the jump of an atom from the a- to the c- side of the interface. Thus v_g is independent of the DB concentration, and this case will never represent the data of Figures 1 and 2.

b) For $\nu' \gg \nu_{DB}$, it must be the case that $v_g = \nu_{DB}d$ and v_g takes the form $v_g = d\nu_{DB_0} \exp(-E_{DB}/kT)$. This may be compared to Equations (1) and (2) and the data of Figures 1 and 2 where E_{DB} represents the activation energy for the diffusion of DBs and/or the electrostatic barrier of their capture at the interface. In order to allow comparison with Equation (1), E_{DB} must be independent of ϕ and should depend slightly on C_D. Furthermore, in order to describe the data, ν_{DB_0} should be an increasing function of the DB concentration, which is reasonable.

c) If ν' and ν_{DB} are roughly the same order of magnitude and have roughly the same activation energy, we may write

$$\nu' = \nu'_0 \exp(-E'/kT) = \nu'_0 \exp(-E_{DB}/kT) \tag{7}$$

and v_g becomes

$$v_g = \left[\frac{\nu_{DB_0} \nu'_0}{\nu'_0 + \nu_{DB_0}} d\right] \exp(-E'/kT)$$

where the factor in square brackets represents the prefactors of Equations 1 and 2 and the intercepts in Figures 1a and 1b. If E' and E_{DB} are only approximately equal, the activation energy E' may vary slightly with doping concentration. In conclusion, we feel that cases b) or c) best describes the data.

140

REFERENCES

1. Laser and Electron-Beam Solid Interactions and Materials Processing, J. F. Gibbons, L. D. Hess and T. W. Sigmon, eds., North-Holland, NY 1980.

2. Laser and Electron-Beam Interactions with Solids, B. R. Appleton and G. K. Celler, eds., North-Holland, NY 1981.

3. M. Wautelet, L. D. Laude and R. Andrews, Physics Letters 77A, 274 (1980).

4. A Barna, P. Barna and J. Pocze, J. Non-Cryst. Solids 8, 36 (1972)

5. P. Germain, S. Squelard and J. Bourgoin, J. Non-Cryst. Solids 23, 159 (1977) and Radiation Effects in Semiconductors, Dubrovnik 1976, N. Urli and J. Corbett eds.

6. L. Laude and R. Willis, AIP Conf. Proc. No. 20, P. 65 (1975).

7. P. Germain, K. Zellama, S. Squelard, J. Bourgoin and A. Gheorghiu, J. Appl. Phys. 50,6986 (1979).

8. P. Germain and S. Squelard, to be published.

9. L. Csepregi, E. Kennedy, T. Gallagher, J. Mayer and T. Sigmon, J. Appl. Phys. 48, 4241 (1977).

10. The error in activation energies from reference 9 stems principally from the RBS instrument resolution of 20 keV (434 Å in depth), which results in an error in the determination of the activation energy of ∿0.2 eV based on a leastsquares fit to a straight line on the Arrhenius plot.

11. A. Lietoila,R. B. Gold and J. F. Gibbons, Appl. Phys. Lett. 39, 810 (1981).

12. S. Kokorowski, G. Olson and L. Hess. J. Appl. Phys. 53, 921 (1982) and references to earlier work sited therein.

13. P. Germain and M. A. Paesler, to be published.

14. N. Mott and E. A. Davis, Electronic Processes in Non-Crystalline Materials, 2nd edition (Clarendon Press, Oxford) 1979.

15. Applying error bars of earlier Hughes work to the data of ref. 11, one cannot make firm conclusions about the combined effects of doping and irradiation.

16. P. Thomas, M. Brodsky, D.Kaplan and D. Lepine, Phys. Rev. B 18, 3059 (1978).

17. D. Kaplan, private communication.

18. J. Mayer in Thin Films, Preparation and Properties, K. Rosenberg, ed., Pasadena, CA (1981).

19. K. Zellama, P. Germain, S. Squelard, J. Bourgoin and P. Thomas, J. Appl. Phys. 50, 6995 (1979).

20. W. A. Harrison, Surface Science 55, 1 (1976).

21. D. Chadi, Phys. Rev. Letters 43, 43 (1979).

22. L. Csepregi, E. F. Kennedy and J. W. Mayer, J. Appl. Phys. 49, 3906 (1978).

LASER-INDUCED SOLID PHASE CRYSTALLIZATION IN AMORPHOUS SILICON FILMS

G.L. OLSON, S.A. KOKOROWSKI,* J.A. ROTH AND L.D. HESS
Hughes Research Laboratories, 3011 Malibu Canyon Rd., Malibu, CA 90265

ABSTRACT

We review recent work on the kinetics of laser-induced solid phase epitaxial crystallization of silicon as determined from time-resolved reflectivity measurements. Specific topics which are addressed include: the intrinsic kinetics of solid phase epitaxy (SPE) in ion-implanted and UHV-deposited films; SPE rate enhancement by implanted dopant atoms and the effects of electrical compensation on the SPE rate; and the temperature dependence of SPE and competing processes in samples containing impurity atoms at concentrations exceeding the solid solubility limit. The high temperature kinetics results are compared with predictions from transition state theory and are discussed with respect to a proposed depression in the amorphous Si melting temperature.

INTRODUCTION

Studies of solid phase crystallization kinetics in ion-implanted and vacuum-deposited amorphous films have contributed significantly to the understanding of crystallization mechanisms, competitive crystallization processes, and the effects of impurity atoms on the amorphous-to-crystalline transition. Since the pioneering work of Csepregi and co-workers [1-3] on the kinetics of solid phase epitaxy (SPE) in silicon, numerous investigations have been directed toward determining kinetic parameters and elucidating the microscopic details of the SPE process. That work has complemented earlier as well as more recent studies of random nucleation and crystallite growth which can occur in parallel with SPE in amorphous films. Most of the research that has been conducted in these areas has, of necessity, been limited to temperatures $\lesssim 600°C$ because of the experimental difficulties associated with (1) rapidly raising the temperature of an amorphous film to a desired level without effecting substantial crystallization at lower temperatures, and (2) monitoring the rapid rate of crystallization occurring at high temperatures. The former problem has been alleviated by application of directed beam heating methods, and the development of capabilities for in-situ measurement of SPE and random nucleation and growth rates has essentially removed the second impediment to the measurement of high temperature solid phase crystallization kinetics. Consequently, it has now become possible to examine in considerable detail such issues as: the temperature dependence of solid phase crystallization in various systems such as ion-implanted and vacuum-deposited films on bulk silicon or silicon-on-sapphire; the influence of impurity atoms and the effects of dopant compensation on the rate of SPE at high temperatures; the temperature dependence of epitaxial growth in systems containing impurity atom concentrations in excess of the solid solubility limit; and the energetics of phase transitions which occur between metastable and stable phases in silicon during directed beam heating. In this paper we present an overview of recent studies of high temperature solid phase crystallization kinetics which have been performed in these areas using cw laser heating and in-situ time-resolved

---

*Current Address: Dikewood, A Division of Kaman Sciences Corp.,
2716 Ocean Park Blvd., Santa Monica, CA 90405

Mat. Res. Soc. Symp. Proc. Vol. 13 (1983) © Elsevier Science Publishing Co., Inc.

optical reflectivity techniques. Following a description of experimental
methods for determining fast crystallization rates and temperatures during the
crystallization process, we present and discuss data on the intrinsic rate of
SPE in ion-implanted and vacuum-deposited films over a wide temperature range,
the influence of boron and phosphorus on the SPE process in Si, the effect of
temperature on solid phase crystallization in Si(100) samples implanted with
impurities to levels beyond the solid solubility limit, and implications of the
observation of solid phase epitaxy near the crystalline Si melting point on the
issue of the amorphous-to-liquid phase transition in silicon.

EXPERIMENTAL

A variety of experimental techniques have been employed in the past for
monitoring solid phase crystallization dynamics in thin films. These include,
for example, Rutherford backscattering and ion channeling [1-3], dynamic
microscopy [4], and electrical conductivity [5]. Although these techniques can
be extremely useful diagnostic tools for specific regimes of temperature and
crystallization rate, they are of limited utility for probing rapid crystallization processes at high temperatures (>550°C). A useful method for
obtaining such data employs time-resolved reflectivity (TRR) of the sample
during cw laser heating [6]. As epitaxial crystallization proceeds in a sample
containing a thin amorphous layer on an optically thick substrate, the net
reflectivity of the sample at a wavelength λ will oscillate due to constructive
and destructive interference occurring between light reflected from the surface
and from the advancing epitaxial growth interface. Since each consecutive
interference maximum corresponds to a change in amorphous layer thickness of
$\lambda/2n$ (652 Å in Si for 6328 Å light), it is possible to directly extract the
rate of interface movement by monitoring the temporal change in the net
reflected light intensity.

The experimental configuration employed for such experiments is shown in
Figure 1. A spatially filtered and focused HeNe laser serves as the probe
source; an argon laser beam (all visible lines) heats the sample and the
reflected Ar laser beam triggers the detection electronics. The spatial intensity profiles of the heating and probe beams are shown in Figure 2 for two
different heating beam diameters. The HeNe beam has a narrow profile and is
centered within the larger heating beam as shown in both displays. The beam
profiles are experimentally measured by chopping the beam at the sample plane
and differentiating the output of a photodiode which monitors the chopped signal [7]. Knowledge of the chopping frequency, radial distance from the chopper
axis to beam center, and temporal width of the differentiated signal allows the

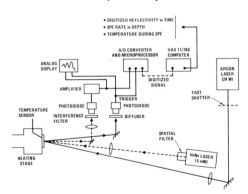

Fig. 1. Experimental configuration employed for time-
resolved reflectivity
measurements.

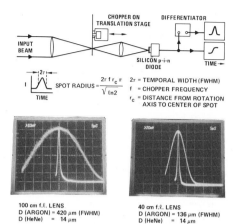

Fig. 2. Determination of heating beam and probe laser spot sizes. Spatial intensity profiles of the heating and probe beams are compared for two different heating beam diameters.

spot size to be directly determined (see Figure 2). These measurements are regularly performed to monitor both spot size and mode quality of the focused beams. As shown in Figure 1, the TRR data are displayed in analog form and are also digitized by an on-line A/D conversion system. The computer data acquisition and analysis system converts the data to absolute reflectivity versus time and SPE rate versus depth for each experiment.

Although the accurate measurement of fast crystallization rates is crucial in these studies, an accurate determination of temperature during crystallization is equally important when temperature dependent effects are to be investigated. The temperature measurement technique we employed exploits the fact that the refractive index (and hence the reflectivity) of silicon is a function of temperature and the temperature dependence has been accurately measured over a wide temperature range [8,9] (see Figure 3). Following SPE, the heating beam is removed from the sample by closure of a fast shutter. When the heating beam is blocked, the temperature decreases abruptly, and the reflectivity of the crystalline region probed by the HeNe beam drops to the reflectivity of single crystal Si at the substrate temperature. Since the

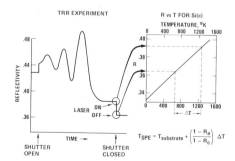

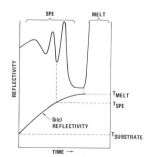

Fig. 3. Experimental determination of temperature in time resolved reflectivity measurements of laser-induced SPE.

Fig. 4. Measurement of temperature during SPE for temperatures approaching T_{melt}.

dependence of reflectivity on temperature is known, and the substrate temperature is accurately measured, the temperature during laser heating can be straightforwardly determined as shown in Figure 3. The temperature in the amorphous material during SPE is slightly lower than that of the crystalline solid after completion of SPE because a smaller fraction of the heating beam power is absorbed by amorphous Si than by crystalline Si ($R_a > R_c$). Since the temperature increase produced by the heating beam is proportional to $(1-R)$ [10], the temperature change of the amorphous material during SPE is given by $[(1-R_a)/(1-R_c)]\Delta T$. This value when added to the measured substrate temperature, gives the temperature during SPE.

When cw laser heating is used to raise the temperature of the amorphous material to levels near the crystalline melting point, the determination of temperature during SPE is complicated by the fact that temperature varies with time. However, it is possible to determine this time dependence by monitoring the temporal variation of the reflectivity during irradiation of a crystallized region. This measured variation of reflectivity (and therefore temperature) with time along with the known substrate temperature and the time at which melting occurs (abrupt increase in reflectivity) allow the temperature at any time between t_0 and t_{melt} to be determined. This approach is illustrated in Figure 4 and is discussed in greater detail in Refs. 11 and 12.

The result of using time-resolved reflectivity techniques for the determination of solid phase epitaxial crystallization rates and temperatures during crystallization of an ion-implanted silicon film is illustrated by the

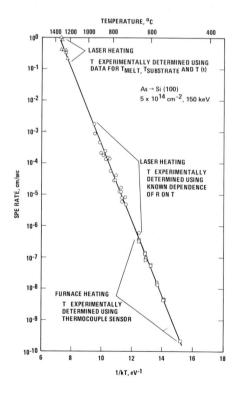

Fig. 5. SPE rate and temperature data obtained by time-resolved reflectivity measurements during crystallization of arsenic-implanted Si(100).

data shown in Figure 5. Data above ~ 650°C were obtained from TRR measurements of both rate and temperature as discussed above. The lower temperature data (squares) were obtained by heating the sample on a stage fitted with an enclosure having an AR-coated window to permit TRR measurement of SPE rates; the temperature of the resistively heated stage was measured with a thermocouple. These data are in excellent agreement with those obtained earlier by TRR measurement of rate and calculation of temperature [13] and serve to illustrate the wide range over which experimental data can be obtained with these techniques.

RESULTS AND DISCUSSION

Intrinsic SPE Kinetics
    The determination of the intrinsic rate of SPE in silicon is of central importance in the development of a comprehensive description of the epitaxial crystallization process. Currently available data on intrinsic SPE kinetics have for the most part been obtained in self-implanted silicon films and at temperatures $\lesssim$ 575°C [1,14]. However, there is considerable discrepancy in the values for kinetics parameters which have been reported in those studies. Moreover, only limited data exists [15-17] for intrinsic SPE kinetics in amorphous films produced by vacuum deposition techniques. In this section we present results of TRR measurements which significantly extend the temperature range of SPE kinetics data in both self-implanted and UHV-deposited amorphous Si films.
    The temperature dependence of SPE growth rates in silicon-implanted Si(100) wafers and in thin amorphous films prepared by deposition of Si onto clean Si(100) substrates in an ultra-high vacuum (UHV) environment are given in Figure 6. The ion-implanted films were prepared by multiple dose $^{28}Si^+$ implantation into Si(100) substrates maintained near 77°K. Samples prepared both at the Stanford Electronics Laboratories and in our laboratory were studied and the rate versus temperature results were found to be identical.

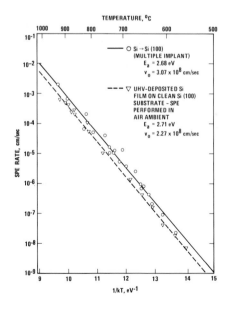

Fig. 6. Intrinsic SPE kinetics in amorphous Si films prepared by Si implantation and UHV deposition onto Si(100) substrates.

The data are accurately described by an Arrhenius expression,
$v = v_0 e^{-E_a/kT}$, where v is the SPE rate and E_a is the activation energy
for the process. A least squares fit to the data in Figure 6 gives an activation
energy of $2.68(\pm.05)$eV for the SPE process in the ion-implanted films (solid
line) and an activation energy of $2.71(\pm.05)$eV for the UHV-deposited films
(dashed line). The observation of nearly identical activation energies for
intrinsic SPE in Si-implanted and UHV-deposited films shows that although there
may be differences in the microstructure in the two types of films due to the
method of producing the amorphous layer, the intrinsic SPE growth dynamics are
essentially the same.

The earlier work of Csepregi et al. [1] on intrinsic SPE kinetics at
temperatures from 475 to ~575°C yielded a considerably lower activation energy,
2.3 eV, whereas a recent study by Lietoila, et al. [14] gave an activation
energy of 2.85 eV over the same temperature range. Given the much wider
temperature range of the present rate data obtained by TRR and the consistency
of activation energies obtained in samples prepared by different techniques, we
feel that the activation energies deduced from the reflectivity measurements
more accurately describe the intrinsic SPE process.

<u>SPE Rate Enhancement By Implanted Dopant Atoms</u>
An area of active research in the field of solid phase crystallization is
the development of an accurate description for the effects of impurities on the
rate of solid phase epitaxy. In the past it was shown [2,18] that large
enhancements or reductions in SPE rate relative to the intrinsic rate can
result from the introduction of dopants or other impurities at concentrations
$\gtrsim 10^{20}$cm^{-3}. An example of this effect is given in Figure 7 where TRR data
obtained during epitaxial growth in a boron-implanted Si(100) film previously
amorphized by Si implantation is presented. As indicated by the nonuniform
spacing of the interference features in the TRR signature, the SPE growth rate
varies markedly with position (depth) of the crystal/amorphous interface in
the sample. The SPE rate as a function of depth can be determined directly
from the TRR data and is given in Figure 7b. Also plotted in Figure 7b is the
boron concentration profile for this sample calculated from the Pearson IV

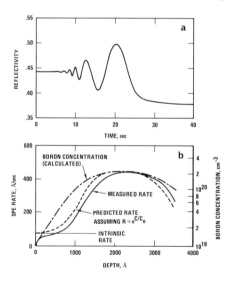

Fig. 7. TRR measurements of
epitaxial growth in B-implanted
Si(100) previously amorphized by Si
implantation. (a) Reflectivity as
a function of time for SPE at 653°C.
(b) SPE rate as a function of depth
obtained from data in (a) compared
to the rate predicted from the boron
concentration profile.

coefficients [14]. Note that at the peak of the boron concentration profile
the SPE rate is approximately six times the intrinsic value. Previously it was
shown [18] that for amorphous films containing either phosphorus or oxygen the
SPE rate depends exponentially on concentration. Assuming that the same
dependence holds for boron as well, we can calculate the expected SPE rate
versus depth for the present sample. The resulting predicted rate, shown as a
dashed line in Figure 7b is in good agreement with the measured rate
variation, and can be taken as evidence that an exponential dependence of rate
on concentration is appropriate for boron.

Important insight into a description of the SPE rate enhancement caused by
the presence of certain dopants has been provided recently by Suni et al.
[19,20] and Lietoila et al. [14]. They hypothesized that the SPE rate is
controlled by the number of vacancies at the amorphous/crystalline interface as
first suggested by Csepregi et al. [21]. Since the vacancy concentration is
dominated by charged vacancies in heavily doped Si [22], and the Fermi level
position controls the concentration of charged species, then insofar as
changes in dopant concentration affect the Fermi level, the SPE rate should be
affected as well. Correspondingly, if the Fermi level is driven to near midgap
by the addition of electrically compensating dopant atoms, then the vacancy
concentration and the SPE rate should revert to their intrinsic values.

Data supporting this conjecture have been obtained over a limited
temperature range [14,19,20] but considerable uncertainties remained about the
behavior at high temperatures. However, using the laser heating and
reflectivity techniques described here, we have now investigated these effects
over a wide range of temperatures. The results are given in Figure 8 for SPE
in previously amorphized films containing nearly uniform concentrations of:
(1) boron only (diamonds); (2) phosphorus only (triangles); and (3) boron plus
phosphorus (compensating concentrations – squares). Activation energies for

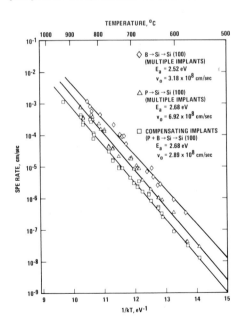

Fig. 8. SPE rate vs 1/kT for
samples containing boron,
phosphorus, and boron plus
phosphorus in compensating
concentrations in Si(100)
samples previously amorphized by
silicon implantation. (Sample
preparation is described in detail
in Ref. 14).

samples containing boron, phosphorus and compensating implants are $2.52(\pm.05)$eV, $2.68(\pm.05)$eV and $2.68(\pm.05)$eV, respectively. The temperature dependence of the SPE growth rate in the compensated samples is seen to be identical to that measured in self-implanted samples (Figure 6). These results strongly support the conclusions of Suni et al. and Lietoila et al. that the rate enhancement caused by dopant atoms is an electronic effect and show that this effect exists over a wide range of temperatures. A more detailed description of these experiments and a detailed discussion of the microscopic aspects of vacancy-mediated growth will be presented in a future publication.

SPE in Amorphous Films Containing Impurity Concentrations Greater than the Solid Solubility Limit

It has been shown that thin amorphous silicon films containing ion-implanted impurity atoms at concentrations considerably in excess of equilibrium solid solubility limits can be crystallized epitaxially in the solid phase [23-31]. The maximum impurity concentration which can be incorporated into substitional lattice sites by solid phase epitaxy (SPE) depends strongly on the implant species, crystallographic orientation, and annealing conditions (temperature and time). Recent investigations have shown that during SPE at low temperatures ($<600°C$) in films containing high doses of certain implants, the epitaxial crystallization process is retarded as the crystal/amorphous interface approaches the surface of the film [26-28]. Growth retardation is accompanied by impurity segregation at the interface and a limitation in the number of impurity atoms which are substitutionally incorporated into lattice sites. It was suggested [27] that local strain effects at the growth interface due to size differences between the implanted atoms and silicon are responsible for the retardation of epitaxial growth in samples containing high dose implants. It was also suggested that at higher temperatures the probability for polycrystallite nucleation in these films is greater and might effectively prevent crystallization of the entire film by SPE.

In work presented at this symposium last year [28] it was shown by using cw laser heating, TRR, transmission electron microscopy and Rutherford backscattering that although significant SPE retardation occurs at low temperatures, high quality epitaxy without interface retardation or poly-crystallite nucleation can be produced in indium-implanted films at temperatures $\gtrsim750°C$. In addition to the studies on indium we have now also investigated the recrystallization kinetics as a function of temperature and concentration in films containing bismuth, antimony, and arsenic, and in each case have found the competition between SPE and inhibiting processes such as precipitate formation to be highly dependent on temperature.

An example of temperature dependent changes which can occur in Si(100) samples containing high impurity atom concentrations is shown in Figure 9. An arsenic-implanted sample (1.1×10^{16} cm^{-2}, 100 keV) with a peak impurity concentration of 2.4×10^{21} cm^{-3} was heated at various laser powers and the crystallization rates were monitored by TRR. For the lowest temperature shown (T = 574°C, upper plot) severe retardation of the epitaxial growth interface is observed. This is evident from the fact that the last interference peak is much broader than preceding peaks and from the observation that the final reflectivity is much higher than the reflectivity of crystalline silicon. However, as indicated by the sequence of data in the figure, as the crystallization temperature is increased by raising the incident laser power, progressively less interface retardation occurs. This is consistent with the results we previously reported for indium-implanted layers [28] and suggests that (1) at low temperatures, precipitate formation at the epitaxial interface is a dominant process; (2) at high temperatures the SPE process is much faster than the formation rate of precipitates or defect clusters which would act to inhibit epitaxial growth; and (3) random nucleation and

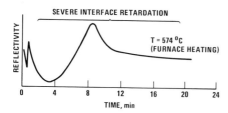

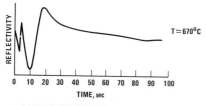

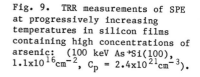

Fig. 9. TRR measurements of SPE at progressively increasing temperatures in silicon films containing high concentrations of arsenic: (100 keV As$\rightarrow$Si(100), 1.1×10^{16}cm^{-2}, C_p = 2.4×10^{21}cm^{-3}).

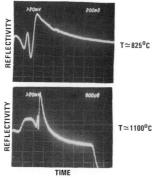

crystallite growth are <u>not</u> dominant processes at high temperatures either in these samples or in the indium-implanted samples discussed previously.

High Temperature SPE Kinetics.
An interesting issue which has recently been the subject of numerous investigations is the determination of the melting temperature of amorphous silicon. It has been suggested [32,33] that the amorphous silicon melting temperature $T_m(a)$ is significantly lower than that of crystalline material, $T_m(c)$. Bagley and Chen [32] and Spaepen and Turnbull [33] estimated a depression of more than 300°K from theoretical considerations, and Baeri et al. [34] reported a reduction of more than 500°K in $T_m(a)$ relative to the crystalline state melting point based on interpretation of structural changes produced by pulsed electron beam irradiation of amorphous Si films. In contrast to those reports, we have observed solid phase epitaxial growth in ion-implanted amorphous layers at temperatures very close to $T_m(c)$. (See Figure 5 and Refs. 11 and 12). More precisely, the temperature of an amorphous film was raised to within 40°K of $T_m(c)$, yet melting of the amorphous material was not observed.

We have repeated those experiments using very thick (4000 Å) ion-implanted
layers as well as UHV-deposited films on both clean and oxidized substrates and
in no cases have we observed melting below $T_m(c)$.
   In considering these apparently conflicting results, it is important to note
that there are two fundamentally distinct issues to be addressed with regard to
the melting of amorphous Si. The first is the establishment of the static
thermodynamic parameters of the phases involved, as is typically expressed by a
plot of free energy versus temperature. Within this framework the effective
melting point of amorphous Si is defined as the temperature at which the amorphous and liquid Si free energy curves cross. The second issue is one of the
dynamics of transitions from one phase to another, particularly the case where
$T_m(a)$ lies below $T_m(c)$. The central question is whether amorphous Si
(which is of course not a stable, equilibrium phase), when heated to
temperatures between $T_m(a)$ and $T_m(c)$, would actually be observed to convert
directly to the lowest free energy state available, namely solid crystalline
Si, or would first make a transition into the energetically less favorable
liquid state. In the context of this discussion, it is evident that our
failure to observe melting during high temperature SPE does not necessarily
rule out the possibility of a reduced melting point, but rather should be
interpreted more as a statement on the kinetics of melting relative to solid
phase crystallization than on the thermodynamics of the phase transition.
   We have found, however, that the SPE rates we have measured can be used to
place upper and lower limits on the melting temperature of amorphous Si because
theoretically the temperature dependence of the SPE growth rate in the high
temperature regime is quite sensitive to the value of $T_m(a)$. This
sensitivity can be understood within the framework of a transition state
description of the thermally activated SPE growth process [35] by noting that
the amorphous/crystal interface velocity, v, can be expressed as:

$$v = \frac{kT}{h} \, \delta e^{-g*/kT}(1 - e^{-\Delta g/kT}) \quad . \tag{1}$$

In this expression, Δg is the free energy difference between amorphous and
crystalline Si, g* is the free energy relative to amorphous Si of the hypothetical "activated complex" which appears in the transition state descrip-
tion, δ is the $\langle 100 \rangle$ interplanar distance in Si, h is Planck's constant, and T
is the temperature. If Δg and g* are decomposed into their entropic and
enthalpic parts, i.e., $\Delta g = \Delta h - T\Delta s$, and $g* = h* - Ts*$, then, given values
for Δh, Δs, h* and s*, the temperature dependence of v can be calculated and
compared with our experimental measurements.
   Using the value of $\Delta h = 0.1$ eV based on previous measurements [36,37]
we have calculated curves of interface velocity versus temperature for three
combinations of Δs and g*, and these are presented in Figure 10. We have also
plotted the free energies of the phases (relative to crystalline silicon) as a
function of temperature so that the impact of Δs on the assumed transition
temperature can be seen. For a particular choice of Δs, g* is uniquely determined by requiring a good fit to the lower temperature rate data. Thus Δs is
effectively the only remaining free parameter and its effect on the high temperature end of the curve is quite dramatic, permitting the range of allowable
values to be determined quite straightforwardly. We note from the figure that
values of Δs greater than about 0.5k cause the theoretical SPE rate to roll over
at temperatures below the crystal melting point, in marked disagreement with our
measured data. Of particular interest is the case of $\Delta s = 0.68$k (dashed line)
since this value causes the melting temperatures of amorphous and crystalline Si
to be equal on the free energy diagram, and thus also forces $v \rightarrow 0$ at T_{melt}.
From Figure 10 it is clear that the experimental rates contradict this choice of
equal melting points. It has also been noted by Turnbull [38] that since the
measured SPE rates do not roll over at high temperatures the assumption of equal

151

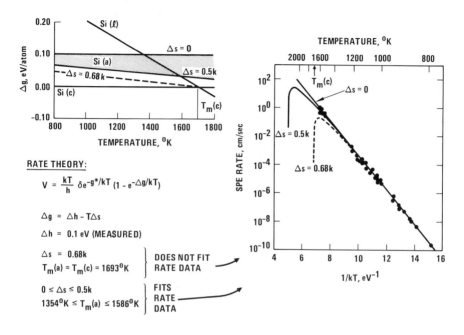

Fig. 10. Determination of the range of allowable melting point values for amorphous silicon from a comparison of experimental and theoretically predicted SPE rates.

melting points is inconsistent with the data. However, the degree of melting point suppression required to obtain agreement between theory and experiment need not be as great as suggested in Ref. 38. The range of acceptable Δs values which permit a reasonable description of the high temperature rate data is found to be 0 ≤ Δs ≤ 0.5 k, as depicted by the shaded region on the free energy diagram. That is, assuming that the value of Δh = 0.1 eV is correct over the entire temperature range, values of Δs in the range 0 to 0.5k give agreement between rates predicted by Eq. (1) and our experimental results. This treatment therefore implies that the amorphous to liquid transition temperature should lie in the range between 1354°K and 1586°K in order for the transition state formulation to be consistent with our rate versus temperature data. The fact that melting was not observed in our experiments even at temperatures in excess of 1600°K [11,12] then suggests either that kinetic aspects of melt nucleation dominate at high temperatures or that the thermodynamic interpretation must be changed. Since the rate of energy deposition and the intrinsic rate of melting are much faster than SPE in our experiments [11,12], sufficient time exists for an amorphous-to-liquid phase transition to occur before the amorphous solid crystallizes. However, if the amorphous material is continually changing structure during heating, as suggested by Turnbull [38,39], then the melt nucleation kinetics might be altered sufficiently to make the amorphous-to-liquid phase transition unobservable in cw heating experiments of the type we have reported.

SUMMARY

The dynamics of solid phase crystallization processes in silicon were studied experimentally over a wide range of temperatures and dopant concentrations using techniques of cw laser heating and time-resolved reflectivity. Solid phase epitaxial crystallization rates were determined as a function of position within the amorphous films as well as the temperature during SPE from optical reflectivity measurements. On-line computer analysis of the data provides near real-time SPE rate and temperature information over the temperature range of 500 to 1400°C.

In general, the temperature dependence of the SPE rate can be accurately described by an Arrhenius expression, $v = v_0 \exp(-E_a/kT)$, over this entire temperature range. The activation energies for undoped deposited and ion-implanted films were determined to be 2.71 eV and 2.68 eV, respectively. Most dopants increase the rate of SPE relative to the intrinsic rate. However, when equal concentrations of donor and acceptor dopants such as P and B are present in the film, their enhancement effect is negated and the SPE rate reverts to the intrinsic value. For samples containing impurity concentrations in excess of the solid solubility limit, retardation of SPE by impurity precipitation is a dominant effect at low temperatures. However, at high temperatures the same samples crystallize in a regular epitaxial manner.

Measurements of amorphous silicon SPE rates at temperatures near the crystalline melting point provide important information concerning the amorphous silicon melting temperature. Analysis of the observed rate data in terms of transition state rate theory leads to the conclusion that the transition temperature, $T_m(a)$, for the conversion between amorphous and liquid silicon can be no higher than 1586°K; a lower limit of 1354°K for $T_m(a)$ is also deduced. Since we did not observe melting in cw laser heating experiments at temperatures in excess of those values we conclude that for temperatures above the inferred crossover point of the amorphous Si and liquid Si free energy curves (i.e., the "melting point" of amorphous Si) and below $T_m(c)$, transition directly to the crystalline state is kinetically favored over the two step amorphous-to-liquid-to-crystalline solid transition which is also allowed by the free energy diagram. At the same time we acknowledge the possibility that under certain transient conditions (e.g., explosive crystallization [40]), it may be possible to achieve the transition from amorphous solid to supercooled liquid.

The use of laser heating and optical diagnostic techniques offers the potential for investigating other important issues in solid phase crystallization that have not been directly addressed in this paper. For example, the kinetics of random nucleation and crystallite growth in ion-implanted layers at high temperatures has not been investigated in detail; this will be necessary before the competition between this process and epitaxial growth can be quantified. These studies should provide a basis for the development of a theoretical model of SPE and random nucleation and growth in silicon, and it is expected that this will be a useful framework for understanding the recrystallization behavior of compound semiconductor materials as well.

ACKNOWLEDGMENTS

The authors wish to gratefully acknowledge J. Narayan and T.W. Sigmon for their helpful discussions regarding this work and for supplying well characterized samples of ion-implanted Si. We also wish to thank R.S. Turley for his contributions to the development of the data acquisition and analysis system and acknowledge R.F. Scholl and L.M. Lewis for their expert technical assistance.

153

REFERENCES

1. L. Csepregi, J.W. Mayer, and T.W. Sigmon, Phys. Lett 54A, 157 (1975).

2. L. Csepregi, E.F. Kennedy, T.J. Gallagher, J.W. Mayer and T.W. Sigmon, J. Appl. Phys. 48, 4234 (1977).

3. L. Csepregi, E.F. Kennedy, J.W. Mayer, and T.W. Sigmon, J. Appl. Phys. 49, 3906 (1978).

4. B. Drosd and J. Washburn, J. Appl. Phys. 51, 4106 (1980).

5. K. Zellama, P. Germain, S. Squelard, J.C. Bourgoin, and P.A. Thomas, J. Appl. Phys. 50, 6995 (1979).

6. G.L. Olson, S.A. Kokorowski, R.A. McFarlane, and L.D. Hess, Appl. Phys. Lett. 37, 1019 (1980).

7. J.A. Arnaud, W.M. Hubbard, G.D. Mandeville, B. de la Clavière, E.A. Franke, and J.M. Franke, Appl. Opt. 10, 2775 (1971).

8. M.A. Hopper, R.A. Clarke, and L. Young, J. Electrochem. Soc. 122, 1216 (1975).

9. Y.J. van der Meulen and N.C. Hien, J. Opt. Soc. Am. 64, 804 (1974).

10. See for example, S.A. Kokorowski, G.L. Olson, and L.D. Hess, in Laser and Electron-Beam Solid Interactions and Materials Processing, J.F. Gibbons, L.D. Hess, and T.W. Sigmon, eds. (North Holland, New York, 1981), p. 139.

11. S.A. Kokorowski, G.L. Olson, J.A. Roth, and L.D. Hess, Phys. Rev. Lett. 48, 498 (1982).

12. S.A. Kokorowski, G.L. Olson, J.A. Roth, and L.D. Hess, in Laser and Electron Beam Interactions with Solids, B.R. Appleton, G.K. Celler, eds. (North Holland, New York, 1982), p. 195.

13. S.A. Kokorowski, G.L. Olson, and L.D. Hess, J. Appl. Phys. 53, 921 (1982).

14. A. Lietoila, A. Wakita, T.W. Sigmon, and J.F. Gibbons, J. Appl. Phys. 53, 4399 (1982).

15. L.S. Hung, S.S. Lau, M. von Allmen, J.W. Mayer, B.M. Ullrich, J.E. Baker, P. Williams, and W.F. Tseng, Appl. Phys. Lett. 37, 909 (1980).

16. J.C. Bean and J.M. Poate, Appl. Phys. Lett. 36, 59 (1980).

17. J.A. Roth, G.L. Olson, S.A. Kokorowski, and L.D. Hess, in Laser and Electron-Beam Solid Interactions and Materials Processing, J.F. Gibbons, L.D. Hess, and T.W. Sigmon, eds. (North Holland, New York, 1981), p. 413.

18. E.F. Kennedy, L. Csepregi, J.W. Mayer, and T.W. Sigmon, J. Appl. Phys. 48, 4241 (1977).

19. I. Suni, G. Göltz, M.G. Grimaldi, M.A. Nicolet, and S.S. Lau, Appl. Phys. Lett. 40, 269 (1982).

154

20. I. Suni, G. Göltz, M.A. Nicolet, and S.S. Lau, Thin Solid Films 93, 171 (1982).

21. L. Csepregi, R.P. Küllen, J.W. Mayer, and T.W. Sigmon, Sol. State Commun. 21, 1019 (1977).

22. J.A. Van Vechten and L.D. Thurmond, Phys. Rev. B 14, 3539 (1976).

23. P. Blood, W.L. Brown, and G.L. Miller, J. Appl. Phys. 50, 173 (1979).

24. A. Lietoila, J.F. Gibbons, T.J. Magee, J. Peng, and J.D. Hong, Appl. Phys. Lett. 35, 532 (1979).

25. S.U. Campisano, E. Rimini, P. Baeri, and G. Foti, Appl. Phys. Lett. 37, 170 (1980).

26. J.S. Williams and R.G. Elliman, Appl. Phys. Lett. 37, 829 (1980).

27. J.S. Williams and R.G. Elliman, Nucl. Instr. Meth. 183, 389 (1981).

28. J. Narayan, G.L. Olson, and O.W. Holland in Laser and Electron Beam Interactions with Solids, B.R. Appleton, G.K. Celler, Eds. (North Holland, New York, 1982), p. 183.

29. A. Lietoila, R.B. Gold, J.F. Gibbons, T.W. Sigmon, P.D. Scovell, and J.M. Young, J. Appl. Phys. 52, 230 (1981).

30. J.S. Williams and R.G. Elliman, Appl. Phys. Lett. 40, 266 (1982).

31. J. Narayan and O.W. Holland, Appl. Phys. Lett. 41, 239 (1982).

32. B.G. Bagley and H.S. Chen, Laser-Solid Interactions and Laser Processing, S.D. Ferris, H.J. Leamy, and J.M. Poate, eds., Am. Inst. Phys. Conf. Proc. 50, 97 (1979).

33. F. Spaepen and D. Turnbull, ibid p. 73 (1979).

34. P. Baeri, G. Foti, J.M. Poate, and A.G. Cullis, Phys. Rev. Lett. 45, 2036 (1980).

35. See for example, J.W. Christian, The Theory of Transformations in Metals and Alloys (Pergamon Press, 1965) Chap. XI.

36. J.C.C. Fan and C.H. Anderson, Jr., J. Appl. Phys. 52, 4003 (1981).

37. E.P. Donovan, private communication.

38. D. Turnbull in Metastable Materials Formation by Ion Implantation, S.T. Picraux and W.J. Choyke, eds. (North Holland, New York, 1982) p. 103.

39. D. Turnbull, "Laser Annealing," paper presented at APS meeting, March 11, 1982, Dallas, TX.

40. See for example, H.J. Leamy, W.L. Brown, G.K. Celler, G. Foti, G.H. Gilmer, and J.C.C. Fan in Laser and Electron-Beam Solid Interactions and Materials Processing, J.F. Gibbons, L.D. Hess, and T.W. Sigmon, eds. (North Holland, New York, 1981), p. 89.

SOLID-PHASE-EPITAXIAL GROWTH OF ION IMPLANTED SILICON USING CW LASER AND ELECTRON BEAM ANNEALING*

J. NARAYAN, O. W. HOLLAND
Solid State Division, Oak Ridge National Laboratory, Oak Ridge, TN 37830

and

G. L. OLSON
Hughes Research Laboratories, 3011 Malibu Canyon Road, Malibu, CA 90265

ABSTRACT

The nature of residual damage in As^+, Sb^+, and In^+ implanted silicon after CW laser and e^- beam annealing has been studied using plan-view and cross-section electron microscopy. Lattice location of implanted atoms and their concentrations were determined by Rutherford backscattering and channeling techniques. Maximum substitutional concentrations achieved by furnace annealing in a temperature range of 500-600°C have been previously reported [1] and greatly exceeded the retrograde solubility limits for all dopants studied. Higher temperatures and SPE growth rates characteristic of electron or cw laser annealing did not lead to greater incorporation of dopant within the lattice and often resulted in dopant precipitation. Dopant segregation at the surface was sometimes observed at higher temperatures.

INTRODUCTION

Ion implantation damage in silicon caused by moderate fluences of heavy ions generally consists of an amorphous layer followed by a band of dislocation loops. These loops below the amorphous layer are primarily created by ion range straggling. The number density of loops decreases with increasing mass of the incident ions, and decreasing substrate temperature and dose of the implanted ions. These amorphous layers on semiconductor substrates can be recrystallized either by solid-phase-epitaxial (SPE) or by liquid-phase-epitaxial (LPE) growth [1,2]. The SPE growth can be achieved in a furnace by heating the whole wafer, or by using CW laser and e^- beams which provide spatial and temporal selectivity. A characteristic feature following SPE growth is the presence of a band of dislocation loops below the original amorphous layer. This band is essentially the same band of dislocations that is present after implantation although the loops may coarsen and the number density decrease depending upon the details of the thermal treatment. These loops can be completely removed under LPE growth conditions, if the melt-depth, for example, induced by pulsed laser irradiation penetrates the dislocation band [2]. In this investigation, we have studied the residual damage using cross-section and plan-view electron microscopy. Dynamics of the crystallizing interface was monitored by time-resolved reflectivity techniques [3]. The dopant concentrations as a function of depth were studied by backscattering and channeling techniques.

*Research sponsored by the Division of Materials Sciences, U. S. Department of Energy under contract W-7405-eng-26 with Union Carbide Corporation.

Mat. Res. Soc. Symp. Proc. Vol. 13 (1983) Published by Elsevier Science Publishing Co., Inc.

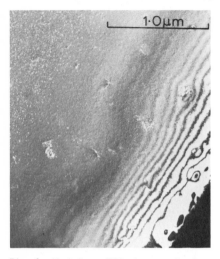

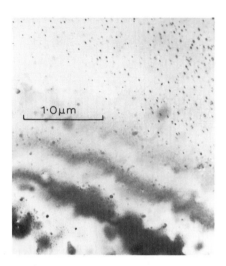

Fig. 1. Weak-beam TEM micrograph showing residual damage in arsenic implant-
ed (100 keV As$^+$, 5.0x10^{14} cm^{-2}) and CW
laser (Ar$^+$ ion) annealing (10W) silicon.

Fig. 2. Bright-field micrograph similar
to Fig.1 after annealing at 900°C/20
min. Note no loop-type damage in the
top layer except some polishing
contamination.

EXPERIMENTAL DETAILS

Both <100> and <111> orientations of silicon single crystals were implanted
with fluences of As$^+$, Sb$^+$, Ga$^+$, In$^+$, and Bi$^+$ ions ranging from 5.0 x 10^{14} to
1.0 x 10^{17} ions cm^{-2} in an energy range of 100 to 250 keV. The laser annealing
was performed by CW Ar$^+$ scanning laser: scanning speed 3.4 to 34 cms^{-1}, spot
diameter 80 µm (1/e intensity), and laser power 6 to 20 watts. The electron-
beam annealing was performed using 30 KV beam at 80 µA and spot size 8 µm.
Specimens were ion thinned for cross-section electron microscopy or chemically
thinned for plan-view electron microscopy. Rutherford backscattering and channeling analysis were performed using a 2.0 MeV He$^+$ beam to obtain both the lattice location and concentration of dopants. The solid-phase crystallization
kinetics in the amorphous silicon films was obtained by in-situ time-resolved
reflectivity measurements.

RESULTS AND DISCUSSION

TEM

Figure 1 shows plan-view TEM micrograph from an arsenic implanted (100 keV
As$^+$, 5.0 x 10^{14} cm^{-2}) specimen after laser annealing with a 10 watt CW laser
beam. The as-implanted specimens contained an amorphous layer 1300 Å thick
followed by 200 Å wide band of dislocation loops. After the laser treatment,
the top SPE grown layer contained dislocation loops ranging in size from 10 to
50 Å. We also observed tangles of dislocations in the near surface layer. The
dislocation loops in the underlying dislocation band are shown in the top left
corner of the micrograph. The average size and number density of these loops is
much higher than those present in the SPE grown region. It should be mentioned
that the specimens for plan-view microscopy were thinned from the backside with
a wedge near the edge. Therefore, the area near the edge contains information

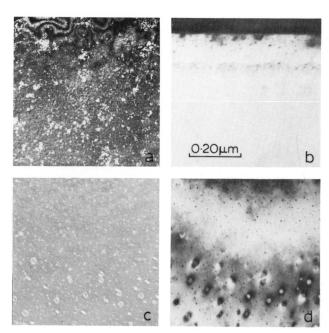

Fig. 3. Residual damage in indium implanted (125 keV In$^+$, 1.0x10^{15} cm^{-2}) and CW laser (10 W) annealed silicon: (a) plan-view, weak beam micrograph, (b) corresponding cross-section micrograph, (c) plan-view, weak-beam after thermally annealing (800°C/20 min) these specimens, (d) corresponding bright-field.

about the top layer and the area away from the edge contains microstructures corresponding to the underlying dislocation band and the defects in the top layer. When some of the laser annealed specimens were subsequently heated in the furnace at 900°C, the dislocation loops in the underlying dislocation band coarsened by loop coalescence involving conservative climb and glide, whereas the residual defects in the top SPE grown layer (small loops and dislocation tangles) annealed out, as shown in Fig. 2. Figure 3 shows TEM results for indium implanted (100) silicon. The as-implanted (125 keV ^{115}In$^+$, 1.0 x 10^{15} cm^{-2}) specimens contained a 1120 Å thick amorphous layer followed by a band of dislocation loops. Plan-view and cross-section micrographs of this specimen after annealing with a 10 W CW laser are shown in Fig. 3(a) and 3(b) respectively. Indium precipitates in addition to loops and tangles in the top SPE

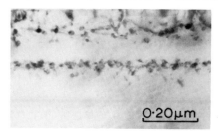

Fig. 4. Cross-section, bright-field micrograph showing residual damage in antimony implanted (200 keV, 4.4x10^{15} cm^{-2}) and CW laser (12W) annealed silicon.

158

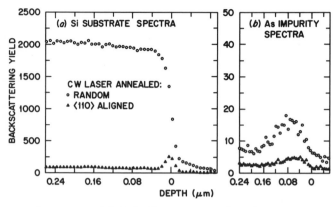

Fig. 5. CW laser annealed (100)Si implanted with 100 keV As$^+$, 5.0x10^{14} cm^{-2}.

grown layer are evident in the cross-section micrograph with only dislocation loops in the underlying dislocation band. The plan-view micrograph in Fig. 3(a) contains dislocation tangles, loops and precipitates. The top left corner is near the edge of the specimen and it contains microstructure corresponding to the top SPE grown layer. Weak beam dark-field and bright-field micrographs of identically laser annealed specimens which were subsequently heated to 800°C for 20 minutes are shown in Figs. 3(c) and 3(d) respectively. These micrographs show mostly precipitates in top layer (top right corner) and large loops in the underlying dislocation band. Cross-section TEM results from antimony implanted (200 keV, 4.4 x 10^{15} cm^{-2}) specimens are shown in Fig. 4 after annealing with a 12 Watt laser beam. The as-implanted specimens contained a 1580 Å thick amorphous layer followed by a dislocation band. A considerable coarsening of loops is observed in the underlying dislocation band after laser annealing. The SPE grown layer contains antimony precipitates and dislocation tangles. It should be mentioned that in the case of both indium (Fig. 3) and antimony (Fig. 4) implanted specimens, the doses were considerably higher than equilibrium values. The dislocation tangles in these indium and antimony implanted specimens are formed because of precipitation of dopants ahead of the crystallizing interface. These dislocations are generated around the precipitates as the crystallizing interface passes.

Backscattering and Channeling
    Si-As system, CW laser annealing. Figure 5 shows backscattering yield spectra for Si and As after CW laser annealing with 10 Watt beam. The specimens were implanted with 100 keV As$^+$ to a dose of 5.6 x 10^{14} cm^{-2}. The random and <110> aligned yields as a function of depths for both Si and As are shown in Fig. 5. The aligned spectrum for scattering from the Si matrix is nearly identical to one from a virgin crystal (not shown) which is a clear evidence that the recovery of the lattice was good. The reduced scattering yield in the aligned spectrum from the As atoms (seen in Fig. 5b) indicates that arsenic has been incorporated into the lattice during the recovery. From the calculated profiles of both total and substitutional concentrations in Fig. 6, it is obvious that the incorporation was good over the entire SPE grown layer with a ~0.9 substitutional fraction at the peak concentration.
    Si-As system, e$^-$ beam annealing. Silicon (100) specimens were implanted with 100 keV As$^+$ (dose = 5.8 x 10^{14} cm^{-2}) and annealed with e$^-$ beam (current 80 μA, spot size 8 μm, energy 30 KV, scanning velocity 150 cms^{-1}) using 95% overlap between traces. Figure 7 shows backscattering yield spectra of Si and As before and after e$^-$-beam annealing. From the aligned, as-implanted spectrum it was determined that the thickness of the amorphous plus dislocation band layer to be

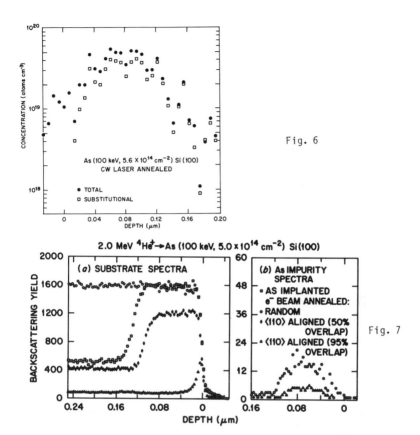

Fig. 6

Fig. 7

~1500 Å, which is in good agreement with the cross-section TEM results. The aligned spectrum from the sample scanned for 50% overlap is rather curious. The depth over which damage is distributed is very similar in this sample to that in the as-implanted sample. However, the scattering yield from the damage region is not as great as would be expected from a random solid. It was suspected that this behavior was due to a nonhomogeneous annealing over the face of the sample. If such nonhomogenities persisted over the spot size of the analysis (ion) beam, 1 mm, it could easily account for this behavior. Our suspisions were confirmed by TEM cross-section results which showed large regions between the traces where solid-phase-epitaxial did not occur up to the surface. In contrast, the aligned Si spectrum from the sample scanned for 95% overlap was very similar to a virgin spectrum indicating almost a complete recovery of silicon lattice. A comparison of random and aligned As spectra for these specimens shows a large fraction of substitutional arsenic. The arsenic concentration (total and substitutional) profiles extracted from these spectra are plotted in Fig. 8 and show a good incorporation of arsenic into substitutional sites.

Si-Sb System
    In this investigation, Si(100) specimens were implanted with 200 keV, $^{121}Sb^+$ to a dose of 0.86 x 10^{16} cm^{-2} and annealed with 11 watts cw Ar$^+$ laser beam (scanning speed 34 mm/s, scan distance ~6 mm, substrate temperature 330°C,

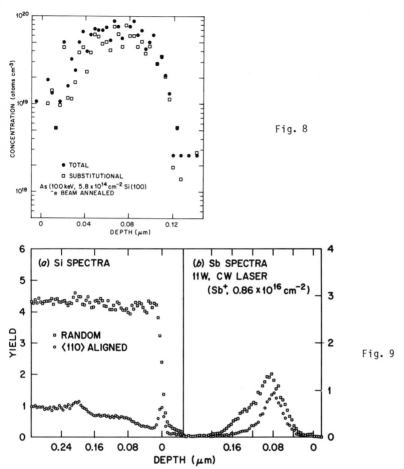

Fig. 8

Fig. 9

overlap > 60%). The RBS spectra for Si and Sb are plotted in Fig. 9. The peak at 0.2 μm in the aligned Si spectrum indicates the presence of a rather large density of extended defects. A comparison between aligned and random Sb spectra shows good dopant incorporation up to 800 Å and after that only a fraction of Sb in is substitutional sites. This can be seen more clearly in Fig. 10 which shows substitutional and total concentration of antimony as a function of depth. Figure 10 also contains a small Sb peak near the surface, indicating segregation of antimony in the near surface regions. The TEM results in Fig. 4 are in qualitative agreement with the RBS results.

Si-In System

The (100) Si specimens were implanted with 125 keV, $^{115}In^+$ ions to a dose of 1.0×10^{15} cm^{-2}. The laser annealing was performed with 12 CW Ar^+ laser (scanning velocity 34 mm/s, scan length 6 mm, overlap > 60%, substrate temperature 338°C). The Si and In spectra in Fig. 11, after CW laser annealing, show that the silicon lattice has recovered and that only a small fraction of In is in substitutional sites. The plot of indium concentrations (substitutional and total) versus depth are shown in Fig. 12. It shows a good incorporation up to

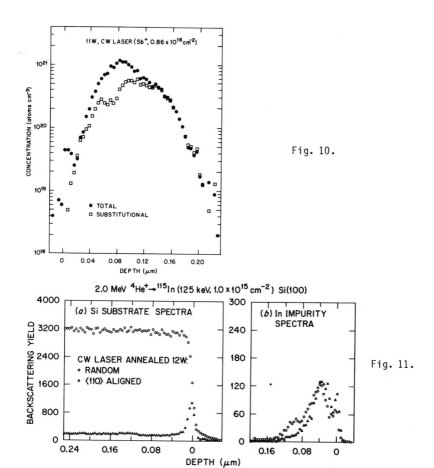

Fig. 10.

Fig. 11.

800 Å after which only a small fraction is substitutionally located. Additionally considerable redistribution of indium after laser annealing has occurred as well as surface segregation of dopant. The peak concentration of In is greatly decreased from that in the as-implanted profile (not shown).

Metastable Alloying

It has been shown previously [4] that by solid-phase-epitaxial growth in the temperature range 500-600°C in a furnace, supersaturated silicon alloys can be formed. Table I summarizes these results, showing that for Si-Bi system, the substitutional concentration of bismuth can exceed retrograde solubility limit as much as a factor of 560. One of the goals of this investigation was to see if the substitutional concentration could be increased further by SPE growth at higher temperature where the velocity of the crystallizing interface is higher. However at higher temperatures, which are obtained by CW laser annealing, the precipitation of dopants leads to a decrease in the observed substitutional concentrations. There were no cases using CW laser annealing which exceeded the

TABLE I
Comparison of observed and calculated solubility limits in Si-In, Si-Sb, Si-Bi, Si-As, and Si-Ga systems

Dopants	Retrograde Maximum Solubility	Limiting Conc. Under SPE Growth Conditions	Predicted Maximum Solubility Limits
Sb	7×10^{19} cm^{-3}	1.3×10^{21} cm^{-3} 1.6×10^{21} cm^{-3}(1) 2.5×10^{21} cm^{-3}(2)	3.0×10^{21} cm^{-3}
In	8×10^{17}	5.5×10^{19}	1.5×10^{21}
Bi	8×10^{17}	4.5×10^{20}	1.0×10^{21}
Ga	4.5×10^{19}	2.5×10^{20}	6.0×10^{21}
As	1.5×10^{21}	9.0×10^{21}	$1.5 \pm 0.5 \times 10^{22}$

(1) and (2) represent lower free energy states.
(1) Low-current implant at room temperature.
(2) Liquid nitrogen temperature implant.

maximum substitutional concentrations achieved by furnace annealing. The maximum limit of solute concentration is reached when gain in free energy due to amorphous-crystalline transformation is equal to the increase in strain free energy associated with crystalline silicon as a result of difference in covalent radius of the dopant and the host [4]. The calculated maximum values for various dopants are summarized in Table I. It can be seen in the table that the observed concentrations for furnace annealed samples approached the calculated maxima for the Si-Sb and Si-As systems. For other silicon/impurity systems, the formation of defects and solute segregation at the interface during SPE growth prevented the maximum achievable concentrations.

SUMMARY AND CONCLUSIONS

Ion implantation damage generally consists of amorphous layer followed by a band of dislocation loops. During CW laser or e$^-$ beam annealing of ion implanted amorphous layers, the underlying amorphous-crystalline interface moves to the surface leaving behind a band of dislocation loops. Residual damage in the SPE grown layers consist of small dislocation loops ranging from 10-50 Å. By subsequent thermal annealing in a furnace, the loops in the SPE grown layers can be annealed out. However, the loops in the underlying dislocation, depending upon their size and number density, coarsen by dislocation climb and glide mechanisms. The results described in this paper were obtained from specimens which held at 300 to 350°C during CW laser or e$^-$ beam irradiation to avoid large temperature gradients in the samples. If the samples are held at room temperature slip dislocations are generated from the surface as shown in Fig. 13. The dislocations are confined in the irradiated area and they are distributed from the surface to a depth ~7.0 μm. The coarsening of dislocation loops occurs in the underlying dislocation band at these temperatures.

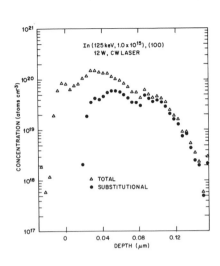

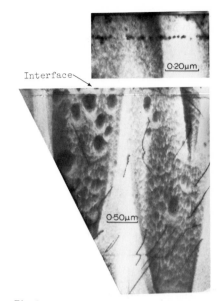

Fig. 12.

Fig.13 Slip dislocations generated in laser annealed Si.

Supersaturated solutions can be formed by SPE growth, the maximum concentrations are obtained in the temperature range 500 to 600°C. The SPE growth at higher temperatures does not result in higher concentrations. The precipitation of dopants at high temperatures probably limits the maximum achievable concentrations.

REFERENCES

1. J. Narayan and O. W. Holland, Phys. Stat. Sol. (a) 73, 225 (1982).
2. C. W. White, J. Narayan and R. T. Young, Science 204, 461 (1979).
3. G. L. Olson, S. A. Kokorowski, R. A. McFarlane and L. D. Hess, Appl. Phys. Lett. 37, 1019 (1980).
4. J. Narayan and O. W. Holland, Appl. Phy. Lett. 41, 239 (1982).

EXPLOSIVE CRYSTALLIZATION IN a-Ge AND a-Si : A REVIEW

D. BENSAHEL, G. AUVERT

CNET-CNS, BP.42, 38240, MEYLAN, France

ABSTRACT :

Self-sustained or explosive crystallization results on a-Si and a-Ge are reviewed and compared. cw laser experiments on a-Si allow to follow all the phenomena encountered, and to experimentally measure the crystallization front velocity. The published theories agree well with the experimental results on a-Si. Moreover, a-Ge presents the same behaviour as a-Si, but for a-Si, the high nucleation rate can induce a specific explosive growth.

I. INTRODUCTION :

Explosive crystallization (XCR) is a generic term which has been used for over a century to describe the rapid crystallization of amorphous materials due to the free energy release during the amorphous to crystalline transition (1). This energy release heats the adjacent material which in turn crystallizes and drives the reaction outwards. The material properties of the film either macroscopic (thickness, preparation of the samples ...) or microscopic (crystallinity, thermal history, nucleation and growth processes ...) as well as its underlying substrate modify the features observed after XCR. In all the cases however, high crystallization front speeds which can vary from 2 to 16 m/sec. have been obtained. These high speeds seemed to be difficult to imagine taking place in the solid phase regime, and theories have concluded that a thin supercooled liquid layer must be present during the crystallization process.

The aim of this paper is to review and compare the different results obtained in the XCR field in the case of a-Ge and a-Si since it appears that this kind of crystallization can have some technological interest for low temperature processes. All along this paper, we discuss more in detail the XCR of a-Si, that we will compare to the a-Ge behaviour when necessary. The differences between a-Si and a-Ge are explained as due to differences in their nucleation rates. We will show that XCR is a pure thermal process with three defined steps : initiation, propagation and quenching that we will describe using the two main published theoretical approaches to cw laser induced rapid crystallization.

All the features observed in the following have been obtained after XCR of a-Si by cw Argon laser irradiation. The laser scan speed is then the parameter that we will use to describe all the phenomena encountered in the crystallization of a-Si from the furnace annealing to the Q-switched laser experiments. In particular, for slow scans, we observe isolated explosive phenomena that turn into cooperative movement leading to the formation of a crystallization front when the laser scan speed is increased. This results in periodic crescent shaped macroscopic features. The discussion is then based upon the propagation and the quenching of these events. On the other hand , in the high scan speed regime, we obtain continuous crystallized sheets useful for technological applications. Finally, since all the mechanisms involve the nucleation and growth properties of the material, the impurities are supposed and shown to play a key role in the different behaviours encountered.

II. FURNACE-LIKE CRYSTALLIZATION :

Since crystallization starts by nucleation and proceeds by formation of new nuclei and crystal growth, differences between a-Si and a-Ge become apparent when microscopic analysis is performed to determine the grain size. A comparison of experimental results found by Barna et al. (2) for a-Ge and by Koster (3) for a-Si, shows that the nucleation rate of Si is higher than that of Ge. This difference in crystallization process is even more marked at high temperature and for short annealing times such as those encountered in cw laser irradiation. The furnace crystallization results of Blum et al. (4,5) can be reproduced in cw laser experiments under slow scan conditions when the dwell time is long enough to crystallize all the amorphous material (e.g. 10 cm/sec. with a spot diameter of 60 microns). In that case, like in furnace annealing, the latent heat of crystallization is lost to the surrounding.

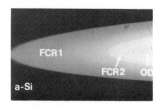

Fig.1 : Optical transmission : FCR-1 and FCR-2 occur before the appearance of the Optical Damage zone. Line width: 40 μm

But when the dwell time shortens (e.g. scan speed between 20 and 30 cm/sec.) the use of the latent heat is responsible for the existence of two modes of crystallization (labelled by analogy furnace-like crystallization FCR-1 and FCR-2) below the crystalline melting point of Si (6). In Fig.1, we show an optical transmission micrograph of the beginning of a single cw laser crystallized line in a-Si deposited on quartz substrate, showing how the degree of crystallinity affects the absorption coefficient : the more transparent the sample, the more crystallized it is. We see, as the laser power increases from left to right, the two FCR-1 and FCR-2 regimes before the appearance of the so-called optical damage zone characterizing local melting of the film. These changes in optical transmission are correlated with the different grain size of FCR-1 and FCR-2 as shown in Fig.2. A steep transition exists between the a-Si and FCR-1 regions and between the FCR-1 and FCR-2 ones. The diffraction pattern in FCR-1 shows that this zone is a mixture of polycrystalline and amorphous material, while all the FCR-2 region is crystallized. The abrupt increase in grain size between FCR-1 and FCR-2 in Fig. 2b suggests that, in addition to the gaussian temperature profile induced by the laser spot, another heat source contributes to the local temperature. Since the scan speed is around 20 cm/sec., the temperature rises rapidly and, from the amorphous phase, grains begin to grow. This growth may be due to nuclei either present before laser treatment or formed during heating. The latent heat released during this crystallization increases the local temperature further in a self-enhanced process which appears if the following conditions are simultaneously met : (i) the purely laser-induced crystallization involves a significant fraction of the amorphous layer, so that the released heat is sufficient to initiate the self-enhancement ; (ii) the released heat does not diffuse away from the layer during laser treatment.

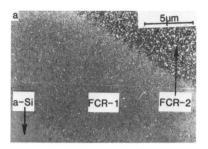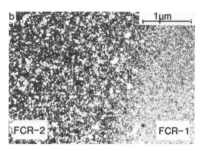

Fig.2 : Electron transmission micrograph of the half width of a scanned line showing the two FCR regimes. (b) Blow-up of the transition zone. Note the correspondance between the absorption coefficient of Fig.1 and the grain size (7)

The first condition is clearly met since the FCR-2 crystallization process starts above that of FCR-1 in which state the crystallized fraction is already significant. As for the second condition mentioned, it can be explained by one of the two possibilities : either each growing crystallite makes use of the latent heat, or it is put to "collective" use on the scale of the layer thickness. Under our conditions, the dwell time is about 100μsec., leading to heat diffusion lengths of 20 microns in Si or 2 microns in silica. Clearly then, the latent heat can only be used on the scale of the complete layer. It thus becomes clear that FCR-2 is strongly dependent on dwell time. For longer times, as in furnace annealing, the latent heat will be dissipated without any significant increase in the growth rate, whereas for shorter times, it is completely confined in the Si layer, leading to explosive phenomena. At scan speeds higher than 30 cm/sec., the local and vertical use of the latent heat results in a cooperative movement and generation of a front giving rise to macroscopic explosive phenomena (7).

In a-Ge, since the nucleation rate is low, in cw laser experiments, the range of FCR is very narrow (8) and with short dwell times (< 1msec.), no solid phase crystallization occurs below the melting point. Hence, if an explosive event can be initiated, it must be by a local melting of the film, as shown in Fig.3. In this case, we clearly see the locally melted zone due to the laser probe and, extending outwards, the explosive phenomenon. This behaviour was first encountered in a-Ge when the XCR was initiated either mechanically, optically (9,10) or electrically (11).

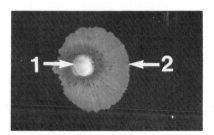

Fig.3 : cw laser induced crystallization at 1 m/sec. in a-Ge. Melting of the film <1> (diameter : 2 μ m) initiates XCR <2>. Scan direction is from left to right

168

III. DYNAMICS OF THE EXPLOSIVE PHENOMENA :

In the laser induced XCR of a-Si, we have shown previously (7) that above a scan speed of 30 cm/sec., a crystallized front can be generated. This runaway front propagates then at a high speed (Vc). Chapman et al. (12) have demonstrated that the competition between the laser spot and the moving crystallized front results in periodic structures. In the case of a gaussian circular laser beam, as shown in Fig.4, the periodicity is constructed between two thresholds temperatures Tn and Tq for the nucleation and the quenching of the event respectively. The runaway front starts at time t1 when Tn is at position X1. The front stops at time t2 at X2. During this time interval, the laser moves a distance "a" and crystallization proceeds on distance d=a+b. "d" is then the spatial period of the crescents induced by the laser. The dependence of the period with the laser power and scan speed is shown in Fig.5 where Po is the laser power needed to induce the first occurrence of a crescent of period do. We see, in Fig.5a, that a minimum laser power is necessary to induce XCR. If the laser power further increases, the period "d" decreases. This is related to the distance between Tn and Tq isotherms which decreases as the maximum laser induced temperature profile is increased. In Fig.5b, the period "d" is an hyperbolic function of the laser scan speed (13). The period becomes infinite when the laser spot moves as fast as the crystallization front Vc, that is to say that we have a direct experimental measure of the front velocity. This front velocity depends on the properties of the starting material. In Fig.6, we show that two kinds of XCR can exist along the same scanned line in plasma-CVD deposited a-Si on quartz substrates (14). These two kinds of XCR, crescent shaped, were labelled LP- and SP- XCR for liquid and solid phase respectively by analogy with the published results in a-Ge as described below.

It is important to recognize that the level of disorder, or degree of crystallinity, plays a crucial role in determining whether a LP- or SP- XCR event will occur : this is due to the dependence of both the Si absorption coefficient, i.e. film color, and the magnitude of the latent heat on the grain size (15).

Fig 8

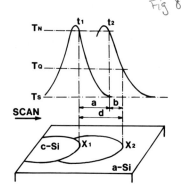

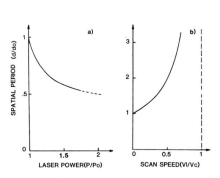

Fig.4 : Dynamics of the competition between the laser scan speed and that of the crystallization front.

Fig.5 : Spatial period of the crescents versus : (a) the laser power, (b) the scan speed. The parameters are normalized.

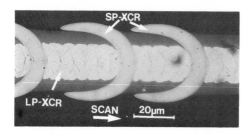

Fig.6 : Two different kinds of explosive event can exist in a-Si along the same scanned line at 85 cm/sec. Optical transmission picture (14)

As the laser power is increased above 30 cm/sec., LP-XCR events develop. Further increase of the speed makes the small crescents wider, but LP-XCR cannot be extended into continuous lines because of competition from the appearance of the SP-XCR event, which begins to develop above 100 cm/sec. However, if the a-Si deposit is initially brought to a FCR-1 structure by a first low speed scan at 20 cm/sec., then SP-XCR will not occur and LP-XCR crescents can be studied as a function of scan velocity. Fig.7 shows the extension of LP-XCR crescents until we obtain the continuous line at 2.2 m/sec. Note that there is a competition between LP-XCR growth and FCR-2 growth at speeds greater than 2.2 m/sec. Scanning the laser at 2.5 m/sec., thereby exceeding the speed of the crystallization front, produces a continuous line of FCR-2 which is bounded on both sides by LP-XCR zones as shown in Fig.7. This is an important observation because there is no doubt that the FCR-2 material has not melted, yet the transformation proceeds at 2.5 m/sec. This dual mode of crystallization is not observed for continuous lines of SP-XCR as shown in Fig.8, because at speeds higher than 10 m/sec., furnace-like crystallizations, which are time dependent, are suppressed.

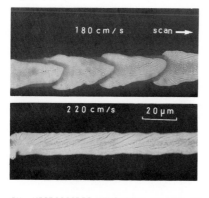

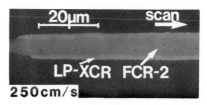

Fig.7 : Optical transmission : influence of the laser scan speed in precrystallized (FCR-1) a-Si. This cannot be obtained directly in a-Si since, above 1m/sec., the other explosive event overtakes the LP-XCR one (15)

IV. MICROSCOPIC BEHAVIOUR OF XCR IN a-Si :

The differences between the two LP- and SP-XCR events in a-Si are more marked when we study their crystallographic behaviour (6). In Fig.9, we show electron transmission micrographs of the two possible explosive growths : the LP-XCR crescent is composed by large grains expanding in a diverging pattern, and showing a surface roughness. The grains initiate from a FCR-2 region and extend then outwards until they stop abruptly. This LP-XCR event is dominated by a growth process.

170

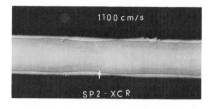

Fig.8 : Optical transmission : influence of the laser scan speed in the SP-XCR mode. Note that this regime is a two step process : SP2-XCR with a front velocity of 11 m/sec. and SP1-XCR which develops above this limit speed (15)

On the other hand, the SP-XCR crescent is initiated from a FCR-1 region and we encounter in all the crescent small homogeneous and well crystallized grains (SP1-region) except in the outer edge of the crescent where the grains become elongated (SP2-region). This event is then composed by two kinds of grains and is dominated by a nucleation mechanism. As the laser scan speed is increased, we observe the following dynamic sequence shown in Fig.8 : opening of the crescents, continuous line of SP2-XCR at 11 m/sec., and above, a continuous line of SP1-XCR (not shown in the picture) (15).

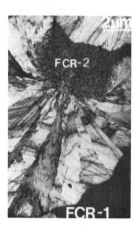

Fig.9 : Transmission electron micrographs of : at left, LP-XCR dominated by grain growth ; at right, SP-XCR dominated by nucleation (SP1-XCR). Note the small SP2-XCR region at the outer edge of the crescent (6)

V. THE a-Ge CASE :

   In the previous sections, we described several phenomena observed in a-Si irradiated with a cw laser. The results are consistent with one another and for instance, we do know that two XCR modes can occur. In the case of a-Ge, the range of published data is less complete. Furthermore, in a-Ge, the substrate temperature is a dominant parameter which is less sensitive in Si. The free energy release during crystallization is used first to sustain the crystallization in the film and secondly to locally heat up the substrate. Fan et al. (16) and Gold et al. (17) have shown that below a threshold substrate temperature, no runaway crystallization can be obtained. This is due to the substrate which consumes too much energy. As a confirmation, Koba et al. (18) have shown that a thick film can explosively crystallize with a lower temperature than a thin one. In Fig. 10a is shown a self-sustained crystallization which extends over a large area. Fig.10b shows the influence of the substrate temperature on the extent of the crystalline region showing that above a threshold temperature, an infinite crystalline region can develop. The hyperbolic behaviour is similar to that observed in the cw laser experiments on a-Si (Fig.5b). In fig.10c, we show that the threshold temperature for infinite crystallization varies as the inverse of the film thickness. This point confirms the existence of an energy balance in the XCR process.

$Fig\ 2$

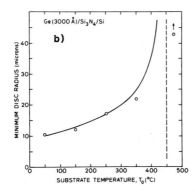

a)

Fig.10 : influence of the substrate temperature on the extension of XCR in a-Ge. (a) optical micrograph : the local initiation is made by a 20 μ m laser spot, (b) extension of the crystallized area versus the substrate temperature (17), c) threshold temperature for infinite propagation versus the inverse of the film thickness $\langle x\ 10^3 cm^{-1}\rangle$ (18)

Ge(3000 Å)/Si$_3$N$_4$/Si

b)

MINIMUM DISC RADIUS (microns)

SUBSTRATE TEMPERATURE, T$_0$(°C)

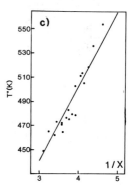

c)

T°(K)

1 / X

However, it is important to note that the self thermal enhancement induces a self-sustained crystallization leading eventually to the formation of a liquid region and to surface damage. Such damage can stop the rapid crystallization and then lower the observed runaway speed. Indeed, runaway velocities of 1 m/sec. have been measured for XCR of thick films by using either an ultra high speed camera (19, 20, 21), or in-situ optical transmission measurements during cw laser irradiation (12). On the other hand, by scanning a 2 microns film up to 6 m/sec., Greene et al. (22) still observed crescents. We confirm this last result in Fig.11 where we show the cw laser induced XCR of a 0.3 micron a-Ge film deposited by vacuum evaporation on quartz substrates (23). A runaway speed near 16 m/sec. is obtained experimentally, and the surface of the crystallized Ge remains very smooth.

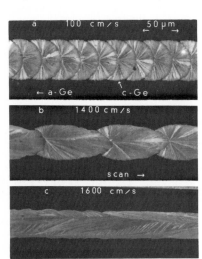

Fig.11 : influence of laser scan speed in XCR of a-Ge, .3 μm thick deposited on quartz substrate. The surface is smooth after crystallization

VI. THEORIES OF SELF-SUSTAINED CRYSTALLIZATION

The discrepancies between the above results in a-Ge or the existence of two kinds of XCR in a-Si can be qualitatively described by the theories of Zeiger et al. (24) and Gilmer et al. (25). Both these theories describe the phase-boundary dynamics during XCR of amorphous films. The two dimensional model of Zeiger et al. (24) supposes that we pass directly from amorphous to crystalline structure and the basic hypotheses are then : (i) the crystallization front is planar, (ii) no melting occur, (iii) the latent heat is totally released in the crystallization plane. It is then a two dimensional heat flow problem, where, in the steady-state case, the runaway velocity is obtained with no undetermined parameters. However, the question of the stability of a steady-state solution is not easy to solve mathematically. In Fig.12, the theoretical thermal profile around the boundary is shown. This kind of result is probably relevant for the SP1-XCR behaviour where we have a high nucleation rate, although it can only be qualitative since the nucleation region certainly has a finite extent, as opposed to a liquid-solid interface.

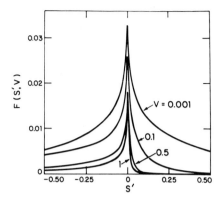

Fig.12 : theoretical normalized thermal profile around a crystallization front moving in Solid Phase mode at various normalized speed (24)

This case of an intermediate liquid phase present during the XCR process has been investigated by Gilmer et al. (25). The corresponding thermal profile is schematically shown in Fig.13a. All the excess internal energy is released in the solidification front and most of it is conducted through the thin liquid layer to drive the liquid-amorphous interface. From the analysis, we obtain the dependence of the liquid layer thickness versus the front velocity as shown in Fig.13b. Near the melting point, the velocity increases when the liquid zone becomes thinner. These results are in agreement with our results on a-Si, if we assume that in the LP-XCR case, we have a large melted zone. On the other hand, the value of 11 m/sec. for the SP-XCR event suggests that we have also a supercooled liquid layer but very thin in that case.

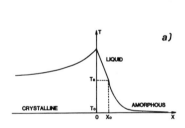

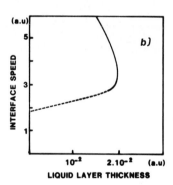

FIG.13 : (a) schematics of the intermediate Liquid Phase. (b) Interface speed versus the liquid layer thickness (25)

For a-Ge, there remains some doubt since experiments are not done together. Moreover, a low runaway speed (around 1 m/sec.) can be explained by a high release of latent heat due to a high film thickness associated with a low energy loss in the substrate. Two types of consequences can be imagined : the first is that the runaway speed is decreased as the liquid layer increases, and a surface roughness is induced due to surface tension and/or thermal stress in an encapsulating oxide layer. The second is that thermal stress in the substrate extends far enough to give rise to mismatch between layer and substrate in cold regions, leading to cracking of the layer, which of course will stop the propagation of the crystallization front. Note also that our result of 16 m/sec. for XCR of a-Ge and the associated smooth surface suggest that we are in the case of a narrow liquid phase involved, like the SP2-XCR case of a-Si (Fig.8).

VII. IMPURITIES

As it can be seen in all the publications concerning the explosive process, preparation modes and impurity content of the film are the key parameters in this kind of crystallization. The most common methods of preparation are either sputtering in an argon atmosphere or high vacuum evaporation in the a-Ge case. These methods give the same explosive behaviour and in that case, argon or oxygen impurities have been found to impede the explosive growth (26). For a-Si, we have found that both kinds of explosive phenomena were dependent on the film growth technique. The two regimes were clearly observed on a-Si obtained by low frequency plasma-CVD (7) having less than 1 % of hydrogen. The liquid phase explosive crystallization appears less and less clearly depending on whether a-Si is deposited by using LPCVD at low temperature, vacuum evaporation, sputtering or high frequency plasma-CVD in that order. It is very likely that both the amount of latent heat available and the impurity content play a role. Implantation of impurities has been used in both a-Ge and a-Si to test the validity of the theoretical model of the liquid phase via impurity segregation. In a-Ge, the broadening of the in-depth Sb and Pb implantation profiles after explosive crystallization was explained as a consequence of the liquid phase diffusivities for the two impurities (27). In a-Si, we have performed such experiments but we were interested in the lateral segregation of the dopants according to the solid and liquid explosive growth. In particular we have shown that the implanted layer explodes separately from the underlying part of the film in the SP-XCR mode (28). As another consequence, we have also shown that, as in classical furnace annealing, the impurity can increase the growth rate and decrease the different threshold temperatures for growth. A one to one correlation also exists between the features encountered in explosive processes on implanted mono-Si (29,30) and the events observed on deposited a-Si.

VIII. CONCLUSIONS :

We have investigated all the crystallization processes that can occur in a-Ge and a-Si from furnace annealing to cw laser experiments involving dwell times around 1 μ sec. Two main parameters appear : the properties of the starting material (i.e. nucleation and growth), and the characteristic time of crystallization. In the .1 msec. range, we can observe isolated microscopic explosive phenomenon, especially in silicon which has a higher nucleation rate that germanium. In the (.1 msec -1 μ sec.) range, extended XCR occurs. Front velocities observed either in a-Ge or in a-Si are in agreement with a theory involving the existence of a narrow liquid phase during the XCR process. The width of this liquid phase is directly related to the front velocity and it is interesting to note that the highest front velocities in a-Ge and a-Si are nearly equivalent and in good agreement with those obtained from Q-switched experiments. However, the high nucleation rate of a-Si makes that below 1 μsec. dwell time, we will only obtain nucleation and hence small grains. Specific details of explosive recrystallization phenomenon in amorphous layers are given in a paper by Narayan et al. in these proceedings (31).

ACKNOWLEDGMENTS :

We wish to thank V.T. Nguyen for continuous support, J.C. Pfister for fruitful discussions, A. Raharijaona and R. Carré for their technical help.

REFERENCES :

(1) G. GORE Phil. Mag. 9, 73 (1855)

(2) A. BARNA, P.B. BARNA, J.F. POCZA, J. Non-Cryst. Sol. 8, 36 (1972)

(3) U. KOSTER Phys. Stat. Sol. (a) 18, 313 (1978)

(4) N.A. BLUM, C. FELDMAN, J. Non-cryst. Sol. 11, 242 (1972)

(5) N.A. BLUM, C. FELDMAN, J. Non-cryst. Sol. 22, 29 (1976)

(6) G. AUVERT, D. BENSAHEL, A. PERIO, F. MORIN, G.A. ROZGONYI, V.T. NGUYEN "Laser and Electron Beam Interactions with Solids" edited by B.R. APPLETON, G.K. CELLER (North Holland press, New York 1982) pp 535

(7) D. BENSAHEL, G. AUVERT, Y. PAULEAU, J.C. PFISTER Revue de Physique Appliquée Dec. 82 pp 21 (in english)

(8) R.L. CHAPMAN, J.C.C. FAN, H.J. ZEIGER, R.P. CALE "Laser and Electron Beam Interactions and Materials Processing" edited by J.F. GIBBONS, L.D. HESS, T.W. SIGMON (North Holland press, New York 1981) pp 81

(9) T. TAKAMORI, R. MESSIER, R. ROY Appl. Phys. Lett. 20, 201 (1972)

(10) G. BADERTSCHER, R.P. SALATHE, H.P. WEBER, Appl. Phys. 25, 91 (1981)

(11) K. J. CALLANAN, A. MATSUDA, A. MINEO, T. KUROSU, M. KIKUCHI Sol. Stat. Comm. 15, 119 (1974)

(12) R.L. CHAPMAN, J.C.C. FAN, H.J. ZEIGER, R.P. CALE Appl. Phys. Lett. 37, 292 (1980)

(13) G. AUVERT, D. BENSAHEL, A. GEORGES, V.T. NGUYEN, P. HENOC, F. MORIN, P. COISSARD Appl. Phys. Lett. 38, 613 (1981)

(14) G. AUVERT, D. BENSAHEL, A. PERIO, V.T. NGUYEN, G.A. ROZGONYI, Appl. Phys. Lett. 39, 724 (1981)

(15) D. BENSAHEL, G. AUVERT, V.T. NGUYEN, G.A. ROZGONYI - id - Ref.6 pp. 541

(16) J.C.C. FAN, H.J. ZEIGER, R.P. CALE, R.L. CHAPMAN Appl. Phys. Lett. 36, 158 (1980)

(17) R.B. GOLD, J.F. GIBBONS, T.J. MAGEE, J. PENG, R. ORMOND, V.R. DELINE, C.J. EVANS, Jr. "Laser and Electron Beam Processing of Materials" edited by C.W. WHITE, P.S. PEERCY (Academic Press, New York 1980) pp 221

176

(18) R. KOBA, C.E. WICKERSHAM Appl. Phys. Lett. 40, 672 (1982)

(19) A. MINEO, A. MATSUDA, T. KUROSU, M. KIKUCHI Sol. Stat. Comm. 13, 329 (1973)

(20) T. TAKAMORI, R. MESSIER, R. ROY J. Mat. Sci. 8, 1809 (1973)

(21) T. TAKAMORI, R. MESSIER, R. ROY J. Mat. Sci. 9, 159 (1974)

(22) J.E. GREENE, K.C. CADIEN, D. LUBBEN, G.A. HAWKINS, G.R. ERIKSON, J.R. CLARKE Appl. Phys. Lett. 39, 232 (1981)

(23) The as-deposited samples were kindly provided by L.D. LAUDE of the University of Mons (Belgium)

(24) H.J. ZEIGER, J.C.C. FAN, B.J. PALM, R.L. CHAPMAN, R.P. CALE Phys. Rev. B 25, 4002 (1982)

(25) G.H. GILMER, H.J. LEAMY - id - Ref. 17 pp. 227

(26) C.E. WICKERSHAM Sol. Stat. Comm. 34, 907 (1980)

(27) H.J. LEAMY, W.L. BROWN, G.K. CELLER, G. FOTI, G.H. GILMER, J.C.C. FAN Appl. Phys. Lett. 38, 137 (1981)

(28) D. BENSAHEL, G. AUVERT J. Appl. Phys. accepted for publication Jan. 83

(29) D. BENSAHEL, G. AUVERT, M. DUPUY J. Appl. Phys. accepted for publication Jan. 83

(30) M. LERME, T. TERNISIEN D'OUVILLE, D.P. VU, A. PERIO, G.A. ROZGONYI, V.T. NGUYEN - id - Ref. 6 pp. 547

(31) J. NARAYAN et al, these proceedings.

EXPLOSIVE RECRYSTALLIZATION OF ION IMPLANTATION AMORPHOUS SILICON LAYERS*

J. NARAYAN,
Solid State Division, Oak Ridge National Laboratory, Oak Ridge, TN 37830

G. R. ROZGONYI,
Materials Engineering Dept., North Carolina State University, Raleigh, NC 27650

D. BENSAHEL, G. AUVERT, V. T. NGUYEN,
CNET, Meylan, Grenoble, France

and A. K. RAI
Universal Energy Systems, Inc., Dayton, OH 45432

ABSTRACT

Explosive recrystallization of amorphous layers, induced by scanned electron or CW Ar^+ laser beams, has been investigated in arsenic implanted (100) specimens. We have studied the microstructural changes using plan-view and cross-section electron microscopy to obtain the mechanism of explosive recrystallization phenomenon. The confinement of heat in the amorphous layers during laser irradiation and the nature of amorphous silicon determine the modes of explosive recrystallization. We indicate that, depending upon the degree of undercooling, two distinct modes of liquid-phase recrystallization are observed: one occurring at a velocity of ~ 2 ms^{-1} and the other at ~ 10 ms^{-1}. Explosively recrystallized grains contain $\langle 110 \rangle$ and $\langle 1\overline{1}0 \rangle$ as surface normal and growth direction respectively. We present a model for unseeded crystallization to explain the $\langle 110 \rangle$ texture observed in diamond cubic lattices.

INTRODUCTION

The phenomenon of explosive recrystallization (XCR) is referred to as a process in which a crystallizing interface advances rapidly or runs away from a thermal, optical or mechanical [1-4] source which was responsible for its initiation. Most of the studies so far have concentrated on amorphous layers on insulating or glass substrates, where two rather distinct modes of crystallization have been reported [3]. In the case of silicon layers or glass substrates, the first mode involves crystallization velocity ~ 2 ms^{-1}, and the second mode involves crystallization velocity ~ 10 ms^{-1}. It has been suggested that in the first mode, where relatively large crystallites are formed, the crystallizing interface contains a thin layer of liquid that moves along the scanned beam [5]. The second mode has been proposed to involve solid-phase crystallization [3,4]. In this investigation, we have studied explosive recrystallization of thin (~ 1300 Å) amorphous layers produced by arsenic ion implantation of (100) Si. This mode of XCR is similar to the second one described above. In this paper, we indicate that both modes of XCR involve liquid-phase crystallization. The two modes are related to each other by different degrees of undercooling. We show that surface normal of recrystallized film and lateral crystallization directions are $\langle 110 \rangle$, $\langle 1\overline{1}0 \rangle$ instead of $\langle 110 \rangle$ and $\langle 001 \rangle$ respectively as reported earlier [2]. Fundamental aspects of this phenomenon related to growth directions in unseeded crystallization are discussed.

*Research sponsored by the Division of Materials Sciences, U. S. Department of Energy under contract W-7405-eng-26 with Union Carbide Corporation.

Mat. Res. Soc. Symp. Proc. Vol. 13 (1983) Published by Elsevier Science Publishing Co., Inc.

178

EXPERIMENTAL

Silicon single crystals (having <001> orientations, resistivity 2-6 Ω-cm, 0.5 mm thick) were amorphized by $^{75}As^+$ ion implantation (100 keV, dose = 5.0 x 10^{14} cm^{-2}). This leads to the formation of 1300 Å thick amorphous layer followed by a ~200 Å wide band containing dislocation loops. These specimens were treated with CW Ar$^+$ laser operating in multimode. The diameter of the Gaussian shaped laser spot was 60-70 μm at 1/e^2. The power of the laser beam varied from 15 to 20 watts and the scanning velocity from 150 to 300 cms^{-1}. The electron beam irradiation was performed with a 8 μm (at 1/e^2) diameter beam, which was scanned at 150 to 200 cms^{-1}. The maximum power of the electron beam was 6 watts.

RESULTS AND DISCUSSION

The nature of explosive crystallization induced by electron or laser beams is basically similar, although the power and scanning velocity required to induce XCR are different in the two cases due to differences in the heat deposition. The absorption of laser light depends upon the optical properties of near surface layers, but usually the energy is deposited in ~100 Å thick surface region. The energy deposition in the case of electron beams is determined by electronic stopping power. The width of e$^-$-beam recrystallized region in Fig. 1 is considerably smaller than that due to the laser beam because of corresponding differences in the beam diameter. Figure 1 contains explosively recrystallized crescents with average periodicity of 3.0 μm. It is interesting to note that the next explosive event starts immediately after the end of the preceding one. The surface morphology of recrystallized areas was found to be smooth. Figure 2 shows ending and beginning of an explosive event or explosion at a higher magnification. At the beginning of an explosion, the region is dominated by high number density of nuclei, as shown at n. It is envisaged that a highly undercooled state of molten silicon in the region provides a large driving force for nucleation. Some of these nuclei can develop a crystallization front and sustain the growth which is derived by the release of heat of crystallization. The top left corner of the micrograph contains termination of an explosive event, where the average grain size is considerably smaller than that between the source and the termination point.

Figure 3 shows termination or quenching of explosive recrystallization in the amorphous region. The boundary between the recrystallized and amorphous regions is sharp. Although the regions near the end have finer grains but contain no amorphous zones. This point was confirmed by detailed STEM microdiffraction studies where diffraction information from the areas ≤ 400 Å diameter could be obtained.

The texture which is characterized by orientation normals to the recrystallized layers and the crystallizing interface, is of particular interest during XCR because it provides information on unseeded crystallization. Figure 4(b) shows a microdiffraction pattern from an explosively recrystallized region (encircled) in Fig. 4(a). Surface normal and crystallizing direction were determined to be [110] and [1$\bar{1}$0] respectively. Previous studies [2] had reported <110> surface normals for the recrystallized grains of Si and Ge on glassy substrates, but the crystallizing directions were found to be <001>. This error may be due to the lack of microdiffraction studies. A model has been proposed (to be described later) to explain <110>/<1$\bar{1}$0> texture during unseeded crystallization in diamond cubic lattices.

Electron beam recrystallized specimens were investigated by cross-section electron microscopy and the results are shown in Fig. 5. The micrographs [bright-field (a), dark field (b)] show that ~300 Å thick layer above the original dislocation band grows by normal solid-phase-epitaxial (SPE) growth in conjunction with (001) substrate. After that the crystallizing interface breaks

179

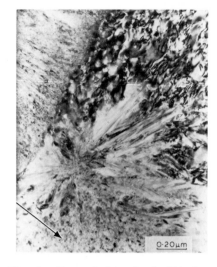

Fig. 1. Formation of periodic crescent in explosively recrystallized layer (Period 3 μm). The arrow shows the direction of scanning.

Fig. 2. Part of the area in Fig. 1 at a higher magnification, the arrow shows the scanning direction and 'n' indicates the point of explosive event.

Fig. 3. The boundary between recrystallized and amorphous region.

180

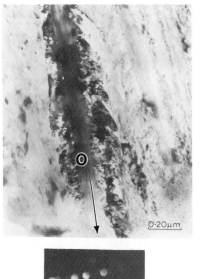

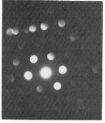

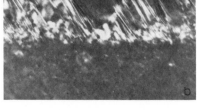

Fig. 4. Explosively recrystallized grain and a microdiffraction pattern from the grain. The arrow represents the direction of grain crystallization.

Fig. 5. (110) cross-section micrographs: (a) bright-field, (b) dark-field. The arrow indicates the original dislocation band.

into an explosive mode of crystallization. Figure 5 contains edge-on view of recrystallized grains which were found to have [1$\bar{1}$0] orientation, in agreement with the plan-view results of Fig. 4. The micrograph in Fig. 5 also shows the smoothness of outer surface, which is a characteristic feature of XCR occurring ~10 ms^{-1} in silicon.

CW Ar$^+$ LASER INDUCED RECRYSTALLIZATION

Explosive recrystallization of amorphous layers in ion implanted silicon was obtained for a particular combination of laser power and scanning velocity. The XCR crescents (A) were separated by furnace-like crystallized regions (B), as shown in an optical micrograph (Fig. 6). The separation between the crescents in laser crystallized specimens near the threshold was found to be irregular, probably reflecting spatial inhomogenieties in the laser beam. Figure 7 shows a nucleus of an explosion, which was present near the boundary between A and B regions. It is interesting to note that the crystallization front did not develop in the furnace-like crystallized regions (B), which contain small size

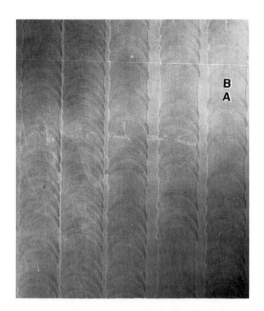

Fig. 6. Optical micrograph showing explosively (laser-induced) recrystallized crescents (A) and furnace-like crystallized regions (B).

Fig. 7. Dark-field micrograph showing the nucleus of an explosion 'n'.

grains 100-300 Å. Thus, the explosive recrystallization front develops ahead of the scanning beam in the amorphous region. The texture of large grains in Fig. 7 was determined to <110> and <1$\bar{1}$0>, similar to e⁻-beam recrystallized grains shown in Fig. 4.

The microstructures of regions A and B were investigated using cross-section electron microscopy and the results summarized in Figs. 8 and 9. The region 'B' between the crescents contain SPE grown region ∿300-400 Å wide above the dislocation band. The interface region contains twins in {111} planes, above which polycrystalline regions (average crystallite size ∿100 Å) are observed as shown in Fig. 8. Figure 9 shows cross-section micrographs of regions near the boundary of A and B [Fig. 9(a)] and A regions [Fig. 9(b)]. Figure 9(a) shows 111 facets near the interface, which contain twins. Explosive events are beginning as shown at n in Fig. 9(a). In Fig. 9(b) the crystallizing front has developed leading to XCR.

From the above observations, we propose a following mechanism for explosive recrystallization of ion implanted amorphous layers. During laser irradiation, the photon energy from the laser beams is ultimately transferred to the lattice in times less than a nanosecond, which heats the amorphous layer as well as the substrate. By scanning the CW laser beam, the laser-solid interaction time as well as the confinement of heat can be controlled. From cross-section micrograph we determined ∿300 Å wide SPE grown layer. Beam interaction time was obtained from the scanning velocity. From the above thickness, we obtained SPE growth velocity ∿0.1 cms⁻¹, from which specimen temperature was deduced to be 1150 ± 50°C. The crystallizing (001) substrate interface tends to become unstable and

182

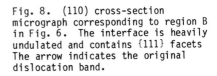

Fig. 8. (110) cross-section micrograph corresponding to region B in Fig. 6. The interface is heavily undulated and contains {111} facets The arrow indicates the original dislocation band.

Fig. 9. (a) (110) cross-section micrograph corresponding to the boundary of A and B regions; (b) (110) cross-section corresponding to regions A in Fig. 6.

develops facets along {111} planes. This slows down the advancement of the growing interface because the growth on {111} planes is slower by a factor of 1/24 compared to (100) growth rate. The SPE growth on {111} also leads to the formation of twins. During this time when the interface advancement is slowed down, the temperature of the remaining amorphous layer is reached to 0.80 to 0.95 T_m, where T_m is the melting point of silicon. The recrystallization in the highly supercooled layer ($\sim$0.80 T_m) occurs with velocities $\sim$10 ms^{-1}, whereas in layers ($\sim$0.95 T_m) the crystallization velocity is $\sim$2 ms^{-1}. It is proposed that some of the twins or other defects which are present near the interface may provide a nucleus for the development of crystallization front. The heat of crystallization is utilized to maintain highly supercooled thin liquid layer. We have shown that crystallization occurs along the <110> directions. If the explosive crystallization cannot be initiated, for example, due to lower laser power as a result of beam power fluctuation, then the nucleation of crystallites occurs in the amorphous layer by solid-phase crystallization, as shown in region B in Fig. 8. Once the layer has developed the crystallites (B region in Fig. 6), it can no longer be highly supercooled and crystallization can be induced only near $\sim$0.95 T_m with a growth velocity $\sim$2 ms^{-1}. The growth front velocity in explosively recrystallized regions such as A is estimated to be $\sim$10 ms^{-1}. This has been shown by obtaining continuous recrystallized crescents by matching the growth-front velocity with laser scanning velocity. We have found that microstructures under pulsed laser melting conditions (such as shown in Fig. 10) are similar to the present explosively recrystallized microstructures. The crystallization velocity under laser-melted, liquid-phase growth was estimated from arsenic-profile broadening, and it was found to be $\sim$10 ms^{-1}. Therefore, it is suggested that explosive crystallization by rapid CW laser or e$^-$-beam scanning is similar to that obtained during pulsed laser melting of thin amorphous layers and crystallization under highly supercooled state.

Fig. 10. (110) cross-section micrograph from pulsed-laser melted (ruby laser λ=0.693 µm, τ=15x10^{-9} s, E=0.5 J/cm^{-2}) and recrystallized region.

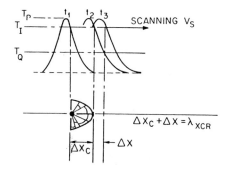

Fig. 11. Schematic representation
of explosive recrystallization and
associated temperature profiles.

Figure 11 is a schematic representation of the XCR phenomenon during scanning.
The temperature profile induced by the beam is a critical function of laser
parameters (wavelength, power, beam shape, etc.) and scanning velocity. The
temperature profile at time t_1 produces a peak temperature T_p, the explosive
crystallization starts at T_I in the leading edge of the beam. The crystallization velocity V_s is usually faster than the scanning velocity and the crystallization front tries to move ahead but after a distance ΔX_c the front is quenched
or terminated at T_Q when the heat supplied by the laser beam plus the heat of
recrystallization are not enough to sustain liquid-phase growth. Up to time t_2,
the beam energy is simply utilized to heat the already recrystallized layer.
Until time t_3 another explosive event starts at distance 'ΔX' which depends on
the heating beam parameters and the scanning velocity. The distance between the
XCR crescents $\lambda_{XCR} = \Delta X_c + \Delta X$.
    In the following we deal with a fundamental problem of formation of texture.
The formation of texture during explosive crystallization provides information
on the nature of unseeded crystallization. In other words, if the seed is not
present during crystal growth, then, is there a preferred direction of crystal
growth? In previous studies, <110> surface normals for recrystallized Si and Ge
layers on glass substrates were reported [2,3], however, the lateral growth
direction was erroneously identified to be <001>. Present studies, using
detailed microdiffraction techniques, have clearly identified that surface normals and lateral directions of unseeded XCR recrystallized layers are <110> and
<1̄10> respectively. A crystal growth model to explain <110> texture is pro-
posed. In this model it is recognized that the basic building unit in the
diamond cubic lattice is <110> chain of atoms, as shown in Fig. 12 of ref. 6.
The bonds in this chain are along <111> directions, their length and angle are
respectively $\sqrt{3}\,a_0/4$ and 109.5°, where a_0 is the lattice constant. During
crystallization, these basic chains of atoms are formed along with the requirement of completing a sixfold or a hexaring in {111} planes. The hexaring structure is characteristic of crystalline structure in diamond cubic lattices. It
should be noted that the requirement of completing a hexaring is similar to
Spaepen's model of crystallization [7], and it is equivalent to forming two
bonds with the crystalline phase as proposed by Drosd and Washburn [8]. During
crystallization when these basic chains are completed, steps at the interface
are created. The steps grow by further incorporation of atoms or new steps may
be created by completing additional chains until the interface advances by one
plane spacing. The formation of <110> || <1̄10> texture requires the development
of two sets of <110> chains in these two directions as illustrated in Fig. 13 of
ref. 6 for (110) growth.

184

SUMMARY AND CONCLUSIONS

During explosive recrystallization of ion implanted amorphous layers using laser or electron beams, we observe about 300 Å wide SPE grown layer above the original band of dislocation loops. Using the thickness and the specimen-beam interaction times, the temperature of the specimen was estimated to be 1150 $\pm$ 50°C. The crystallizing interface tends to become unstable resulting facets in the {111} planes. The SPE growth on {111} planes leads to the formation of twins, and it slows down the growth rate by a factor of about 1/24 compared to that along {100} planes. During this time, the temperature of the remaining amorphous layer is raised from 0.80 to 0.95 T_m, where T_m is the melting point of crystalline silicon. The crystallization velocities corresponding to 0.80 and 0.95 T_m supercooled states are estimated to be 10 and 2 ms^{-1} respectively. Crystallization starts from twins and interfacial defects protruding in the undercooled layer. From a comparison of these explosively recrystallized microstructures with those obtained by pulsed laser irradiation, we estimate recrystallization velocity $\sim$10 ms^{-1}. This rapid crystallization leaves high density of residual defects. The explosive recrystallization $\sim$2 ms^{-1} results in relatively defect-free large grains. The crystallization ends sharply with no evidence of the presence of amorphous zones near the end of the crescent. However, the average grain size is considerably smaller near the end than that between the explosion nucleus and the termination point. We have experimentally shown that unseeded crystallization, which is involved during explosive recrystallization, results in the formation of <110> texture, which is consistent with the proposed model of crystallization in diamond cubic lattices.

REFERENCES

1. T. Takamori, R. Messier and R. Roy, Appl. Phys. Lett. 20, 201 (1972).

2. R. L. Chapman, J.C.C. Fan, H. J. Zeiger and R. P. Gale, Appl. Phys. Lett. 37, 292 (1980); R. P. Gale, J.C.C. Fan, R. L. Chapman and H. J. Zeiger, p. 439 in Defects in Semiconductors, ed. by J. Narayan and T. Y. Tan, North Holland, New York, 1980.

3. G. Auvert, D. Bensahel, A. Perio, V. T. Nguyen and G. A. Rozgonyi, Appl. Phys. Lett. 39, 724 (1981).

4. M. Lerme, T. Ternisien D'ouville, Duy-Phack Vu, A. Perio, G. A. Rozgonyi and V. T. Nguyen, p. 547 in Laser and Electron Beam Interactions with Solids, ed. by B. R. Appleton and G. K. Celler, North Holland, New York, 1982.

5. H. J. Leamy, W. L. Brown, G. K. Celler, G. Foti, G. H. Gilmer and J.C.C. Fan, Appl. Phys. Lett. 38, 137 (1981).

6. J. Narayan, J. Appl. Phys. (Dec. 1982).

7. F. Spaepen, Phil. Mag. 30, 417 (1974).

8. R. Drosd and J. Washburn, J. Appl. Phys. 53, 397 (1982).

VISUALIZATION OF TEMPERATURE PROFILES IN a-Si CRYSTALLIZED IN THE EXPLOSIVE MODE BY CW LASER.

MICHEL RIVIER, ANDRAS CSERHATI AND GEORGE GOETZ
Bendix Advanced Technology Center, Columbia MD. 21045   USA

ABSTRACT

We present an experimental method, based on the isothermal nature of the boundaries of explosively crystallized silicon, to display instantaneous isotherms produced by fast scanning laser beams. This method is especially useful for the case of multilayer structures or complex geometries, where thermal profiles simulation is a difficult problem. We can also obtain directly a measure of the spatial intensity distribution in the laser beam.

INTRODUCTION

The temperature profiles induced by a scanning CW laser beam control the crystallization processes, in the liquid or the solid state. When the silicon film is melted, thermal gradients at the trailing edge of the molten area determine size and direction of elongated grains. Appropriate beam shaping [1] leading to temperature profile shaping has been found succesful in controlling the grain direction. We show that explosive crystallization obeys similar relations, except that the leading edge is important.

The simulation of temperature profiles requires the solution of a heat flow equation.Analytical solutions are only used in simple cases such as steady state, continuous homogeneous films, or a laser beam with circular or cylindrical symmetry [2-6]. The cases of complex geometries, multilayered samples, fast scanning beams are only solved numerically by a finite difference analysis. This often requires extensive computer time and leads to meaningful solutions only if the temperature variation of the properties of the layers are known.

EXPLOSIVE CRYSTALLIZATION OF AMORPHOUS SILICON

The samples used in this study are thin films (3000-5000A) of electron beam evaporated silicon deposited at a base pressure of 10^{-7} torr at a rate of 50 A/s on fused quartz substrates at a temperature of 350 C.No evidence of crystallinity was found in these films by TEM in the diffraction mode. Laser annealing was performed in air with a modified Coherent System 5000 annealer.

A laser beam scanned at high speeds (0.5-10 m/s), on amorphous silicon layers results on explosive crystallization of the material [7-12]. A typical XCR periodic structure created by a circular laser beam operating in the TEM00 mode, swept at a speed of 1m/s is shown in Figure 1. The XCR phenomenon starts in the hottest region of the spot, where the nucleation temperature is exceeded. Grain growth, originating from these nucleation sites, occurs at high velocities, of the order of 14 m/s [12]. For the scanning speed used (1m/s), the growth front, aided by the release of the free energy of crystallization, moves ahead of the beam and stops where a critical quenching temperature T_q is reached or where it encounters crystalline material. Each

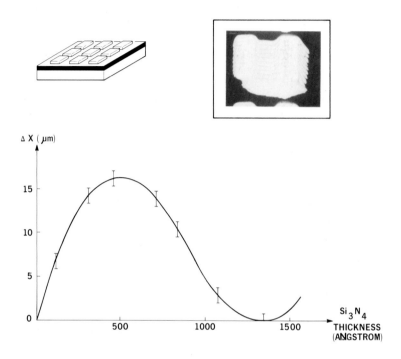

Fig. 1. Optical micrograph of a periodic XCR pattern and corresponding temperature profile induced by a laser beam scanned along x.

of the periodic fronts therefore represents an instantaneous isotherm. If the laser speed is much slower than the crystallization front, the temperature profile created by the scanning beam can be considered as "frozen in". The XCR front will therefore represent an accurate replica of the position of the crystallization isotherm at the instant of the nucleation event. On Figure 1, the computer simulation of the temperature profile obtained by scanning a circular gaussian beam fits well with the shape of the resulting XCR pattern. The Tq isotherm extends completely around the scanning spot but only the portion in the amorphous material can be seen in the pattern. This portion of the XCR pattern gives the best information on the beam profile since its shape is only weakly influenced by the beam velocity. At high scan speeds, the velocity of the beam is not negligible with respect to the crystallization speed. The XCR front still is an isotherm, but it does not represent one instantaneous isothermal section of the beam. Bensahel et al. [12] have shown that if the laser beam speed equals exactly the crystallization speed, the XCR pattern does not show the periodicity seen in Figure 1, but rather propagates to infinity.

DIRECT IMAGING OF INSTANTANEOUS ISOTHERMS

At the velocities of interest (Vlaser < Vcrystal.), it is possible to image directly one instantaneous isotherm created by the laser beam on the sample. On an homogeneous material,the shape and size of the isotherm is only function of the laser beam characteristics. It depends on the size of the laser spot, the laser power and the scan velocity. On the other hand, with fixed laser parameters, the shape of the isotherm depends on the local properties of the material such as optical absorption and thermal conductivity. The shape of the XCR front gives direct information on the laser spatial intensity distribution as well as material homogeneity. As an example, on Figure 2, we show optical transmission micrographs of a single XCR event where the beam had different intensity distribution profiles. In the case of a circular gaussian beam, we have already shown that the XCR

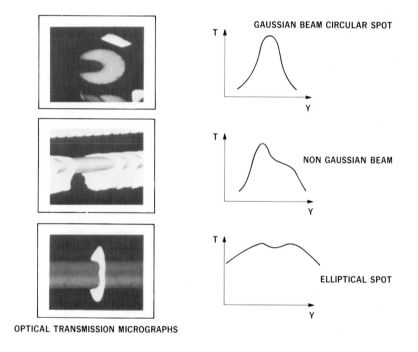

OPTICAL TRANSMISSION MICROGRAPHS

GAUSSIAN BEAM CIRCULAR SPOT

NON GAUSSIAN BEAM

ELLIPTICAL SPOT

Fig. 2. Optical transmission micrographs of XCR events created by three different beam profiles.

front is elliptical. If the laser operates in a higher order mode or if the optics is modified, the intensity distribution profile is no longer gaussian. In Figure 2b, the XCR reflects an asymetrical power distribution created by misaligned optics. With an elliptical beam geometry and with a "doughnut" mode we can obtain a flat thermal profile in the direction perpendicular to the scanning. This is observed as a flat XCR front in Figure 2. The dimensions of the flat region can be extended up to several hundreds of microns by varying beam power and excentricity. Any thermal perturbation can be direcly visualized as a perturbation in this flat profile.

We show on Figure 3 the influence of a thin anti- reflecting layer such as Si3N4 on the thermal profiles. More energy is coupled to the silicon film coated with the nitride layer, resulting locally in a higher temperature. The instantaneous quenching isotherm (Tq) is therefore located further ahead of the beam under the nitride pads than on the uncoated areas. A comparison of

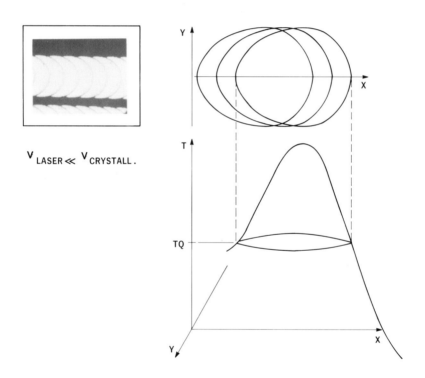

$$V_{LASER} \ll V_{CRYSTALL}.$$

Fig. 3. Displacement of the position of the XCR front as a function of the dielectric thickness.

the distance of the XCR fronts on bare Si and under Si3N4 indicate the effect of the antireflecting coating, which obeys a sine dependence on the layer thickness. The best fit to this curve gives a value of the index of refraction of our sputtered nitride film at the average temperature reached during the process.

CONCLUSIONS

We have shown how XCR patterns in a-Si films can be used to obtain information laser beam intensity profiles and surface temperature profiles induced in laser heated surfaces. The latter can be particularly important when the heated surface has a complex structure, and can be used for the optimization of temperature profiles for recrystallization.

ACKNOWLEDGEMENTS

We would like to thank Drs. G. Auvert and D. Bensahel of CNET Grenoble for fruitful discussions, R. Diehl, T. Loughran and V. Lucian for their skilled technical assistance.

REFERENCES

1. T. J. Stultz, J. F. Gibbons, Appl. Phys. Lett. 39, 6 (1981).

2. M. Lax, J. Appl. Phys. 48, 3919 (1977).

3. M. Lax, Appl. Phys. Lett. 33, 786 (1978).

4. Y. I. Nissim et al., J. Appl. Phys. 51, 274 (1980).

5. M. L. Burgener, R. E. Reedy, J. Appl. Phys. 53, 4357 (1982).

6. J. E. Moody, R. H. Hendel, J. Appl. Phys. 53, 4364 (1982).

7. J. C. C. Fan et al., Appl. Phys. Lett. 36, 158 (1980).

8. H. J. Leamy et al., Appl. Phys. Lett. 38, 137 (1981).

9. G. Auvert et al., Appl. Phys. Lett. 39, 724 (1981).

10. G. Auvert et al., Appl. Phys. Lett. 38, 613 (1981).

11. G. Auvert et al. in: Laser and electron-beam interactions with solids, B. R. Appleton, G. R. Celler eds. (North Holland, New York 1982)pp 535-546.

12. D. Bensahel et al. in: Laser and electron-beam interactions with solids, B. R. Appleton, G. R. Celler eds. (North Holland, New York 1982)pp 541-546.

SURFACE ELECTROMAGNETIC WAVES IN LASER MATERIAL INTERACTIONS*

D. J. EHRLICH, S. R. J. BRUECK AND J. Y. TSAO
Lincoln Laboratory, Massachusetts Institute of Technology
Lexington, Massachusetts 02173

ABSTRACT

Transient ripple formation in pulsed laser annealing of semiconductor materials is shown to arise from amplified surface polariton scattering. The permanent ripple patterns are frozen in after transverse thermal and material diffusion during the laser-annealing process. The model is in good agreement with time-resolved measurements of the ripple formation in 1.06-μm laser annealing of Ge. Dispersion of the ripple wavevector in the UV spectral region is discussed.

INTRODUCTION

The interactions of laser beams with solid surfaces produce a variety of surface morphology changes, many of which show ripple structure with periods comparable to optical wavelengths. Such structure has been seen widely in laser annealing experiments, [1-4] as well as in laser-induced chemical vapor deposition [5] and laser photodeposition [6].

In recent publications, [7,8] we have described a complete mechanism for the growth of periodic fringes at or near the laser wavelength (i.e., at ~ $\lambda/1.1$-$\lambda/1.8$) during low intensity UV laser photodeposition of metal films. An extension to the more complex case of pulsed-laser annealing of Si and Ge has also been outlined.

OBSERVATIONS

The most important qualitative observations on ripple formation in laser annealed Si, Ge, and GaAs are (1) for visible and IR wavelengths, the ripple period is approximately λ, (2) the dominant ripple orientation is perpendicular to the (linear) laser polarization, (3) ripples can be formed using laser pulses at least as short as ~ 100 ps (ref. 3), (4) during an initial formation period the modulation depth and degree of organization in the ripple pattern increases rapidly with integrated laser exposure. Of these observations, perhaps the most difficult to explain are, first, the polarization dependence, which is perpendicular to interference between the incident and any scattered transverse electromagnetic wave and, second, the transverse material motion required to form a ripple if this is to occur <u>during</u> a 100 ps pulse.

In order to add to the information available about the dynamics of the ripple formation, we have recently carried out [8] a direct 1-ns-resolution optical diffraction measurement, in which the ripple formation was correlated with the well established reflectivity change occurring during pulsed laser annealing; 7-ns pulses were used to anneal Ge. The important features of the measurement can be summarized as follows: (1) no diffraction signals, and therefore no ripple structures, are observed below the melt threshold; (2) very large <u>transient</u>

*This work was supported by the Department of the Air Force, in part with specific support from the Air Force Office of Scientific Research, by the Defense Advanced Research Projects Agency, and by the Army Research Office.

diffraction signals, characteristic of strong transient ripples, are seen just above threshold but these relax nearly completely to the original, nondiffracting, surface during recrystallization; (3) weaker transients but stronger permanent ripples are formed on single pulses as the pulse energy is increased; and (4) no ripples are formed at high energies when the melt lasts longer than ~ 200 ns. The details of these measurements can be found in Ref. 8.

MECHANISM

Our conclusion, based on the sum of these observations, is that ripple formation is by two processes. The first process is an exponential growth of ripples at the air/liquid/solid interfaces by stimulated light scattering. This process has been traced specifically to an interaction with a surface polariton wave (SPW) propagating along the interfaces. The second process is the dynamic evolution of these ripples during the melting and regrowth. The second process can completely reverse the first; permanent ripples occur only with the incomplete mutual cancellation of the modulation at the two interfaces on regrowth. In turn this can only occur if the processes are accompanied by transverse material flow. The first process has been discussed in some detail in previous papers; here we summarize that discussion and make some comments on the use of short wavelength light, where the SPW dispersion relation becomes nonlinear. We will elaborate in more detail on the second, material aspect, of ripple formation.

Stimulated Scattering

The exponential growth begins when the laser melts the semiconductor surface, forming a thin liquid layer with the metallic optical properties [10,11] necessary to support the SPW. Laser light is scattered into a SPW, initially by surface irregularities. The interference between this wave and the incident laser field modulates the light intensity above the surface, which in turn modulates the propagation velocity of the liquid-solid interface. The resulting depth and index-of-refraction modulations increase the SPW scattering efficiency, producing positive feedback and exponential gain for the ripples and the SPW. Since the laser beam is scattered most strongly at the liquid/solid interface, this light scattering process can only persist during the short period during which the molten layer is less than ~ 30 nm thick.

The exponential gain coefficient for growth of the ripple amplitude is a rapidly varying function not only of the liquid layer depth but also of the ripple spatial frequency, q. It peaks near $q \sim k_o = \lambda^{-1}$ for typical visible- and IR-laser annealing experiments of semiconductors, although substantial deviations from λ^{-1} will occur at shorter wavelengths as discussed below. The theoretical calculations are analytic, based on perturbation theory; no free parameters are needed to predict the absolute gain coefficient. Representative results for Ge with $\lambda = 1.06$ µm are shown in Figs. 1 and 2. Note that the absolute gain coefficient peaks at q slightly greater than λ^{-1} with a value that averages ~ 30 cm^2/W-s for a melt depth in the range 2-20 nm, where strong scattering occurs. From previous experiments [9] it is known that at near threshold melting the liquid/solid interface typically sweeps through this range in ~ 10 ns, which approximates the duration of our laser pulse. During this period, at near threshold intensities, (~ 100 mJ/cm^2), this value of the gain coefficient corresponds to an exponential growth in ripple amplitude with a exponential period ~ 2 ns. This implies a rapid growth from random noise in the form of scratches or dust to a well formed ripple.

The results of ripple gain are shown in the electron micrographs in Fig. 3. After a single laser pulse at near threshold melting conditions, ripples formed in various regions are not well phase-related due to the randomness of the initial noise source. As shown in Fig. 3(b), the ripples intensify nonlinearly with the number of laser pulses and become more coherent. The increased phase coherency of the ripples after multiple pulses is analogous to the line narrowing which occurs in homogeneously broadened lasers.

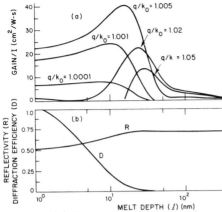

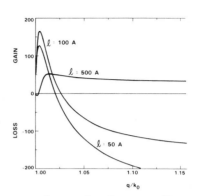

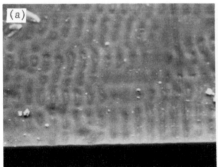

Fig. 1. (a) Calculated small-signal
gain coefficients for the depth modulation of the liquid/solid interface
as a function of melt depth ℓ for 1.06-µm
irradiation of Ge at near threshold melting intensities. The normalized ripple
spatial frequency, q/k_0, is treated as a
parameter. (b) The variation of the
488-nm probe laser reflectivity and relative diffraction efficiency for a small
fixed-amplitude ripple.

Fig. 2. Results from a small-
signal gain calculation (similar to
those of Fig. 1) for 1.06-µm irradiation of Ge. Relative gain (+)
or loss (−) is plotted as a function
of the normalized ripple spatial
frequency, q/k_0, for several liquid
layer depths, ℓ. Note that there
can be either loss or gain depending on the melt depth and ripple
frequency.

Fig. 3. Scanning electron micrographs of a Ge surface, (a) after one and
(b) after 20 pulses at near threshold melting using a 0.53 µm laser. Note the
increased phase coherency and modulation depth, indicative of ripple gain, with
repeated shallow melting.

An interesting aspect of the amplified surface polariton model of ripple
formation is the predicted dispersion of the spatial wavevector as laser frequencies in the UV spectral region are used to generate ripple structures.
There is a substantial deviation of the predicted spatial wave-vector from the

194

free space wavevector in this spectral region. This is shown in Fig. 4 where
the dispersion curves for surface polaritons in liquid Ge and in Au and Ag are
shown. For these latter two metals, literature values [12] of the optical constants have been used. For liquid Ge, an extrapolation of the optical constants
in the IR-Vis spectral region based on a free-electron model is used [10,11].
These curves show several interesting features. For Ag, the familiar surface
polariton dispersion relation is obtained; for Au the usual bending of the curve
away from the light line is modified by the strong interband transitions that
arise in the visible. For liquid Ge, there is a significant deviation from the
light line in the region above ~ 5 eV; we have carried out preliminary experiments in this short wavelength regime using ArF and F_2 lasers and have confirmed
this behavior, but with somewhat different values of the plasma frequency ω_p and
the electron damping time τ than those extrapolated from the visible spectral
region. Details of these measurements will be presented elsewhere. Similar
dispersion behavior, consistent with the polariton model, has also been confirmed for laser photodeposition of Cd films [7].

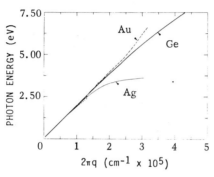

Fig. 4. Calculated variation of the stimulated-surface-polariton-wave ripple
frequency vs laser photon energy for Ge, Ag and Au. The calculation is for normal incidence irradiation.

Ripple Formation

    The dynamic material processes occurring after amplified polariton scattering
are equally important to a full explanation of ripple growth. This importance
becomes obvious, for example, upon considering the results of the diffraction
probe experiment, which show that the conditions of optimum transient ripples do
not correspond to optimum (single pulse) permanent ripple formation. Similar
questions are also salient for 100 ps experiments; in this case, the laser pulse
has ended well before the transverse material flow required to form a permanent
ripple can occur. These results and others cited above have been used to con-
struct the preliminary model of the material kinetics discussed below. The key
aspect to explain is the mediating process which transforms an efficient transient ripple, which forms during the short period of optical scattering, into
transverse flow and permanent ripples.
    As noted above, the transient ripple growth due to amplified surface polari-
ton scattering occurs predominantly during the initial phases of the laser-
material interaction when the melt depth is less than ~ 30 nm. The effect of
the spatially modulated laser intensity distribution, due to the interference
between the incident laser beam and the surface polariton, is to spatially modulate the depth of the melt. This in turn amplifies the scattering into the surface polariton and increases the intensity modulation. Note that this process
does not require any transverse material flow during the growth of the transient
ripple. Nevertheless, such flow will inevitably occur, due to a geometrical
redistribution accompanying the ~ 5% volume contraction of the Ge on melting.

This is discussed in more detail below. Because of this geometrical effect, the ripple amplitude in the air/liquid interface is not quite 5% of the ripple in the liquid/solid interface.

As the laser-material interaction progresses and the melt front propagates further into the material, the laser beam can no longer sense the modulation of the liquid/solid interface through the optically dense liquid and the scattering into the surface polariton and hence the growth phase of the ripples ends (see Fig. 1). During the growth phase, however, the isotherms in the liquid are also rippled; in fact, it is the modulated temperature profile that drives the melt depth modulation. After the ripple growth phase has ended, the melt front can still be driven by the laser further into the substrate before returning to the surface. This drive-in and regrowth phase can take from ∼ 10 ns to ∼ 500 ns, depending on the laser intensity and the resultant melt depth [9]. During this period, transverse heat diffusion will lead to a straightening of the spatial isotherms and subsequently of the liquid/solid interface. Because of the Ge density change on resolidification, there is a corresponding straightening effect on the front surface ripple. However, because of the geometrical displacement of this density change the effect can be smaller than the full ∼ 5% volume contraction. The key point to appreciate is that there is a permanent ripple structure only if the depth at which the liquid/solid interface straightens is different (larger) than the depth at which it originally became rippled. If the two processes occur at approximately the same depth in the material, there will be an almost complete cancellation of the ripple structure on regrowth – this is the condition (discussed above) for a strong transient diffraction signal but no permanent ripple structure.

This model can be made quantitative by considering the effect of melting the liquid/solid interface, at an effective depth for growth, ℓ_g, on the profile of the air/liquid interface. If the liquid/solid interface is melted an incremental amount $\Delta z_{\ell s}(x)$, where x is the transverse coordinate, the incremental change in the height of the air/liquid interface is given by

$$\Delta z_{a\ell}(x) = \int \Delta z_{\ell s}(x') \, \ell_g \, (\rho_\ell/\rho_s - 1)/\pi(\ell_g^2 + (x - x')^2) dx' \quad , \tag{1}$$

where ρ_s, ρ_ℓ are the solid and liquid mass densities, respectively. In deriving this expression, we have assumed that the flow of material due to local melting of the liquid/solid interface is radially inwards, which is a geometrical divergence problem whose point impulse response function is $h(x-x') = \ell_g(\rho_\ell/\ell_s - 1)/\pi(\ell_g^2 + (x-x')^2)$. For a sinusoidal incremental change in the liquid/solid interface, with amplitude $u_{\ell s}$, the incremental change in the air/liquid interface will also be sinusoidal, but with a reduced amplitude $u_{a\ell} = u_{\ell s}(\rho_s/\rho_\ell - 1)e^{-q\ell_g}$. If the liquid/solid interface is now straightened out at a melt depth ℓ_r, the net ripple in the air/liquid interface is

$$u_{a\ell} = u_{\ell s} \, (\rho_s/\rho_\ell - 1) \, [e^{-q\langle\ell_g\rangle} - e^{-q\langle\ell_r\rangle}] \quad , \tag{2}$$

where the angular brackets have been used to indicate that averages must be taken over the growth and decay depths during the dynamic laser-annealing process. Note that if $\ell_g \sim \ell_r$, then $u_{a\ell} \sim 0$ and there is no net effect.

This model gives a good qualitative explanation of all of the experimental features. For shallow melts where the growth and decay of the liquid/solid ripple occur at comparable liquid depths there is a strong transient ripple and almost complete cancellation on regrowth. At higher laser intensities, the measured transient ripple is lower due to the rapid evolution of the liquid layer past the ∼ 20-30 nm depth on a time scale short compared to the detector response time, but there is a much stronger permanent ripple due to an incomplete cancellation on regrowth.

SUMMARY

In summary, we have presented a complete model for the ripple formation process in laser annealing of semiconductor materials. Absolute numbers for the rate of amplitude growth and for the dependence of the ripple period on laser wavelength have been calculated. This model is in substantial accord with previous observations and recent dynamic measurements of the ripple formation. For simplicity, the model divides the overall rippling process into two steps, the growth of a ripple in the liquid/solid inter-face, and then its decay. The two steps need not be temporally distinct: the forces driving the decay should be regarded as always present. The effect of the growth/decay of a ripple in the liquid/solid interface on the growth/decay of a ripple in the air/liquid inter-face is more subtle, and depends on the depth of the melt during the growth/decay process. A ripple is frozen into the solid due to the incomplete cancellation of these effects.

ACKNOWLEDGMENT

We would like to thank Paul Nitishin for extensive electron microscopy, and P. L. Kelley for helpful discussions.

REFERENCES

1. J. C. Koo and R. E. Slusher, Appl. Phys. Lett. 28, 614 (1976).

2. M. Oron and G. Sorenson, Appl. Phys. Lett. 35, 782 (1978).

3. P. Fouchet and A. E. Siegman, Appl. Phys. Lett. 40, 824 (1982).

4. J. F. Young, J. E. Sipe, J. S. Preston, and H. M. Van Driel, Appl. Phys. Lett. 41, 261 (1982).

5. D. J. Ehrlich, R. M. Osgood, Jr., and T. F. Deutsch, Appl. Phys. Lett. 39, 957 (1981).

6. R. M. Osgood, Jr. and D. J. Ehrlich, Optics Lett. 7, 385 (1982).

7. S. R. J. Brueck and D. J. Ehrlich, Phys. Rev. Lett. 48, 1678 (1982).

8. D. J. Ehrlich, S. R. J. Brueck, and J. Y. Tsao, Appl. Phys. Lett. 41, 630 (1982).

9. C. M. Surko, A. L. Simons, D. H. Auston, J. A. Golovchenko, R. E. Slusher, and T. N. C. Venkatesan, Appl. Phys. Lett. 34, 635 (1979).

10. J. N. Hodgson, Phil. Mag. 6, 509 (1961).

11. A. Abraham, J. Tauc, and B. Velicky, Phys. Stat. Sol. 3, 767 (1963).

12. P. B. Johnson and R. W. Christy, Phys. Rev. B 6, 4370 (1972).

LASER INDUCED PERIODIC SURFACE STRUCTURE

H.M. VAN DRIEL, JEFF F. YOUNG AND J.E. SIPE
Department of Physics and Erindale College, University of Toronto, Toronto, Ontario, Canada M5S 1A7

ABSTRACT

Laser induced periodic surface structure can be understood as a universal phenomenon which occurs when high intensity pulses are absorbed near the surface of solids or liquids. The phenomenon occurs on metals, semiconductors and insulators because of the interference between the incident pulse and an induced "radiation remnant". This scattered field may be enhanced by the existence of true surface modes such as surface plasmons or phonon-polaritons but this is not essential. The universality characteristics include beam polarization, since we show that circularly polarized light can induce surface ripples, with the damage structure showing a dependence on the sense of rotation. We also present time resolved results of the formation of the ripples to illustrate the essential dynamical processes that occur.

INTRODUCTION

Single, high intensity laser pulses can induce permanent or transient periodic structures on the surfaces of solid or liquid metals, semiconductors and insulators[1-16]. This effect has been observed for s and p polarized light, of different wavelengths (0.15 -11.3 µm), pulse widths and for various angles of incidence. This curious symmetry breaking phenomenon is of importance in various aspects of laser-material processing and, at a more fundamental level, is of interest in revealing how coherent radiation interacts with a surface. It is generally acknowledged that the periodic structures are the result of interference between the incident beam and a "surface scattered field" induced by microroughness. Although various authors have suggested different forms for the "surface wave", including pure transverse electromagnetic,[2,12] surface plasmon polaritons[3,16] or surface phonon polaritons[15], the universality of the phenomenon requires an equally universal explanation. At last year's (1981) Materials Research Society Meeting we[4] suggested a new non-radiative surface field structure, which we called a radiation remnant, that could account for the universal characteristics. These fields, which in general have a longitudinal component, propagate along the surface while their amplitude falls off as $x^{-3/2}$, where x is the distance from a line source.[5,9] They are associated with discontinuities in the derivatives of a response function rather than with the pole singularities characteristics of true propagating electromagnetic modes. These new field structures exist on materials of arbitrary dielectric constant and occur in addition to any true surface modes which may exist such as surface polaritons. These radiation remnants show a formal similarity to "lateral waves" in such diverse areas as total internal reflection of light[17] and "ground radio waves"[18]. We have developed a detailed universal theory[9] which takes into account the radiation remnant fields using a new variational principle to calculate the surface field structures. This theory accounts for all the ripple spacings, orientations, material and polarization dependences that have been observed to date in a wide variety of materials. We now show that the universality of the phenomenon can be extended to include circularly as well as linearly polarized light. We also report time resolved data on the ripple formation which indicates that the growth of the patterns on semiconductors, at

Mat. Res. Soc. Symp. Proc. Vol. 13 (1983) ©Elsevier Science Publishing Co., Inc.

least, is consistent with the radiation remnant picture.

SPATIAL SPECTRA OF RIPPLES

Because of the spatial periodicity of the ripples, we have found it most convenient to study the Fourier transform of the surface as revealed by the diffraction pattern of a weak cw probe beam, rather than to view the ripples directly. For λ = 0.53 and 1.06 μm, 15 nsec damaging pulses, the patterns produced from Ge, Si, Ga As, Al etc. lie on two overlapping circles, the Fourier co-ordinates, $\vec{K}$, of which satisfy

$$\left| \vec{K} \pm \vec{K}_i \right| = 2\pi/\lambda \qquad (1)$$

where $\vec{K}_i$ is the component of the incident beam wavevector which is parallel to the surface ($|\vec{K}_i| = 2\pi\lambda^{-1}\sin\theta$). The _extent_ to which these circles are filled out depends on beam polarization, angle of incidence and material. It should be noted that the surface patterns evolve on a shot-to-shot basis and it may take as many as 20 shots before a "steady-state" pattern emerges which is independent of the initial surface conditions. Figure 1 shows, for various angles of incidence and polarizations of the damaging beam, a summary of diffraction patterns observed from the metals and semiconductors.

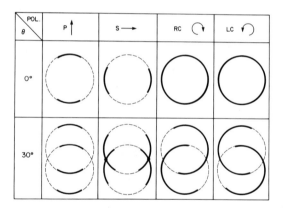

Fig. 1. Summary of diffraction patterns (darkened regions) observed from metals and semiconductors which were damaged by laser pulses for different polarizations and angles of incidence.

The metals tend to fill out more of the circles than do the semiconductors. The information contained in these patterns reveal the orientation, spacings and relative amplitudes of the ripple patterns which are simultaneously present. The two circles also contain, as special cases, the $\lambda/(1 \pm \sin\theta)$ fringes which are oriented perpendicular to a linearly polarized field as reported by several authors[2,3,4,12] and the $\lambda/\cos\theta$ fringes, oriented parallel to the field as first reported by us last year[4].

We report for the first time, that circularly polarized light can induce periodic surface structure in metals and semiconductors. Figure 1 shows that for normal incidence, the diffraction pattern is a uniformly illuminated circle, revealing that fringes of spacing λ exist for all orientations along the surface. Figure 2 reveals the corresponding real space picture from λ = 1.06μm

induced damage on germanium. No obvious indications of periodicity are seen
because of the interference of the randomly oriented fringes. For non-zero
angle of incidence circularly polarized light induces interesting asymmetric
diffraction patterns which depend on the sense of rotation.

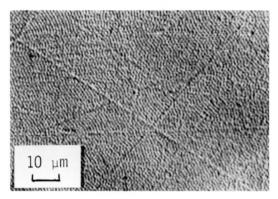

10 μm

Fig. 2. Surface of Ge damaged by 1.06μm, circularly polarized light at normal
incidence.

Diffraction patterns for right and left handed circular polarization are mirror
images of each other in the plane of incidence as indeed right and left circularly polarized light are. All of these results are in agreement with our detailed theory[9],[19]. This interesting feature serves as a simple way of determining the sense of rotation of an intense light beam.

Equation 1 is basically a statement[9] of the fact that the dominant compon-
ent of the radiation remnant wave has wavelength λ when $|\varepsilon| >> 1$ where ε is the
complex dielectric constant. In the care of $Re(\varepsilon) \lesssim -1$ (e.g. metals) the sur-
face polaritons dominate the radiation remnant peaks. For $|\varepsilon| \sim 1$ (e.g. diel-
ectrics) the right hand side of equation 1 is replaced by $2\pi n/\lambda$ where n is the
real part of the refractive index. This reflects the fact that the "material"
radiation remnant dominates the scattered field.[5] We have observed the appropriate ripple patterns on quartz, NaCl, BK7 glass and KZF2 glass at 10.6 m for
which $\varepsilon \gtrsim 1$ and for which surface polaritons are not present. Similar results
have been observed by Soileau et al.[8],[20] in NaCl, MgF, KCl and ZnSe. In none
of these materials are surface polaritons present at 10.6μm. Indeed, the ripples
reported by Keilmann and Bai[15] for quartz at 9-11μm may be identified with radiation remnants for $\lambda \gtrsim 9.5$μm with only results near 9.3μm possibly due to the
presence of a surface polariton at that wavelength.

GROWTH KINETICS

In order to look at the growth kinetics of the ripples we have studied
their evolution on a Ge surface by simultaneously time resolving the specular
and first order diffraction of a 0.52μm cw probe beam which overlaps with a
1.06μm damaging laser beam incident at 30°, s-polarized. We have previously
noted[21] that the transient diffraction is very large when the specular reflectivity is less than the 76% value characteristic of uniformly molten Ge,
whereas very little transient diffracted signal is observed when the reflectivity is 76%. No transient or permanent diffraction is observed if the reflectivity at no time rises above that of solid Ge. We suggested that these results
are consistent with a model in which the Ge is inhomogeneously melted while the

reflectivity is less than 76% and that the large transient diffraction results
from the alternating regions of high and low reflectivity. When the reflectivity is 76%, the uniformly molten surface has little periodic structure and hence
the diffraction is weak. Here we report further experimental results that confirm that the large transient diffraction is due to an alternating solid/liquid
region and further that at higher fluences oscillatory transient ripple structure is induced on the uniformly molten surface.

Figure 3 shows the calculated[22] first order diffraction from an alternating solid/liquid surface as a function of the polarization and angle of incidence of the probe beam. Note that at incident angles between 30° and 80° the
ratio of p to s-polarized diffraction is significantly larger than unity.

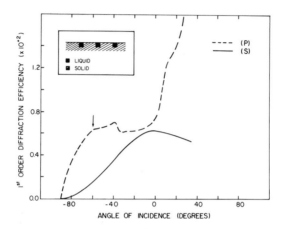

Fig. 3. A plot of the calculated first order diffraction efficiency of a 0.52
μm probe beam incident perpedendicular to strips of molten Ge on a solid Ge
surface as a function of angle of incidence and polarization. The dielectric
constants were taken as 22 + 23i for the solid and -19.8 + 23i for the liquid regions respectively. The spacing of the metallic strips was 1.2μm and the molt-
en regions were 0.4μm wide and 160 Å deep. The calculation employs a plane
wave approximation.

Similar calculations for corrugated solid or liquid surfaces indicate that the
p to s ratio is unity to within 5% for all angles of incidence. By placing
Glan-Taylor prisms in the probe beam path just before and just after reflection
from the Ge surface, both polarizations of the transiently diffracted beams
were measured for each damaging laser pulse. Figure 4 summarizes the typical
diffraction signals observed at three different fluences. The specular reflectivity at 140 mJ/cm^2 reaches a maximum value of 60% while at the other two fluences the reflectivity reaches a flat-topped value of 76%. The large transient
diffracted signals at each fluence level occur only when the specular reflectivity is less than 76% and their amplitudes are 3.5 ± .5 times larges for p as
opposed to s-polarizations at an angle of incidence of 60°. The ratio of peak
amplitudes is essentially unity at 8° angle of incidence, while the reflectivity is 76% the diffracted signals are the same amplitude for p and s polarizations at both 60° and 8° as are the dc (permanent) level shifts. We therefore
conclude that the large transient diffracted signals are due to an alternating
solid/liquid region at the surface.

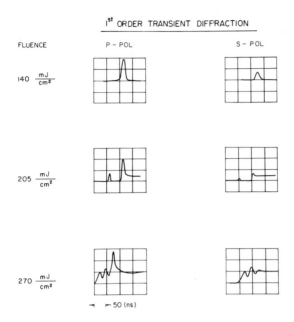

Fig. 4. Typical first order diffraction signals observed with the 0.52 μm
probe beam incident at 60° and perpendicular to the dominant fringes formed on
the Ge surface by a 1.06 μm s-polarized damaging beam incident at 30°. The s
and p-polarized components of the diffracted beam were recorded separately by
different photo-multipliers of equal gain at three different fluence levels of
the 1.06 μm beam.

The oscillatory nature of the diffraction at 270 mJ/cm^2 is quite interesting. The period of oscillation is 25 nsec and the diffraction efficiency is
the same for p and s polarizations. Keilmann[10] has also studied laser induced
ripples on liquid mercury surfaces and we have developed a theory[22] based on
inhomogeneous thermal coupling to surface capillary waves in order to explain
this phenomenon. The calculated natural period for oscillation of a standing
capillary wave of wavelength 1 μm on molten Ge is 25 nsec and hence we conclude
that at fluences such that the Ge surface melts very quickly to a significant
depth (> 500 Å), the incident laser beam can scatter from surface roughness present on the liquid and produce capillary waves with the same wavelength as the
induced permanent damage.

We close with some comments on a recent proposal[23] that the Ge surface
must melt uniformly prior to the occurence of ripple structure in order that
surface plasmons might be produced. Two experimental facts indicate that at
least for fluences ≲ 175 mJ/cm^2 the maximum possible depth of any such
molten layer that might form is < 20Å. As reported previously[21], the transient
diffraction occurs when and only when the specular reflectivity rises above its
solid value. If one interprets any rise in reflectivity purely in terms of
skin depth effects, this implies that, within experimental error, transient diffraction occurs at all melt depths. If one allows for experimental error, the

rise in reflectivity might precede the diffraction signal by enough that one could claim that the surface was uniformly molten for a depth of no more than 15 Å. Also, the absolute magnitude of the transient diffraction signal can be as large as 3×10^{-3} times the incident probe intensity. Using the same exact diffraction calculation procedure as above[22], assuming that a corrugated solid/ liquid interface is buried beneath a uniform depth of melt, this melt depth cannot exceed 20 Å and still be able to account for the absolute diffraction efficiency. Given these facts, we conclude that surface plasmons cannot be invoked to explain the induced periodic structure. Molten layers of 20 Å are not sufficient to produce pole-like structure in the Fresnel coefficient and a more general treatment of the electrodynamics would be required in the event that such thin layers existed. We believe that radiation remnants are responsible for the initial inhomogeneous energy deposition since they exist on the surface of solid Ge.

To illustrate that there is no overwhelming effect due to plasmons at uniform melt depth even as large as 40 Å we show in figure 5 the first order inhomogeneous intensity calculated exactly for three different cases in which 40 Å thick corrugations exist at or near the surface.

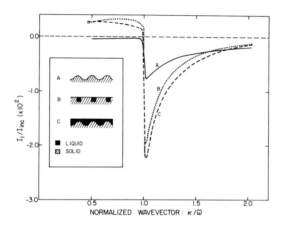

Fig. 5. Plots of the calculated normalized first order inhomogeneous intensity (I_1/Inc) just beneath the selvedge region for a 1.06 μm damaging beam normally incident on three different periodic surface structures as a function of the normalized wavevector of the surface roughness. The peak to peak amplitude of the corrugations was 40Å in each case; the width of the molten regions in the case of the alternating solid/liquid surface was always taken as 0.25 x the period. The dielectric constants were taken as 16 + 0.8i and -30 + 72i for the solid and liquid regions respectively. For an incident intensity I_{inc} the field is written as $I_0 + I_1 \cos kx$; $\omega = 2\pi/1.06 \times 10^{-6} m^{-1}$.

The shapes of all three curves are basically the same, showing a large peak at the period at which surface damage is induced. The amplitude of the inhomogeneity is greater for a solid ripple submerged beneath a 40 Å molten layer as opposed to a simple air/solid corrugation, however, there is no reason to suspect, without doing the detailed inhomogeneous thermal diffusion and phase transistion calculation, that the field structure in the air/solid case is incapable of causing inhomogeneous melting. Also since the amplitude of the inhomogeneous field intensity is essentially the same for the locally melted case as the uniformly melted case there is no reason, a priori, to conclude that the feedback should be any stronger for the case of a uniformly molten molt front as opposed to an inhomogeneous one. Note, in our model strong feedback commences from the <u>solid</u> phase at the onset of inhomogeneous melting (see Figure 5) whereas in the model proposed by Brueck et. al.[16,23] the feedback via "stimulated" scattering occurs only <u>after</u> a uniform molten region forms. As indicated above, the experimental evidence demonstrates that no such uniformly molten layer precedes the growth of the ripple structure.

In conclusion we have demonstrated that the phenomenon of laser induced periodic surface structure is a very general one which can be understood in equally general terms as due to scattering of polarized, coherent radiation at randomly rough solid of liquid surfaces. The exact nature of the scattered fields depends on the dielectric constant of the medium and can in general, but need not be, influenced by surface excitations. The effect of the inhomogeneous intensity distribution on the surface also depends on the particular material; both localized melting, in the case of solid Ge, and excited capillary waves in the case of molten Ge have been identified.

ACKNOWLEDGEMENTS

We gratefully acknowledge the financial support of the Natural Sciences and Engineering Research Council of Canada and the Atkinson Foundation.

REFERENCES

1. M. Birnbaum, J. Appl. Phys. <u>36</u>, 3688 (1965).

2. H.J. Leamy, G.A. Rozgonyi, T.T. Sheng and G.K. Celler, Appl. Phys. Lett. <u>32</u>, 535 (1978).

3. N.R. Isenor, Appl. Phys. Lett. <u>31</u>, 148 (1977).

4. J.F. Young, J.E. Sipe, M.I. Gallant, J.S. Preston and H.M. van Driel, in "Laser and Electron Beam Interactions with Solids" B.R. Appleton and G.K. Celler, eds. (North Holland, 1982).

5. J.F. Young, J.E. Sipe, J.S. Preston and H.M. van Driel, Appl. Phys. Lett. <u>41</u>, 261 (1982).

6. J.A. van Vechten, Solid State Commun. <u>39</u>, 1285 (1981).

7. C.T. Walters, Appl Phys. Lett. <u>25</u>, 696 (1974); D.C. Emmony, R.P. Howson, and L.J. Willis, Appl. Phys. Lett. <u>23</u>, 598 (1973).

8. P.A. Temple and M.J. Soileau, IEEE J. Quant. Elec. <u>17</u>, 2067 (1981).

9. J.E. Sipe, J.F. Young, J.S. Preston and H.M. van Driel, (Phys. Rev. B, in press).

10. F. Keilmann and Y.H. Bai, Conference on Lasers and Electro-optics, Phoenix, Az. 1982 (paper WK5); F. Keilmann XIII Int. Conf. on Quantum Electronics, Munich, West Germany (1982).

11. J.F. Young, J.S. Preston, H.M. van Driel and J.E. Sipe (Phys. Rev. B, in press).

12. M. Oron and G. Sorenson, App. Phys. Lett. <u>35</u>, 782 (1979).

204

13. M.F. Becker, R.M. Walser, Y.K. Jhee and P.Y. Sheng, SPIE Conference on "Picosecond Lasers and Applications" L. Goldberg Ed. P; R.M. Walser, M.F. Becker, D.H. Sheng, and J.G. Ambrose, in "Laser and Electron Beam Interactions and Material Processing," T.J. Gibbons, W. Hess, and T. Sigmon, eds. (Elsevier, New York, 1981), 177.

14. D.J. Ehrlich, S.R.J. Brueck, and J.Y. Tsao, XII Int. Conference on Quantum Electronics, Munich, West Germany, 1982.

15. F. Keilmann and Y.H. Bai, Appl. Phys. A, $\underline{29}$, 9 (1982).

16. S.R.J. Brueck and D.J. Ehrlich, Phys. Rev. Lett. $\underline{48}$, 1678 (1982).

17. T. Tamir and A. Oliner, J. Opt. Soc. Am. $\underline{59}$, 942 (1969) H. Osterberger and L.W. Smith. J.O.S.A. $\underline{54}$, 1973 (1964).

18. L.M. Brekhouskikh, "Waves in Layered Media" (Academic Press Inc., New York, 1960) ChIV, p 234.

19. H.M. van Driel, J.E. Sipe and Jeff F. Young, Phys. Rev. Lett. (in press).

20. M.J. Soileau and E.W. Van Stryland, Optical Society of America 1982 Annual Meeting, paper WE7.

21. Jeff F. Young, J.E. Sipe and H.M. van Driel, Phys. Rev. B, (in press).

22. J.E. Sipe, H.M. van Driel and Jeff F. Young (unpublished).

23. S.R.J. Brueck and D.J. Ehrlich, App. Phys. Lett. $\underline{41}$, 630 (1982).

PICOSECOND LASER-INDUCED SURFACE TRANSFORMATIONS IN SOLIDS

P.M. FAUCHET, ZHOU GUOSHENG AND A.E. SIEGMAN
Edward L. Ginzton Laboratory, Stanford University, Stanford, CA 94305, USA

ABSTRACT

We report the results of a study of surface transformations in semiconductors (Si, GaAs) and metals (Cu) produced by
100 ps long pulses from a Nd:YAG laser system. We present and
outline the conclusions of our model for growth of spontaneous
periodic surface structures or ripples often associated with
laser damage and annealing. The effect of repetitive subthreshold illumination is also examined. Theoretical analysis
is presented to support time-resolved experiments performed
with two excitation pulses at different wavelengths.

1. INTRODUCTION

Picosecond laser sources as a means of inducing surface transitions on semiconductors and other solids are of scientific and technological interest.
Ultrashort pulses in the picosecond (and more recently femtosecond) range have a
duration comparable to relaxation, recombination and diffusion times of interest
and therefore are capable of yielding meaningful time-resolved information about
the dynamics of highly excited systems [1]. Some technological applications of
these ultrashort pulses are also possible, using their unique property to
locally induce remarkable phases [2]. We present here three results of our
investigation of laser-induced phase transformations in solids, with emphasis on
semiconductors such as Si and GaAs.

2. SPONTANEOUS PERIODIC SURFACE STRUCTURES

It has been known for several years that laser annealing and damage may be
accompanied by formation of surface structures. We restrict our attention to
the case of spontaneous periodic surface structures or "ripples" [3] with the
well-characterized properties summarized as follows: For p-polarized light, the
ripple spacing Λ is given by $\Lambda = \lambda/n(1 \pm \sin \theta)$ where λ is the free space
wavelength of the laser, θ the angle of incidence and n the refractive index
of the medium in contact with the surface; for s-polarization, $\Lambda = \lambda$. The
ripples can extend coherently over many successive laser shots if the raster
scan direction is perpendicular to the ripples. Ripples can also be created by
multiple subthreshold illumination, though in general they usually appear around
melting threshold.

To us, this indicates that these particular periodic structures result from
an optical interference between the incident light beam and a diffracted surface
wave that comes from scattering of the incident wave off a grating-like structure
on the surface. The modulation of the incident power - and in turn, the strength
of the grating - may grow, depending upon the exact nature of the grating, leading to an exponential growth or positive feedback regime similar to small-scale
self-focusing in dielectrics. Only the Fourier component of the initial random
disturbance which diffracts light almost exactly along the surface grows, and
the strength of the initial grating is unimportant since exponential growth processes can start from noise. We note that significant positive feedback has to

occur in less than the pulse duration.

To check these ideas, we have solved Maxwell's equations with appropriate boundary conditions to first order; i.e., the modulation remains proportional to the grating strength [4]. Three cases have been investigated, a height grating, a temperature grating, and an electron-hole grating, the last two cases being variations of a complex dielectric function grating. Our calculations show that electron-hole density fluctuations give rise to negative feedback (the fluctuations die out), while a temperature grating leads only to small positive or negative growth coefficient, depending upon the form of the permittivity. When we use Chuang and Kong's method [5] to compute the modulated power for a height grating, we obtain a large modulation for a molten or metallic-like surface and a small modulation for unmelted Si or GaAs (Fig. 1). From this, we hypothesize that at least partial melting is necessary: this is consistent with previous observations [6] and the very recent findings of the Xerox group [7] who has shown that a laser can induce lamellar melting with alternating regions of solid and liquid Si. This situation corresponds to the maximum grating strength (largest permittivity difference). Note that for ultrashort illumination followed by very long melt duration, the ripples should be washed out because of mass movement, as we have observed in Si.

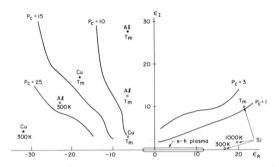

Fig. 1. Modulation of absorbed power, P_c (in %), for surface corrugation equal to $0.01 \times \lambda$. The full and open circles are the complex permittivity $\mathcal{E}_R + j \mathcal{E}_I$, for materials illuminated at 1.06 μm and 532 nm, respectively.

Our theoretical results reproduce the wavelength and polarization properties of the ripples, and explain the raster scan results in terms of a simple scattering phenomenon. Another proposed mechanism for ripple formation suggests however that surface plasmons or polaritons [8] play the central role in ripple formation. Van Vechten has further proposed that by applying a weak second pulse of different linear polarization after a threshold pulse, it should be possible to rotate the rippled pattern. We have performed such an experiment in Si and GaAs with delays from 0 to 5 ns [9] but so far we have been unable to confirm that hypothesis.

3. GRADUAL TRANSITIONS UNDER REPETITIVE SUBTHRESHOLD ILLUMINATION

Single (picosecond) pulse illumination can produce a host of different surface transitions with various threshold energies. Multiple-pulse illumination with pulse energy below these thresholds can, however, also produce similar results [10]. We have investigated these multi-pulse transitions on Si, GaAs and Cu surfaces of different crystalline structure illuminated at 1.06 μm or 532 nm. The diagnostic tools are optical and scanning electron microscopes.

Each transformation, from flat to rippled surface, from amorphous to crystalline material, or from single crystal to damage, exhibits a common behavior displayed in figure 2.

For intensities of the order of the single-shot threshold I_T^1 the transformed area first increases dramatically and finally becomes constant, while for $I \ll I_T^1$, there is an incubation period before the transformation starts. As I decreases, the incubation period becomes longer, and finally a safe level is reached (no transformation for an arbitrarily large number of shots). This can be interpreted as a decrease of the effective threshold I_T^n with increasing number of shots.

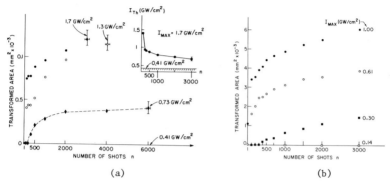

(a)   (b)

Fig. 2(a). Damage curves for c-Si irradiated at 532 nm. The inset illustrates the decrease in effective threshold with repetitive illumination.
Fig. 2(b). Annealing curves for a-Si irradiated at 1.06 μm.

Avrami's equation (which usually describes nucleation and growth mechanisms) has been applied previously [11] to explain similar results obtained for the damage transition on thin Si films irradiated at 1.06 μm. However, no general microscopic description has been proposed. For the amorphous to crystalline transition for example, the evidence suggests that microcrystallites [12] are formed during the incubation period and later grow to form large grains. For other transformations, however, there is not much experimental evidence. We are still trying to elucidate this problem, for example by employing more sensitive diagnostic tools.

An obvious and important practical consequence of this work is that knowledge of the single shot damage threshold is hardly adequate, since the performances of optical components severely degrade with repetitive subthreshold illumination. Moreover, if the substrate is also brought to a temperature higher than ambient as a result of repetitive illumination, the safe level for permanent operation is further lowered. This additional mechanism does not play a role in our experiments but must also be taken into consideration when studying exit facet damage in laser diodes.

4. DYNAMICS OF HIGHLY EXCITED SOLIDS AND MELTING

Under intense, ultrashort pulse illumination, a dense electron-hole plasma (EHP) is formed at the surface layer. Melting of that layer becomes strongly dependent upon the dynamics of the EHP. Lietoila and Gibbons have developed a model [13] that successfully accounts for results on Si under picosecond laser illumination at 532 nm [14] and under nanosecond illumination at 1.06 μm and

208

532 nm. An important feature of their model is the influence of free carrier absorption (FCA) in the infrared. To test their ideas on FCA and other parameters, we have used two ultrashort pulses, one at 1.06 μm, the other at 532 nm, with a variable delay between them. In our experiment, a first, weak pulse at 532 nm "prepares" the sample, i.e., creates a dense EHP while increasing only slightly the temperature; the second pulse at 1.06 μm is delayed by variable increments between 0 and several ns, and its intensity adjusted so that melting point is just reached on the surface. As the delay is increased, the infrared intensity required increases, under the combined action of recombination and electronic and heat diffusion.

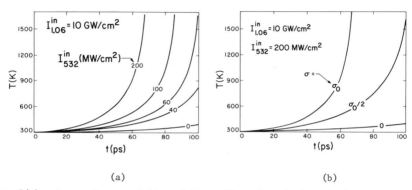

(a)   (b)

Fig. 3(a). Temperature evolution of the c-Si surface during simultaneous illumination by 100 ps pulses at 1.06 μm and 532 nm ($\sigma = \sigma_0$);
Fig. 3(b). Influence of the FCA cross-section σ on temperature evolution for same conditions as in Fig. 3(a).

To describe FCA, we write $\alpha = n\sigma$ where n is the carrier density and σ is the FCA cross-section. To predict the temporal evolution of the carrier density, and electron and lattice temperatures, we use a computer code which does not include the secondary parameters in Lietoila and Gibbons' model and which neglects diffusion during the 100 ps pulse. We have verified that our results coincide with theirs when checked for the conditions of reference 14. Figure 3a shows the temperature on a crystalline Si surface as a function of time for simultaneous illumination by two pulses, and figure 3b indicates the sensitivity of the results to the exact value of σ (here, $\sigma_0 = 5.1\times10^{-18}(T/300)(1.17/\hbar\omega)^2$ cm^2 [15]). In figure 4, we show the 1.06 μm intensity just needed to reach melting point as a function of the intensity at 532 nm. The broken lines correspond to the case where the infrared pulse arrives just before the visible pulse (negative delay), while the full curves are for simultaneous illumination. The synergistic effect for zero delay is evident: for example, for $I_{532} = 1$ GW/cm^2, the amount of 1.06 μm radiation needed decreases by a factor of six as the delay becomes positive. For increasing delays, the curve in figure 4 undergoes a continuous deformation from positive to negative concavity until it should finally approximate a rectangle for extremely long delays.

For simplicity of diagnostics, we have so far performed such experiments on ion-implanted <100> Si substrates (10^{15} cm^{-2} As$^+$ at 100 keV), the results of which are shown in figure 5. The synergistic effect is also present in ion-implanted Si but as expected, there are differences with the c-Si predictions. For negative delays, the infrared absorption is stronger due to damage and

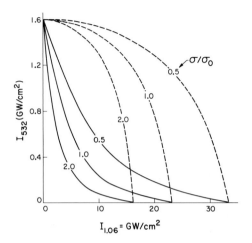

Fig. 4. Threshold intensity required to melt the c-Si surface by simultaneous illumination (full curves) and successive illumination with negative delay (broken curves).

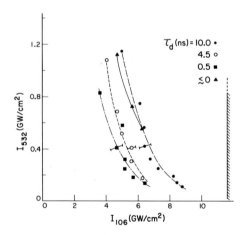

Fig. 5. Experimental threshold intensity required to melt a-Si for various delays. For $\tau_d < 0$, the hatched boundary is a lower bound of $I_{1.06}$ required to melt given $I_{532} \leq 1$ GW/cm^2.

doping levels, while for large positive delays, the influence of diffusion is reduced. Nevertheless, at $I_{1.06} = 6GW/cm^2$, the amount of visible light required to melt decreases by a factor of at least 5 as τ_d becomes positive and recovers to about half its initial value after 10 ns.

ACKNOWLEDGEMENTS

We wish to thank Professor Gibbons for bringing some of these problems to our attention two years ago. This work was supported by the Air Force Office of Scientific Research.

REFERENCES

1. P.M. Fauchet, Phys. Stat. Sol.(b) 110, K11 (1982).

2. C-J. Lin and F. Spaepen, Appl. Phys. Lett. 41, 721 (1982).

3. P.M. Fauchet and A.E. Siegman, Appl. Phys. Lett. 40, 824 (1982).

4. Zhou Guosheng, P.M. Fauchet and A.E. Siegman, Phys. Rev. B. (15 Nov. 1982).

5. S-L. Chuang and J.A. Kong, Proc. IEEE 69, 1132 (1981).

6. H.J. Leamy, G.A. Rozgonyi, T.T. Sheng and G.K. Celler, Appl. Phys. Lett. 32, 535 (1978).

7. D.K. Biegelsen, R.J. Nemanich and W.G. Hawkins, these Proceedings. The surface polariton model [D.J. Ehrlich, S.R.J. Brueck and J.Y. Tsao, Appl. Phys. Lett. 41, 630 (1982); F. Keilmann and Y.H. Bai, Appl. Phys. A29, 9 (1982)] does not appear to apply here since it would require complete surface melting ($\varepsilon_R < -1$). Note however that our model agrees with the surface polariton's results for $\varepsilon_R < -1$.

8. J.A. Van Vechten, Solid State Commun. 39, 1285 (1981).

9. P.M. Fauchet, Zhou Guosheng and A.E. Siegman, in Picosecond Phenomena III, K.B. Eisenthal, R.M. Hochstrasser, W. Kaiser and A. Laubereau, editors, Springer-Verlag, 1982, pp 376-379.

10. P.M. Fauchet, "Gradual surface transitions on semiconductors induced by multiple picosecond laser pulses," Phys. Lett. A, in press.

11. D.Y. Sheng, R.M. Walser, M.F. Becker, and J.F. Ambrose, Appl. Phys. Lett. 39, 99 (1981).

12. J-F. Morhange, G. Kanellis and M. Balkanski, Solid State Commun. 31, 805 (1979).

13. A. Lietoila and J.F. Gibbons, Appl. Phys. Lett. 40, 624 (1982).

14. J.M. Liu, R. Yen, H. Kurz, and N. Bloembergen, Appl. Phys. Lett. 39, 755 (1981).

15. J.R. Meyer, M.R. Kruger and F.J. Bertoli, J. Appl. Phys. 51, 5513 (1980).

ALIGNED, COEXISTING LIQUID AND SOLID REGIONS IN PULSED AND CW LASER ANNEALING OF Si

R. J. Nemanich, D. K. Biegelsen, and W. G. Hawkins,*
Xerox Palo Alto Research Center, Palo Alto, CA 94304

ABSTRACT

Aligned, coexisting liquid and solid regions are observed in cw laser annealing of polycrystalline Si films on quartz substrates. These stripe patterns are the precursors of surface topography that exists after cooling. It is proposed that a similar situation exists in the pulse annealing process. A calculation of the temperature evolution which assumes stripe symmetry and kinetic restraints of the crystallization process has been carried out. These calculations indicate a lattice temperature of between 1100 and 1300 K, 10 nsec after the sample has fully solidified.

INTRODUCTION

In the studies of pulsed laser annealing of semiconductors, two areas of special interest are the observation of oriented surface topography (i.e. ripples) [1,2,3] and the Raman measurements of a low lattice temperature shortly after the annealing pulse [4]. By and large the experimental and theoretical investigations have treated these aspects separately. In this study it is shown that aligned liquid and solid regions are responsible for the surface topography of cw laser annealed Si. Furthermore, it is suggested that this configuration also exists in pulse laser annealing and can account for the Raman observation through a purely thermal analysis.

The occurrence of surface topography has been observed for pulse laser annealing with pulse energies from ~ .4 to 2 J/cm^2. When higher pulse energies are used, the surface is meniscus like, indicating clearly that the annealed region of the sample has totally melted. An unusual characteristic of the surface topography is the formation of parallel linear ridges or ripples. While several spacings have been reported [3], for annealing with normally incident polarized light, the predominant pattern exhibits ripples aligned perpendicular to the polarization which are separated by a distance equal to the free space wavelength of the incident laser. Several analyses have indicated that the ripples are due to a standing electromagnetic wave at the surface and this pattern arises from an interference between the incident laser and a surface wave [1,2]. The origin of the surface wave and the coupling to the semiconductor are aspects yet to be defined.

With respect to a purely thermal analysis of the laser annealing process, it is difficult to understand how the ripple structures could persist in the molten state after the incident radiation is removed. Of course, the low lattice temperature of ~ 700 K deduced from Raman measurements [4] is also difficult to reconcile with the purely thermal analysis.

*Permanent address: Xerox, Webster Research Center, Webster, NY 14580

Mat. Res. Soc. Symp. Proc. Vol. 13 (1983) ©Elsevier Science Publishing Co., Inc.

EXPERIMENTAL

To understand the formation of the surface ripples, cw laser annealing of Si has been monitored *in situ*. This has been accomplished by annealing polycrystalline Si films deposited on quartz substrates. The laser annealing was carried out with 10.6 μm radiation from a CO_2 laser which was focused cylindrically or to a broad spot. Since surface topography is related to the laser wavelength, the 10.6 μm radiation allows easy identification of the precursor of the surface structures. In the annealing configuration used, the black–body radiation or reflected light could be easily monitored from the back side of the quartz substrate. This was accomplished by positioning a microscope objective near the substrate and imaged on a television camera.

RESULTS AND DISCUSSION

A) The Origin of Ripples in CW Annealing
With the capability of monitoring *in situ* the processes involved in the cw laser annealing, the precursors of the surface ripples can be identified. An image of the annealed spot in reflected illumination which is shown in Fig. 1a indicates how the surface structures are formed. In the figure the bright regions are liquid while the dark regions are solid. Shown in Fig. 1b is a video micrograph of the same region after the sample has cooled. It is clear that the aligned, liquid and solid regions are the precursors of the surface ripples. We note here that the grating structure is stationary under uniform illumination. However, when a distorted wave front is used, the liquid lines "sweep" through the irradiated area.

It should be noted that there have been several reports of coexisting liquid and solid regions in cw laser annealing [5,6], but the alignment of the lamellae have not previously been observed. To achieve the alignment it was necessary to obtain a relatively uniform beam profile to minimize temperature gradients. This was achieved with a cylindrical lens or with a long focal length spherical lens. The formation of the liquid regions occurred in all cases first by nucleation of stable circular regions of ~ 3 μm diameter [6] which then coalesced into the stripes as the laser power was increased.

Although initially the radiation may be absorbed in the fused silica substrates, nevertheless, when the Si is heated to near the melt temperature, the absorption length for 10 μm radiation due to free carriers is ~ .25 μm which is much less than the thickness of the Si films (0.5 – 2.0 μm). However, to assure that the radiative interactions are due to the Si, the annealed (rippled) Si film was etched from the quartz substrate. Examination of the quartz surface using Nomarski microscopy showed no evidence of ordered surface topography. Furthermore with a 2 μm SiO_2 encapsulating layer, aligned lamellae are not observed whereas they are observed for encapsulating layers $\leq$ 400 nm. These results indicate that the formation of the aligned coexisting liquid and solid phases is due to a coherent interaction of the incident beam and a scattered wave radiated solely by the Si film.

B) Considerations for Pulse Laser Annealing
The question now arises whether liquid and solid regions also coexist in the pulse annealing process. Two basic differences between the cw and pulse laser annealing processes are that fields due to the incident radiation are much higher and a steady state condition never is reached in the pulse annealing process. Since it is known that both the surface ripple structures of pulse annealing and the liquid solid regions in the cw process are due to the radiation pattern, then the higher fields of pulse irradiation should amplify this effect. Furthermore the fact that a steady state condition is never reached during pulses allows large temperature

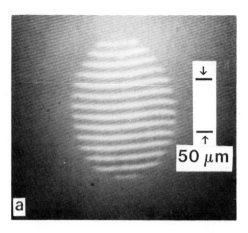

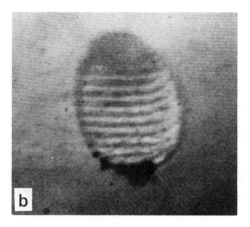

Fig. 1.   A reflected light image of the laser annealed region of a .5 μm Si film (a). The bright regions are liquid. Also shown is the video image of the surface topography after cooling (b).

variations over small distances.   Thus both of these properties of the pulse annealing process seem to favor the formation of aligned coexisting liquid and solid regions.

Two aspects of the silicon melting transition which could contribute to the surface scattering mechanisms are (1) the higher reflectivity, and (2) the density increase of the molten regions. Both of these properties could supply a positive feedback mechanism for the standing electromagnetic surface wave [7]. Either the metallic character or the surface contraction of the molten regions will create a diffraction grating that will diffract the incident light along the surface.   This combined with the incident light will set up a standing electromagnetic wave at the surface, the wavelength of which is the same as the spacing of the liquid and solid regions.

C) Calculation of Temperature Variation

To address the questions involving the temperature achieved during and after the pulse annealing process, a calculation based on thermal conduction to the substrate has been caried out. The calculation that is described here contains two noteworthy aspects. The first is that the lateral gradients due to an assumed stripe symmetry are obtained, and the second is that kinetic restraints which relate the solidification rate to the undercooling of the melt are included. The computer calculation method is similar to that used in other investigations [8], but because of the stripe symmetry, the problem was reduced to a two–dimensional heat flow calculation. To compare with typical experiments it was assumed that the melt regions were 260 nm wide, extended 200 nm into the sample and were separated by 260 nm of solid regions. Because of the periodicity of the stripe arrangement, the lateral extent of the calculation could be confined between the center of an initially solid region and the center of an initially liquid region (both represent mirror planes). To obtain initial conditions corresponding to an absorbed energy of ~ .4 J/cm^2 (which would result from a .8 J/cm^2 pulse), all the solid and liquid regions at less than 400 nm depth were at the melt temperature (1685 K), and the temperature profile decayed with a gaussian form with a 300 nm width. The heat flow was calculated for (10 nm)^3 regions at every 2.5 psec time interval. The energy flow between cells was approximated by the product of the temperature difference and the thermal conductivity. A temperature independent thermal conductivity (.05 cal/sec. cm.K) and heat capacity (.24 cal/gm–K) of solid Si at the melt temperature were used. The value of the latent heat of the liquid used was 430 cal/gr. The initial conditions of solid and liquid at the melt temperature would evolve ~ 20 nsec after the beginning of a 5 nsec pulse. Shown in Fig. 2 is the temperature profile that evolved 25 nsec later. A small region of melt remains. It should be noted,

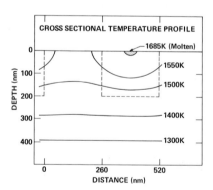

Fig. 2.   A calculation of the cross sectional temperature profile for the parallel stripe configuration just before complete solidification.

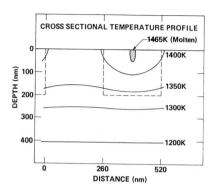

Fig. 3. A calculation of the cross sectional temperature profile as in Fig. 2 but with kinetic restraints of the recrystallization process included.

however, that the center of the initial solid regions exhibit a temperature of ~ 1520 K averaged over the top 200 nm (which is probed by visible radiation). At 5 nsec later, the lateral variations of the temperature profile have largely disappeared and the average temperature of the top 200 nm is ~ 1415 K.

An intriguing result that the calculation yielded is that a melt front velocity of ~ 800 cm/sec is obtained. It has been shown, however, that for uniform interface velocity, the undercooling is linearly related to the interface velocity [9]. While detailed measurements are not available, a reasonable estimate for the interface rate constant is between 1 and 10 cm/sec–K [10]. This factor has been included in the calculation. The resulting temperature profile for the interface rate constant equal to 5 cm/sec–K is shown in Fig. 3. The profile which evolved in 35 nsec, shows a temperature of ~ 1360 K averaged over the top 200 nm of the center of the initially solid regions. The measurements indicate a temperature drop of ~ 120 K in 10 nsec after complete solidification.

D) Relation to Raman Scattering Measurements

The calculation described in the previous section is a description of the situation that may exist for the time–dependent Raman scattering measurements [4]. These measurements used a pulse of .4 to 1.0 J/cm^2 of visible radiation which could cause aligned–liquid solid regions with a periodicity of ~ .5 μm. Because of the metallic character of the liquid and the resulting grating effect, when the melt is present, there would be little or no Raman signal. However, measurements which do not rely on visible or longer wavelength light could measure the recrystallization properties. Furthermore when the kinetic restraints are included in the calculation, for 10 nsec after solidification surface temperatures of between 1100 to 1300 K are calculated for rate constants of 1 and 10 cm/sec–K respectively. This is to be contrasted with recent Raman measurements which show a temperature of ~ 700 K, 100 nsec after the annealing pulse. There is clearly disagreement remaining, but this may be accounted for by somewhat different initial conditions or constants in the calculation or the time after the pulse when the two results are compared.

CONCLUDING REMARKS

One aspect that remains to be addressed is the recrystallization of the solid regions between the molten stripes. For samples amorphized by implant, it is feasible that these regions recrystallize by solid state regrowth. There should be an observable difference, then, in the dopant redistribution between the initially solid and molten regions.

In this study we have demonstrated that the precursor of surface ripple structures in cw laser annealing of Si is aligned, coexisting liquid and solid regions. It is proposed that similar structures exist in the pulsed laser annealing process. The high reflectivity of the molten state will establish a diffraction grating which will cause a standing electromagnetic wave at the surface. While the existence of aligned liquid and solid regions explain the persistence of the surface ripples through the time when the melt exists, it can also account for the Raman results.

ACKNOWLEDGMENTS

We acknowledge helpful discussions with R. M. White, C. Herring, K. A. Jackson, P. M. Fauchet and A. E. Siegman.

REFERENCES

1. H. J. Leamy, G. A. Rozgonyi, T. T. Sheng and G. K. Celler, Appl. Phys. Lett. 32, 535 (1978).

2. P. M. Fauchet and A. E. Siegman, Appl. Phys. Lett. 40, 824 (1982).

3. D. Haneman and R. J. Nemanich, Solid State Commun., 43, 203 (1982).

4. H. W. Lo and A. Compaan, Phys. Rev. Lett. 44, 1604 (1980); A. Compaan, H. W. Lo, A. Aydinli and M. C. Lee, in *Laser and Electron-Beam Solid Interactions and Materials Processing*, edited by J. F. Gibbons, L. D. Hess and T. W. Sigmon (North Holland, New York 1981) p. 15; A. Compaan, H. W. Lo, M. C. Lee and A. Aydinli, J. de Phys. 42, C6–453 (1981).

5. M. A. Bosch and R. A. Lemons, Phys. Rev. Lett. 47, 1151 (1981).

6. W. G. Hawkins and D. K. Biegelsen, to be published.

7. Z. Guosheng, P. M. Fauchet and A. E. Siegman, Phys. Rev. B (in press).

8. R. F. Wood and G. E. Giles, Phys. Rev. B 23, 2923 (1981).

9. For a review see W. A Tiller, in *The Art and Science of Growing Crystals*, edited by J. J. Gilman (Wiley, New York) p. 276.

10. K. A. Jackson, private communication.

INTERFEROMETRIC MEASUREMENTS OF SURFACE PROPERTIES DURING LASER ANNEALING

MARIO BERTOLOTTI, ALDO FERRARI, CONCETTA SIBILIA
Istituto di Fisica,Ingegneria,Università Roma, Italy and GNEQP of CNR,Italy
and
PETER JANI, Central Research Institute for Physics of the Hungarian
Academy of Sciences, Budapest, Hungary.

ABSTRACT

A simple interferometric method is presented which allows measurement of small vertical displacement of a surface heated by a laser beam. Calculations applied to a silicon crystal in the case of a c.w. laser show reasonable agreement with experiment. The method can be applied to assess surface temperature and thermal constants.

INTRODUCTION

The problem of the determination of the temperature reached by a semiconductor surface irradiated with a laser is considered by using an interferometric method to measure the surface thermal dilatation.

Application to the case of slow heating with a cw laser is here treated.

The basic point is the fact that under a temperature increase the illuminated surface suffers a vertical displacement (here denoted as dilatation) which is the measured quantity. Unfortunately the relation between temperature increase and dilatation is rather involved; however with some approximation a solution in closed form can be given which is rather good near equilibrium. The method can be used to derive also other interesting parameters of the material under study.

THEORETICAL RELATION BETWEEN DILATATION, TIME AND TEMPERATURE

The problem of heating and subsequent displacing of a thin semiconductor slab of thickness l illuminated on one side by a gaussian cw laser is treated. The laser intensity is written as

$$I(r) = I_0 \exp \left(- 2r^2/w_0^2 \right) \tag{1}$$

where w_0 is the spot.

A thermoelastic quasistatic formulation is used (1). The starting equation for the displacement $\vec{u}$ is taken as

$$\nabla^2 \vec{u} + \frac{1}{1-2\nu} \operatorname{rot} \vec{u} = \alpha^* \operatorname{grad} T \tag{2}$$

where

$$\alpha^* = 2\alpha \ \frac{1 + \nu}{1 - 2\nu}$$

being ν the Poisson ratio, α the thermal dilatation coefficient, and T the

Mat. Res. Soc. Symp. Proc. Vol. 13 (1983) ⓒElsevier Science Publishing Co., Inc.

temperature.

A solution is found by using a method proposed by Goodier (1) which makes use of the thermoelastic potential Φ (x,y,z,t) related to u by

$$\overline{u}_1 = \frac{\partial \Phi}{\partial x} + \frac{\partial \Phi}{\partial y} + \frac{\partial \Phi}{\partial z} \tag{3}$$

being $\overline{u}_1$ a particular integral of (2). Eq.(2) then writes

$$\nabla^2 \Phi = (\frac{1+\nu}{1-\nu})\ T \tag{4}$$

being T the temperature distribution which in our case is taken as

$$T_{ist} = \frac{Q_o(1-R)w_o^2\ \exp(-at)\ \exp(-r^2/(4\chi\ t+w_o^2))}{1_o^2 c(4\chi t + w_o^2)} \int_{-1_o}^0 dz'\exp(-(z-z')^2/4\chi t) \tag{5}$$

where Q_o is the energy incident per unit area at the center of the gaussian profile of radius w_o, a = $2H\chi/1_o k$, being H the heat transfer by free convection (which is a function of the geometry, the orientation and the temperature difference; for surfaces in air H $\sim$.5.10^{-4} W cm^{-2} C (2)), R is the reflectivity, χ is thermal diffusivity, c is thermal capacity, and k is thermal conductivity.

If $T-T_o = \Delta T << \frac{1-\nu}{1+\nu}\ \frac{kl_o}{2\alpha}\ \frac{1}{H}$

Eq.(4) can finally be put into the form

$$\nabla^2\ (\frac{\partial \Phi}{\partial t} - \chi\alpha\ \frac{1+\nu}{1-\nu}\ T_{ist}) = 0 \tag{6}$$

from which the expression of dilatation S(t) at the spot center as a function of time can be derived after some lengthy calculations as

$$S(t) = \frac{\alpha(1+\nu)\ Q_o(1-R)\ w_o^2}{(1-\nu)\ 1_c^2\ c} \int_0^t \frac{dt\ \exp(-at)}{4\chi t+w_o^2}\ -\alpha(1+\nu)I_o\frac{(1-R)w_o^2}{k}$$

$$\{\exp(\overline{k}^2 w_c^2)\left[\ E_i(-\chi\overline{k}^2 t-\overline{k}^2 w_c^2\ )- E_i(-\overline{k}^2 w_o^2\ )\right]\}\ , \tag{7}$$

where $\overline{k}^2$= $2H/1_o k$ and E_i is the exponential integral.

EXPERIMENTAL SET-UP

The surface dilatation is measured via an interferometric method as shown in Fig.1. A beam from a He-Ne laser is made parallel by a telescope T, and split into two beams 1 and 2 at the semi-reflecting surface Su. Beam 2 is reversed by a corner prism P while beam 1 is focused on the semiconductor surface

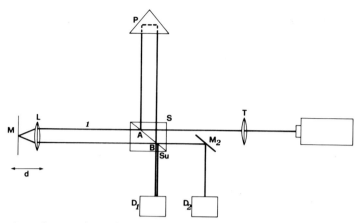

Fig. 1. Interferometric system

Se and superposed to beam 2 on the detector D. If the light intensity in the
two split beams is the same I, the intensity on the detector I_D can be written
as

$$I_D = I \left[1+\cos \Delta\phi\right];\qquad(8)$$

where $\Delta\phi$ is the phase difference between the beams, which depends on the distances of the corner prism and surface Se from the surface Su

$$\Delta\phi = \frac{2\pi}{\lambda}(d_1 - d_2) + \delta\phi\qquad(9)$$

where d_1 and d_2 are the distances traveled from point A to B through the two
paths respectively, and $\delta\phi$ takes into account phase shifts at reflections.
     The heating laser beam impinges on the surface Se in the same point which is
inspected via the interferometer. The dilatation S of the surface produces a
phase shift

$$\delta\ (\Delta\phi) = \frac{2\pi}{\lambda}\ 2S\ .\qquad(10)$$

The number of fringes detected by D and plotted on an X-Y writer gives there
fore a measurement of S, the absolute accuracy of which was better than a
quarter of a fringe, i.e. ~ 0.1 μm.

EXPERIMENTAL RESULTS

     The dilatation S(t) as a function of time obtained from a crystalline sili-
con sample irradiated with a cw Ar laser is shown in Fig.2. For comparison the
theoretical values as derived from Eq.(7) are shown as a continuous line. For
small times the agreement is not very good; this probably is due to a spatial
integration of the probing beam over the heated surface. At equilibrium a large

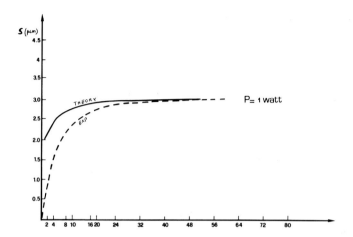

Fig. 2. Experimental results on crystalline silicon.

surface region suffers the same dilatation and the spatial integration effect is no more present, which explains the good agreement between experimental and theoretical saturation values. The method seems therefore very useful as a noncontact probe of surface temperature during laser irradiation, which can be used in routine processing upon calibration. It can also be used to determine some material property like reflectivity.

Application of the method to the case of pulsed irradiation is now being considered. The interferometric method has an inherently fast response and is therefore able to detect very sudden displacements. Thermoelastic theory has however to be refined when going to very short times.

REFERENCES

1. J. L. Nowinski, Theory of thermoelasticity with applications - Sijthoff and Noordhoff Intern. publishers, Alphen Aan Dem Rijn, The Netherlands, 1978.

2. J. F. Ready, Effect of high power laser radiation, Academic Press Inc., New York, 1971.
   H. S. Carslaw and J. C. Jaeger, Conduction of Heat in Solids, Oxford Univ. Press, London 1959.
   M. Bass in Physical Processes in Laser-Materials Interactions, ed.M. Bertolotti, Plenum Pu. Co. New York 1982.

IMPURITY DIFFUSION IN SILICON ANNEALED BY SEMI-CONTINUOUS LASER

Jean-Eric BOUREE, Claude LERAY and Michel RODOT
CNRS, 1 Place A. Briand, 92190 - Meudon, France

ABSTRACT

The diffusion of an impurity into a solid which is irradiated by laser pulses of some milliseconds duration has
been analysed in terms of effective diffusion time and temperature. In the case of Fe in Si, the diffusion coefficient
is found to be similar to that measured by COLLINS and
CARLSON, and independent of boron doping. In the case of
Al in Si, a value of 2.10^{-10} cm^2/s at 1400°C has been found
both for single crystals and for fine-grained polycrystals.

INTRODUCTION

Solid state laser annealing may be a way to measure atomic diffusion coefficients, as shown by MATSUMOTO [1] for the cases of As and P in Si. This
technique is in principle well adapted to the study of fast-diffusing impurities
since it allows to choose short effective diffusion times t_{eff}. For the case of
Fe in Si, first studied by STRUTHERS [2], interstitial diffusion is very fast
($D = 10^{-6}$cm^{2}s^{-1} near 900°C), so that Fe diffusion in an oven for some minutes
is currently used to obtain homogeneous Fe-doped wafers. Another analogous case
might be that of intergranular diffusion, into polycrystals, of normally slow-
diffusing impurities like Al, at least if we envisage a possible extrapolation
at high temperatures of results recently obtained near 400°C [3]. We have made
measurements of D for Fe and Al in Si using semi-continuous laser annealing,
i.e. a continuous laser chopped to give pulses of typically some milliseconds;
impurity profiles were measured by SIMS and interpreted using a model that we
derived following the line of GOLD and GIBBON's model [4].
An important feature sought from these experiments is the behaviour of electron lifetime killing impurities like Fe or Al in the growth of epitaxial solar
cells. When using a cheap impure substrate, the quality of the CVD-grown active
layer is determined by the presence of such impurities. Conversely, one can
tolerate in the substrate an impurity content 10 times larger than the one
acceptable near the surface of the active layer if the latter is grown in 20
minutes, for a diffusion coefficient of 10^{-8}cm^{2}s^{-1} : the exact knowledge of D as
a function of temperature T is a prerequisite to define an acceptable purity for
the substrate. It can also help tailoring the gettering step which may be used
before CVD [5]. This goal explains why we were particularly interested in the
possible dependence of D_{Fe} on the boron content in the diffusion sample : large
boron doping of the back of the CVD layer has been proposed to improve the cell
[5]; on the other hand it is known that the diffusion kinetics may be affected
by donor-acceptor interactions or by the formation of complexes [6]. The diffusion of Fe in Si raises a special problem, since two diffusion kinetics have
been observed [7].

MODELLIZATION OF LASER ANNEALING

A knowledge of the temperature distribution inside the illuminated solid is

Mat. Res. Soc. Symp. Proc. Vol. 13 (1983) ©Elsevier Science Publishing Co., Inc.

necessary to account for laser annealing |8| |4| |1|. The thermal effects indu-
ced by a CW laser are well described by the usual heat diffusion equation :

$$\rho C(\partial T/\partial t) - \nabla(K.\nabla T) = Q \qquad (1)$$

Where Q is the absorbed radiant energy and K, ρ, C are the heat conductivity,
volumic mass and thermal capacity of the solid. Equation (1) is solved after
being linearized |9| |10| and integrated using the Green source function. To
write the analytic solution we assume the silicon reflectivity R to be constant-
ly equal to 0.4 |11|. Furthermore, like NISSIM |10| we assume the thermal conduc-
tivity to vary with temperature as $A/(T-T_k)$, with A = 299 W.cm^{-1} and T_k= 99 K.
The solid is viewed as semi-infinite, since its dimensions are large compared to
the diameter 2 ω of the laser spot (50-100 μm) and the thermal diffusion length.
The heat source is assumed to be plane, because ω is much larger than the absorption length (0.5 μm for a laser with ca. 5000 Å wavelength). Finally the
laser beam intensity is supposed to have a gaussian distribution :

$$I(r) = \frac{P(1-R)}{\pi\omega^2} \exp - (\frac{r^2}{\omega^2}) \qquad (2)$$

where r is the distance to the spot center and P the beam power. ω is a parameter which can be calculated from the power Pf necessary to melt the center of
the laser spot, for a given pulse duration τ (Fig.1); the maximal temperature T_m
at this point is a function of P shown by Fig.2, while Fig.3 illustrates the
thermal inertia of the sample during a short annealing.
    For a thermally activated phenomenon like the diffusion of an impurity, the
diffusion length follows an Arrhenius law. For laser annealing, we show that :

$$\ell^2 = D_o\ t_{eff} \exp (-E_a/kT_{eff}) \qquad (3)$$

where D_o and E_a are constants, t_{eff} and T_{eff} the effective diffusion time and
temperature. The temperature induced by the laser pulse being inhomogeneous, we
consider the average $\overline{\ell^2}$ of ℓ^2 over the annealed area and show that :

$$\overline{\ell^2} = D_o.\tau.f_i.f_g \exp (-E_a/kT_m) \qquad (4)$$

This means that T_{eff} is equal to the maximum temperature reached at the laser
spot center and that $t_{eff} = \tau.f_i.f_g$, where f_i illustrates the thermal inertia
and f_g the temperature distribution of the sample. In order to get an extended
homogeneous annealed area, we can cover this area with several series of superposed spots, so that t_{eff} is multiplied by a factor f_r describing the spot over-
lap : $t_{eff} = \tau f_i f_g f_r$ (Fig.4). By an iterative process indicated by SHIBATA |12|,
we can deduce from the experiments the values of D_o and E_a characterizing the
diffusion.

EXPERIMENTAL CONDITIONS

Sample preparation
    Several samples have been used :
- Wacker-Heliotronic single crystals, slightly boron doped (1 Ωcm),
- large-grained polycrystals with different B concentrations, from the same
origin,
- very fine-grained polycrystals from the same origin. These grains are elongated with a mean width of 3 μm.
- pure single crystals on which an epitaxial layer of ca. 50 μm has been grown
from vapour phase at 1150°C under 75 torrs (by Comurhex).
    All samples were mechanically polished, degreased and etched (chemical or

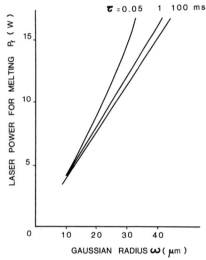

Fig. 1. Minimum laser power for melting P_f, as a function of gaussian radius and pulse duration.

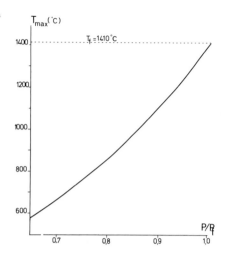

Fig. 2. Maximum temperature vs. reduced power P/P_f.

	P_f (W)	τ (ms)	d (μm)	Ea (eV)
Fe :	15.5	15	20	0.87
Al :	15.5	100	10	3.3

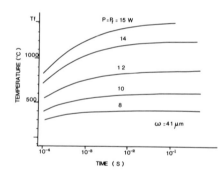

Fig. 3. Temperature rise at the spot center vs. time, for different laser powers (here P_f = 15 W, ω = 41 μm).

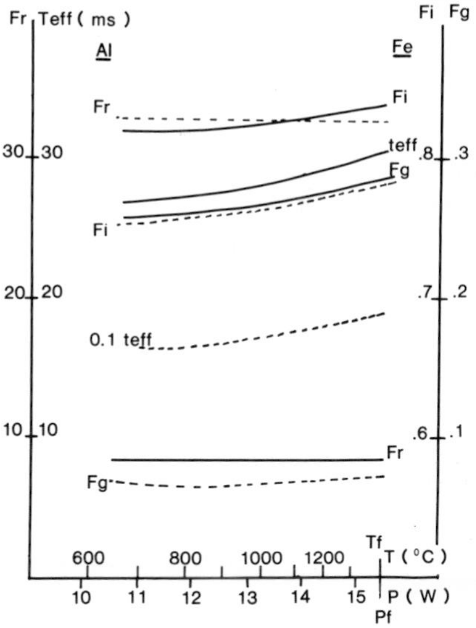

Fig. 4. Reduction factors and effective annealing time for the diffusion of Fe and Al into Si (d is the distance between consecutive laser spots in a hexagonal lattice).

mechano-chemical etching).
   The impurity Fe or Al was deposited on the sample surface :
- either by electron-beam evaporation under ultra-high vacuum (better than
10^{-8} torr). The evaporated layer is 40 Å to 200 Å thick and is covered by a 200
to 500 Å thick Si layer which is both antireflecting for light and avoiding
oxidation.
- or by ion implantation at doses between 10^{11} and 10^{13} cm^{-2} under an energy of
100 keV.

<u>Semi-continuous laser annealing</u>
   A continuous Ar$^+$ laser of 29 W maximum power emits, in the spectral range 4880-
5140 Å, a beam which is chopped by a rotating disk and focused on the sample |13|.
Between two pulses the table on which the sample is fixed may be automatically
moved in two perpendicular directions, so that an area of the sample, about
1 mm^2, is covered by laser spots drawing a pattern designed in advance. The
sample is kept under an argon atmosphere. We control four experimental parameters:
the beam power P, the pulse duration   (10 µs to 100 ms), the gaussian radius ω
of the laser spot and the distance d between two consecutive spots.

<u>Method of analysis</u>
   A SIMS apparatus, of either the
SMI 300 or the IMS 3F model of Cameca,
allows to analyze the center (60 µm
diameter) of a 0.5 mm diameter area,
eroded at a constant rate, with a
resolution of 100 to 200 Å in depth.
We choose O_2^+ as the primary ion
(5.5 or 12.5 keV) and $^{11}B^+$, $^{27}Al^+$,
$^{54}Fe^+$, $^{52}Si_2^+$ (or $^{29}Si^+$) as the
secondary ions, in imaging mode (the
contribution of $^{54}Al_2^+$ is verified to
be negligible). We thus obtain
profiles on 1-2 µm depths, like
those of Fig. 5.

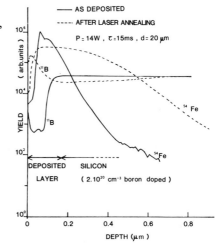

<u>Fig.5</u>. Depth profiles in Si, as
       measured with Cameca IMS 3F
       microprobe.

RESULTS AND DISCUSSION

   Fig. 6 shows Fe profiles after laser annealing at different laser powers, as
compared to the profile in non-annealed Si. The annealing temperature T_{eff} is
deduced from the laser power using Fig. 2. The curves of Fig. 6 are interpreted
as due to a diffusion from a plane source, which gives the value of the product
$D.t_{eff}$ Fig. 4 is then used to deduce D (T_{eff}). The results are plotted on Fig.8.
Our best value for D_o and E_a are : $D_o = 3.10^{-3}cm^2s^{-1}$; $E_a = 1.1$ eV.
   A first observation is that our measurements agree rather well with the
results of "slow diffusion" obtained by COLLINS and CARLSON |7|. They differ
largely from those of the "fast diffusion" of the latter authors which coincide
with STRUTHER's |2| results. A reasonable explanation is the following. In our
experiments and in those of COLLINS and CARLSON, the diffusion source is pure
iron; starting from a 20 Å Fe layer, the mean Fe concentration in the 1 µm thick
layer where Fe has diffused is larger than 10^{19} cm^{-3}, while the solubility
limit is only 9.10^{15} cm^{-3} at 1100°C : the Fe diffusion should then take place
via a mechanism of Fe-Si reaction. On the contrary STRUTHERS used a gaseous

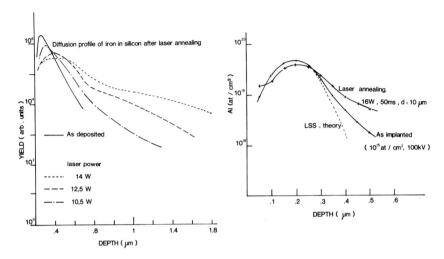

Fig.6. Iron profiles in CVD-Si coated with an Fe layer; full line : before laser annealing; dashed lines : after laser annealing for different beam powers (pulse duration 15 ms).

Fig.7. Examples of Al profiles in Al-implanted Si single crystals, as implanted and after laser annealing. The theoretical profile after implantation is also shown.

source, so that his Fe content could be permanently smaller than the solubility limit, and diffusion could occur inside the homogeneous Si phase. We verified, by SIMS imaging, the presence of tiny precipitates in the diffusion region of our samples.

A second conclusion which can be drawn is that there is no significant difference of D_{Fe} in samples very rich and very poor in boron. In view of the recent papers |14| studying Fe-B interactions by DLTS measurements and showing that the Fe-B pairs are dissociated above 150°C, this result is not surprising. As a consequence, a boron-doped layer cannot play the role of diffusion barrier during the growth of a CVD epitaxial layer.

Fig. 7 shows a typical result about Al diffusion. The Al profile after implantation follows the theoretical profile except for a rather large tail which may be due to partial channelling and diffusion during implantation (see e.g. |15|). Laser annealing causes a further diffusion, which is analysed using a finite element calculation : the diffusion process is simulated, starting from the initial profile, to obtain a final profile approaching the real one as much as possible. The results are shown on Fig. 8 for two samples : a single crystal and a fine-grained polycrystal. Both diffusion coefficients (2.10^{-10} cm^2s^{-1} at 1400°C) are almost equal to the one already published for single crystals |16|. This finding casts a doubt on a possible extrapolation of HWANG's results|3|to higher temperatures; this author effectively remarked that his pre-exponential factor was abnormally high; his results (obtained in a very narrow temperature range) and ours can be reconciled if we admit an activation energy of intergranular diffusion lower than that of 2.64 eV given by HWANG's paper. Like in the Fe case, the absolute value of Al concentrations overpasses somewhat the

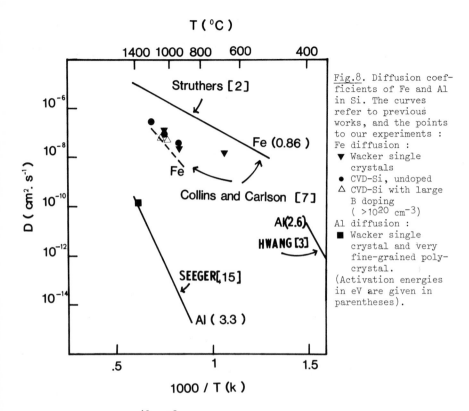

T (°C)

Fig.8. Diffusion coefficients of Fe and Al in Si. The curves refer to previous works, and the points to our experiments :
Fe diffusion :
▼ Wacker single crystals
● CVD-Si, undoped
△ CVD-Si with large B doping (>10²⁰ cm⁻³)
Al diffusion :
■ Wacker single crystal and very fine-grained poly-crystal.
(Activation energies in eV are given in parentheses).

solubility limit (2.10^{19} cm^{-3} at 1100°C); other samples which were implanted at lower doses did not allow accurate interpretations in this case where D is very small. Our result is a first indication that intergranular diffusion of Al could be non-dangerous when CVD-depositing an epitaxial layer to build solar cells. However this experiment has to be improved, since in this case (τ = 100ms) the sample temperature was not strictly controlled.

CONCLUSIONS

Semi-continuous laser annealing allows to perform experiments from which diffusion coefficients may be confidently deduced, provided that appropriate values are chosen for pulse duration, laser shot overlapping and laser power. Fig. 1 to 4 can help to select these conditions.

Compared to scanning of a continuous laser, our method takes a little more time, but gives more uniform annealing conditions because no preferential direction of the sample is implied.

Iron is confirmed to be a rapidly diffusing impurity in Si with a value of D near that of COLLINS and CARLSON, which is attributed to the silicide formation kinetics. Boron concentration does not change the diffusion coefficient.

The Al diffusion coefficient is smaller but still measurable; intergranular diffusion is negligible at 1000°C.

ACKNOWLEDGMENTS

We are indebted to R. LEGROS and J. PLASSARD for using the laser annealing apparatus that they have built, to M. TESSIER, J. CHAUMONT and J. RUSTE for respectively making available their ultra-vacuum evaporation, implantation and SIMS apparatus, M. BARGUES for depositing the CVD layers and D. HELMREICH for delivering Si crystals.

REFERENCES

11. S. Matsumoto, J.F. Gibbons, V. Deline, C.A. Evans Jr., Appl. Phys. Lett. 37, 821 (1980).

2. J.D. Struthers, J. Appl. Phys. 27, 1560 (1956).

3. J.C.M. Hwang, P.S. Ho, J.E. Lewis, D.R. Campbell, J. Appl. Phys. 51, 1576 (1980).

4. R.B. Gold, J.F. Gibbons, J. Appl. Phys. 51, 1256 (1980).

5. R.G. Wolfson, R.G. Little, 15th IEEE Photovoltaic Spec. Conf. Proc. (1981) pp. 595-597.

6. R.B. Fair, G.R. Weber, J. Appl. Phys. 44, 273 (1973); R.B. Fair, J.C.C. Tsai, J. Electrochem. Soc. 122, 1689 (1975).

7. C.B. Collins, R.O. Carlson, Phys. Rev. 108, 1409 (1957).

8. Z.L. Liau, B.Y. Tsaur, J.W. Mayer, Appl. Phys. Lett. 34, 221 (1979).

9. H.S. Carlslaw, The conduction of heat in solids (Oxford Univ. Press, 1959) p. 89.

10. Y.I. Nissim et al.J.Appl. Phys. 51, 274 (1980).

11. M.O. Lampert, J.M. Koebel, P. Siffert, J. Appl. Phys. 52, 4975 (1981).

12. T. Shibata, T.W. Sigmon, J.F. Gibbons, Laser and electron beam processing of materials, C.W. White and P.S. Peercy ed. (Acad. Press 1980), p. 530.

13. H. Tews, M. Schneider, C. An, Appl. Phys. Lett. 40, 41 (1982).

14. K. Graff, H. Pieper, J. Electrochem. Soc. 128, 669 (1981); K. Wünstel, P. Wagner, Appl. Phys. A 27, 207 (1982).

15. F.N. Schwettmann, Appl. Phys. Lett. 22, 570 (1973).

16. A. Seeger, K.P. Chik, Phys. Stat. Sol. 29, 455 (1968).

SEARCH FOR VACANCIES IN LASER ANNEALED SILICON*

H. J. STEIN AND P. S. PEERCY
Sandia National Laboratories, P. O. Box 5800, Albuquerque, New Mexico

ABSTRACT

In an effort to enhance vacancy trapping and detection in laser-annealed Si, float zone Si was implanted with oxygen to achieve concentrations between one and two orders of magnitude greater than the equilibrium saturation limit. Oxygen, which is known to be an effective trap for vacancies, was found to incorporate efficiently into Si regrown with Q-switched laser annealing Interstitial oxygen, oxygen-vacancy defects and divacancies were observed after implantation. The laser-regrown layers, however, were free of detectable vacancy-associated defects.

INTRODUCTION

Previous studies have shown that laser annealing can produce Si layers which are free from extended defects detectable by TEM [1]. Small defects, however, such as vacancies, vacancy impurity complexes, isolated interstitials and divacancies, could remain in the layers and would not be resolved in the TEM measurements. Electrically active states associated with small defects have been reported by Kimerling and Benton [2] for laser-annealed layers which exhibited perfect microstructures. They found an energy level for one electrically detectable defect to be different in oxygen-containing from that in oxygen-free Si. In addition, a recent x-ray diffractometry study by Servidori, et al., [3] showed residual disorder not detectable by TEM in ion implanted and laser annealed Si. The disorder was ascribed to carbon and oxygen impurities associated with vacancy-type defects.

Infrared absorption has been used extensively for measurements of interstitial oxygen, O_I, and of irradiation-produced oxygen-vacancy, O-V, and divacancy, V-V, defects in Si [4]. Infrared absorption, however, would not be sufficiently sensitive to observe O-V defects in thin layers with oxygen at the saturation concentration [5] of molten Si. On the other hand, if trapping of implanted oxygen at the resolidification interface occurs for O in Si, then vacancy trapping by > 5 percent of the implanted oxygen should be observable by infrared absorption methods. Oxygen also enhances vacancy trapping by other vacancies to form V-V defects [6].

We have used infrared absorption to search for vacancy trapping in oxygen implanted and laser-annealed Si, and to obtain information on O retention above the equilibrium saturation concentration in laser-melted Si.

*This work performed at Sandia National Laboratories supported by the U.S. Department of Energy under contract number DE-AC04-76DP00789.

EXPERIMENTAL DETAILS

Silicon samples 6 x 6 x ~ 0.6 mm were prepared from (111) and (100)
float zone Si. A syton polish followed by a chemical etch polish was found
to yield surfaces that were best suited for the experiments. The crystalline
defects after implantation were more clearly resolved when the surfaces were
etched, particularly for the lower energy implants, and the etch gave a
slightly uneven surface which, combined with bevel polishing, helped to
suppress interference fringes in the optical absorption measurements. No
differences in oxygen-related effects were observed for (111) and (100) Si.

Infrared absorption measurements for O_I and O-V defects were made at
~ 300 K on a Nicolet 7199 FTIR spectrometer which incorporated a cooled HgCdTe
detector, and measurements for V-V defects were made on an upgraded model
221 Perkin Elmer CaF$_2$ prism spectrophotometer.

Implantations were performed 5° off axis in an ion-pumped sample chamber
using an Accelerators Inc. implanter. Temperatures during implantation were
300 K and were monitored with a thermocouple on the sample holder. Laser
annealing was performed in air with a Q-switched ruby laser beam passed
through a homogenizer. The pulse energy was measured by a calibrated photodiode which monitored a split-off portion of the pulse energy. The reflectance changes during annealing were also monitored using the 4880 A line of
an argon laser.

BACKGROUND

Figure 1 shows absorption bands in 1 mm thick crucible-grown Si after 10^{17}
neutrons/cm^2 (E > 10 keV) to illustrate O-V, O_I, V-V absorption bands in Si.
Absorbance by a low O concentration (< 10^{16} O/cm^3) float zone reference sample
was subtracted to cancel absorption by phonons in the Si substrate. The intensity of the 1106 cm^{-1} O_I band and the ASTM-recommended [7] conversion factor
of 2.45 x 10^{17} O atoms/cm^2 per unit peak absorption coefficient, indicate

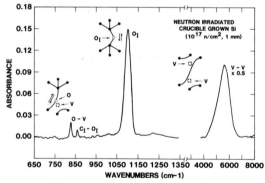

Figure 1. Illustration of
infrared absorption bands
for interstitial oxygen, O_I,
and trapped-vacancy (oxygen
vacancy, O-V, and divacancy,
V-V) defects in neutron-
irradiated crucible-grown
(8.5 x 10^{17} O/cm^3) Si.

8.5 x 10^{17} cm^{-3} dispersed O_I. The small peak near 1230 cm^{-1} indicates an
additional concentration of precipitated O. Neutron irradiation produces
vacancies which are trapped to induce the 829 and 5550 cm^{-1} bands assigned
to the O-V and V-V defects, respectively. Models for these defects and O_I
are sketched adjacent to the respective absorption bands. The neutron-induced
862 cm^{-1} band has been assigned [8] to interstitial carbon-O_I (C_I-O_I) defects and indicates carbon, in addition to O_I, is present in the crucible
grown Si. The small absorption band near 935 cm^{-1} may also be neutron induced

since absorption at this wavelength has been ascribed [9] to interstitial Si trapped by O_I.

RESULTS AND DISCUSSION

The present investigation is concerned primarily with O_I, O-V and V-V defects in oxygen-implanted and Q-switched laser annealed Si. Absorption bands for O-V, O_I and V-V defects produced by implantation of 3 x 10^{15} 220 keV ^{16}O/cm^2 into two surfaces of (111) Si are shown by the solid line in Fig. 2. Assuming the ASTM conversion factor and an implantation depth of 5500 A gives an average dispersed O_I of 2 x 10^{19} cm^{-3} which is approximately a factor of 10 greater than the equilibrium saturation limit for O_I in Si. If the same conversion factor is applied to the O-V band, then a total of 3.6 x 10^{15} O atoms/cm^2 is obtained compared to a total implantation fluence of 6 x 10^{15}/cm^2. The efficiency of O incorporation into these defects may be even higher than inferred from these relative numbers since the value of the conversion factor has been a subject of discussion for many years, and a factor as large as 4.81 x 10^{17}/cm^{-2} per unit absorption coefficient has been reported [7]. Divacancies are produced by the O-ion implantation and the band intensity, based upon a previous calibration, indicates ~ 3 x 10^{15}V-V/cm^2.

The dashed line in Fig. 2 shows the absorption measured after ruby laser annealing the O-implanted layers with 2.9 J/cm^2. This energy is sufficient to melt through the depth of the O-implanted layer. The data show that the laser annealed layers are free from trapped vacancies detectable by either O-V or V-V absorption bands.

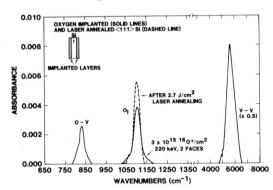

Figure 2. Infrared spectra for ^{16}O-implanted oxygen-free Si (solid line) showing interstitial oxygen, oxygen-vacancy, and divacancy defects, and for the same sample after laser annealing. The laser annealed layers (2 layers) are free of resolved trapped-vacancy defects.

An O_I band intensity indicative of a concentration significantly above the equilibrium saturation limit, and even increased above that for the implanted layer, is quenched into the laser annealed layers. This result is in contrast to other studies at higher implant fluences where oxide formation was observed in laser annealed oxygen-implanted layers [10].

To assure that we are measuring implanted oxygen rather than O from some other source such as O_I in the substrate, O incorporated from a thin oxide layer, or O from the ambient [11], an implantation was performed with the ^{18}O isotope. The spectra after implantation and after laser annealing are shown in Fig. 3. The spectra shown are for two stacked samples, each of which was implanted with 2 x 10^{15} 200 keV ^{18}O/cm^2 into two faces. The isotopically shifted O_I and O-V bands appear at 1052 and 793 cm^{-1}, respectively, as expected for the ^{18}O isotope in these defects. The divacancy band is also produced by the implantation. After laser annealing with 2.7 J/cm^2, the spectrum shows no evidence for trapped vacancies consistent with the laser-annealed ^{16}O implanted Si data. The isotopically shifted O_I band after laser annealing confirms that implanted oxygen is retained under laser annealing. The spectra are not shown, but we have noted that absorption changes in the 600 cm^{-1} Si multiphonon absorption region of the spectrum are produced by implantation and laser annealing. Additional work is necessary to determine if the changes are caused by oxygen which is known to increase the lattice parameter, or if the spectral changes are due in part to carbon introduced during implantation and annealing.

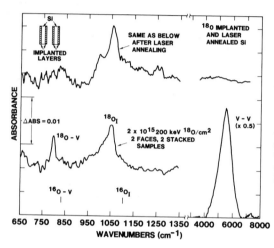

Figure 3. Infrared spectra before and after laser annealing of 4 layers of ^{18}O implantation of oxygen free Si. The spectra are shifted because of the difference in isotopic mass relative to ^{16}O in Si. These data confirm retention of implanted oxygen under laser annealing.

Since the O_I concentration was increased by laser annealing, and because the possibility of observing trapped vacancies increases with oxygen concentration, experiments were performed at larger O_I concentrations. These experiments utilized two-energy implants and reduced ion energies. With the reduced implant energy, a lower laser exposure could be used to obtain a higher regrowth velocity which may further enhance oxygen trapping. Data obtained after implanting 2 x $10^{15}/cm^2$ 50 and 100 keV O ions/cm^2 into (100) Si are shown by the lower trace in Fig. 4. The intensities of the 1106 and 829 cm^{-1} bands indicate 7.9 x $10^{15}/cm^2$ O_I and 4.5 x $10^{15}/cm^2$ O-V defects compared to a total implantation fluence of 1.6 x $10^{16}/cm^2$. The implanted O is still efficiently incorporated into crystalline Si at these higher concentrations; however, the divacancy band is distorted and has a smaller intensity relative to the O_I band than in Fig. 2 indicating a high level of disorder is present in the implanted layer. The broad feature in the absorption curve near 950 cm^{-1} is probably associated in part with interference effects caused by an implantation-induced increase in the refractive index.

The upper spectrum in Fig. 4 shows the absorption for the implanted samples after ruby laser annealing at 1.88 J/cm^2. The laser-annealed layers

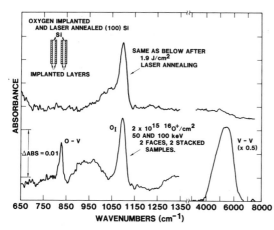

Figure 4. Infrared spectra before and after laser annealing of 4 layers of high fluence ^{16}O-implanted oxygen-free Si. Interstitial oxygen concentration is 1.4×10^{20} cm^{-3} with no resolved trapped vacancies in the laser-annealed layers.

are still free from detectable trapped vacancies even for the higher O_I concentrations. Although the O-V band has been annealed out, the intensity of the O_I band at 1106 cm^{-1} remains essentially unchanged by the laser annealing. Consequently, the estimated average O_I concentration of 1.4×10^{20}/cm^{-3} obtained from the intensity of the 1106 cm^{-1} band is quite likely the limit of O_I incorporation under the conditions of these experiments. A weak, broad, absorption near 1225 cm^{-1} is indicative of oxygen precipitation in crystalline Si and the increased absorbance in the 1000 cm^{-1} region observed after laser annealing may signify the onset of oxide formation.

SUMMARY AND CONCLUSIONS

Silicon layers were implanted with oxygen to increase the probability for vacancy-trapping by oxygen and by other vacancies. Trapping of implanted O is shown to occur for Si layers regrown from the melt. Implanted O atoms are effectively incorporated into dispersed O_I sites in laser-annealed layers to average concentrations of $\sim 10^{20}$/cm^3, which is nearly two orders of magnitude higher than the equilibrium saturation limit of O at the melting temperature of Si. Laser-annealed oxygen-implanted layers are found to be free from trapped vacancies within the detection limits for O-V and V-V defects by infrared absorption. These studies place an upper limit on trapped single vacancies in laser-annealed Si of 5×10^{18} V/cm^3. While there have not been any direct measurements of vacancy concentrations at the melting temperature of Si, an estimate of 2.5×10^{18} cm^{-3} has been made from quenching experiments where donor formation in float zone Si was assumed to be a measure of the vacancy concentration.[12] On the other hand, thermodynamic considerations utilizing the 0.33 eV migration energy for vacancies determined from irradiation damage studies predicts 10^{15} vacancies/cm^3 at the melting temperature.[12]

234

1. See for example, G. Foti in <u>Laser and Electron Beam Processing of Materials</u>, ed. by C. W. White and P. S. Peercy, Academic Press, 168 (1980).

2. L. C. Kimerling and J. L. Benton, ibid, p. 385.

3. M. Servidori, A. Zani and G. Garulli, Phys. Stat. Sol. (a) <u>70</u>, 691 (1982).

4. For a review see, H. J. Stein, <u>Radiation Effects in Semiconductors</u>, ed. by J. W. Corbett and G. D. Watkins, Gordon and Breach Science Publishers, 125 (1971).

5. Y. Yatsurugi, N. Akiyama and Y. Endo, J. Electrochem. Soc. <u>120</u>, 975 (1973).

6. H. J. Stein and W. Beezhold, Appl. Phys. Let. <u>17</u>, 442 (1970).

7. ASTM F121-80, Annual Book of ASTM Standards.

8. R. C. Newman and A. R. Bean, <u>Radiation Effects in Semiconductors</u>, ed. by J. W. Corbett and G. D. Watkins, Gordon and Breach Science Publishers, 155 (1971).

9. A. Brelot, <u>Radiation Damage and Defects in Semiconductors</u>, ed. by J. E. Whitehouse, Institute of Physics, London, 191 (1972).

10. S. W. Chiang, Y. S. Liu and R. F. Reihl, Appl. Phys. Let. <u>39</u>(9), 752 (1981).

11. Koichiro Hoh, Hiroshi Koyama, Keiichiro Uda and Yoshio Miura, Japanese J. of Appl. Phys. <u>19</u>, L375 (1980).

12. See review by A. Seeger and K. P. Chik, Phys. Sta. Sol. <u>29</u>, 455 (1968).

MICROPROBE RAMAN ANALYSIS OF PICOSECOND LASER ANNEALING OF IMPLANTED SILICON

J. SAPRIEL, Y.I. NISSIM and J.L. OUDAR
Centre National d'Etudes des Télécommunications*, 196 rue de Paris 92220
BAGNEUX - (FRANCE)

ABSTRACT

Picosecond laser annealing has been performed on implantation-
amorphised silicon. A multiannular (up to five rings) recrystallization pattern has been generated by a single~30 psec pulse at
1.06 μm and 0.53 μm wavelength and energy density just below the
damage threshold. The different patterns have been investigated by
scanning the surface with a Raman microprobe with a 1μm spacial resolution. Information are thus given on the different phases (amorphous or crystalline) and on lateral as well as in depth dimensions
of the different rings.

INTRODUCTION

The study of picosecond laser annealing of implanted silicon and the use of
Raman scattering as a versatile and non destructive tool for the characterization of the recrystallization pattern is reported here. A rather complete set
of experiments are presented on silicon samples, at different annealing energies and wavelengths. Each pattern resulting from single pulse annealing is
analysed parallel and perpendicular to the free surface.

When a pulsed beam is used to anneal implanted silicon, it is likely that
melting of a surface layer occurs during irradiation, followed by liquid
phase epitaxial regrowth. The resolidification time is of the order of tens of
nanoseconds. The velocity of the solid-liquid interface is thus slow enough for
the lattice to reconstruct. However disordered material can be formed when this
velocity is increased above the value necessary for epitaxial regrowth to take
place. Such high resolidification rates occur in single crystal silicon for
extremely high absorption (U.V. photons in the nanosecond regime [1]) or with
extremely short pulses (picosecond irradiation [2])resulting in amorphous silicon formation. On ion implanted amorphized silicon similar experiments led to
an annular recrystallization pattern [3,4] formed with single crystal, amorphous and polycrystalline materials.

SAMPLE PREPARATION AND THE PICOSECOND ANNEALING

Single crystal silicon (100) oriented, were implanted with As$^+$ ions at a
dose of $1\times10^{15}/cm^2$ and at an energy of 100 keV creating à 1000 Å thick amorphous layer. The annealing was carried out with a mode locked Nd-YAG laser
described in ref [5]. Single pulses of average duration 27 psec at 1.06μm and
20 psec at 0.53 μm can be extracted from this system. The sample was placed at

---

* Laboratoire associé au CNRS (LA250)

the focal point of a one meter focal length lens. The laser was operated at a
low repetition rate in combination with a shutter, so that the sample could be
translated between each pulse. Each pulse focused through the lens had a gaus-
sien spatial and temporal profile. The spatial profile is visualized on an
array of photodiodes in Fig.1.

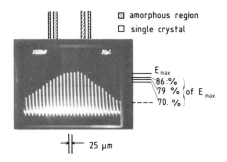

amorphous region

single crystal

E_{max}
86 :%
79 % } of E_{max}
70. %

25 µm

Fig.1 : Spatial distribution of the pi
cosecond laser beam. The energies corresponding to the crystalline-
amorphous transitions are indicated.

The annealing has been first performed by irradiation at 1.06 µm. The incident energies were set to be just below the laser induced damage threshold
($2J/cm^2$) and half of this value ($1J/cm^2$). Under a conventional Nomarski optical
microscope up to five different annular region could be observed (fig.3a) at
$2J/cm^2$. The formation of this pattern is controlled by the incident energy
level. A decrease in energy is responsible for a change from a halo to a dot
pattern of the different areas (fig4a for $E_0=1J/cm^2$). Finally a similar experiment was carried out at 0.53 µm wavelength just below the limit of substrate
deterioration ($E_0 \simeq 1.4J/cm^2$). The results are presented and compared in the
following.

THE RAMAN SCATTERING AND THE SPATIAL ANALYSIS OF THE ANNEALED PATTERN

Raman scattering has been used previously to study nanosecond pulsed annealing of implanted silicon [6]. Single crystal silicon can be easily distingui-
shed from polycrystalline or amorphous silicon as the corresponding Raman spectra are quite different. Crystalline silicon exhibits one first order Raman-
active peak (symmetry F_{2g}) whose frequency is 521 cm^{-1} and whose full width at
half maximum is about 3 cm^{-1} at room temperature. The presence of strain or
microcrystallinity produces a frequency shift. Moreover, microcrystallinity
broadens the Raman peak which becomes asymmetric. The amorphous phase is characterized by very broad, low-intensity Raman bands peaking around 160 and 480
cm^{-1}. It is worthwhile mentioning that we observed single crystal and amorphous
silicon in the annealed spot but no appearance of polycrystalline silicon was
detected in the Raman spectra. Moreover the single crystal which results from
the annealing was stress free in the limit of detection of our experimental set
up (strain of the order of $0.5.10^{-3}$). The scattering efficiency of a crystal
is strongly dependent on the orientation of the sample relative to the incident

and diffracted light polarization $\vec{e}_i$ and $\vec{e}_s$. Our Raman measurements were performed in backscattering arrangement along $[001]$, the direction perpendicular to the crystalline silicon substrate. For $\vec{e}_i$ along $[100]$, the diffracted light is polarized along $[010]$ and for $\vec{e}_i$ along $[110]$ $\vec{e}_s$ is parallel to $\vec{e}_i$. A conventional experimental set up $[7]$ was first used to perform precise polarized Raman measurements which showed that the recrystallization obtained by annealing was oriented in the same crystallographic directions as the substrate. The exciting laser lines were 5145Å and 4880Å (Ar$^+$ ion laser) and the laser beam was focused on a diameter of 15 μm. The spatial resolution was subsequently improved to 1μm by the use of a Raman microprobe described in Fig.2. The incident beam of the microprobe was used to scan the diameter of the whole recrystallization pattern which can be considered as a film of variable thickness x on a crystalline substrate. All our measurements are based on the reduced Raman intensity r of the 521 cm^{-1} peak ; r is ratio of this peak intensity in the annealed area to a single crystal silicon taken as a reference (r = Iannealed/-Icrystal).

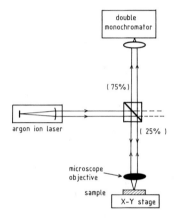

Fig.2 : Schematic diagram of the Raman microprobe. The transmission of the double prism is 75 %.

The value of r is given in Fig.3-b, and 4-b as a function of d, the distance between the center and the investigated point of the recrystallization pattern. The expression of r is determined elsewhere $[5]$. It is worthwile pointing out that assuming that the surface is not damaged by the annealing, the value of r depends only on the absorption coefficient α $[8]$ and the thickness x of the amorphous layer :

$$r = \exp - 2\alpha\, x$$

Thus this value can give the thickness of the amorphous layer. In Table I is collected the information about the phase (amorphous or crystalline) the thickness x of the amorphous layer perpendicular to the surface, the diameter d of the central region, the lateral thicknesses d of the rings which appear on the patterns of Fig. 3a, 4a (λ = 1.06μm) and the pattern obtained at λ=0.53μm.

238

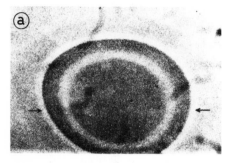

Fig.3 : Picosecond laser annealed spot of ion implanted amorphized silicon for $E=2J/cm^2$ and $\lambda=1.06\mu m$:
a) Nomarski optical micrograph
b) Spatial evolution of the reduced Raman intensity r of the $521cm^{-1}$ peak obtained with the microprobe.

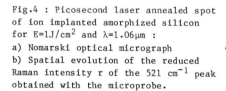

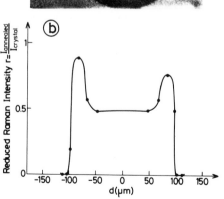

Fig.4 : Picosecond laser annealed spot of ion implanted amorphized silicon for $E=1J/cm^2$ and $\lambda=1.06\mu m$:
a) Nomarski optical micrograph
b) Spatial evolution of the reduced Raman intensity r of the $521 \ cm^{-1}$ peak obtained with the microprobe.

Table I : Analysis of the different regions of the recrystallization pattern obtained by picosecond pulse annealing. The thickness of the amorphous rings are calculated with $\alpha = 3\times10^5$ cm^{-1} [8].

	$\lambda = 1.06\,\mu m$ $E \simeq 2J\,/\,cm^2$	$\lambda = 1.06\,\mu m$ $E \simeq 1J\,/\,cm^2$	$\lambda = 0.53\mu m$ $E \simeq 1.4\,J\,/\,cm^2$
CENTRAL REGION	stress free single crystal $d \simeq 180\,\mu m$	amorphous $x \simeq 120$ A $d \simeq 130\,\mu m$	stress free single crystal $d \simeq 130\,\mu m$
FIRST RING	amorphous $x \simeq 100$ Å $d \simeq 30\,\mu m$	stress free single crystal + amorphous layer ($x \simeq 30$ Å) $d \simeq 30\,\mu m$	amorphous $x \simeq 100$ Å $d \simeq 30\,\mu m$
SECOND RING	stress free single crystal $d \simeq 25\,\mu m$	amorphous $x > 330$ Å $d \simeq 25\,\mu m$	stress free single crystal $d \simeq 7\,\mu m$
THIRD RING	amorphous $x > 330$ Å $d \simeq 20\,\mu m$		amorphous $x > 330$ Å $d \simeq 20\,\mu m$
FOURTH RING	not identified		

DISCUSSION AND CONCLUSION

With the help of the Raman microprobe we have been able to obtain a lateral as well as in depth picture of the multiannular pattern of picosecond laser annealed implanted silicon. This pattern varies as a function of the annealing experimental conditions (energy and wavelength), but in all cases the transitions between crystalline and amorphous regions appear rather sharp. At least two amorphous rings have been observed with the thinner one beeing the centermost. The supercooling of the edge of the molten layer and energy density consideration [9] can explain, at most, the formation of one amorphous ring but fail to account for the multiannular pattern described in this paper. It is likely that fluctuations occur during the solidification process around the threshold of 18 m/s which is considered [9] as the critical velocity for crystalline material formation. The latent heat of crystallization could interfere with the epitaxial process from the liquid-solid interface up to the surface, thus imposing a lateral variation in the resolidification speed.

ACKNOWLEDGEMENT

The authors would like to thank the Grenoble CNET for carrying the implantations and the Yvon Jobin company for kindly letting then use their Raman microprobe.

240

1. R. Tsu, R.T. Hodgson, T.Y. Tan and J.E. Baglin, Phys. Rev. Lett 42, 1356
(1979)

2. P.L. Liu, R. Yen, N. Bloembergen and R.T. Hodgson, Appl. Phys. Lett. 34,
864 (1979)

3. P.L. Liu, R. Yen, N. Blombergen and R.T. Hodgson in "Laser and Electron
Beam Processing of Materials", C.W. white and P.S. Peercy eds.(Academic
Press New York 1980) p.156

4. G.A. Rozgonyi, H. Baumgart, F. Phillip, R. Uebbing and H. Oppolzer in
"Laser and Electron Beam Interaction with Solids" B.R. Appleton and
G.K. Celler eds.(North Holland, New-York 1982) p.177

5. Y.I. Nissim, J. Sapriel and J.L. Oudar, Submitted to Appl. Phys.Lett.

6. J.F. Morhange, G. Kannelis and M. Balkanski, Solid St Comm. 31, 805 (1979)

7. B. Jusserand and J. Sapriel, Phys. Rev. B,24, 7194 (1981)

8. M.H. Brodsky, R.S. Title, K. Weiser and G.D. Pettit, Phys. Rev. B, 6, 2632
(1970)

9. A.G. Cullis, H.C. Webber, N.G. Chew, J.M. Poate and P. Baeri, Phys. Rev.
Lett. 49, 219 (1982)

ANALYSIS OF TEMPERATURE AND STRESS PROFILES INDUCED BY A CW LINE SCANNED
ELECTRON BEAM IN <100> ORIENTED SILICON

G. G. BENTINI AND L. CORRERA
C.N.R. - ISTITUTO LAMEL, Via Castagnoli 1, 40126 Bologna, Italy

ABSTRACT

   Evaluation of thermal profiles along the scan direction of a line shaped e-beam as well as the topographic
distribution of the thermally induced stresses have been
performed by solving the heat diffusion equation for several incident power values. The resulting stresses have
been computed in the "nearly isotropic" approximation but
taking into account that the slip planes, in <100> oriented silicon crystals, are {111} and the slip directions in
the plane are <110>.
   The threshold of damage introduction has been evaluated by comparing the computed stresses with the yield
stress of the material at any annealing temperature.
Experimental observations based on X-Ray topography have
been performed in order to study the damage introduction
on <100> silicon samples annealed in the same conditions
used for stress computation. A very good agreement between
computed and experimental results was found.

INTRODUCTION

   There is a growing interest in the use of a scanning e-beam source to
anneal wide areas in semiconductors [1,2,3] working in solid phase regime. In
fact the recrystallization of thin films as well as the silicide formation or
the recovery of damage consequent to Ion Implantation, can successfully be
achieved. However, severe stresses can be generated in the sample, due to the
thermal gradient induced by the scanning beam; if these stresses become greater
than the elastic limit a severe damage is introduced.
   Aim of this work is to evaluate both the temperature profile and the
topographic distribution of stresses in <100> silicon samples as a function of
the power dissipated by the beam in the case of a line shaped source scanned
across the sample.
   The ratio between the computed stresses and the yield stress of the
material has been evaluated in order to investigate the threshold of damage
introduction and its topographic distribution as a function of the irradiation
conditions.
   The experimental observations of defects introduction as a function of the
beam power have been performed by using a linear energy source obtained by fast
multiscanning an e-beam along a line; the sample was mechanically driven
through the scanned area.

TEMPERATURE PROFILE

   The temperature profile computation has been performed in the geometry
shown in fig.1, and considering a line source oriented along the y direction
and gaussian along the x direction. In order to match the available experimental
conditions, the scan speed and the gaussian width were chosen to be 1 cm/sec
and 0.8 cm respectively.

Mat. Res. Soc. Symp. Proc. Vol. 13 (1983) ©Elsevier Science Publishing Co., Inc.

Under these conditions the dwell time is long enough to obtain a heat diffusion length much longer than the wafer thickness, and a volume heat source uniform in y and z (shadowed area in fig.1) running along the x direction can be considered. In this case the temperature gradient is effective only along the x direction and the one-dimensional heat diffusion approximation can be used.

$$C\varrho \frac{\partial T(x,t)}{\partial t} - \frac{\partial}{\partial x} (K \frac{\partial T(x,t)}{\partial x}) = -Q(x,t) - R(T) \qquad (1)$$

$$Q(x,t)=(I_o/d) \exp(-(x-vt)^2 \ell^2); \quad R(T)=2 \varepsilon' \sigma'(T^4-T_o^4)/d$$

Were: Q(x,t)= strip source term   R(T)= radiative losses term
        K = thermal conductivity   v= translation speed
        2b= sample width   1/ℓ= gaussian half-width
        I_o = peak power density   T = room temperature
        P= beam total power ε'= silicon emissivity
        T= temperature   L= sample length
        C= specific heat   d= sample thickness
        σ'= Stephan constant   ϱ= density

The temperature dependence of thermal conductivity K can be taken into account by introducing the "linear" temperature ϑ defined by:

$$\vartheta = 1/Ko \int_0^T K(T) dT' \qquad (2)$$

and transforming eq.(1) in a linear equation as described in ref.4.

Once the ϑ temperature is computed, the non linear temperature T can be obtained by inversion of eq.(2) using for K(T) an exponential expression [5].

Fig. 2 shows the asymmetric temperature gradient along x axis as a function of power when the beam is positioned at the center of the sample and v=1 cm/sec.

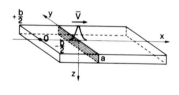

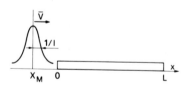

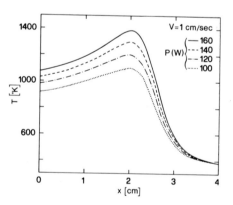

Fig. 1. Geometry of the sample and cross-section of the line-shaped source.

Fig. 2. Temperature profiles as a function of power when the beam is at the center of the sample.

THERMAL STRESS ANALYSIS

Once the temperature profile is known, the thermal stress analysis can be carried out in the two-dimensional case (i.e., supposing the sample thickness much lower than the other dimensions) by using the relationships [6] :

$$\frac{\sigma_{xx}}{\alpha\,E} = -\frac{1}{6}\left(3y^2-\left(\frac{b}{2}\right)^2\right)\frac{d^2T}{dx^2} + \frac{1}{180}\left(15y^4-30y^2\left(\frac{b}{2}\right)^2+7\left(\frac{b}{2}\right)^4\right)\frac{d^4T}{dx^4} + \ldots$$

$$\text{and}\ :\ \sigma_{yy} = f\left(y,b,\frac{d^4T}{dx^4},\frac{d^6T}{dx^6}\right)\ \ ;\ \ \sigma_{xy} = f\left(y,b,\frac{d^3T}{dx^3},\frac{d^5T}{dx^5}\right)$$

(3)

This analysis is strictly valid in the case of isotropic bodies; dealing with crystalline solids both the Young's modulus E and the thermal expansion coefficient α are functions of crystallographic directions. So, generally, the α coefficient is replaced by a tensor $\alpha_{i,k}$ of rank two, whereas the Lamé coefficients λ and μ are repleaced by the components of a tensor (elastic modulus tensor) $\lambda_{i,k,l,m}$ of rank four [7] .

In the case of cubic crystals, the number of independent components of $\alpha_{i,k}$ is 1; this means that in this respect they behave as isotropic bodies, whereas the independent components of elastic modulus tensor are reduced to 3,

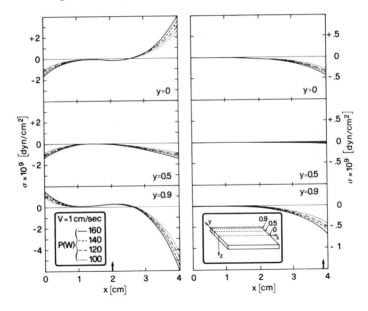

Fig. 3. Values of the thermal stresses as a function of the x coordinate, for three different sections along y and two beam positions (indicated by arrows). The stress distribution is symmetrical with respect to the longitude axis of the sample (y=0).

e.g., λ_{xxxx}, λ_{xxyy}, and λ_{xyxy}. In correspondence to these three elastic moduli there are three Young moduli values, i.e., E<100>, E<111>, E<110>; in silicon crystal, having diamond structure, the values at 300°K are E<100>= 1.3 10^{12} dyn/cm², E<110>= 1.7 10^{12} dyn/cm², E<111>= 1.9 10^{12} dyn/cm² [8].

These values are slightly dependent on temperature; by assuming an average modulus for the three crystalligraphic directions, one gets an approximation better than 20%.

Inside this approximation silicon crystals can be considered as "nearly isotropic" and eqts.(3) can be taken into account by resolving the stresses on the <110> slip directions laying on the {111} planes.

The values of the stress computed as a function of the power, for three different sections of the sample along the y direction, when the e-beam is at the center and at the end of the sample, are shown in fig.3; positive values correspond to tensile stresses, whereas negative values correspond to compressive stresses.

It can be observed that there are relatively small differences among the stress values induced by different beam powers; moreover the temperature reaches the maximum value at the edge of the sample which is irradiated last, so it can be expected that defects will be introduced mainly in this region.

The introduction of defects has been investigated by comparing the computed stresses with the yield stress σ_E of silicon evaluated by using the relationship [9] :

$$\sigma_E(T, \dot{\varepsilon}) = H \ (\dot{\varepsilon})^{1/n} \ \exp(U/nkT)$$

Were: H= 4.4 10^5 dyn cm^{-2} sec$^{1/n}$; n= 2.1; U= 2.3 eV
$\dot{\varepsilon}$ = strain rate (sec^{-1}); k= Boltzmann constant

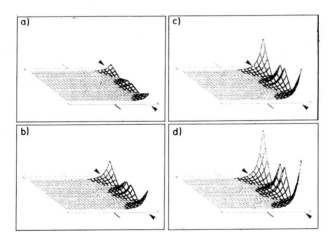

Fig. 4. Map of the ratio R between the thermal stress and the yield stress, at a scan speed of 1cm/sec (see text.). Arrows indicate the beam position.

Fig.4 (a,b,c,d) reports the map of the ratio R between the induced stresses and the yield stress as a function of the dissipated power, when the beam position is near the outer edge of the sample, i.e. the position where the temperature reaches its maximum. The dark regions indicate where R > 1, that is, where the defects are introduced as the elastic limits are exceeded.

It can be seen that in spite of the relatively small differences among the stresses induced at different beam power values, there is a strong effect on the ratio σ / σ_E; this is due to the exponential dependance of the yield stress on the maximum temperature reached by the sample during the cw annealing. For the beam power values investigated, the difference among the temperature maxima is of the order of 300°K (see fig.2).

EXPERIMENTAL

An experimental observation of defects introduced by line scan annealing has been performed by using an e-beam fast multiscanning along a line, already described [2].

Silicon samples <100> oriented have been irradiated with beam powers ranging from 100 W to about 160 W, whereas the translation speed across the scanned region was kept v= 1 cm/sec.

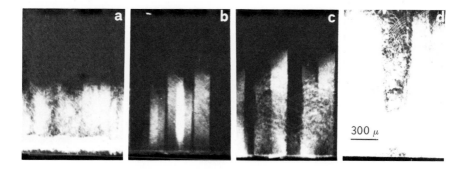

Fig. 5. X-Ray topographs showing the introduction of defects at the hottest edge of the sample. Translation speed v=1 cm/sec. Beam power:
a) 100 W; b) 120 W; c) 140 W; d) 160 W.

X-Ray topographs performed on the samples are reported in fig.4, confirming that in these conditions the introduction of defects starts at the edge of the sample which is irradiated last and that the damaged region extends deeper and deeper inside the sample as a function of the power dissipated by the e-beam.

REFERENCES

1. R. A. Mc Mahon, H. Hamed, Electronic Lett. 15, 45 (1979).

2. G. G. Bentini, R. Galloni, R. Nipoti, Appl. Phys. Lett. 36, 661 (1980).

3. J. A. Knapp, S. T. Picraux, J. Appl. Phys. 53, 1492 (1982).

4. H. S. Carslaw, J. C. Jaeger in: Heat Conduction in Solids (Oxford Univ. Press, London 1971).

5. G. G. Bentini, L. Correra, J. Appl. Phys., in press.

6. B. A. Boley, J. H. Weiner, Theory of Thermal Stresses (John Wiley & Sons New York 1960) pp. 320-323.

246

7. L. D. Landau, E. M. Lifshitz, Theory of Elasticity (Pergamon Press 1959)
    pp. 40-42.

8. Yu. A. Burenkov, S. P. Nikanorov, A. V. Stepanov, Sov. Phys. - Solid State
    13, 2516 (1972).

9. H. Siethoff, P. Haasen in: Lattice Defects in Semiconductors, R. R.
    Hasiguti Ed. (University Tokio Press 1968) p.491.

PART III
RAPID CRYSTALLIZATION, IMPURITY INCORPORATION, AND METASTABLE SURFACES

ISING MODEL SIMULATIONS OF IMPURITY TRAPPING IN SILICON

George H. Gilmer

Bell Laboratories
Murray Hill, New Jersey 07974

ABSTRACT

Laser annealing experiments on silicon have shown that rapid solidification can trap large amounts of certain impurities in the crystal lattice. Concentrations that exceed the equilibrium solubility limits by several orders of magnitude have been obtained. In this paper we discuss the impurity trapping process using Monte Carlo simulation data from the kinetic Ising model. The dependence of the impurity concentration in the crystal on the solidification rate is calculated. The simulation data are compared with recent laser annealing results for bismuth and indium. Excellent agreement between the model and the bismuth experiments is obtained. The larger trapping rate on the (111) relative to the (100) orientation is found to be caused by the slower crystallization kinetics on the (111) face. Similar results are obtained for indium, although the difference in trapping on the (111) and (100) faces is somewhat smaller in the model than in the experiment.

I. INTRODUCTION

Recently the trapping of impurities by rapid solidification has been studied by pulsed laser annealing of silicon[1]. Arsenic, gallium, indium, bismuth and other elements are incorporated in substitutional sites at concentrations that exceed the solid solubility limits[2]. Because the concentration in the solid deviates appreciably from the equilibrium value, these experiments can provide information on the kinetics of crystallization. The structure of the crystal-melt interface and the transition rates between liquid and crystalline configurations will certainly influence the trapping of impurities. In order to examine these effects in detail, we have developed an atomistic model of a diamond cubic (DC) crystal-fluid interface oriented along the (111), (110), and (100) directions. Studies of the trapping process in this model yield values of the parameters that provide a fit to the experimental data on silicon. The effects of different rate constants on the trapping process can be evaluated in a straightforward manner, since the simulation model permits detailed changes in the rates and interaction energies of the atoms at the interface.

A large number of models of trapping have been examined[3-15]. The purpose of many of these studies was to explain inpurity inhomogeneities in crystals grown from the melt by conventional methods[16]. In the case of laser-annealed silicon, diffusion in the liquid phase can be calculated numerically, and the kinetic distribution coefficient between the crystal and the adjacent liquid can therefore be inferred from the depth distribution of the impurity after crystallization. It is this interfacial distribution coefficient that is to be derived from the models. Explanations of trapping based on the idea that the impurity diffusing in the liquid phase is overtaken by the rapidly moving interface are not valid in this case. This effect is included in the calculation of the distribution coefficient. Instead, trapping is determined by the rates of crystallization and melting of atoms at the interface. Rate-theory models and those that employ the one-dimensional motion of a kink site along the interface tend to predict less trapping than that observed experimentally. Several modifications of these models have been suggested recently. Jackson et. al.[12] obtained agreement with experiment by including a "trapping" term that accounts for the immuring of small clusters of material at the liquid composition in the growing crystal. A similar idea postulates the initial crystallization of each layer of atoms at the liquid composition, and subsequent redistribution of material by diffusion[15]. Another approach is to assume a specific dependence of the rate constants on the interface speed[13]. A possible conclusion from these studies is that the mobility of the impurities is somehow reduced before their potential energy increases to the bulk crystal value. This contrasts with the classical rate theory model[5], in which the full potential energy is achieved in one step when the impurity crystallizes. However, the exact mechanism may not be accessible to such simplified models. Cooperative interactions between adjacent sites may play an important role, since the lateral bonds will tend to stabilize the impurity until a new layer crystallizes.

Several details relating to the interaction of the impurity with atoms in the interfacial region may influence trapping. Segregation of the impurity at the interface may have a large effect[17]. The concentration in this region may be considerably higher than in either bulk phase. Since this is the region where crystallization occurs, the chance of trapping impurities is increased proportionally. Also, atomic size may be important. A large impurity can be accomodated at the interface by relaxation into the fluid. The full potential energy will be attained only after additional layers are deposited and the impurity is constrained to fit into the host lattice.

Particularly revealing are the recent experiments on different crystallographic orientations by Baeri et. al.[18] and Poate[19]. Large differences in trapping on the (111) and (100) interfaces were measured for bismuth and indium, with the larger concentration of impurity resulting from (111) growth. The (111) face is a singular orientation and exhibits facets during Czochralski growth, whereas all other orientations are rounded[20]. Comparisons between model and experiment will provide insight into the different mechanisms of crystal growth on these "rough" and "smooth" faces[21] and the consequences for trapping. For this purpose it is essential that the model contain sufficient detail to represent the different surface structures.

Section II contains a description of the crystal-melt interface model, and the relations that give the effective bond energies in the equivalent crystal-vapor interface. Kinetic rate constants consistent with the equilibrium statistical mechanics are also discussed. Section III is a treatment of the growth of the (111), (110) and (100) faces of pure silicon crystals by Monte Carlo simulations of the model. Section IV contains simulation data for the distribution coefficients during rapid crystal growth. Various factors that may influence the trapping of impurities are evaluated. Conclusions are contained in Section V.

II. ISING MODEL OF THE CRYSTAL-MELT INTERFACE

Lattice models of a crystal and its melt provide a good representation of the equilibrium phase diagrams of many systems. Thurmond et. al.[22] have used the mean-field approximation of this model to match the phase diagrams of some of the alloys of silicon. These models are also relatively simple and provide a good basis for simulation studies of trapping.

Although these calculations required only bulk liquid and crystalline phases, we we are primarily interested in the interface between the phases. Since the structure of this interface plays a crucial role in trapping, we have chosen to represent it in atomistic detail, using the Monte Carlo method to simulate the "crystallization" and "melting" of atoms in this region. Our model is equivalent to the spin-1 Ising model; the lattice sites may be in one of three states: crystalline type A, crystalline type B, or liquid. The occupancy of every site in the crystalline region is specified, whereas the liquid is represented by mean field theory. The model excludes a gradual transition from the liquid to the crystalline state, although a layer of atoms in the vicinity of the interface may have properties intermediate between the two bulk phases simply because it contains a mixture of liquid and crystalline sites. Recent molecular dynamics studies provide evidence in support of a sharp distinction between the liquid and crystalline states; trajectory plots indicate that liquid and crystalline regions coexist within a single layer of atoms[23]. Since the Lennard-Jones crystal used in these studies has a relatively small entropy of fusion, one would expect an even more abrupt transition from crystal to liquid in the silicon system.

Mean-field theory can be used to calculate an approximate equilibrium distribution coefficient between the crystal and its melt. Although our model of the crystal gives results that are exact (in principle), the mean-field approximation is quite accurate for the small concentrations of B atoms that are treated. The mean-field canonical partition function for one of the bulk phases is[24]

$$Q = (q_A)^{N_A}(q_B)^{N_B}\frac{(N_A+N_B)!}{N_A!N_B!}\exp[\frac{z\,(N_A^2\phi_{AA}+2N_AN_B\phi_{AB}+N_B^2\phi_{BB})}{2kT\,(N_A+N_B)}],\tag{1}$$

where q_A and q_B are the individual partition functions of A and B atoms at their sites (excluding interactions with neighbors), N_A and N_B are the numbers of A and B atoms, z is the coordination number of the lattice, kT is the product of Boltzmann's constant times the temperature, and ϕ_{ij} is the energy of the bond between a nearest neighbor pair of atoms of types i and j. The chemical potential of the A species is

$$\mu_A = -\frac{\partial \ln Q}{\partial N_A}\Big|_{T,N_B} = -\ln q_A + \ln C_A - \frac{z}{2kT}(\phi_{AA}+C_B^2w),\tag{2}$$

where $w = 2\phi_{AB} - \phi_{AA} - \phi_{BB}$, and the Stirling approximation has been used to simplify the factorial terms. Equating the chemical potentials in the two bulk phases, we obtain

$$\frac{C_A^C}{C_A^L} = \frac{q_A^C}{q_A^L}\exp\{\frac{z}{2kT}[\phi_{AA}^C+(C_B^C)^2w^C-\phi_{AA}^L-(C_B^L)^2w^L]\},\tag{3}$$

and in the case where $C_A^C \approx C_A^L \approx 1$, the temperature is approximately equal to the melting point T_m of the pure A crystal

$$\frac{q_A^C}{q_A^L}\approx\exp\{-\frac{z}{2kT_m}(\phi_{AA}^C-\phi_{AA}^L)\}.\tag{4}$$

The concentrations of the B species are given by an equation analogous to (3), and using (4) the distribution coefficient at T_m is

$$K^e \equiv \frac{C_B^C}{C_B^L} = \frac{q_B^C q_A^L}{q_B^L q_A^C}\exp\{\frac{z}{kT_m}[(\phi_{AB}^C-\phi_{AB}^L)-(\phi_{AA}^C-\phi_{AA}^L)]\}.\tag{5}$$

Note that the exponential contains the net internal energy change resulting from a transfer of a B atom from the liquid to the crystal, and an A atom from the crystal to the liquid phase.

The canonical partition function of the two-phase system with an interface is a sum over configurations with a total of N_A A atoms and N_B B atoms in both phases. But the number of atoms in the crystalline phase is variable, with $0 \leqslant N_A^C \leqslant N_A$ and $0 \leqslant N_B^C \leqslant N_B$. The partition function is

$$Q_I = \sum_i (q_A^C)^{N_A^C(i)}(\frac{q_A^L}{C_A^L})^{[N_A-N_A^C(i)]}(q_B^C)^{N_B^C(i)}(\frac{q_B^L}{C_B^L})^{[N_B-N_B^C(i)]}\exp[-E\,(i)/kT],\tag{6}$$

where configuration i has $N_A^C(i)$ and $N_B^C(i)$ atoms in the crystal of types A and B, and energy $E\,(i)$. Included in the sum are distinct configurations of the crystal only; from eq. (1) and Stirling's approximation the number of configurations in the mean-field liquid is $(C_A^L)^{-N_A^L}\times(C_B^L)^{-N_B^L}$ for N_A^L and N_B^L liquid A and B atoms, and this term is included in (6). But this can also be expressed in a form resembling a grand canonical ensemble,

$$Q_I = F(T)\sum_i\exp[\mu_A^0 N_A^C(i)+\mu_B^0 N_B^C(i)-E\,(i)/kT],\tag{7}$$

where $F(T) = (\frac{q_A^L}{C_A^L})^{N_A}(\frac{q_B^L}{C_B^L})^{N_B}$, and

$$\mu_A^0 = \ln(q_A^C/q_A^L)+\ln C_A^L\tag{8}$$

$$\mu_B^0 = \ln(q_B^C/q_B^L) + \ln C_B^L.$$

The energy of the bulk liquid is calculated from the average bond energy ϕ^L,

$$E_L = -\frac{Nz}{2}\phi^L \equiv -\frac{Nz}{2}[(C_A^L)^2\phi_{AA}^L + 2C_A^L C_B^L \phi_{AB}^L + (C_B^L)^2\phi_{BB}^L]. \tag{9}$$

When an A atom in the liquid phase crystallizes, the concentration of the remaining liquid changes by an amount $\Delta C_A = -\dfrac{C_B}{N}$, and from (9) the resulting change in energy is

$$\Delta E_A^L = zC_B^L[C_A^L\phi_{AA}^L + (C_B^L - C_A^L)\phi_{AB}^L - C_B^L\phi_{BB}^L], \tag{10}$$

and when a B atom crystallizes the change in energy is given by the analogous expression.

The total change in energy when an A atom crystallizes at a site with n_A and n_B nearest neighbors of types A and B is

$$\Delta E_A = \Delta E_A^L + z(\phi^L - \phi_{AL}^I) + n_A(2\phi_{AL}^I - \phi^L - \phi_{AA}^C) + n_B(\phi_{BL}^I + \phi_{AL}^I - \phi^L - \phi_{AB}^C), \tag{11}$$

and similarly

$$\Delta E_B = \Delta E_B^L + z(\phi^L - \phi_{BL}^I) + n_A(\phi_{BL}^I + \phi_{AL}^I - \phi^L - \phi_{AB}^C) + n_B(2\phi_{BL}^I - \phi^L - \phi_{BB}^C). \tag{12}$$

Here ϕ_{AL}^I and ϕ_{BL}^I are the energies of bonds between crystalline A and B atoms in the interface with nearest neighbor liquid sites.

Equations (11) and (12) are analogous to the corresponding expressions for a crystal-vapor system, but with effective bond energies

$$\Phi_{AA} = \phi_{AA}^C + \phi^L - 2\phi_{AL}^I \tag{13}$$

$$\Phi_{AB} = \phi_{AB}^C + \phi^L - \phi_{BL}^I - \phi_{AL}^I \tag{14}$$

$$\Phi_{BB} = \phi_{BB}^C + \phi^L - 2\phi_{BL}^I \tag{15}$$

Then eqs. (11) and (12) can be written as follows

$$\Delta E_A = E_A^0 - n_A\Phi_{AA} - n_B\Phi_{AB} \tag{16}$$

$$\Delta E_B = E_B^0 - n_A\Phi_{AB} - n_B\Phi_{BB}. \tag{17}$$

In the limit $C_B^L \approx 0$ which will be treated below,

$$E_A^0 = z(\phi_{AA}^L - \phi_{AL}^I) \tag{18}$$

$$E_B^0 = z(\phi_{AB}^L - \phi_{BL}^I) \tag{19}$$

since in this case from (9) $\phi^L \approx \phi_{AA}^L$. The one important difference between the equations describing a crystal-vapor system and eqs. (16) and (17) is the presence of the terms E_A^0 and E_B^0. An atom that condenses from the vapor experiences a change in energy as a result of interactions with neighboring atoms in the crystal, whereas (16) and (17) imply an additional energy change that is associated purely with the change of state. The magnitude of these terms can influence the surface structure, the roughening transition temperature, and surface segregation of impurities. Thus, the system can be represented as the equivalent crystal-vapor ensemble, with the distribution coefficient

$$K^e = \frac{q_B^C q_A^L}{q_B^L q_A^C}\exp\{[z(\Phi_{AB} - \Phi_{AA}) + (E_A^0 - E_B^0)]/kT\}. \tag{20}$$

The kinetics of crystal growth are determined by the transition rates between the liquid and crystalline states of atoms in the interface. In equilibrium, the transition rates must be consistent with eq. (6) by the principle of microscopic reversibility. Following Jackson[5], we assume that the rate at which A atoms of the liquid crystallize at a site, k_A^+, is independent of the state of the nearest neighbor sites, and is proportional to the concentration of A atoms in the liquid, and a similar assumption applies to the B atoms:

$$k_A^+ = C_A^L k^0 \tag{21}$$

$$k_B^+ = C_B^L k^{0.}$$

Under equilibrium conditions, the rates at which crystalline atoms make transitions to the liquid state is

$$k_A^-(n_A,n_B) = k^0 \left(\frac{q_A^L}{q_A^C}\right) \exp[(E_A^0 - n_A\Phi_{AA} - n_B\Phi_{AB})/kT] \tag{22}$$

$$k_B^-(n_A,n_B) = k^0 \left(\frac{q_B^L}{q_B^C}\right) \exp[(E_B^0 - n_A\Phi_{AB} - n_B\Phi_{BB})/kT].$$

These expressions are derived from the partition function (6) using eqs. (21) and the principle of microscopic reversability.

The driving force for the growth of the crystal resulting from an undercooling is a consequence of the temperature dependence of the terms in the exponent of (22). Since the interaction energies are excluded from the site partition functions, we can assume that the ratio is relatively insensitive to temperature, and approximate this term by a constant. Then, the melting rate at a kink site is related to the crystallization rate by the expression

$$k_A^-(z/2,0) = k^0 \exp[-(1/kT - 1/kT_m)L] = k^0 \exp[-\frac{L\Delta T}{kTT_m}], \tag{23}$$

where the heat of fusion is

$$L = z\Phi_{AA}/2 - E_A^0. \tag{24}$$

The difference of the chemical potential between the A species in the crystal and in the supercooled melt from eq. (2) is the usual expression for the driving force for crystal growth from the melt[5],

$$\Delta\mu = \mu_A^C - \mu_A^L = L\Delta T/T_m, \tag{25}$$

which is consistent with (23).

III. GROWTH RATES OF DC (111), (110) AND (100) FACES

The Monte Carlo simulations are performed using a procedure similar to those described previously[25]. Sites in the interfacial region are represented by a square (48x48) array of integers. Elements $S(I,J)$ of the array specify the state of occupation of all sites in a column that traverses the interfacial region, extending from the bulk crystal below to the bulk liquid above. Each element is determined by the M sites in its column as follows

$$S(I,J) = \sum_{K=0}^{M} 3^K \alpha_{IJK}, \tag{26}$$

where I and J specify the position of a site in layer K; and $\alpha_{IJK} = 0$, 1, or 2 if the site is in the liquid state, or crystalline A or B, respectively. (Multiplication by 3^K permits information for the entire column to be stored in a single element of the array.) The solid-on-solid (SOS) restriction employed in earlier simulations is relaxed in this case, although isolated atoms of either phase are not permitted. The region includes $M = 15$ to 25 layers, and it is translated with the interface as crystallization proceeds. Information on the location, occupancy, and coordination of each site is stored in two additional arrays; these facilitate the Monte Carlo procedure used to select sites for transitions between states. Sites are selected using a random number generator, but with a probability that is proportional to the rate constant k^0, k_A^-, or k_B^- for the site. The net rate of accumu-

254

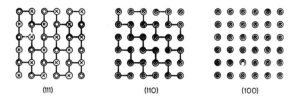

Fig. 1.   Schematic diagram of nearest neighbor bond configurations for the three DC faces.
Circles and crosses indicate single bonds to atoms in the layer above or below, respectively, except for the (100) diagram where each atom has two bonds in each direction.

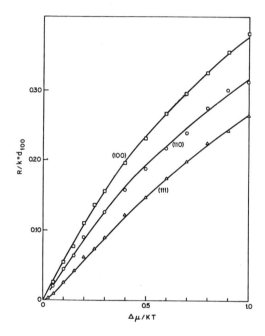

Fig. 2.   Monte Carlo growth rates on the (111), (110) and (100) faces. All rates are normalized by $k^0 d_{100}$.

lation of crystalline A and B atoms is monitored during the simulation and used to obtain the crystal growth rate and composition.

The bonds within the (111), (110) and (100) layers of atoms are illustrated schematically in
Fig. 1. Only the (111) face has a connected network of bonds between nearest neighbors. The
other two faces can not exhibit a roughening transition since they contain non-interacting rows of
atoms; they are "rough" under all conditions. Even in the presence of weak second-neighbor forces,
these interfaces could also be expected to roughen at temperatures well below the melting point of
silicon. The weak lateral bonding would produce low transition temperatures.

The effective bond energy Φ_{AA} for the silicon crystal-melt system can be calculated by comparing experimental observations with simulations. The small (111) facet observed during the Czochralski growth of dislocation-free silicon suggests that the face is quite close to its transition temperature, and estimates of the undercooling at the center of the facet are in the vicinity of 1.5-$5^0C^{(26)}$. Van Enckevort and van der Eerden[27] have simulated growth on the Ising DC(111) face, and have calculated a roughening transition temperature $kT_R^{(111)}=0.37\Phi_{AA}$.

Fig. 2 shows our simulation data for growth rates on the three faces with the melting temperature $kT_m=0.35\Phi_{AA}$. We show in a later discussion that the (111) kinetics are consistent with the calculations of the experimental undercooling. The chemical potential driving force is given in the figure, but the corresponding undercooling can be obtained from eq. (25) with $L=3.24kT_m$ for the latent heat of fusion per atom of silicon, and $T_m=1700^0K$. With these values an undercooling of 50^0C yields a driving force $\Delta\mu\approx0.1kT$. Note that these values for T_m and L give $E_A^0=2.47kT_m$, the positive value indicating that this component of the energy increases on crystallization, whereas the energy contributed by Φ_{AA} decreases.

The velocities of the (100) and (110) faces are approximately proportional to $\Delta\mu$ for small values of the driving force, whereas the (111) data contains a non-linear region near the origin as expected for growth limited by two-dimensional nucleation. The kinetic data of van Enkenfort and van der Eerden at $kT=0.34\Phi_{AA}$ exhibit a slightly larger non-linear region, consistent with the lower temperature. The growth rates of Fig. 2 are highly anisotropic at small $\Delta\mu$ with $R_{100}/R_{111}\approx2$ at $\Delta\mu=0.1$. A surprising feature of these results is the persistence of the anisotropy at large values of the driving force. Although the ratio $R_{100}/R_{111}\approx1.6$ at $\Delta\mu=0.5$ is smaller, the difference $R_{100}-R_{111}$ increases with $\Delta\mu$. This contrasts with data for the face centered cubic lattice[28]. In that case the difference in rates diminishes at large values of $\Delta\mu$ and growth becomes almost isotropic. Apparently lattices with a high coordination number are more susceptible to kinetic surface roughening, and this reduces the anisotropy in growth rates. All velocities R in Fig. 2 are normalized to k^0d_{100}, where d_{100} is the (100) layer spacing. In the case of the (100) face, this dimensionless ratio is simply

$$R_{100} = \Delta N_A/\Delta N_A^+. \tag{27}$$

Here ΔN_A is the net increase in the number of atoms in the crystal during the course of the simulation, and ΔN_A^+ is the number of crystallization events at selected liquid sites and it provides a measure of the rate constant k^0. These liquid sites are selected, one in each column, from those with crystalline neighbors. (Crystallization on liquid sites with no crystalline neighbors is excluded.) The value of R/k^0d_{100} for the (100) face can exceed unity, since the absence of a SOS restriction permits crystallization on several sites in the same column. The growth rates of the (111) and (110) faces were calculated in a similar manner and scaled by the ratio of layer spacings: $d_{111}/d_{100} = 4/\sqrt{3}$, and $d_{110}/d_{100} = \sqrt{2}$.

The rate constant k^0 is not known for this system, and therefore the calculated rates can not be directly compared with those measured during laser annealing. In Section IV we compare the calculated distribution coefficients with those measured experimentally, and from that analysis we estimate that $k^0d_{100}\approx13.5m/s$. Using this value we can make a rough calculation of the magnitude of the undercooling during Czochralski growth implied by the (111) kinetics of Fig. 2. The growth rate resulting from two-dimensional nucleation has the form[29]

$$\frac{R}{k^0d} = A(\Delta\mu/kT)^{5/6}\exp(-B\epsilon^2/\Delta\mu kT), \tag{28}$$

where ϵ is the edge free energy per site of a step on the surface, and A and B are constants of order unity. Equation (28) provides a good fit to the data for $\Delta\mu<0.5$, using the unknown step free energy as a fitting parameter. From this expression we calculate an undercooling of $\Delta T\approx2^0$ corresponding to a typical Czochralski rate of $10^{-5}m/s$. This calculation confirms that $kT_m=0.35\Phi_{AA}$ is an appropriate value for the melting temperature in the model.

IV DISTRIBUTION COEFFICIENTS

In this section we consider the effect on the distribution coefficient K of several factors relating to the interaction of impurities with the crystal surface. We compare our results with laser annealing experiments on bismuth[18] and indium[19]. These two impurities have approximately equal values of the equilibrium distribution coefficient and both can be accurately represented using a common value of $K^e = 4\times10^{-4}$ ($K^e \approx 7\times10^{-4}$ and 4×10^{-4} for bismuth and indium, respectively). Test simulations show that the small difference in K^e for the two systems has little effect on the kinetic value of K for interfacial conditions that match the experiments. Several additional simplifying assumptions are employed. First, in all cases the ratios of the site partition functions of the A and B species are equated, i. e., $q_A^C/q_A^L = q_B^C/q_B^L$. Second, the effects of the impurity concentration on K are minimized by limiting C_B^C to values less than 0.02. Experimental concentrations include values of C_B^C in the range of 10^{-4} at the point where crystallization is initiated, to 5×10^{-4} near the crystal surface. Since a constant value of K provides a fit to the experimental data over the entire range of concentrations, the neglect of these effects is perhaps justified. A third simplification is the exclusion of impurity migration in the crystal. Our concern here is purely with the interface distribution coefficient and it is natural to exclude effects of diffusion in both bulk phases. Effects due to bulk diffusion can be added by a numerical analysis after the interface coefficient is calculated, but diffusion in the crystal is slow and should have little effect in any case. In addition, as noted previously, a fully coordinated atom of either type is not permitted to melt.

Simulation data for the distribution coefficients are shown in Fig. 3. The hexagons indicate the values of K measured for (100) simulations in the absence of surface segregation, i.e., $E_B^0 = E_A^0 = 2.47kT_m$. As the undercooling increases, K also increases on both faces, with the (100) face having the largest distribution coefficient. This data contrasts sharply with the experiments which give values of $K \approx 0.3$ at the highest growth rates. Furthermore, it is not consistent with the classical one-dimensional models, which predict a decrease in K with increasing $\Delta\mu$. The difference with the classical models is a consequence of the larger value of Φ_{AB} employed in the simulations. The usual assumption that $E_B^0 = E_A^0 = 0$ produces the negative value of $\Phi_{AB} = -0.24kT_m$, according to eq.(20). An impurity crystallizing at any site on the crystal surface must experience an increase in potential energy and at low temperatures the probability of this transition becomes relatively small. However, with the values of E_B^0 and E_A^0 listed above, $\Phi_{AB} = 0.90kT_m$, and in this case impurities crystallizing at certain sites cause a reduction in the energy. Since a variety of sites are always present during crystal growth, K should approach a finite value as $T \to 0$. (In all cases we have chosen $\Phi_{BB} = 2\Phi_{AB} - \Phi_{AA}$, the ideal alloy value. This has the effect of minimizing the consequences of finite impurity concentrations, although the number of BB bonds is usually negligible for such small values of C_B^C)

The reconcilliation of these results with experiment is not so simple. The data at large $\Delta\mu$ are orders of magnitude too small. Although these results are an improvement over the classical theory, it is apparently necessary to include surface segregation or other such phenomena present in real crystal-liquid interfaces in order to obtain quantitative agreement with experiments.

The energy of crystallization of the impurities E_B^0 is probably less than the rather large value of E_A^0 for the silicon atoms. The silicon atom in the interface could be expected to have a large potential energy, since it is in a state intermediate between the covalently bonded crystal and the metallic liquid. Bonding to the crystal would be disrupted by the disordered environment, and the density is not suited for the metallic melt. The impurities should not be greatly affected by the change in structure, and we expect that for most materials $E_B^0 < E_A^0$. This causes segregation of the impurities at the interface, and enhanced trapping.

Stress is produced by impurities in the crystal which have atomic diameters appreciably different from that of the host system, and this may also influence trapping. It is well known that surface segregation is also affected by atomic size effects[30]. A large impurity at the interface with few crystalline neighbors will produce little stress, since the crystal can relax into the liquid.

During crystal growth the stress can be expected to increase gradually as more material is crystallized, and eventually the energy associated with the strain field will approach the bulk value. A gradual increase of this type can greatly enhance trapping, since the excess energy associated with the impurity is distributed over a large number of crystallizing atoms. A single crystallization event associated with a large increase in potential energy is rejected with a high probability, but a large number of events with only a small perturbation from the normal energy for a host atom are not as likely to be rejected.

A gradual increase in the impurity energy is not easily added to the Monte Carlo simulation, but this effect can be partially included by delaying the point at which the potential energy of the impurity reaches its bulk value. In addition to the crystallization energy E_B^0, we include a second term E_B^1 only when the impurity is fully coordinated. The change in energy when an impurity crystallizes at a site with four crystalline nearest neighbors is

$$\Delta E_B = E_B^0 + E_B^1 - n_A \Phi_{AB} - n_B \Phi_{BB} \qquad n_A + n_B = z, \qquad (29)$$

and eq. (20) is replaced by

$$K^e = \frac{q_B^C q_A^L}{q_B^L q_A^C} \exp\{z\, (\Phi_{AB} - \Phi_{AA})/kT + (E_A^0 - E_B^0 - E_B^1)/kT \}. \qquad (30)$$

For a given value of K^e a large value of the sum $E_B^0 + E_B^1$ corresponds to large values of Φ_{AB}, and such an impurity will be susceptible to trapping in large quantities unless E_B^0 is sufficiently large to oppose the strong AB bonds.

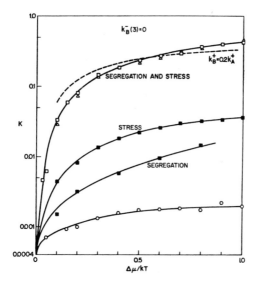

Fig. 3   Distribution coefficient K vs. driving force $\Delta\mu/kT$. The open squares and triangles correspond to $E_A^0 - E_B^0 = 1.80kT_m$, $E_B^1 = 5.71kT_m$, and $\Phi_{AB} = 1.88kT_m$ (the dashed curve indicates data for identical bond parameters, but with the B crystallization rate reduced by a factor of 5). The solid squares correspond to $E_A^0 = E_B^0$, $E_B^1 = 3.91kT_m$, and $\Phi_{AB} = 1.88kT_m$; the squares with crosses to $E_A^0 - E_B^0 = 1.80kT_m$, $E_B^1 = 1.80kT_m$, and $\Phi_{AB} = 0.90kT_m$; and the open hexagons to $E_A^0 = E_B^0$, $E_B^1 = 0$, and $\Phi_{AB} = 0.90kT_m$.

The open squares and triangles are data obtained for (100) and (111) faces with an energy $E_A^0 > E_B^0$ favoring segregation, and also an energy E_B^1 to account for stress ($E_A^0 - E_B^0 = 1.80kT_m$, $E_B^1 = 5.71kT_m$, $\Phi_{AB} = 1.88kT_m$). These data are in the correct range to fit the laser annealing results. The value of E_B^1 is quite large, but this is probably required to offset the more gradual increase in strain energy that occurs in reality. The difference between the values of K measured on the two faces is within the statistical error of the simulation data.

The rate of crystallization of the B atoms was reduced to $k_B^+/C_B^L = 0.2k_A^+/C_A^L$ in order to assess the effect of these rate constants on K. The dashed line represents these results. Although the differences are relatively small, clearly the reduced kinetics for the impurity accentuates the features mentioned above, i.e., the initial rise is even steeper and at high driving force K reaches a plateau at $K \approx 0.35$. The increase in trapping at low driving force is contributed by impurities in sites of low coordination number that are engulfed by advancing clusters and steps; these atoms are less likely to escape because of the slower kinetics. At faster growth rates, k_B^+ is comparable with the time required to grow a monolayer. Less time is available to populate the sites with impurity, and the trapping is reduced.

Other data in Fig. 3 shows the effect on K of reducing the influence of segregation and stress. The data corresponding to stress show a tendency to approach an asymptote at large $\Delta\mu$, although without segregation this value is smaller than that observed experimentally. Agreement with experiment could probably be obtained with a larger value of E_B^1.

In all cases we have excluded the melting of impurities with three bonds to the crystal. Crystallization of impurities on the corresponding liquid sites is also excluded, of course, as required for microscopic reversibility. This accounts for the reduced mobility of impurities in crevices and

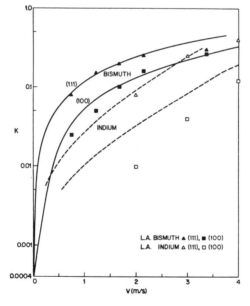

Fig. 4.   Distribution coefficient K vs. interface speed V (meters/second). Experimental data for bismuth (solid triangles and squares) are compared with simulation data from Fig. 3. Experimental data for indium are compared with simulation data for $E_A^0 - E_B^0 = 2.16kT_m$, $E_B^1 = 4.51kT_m$, and $\Phi_{AB} = 1.49kT_m$.

other sites of high coordination. Simulations without this restriction give simular results, although the values of K for the (111) face are appreciably lower than those for the (100) face.

Fig. 4 provides a direct comparison of the simulation data of Fig. 3 with the experimental results. The bismuth curves are taken from the data of Fig. 3 representing simulations with segregation and stress. The growth rates of Fig. 2 have been used to convert $\Delta\mu$ into interface velocity. corresponding to the values of $\Delta\mu$. The unknown value of $k^0 d_{100}$ was used as a fitting parameter to optimize agreement between the two sets of data. In Fig. 4 we have used the value $k^0 d_{100} = 13.5 m/s$. Although the (111) and (100) faces had approximately equal distribution coefficients for equal values of $\Delta\mu$, the (111) face has a larger value of K when the interface speeds are equal. The slower kinetics on the (111) face require a larger $\Delta\mu$ to achieve the same velocity. An analysis of the heat conduction equations for pulsed laser melting shows that the interface speed is approximately equal for the two faces under identical heating conditions in spite of the slower kinetics on the (111) orientation[31]. That is, the conduction of latent heat from the advancing interface is the primary rate limiting factor.

An interesting feature of these results is the steep rise in the (111) distribution coefficient K_{111} for small velocities. Because of this, the ratio K_{111}/K_{100} reaches a maximum of ≈ 10 near $V = 0.1 m/s$. Also, it helps to explain the frequently observed core of impurities behind the (111) facet during slow Czochralski growth[20] and the presence of striations resulting from growth rate changes at small values of V[16,32].

Some significant differences between the simulation and the bismuth data are observed at large velocities. Better agreement in this region is obtained using the dashed curve of Fig. 3 corresponding to smaller bismuth crystallization rates. Slower rates for the bismuth atom may result from its larger atomic radius.

Distribution coefficients measured for indium are also shown in Fig. 4, together with simulation data for a smaller E_B^1. In this case the orientation effect is appreciably larger than that predicted by the simulation model, but reasonably good agreement is possible by appropriate choice of the interaction energies.

V CONCLUSIONS

The simulation results demonstrate that lattice models of the crystal-melt interface can explain the large distribution coefficients observed during rapid solidification. Furthermore, information on the different atomic processes and their contribution to trapping has been obtained. The model is uniquely suited to such a study since it permits detailed control of the energy levels and kinetics of atoms at the interface.

Three different conditions are apparently required to achieve the level of trapping observed in the experiments. (i) Segregation of the impurities in the interface. (ii) A delay in the approach to the bulk potential energy of an impurity, such as would be expected if the impurity is larger than the host species and produces a stress field. (iii) An inhibition of diffusion of impurities from sites of high crystalline coordination. All of these factors can be expected to be present to some degree in crystal-melt systems. They can also be included in the model by relatively simple modifications of the simulation procedure, although they are not usually present in Ising model simulations of crystal growth.

Strong segregation of impurities at the interface is expected for systems such as silicon with large values of the crystallization energy E_A^0. This component of the crystallization energy is an essential part of the equivalence relations between the crystal-liquid and the crystal-vapor systems. The large value of E_A^0 is required for consistency with (111) facets observed experimentally.

The orientation dependence of trapping in silicon is of great interest, since it provides a direct comparison of the growth of the singular (111) and the non-singular or rough (100) faces. The model predicts the correct order for these faces provided diffusion of impurities from sites of high coordination is inhibited. The higher impurity concentration obtained by (111) growth is a consequence of the larger driving force required for crystallization on this face. The simulation data show that the (111) kinetics are appreciably slower than those for the (100) orientation, and this anisotropy persists even at large values of the driving force. Without the difference in kinetics, the larger trapping rate on the (111) face could not be explained. At equal values of the driving force, the trapping rate on the (100) face is equal to or greater than that on the (111) for all conditions simulated. This effect appears to be inherent in the crystal growth process, since the faster growth permits less time for the impurity to escape.

Acknowledgments

The author is grateful for many useful discussions with K. A. Jackson and J. M. Poate.

REFERENCES

(1) *Laser and Electron-Beam Interactions with Solids*, B. R. Appleton and G. K. Celler, Eds. (Elsevier, Amsterdam, 1982)

(2) *Laser and Electron Beam Processing of Materials*, C. W. White and P. S. Peercy, Eds. (Academic Press, New York, 1980)

(3) A. G. Cullis, H. C. Webber, J. M. Poate, and A. L. Simons, Appl. Phys. Lett. **36**, 320 (1980); C. W. White, S. R. Wilson, B. R. Appleton, and F. W. Young, Jr., J. Appl. Phys. **51**, 738 (1980); A. G. Cullis, D. T. J. Hurle, H. C. Webber, N. G. Chew, J. M. Poate, P. Baeri, and G. Foti, Appl. Phys. Lett. **38**, 642 (1981).

(4) R. N. Hall, J. Phys. Chem. Solids **3**, 63 (1957).

(5) K. A. Jackson, Can. J. of Phys. **36**, 683 (1958).

(6) A. Trainor and B. E. Bartlett, Solid State Electron. **2**, 106 (1961).

(7) P. J. Holmes, J. Phys. Chem. Solids **24**, 1239 (1963).

(8) V. V. Voronkov and A. A. Chernov, Sov. Phys.-Cryst. **12**, 186 (1967).

(9) J. C. Baker and J. W. Cahn, Acta. Met. **17**, 575 (1969).

(10) A. A. Chernov, Sov. Phys.-Uspekhi **13**, 101 (1970).

(11) D. E. Temkin, Sov. Phys.-Cryst. **17**, 405 (1972).

(12) K. A. Jackson, G. H. Gilmer, and H. J. Leamy, in: Ref. 2, p.104.

(13) R. F. Wood, Appl. Phys. Lett. **37**, 302 (1980); and Phys. Rev. **B25**, 2786 (1982).

(14) H. E. Cline, J. Appl. Phys. **53** (1982).

(15) M. Aziz, J. Appl. Phys. **53**, 1158 (1982).

(16) A. J. R. de Kock, in: *Crystal Growth and Materials*, E. Kaldis and H. J. Scheel, Eds. (North-Holland, Amsterdam, 1977).

(17) J. W. Cahn, S. R. Coriell and W. J. Boettinger, in: Ref. 2, p.81.

(18) P. Baeri, G. Foti, J. M. Poate, S. U. Campisano and A. G. Cullis, Appl. Phys. Lett. **38**, 800 (1981).

(19) J. M. Poate, in: Ref. 1, p.121.

(20) D. J. T. Hurle, in: *Crystal Growth: An Introduction*, P. Hartman, Ed. (North-Holland, Amsterdam, 1973), p.210.

(21) Several reviews treat the roughening transition and its effect on kinetics. See: J. D. Weeks, in: *Ordering in Strongly Fluctuating Condensed Matter Systems*, T. Riste, Ed. (Plenum, 1980), p. 293; J. P. van der Eerden, P. Bennema, and T. A. Cherepanova, in: *Progress in Crystal Growth and Characterization*, B. R. Pamplin, Ed. (Pergamon, Oxford, 1979), Vol. 3, p. 219; and J. D. Weeks and G. H. Gilmer, Adv. Chem. Phys. **40**, 157 (1979).

(22) C. D. Thurmond and M. Kowalchik, Bell Syst. Technical J. **39**, 169 (1960); C. D. Thurmond and J. D. Struthers, J. Phys. Chem. **57**, 831 (1953).

(23) J. Q. Broughton, A. Bonissent and F. F. Abraham, J. Chem. Phys. **74**, 4029 (1981); J. Q.

Broughton and F. F. Abraham, Chem. Phys. Lett. **71**, 456 (1980).

(24) T. L. Hill, *Introduction to Statistical Thermodynamics* (Addison-Wesley, Reading, 1960) p.371.

(25) G. H. Gilmer, J. Crystal Growth **35**, 15 (1976); and J. Crystal Growth **49**, 465 (1980).

(26) T. F. Ciszek, J. Cryst. Growth **10**, 263 (1971); W. D. Edwards, Can. J. Phys. **38**, 439 (1960); T. Abe, J. Cryst. Growth **24/25**, 463 (1974).

(27) W. J. P. van Enckevort and J. P. van der Eerden, J. Cryst. Growth **47**, 501 (1979).

(28) G. H. Gilmer and K. A. Jackson, in: *Crystal Growth and Materials*, E. Kaldis and H. J. Scheel, Eds. (North-Holland, Amsterdam, 1977), p. 79.

(29) W. B. Hillig, Acta. Met. **14**, 1968 (1966).

(30) F. F. Abraham, N.-H. Tsai, and G. M. Pound, Surf. Sci. **83**, 406 (1979).

(31) G. H. Gilmer, unpublished.

(32) A. Murgai, H. C. Gatos, and A. F. Witt, J. Electrochem. Soc. **123**, 224 (1976).

AMORPHOUS Si, CRYSTALLIZATION AND MELTING

J. M. POATE
Bell Laboratories, Murray Hill, New Jersey 07974

ABSTRACT

Recent experiments dealing with the thermodynamics of crystallization and melting of amorphous Si are reviewed. Differential scanning calorimetry measurements give the heat of crystallization of implanted, amorphous Si to be $11.3 \pm .8$ kJ/mole. Gibbs free energy calculation based on these measurements indicate that amorphous Si melts at a temperature of $1460°$K compared to the crystalline value of $1685°$K. Evidence for this reduced melting temperature also comes from rapid heating measurements using a) structural information after solidification and b) dynamic conductance measurements during the melt. Solid phase epitaxial regrowth experiments which apparently do not show such a depression in amorphous melting temperature will be discussed.

INTRODUCTION

The advent of rapid heating techniques has created much interest in the basic thermodynamic properties of Si. The condensed phases of Si that take part in these rapid-heating transformations are the solid (amorphous and crystalline) phases and the liquid phase. The thermodynamic relationships of the liquid and crystal phases are quite well established but there is some uncertainty regarding the amorphous phase. The fact, however, that amorphous Ge and Si are in a higher free energy state than the crystalline state led Bagley and Chen[1] and Spaepen and Turnbull[2] to propose that the amorphous phase melts at a discrete temperature $(T_{a\ell})$ which is lower than the crystalline melting temperature $(T_{c\ell})$. These initial predictions were based on calorimetric measurements of the heat of crystallization (ΔH_{ac}) of amorphous Ge as no measurements existed for Si. The Gibbs free energies of the amorphous and liquid phases were calculated with respect to the crystalline phase and the intersection of the liquid and amorphous free energy curves gave $T_{a\ell} \approx 0.8 T_{c\ell}$.

The argument regarding this remarkable depression in melting temperature is principally based on the premise that the phase transition from the amorphous to liquid phase is discontinuous and first order. The reasoning behind this premise is quite plausible when it is realized that the transition involves a fundamental change in bonding from the covalent, amorphous phase with four-fold coordination to the metallic liquid with 11-12 fold coordination. Clearly this transition will only occur in very rapid heating measurements otherwise the amorphous phase will recrystallize directly in the solid phase as observed, for example, in furnace heating measurements. The transition should be reversible so that undercooling or supercooling the melt beneath $T_{a\ell}$ should lead to growth of amorphous Si in the melt.

This paper will review recent experiments addressing some of these issues. Firstly, measurements of the heat of crystallization of implanted, amorphous Si layers will be discussed along with calculations of the Gibbs free energies and amorphous melting temperatures. Experiments on the melting of amorphous Si will then be reviewed. These measurements fall into two classes; structural observations following the melting and solidification and dynamic observations using the transient conductance technique. Such observations indicate a considerable depression in the amorphous melting temperature in contradiction to the recent solid phase measurements[3] of the Hughes group. This controversy will be discussed in terms of the kinetics of the solidification and crystallization process.

Mat. Res. Soc. Symp. Proc. Vol. 13 (1983) ©Elsevier Science Publishing Co., Inc.

AMORPHIZATION AND CALORIMETRY MEASUREMENTS

The experiments of Donovan et al.[4] to be discussed here are the first direct measurements of the heat of crystallization of implanted, amorphous Si. One of the reasons that no measurements had been made previously is simply due to the fact that the amount of heat released from typical implanted layers of 1000Å thickness would only be approximately a hundredth of a joule; this amount of heat is barely detectable in scanning calorimeters. Therefore to utilize the advantages of implantation, layers must be fabricated with thicknesses exceeding a micron. To achieve this thickness, Si was bombarded with MeV inert gas ions. In order to minimize the concentration of implanted gas, bombardments were carried out with the Si substrates held at 77°K where the damaging and amorphization processes are much more efficient than, for example, room temperature bombardment. The threshold doses for amorphization for Ar and Xe bombardment at 77°K were determined and typical bombardments were carried out at doses twice or greater than the amorphization thresholds. In order to produce a uniform damage layer with approximately constant impurity incorporation, 3 or 4 implantations were employed as shown in Figure 1. In this case the Xe concentration is about 1 part in 10^4. The areal density (g/cm^2) of the amorphous layer was obtained from the energy loss[5] of 1.0 MeV H^+.

Most differential scanning calorimeter (DSC) measurements were made on (100)Si wafers cut to 7.6 mm diameter and implanted on both sides. Temperature scan rates of 10,20 or 40°K/min were used and the differential power needed to keep the implanted wafer and a reference wafer at the same temperature was determined. Figure 2 shows the DSC trace from a (100) wafer amorphized with Xe at the energies and doses given in Fig. 1. The scan rate was 40°K/min. The rate at which heat is released is exponential and at 970°K solid phase epitaxy has ceased. The samples were examined by channeling, following the heating, to ensure that single crystal growth had resulted. The power was integrated with time to obtain ΔH_{ac}. Calorimetric measurements on 20 samples amorphized under different conditions gave a value of ΔH_{ac} of 11.3±0.8 kJ/mole. Similar measurements on Ar amorphized Ge wafers gave ΔH_{ac} (Ge) of 12±1kJ/mole in agreement with previous measurements for deposited films.[6]

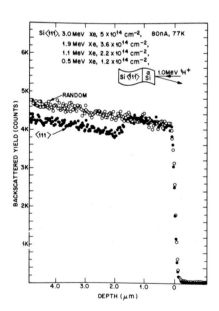

Fig. 1 Rutherford backscattering and channeling spectra of typical amorphous layers used in the calorimeter measurements. From Donovan et al.[4]

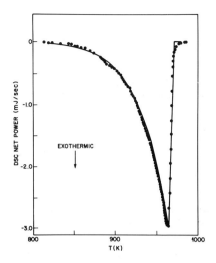

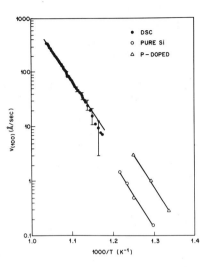

Fig. 2 Differential scanning calorimeter (DSC) data for the heat released from a 1.8 μm Si <100> amorphous layer. The Xe implantation conditions are the same as those shown in Figure 1. The smooth curve is from the fit to the velocity data of Figure 3. From Donovan et al.[4]

Fig. 3 Epitaxial Si <100> regrowth rate obtained from the DSC data in Figure 2. The data at T<850°K of Csepregi et al.[7] are shown for comparison. From Donovan et al.[4]

The rate of heat release by the advancing interface should only be dependent upon the velocity of the interface, $v_{<100>}$. The power release $\dfrac{dH}{dt}$ can therefore be written as

$$\frac{dH}{dt} = v_{<100>} \frac{A}{\overline{V}_a} \Delta H_{ac} \tag{1}$$

where A is the amorphized area and $\overline{V}_a$ is the molar volume of amorphous Si which is assumed to be the same as the crystalline value of 12.05 cm³/mole. The velocity can therefore be determined as a function of temperature as shown in the Arrhenius plot of Fig. 3. The straight line is a fit to the data with $v_{<100>} = v_o \exp(-\Delta H'/RT)$ with $v_o = 1.77 \times 10^{14}$ Å/sec and $\Delta H' = 216$ kJ/mole (2.24 eV). Csepregi et al[7] had originally demonstrated an Arrhenius temperature dependence of the crystallization rate with $v_o = 3.5 \times 10^{14}$ Å/sec and $\Delta H' = 227$ kJ/mole for Si implants. It is well known[3] that some implants (eq Ar or Xe) will retard regrowth while others (see Fig. 3) can enhance regrowth. The object of the present work was not to establish kinetics of epitaxy but rather to determine ΔH_{ac}. Nevertheless the Arrhenius behavior of the heat release gives considerable confidence in the cleanliness of the films and the accuracy of the ΔH_{ac} measurements. Recently Fan and Anderson[8] have measured a ΔH_{ac} of 9.9 ± 1.2 kJ/mole for Si from sputter deposited films. There is considerable scatter in their results probably because of entrapped impurities. The DSC traces do not have exponential dependences because of polycrystal formation. The present results indicate that high energy implantation and the DSC techniques could provide interesting kinetic data to complement lower temperature furnace.

GIBBS FREE ENERGY DIAGRAMS AND AMORPHOUS MELTING TEMPERATURES

Figure 4 shows the calculation by Donovan et al.[4] of the Gibbs free energy (ΔG) of liquid and amorphous Si referenced to crystalline Si. The liquid line was calculated from existing thermodynamic data. The amorphous curves were calculated from the following equations

$$\Delta S_{ac} = \Delta S_{ac}^{o} + \int_{0}^{T} (\Delta C_p/T) dT \tag{2}$$

$$\Delta H_{ac} = \Delta H_{ac}(T_1) + \int_{T_1}^{T} \Delta C_p dT \tag{3}$$

$$\Delta G_{ac} = \Delta H_{ac} - T\Delta S_{ac} \tag{4}$$

The residual entropy difference ΔS_{ac}^{o} was taken to be 0.2R, the upper limiting value for the excess configurational entropy of the ideal four-coordinated random network calculated by Spaepen.[9] As the heat capacity differences (ΔC_p) are not known, the following assumptions were made. For the line labelled a(1), $\Delta C_p = 0$ for all temperatures. For the line labelled a(2) a more realistic assumption was made, $\Delta C_p = 0$ for T<80K otherwise $\Delta C_p = -0.224 + 4.8(T/1685)$ J/mole K. This latter assumption uses the analytical fit to the ΔC_p data for Ge measured by Chen and Turnbull[6] over the range 200–500K and replaces the value of the melting point $T_{c\ell}$ of crystalline Ge (1210K) by that for Si (1685K).

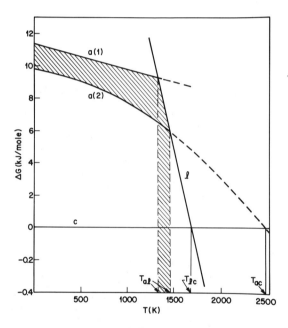

Fig. 4 Calculation of the isobaric Gibbs free energy, relative to crystalline Si(c), of the metallic liquid(ℓ) and amorphous(a) phases. The lines a(1) and a(2) represent different limits to the amorphous free energy. The hatched region therefore gives an estimate of the uncertainty in $T_{a\ell}$. From Donovan et al.[4]

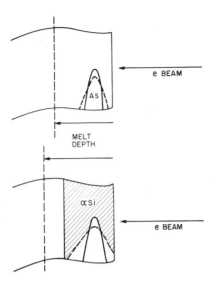

e BEAM

AS

MELT
DEPTH

Fig. 5 Schematic of e-beam
experiment to determine ΔH_{ac}.
The As marker profile is shown
before and after irradiation.

As discussed previously the intersection of the amorphous and liquid lines in Figure 5 represents a first-order phase change, so that $T_{a\ell}$ can be considered as the melting temperature of amorphous Si. The hatched region represents possible limits to the values of $T_{a\ell}$. Line a(1) gives a lower bound to $T_{a\ell}$ of 1335K, with $\Delta H_{a\ell} = 37.9$ kJ/mole. Line a(2) gives an estimate for $T_{a\ell}$ of 1460K, with a heat of fusion $\Delta H_{a\ell} = 37.8$ kJ/mole. These values of $\Delta H_{a\ell}$ are in good agreement with the results of Baeri et al.[10], who found $\Delta H_{a\ell} = 34.3 \pm 4.2$ kJ/mole from electron beam heating measurements which will be discussed in the following section.

These Gibbs free energy calculations based on the measured heat of crystallization of amorphous Si, indicate that $T_{a\ell}$ is depressed some 250°K beneath $T_{c\ell}$. The validity of this concept rests on the premise that the transition from the amorphous phase to the liquid is thermodynamically discontinuous and first order. Experimental evidence for this premise will be reviewed in the following section.

THE MELTING OF AMORPHOUS Si

There is no doubt that amorphous Si can be melted using rapid heating techniques. The questions lie in the nature of the melting and the actual melting temperature. One of the first experiments to try and explore some of these thermodynamic questions was the electron beam heating measurement of Baeri et al.[10] Electron pulses (50 nsec) were used to heat the surface layers because the coupling between the incident beam and irradiated samples is independent of the physical state (amorphous, crystalline or liquid) of the Si. The concept of the experiment is shown schematically in Figure 5. Arsenic markers are implanted to exactly the same depths in amorphous (upper) and single crystal (lower) Si. The samples are irradiated with sufficient and equal energies so that amorphous Si is not only melted but raised to a temperature greater than $T_{c\ell}$. In this way the single crystal Si will melt underneath the amorphous Si and liquid phase epitaxy will ensue on cooling. The fact that energy is stored in the amorphous layers implies that the melt depths will be greater in the amorphized samples and that diffusion of the As markers will be greater than in the crystalline samples.

Indeed the experimental data showed precisely this sort of behavior and the enthalpy of melting of amorphous Si was calculated to be 34 ± 4 kJ/mole implying a heat of crystallization of amorphous Si to be 16 ± 4 kJ/mole. The large errors were due to uncertainties in the electron beam parameters. The agreement in enthalpy values between these early melting measurements and the more recent solid phase calorimetry measurements are significant. It would appear that the thermodynamic concepts are sound. Moreover the fact that, in the melting experiments, constant values of ΔH_{ac} were obtained for different heating conditions supports the contention that the transition is first order. For example in a glass transition, ΔH would be a function of the heating conditions.

Unusual microstructures were observed in the e-beam experiments for irradiation of the amorphous samples at energies $\leqslant 0.55$ J/cm² which were not sufficient to melt the single crystal samples (the signature of melting of the crystal samples was taken to be measurable diffusion of the As markers). The microstructures are shown in Figure 6 for transmission electron microscopy (TEM) micrographs of cross-sectioned samples. The upper micrograph shows the 1900Å amorphous layer (A) before heating. The middle micrograph shows the unusual layered structure resulting from an irradiation of 0.5 J/cm². Such structures are observed over the energy range 0.42-0.5 J/cm². They consist of approximately 800Å of highly defective, epitaxial Si (B) at the interface with the outermost material consisting of large grain polycrystallites (C). The lower micrograph of the 0.55 J/cm² irradiation shows complete epitaxial recrystallization (D) with some defects at the interface.

The unusual layered microstructure of the middle micrograph was interpreted in terms of the melting of amorphous Si at temperatures hundreds of degrees beneath $T_{c\ell}$. The greatly undercooled melt then will recrystallize from both the single-crystal interface and random nucleation centers at the free surface to produce the layered structure. This layered structure is not an artifact of e-beam heating as Baeri[11] has been able to reproduce it using laser heating with approximately the same temperature profiles as those generated by electron deposition. There is one other significant fact regarding the properties of the layered structure; no redistribution or diffusion (>100Å) of the As marker was observed. This lack of motion was interpreted in terms of recrystallization velocities exceeding 10 m/sec.

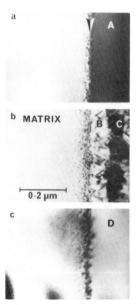

Fig. 6 TEM micrographs of cross-sectioned samples. Upper, 1900Å — amorphous layer before irradiation. Middle, irradiation of 0.5 J/cm², note unusual layered structure. Bottom, irradiation of 0.55 J/cm². From Baeri et al.[10]

The above interpretation of the e-beam heating experiment was disputed by Kokorowski et al.[12] in light of their apparent findings that amorphous Si can be heated to within 50°K of T_{cl} without melting. The solid phase results are reviewed by Olson et al.[3] in these proceedings. The main criticisms by Kokorowski et al.[12] of the interpretation of the e-beam experiment were that the layered structure was not sufficient evidence for melting and that, if melting did occur, the lack of diffusion of As must mean that the amorphous layer melts and solidifies within the time of the heating pulse. The phenomena of melting and solidification during the heating pulse had not been considered by Baeri et al[10]

These valid criticisms have been addressed to some extent in recent transient conductance measurements of the heating of 1000Å amorphous Si layers[13] by 30 nsec ruby laser irradiation. This technique is reviewed by Thompson and Galvin[14] in these proceedings. The samples consisted of 1000Å thick implanted, amorphous layers on 5000Å thick silicon on sapphire. Figure 7 shows the current traces from the low energy transient conductance measurements which have been converted to melt depths versus time. The dashed curve at 0.08 J/cm² is the photoconductivity response without melting and indicates the time duration of the laser pulse. The curve should not be read as melt depth versus time. The onset of melting is quite obvious with the trace rising sharply from the photoconductive response. This behavior is illustrated for a low energy irradiation of 0.25 J/cm². Melt is seen to start at approximately 30 nsec with the sharp increase in current and propagates approximately 200Å in the amorphous Si. The melt then recrystallizes within the time span of the laser pulse at a velocity of 1 m/sec. This result refutes, to some extent, the criticism of Kokorowski et al.[12] of the original e-beam measurements. Firstly the amorphous layers does melt and resolidify at very low energies and does not recrystallize by some unknown mechanism (e.g., explosive crystallization). Moreover once molten, the layer can resolidify while the heating pulse is on. The lack of As marker diffusion therefore results, not as originally thought, from high regrowth

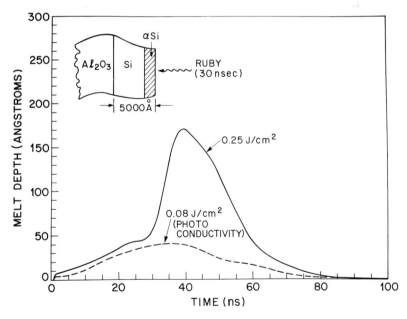

Fig. 7 Transient conductance measurement of 30 nsec, ruby laser irradiation of a 1000Å amorphous layer on 5000Å SOS. (Dashed curve is the photoconductivity transient) From Thompson et al.[13]

velocities ~10m/sec but probably from the fact that the melt only exists for exceedingly short times ~10 nsec. Of course these results say nothing about $T_{a\ell}$. Higher energy irradiation of 0.78 J/cm² (Figure 8) shows the melt-in front penetrating the amorphous layer into the underlying Si single crystal substrate. The intriguing feature is the plateau at 1000Å during the melt-in process. Heat flow calculations[15] indicate that such a feature can only result from a difference of $T_{c\ell}-T_{a\ell}$ of several hundreds of degrees. Differences between the thermal conductivities or specific heats of amorphous or crystalline Si can only produce a change in slope of the melt-in front.

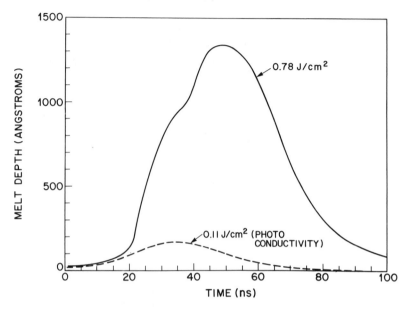

Fig. 8 Transient conductance measurement of 30 nsec, ruby laser irradiation of 1000Å amorphous layer on 5000Å SOS. Note plateau at 1000Å in melt-in kinetics. From Thompson et al.[13]

More transient conductance experiments are required to pin down definitively the phenomena associated with the melting of amorphous Si. Nevertheless all the various pulsed heating measurements do seem to show fairly convincing evidence for $T_{c\ell}-T_{a\ell}$ being of the order of several hundreds of degrees. In addition, as discussed by Cullis[16] in these proceedings, there is considerable evidence for the reverse transition when amorphous Si is quenched directly from the undercooled melt. The solid phase measurements of Olson et al.[3] are therefore very puzzling if amorphous Si, using slow cw laser heating, can be raised to within 50°K of $T_{c\ell}$ without melting. Moreover, as pointed out by Turnbull,[17] the fact that the solid phase regrowth rate appears[3] to have a single activation energy at these high temperatures poses another puzzle. If $T_{c\ell} \approx T_{a\ell}$ as originally proposed by Kokorowski et al.[3], the regrowth rate should fall off sharply at temperatures approaching $T_{c\ell}$ because there will be no differences in free energies and no driving force for crystallization. The regrowth rate does not fall off in the vicinity of $T_{c\ell}$ implying that the free energy diagram of Figure 4, with large differences in free energy at T_{ascrl} is essentially correct. The amorphous Si must therefore be superheated above $T_{a\ell}$ if the Hughes temperature estimates are not in error. In other words the cw measurements appear to provide additional evidence for $T_{c\ell}-T_{a\ell}$ being considerably greater than 50°K. However, there is an inconsistency in these arguments. To rationalize the results of the different experiments, superheating of amorphous Si is invoked in the slow, cw heating but not in the rapid, pulsed heating. The reverse behavior would seem more logical.

Acknowledgements

I am indebted to my colleagues at Bell (D. C. Jacobson), Harvard (E. P. Dovovan, F. Spaepen and D. Turnbull), Malvern (A. G. Cullis, H. C. Webber and N. G. Chew), Cornell (M. O. Thompson and J. W. Mayer) and Sandia (P. S. Peercy) for discussion and collaboration on the experiments reviewed here.

REFERENCES

[1] B. G. Bagley and H. S. Chen in "Laser Solid Interactions and Laser Processing" ed. S. D. Ferris, H. J. Leamy and J. M. Poate (AIP Conf. Proc. No 50) p. 97 (1978).

[2] F. Spaepen and D. Turnbull in "Laser Solid Interactions and Laser Processing" ed. S. D. Ferris, H. J. Leamy and J. M. Poate (AIP Conf. Proc. No 50) p. 73 (1978).

[3] G. L. Olson, S. A. Kokorowski, J. A. Roth and L. D. Hess, these proceedings and S. A. Kokorowski, G. L. Olson, and L. D. Hess, J. Appl. Phys. *53*, (1982) 921.

[4] E. P. Donovan, F. Spaepen, D. Turnbull, J. M. Poate and D. C. Jacobson, to be published.

[5] H. H. Anderson and J. F. Ziegler, "Hydrogen Stopping Powers and Ranges in All Elements" Pergamon, Oxford 1977.

[6] H. S. Chen and D. Turnbull, J. Appl. Phys. *40* (1969) 4212.

[7] L. Csepregi, E. F. Kennedy, J. W. Mayer and T. W. Sigmon, J. Appl. Phys. *49* (1978) 3906.

[8] J. C. C. Fan and H. Anderson, J. Appl. Phys. *52* 91981) 4003.

[9] F. Spaepen, Phil. Mag. *30* (1974) 417.

[10] P. Baeri, G. Foti, J. M. Poate and A. G. Cullis, Phys. Rev. Lett. *45* 2036 (1980).
     P. Baeri, G. Foti, J. M. Poate and A. G. Cullis in Materials Research Society Symposia Proc. ed. by J. F. Gibbons, L. D. Hess and T. W. Sigmon *1* (1981) 39.

[11] P. Baeri in Materials Research Society Symposia Proc. ed. by B. R. Appleton and G. K. Celler, *4* (1982) 151.

[12] S. A. Kokorowski, G. L. Olson, J. A. Roth and L. D. Hess, Phys. Rev. Lett. *48* (1982) 498.
     S. A. Kokorowski, G. L. Olson, J. A. Roth and L. D. Hess, Materials Research Society Proc. *4* (1982) 195.

[13] M. O. Thompson, G. J. Galvin, J. W. Mayer, P. S. Peercy, A. G. Cullis, N. G. Chew, H. C. Webber, D. C. Jacobson and J. M. Poate, to be published.

[14] M. O. Thompson and G. J. Galvin, these proceedings.

[15] A. G. Cullis, H. C. Webber N. G. Chew, private communications.

[16] A. G. Cullis these proceedings.

[17] D. Turnbull in Materials Research Society Symposium, Proc. ed. S. T. Picraux and W. J. Choyke, North Holland, Amsterdam, *7* (1982) 103.

IMPURITY REDISTRIBUTION IN ION IMPLANTED Si AFTER PICOSECOND
Nd LASER PULSE IRRADIATION

S.U.CAMPISANO,P.BAERI,E.RIMINI,G.RUSSO
Istituto di Fisica, 57 Corso Italia I 95129 Catania (Italy)

A.M.MALVEZZI
CISE SpA-P.O.Box 12081 I 20100 Milano (Italy)

ABSTRACT

Impurity redistribution in Bi-implanted Si and in As-
implanted Si has been investigated after irradiation with
25 ps Nd(λ=1.06 μm) laser pulse in the energy range 0.1-1.5
J/cm^2. Channeling effect in combination with 2.0 MeV He^+
backscattering in glancing detection has been used to characterize the epitaxial crystallization, the impurity location and its depth distribution. The amorphous to single
crystal transition occurs at an energy density of about
0.4 J/cm^2. Bi atoms are located after crystallization
in substitutional lattice sites for the in depth part of
the distribution. Part of the Bi atoms accumulated at
the sample surface and the amount of segregation increases
with the pulse energy density and depends on the substrate
orientation. A computer model has been also developed to
calculate several parameters of interest, as the melt threshold, the melt duration, the carrier temperature etc including a detailed description of the absorption and of the
energy relaxation processes. The calculations indicate that
the simple thermal description accounts quantitatively for
the experimental data on melt duration and impurity segregation.

INTRODUCTION

Several phenomena caused by irradiation of semiconductors with high power
picosecond laser pulses have been recently investigated[1,2]. Crystal to amor
phous transition occurs in silicon irradiated at 0.53 μm wavelength in the ener
gy density range 0.20-0.25 J/cm^2 [3]. Detailed analysis of the number of emitted charged particles (electrons and ions) during irradiation has shown that
above a well defined energy density value, positive and negative particles are
detected in equal amounts. Electron and lattice temperature remain then practically equal during irradiation and the energy relaxation time between electrons and phonons should be then lower than 10^{-11}s. This estimate is also
supported by time-resolved reflectivity measurements performed on Si single
crystals both on nanosecond and on picosecond time scale. It has been found
[4] for instance that the reflectivity measured with a probe delay of 100 ps

and with 1.06 μm wavelength increases abruptely for an energy density sligtly larger than 0.2 J/cm^2-0.53 μm pump pulse. The measured value of 0.76±0.03 is characteristic of molten silicon at the probe wavelength and the rise by itself indicates the occurrence of a first order transition. Effects associated to the overheating of the silicon are seen at larger energy density value but disappear with time due to the cool down of the liquid silicon layer. All these experiments point out that a thermal model remains valid also on this time scale, in spite of the large electron-hole concentration and of plasma effects.

Other phenomena, mainly investigated in the nanosecond irradiation regime, are related to the impurity behaviour. Solidification of the molten layer occurs at a solid-liquid interface velocity of a few m/s, several orders of magnitude higher than conventional values [5]. The segregation coefficient for many impurities in Si increases by several orders of magnitude. Its dependence upon velocity has been investigated by changing either the Si thermal conductivity or the laser pulse length and the energy. Unique segregation coeffi cients were measured independent of the way the velocity was obtained. For example the interfacial distribution coefficient of Bi in Si varies from 0.01 to 0.3 for (100) oriented substrates in the velocity range 0.8-4 m/s[6]. This last parameter has been determined directly by time resolved conductance measurements in Si single crystals [7]. The measured interface velocities are in quite good agreement with those evaluated within the melting model framework.

Only few experiments are reported on the dopant redistribution after pico-second irradiation[8]. In a previous work we reported some measurements on the Bi-implanted Si system after irradiation with 25 ps or 35 ns Nd(1.06 μm) laser pulses. The impurity distribution was similar for both duration pulses. In the present work we report a more detailed investigation of the Bi dopant redistribution in Si and some results on the dynamic of the energy transfer from light to lattice.

EXPERIMENTAL

Si wafers of <100> orientation were implanted with 120 keV Bi-10^{15}ion/cm^2. The irradiation was performed in air using Nd:YAG laser pulses of 30 ps duration. The energy density was measured for each pulse using a beam splitter and a calibrated photodiode. It ranged between 0.1 and 2.0 J/cm^2 at the center of the TEM$_{oo}$ o Gaussian beam spot, of 3 mm full width at half-maximum. The spatial distribution of the energy in the cross plane was tested by a pho todiode array with 18 μm resolution.

The irradiated samples were analyzed with channeling effect in combination with 2.0 MeV He$^+$ backscattering technique. To enhance depth resolution the scattered ion projectiles were detected at glancing geometry. The analyzing beam was collimated to a spot of 0.2 mm×0.2mm to probe only the center of the irradiated area, i.e. of the maximum power density of the Gaussian picosecond pulse.

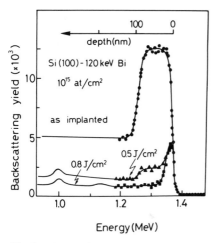

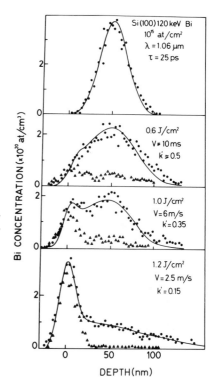

Fig.1. Channeling analysis of a Si sample implanted with 120 keV Bi-10^{15} at/cm^2 and irradiated with 25 ps-1.06 µm laser pulses of different energy density.

Fig.2. Bi concentration profiles in as-implanted sample and after irradiation at 1.06 µm 25 ps laser pulse for different energy density. The full circles and the open triangles refer respectively to random and aligned incidence of the analyzing beam. The full curves are calculated values assuming the couple of K-v values.

RESULTS AND DISCUSSION

Irradiation of ion implanted Si samples with 1.06 µm-picosend laser pulse causes the amorphous to single crystal transition above a threshold energy density value. Visual inspection of the samples indicates that at 0.1-0.2 J/cm^2 the center of the irradiated spot changed its initial milky appearance into the metallic reflectance characteristic of crystalline material. At these energy density values the aligned yield does not change from the as-implanted case and indicates a transition to polycrystalline material. Single crystal transition occurs above 0.4 J/cm^2 as shown by the drastic decrease of the aligned yield reported in Fig.1 . At 0.5 J/cm^2 irradiation the channeling analysis evidences residual disorder at the initial amorphous-single crystal interface. Free of defects single crystal layers are formed at higher energy density values.

The crystallization of amorphous Si layer under picosecond irradiation is then similar to that produced by nanosecond pulse. The process is characterized by an energy density threshold value above which the layer is single crystal and below which it is polycrystalline.

Broadening and surface accumulation of Bi implanted atoms occurs after laser irradiation. The depth distribution as measured by aligned (open triangles) and random (full circles) yield is shown in Fig.2 as a function of the energy density value. At 0.6 J/cm^2 only a broadening of the initial Gaussian profile is evidenced. The large attenuation of the yield for beam incidence along the (100) axis direction indicates that about 90% of the implanted Bi atoms are located in substitutional lattice sites. The negligible surface peak in the aligned yield shows also a negligible surface accumulation of dopant.

With increasing the energy density value broadening and surface accumulation becomes more and more relevant. At 1.2 J/cm^2 about 35% of the implanted Bi atoms is rejected at the sample surface while the in-depth dopant is substitutionally located. A comparison of the aligned with the random yield provides an accurate estimate of the surface accumulation when incorporation of the dopant in cell structure inside the sample is negligible[9]. Broadening and surface accumulation are described quite well in terms of a transient liquid layer whose solidification velocity depends on the energy density value. As a general trend with increasing the energy density and then the thickness of the melted layer there is a decrease of the solidification velocity.

The dopant profiles are fitted by a numerical solution of the mass transport equation[10]. For instance the full lines of Fig.2 represent calculated values assuming a diffusion coefficient of the impurity in liquid silicon equal to 2.10^{-4}cm^2/s and an interfacial distribution coefficient, K′ dependent on the solidification velocity. Several K′-v combination can fit the same experimental profiles. For nanosecond irradiation the solidification velocity can be computed by heat-flow calculations so that the K′-v relation is unique. It must be point out that the computed velocity values agree within few percents, with the experimental ones determined by time-resolved conductance measurements.

To fit the profiles of Fig.2 for the picosecond case we have chosen the couples of K′-v values previously determined on the same Bi-Si system after nanosecond irradiation. The K′-v used curve is reported in Fig.3, full line covers the range of K′-v previously measured, the dashed line is instead an extrapolation. At 1.2 J/cm^2 the profile is fitted by K′=0.15 and v=2.5 m/s. At 1.0 and at 0.6 J/cm^2 the couples of values are v=6 m/s, K′=0.35 and v≥10 m/s,K′≥0.5 respectively. These velocity values should be compared with those obtained by calculations.

The simple thermal model used in the nanosecond regime might be modified to account for the processes occurring during picosecond irradiation. The extension is then not trivial at all. As a starting point for calculations we have considered Si single crystals irradiated with 20 ps-0.53 μm. For this case many "in situ" measurements are available so that a comparison with calculation is feasible. The adopted model follows closely that of ref.[11] until the melting is reached; as soon as a layer starts to melt all the energy stored in the electronic system is converted into heat and the model becomes simi-

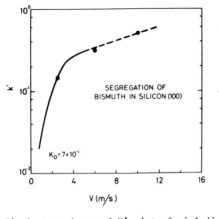

Fig.3. Dependence of K', interfacial distribution coefficient, on the solidification velocity for Bi implanted Si. The full points represent the K'-v values used to fit the data of Fig.2.

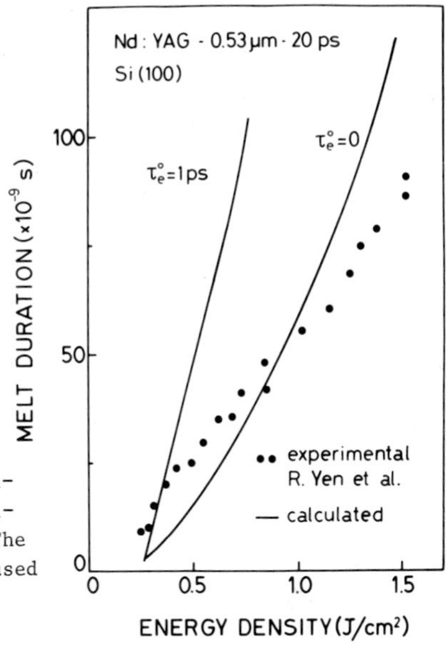

Fig.4. Melt duration vs energy density of 0.53 μm-20 ps laser pulse. The dots are measured values from ref.[3]. The lines are calculated assuming a relaxation energy time constant τ°_e (see eq.1) equal to 10^{-12}s and to 0 s.

lar to that of ref.[10], accounting for the computation of the melt front kinetics. As a parameter of main interest we have considered the rate of energy release of carriers to lattice and we have used the relation|11|.

$$\tau_e = \tau^{\circ}_e \left[1+(p/p_{crit})^2\right] \qquad (1)$$

with p_{crit}=2×10^{21} cm^{-3} and p carrier concentrations. The parameter τ°_e assumed different values, 10^{-11}-10^{-12} and 0. The last choice gives the description used for the nanosecond regime where the energy is assumed to be released instantaneously to the lattice.

Among the several quantities obtained from the calculations we report in Fig.4 the melt duration as a function of the energy density for τ°_e =10^{-12}s and for τ°_e=0 respectively. Experimental data from ref.[3] are also reported for comparison. The best overall agreement with experiments is obtained for τ°_e=0, although the parabolic shape of the calculated trend is quite different from that of measured values. The region at low energy density value requires a τ°_e of about 0.5 ps for a better agreement. At higher energy density value some other effects as overheating, evaporation, etc may change the coupling of the incident beam with the target. All these phenomena are not taken into account in the calculations.

As a first approximation we used the strictly thermal model (τ°_e=0) also for the 1.06 μm irradiation of ion implanted Si samples. The amorphous nature of

278

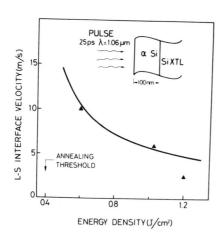

Fig.5. Solidification velocity vs the energy density of 1.06μm-25ps laser pulse. The full line represents calculated value in the strictly thermal model, the triangles are the obtained values from the fit of the data of Fig.2.

the surface layer has been included assuming for the absorption coefficient the value of 25000 cm^{-1}. All the other parameters were assumed the same as for the single crystal. The resulting solidification velocity is reported in Fig.5 as a function of the energy density. With increasing the thickness of the molten layer the velocity decreases due to reduction of the thermal gradient with time. The full triangles represent instead the velocity values obtained by the fits óf the experimental data reported in Fig.2.

The partial disagreement reflects that already shown in Fig.4. The strictly thermal model seems to underestimate the melt duration for energy density value lower than 1 J/cm^2 and to overestimate for energy density value above 1. J/cm^2. By analogy we probably also underestimate the melt duration and then overestimate the solidification velocity in the calculations performed at 1.06 μm irradiation.

In conclusion the overall trend of the experimental data obtained by picosecond irradiation is quite well described by a thermal model. The energy release time constant is less than 10^{-12}s. A more quantitative comparison requires a more detailed description of the several process which occcur at this very power densities as overheating, plasma screening, evaporation etc.

REFERENCES

1. N.Bloembergen,H.Kurz,J.M.Liu and R.Yen in Laser and Eelectron Beam Interaction with Solids, B.R.Appleton and G.K.Celler eds. (North Holland,N.Y. 1982) p.3

2. D.M.Kim,R.R.Shah,D.von der Linde and D.L.Crostwait in Laser and Electron Beam Interaction with Solids,B.R.Appleton and G.K.Celler eds. (North Holland, N.Y. 1982) p.85

3. R.Yen,J.M.Liu,H.Kurz and N.Bloembergen; Appl.Phys. A27,153(1982)

4. J.M.Liu,H.Kurz and N.Bloembergen, Appl.Phys.Lett. $\underline{41}$,643 (1982)

5. P.Baeri,S.U.Campisano,G.Foti and E.Rimini, J.Appl.Phys.$\underline{50}$,788 (1979)

6. P.Baeri,G.Foti,J.M.Poate,S.U.Campisano and A.G.Cullis, Appl.Phys.Lett. $\underline{38}$,800(1981)

7. G.J.Galvin, M.O.Thompson,J.W.Mayer,R.B.Hammond, N.Paulter and P.S.Peercy, Phys.Rev.Lett. $\underline{48}$,33(1982)

8. S.U.Campisano,P.Baeri,E.Rimini,A.M.Malvezzi and G.Russo, Appl.Phys.Lett. $\underline{41}$,456(1982)

9. C.W.White,D.M.Zehner,S.U.Campisano and A.G.Cullis, Chap.4 of Surface Modification and Alloying, J.M.Poate and G.Foti eds. (Plenum Press,N.Y. 1982)

10. P.Baeri and S.U.Campisano, Chap. 4 of Laser Annealing of Semiconductors J.M.Poate and J.W.Mayer eds. (Academic Press, N.Y.1982).

11. A.Lietola and J.G.Gibbons; Appl.Phys.Lett. $\underline{40}$,624(1982).

COHERENT PRECIPITATE FORMATION IN PULSED-LASER AND THERMALLY-ANNEALED,
ION-IMPLANTED Si.*

B. R. APPLETON, J. NARAYAN, O. W. HOLLAND AND S. J. PENNYCOOK
Solid State Division, Oak Ridge National Laboratory, Oak Ridge, TN 37830

ABSTRACT

It will be shown that under suitable conditions ion
implanted impurities in Si can precipitate and grow
coherently within the single crystal lattice during
recrystallization induced by pulsed laser or thermal
annealing. Ion channeling and transmission electron
microscopy (TEM) were used to characterize such precipitates
in Si implanted with Sb and B and thermally annealed, and in
Si implanted with Tl and annealed with a pulsed ruby laser.
The orientations of these precipitates were determined from
TEM and detailed angular scans using ion scattering channeling. The nucleation and precipitation processes will be
discussed in terms of differences in the liquid and solid
phase regrowth mechanisms.

INTRODUCTION

Careful examination of the formation or dissolution of precipitates during
thermal or laser annealing can provide important information on solute atom
interactions, defect-solute interactions, metastable alloy formation, and
crystal growth [1-7]. It is well established that supersaturated substitutional
alloys can be formed in Si by ion implantation doping followed by thermal
annealing [4], or pulsed laser annealing [5]. In the case of thermal (furnace
or cw laser) annealing the near-surface region of the single crystal is turned
amorphous by ion implantation and annealing in the range 450-600°C initiates
solid phase epitaxial (SPE) regrowth. Solute concentrations in these
"defect-free" regrown layers can exceed retrograde solubility limits by more
than a factor of 10^2 [4]. Trapping of the implanted impurities into substitutional sites in excess concentrations is possible because of the relatively
rapid motion of the crystalline/amorphous interface compared to dopant diffusion
in the solid state. If these SPE grown layers are subsequently heated solute
concentrations in excess of equilibrium solubility at this temperature precipitate out [4].

In the case of pulsed laser annealing the near surface of Si melts, the
dopant diffuses in the liquid, and during the rapid liquid phase epitaxial (LPE)
regrowth of the surface the dopants can acquire substitutional sites by solute
trapping at concentrations which greatly exceed equilibrium solid solubility
limits [5]. If these samples are subsequently heated solute concentrations in
excess of the solid solubility limits at that temperature can form precipitates.
Provided the dopant is soluble in liquid Si, (i.e., such as B, P, As, Sb, Ga or
In) precipitates can be redissolved into solution at the previous supersaturated
concentrations by pulsed laser irradiation [6]. A notable exception is Bi which
is immiscible in liquid silicon. In this case, provided the velocity of the
liquid-solid interface is sufficiently slow and the concentration of Bi

*Research sponsored by the Division of Materials Sciences, U. S. Department of
Energy under contract W-7405-eng-26 with Union Carbide Corporation.

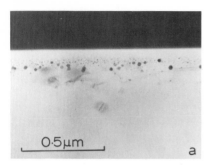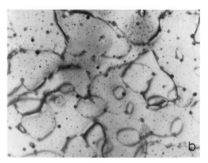

Fig. 1. Cross section (a) and plan view (b) electron micrographs of Sb
implanted (6 x 10^{15} Sb/cm^2, 200 keV) thermally annealed Si.

sufficiently high, precipitates will form in the liquid state and be frozen into
the solid state on solidification [7].

In this paper we extend such studies to show that under suitable conditions
coherent precipitates of implanted dopants can be formed in Si during both LPE
and SPE. Systematics of the effect will be presented as studied by ion
scattering/channeling and TEM. Formation mechanisms are proposed and applications for materials research discussed.

RESULTS AND ANALYSIS

First consider the case of coherent precipitate formation which occurs during
SPE. Samples of (111) Si were implanted at room temperature with 6 x 10^{15}
Sb$^+$/cm^2 at 200 keV. Ion scattering/channeling and TEM showed that as a result
of implantation the first 1580 Å near the surface was turned amorphous and was
followed by a band of dislocations which separated the amorphous surface layer
from the single crystal substrate. This sample was furnace annealed in flowing
dry nitrogen at 550°C for 15 min to initiate SPE. Previous analyses of similarly treated samples [4] shows that SPE starts from the dislocation band with
the single crystal substrate acting as a seed for crystal growth. The amorphous
region crystallizes free from extended defects and the dopant is incorporated
into substitutional sites. Under the present conditions it was observed that Sb
was incorporated into substitutional sites at concentrations ~1.6 x 10^{21} cm^{-3}
which is in excess of the retrograde maximum solubility, 7 x 10^{19} cm^{-3}, by more
than a factor of 20 [4]. This SPE process thus leads to the formation of
metastable supersaturated solid solutions.

Following the SPE process, the Si(Sb) sample was annealed again at 900°C for
30 min. Figure 1(a) is a cross-section micrograph which shows that precipitates
have formed as a result of the metastability of the solid solution. The precipitate distribution is observed to be peaked at the mean projected range (810 Å)
of the implanted ions. The smaller size precipitates observed in the
micrographs are probably formed by a homogeneous nucleation and growth process.
The large size precipitates were found to be connected to the dislocations, as
shown in the plan-view micrograph of Fig. 1(b). The dislocations near some of
the large precipitates are out of contrast because the displacement vector associated with the dislocations is normal to the Burgers vector of the dislocation.

An ion channeling analysis of this sample is shown in Fig. 2 for 2.0 MeV He$^+$
incident parallel to [101]. The scattering yield from the Si region of the
aligned spectra in Fig. 2 for [101] shows a minimum yield ~4% compared to the
random reference spectrum. This indicates that the Si host is relatively good,
unstrained single crystal despite the presence of precipitates. The small peak

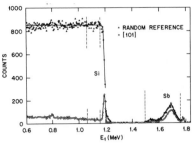

Fig. 2. 2.0 MeV He ion
scattering/channeling analysis of Si ion
implanted (6×10^{15} Sb/cm^2, 200 keV) and
thermally annealed.

in the scattering yield near 0.8 MeV for the [101] spectrum corresponds to scattering from displaced atoms in the dislocation band located at the original amorphous/crystalline interface before SPE. The Sb region of the random spectrum in Fig. 2 shows the distribution in energy (depth) of the precipitate distribution; and the aligned spectrum indicates that some of the Sb has a perferential lattice location. To investigate this further detailed angular scans were performed by monitoring the scattering yield of both Si and Sb within the energy windows shown as dashed lines in Fig. 2, as a function of the angle of incidence to [101]. The results of this angular scan are shown in the extreme right hand panel of Fig. 3. Similar angular scans were measured along the [111] adn [001] axial directions and these are also shown in Fig. 3. These angular scan results are consistent with the behavior expected for coherent Sb precipitates in Si. The minimum yields for Sb are high compared to Si because all the precipitates are not aligned with the Si lattice. Since the Sb and Si angular widths are comparable for the [111] and [001] scans these results could be interpreted as arising from isolated Sb interstitials as well as from aligned precipitates with similar interplanar spacings. However, the only reasonable explanation for the relatively narrower half width for Sb compared to Si along [101] is that is is due to some coherent precipitates with narrower interplanar spacings aligned along [101].

Electron microscope observations showed these precipitates to be rhombododecahedral in shape with 12 faces formed by the {110} planes of Si. Precipitates within the Si matrix were determined from microdiffraction studies (Fig. 4) to have transformed from the normal structure of Sb (the arsenic structure) to a

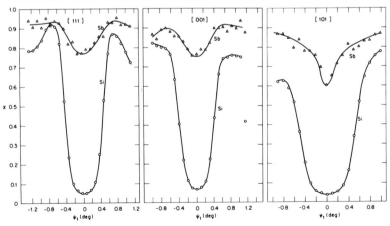

Fig. 3. Detailed angular scans for Si implanted with Sb and thermally annealed.

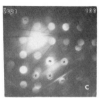

Fig. 4. (a) Microdiffraction pattern from precipitate in thin region showing 0$\overline{1}$11 pole of normal Sb. (b) Same pole from precipitate in thicker region shows absences and can be indexed as 101 pole of new primitive rhombohedral phase. (c) ($\overline{2}$021) pole of normal Sb is identical to (100) pole of the new phase. n=spots from Si matrix; d=double diffraction spots.

new phase having a primitive rhombohedral form. The orientation was determined to be Sb{001} || Si{111} and Sb<010> || Si<121>. Although only small matrix strains were visible under dynamical diffraction conditions, the precipitates are thought to be coherent since no interface dislocations were seen, but Moiré fringes were observed. In addition, precipitates in the thinnest regions of the Si had the normal Sb structure, suggesting that the new phase is stabilized by coherency with the Si lattice. A more detailed description of the identification of the primitive rhombohedral Sb phase will be reported elsewhere.

The second case where coherent precipitates have been observed involved pulsed laser annealing and LPE [8]. Samples of Si {100} single crystals were implanted near room temperature with ^{205}Tl$^+$ to doses of 2.9 x 10^{15} cm^{-2} and 1 x 10^{16} cm^{-2} at 90 keV. Portions of the implanted samples were annealed using a Q-switched ruby laser (15 ns pulse duration). Typical ion scattering/channeling results are shown in Fig. 5 for a Si single crystal implanted with 1 x 10^{16} ^{205}Tl/cm^2 and annealed with a single 1.6 J/cm^2, 15 ns pulse from the ruby laser. The channeling spectrum in Fig. 5 (open triangles) was obtained with the 1.5 MeV He$^+$ beam incident parallel to <111>, while the reference random spectrum (open circles) was obtained by continuously rotating the crystal to average over many crystallographic orientations. Comparison of the implanted impurity profiles in Fig. 5 shows that the fraction of Tl which is substitutional along <111> following pulsed laser annealing is quite small. Despite this, and the fact that the implanted Tl concentration is quite large, comparison of the channeled and random Si spectra shows that the silicon host is relatively good, unstrained, single crystalline material. The Si crystal implanted with 2.9 x 10^{15} ^{205}Tl/cm^2 and annealed (1.6 J/cm^2, 15 ns) showed features similar to those for the sample of Fig. 5. Detailed angular scans were made across the major axial and planar directions of both samples to determine the relative Tl lattice sites. Measurements from the 2.9 x 10^{15} ^{205}Tl/cm^2 sample are shown in Fig. 6 for scans across <110> tracking in the {100} planes, and across <111> tracking

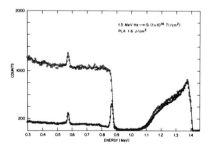

Fig. 5. 1.5 MeV He ion scattering/channeling spectra from a Si {100} single crystal ion implanted (1x10^{16} Tl cm^2) and laser annealed (1 pulse, 1.6 J/cm^2), for incidence-parallel to <111> (open triangles) and in a random (open circles) orientation.

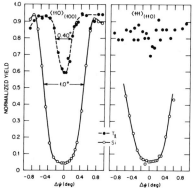

Fig. 6. Detailed angular scans showing
the near surface yields from Tl and Si
for 1.5 MeV He ions incident near <110>
and <111> of a Tl implanted and laser
annealed Si single crystal.

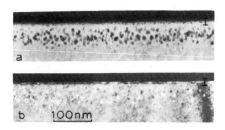

Fig. 7. Bright field cross section
electron micrographs of a Tl implanted
(2.9x10^15/cm^2) pulse laser annealed
(1.6 J/cm^2) Si single crystal under (a)
kinematical and (b) dynamical diffraction conditions.

in the 110 planes. The <110> scan clearly shows that Tl atoms are aligned with
the Si lattice along this direction, but the Tl minimum yield is much higher and
its angular half width much narrower than Si. Although not as pronounced, the
<111> data in Fig. 6 also show a preferential alignment with Si. Scans taken
along the <100> axial direction were comparable to the <110> results in Fig. 6
and 100 planar scans also showed Tl alignment. Angular scans for the higher
dose sample were qualitatively the same. Analysis of the as-implanted portions
of both samples showed that the near surface regions had been made amorphous by
the Tl implantations and consequently no preferential lattice sites could be
detected. This is not inconsistent with early results where large interstitial
Tl concentrations were reported since these results were either obtained from
lower dose implantations (10^{13}-10^{15} cm^{-2}), or were made on crystals that were
heated during implantation and/or annealed afterwards [9-13].

Cross-section electron microscopy was performed on the low dose sample (2.9 x
10^{15} Tl/cm^2, 1.6 J/cm^2) and this analysis verified that coherent Tl precipitates
were present in the Si. Figure 7(a) shows a cross-section electron micrograph
taken under bright-field, kinematical diffraction conditions. Thallium precipitates exhibit black contrast under these diffraction conditions. The average
size of the precipitates was determined to be 100 A with the precipitate distribution extending up to 600 A from the surface. The same area of a specimen
imaged under dynamical conditions is shown in Fig. 7(b). The precipitates
exhibit black-white contrast indicating the presence of unrelaxed strains around
the precipitates. This contrast behavior under dynamical diffraction conditions,
is consistent with the precipitates being coherent with the silicon matrix.

DISCUSSION

The mechanisms by which coherent precipitates form in these two cases are
quite different but both cases provide interesting insights into solute interactions and illustrate how these techniques can be manipulated to study precipitation growth in the solid and liquid state. In the case of Sb in Si a
supersaturated substitutional alloy is first formed by SPE. The Si(Sb) crystallization rate at 550°C is ~100 Å/min so the transit time for the amor-
phous/crystalline interface to pass any particular Sb atom is ~1 s. Assuming a
solid state diffusion D_s ~ 3 x 10^{-24} cm^{2}s^{-1}, an Sb atom can move a distance
δ_1 ~ $\sqrt{D_s\ 1_s}$ = 2 x 10^{-4} Å in the time it takes the interface to pass, and a

distance $\delta_2 \sim \sqrt{D_s \times (9 \times 10^2 s)} = 5 \times 10^{-3}$ Å during the entire SPE process. It is the relatively rapid motion of the interface compared to diffusion distances that is responsible for solute trapping at supersaturated concentrations [4], and with such short diffusion distances during SPE the supersaturated solution doesn't have time to form precipitates. It is important to note at this point that we are dealing with a defect free supersaturated region and that the presence of extended defects can significantly alter precipitation processes [14]. Subsequent annealing of this supersaturated solid solution to 900°C for 30 min is required to initiate the homogeneous nucleation of Sb precipitates shown in Fig. 1. The increased time and temperature provides the mobility required to form precipitates in the solid, and the lower melting temperature of Sb (630°C) allows the Sb precipitates, once formed, to grow coherently in the Si lattice on cooling.

In contrast, the coherent precipitates of Tl formed in Si by pulsed laser annealing nucleate in the liquid state. The absorbed laser light melts the near surface for times $\tau \sim 200$ ns [5,15]. During the time the silicon is molten, assuming a diffusion constant for Tl in molten silicon $D_\ell \sim 5 \times 10^{-4}$ cm^2/s [16], the Tl can move distances $\sim \sqrt{D_\ell \tau} \sim 10^3$ Å. Calculations show that such mobility is sufficient to nucleate precipitates of the dimensions observed while the surface is molten. This process is favored because Tl is immiscible in liquid silicon [17]. Solid state diffusion of Tl ($D_s = 8 \times 10^{-14}$ cm^2/s at 1100°C) [18] would result in mobilities much too low ($\sqrt{D_s \tau} \sim 10^{-2}$ Å) to nucleate such precipitates. However, as in the Sb case, because of the much lower melting temperature of Tl (303°C) compared to Si (1410°C) coherence may be established when the still-molten Tl precipitates solidify in the crystalline Si matrix.

REFERENCES

1. Laser and Electron-Beam Interactions with Solids, ed. by B. R. Appleton and G. K. Celler, North Holland, New York, 1982.
2. Laser and Electron-Beam Solid Interactions and Materials Processing, ed. by J. F. Gibbons, L. D. Hess and T. W. Sigmon, North Holland, New York, 1981.
3. Laser and Electron Beam Processing of Materials, ed. by C. W. White and P. S. Peercy, Academic Press, New York, 1980.
4. J. Narayan and O. W. Holland, Appl. Phys. Lett. 41, 239 (1982) and references therein.
5. C. W. White, S. R. Wilson, B. R. Appleton, and F. W. Young, Jr., J. Appl. Phys. 51, 738 (1980) and references therein.
6. J. Narayan, Appl. Phys. Lett. 34, 312 (1979).
7. C. W. White, p. 109, ibid 1.
8. B. R. Appleton, J. Narayan, C. W. White, J. S. Williams and K. T. Short, submitted for publication in the Proceedings of the 1982 Ion Beam Modification of Materials Conference, Grenoble, France, September 6-10, 1982.
9. L. Eriksson, G. R. Bellavance, and J. A. Davies, Rad. Eff. 1, 71 (1969).
10. L. Eriksson, J. A. Davies, N.G.E. Johansson, and J. W. Mayer, J. Appl. Phys. 40, 843 (1969).
11. G. Fladda, P. Mazzoldi, E. Rimini, D. Sigurd, and L. Eriksson, Rad. Eff. 1, 249 (1969).
12. L. Eriksson, G. Fladda, and K. Bjorkqvist, Appl. Phys. Lett. 14, 195 (1969).
13. B. Dormeij, G. Fladda, and N.G.E. Johansson, Rad. Eff. 6, 155 (1970).
14. J. Narayan, O. W. Holland and B. R. Appleton, submitted for publication.
15. See for example papers and references in Laser and Electron-Beam Interactions with Solids, ed. by B. R. Appleton and G. K. Celler, North Holland, New York, 1982.
16. By analogy with data from K. J. Bachman, p. 502 in Current Topics in Materials Science, Vol. 3, ed. by E. Kaldis, North Holland, New York, 1979.
17. Constitution of Binary Alloys by M. Hansen, McGraw-Hill, New York, 1958.
18. Diffusion in Semiconductors by N. I. Boltaks, Academic Press, New York, 1963.

DOPANT INCORPORATION DURING RAPID SOLIDIFICATION*

C. W. WHITE, D. M. ZEHNER, J. NARAYAN, O. W. HOLLAND, B. R. APPLETON
Solid State Division, Oak Ridge National Laboratory, Oak Ridge, TN 37830

and

S. R. WILSON
Motorola Inc., Phoenix, AZ

ABSTRACT

Incorporation of Group III, IV, V dopants in silicon
occurs as a result of solute trapping during laser
annealing. Distribution coefficients and substitutional
solubilities are far greater than equilibrium values, and
can be functions of growth velocity and crystal orientation.
Mechanisms limiting dopant incorporation at high concentrations are identified and discussed.

INTRODUCTION

In pulsed laser annealing of ion implanted silicon, the rapid deposition of
energy into the near surface region leads to melting, followed by liquid phase
epitaxial regrowth from the underlying substrate [1-4]. During recrystallization the velocity of the liquid-solid interface can be several meters per second
[5] which leads to solidification conditions that are far from equilibrium at
the interface. At these velocities, dopant incorporation occurs by means of
solute trapping and dopant concentrations far in excess of the equilibrium solubility limit can be achieved [6]. Here we review some of the systematic studies
of the incorporation of Group III, IV and V dopants in silicon during rapid
solidification. The behavior of these impurities has been studied at both low
and high dopant concentrations. At low dopant concentrations, the dopant
profiles measured after laser annealing can be compared to model calculations of
dopant redistribution in order to determine the distribution coefficient (k')
from the liquid during solidification [6]. For each impurity there is a maximum
concentration (C_s^{max}) which can be incorporated substitutionally in the lattice
during pulsed laser annealing [6]. The mechanisms which limit the incorporation
of dopant into the crystal lattice are discussed as well as the dependence of
growth kinetics upon interface velocity [7,8] and crystal orientation [9].

EXPERIMENTAL

Group III, IV and V impurities were implanted into both (100) and (111) silicon to doses in the range of 10^{15}-10^{17}/cm^2. Implantation energies were chosen
to give a projected range of $\sim$800 Å. Laser annealing was carried out using a
pulsed ruby laser (6943 Å, $\sim$12 x 10^{-9} s) or a pulsed XeCl laser (3080 Å, $\sim$35 x
10^{-9} s). By changing the energy density obtained from these lasers one can vary
the regrowth velocity in the range of 2-6 m/s as determined from heat flow
calculations. Analysis of the implanted region was carried out using Rutherford

*Research sponsored by the Division of Materials Sciences, U.S. Department of
Energy under contract W-7405-eng-26 with Union Carbide Corporation.

288

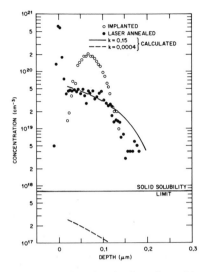

Fig. 1. Dopant profiles for ^{115}In (125 keV, 1.2×10^{15}/cm^2) in (100) Si compared to model calculations. The dashed profile is calculated assuming $k'=k_o$. From Ref. 6.

backscattering (RBS) -ion channeling techniques. Selected crystals were examined by transmission electron microscopy (TEM) in order to determine the microstructure in the near surface region.

DETERMINATION OF DISTRIBUTION COEFFICIENTS

At low dopant concentrations, interfacial distribution coefficients (defined as $k' = C_S/C_L$ where C_S and C_L are dopant concentrations in the solid and liquid phase at the interface) can be determined by comparing measured dopant profiles to model calculations for dopant redistribution by liquid phase diffusion during solidification [6]. In these model calculations, k' is determined by fitting the calculated profile to experimental results using only k' as an adjustable parameter. Figure 1 shows such a fit to the experimental data for the case of In in (100) Si. As a result of laser annealing (Ruby, 1.3 J/cm^2, 4.5 m/s) approximately 60% of the In is zone refined to the surface. The In remaining in the bulk is highly substitutional even though the concentration greatly exceeds the equilibrium solubility limit. The concentration profile can be fit with a value for k' of 0.15, which is much higher than the equilibrium value (k_o = 0.0004). If the recrystallization occurred under conditions of local equilibrium at the interface, the dotted profile in Fig. 1 would be expected and essentially all of the In would have been zone refined to the surface.

Using similar procedures, values for k' have been determined for many impurities in silicon. These are summarized in Table I and compared with equilibrium values (k_o) [10]. Values for k' were determined at a growth velocity of 4.5 m/s except in the cases of B, P and Sb where a velocity of 2.7 m/s was used. In every case $k' > k_o$ which reflects the nonequilibrium nature of this high speed liquid phase epitaxial regrowth process. The increased values for k' relative to k_o show that these dopants do not exchange a sufficient number of times between the solid and liquid phases at the interface to establish their equilibrium concentrations before being permanently incorporated into the solid.

Values for k' are functions of both growth velocity [7,8] and crystal orientation [9]. One can intuitively understand the dependence on growth velocity since as $V \rightarrow 0$, the value for k' must approach k_o. The dependence on crystal orientation is not so obvious and an example is shown in Fig. 2 for (100) and

TABLE I
Comparison of distribution coefficients under equilibrium (k_o) and laser
annealed (k') regrowth conditions.

Dopant	k_o	k'
B	0.80	∿1.0
P	0.35	∿1.0
As	0.30	∿1.0
Sb	0.023	0.7
Ga	0.008	0.2
In	0.0004	0.15
Bi	0.0007	0.4

(111) crystals implanted with [115]In (125 keV, 1.2 x 10^{15}/cm^2) and annealed under
identical conditions. Considerably more In is trapped in the bulk of the (111)
crystal, implying that the value for k' is larger than that for the (100) orientation. Figure 3 summarizes the reported orientation and velocity dependence
for the case of In in silicon [11]. As shown in the figure, up to ∿4 m/s the
value for k' is systematically larger for the (111) case. The increased value
for k' in the (111) case compared to the (100) case at the same growth velocity
has been attributed to a larger cooling on the (111) surface. This is discussed
by Gilmer in these proceedings [12].

MAXIMUM SUBSTITUTIONAL SOLUBILITIES: ORIENTATION DEPENDENCE

As the dopant concentration is increased, ultimately a maximum substitutional
concentration (C_s^{max}) is reached above which the dopant can no longer be incorporated into substitutional lattice sites for a given regrowth velocity [6].

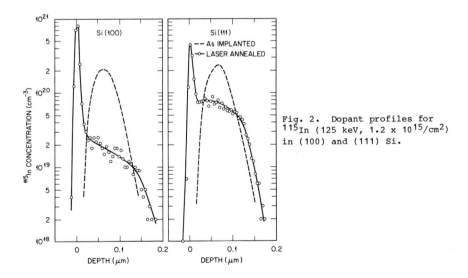

Fig. 2. Dopant profiles for
[115]In (125 keV, 1.2 x 10^{15}/cm^2)
in (100) and (111) Si.

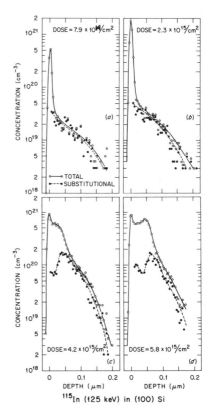

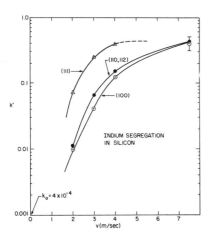

Fig. 3. Dependence of k' on growth velocity and crystal orientation for In in Si. From Ref. 11.

Fig. 4. Dose dependence of solute trapping for In in Si. At low doses the total and substitutional concentrations in the bulk are indistinguishable while at high doses they differ considerably when the concentration exceeds C_s^{max}. From Ref. 22.

This is demonstrated in Fig. 4 for four different doses of In (125 keV) in (100) Si, annealed with a ruby laser at 1.3 J/cm^2. In this example RBS-channeling measurements were used to measure both the total concentration and the substitutional concentrations as a function of depth. At the two lower doses, in the bulk of the crystal the total and substitutional concentrations are virtually identical and the profiles scale with the implanted dose. For the two higher doses, the total and substitutional concentrations are nearly the same up to a concentration of 1.5-2.0 x 10^{20}/cm^3; as the total concentration increases above this value, the substitutional concentration remains the same or decreases somewhat. For these regrowth conditions (V = 4.5 m/s) this is the maximum concentration (C_s^{max}) which can be incorporated into substitutional lattice sites. By similar techniques, values for C_s^{max} have been determined for nine Group III, IV or V impurities in both (100) and (111) silicon at a growth velocity of 4.5

TABLE II
Substitutional solubilties in silicon achieved by recrystallization at 4.5 m/s

DOPANT	c_s^o (cm^{-3})	c_s^{max} (cm^{-3}) (100) Si	c_s^{max} (cm^{-3}) (111) Si	COMMENT
As	1.5×10^{21}	6.0×10^{21}	6.0×10^{21}	Thermodynamic limit
Sb	7.0×10^{19}	2.0×10^{21}	2.0×10^{21}	Cell formation
Bi	8.0×10^{17}	4.0×10^{20}	8.6×10^{20}	Precipitation Cell formation
Ge	5.0×10^{22}	6.0×10^{21}	$>1.2 \times 10^{22}$	Cell formation on (100)
Sn	5.5×10^{19}	9.8×10^{20}	1.4×10^{21}	Cell formation
Pb	---	1.0×10^{20}	3.0×10^{20}	Precipitation Cell formation
B	6.0×10^{20}	2.0×10^{21}	2.0×10^{21}	Mechanical strain
Ga	4.5×10^{19}	4.5×10^{20}	7.2×10^{20}	Cell formation
In	8.0×10^{17}	1.5×10^{20}	4.5×10^{20}	Cell formation
Tl	---	---	---	Coherent precipitation

m/s. These values are summarized in Table II and compared with equilibrium solubility limits (c_s^o). The last column in Table II indicates the mechanism that limited the substitutional solubility under the growth conditions studied. For the remainder of this paper we will discuss these limitations to substitutional solubility.

Substitutional solubilities which can be achieved during pulsed laser annealing appear to be limited by four mechanisms. The most fundamental of these is a thermodynamic limit to dopant incorporation which restricts substitutional solubility even at infinite growth velocity [13]. In addition to this fundamental limit, practical limitations are imposed by interfacial instability during regrowth, by mechanical strain in the recrystallized region, and by dopant precipitation in the liquid for those impurities which phase separate in the liquid. Values for c_s^{max} can be functions of both growth velocity [14] and crystal orientation if interface instability or dopant precipitation is the dominant limitation.

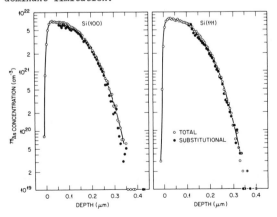

Fig. 5. Limitations to the incorporation of As in (100) and (111) Si. Total and substitutional concentrations are plotted as a function of depth after laser annealing.

Fig. 6. Cross-section TEM micrograph for ^{75}As (100 keV, 1.2×10^{17}/cm^2) in (100) Si after laser annealing.

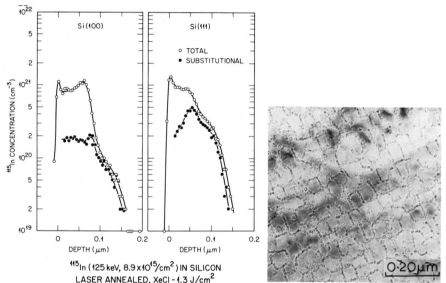

^{115}In (125 keV, 8.9×10^{15}/cm^2) IN SILICON LASER ANNEALED, XeCl ∼1.3 J/cm^2

Fig. 7. Limitations to the incorporation of In in (100) and (111) Si. Regrowth velocity is ∼4.5 m/s.

Fig. 8. Plan-view microstructure for ^{115}In (125 keV, 8.9×10^{15}/cm^2) in (100) Si after laser annealing. Cell walls contain the pure phase of In.

Predictions of thermodynamic limits to dopant incorporation in silicon at infinite growth velocity have been made by Cahn [13]. The basic idea behind these predictions is that on a plot of the Gibbs free energy versus composition at fixed temperature, the solidus and liquidus lines intersect at one point. This is the limit to the solid composition which can be formed from the liquid at any composition since there is no barrier to nucleation of the solid phase. The locus of these points plotted on the phase diagram then determines the T_0 curve, the limit to dopant incorporation during solidification at infinite growth velocity.

Although substitutional solubilities achieved by pulsed laser annealing approach Cahn's predicted limits, in only one case (As in Si) do we appear to have reached this limit [14]. For the case of As in Si, the value for C_s^{max} is independent of growth velocity (in the range 2-6 m/s) and is independent of crystal orientation. Figure 5 shows results obtained at 6.0 m/s (as determined from heat flow calculations) in (100) and (111) Si. In each case the As is measured to be substitutional up to a concentration of 6 x 10^{21}/cm^3. When the total concentration reaches this value, epitaxial growth stops. To ion channeling the near-surface region appears to be amorphous. However, TEM results shown in Fig. 6 reveal that the near surface contains polycrystallites, As precipitates, and even small regions of amorphous material. The line of demarkation between the epitaxially recrystallized region and the disordered near surface region is relatively sharp. It appears as if epitaxial regrowth proceeded normally until a concentration of 6 x 10^{21}/cm^3 was reached at which point the advancing interface slowed considerably and thereafter polycrystallites nucleated and As precipitates formed. A similar microstructure was observed also in the case of (111) Si. Based on the fact that the value for c_s^{max} is independent of velocity (in the range 2-6 m/s) and is independent of crystal orientation we conclude that the thermodynamic limit had been reached. This result is also in reasonable agreement with Cahn's prediction of 5 x 10^{21}/cm^3 for this alloy system [13].

For several of the impurities in Table II, values for c_s^{max} are limited by interfacial instability which develops during regrowth and leads to the formation of a well defined cell structure in the near surface region. These impurities include Ga, In, Sn, Sb and to a certain extent Bi. Interface instability is caused by constitutional supercooling which occurs only when k' < 1. For these impurities it has been found that k' is greater for the (111) case compared to the (100) case and this implies that greater concentrations can be trapped in the lattice for the (111) case before interface instability develops. Figure 7 shows this to be the case for In in silicon. For the (100) case, instability develops when the concentration in the solid reaches 1.5-2.0 x 10^{20} while for the (111) case up to 5 x 10^{20}/cm^3 can be incorporated before the interface becomes unstable. Figure 8 shows the microstructure in the near-surface region of the (100) crystal. This is the well known cell structure that develops as a result of interface instability. The interior is an epitaxial column of silicon extending to the surface and containing the substitutional In. Surrounding each cell is a thin wall containing the nonsubstitutional In in the pure phase. Both the cell size and the concentration at which interface instability develops can be predicted [15,16] with remarkable accuracy using the Mullins and Sekurka theory of interface instability [17] modified to account for the large departures from equilbrium during regrowth.

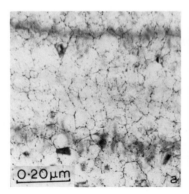

Fig. 9. Plan view microstructure (a) and diffraction pattern (b) for [115]In (125 keV, 8.9 x 10^{15}/cm^2) in (111) Si after laser annealing. The diffraction pattern shows the presence of twins in the near-surface region.

294

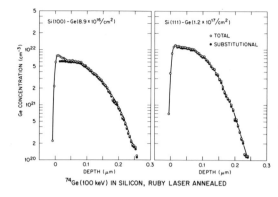

Fig. 10. Limitations to the incorporation of Ge in (100) and (111) Si. Regrowth velocity is ∿4.5 m/s. In the (111) case, Ge introduction by ion implantation is sputter limited.

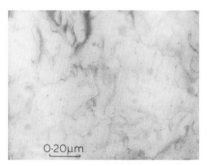

Fig. 11. Plan view micrograph for ^{74}Ge (100 keV, 8.9 x 10^{16}/cm^2) in (100) Si after laser annealing.

Fig. 12. Plan-view micrograph (a) and diffraction pattern (b) for ^{74}Ge (100 keV, 1.2 x 10^{17}/cm^2) in (111) Si. The near-surface region contains twins but no cells after laser annealing.

Figure 9 shows the microstructure in the near surface region for the (111) case. A similar cell structure, although less well defined, is observed in this case. However in the (111) case, in addition to the cells, twins are observed to lie in all (111) planes as demonstrated by the diffraction pattern. During solidification, twins develop due to interface instabilities [18]. Consequently it is not surprising to observe the formation of twins at dopant concentrations comparable to those required for cell formation. At these growth velocities, twins are observed in (111) crystals only for those impurities where k' < 1.

Results for the case of Ge are presented in Fig. 10. In the (100) case the limit to substitutional solubility is ∿6 x 10^{21}/cm^3 while for the (111) case the total and substitutional concentrations are undistinguishable up to concentrations in excess of 1.2 x 10^{22}/cm^2. Figure 11 shows the cell structure formed in the near surface region for the (100) case. This is similar to that observed in previous studies for Ge films deposited on (100) Si [19]. The microstructure in the near-surface region of the (111) crystal is shown in Fig. 12. In this case there is no indication of a cell structure such as that shown in Fig. 11, but defects in the form of twins are observed in the near-surface region. The twins result from interface instability and their presence indicates that at higher Ge concentrations it may be possible to form the cell structure resulting from constitutional supercooling. To investigate this further will require the use of deposited layers because Ge introduction by ion implantation for the (111) case is limited by sputtering.

For several of the dopants listed in Table II, precipitation in the liquid during laser annealing limits the values for C_s^{max}. These impurities include Bi, Pb and Tl. Each of these impurities can phase separate in the liquid. This will occur if the dopant concentration is high enough and if the time available for nucleation and growth of precipitates in the liquid is long enough. Precipitates formed in the liquid will be frozen into the solid during solidification. The three impurities which exhibit this behavior in Table II will also undergo constitutional supercooling because in each case k' < 1. Consequently at low growth velocities precipitation in the liquid will dominate while at high growth velocities interface instability will lead to cell formation. This behavior has been clearly demonstrated in the case of Bi in silicon [14]. For the case of Tl in Si, the precipitates are observed to be coherent with the silicon lattice [21].

In the case of B in Si, the mechanical properties of the implanted region after laser annealing provides the limit to dopant incorporation [14]. When B is incorporated substitutionally in the lattice during laser annealing, the lattice undergoes a one-dimensional contraction [20]. The resulting strain in the implanted region can lead to cracks when the strain exceeds the fracture strength of silicon. For B in silicon this occurs at a B concentration of ∿4 at.%. for either (100) or (111) silicon. Higher concentrations of B could possibly be achieved if a dopant which gives rise to a lattice expansion was simultaneously incorporated.

CONCLUSIONS

Studies of dopant incorporation during pulsed laser annealing have shown that recrystallization of the melted region occurs under conditions that are far from equilibrium at the interface. At low concentrations, model calculations can be used to determine the distribution coefficient during regrowth. At high concentration, maximum substitutional solubilities are measured. These maximum substitutional solubilities appear to be limited by four mechanisms. The thermodynamic limit to dopant incorporation appears to have been reached in the As-Si system. Incorporation of impurities such as Ga, In, Sn, Ge and Sb is limited by interface instability during solidification. Interface instability leads to cell formation in the near-surface region. For (111) crystals, twin formation resulting from interfacial instability is observed when the impurity concentration is comparable to that required for cell formation. In these cases values for C_s^{max} are dependent on crystal orientation due to the dependence of k' on orientation. Impurities such as Bi, Pb and Tl will precipitate in the liquid if the concentration is high enough and if the time available for nucleation and growth of precipitates is long enough. At higher growth velocities, interfacial instability will lead to cell formation. In one case, B in Si, the mechanical strain in the recrystallized regions provides a practical limit to dopant incorporation.

REFERENCES

1. Laser-Solid Interactions and Laser Processing-1978, (ed. by S. D.
    Ferris, H. J. Leamy and J. M. Poate, American Institute of Physics,
    New York), 1979.
2. Laser and Electron Beam Processing of Materials, (ed. by C. W. White
    and P. S. Peercy, Academic Press, New York), 1980.
3. Laser and Electron-Beam Solid Interactions and Materials Processing,
    (ed. by J. F. Gibbons, L. D. Hess and T. W. Sigmon, North Holland, New
    York), 1981.
4. Laser and Electron Beam Interactions with Solids, (ed. by B. R. Appleton
    and G. K. Celler, North Holland, New York), 1982.
5. G. J. Galvin, M. O. Thompson, J. W. Mayer, R. B. Hammond, N. Paulter
    and P. S. Peercy, Phys. Rev. Lett. $\underline{48}$, 33 (1982).
6. C. W. White, S. R. Wilson, B. R. Appleton and F. W. Young, Jr., J.
    Appl. Phys. $\underline{51}$, 738 (1980).
7. A. G. Cullis, H. C. Webber, J. M. Poate and A. L. Simons, Appl. Phys.
    Lett. $\underline{36}$, 320 (1980).
8. P. Baeri, J. M. Poate, S. U. Campisano, G. Foti, E. Rimini, and A. G.
    Cullis, Appl. Phys. Lett. $\underline{37}$, 912 (1981).
9. P. Baeri, G. Foti, J. M. Poate, S. U. Campisano and A. G. Cullis,
    Appl. Phys. Lett. $\underline{38}$, 800 (1981).
10. F. Trumbore, Bell Syst. Tech. Jour. $\underline{39}$, 205 (1960).
11. J. M. Poate, Ref. 4, p. 121.
12. G. H. Gilmer, these proceedings.
13. J. W. Cahn, S. R. Coriell and W. J. Boettinger, Ref. 2, p. 89.
14. C. W. White, Ref. 4, p. 109.
15. J. Narayan, J. Appl. Phys. $\underline{52}$, 1289 (1981).
16. A. G. Cullis, D.T.J. Hurle, H. C. Webber, N. G. Chew, J. M. Poate,
    P. Baeri, and G. Foti, Appl. Phys. Lett. $\underline{38}$, 642 (1981).
17. W. W. Mullins and R. F. Sekerka, J. Appl. Phys. $\underline{35}$, 444 (1964).
18. J. Narayan, J. Appl. Phys., December 1982, in press.
19. H. J. Leamy, C. J. Doherty, K.C.R. Chiu, J. M. Poate, T. T. Sheng and G. K.
    Celler, Ref. 2, p. 581.
20. B. C. Larson, C. W. White and B. R. Appleton, Appl. Phys. Lett. $\underline{32}$, 801
    (1978).
21. B. R. Appleton, J. Narayan, O. W. Holland and S. J. Pennycook, these pro-
    ceeedings.
22. J. Narayan, H. Naramoto and C. W. White, J. Appl. Phys. $\underline{53}$, 912 (1982).

PULSED LASER ANNEALING OF ION IMPLANTED Ge*

O. W. HOLLAND, J. NARAYAN, C. W. WHITE, B. R. APPLETON
Solid State Division, Oak Ridge National Laboratory, Oak Ridge, TN 37830

ABSTRACT

Ion backscattering/channeling and transmission electron
microscopy (TEM) were used to investigate the annealing behavior
of ion implanted Ge single crystals using a Q-switched ruby
laser. The impurities studied were Bi, In, Sb and Pb, which were
implanted at liquid nitrogen temperature into both (100) and
(111) crystal orientations. A rather unique damage structure
which can form during room temperature implantation of Ge is
discussed. Maximum substitutional concentrations, which far
exceed the retrograde maxima, are reported for all the dopants
studied in (100)Ge. The maximum concentrations were limited by
an interfacial instability during epitaxial growth following
laser irradiation which led to the formation of a well-defined
cellular structure.

INTRODUCTION

Pulsed laser annealing has been actively studied in Si. In a simple thermal
melting model, the near surface region, < 1 μm thick, is envisioned to turn
molten as a result of the deposited energy of the laser light. As this layer
cools, epitaxial growth occurs on the crystalline substrate. Growth velocities
between 1-6 m/sec, useful for device applications, have been achieved using Q-
switched lasers. Crystalline material, free from extended defects, results from
this rapid growth [1]. Phenomena such as dopant redistribution [2,3] and surface segregation [4], solute trapping and metastable alloying [5,7],
cell formation [5,7] and resolution of precipitated phases [8] have been
observed after laser irradiation and have been cited as evidence in support of
the melting model. Our work extends the study of such phenomena to another
material germanium. The annealing behavior of ion implanted, group(III,V)
dopants -In and Bi, as well as Sn and Pb are investigated using a pulsed ruby
laser with energy density = 1.3 J/cm^2 and pulse duration = 15 nsec.

One of the most important aspects of our study was the characterization
of a unique damage structure which forms within the amorphous phase of Ge
during implantation [9]. Figure 1 shows a sequence of cross-section, TEM
micrographs from (111)Ge samples implanted with 120 keV In ions at room temperature (RT) to progressively higher fluences. Part a) of the figure shows that
a completely amorphous overlayer has been formed by an implantation dose of
1 x 10^{15} cm^{-2}. An electron, micro-diffraction pattern from this overlayer in
Fig. 1b shows that the layer is indeed amorphous. At a higher implant dose,
surface craters begin to form; craters as deep as 40 nm are evident in Fig 1c
for a 2 x 10^{15} cm^{-2} implant. At a dose of 5 x 10^{15} cm^{-2} (Fig. 1d), a very unique
structure has evolved which consists of regularly spaced pits or craters in the
surface. The material surrounding these deep craters is amorphous as shown by
diffraction techniques. Additionally, we found on exposure to ambient air that
large amounts of carbon and oxygen are absorbed on the surface of the implanted

*Research sponsored by the Division of Materials Sciences, U.S. Department of
Energy under contract W-7405-eng-26 with Union Carbide Corporation.

Mat. Res. Soc. Symp. Proc. Vol. 13 (1983) Published by Elsevier Science Publishing Co., Inc.

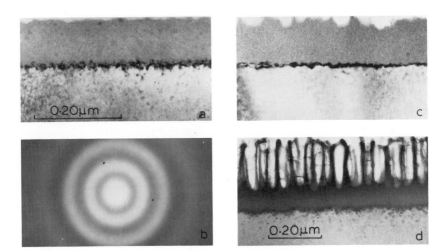

Fig. 1. TEM micrographs from (111) Ge crystals implanted at RT with 120 keV.
In ions at a fluence of (a) $1x10^{15}$, (c) $2x10^{15}$ and (d) $5x10^{15}$ cm^{-2}; and (b) an
electron microdiffraction pattern from the implanted region of sample in (a).

sample which is greatly enhanced by the deep pitting [9]. This surface structure and the adsorbed impurities can affect the annealing behavior of such
samples. In furnace annealing studies whereby ion implantation damage anneals
by epitaxial growth in the solid phase, it was found that this damage was stable
and would not anneal. During laser annealing, the impurities are trapped in the
bulk and can inhibit liquid phase epitaxy. Also, the dopant profiles in as-
implanted samples determined by ion backscattering are difficult, if not
impossible to interpret due to the irregularity of the surface. For these
reasons, the implantations in our study were done at liquid nitrogen (LN_2) temperature, where the critical dose for forming such structure was found to
increase by more than an order of magnitude (i.e. from $2 x 10^{15}$ cm^{-2} for RT
implanted In ions at 120 keV to $3 x 10^{16}$ cm^{-2} for implants at LN_2).

RESULTS AND DISCUSSION

The results of a ^{120}Sn(120 keV, 10^{16} cm^{-2}) implanted and laser annealed
(111)Ge crystal are presented in Fig. 2. An amorphous layer, 140 nm wide, is
formed as a result of the implantation. This is determined from the width of
the damage layer in the aligned spectrum from the as-implanted sample in Fig.
2a. Also, it can be seen in the figure that the aligned spectra from annealed
and virgin (nonimplanted) crystals are almost indistinguishable. This clearly
indicates that essentially all of the implantation damage was removed by the
laser annealing process. This was confirmed by TEM analysis which showed, in
cross-section, that the implanted region was single crystal with no dislocations
or other extended defects present. The micrograph from this sample is displayed
in Fig. 3c. After annealing, nearly all the implanted atoms were incorporated
onto lattice sites as shown by the substitutional dopant profile in Fig. 2b.
Additionally, it is obvious from a comparison of the dopant distribution in the
as-implanted sample with that in the annealed sample, that considerable
redistribution of the dopant occurred as a result of the laser irradiation.
Similar redistribution has been reported in Si and is consistent with diffusion

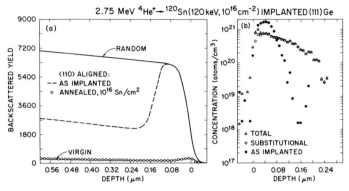

Fig. 2. (a) Backscattering spectra and (b) calculated dopant profiles from a (111) Ge crystal implanted with Sn ions and laser annealed.

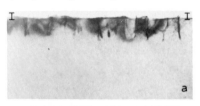

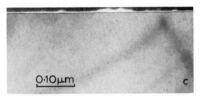

Fig. 3. TEM micrographs showing cell formation following laser irradiation in a (a) (100)Ge crystal implanted with Bi and (b) implanted with Pb ions; and (c) micrograph showing defect-free crystallization of a laser annealed, Sn implanted (111) crystal.

of the dopant in the liquid phase during the time in which the layer remains molten. However little, if any, of the dopant has been segregated to the surface. If the growth were an equilibrium process then the partition of the dopant at the growth interface would be given by the equilbrium distribution coefficient (which gives the ratio of the solute in the solid with that in the liquid in contact with it), whose value is 0.02 [10]. For such a small value of the distribution coefficent, much of the dopant would be zone refined to the surface. The absence of such surface segregation strongly suggests that the growth process does not occur under conditions of local equilibrium and that the effective value of the distribution coefficent is greatly increased over its equilibrium value.

While nonequilibrium distribution coefficients have been shown to be dependent upon the crystal growth velocity [11], orientation dependencies have also been noted [12]. Our results which compare growth in (100) and (111) orientations of Ge are consistent with these previous observations. Figure 4 shows a comparison of the dopant profiles (both total and substitutional) after laser annealing of samples implanted identically but of different orientations. It is clear that there is a significant variation in the dopant distributions. In both cases, the solute that remains in the bulk is highly substitutional. However, a much larger amount of solute (more than twice the amount at the peak concentration) is trapped in the (111) sample. This is consistent with an

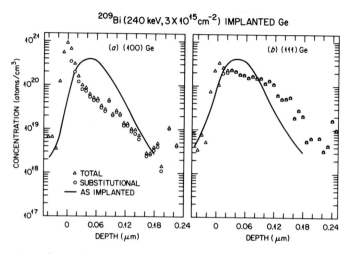

Fig. 4. Comparison of dopant profiles in (100) and (111) orientations of Ge, implanted identically with Bi ions, following laser annealing.

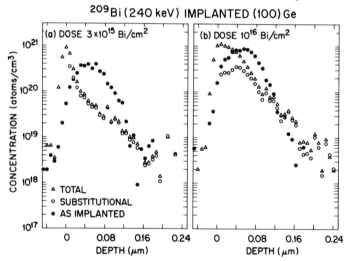

Fig. 5. Comparison of dopant profiles in (100) Ge crystals implanted with Bi ions with a total fluence of (a) 3×10^{15} and (b) 10^{16} cm^{-2}.

effective distribution coefficient which is larger for the (111) growth direction. Since the growth velocity normal to the interface must be identical for the two orientations, the differences have been attributed to a larger undercooling during growth on the (111) substrate compared to (100) [13].

The formation of a well-defined cellular structure after laser annealing was also observed. This structure results from an instability which occurs in the

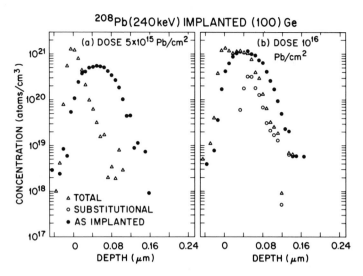

^{208}Pb(240keV) IMPLANTED (100)Ge

Fig. 6. Comparison of dopant profiles in (100)Ge crystals implanted with Pb ions with a total fluence of (a) 5×10^{15} and (b) 10^{16} cm^{-2}.

growth interface when a gradient in the undercooling of the liquid ahead of the interface (as a result of constitutional variations) exists. If the actual temperature gradient in the liquid is less than that of the liquidus temperature of the constitutionally enriched liquid, cell formation can result. Figure 5 shows dopant profiles for two different implant fluences of 240 keV Bi ions implanted into (100)Ge. In the low dose implant, it can be seen that part of the dopant has been segregated to the surface and that which remains in the bulk is highly substitutional. Ion channeling results indicate that the epitaxial growth was essentially defect-free. However, in the higher dose implant (Fig. 5b), a significant fraction of the dopant is in substitutional sites up to 60 nm from the surface where a significant drop occurs. The dopant in the top 60 nm has not been segregated to the surface and a large portion is not substitutionally located. A very large dechanneling rate was observed in the aligned spectrum from this region suggesting that a rather high density of extended defects or other structure are present. TEM analysis determined that a well-defined cellular structure had formed in the region and that the non-substitutional fraction of Bi had been laterally segregated to the walls of the cells. The cells are clearly visible in the micrograph from this sample in Fig. 3a.

A rather unusual behavior was observed for 240 keV Pb ions implanted into (100)Ge. As seen in Fig. 6 for a 5×10^{15} cm^{-2} implant, the dopant was completely zone refined to the surface by the laser irradiation. However, increasing the total implant dose by a factor of 2, results in a totally different behavior. Up to 55 nm of the surface, the Pb was incorporated into substitutional lattice sites and thereafter, cells were formed. The cellular structure can be seen in the micrograph in Fig. 3b. These results indicate that there may be a dependence of the distribution coefficient on the solute concentration. However, this was not observed for the other dopants nor was it observed in (111)Ge for implanted Pb ions.

Table I summarizes our results on substitutional solute concentrations for the (100) and (111) orientations of Ge. Maximum solid solubility limits have

TABLE I
Substitutional solubilities achieved by laser at 1.3 J/cm^2

Dopant	c_s^0 (cm^{-3})	Ge(100) c_s^{max} (cm^{-3})	Ge(111) c_s^{max} (cm^{-3})	Calculated c_s^{max} (cm^{-3})
Bi	6.0 x 10^{16}[a]	3.5 x 10^{20}	>5 x 10^{20}	3.0 x 10^{21}
Pb	5.7 x 10^{17}[b]	4.0 x 10^{20}	4 x 10^{20}	3.2 x 10^{21}
In	3.9 x 10^{18}[b]	3.0 x 10^{20}	>4 x 10^{20}	3.0 x 10^{21}
Sn	5.0 x 10^{20}[b]	6.0 x 10^{20}	>7 x 10^{20}	2.0 x 10^{22}

[a]F. Trumbore et al., J. Electrochem. Soc. 109, 734 (1962). [b]From Ref. 10.

been established for all dopants in (100)Ge under the growth conditions studied. Substitutional concentrations of dopants much in excess of retrograde maxima are achieved as a result of the nonequilibrium nature of the growth. The maximum concentration in substitutional sites for all the dopants was limited by the interfacial instability which leads to cell formation. In (111)Ge, only Pb implanted samples formed cells at the implant fluences studied. While the other dopants formed solid solutions in excess of retrograde maximum, no limiting concentration was determined. The table includes calculated thermodynamic solubility limits, which represent solute concentrations under the fastest solidification conditions. These concentrations correspond to the intersection of the liquidus and solidus curves in the free energy diagrams. A similar procedure was used by Cahn et al. [14] to obtain maximum solute concentrations for Si alloys. The measured maxima were considerably smaller than those calculated indicating that much higher growth velocities may be necessary to achieve substitutional concentrations approaching thermodynamic limits.

REFERENCES

1. J. Narayan, R. T. Young and C. W. White, J. Appl. Phys. 49, 3912 (1978).
2. P. Baeri, S. U. Campisano, G. Foti and E. Rimini, Appl. Phys. Lett. 32, 137 (1978).
3. C. W. White, W. H. Christie, B. R. Appleton and P. P. Pronko, Appl. Phys. Lett. 33, 662 (1978).
4. C. W. White, S. R. Wilson, B. R. Appleton and J. Narayan, p. 124 in Laser and Electron Beam Processing of Materials, ed. by C. W. White and P. S. Peercy, Academic Press, New York, 1980.
5. A.G. Cullis, J.M. Poate and G.K. Celler, p. 311 in Laser-Solid Interactions and Laser Processing, ed. by S. D. Ferris, H. J. Leamy and J. M. Poate, AIP, New York, 1979.
6. C. W. White, J. Narayan, B. R. Appleton and S. R. Wilson, J. Appl. Phys. 50, 2967 (1979).
7. C. W. White, S. R. Wilson, B. R. Appleton and F. W. Young. Jr., J. Appl. Phys. 51, 738 (1980).
8. J. Narayan and C. W. White, p. 65 in Laser and Electron Beam Processing of Materials, ed. by C. W. White and P. S. Peercy, Academic Press, New York, 1980.
9. B. R. Appleton, O. W. Holland, J. Narayan, O. E. Schow, III, J. S. Williams, K. T. Short and E. Lawson, Appl. Phys. Lett. 41, 711 (1982) and O. W. Holland, B. R. Appleton and J. Narayan, J. Appl. Phys. (submitted).
10. F. Trumbore, Bell Syst. Tech. Jour. 39, 205 (1960).
11. A. G. Cullis, H. C. Webber, J. M. Poate and A. L. Simons, Appl. Phys. Lett. 36, 320 (1980).
12. P. Baeri, G. Foti, J. M. Poate, S. U. Campisano and A. G. Cullis, Appl. Phys. Lett. 38, 800 (1981).
13. See paper by G. H. Gilmer in these proceedings.
14. J. W. Cahn, S. R. Coriell and W. J. Boettinger, p. 89 in Laser and Electron Beam Processing of Materials, ed. by C. W. White and P. S. Peercy, Academic Press, New York, 1980.

SEGREGATION AND CRYSTALLIZATION PHENOMENA IN GERMANIUM

G. J. CLARK
IBM Thomas J. Watson Research Center, Yorktown Heights, NY 10598 USA
A. G. CULLIS
Royal Signals and Radar Establishment, St. Andrews Road, Malvern, England
D. C. JACOBSON and J. M. POATE
Bell Laboratories, 600 Mountain Avenue, Murray Hill, NJ 07974 USA
MICHAEL O. THOMPSON
Department of Materials Science, Cornell University, Ithaca, NY 14853

ABSTRACT

While many studies have been made of liquid phase epitaxy impurity trapping and segregation in Si little is known about the equivalent processes in Ge. In this paper we have laser annealed Ge <100> and <111> crystals implanted, at liquid nitrogen temperature, with 200 keV ^{210}Bi ions to doses of 2×10^{15} and 10^{16} ions cm^{-2}. The samples were annealed with Q-switched ruby lasers and an XeCl excimer laser. We have observed 1) velocity and orientation dependence of the Bi segregation coefficient 2) interface instability and cell formation resulting from constitutional supercooling and 3) amorphization and defect production at high velocities. The phenomena are shown to be analogous to those seen in Si.

INTRODUCTION

The energy from pulsed laser irradiation can be used to melt, for a very short time, silicon or germanium crystals to a depth of several thousand angstroms. The melting is followed by non-equilibrium liquid phase, epitaxial regrowth at velocities of up to 20 m s^{-1}. There has been considerable recent activity[1] in the use of these experimental conditions to study high velocity non-equilibrium crystal growth in Si. Segregation studies[1] have been made on the movement and trapping of implanted impurities in silicon during non-equilibrium crystal growth, the onset of defect-formation during crystal regrowth at high velocities[2] and the amorphization of the Si at extremely high regrowth velocities. Specifically it has been observed that following pulsed ruby laser annealing of Bi and In implanted Si both velocity and orientation dependence are observed in the segregation coefficients up to velocities of ~ 5 m s^{-1} where the segregation coefficients saturate at approximately the same value for both the <100> and the <111> orientations. Equilibrium solid solubilities are exceeded by many orders of magnitude. At high doses (~ 10^{16} ions cm^{-2}) cellular microstructure is observed. This occurs when there is constitutional supercooling in the resolidifying liquid so that the liquid-solid interface becomes unstable and breaks down. At velocities greater then ~ 5 m s^{-1} defects in the form of twins are observed for the <111> orientation but not for the <100>. At extremely high velocities (> 15 m s^{-1}) amorphous Si is produced.

In this paper, we measure for the first time such segregation and recrystallization phenomena in Ge with the aim of determining the parallel between Si and Ge in this regime of ultra-rapid crystal growth. Specifically we have investigated segregation and crystallization of Bi implanted <100> and <111> Ge with regrowth velocities in the range 2 to 17 m s^{-1}.

EXPERIMENTAL

Ge <100> and <111> polished wafers were implanted with 200 keV ^{210}Bi ions. Implant doses were 2.10^{15} ions cm^{-2} (low dose) and 10^{16} ions cm^{-2} (high dose). All implants were carried out at liquid nitrogen temperature to avoid the surface craters and incorporation of impurities observed at room temperature implants into Ge[3].

The samples were melted using pulses either from Q-switched ruby lasers or a XeCl excimer laser. The velocity of regrowth was controlled by varying the energy and pulse length of the laser pulse. The basic Q-switched ruby laser delivered a 30 ns pulse at 694 nm. Pulses of lengths down to 2 ns duration were obtained by use of a synchronized Pockels cell beam switch. For the amorphization studies the pulses were frequency doubled. The XeCl$_2$ excimer laser delivered a 30 ns pulse at 308 nm. The samples were held at room temperature in a nitrogen atmosphere during the pulsed annealing. The spatial uniformity of the laser pulse was better then ± 5% over the regions studied. Laser pulse energies were calibrated using a calorimeter and considered to be accurate to ± 5% for the ruby lasers and ± 10% for the excimer laser.

The irradiated samples were studied using Rutherford backscattering of 2 MeV He ions (RBS) and transmission electron microscopy (TEM) in both planar and cross-sectional configurations.

Computer heat flow calculations similar to those of Baeri and Campisano[4] were used to estimate the regrowth velocities for various ruby (694 nm) pulse lengths and energies. Temperature dependent values of the specific heat[5] and the thermal conductivity[6] were used. As little data exists for the thermal properties of the amorphous phase we used crystalline values for all the parameters except the latent heat of the melting ($\Delta H_{a-l} = 410$ J/g and $\Delta H_{a-l} = 240$ J/g).[7] An amorphous surface reflectivity of 40% was used with absorption coefficients of 2×10^5 cm^{-1} in the amorphous layer and 1×10^5 cm^{-1} in the crystalline bulk. Reliable data for the liquid absorption and reflectivity were not available so values of 4×10^5 cm^{-1} and 70% were respectively assumed. In Fig. 1 the average regrowth velocity is plotted against energy density for pulse durations of 5, 15 and 30 ns. Since the absorption coefficient is large in both the amorphous and crystalline phases, the regrowth velocity is dominated by heat conduction into the substrate. These velocity estimates allowed the surface accumulations obtained from RBS measurements to be converted to segregation coefficients using the approach of Baeri and Campisano.[4] These calculations involve the liquid state diffusivity of Bi in Ge. As the liquid diffusivity of Bi in Ge has not been measured we have assumed a value of 5×10^{-4} cm^{2}s^{-1}. This value was chosen to make the cell structure data and segregation data compatible. This value is not inconsistent with liquid phase diffusivities for other elements in Ge.[8]

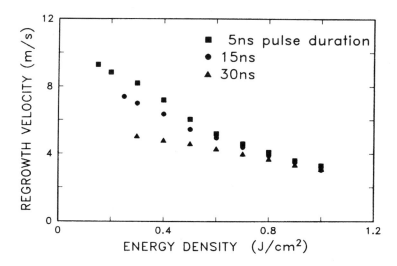

Figure 1. Average liquid-phase regrowth velocity for ruby irradiation of 100 nm of amorphous Ge on crystalline Ge as a function of laser energy density and pulse duration.

RESULTS AND DISCUSSIONS

1. Orientation and Velocity Dependent Segregation.

The orientation and velocity dependence of Bi segregation in Ge following laser annealing was studied using low dose $(2.10^{15}$ ions cm^{-2}) implants.

For low velocities the orientation dependence of the Bi segregation is clearly shown in Fig. 2 with much more Bi trapped on lattice sites in the <111> case then in the <100> case. In that figure random and aligned RBS spectra are shown for both <100> and <111> Ge samples that have been melted with a 30 ns, 1.26 J cm^{-2} pulse from a ruby laser. The velocity of regrowth in this instance is 2.7 m s^{-1}. The aligned spectra gives the amount of Bi zone refined to the surface while the random spectra give the Bi trapped in the bulk of the regrown crystal.

In Fig. 3 we present Bi depth profiles showing a similar orientation dependence for Ge <100> and <111> samples annealed with an excimer laser pulse. The excimer laser pulse was of 30 ns duration and 1.16 J cm^{-2} energy density. The depth profile from the as implanted Bi is also included.

Figure 4 compares Bi depth profiles from Ge <100> and <111> samples following melting at higher velocities. In this case we used a ruby laser pulse of 5 ns duration and 0.5 J cm^{-2} energy density. The calculated regrowth velocity was 6.8 m s^{-1}. The depth profiles are essentially the same and the strong orientation effect at lower velocities (Figures 2 and 3) has disappeared.

For velocities ranging from 2.7 to 6.8 m s^{-1} segregation coefficients were extracted and plotted as a function of velocity in Figure 5. It can be seen that the segregation coefficient (k) in this velocity range is approximately two orders of magnitude greater than the equilibrium segregation coefficient $(k_o = 4.5 \times 10^{-4})$.[9] The maximum equilibrium solid

306

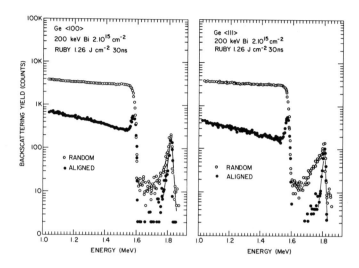

Figure 2. Rutherford backscattering and channeling spectra (2 MeV ⁴He ions) for ruby laser irradiation of Ge <100> and <111) crystals implanted with Bi. The regrowth velocities are 2.7 ms⁻¹.

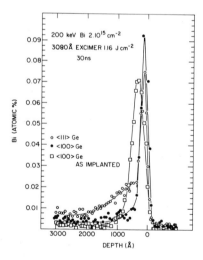

Figure 3. Bi depth profiles for excimer laser irradiation of <100> and <111> Ge crystals. The depth profile for the as implanted Ge is also shown.

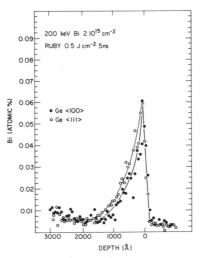

Figure 4. Bi depth profiles for ruby lasers irradiation of <100> and <111> Ge crystals for regrowth velocities of 6.8 ms^{-1}.

solubility is 6.10^{16} Bi cm^{-3} whereas the solid solubility for <111> Ge at 2.7 m s^{-1} is 2.10^{19} Bi cm^{-3} for a Bi implant of 2×10^{15} ions cm^{-2}. It should be noted that data points are not plotted for velocities greater then 5 m s^{-1} because high resolution depth profiles were not available to directly resolve the surface zone refined Bi thus placing some ambiguity in the method of analysis. Nevertheless at the high velocities > 5 m s^{-1} the <100> and <111> depth profiles are virtually indistinguishable. The velocity and orientation dependence of the segregation of Bi in Ge thus appears very similar to that of Bi in Si.

2. Cell Structure.
    The phenomenon of interface instability during crystal regrowth from melts containing high concentrations of low solubility impurities is due to constitutional supercooling in the resolidifying melt. This instability ultimately results in the formation of cellular segregation patterns. The full theoretical treatment of interface breakdown is given by Mullins and Sekerka.[10] That treatment has been extended by Cullis et al.[11] to the high velocity regime encountered during Q-switched laser annealing.

    In the present experiment, we observe for the first time, the phenomenon of cell formation in Ge. Figure 6 shows plan-view TEM images of cellular structure produced by laser irradiation of high dose (10^{16} ions cm^{-2}) Bi implanted <100> Ge. Figures 6a and 6b show the change in the cell dimensions with change in the recrystallization velocity from 2.7 to 5.5 m s^{-1}. The cell dimensions correspondingly change from 100 nm to 60 nm. The square cell morphology exhibited in many cases is consistent with the 4-fold symmetry of the <100> Si crystallographic orientation. The dimensions of the cell structure were observed to be essentially independent of the implanted Bi dose down to a critical value in the region of 3×10^{15} ions cm^{-2}. At doses below this the cells disappeared. It is observed

308

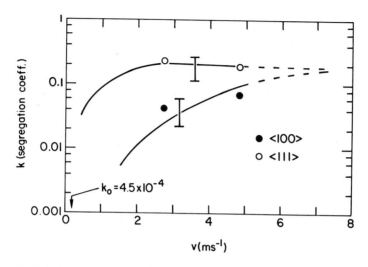

Figure 5. Velocity and orientation dependence of Bi segregation coefficients in Ge.

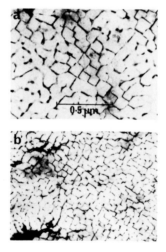

Figure 6. Plan view TEM images of Bi cells in <100> Ge for regrowth velocities of a) 2.7 m s^{-1} and b) 5.5 m s^{-1}.

that cellular breakdown occurs on a scale of the order of $\frac{D}{v}$ where D is the diffusion coefficient of the solute and v is the growth velocity.[11] In this case for $D = 5 \times 10^{-4}$ cm^{2}s^{-1} and $v = 2.7$ m s^{-1} the predicted cell size would be ~ 200 Å, in reasonable agreement with experiment.

3. Amorphization.

Figure 7 shows cross-sectional TEM images of single crystal <111> Ge wafers that have been irradiated with a UV laser pulse of 2.5 ns duration and energy densities of a) 0.23 J cm^{-2} and b) 0.4 J cm^{-2}. Calculations for the 0.23 J cm^{-2}, 2.5 ns case indicate that the velocity of regrowth is $\geq$ 17 m s^{-1}. It is obvious that at these velocities, during resolidification, the melt-solid interface breaks down to form an amorphous Ge layer of ~ 25 nm mean thickness. At somewhat lower velocities, the amorphization ceases but a high density of twins and other extended defects occurs. This is shown in Figure 7b for a regrowth velocity calculated to be 12 m s^{-1}.

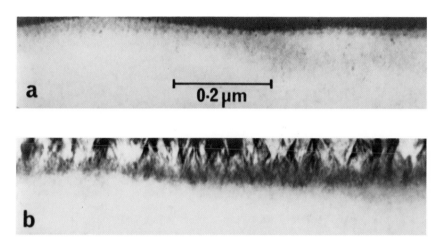

Figure 7. Cross-sectional TEM images of a) amorphous Ge layer grown on <111> Ge at a regrowth velocity of 17 m s^{-1} and b) defects in <111> Ge regrowth velocities of 12 m s^{-1}.

These amorphization and defect producing phenomena are discussed in detail elsewhere in these proceedings.[10] They are analogous to those which occur[2] in Si as discussed by Poate[13] and Turnbull[14] in these proceedings. The <100> Ge did not amorphize under the conditions described above.

SUMMARY

In this paper we have demonstrated for the first time that the phenomena of a) velocity and orientation dependent segregation b) cellular structure due to constitutional supercool-

310

ing and c) amorphization and defect production from the melt all occur in Ge in a similar fashion to that of Si.

This work gives confidence in the experimental and theoretical techniques being used to approach the subject of ultra-fast-solidification in the important elemental semiconductors Si and Ge. Related work is reported in this conference by O. W. Holland et al. [15].

ACKNOWLEDGMENTS

We are indebted to N. G. Chew, R. A. Fiorio, R. F. Marks, P. Saunders and H. C. Webster for considerable experimental assistance.

REFERENCES

1. J. M. Poate in "*Laser and Electron Beam Interactions with Solids* p. 121, ed. B. R. Appleton and G. K. Celler, North Holland, New York (1981).

2. A. G. Cullis, H. C. Webber, N. G. Chew, J. M. Poate and P. Baeri, Phys. Rev. Lett. 49, 219 (1982).

3. B. R. Appleton, O. W. Holland, J. Narayan, O. E. Schow III, J. S. Williams, K. T. Short and E. Lawson, Appl. Phys. Lett. (1982).

4. P. Baeri and S. U. Campisano, Chap. 4, *Laser Annealing of Semiconductors* ed. J. M. Poate and J. W. Mayer, Academic Press, New York (1982).

5. Y. S. Poulovkian and E. H. Buyco, Specific Heat in Metallic Elements and Alloys, Plenum, New York, 1970.

6. C. Y. Ho, R. W. Powell and P. E. Liley, J. Physical and Chemical Reference Data Suppl. No. 1, 288 (1974).

7. J. C. C. Fan and C. H. Anderson, J. Appl. Phys. 52, 4003 (1981).

8. Diffusion Data 1, 48 (May 1967).

9. F. A. Trumbore, W. G. Spitzer, R. A. Logan and C. L. Luke, J. Elect. Chem. Soc. 109, 734 (1962).

10. W. W. Mullins and R. F. Sekerka, J. Appl. Phys. 35, 444 (1964).

11. A. G. Cullis, D. J. J. Hurle, H. C. Webber, N. G. Chew, J. M. Poate, P. Baeri and G. Foti, Appl. Phys. Lett. 38, 642 (1981).

12. A. G. Cullis, these proceedings.

13. J. M. Poate, these proceedings

14. D. Turnbull, these proceedings.

15. O. W. Holland et al, these proceedings.

IMPURITY DISTRIBUTION PROFILES AND SURFACE DISORDER AFTER LASER INDUCED DIFFUSION.

E. FOGARASSY, R. STUCK, P. SIFFERT Centre de Recherches Nucléaires
Laboratoire PHASE , 67037 STRASBOURG CEDEX France
F. BROUTET, J. C. DESOYER Faculté des Sciences POITIERS France

ABSTRACT

A model for laser induced diffusion is proposed, assuming the melting of Si surface under pulsed laser irradiation and the diffusion , in liquid phase, of a thin filmof impurity deposited according to the "limited source" conditions. Depending on the thickness of the film, it results in a dopant distribution profile with a surface disordered layer induced either by segregation effects or precipitation of the dopant in excess of the solubility limit achieved by laser annealing. Experimental results, obtained for a ruby laser irradiation of thin films of different impurities of group III and V, like antimony, bismuth, gallium and indium deposited on silicon substrates are in good agreement with the model. Their effective segregation coefficients have been deduced by fitting the experimental amount of dopant precipitated in the disordered surface layer with the numerical calculations. A cellular structure is seen on surface.

INTRODUCTION

Laser induced diffusion which consists in a thin film deposition on silicon followed by pulsed laser treatment has been demonstrated to be an interesting technique for junction formation especially for solar cell fabrication.[1,2] Like ion implantation combined with laser annealing it allows to tailor the doping profile and to reach doping levels in excess of the thermal equilibrium solubility limit[3] . Whereas several models allowing to calculate the doping concentration profiles have been presented for ion implantation with subsequent laser annealing [4,5], no model exists so far, for laser induced diffusion. Such a model is necessary to determine the concentration of dopant which is not integrated in substitutionnal lattice sites, either because it is accumulated at the surface due to segregation effect, or precipitated because it exceeds the solubility limit achieved by laser treatment. It allows thus to analyze the role of the thickness of the deposited film on the formation of the disordered surface layer. The model presented here is a strictly thermal model assuming diffusion of the dopant in the liquid phase occuring from a limited source located at the surface. A good agreement is obtained between the profiles calculated by this model and those measured by Rutherford Backscattering SPectrometry (R. B. S.). The effective segregation coefficients of Sb, Ga, Bi and In have been deduced from the experimental amount of dopant precipitated in the disordered surface layer, which presents a cellular structure, as demonstrated by transmission electron microscopy.

Mat. Res. Soc. Symp. Proc. Vol. 13 (1983) ©Elsevier Science Publishing Co., Inc.

THE MODEL

     Basically the calculations have been performed for irradiations with a Q-switched ruby laser ($\lambda = 0.69\,\mu m$) single pulses of 20 ns duration. We assume that the dopant films deposited on the surface of silicon are thin enough so that no absorption occurs for the ruby laser light. This appears to be well, verified for thicknesses $\leqslant$ 50 Å.[6].As a consequence, the energy can be considered to be entirely transfered to Si lattice and to lead to superficial melting. The depth of melting can be estimated from heat flow calculations. For laser energy densities ranging between 1 and 2 J/cm^2 the melt front penetrates, typically,to a depth comprised between 1000 and 6000 Å [7]. We assume that during this melting, diffusion of the dopant occurs from the surface according to the basic equation in case of "limited source" conditions. [8].

$$C(x,t) = \frac{N}{\sqrt{\pi . D_L . t}} \times \exp\left(- \frac{x^2}{4 D_L\, t} \right)$$

where N is the initial deposited dopant concentration; D_L is the liquid phase diffusion coefficient ($\simeq 10^{-4}$ cm^2/sec.) [9].

     During the epitaxial regrowth which occurs from the underlying mono-cristalline substrate, the velocity of the solid-liquid interface, as deduced from heat flow calculations, is as high as 2.50 - 3.0 m/sec [7] and the effective segregation coefficient K $= C_S/C_L$ is much higher than the thermal equilibrium segregation coefficient K_o. [5] To calculate the partition of the dopant between the solid and liquid phase, by taking into account the effective segregation coefficient and the diffusion of the dopant, the calculation method of Wang et al. [10]has been used. Finally the dopant is not incorporated substitutionaly in the Si lattice if its concentration in the solid becomes higher than the effective solubility limit C_S for the Laser process. This solubility is much higher than the thermal equilibrium solubility limit C_S^o [11] since due to the very high interface velocity, dopant can be trapped by a solute trapping mechanism.[12, 13] The values of the C_S are assumed to be proportional to the effective segregation coefficient K like at thermal equilibrium :

$$\frac{C_S}{K} = \frac{C_S^o}{K_o}$$

as demonstrated in a previous work. [14]

     It appears,thus that two distinct regimes exist: in the first one the dose deposited is low enough, the solubility limit for laser treatment is never exceeded. Nevertheless,the dopant distribution profiles present a surface accumulation peak due to segregation effects. For higher deposited concentrations, a disordered surface layer, not related to a segregation effect, is formed,which results from the precipitation of dopants in excess of the effective solubility limit. The percent of non substitutional dopant to the total concentration has been calculated as a function of the deposited concentration for 3 different values of K = 1, 0.5 and 0.1 and D_L = 1,2 and 5 x 10^{-4} cm^2/sec which are typical of dopant diffusion coefficients in liquid Si. [9].The two regimes can be clearly seen on Fig.1 for a laser energy density of 1.5 J/cm^2. For low values of the deposited dopant concentration the percent does not depend on the dose. This fraction, which represents the surface accumulated dopant by the segregation increases with decreasing values of K as reported on Fig. 2. These data are in good agreement with the calculated values of Campisano et al.[15]. For higher values of the deposited dopant concentration, as seen on Fig. 1, the concentration of non substitutional dopant increases rapidly with the dose even when there is no segregation as shown on the curve corresponding to K = 1. In these conditions, the distribution profile presents a disordered surface region of several hundred angströms in thickness, which is only due to the in-depth precipitation of dopant atoms in excess of their effective solubility limit. Finally, it is interesting to notice that this effect is very sensitive to the value of K and can , therefore, be used to determine this parameter.

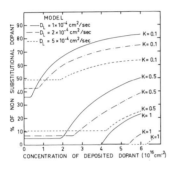

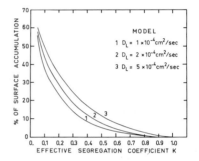

<u>Fig1:</u>calculated % of non substitutional dopant after laser diffusion (1. 5 J/cm²) as a function of the amount of dopant deposited for K= 1,0. 5 and 0. 1, and D_L = 1,2 and 5x10 E-4

<u>Fig. 2:</u>Calculated % of surface accumulated dopant after laser diffusion (1. 5 J/cm²) as a function of K and D_L = 1, 2 and 5x 10 E-4 cm²/s.

COMPARISON WITH EXPERIMENTS

Thin films of high purity dopants as Sb, Ga, Bi and In, were deposited by vacuum evaporation, on monocristalline Si substrate of < 111 > orientation. The deposited layers, as measured by the quartz monitor, ranged in thicknesses between 20 and 150 Å. The amount of dopant deposited was further measured by RBS. The agreement between the two methods was quite satisfactory. The samples covered with a dopant film were irradiated, in air, using the amplified monomode outpout of a 20 ns duration pulsed ruby laser, emitting energy densities in the range 1 to 2 J/cm². The evaporation losses of dopant under laser irradiation, as verified by RBS experiments, are ≤ 10% as reported on Table I for 50 Å dopant films deposited on Si.

The distribution profiles of both total and substitutional dopant were determined by RBS performed under random and channeling conditions with a ⁴He⁺ ion beam of 1 - 2 MeV energy. The backscattered particles were detected by means a surface barrier detector. This arrangement gives an equivalent depth resolution of about 200 Å for Si.

Fig. 3 reports the results obtained for a 20 Å Sb film on Si, after laser irradiation with energy densities E = 1.1, 1.45 and 1.8 J/cm^2. In this case, where we have a low deposited Sb concentration (6.7 × 10^{15}cm^{-2}), the dopant profiles mainly consist of an in-depth distribution with a highly substitional Sb concentration up to 1.2 × 10^{21} cm^{-3} below the solubility limit of Sb achieved by laser annealing (1.3 × 10^{21} cm^{-3}) but largely in excess of the thermal equilibrium solubility (6 × 10^{19} cm^{-3} at 1200°C). The full lines are the calculated profiles according to the approach above described. We notice the good agreement with experimental results, assuming a K value of 0.8, which is much greater than the equilibrium segregation coefficient of Sb in Si (K$_o$ = 0.023).

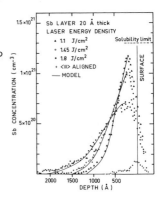

Fig 3: Random and aligned RBS spectra of a 20 Å Sb deposited film after laser irradiation of 1.1, 1.45 and 2 J/cm^2. The full lines are calculated for k=0.8

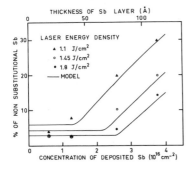

Fig. 4: % of non substitutional Sb, deduced from RBS measurements after ruby irradiation at 1.1, 1.45 and 2 J/cm^2, as a function of the amount of Sb deposited. The full lines are calculated for D$_L$ = 1.5×10 E –4 cm2/sec.

When the thickness of the deposited layer is increased, the Sb distribution profile presents a surface disordered region, in which most of the dopant atoms are non substitutional. The dose of Sb in this surface region has been deduced from RBS measurements, performed respectively before and after removing the disordered layer by using a specific etching, consisting of a H$_2$O(1)–HCl(1)–HNO$_3$(1) solution. We have reported on fig. 4 the percent of non substitutional Sb atoms to the total concentration as measured experimentally, as a function of the

dopant	% loss by evaporation (E=1.8J/cm^2)	%precipited atoms	K$_{effective}$	K$_o$ at equilib.
Sb	0–5	below 5	0.6–0.8	0.023
Ga	0–5	50	0.2	0.008
Bi	10	80	0.1	0.0007
In	8	85	0.05–0.1	0.0004

TABLE I

deposited dopant concentration, for different laser energy densities. We find again the two regimes and the agreement with the calculated curves (full lines) is reasonably good. By increasing the energy density of the pulse, the in-depth profile broadens, leading to a diminution of the maximum sur- face concentration and, therefore, to a better substitutional incorporati on of dopant into the silicon lattice.

As predicted by the numerical calculation (fig. 1), impurities less solu ble in Si should present a high concentration of atoms precipitated at the top surface layer. We have reported on Table I for 50 Å thick films of Sb, Bi, Ga and In deposited on Si, the percentage of precipitated atoms after laser induced diffusion , as deduced from RBS measurements, as shown , for example, for Bi on fig. 5. The deduced values of their effective segrega- tion coefficient are much greater than the equilibrium one and agree reas onably well with previous results obtained on implanted layers. (5)

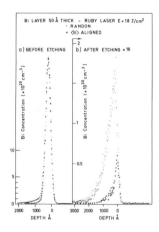

Fig. 5: Random and aligned RBS spectra of a 50 Å Bi deposited film on Si after ruby laser irradiation (1.8 J/cm²), before (a) and after (b) removing the surface diso rdered region

The structure of the disordered surface layer is questionable . As previously demonstrated [16] , an important mechanism which limits the substitutional concentration, achieved by laser annealing, is the interfacial instability which develops during regrowth and leads to lateral segregation of the rejected dopant and to the formation of a well defined cell struc - ture in the near surface region. As for im- planted layers [17] this mechanism dominates in this case of laser induced diffusion of In and Bi, as demonstrated by transmission elec- tron microscopy. An example of these behaviours is shown on micrograph of fig. 6 a for In which gives cells of about 300 Å in diameter, a value which is in good agreement with the characteristic spacing in such structures, given by the diffusion length of rejected dopant $D_L/v \cong 200$ Å, where v is the liquid- solid interface velocity. The non-substitutional In in this region is located in the cell walls. As seen on Fig. 6 b the cell walls penetrate to a depth of $\cong 600$ Å from the surface, in good agreement with the thickness of the disordered surface layer, as deduced from RBS experi- ments on the same sample.

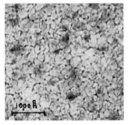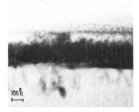

Fig. 6:(a) plan view transmission micrograph and (b) cross-section micrograph showing cell formation after ruby laser induced diffusion of 100 Å In deposited on Si.

CONCLUSION

We have shown that in laser induced diffusion , the experimental profiles , for low dopant deposited concentrations , are well understood by using a model which takes into account the segregation of impurities at the moving solid-liquid interface. However, for increasing dopant deposited concentrations, the segregation effects must be associated to a mechanism of precipitation of dopant atoms in excess of the solubility limit after laser treatment. This leads to the formation of very high disordered surface regions presenting a cellular structure.

REFERENCES.

1. J. Narayan, RT Young, and R. F. Wood, Applied Phys. Lett. 33 (4), 338 (1978)
2. E. Fogarassy, R. Stuck, J. J. Grob and P. Siffert, J. Appl. Phys. 52 (2), 1076 (1981)
3. E. Fogarassy, R. Stuck, J. J. Grob and P. Siffert "Laser and Electron Beam Processing of Materials"edited by C. W. White and P. S. Peercy, Academic Press (1980) p. 117.
4. P. Baeri, S. U. Campisano, G. Foti and E. Rimini, J. Appl. Phys. 50 (2) 788 (1979).
5. C. W. White, S. R. Wilson, B. R. Appleton and F. W. Young Jr., J. Appl. Phys. 51 (1) 738 (1980).
6. Monographie sur les métaux de haute pureté, G. Chaudron, Masson Paris 1972.
7. M. Toulemonde and R. Heddache, Private Communication
8. J. Crank, The mathematics of diffusion - Oxford at the Clarendon Press (1956)
9. H. Kodera Jpn J. Appl. Phys. 2, 212 (1963)
10. J. C. Wang, R. F. Wood, C. W. White, B. R. Appleton, P. P. Proko, S. R. Wilson and W. H. Christie, The proceedings of the Boston Conference on "Laser Solid interactions and Laser processing - 1978" edited by S. D. Ferris, H. J. Leamy, and J. M. Poate . AIP Conf. 50 (1979) p. 123
11. F. A. Trumbore Bell Syst. Techn. J. 39, 205 (1960)
12. J. C. Baker and J. W. Cahn, Acta Met. 17 (1969)
13. K. A. Jackson, G. H. Gilmer and H. J. Leamy "Laser and electron beam processing of materials" edited by C. W. White and P. S. Peercy Academic Press (1980) p. 104.
14. R. Stuck, E. Fogarassy, J. J. Grob and P. Siffert, Appl. Phys. 23, 15 (1980).
15. S. U. Campisano, P. Baeri, M. G. Grimaldi, G. Foti and E. Rimini, J. Appl. Phys. 51 (7), 3968 (1980).
16. G. J. Van Gurp, G. E. Eggermont, Y. Tamminga, W. T. Stacy and J. R. M. Gijsbers Appl. Phys. Lett. 35 (3), 273 (1979).
17. J. Narayan, J. Appl. Phys. 53 (3), 1289 (1981).

SURFACE STUDIES OF LASER ANNEALED SEMICONDUCTORS*

D. M. ZEHNER AND C. W. WHITE
Solid State Division, Oak Ridge National Laboratory, Oak Ridge, Tennessee 37830

ABSTRACT

The surface regions of semiconductor single crystals have been examined following laser annealing in an ultrahigh vacuum environment with the output of a pulsed ruby laser. Atomically clean surfaces with impurity levels below 0.1% of a monolayer can be produced by multiple pulse irradiation. Ordered surface structures are produced on low index oriented crystals as well as crystals slightly misoriented. Metastable surface structures exhibiting (1x1) LEED patterns have been produced on (111) orientations and are believed to be a consequence of the rapid cooling rates of 10^9 degs/sec achieved with the laser irradiation process. The surface and subsurface regions of ion-implanted Si crystals have been examined both before and after laser irradiation and results obtained from Si samples implanted with As are discussed.

INTRODUCTION

The rapid deposition of energy from Q-switched lasers into the near-surface region of semiconductors leads to melting of the crystal to a depth of several thousand angstroms, followed by liquid-phase epitaxial regrowth from the underlying substrate at growth velocities calculated to be of the order of meters per second. This processing technique provides a way of achieving heating and cooling rates (approaching 10^9 deg/sec) which are orders of magnitude faster than those obtained with more conventional techniques. With proper annealing conditions, it has been shown that regions free of extended defects can be formed and substitutional impurities can be incorporated into the lattice far in excess of the equilibrium solubility limits. These observations of the changes that occur in the subsurface region suggest that the surface properties (impurities, structure, electronic energy levels) of laser-annealed semiconductors can be significantly altered by the high heating and cooling rates and rapid velocities of solidification that can be achieved using pulsed lasers.

In this paper we discuss studies concerned with laser irradiation of semiconductor single crystals in an ultrahigh vacuum environment. We show that pulsed laser irradiation can be used to produce atomically clean surfaces. This is accompanied by the observation of ordered surface structures which indicates that the liquid-phase epitaxial regrowth process extends to the outermost monolayers of the crystal. The production of metastable (1x1) surface structures on (111) oriented crystals of Si is discused as well as the geometric and electronic properties of the Si(111) surface. We show that pulsed laser irradiation of vicinal surfaces can be used to produce a well defined stepped surface. Finally, we show that the combination of ion implantation and laser annealing can be used to produce surfaces with electronic properties that cannot be achieved by conventional methods.

*Research sponsored by the Division of Materials Sciences, U. S. Department of Energy under contract W-7405-eng-26 with Union Carbide Corporation.

EXPERIMENTAL

Systematic studies of the surface properties of laser-annealed semi-conductors require that the irradiation of the sample and subsequent analysis must take place in an ultrahigh vacuum (UHV) environment ($< 10^{-9}$ Torr). All measurements to be discussed were made in UHV surface analysis chambers which, after bakeout, had background pressures typically less than 2×10^{-10} Torr. The irradiations were performed in the UHV environment using the output of a pulsed ruby laser and the details of this procedure have been discussed elsewhere [1]. Implanted samples were prepared in a separate ion implantation facility which was also equipped for making Rutherford Backscattering (RBS) measurements. This technique was used to determine the implant profile and to characterize the changes in the subsurface region that occurred with laser annealing.

Because of the desire to examine the surface region, the spectroscopic techniques employed used either electrons or photons as the incident probe. In all cases the detected particle was an electron. As a consequence of the short mean-free path of electrons with energies between 20 and 2000 eV, only the outermost surface region, $\sim$20 Å, was probed. The surface-sensitive spectroscopic techniques employed were Auger electron spectroscopy (AES), low energy electron diffraction (LEED) and photoelectron spectroscopy (PES).

RESULTS

Preparation of atomically clean surfaces. The production of atomically clean surfaces in UHV is a fundamental requirement in areas of basic research directed toward understanding the physical and chemical properties of surfaces and in device technologies where the presence of contaminants introduced either during fabrication or during application can contribute greatly to the degradation of device performance. Recent work has demonstrated that the fastest and most efficient way to obtain an atomically clean surface is by pulsed laser irradiation in UHV.

The application of laser irradiation for the purpose of producing atomically clean Si surfaces is demonstrated by the results shown in Fig. 1. The Auger electron spectrum obtained from an air-exposed Si(100) sample after insertion into a UHV system and following bakeout is shown at the top of Fig. 1. Oxygen (510 eV) and carbon (272 eV) are readily detected on the surface of the as-inserted sample and measurements made by RBS imply a 20-Å oxide thickness, which is typical of air-exposed Si. The features in the region of the Si $L_{2,3}VV$ Auger transition (70-100 eV) are characteristic of Si in silicon dioxide. Irradiation with one laser pulse ($\sim$2.0 J/cm^2) causes a substantial reduction in the levels of O and C present in the surface region as shown in Fig. 1. After exposing the same area to five laser pulses, the Auger electron spectrum shown in Fig. 1 indicates that for the same detection conditions the O and C signals are within the noise level. The Auger electron spectrum obtained from the same area after irradiation with ten laser pulses is shown at the bottom of Fig. 1. Although the O and C signals are not observable in this trace, by increasing the effective sensitivity of the electron detection system, these impurities were determined to be present in surface concentrations of < 0.1% of a monolayer. In addition, the lineshape of the Si $L_{2,3}VV$ transition is that expected from a clean Si surface.

The effect of pulsed laser irradiation on samples that had been sputtered with Ar$^+$ ions (1000 eV, 5 µA, 30 min) was also investigated. Irradiation of the sample with laser pulses of $\sim$2.0 J/cm^2 produced a similar reduction in the contaminant level Auger signals for O and C as observed for the unsputtered surface. Complete elimination, even under increased sensitivty conditions, of the Ar Auger signal occurred after two or three pulses. In addition, if an

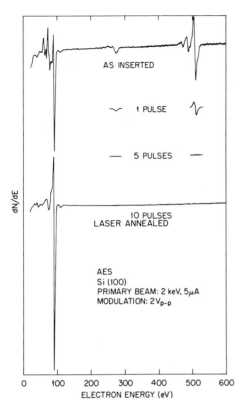

Fig. 1. Auger electron spectra obtained from an uncleaned Si(100) surface and after pulsed laser annealing at ~ 2.0 J/cm^2.

atomically clean surface is exposed to O_2 or CO, the surface can be returned to an atomically clean state by irradiation with five laser pulses.

There are two possibilities as to the ultimate fate of the original oxygen and carbon surface contaminants. The fact that a pronounced pressure rise is observed during the first laser pulse suggests that contaminants are desorbed from the surface during irradiation. Alternatively, since pulsed laser annealing results in the formation of a melted region to a depth of several thousand angstroms, impurities can undergo substantial redistribution by means of liquid phase diffusion during the time the surface region is molten. Whether these impurities are desorbed from the surface or redistributed in depth can be determined by experiments designed to measure the total O and C concentrations in the near surface region before and after laser irradiation.

Such an investigation has recently been performed in which RBS and resonance nuclear reaction $^{16}O(\alpha,\alpha)^{16}O$ techniques in conjunction with AES have been used to determine the oxygen concentration on the surface and in the near surface (3000 Å) region both prior to and after pulsed laser annealing [2]. After irradiation of a Si(100) sample in UHV with eight pulses at an energy density of ~ 1.5 J/cm^2, Auger measurements showed an oxygen concentration at the surface of $\leq 0.3\%$ of a monolayer. The oxygen concentration at a depth of 1100 Å was determined to be $\leq 3.1 \times 10^{18}$ atoms/cm^3 by nuclear reaction analysis. Thus it was concluded that there was no evidence for oxygen interdiffusion during the laser annealing pulse used for cleaning and that the concentration determined in the bulk was less than the oxygen solubility limit.

Although similar results have been obtained in producing atomically clean Ge surfaces [3], problems have been encountered with experiments concerned with compound semiconductors. The production of clean surfaces of GaAs single crystals was investigated using laser annealing only, as well as sputtering with Ar$^+$ ions followed by laser annealing [4]. Similar to results previously observed with Si and Ge, the removal of impurities from the surface region required irradiation with multiple pulses, and complete removal occurred for energy densities > 0.5 J/cm^2. By first sputtering with Ar$^+$ ions, it was found that a surface with no impurities present could be produced by multiple pulse irradiation with energy densities between 0.15 and 0.4 J/cm^2. However, for all cleaning conditions investigated, the lineshapes observed in the Auger electron spectra indicate the presence of Ga in local regions which are nonstoichiometric.

Surface structure. In the previous section it was shown that pulsed laser irradiation could be used to produce atomically clean surfaces in a UHV environment. Since it has been shown that laser irradiation can be used to anneal completely displacement damage in semiconductors, it is of interest then to determine if the crystalline order extends to the outermost monolayer subsequent to irradiation in the UHV environment.

In order to investigate this question we discuss the results obtained from a Si(100) sample using the LEED technique. Examination of the as-inserted sample with LEED showed that diffraction patterns could be observed only at relatively high energies (>250 eV), and they contained an intense background resulting from diffuse scattering. This is consistent with the presence of O and C as determined by AES and shown in Fig. 1. After irradiation with one laser pulse of ∿2.0 J/cm^2, a (2x1) LEED pattern with moderate background intensity due to diffuse scattering was obtained, as shown in Fig. 2. Improvement in the quality of the diffraction pattern occurred with subsequent laser pulses. After five pulses a LEED pattern exhibiting sharp diffraction reflections and very low background intensity was observed, as shown in Fig. 2. The fact that well-defined LEED patterns can be obtained indicates that the crystalline order extends to the outermost monolayers after the liquid-phase epitaxial regrowth process. No detectable change in the LEED patterns was observed with additional pulses, as can be seen by comparing the patterns shown in Fig. 2.

The LEED patterns obtained from the three low-index orientations of Si subsequent to laser annealing with five pulses are shown in Fig. 3 [5]. All patterns show sharp diffraction reflections with low background intensity, and in all cases the quality of the pattern obtained improved with multiple pulse irradiation up to five shots. The (2x1) and (1x2) LEED patterns obtained from the (100) and (110) surfaces are similar to those obtained using conventional thermal treatments and show the presence of reconstructed surfaces. These observations indicate that the atoms in the outermost layers have enough time at a temperature, under the laser annealing conditions used, to reorganize into the reconstructed arrangements from which the LEED patterns are obtained. This is consistent with the proposed surface structure models for the (100) surface that involve only small displacements of the atoms in the outermost monolayers.

In contrast to the patterns obtained from the (100) and (110) surfaces, the (1x1) LEED pattern obtained from the (111) surface and shown in Fig. 3 suggests that as a result of laser annealing the normal surface structure (truncation of the bulk) is obtained, and there is no evidence of any ordered laterial reconstruction [3]. Although a (2x1) LEED pattern can be obtained from a cleaved Si(111) surface, the (7x7) LEED pattern is always observed on a clean thermally annealed crystal surface. Additional investigations showed that the (1x1) LEED pattern could be obtained from the (111) surface when the crystal was held at

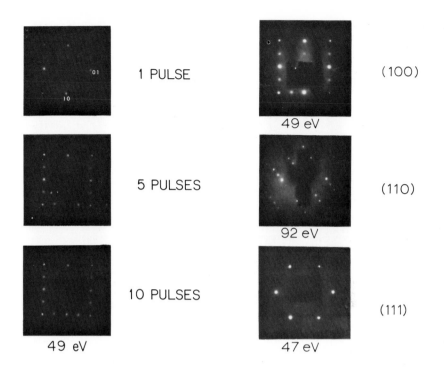

1 PULSE   (100)

49 eV

5 PULSES   (110)

92 eV

10 PULSES   (111)

49 eV   47 eV

Fig. 2. LEED patterns from a Si(100) surface at a primary beam energy of 49 eV. Patterns are shown subsequent to laser annealing at ∿2.0 J/cm^2 for 1, 5, and 10 pulses.

Fig. 3. LEED patterns from clean (100), (110), and (111) Si surfaces at primary beam energies of 49, 92, and 47 eV. Patterns are shown subsequent to laser annealing at ∿2.0 J/cm^2 for 5 pulses.

temperatures in the range 100-700 K during laser annealing. The observation of one-seventh order diffraction reflections, indicative of the reconstructed surface, occurred after either annealing the laser-irradiated surface at temperatures greater than ∿800 K or holding the crystal at this temperature during the laser annealing process. By heating for a sufficient time (>30 min) at temperatures greater than ∿800 K, it was possible to convert the laser annealed surface to one from which a well-defined (7x7) LEED pattern was observed. This result indicates that the structure giving rise to the (1x1) LEED pattern is metastable, and by combining laser annealing with conventional thermal annealing, it is possible to cycle back and forth between the two structures. As with Si, well-defined LEED patterns are obtained from low-index oriented single crystals of Ge subsequent to laser irradiation [3].

Although LEED patterns were obtained for all orientations of GaAs crystals investigated, the quality of the surface structure as reflected in these patterns differed significantly [4]. The highest quality patterns were obtained from the (110) orientation, and reasonable quality patterns were obtained from both A and B type (111) orientations. Very poor quality patterns were obtained from the (100) orientation. In all cases the observation of diffuse background

intensity or streaking indicated the presence of disorder present in the surface region. These observations are consistent with both AES and RBS results
which indicate the existence of excess Ga in local regions which are nonstoichiometric in the near surface.

The observation that a (1x1) LEED pattern is obtained from the (111) surface
of Si after laser irradiation in a UHV environment, coupled with the fact that
this surface is observed to be atomically clean after this treatment, suggests
that it offers the opportunity for investigating a clean semiconductor surface
that exhibits no ordered lateral reconstruction. In particular, it should be
possible to obtain information about surface relaxations. One way to investigate this question is to measure the intensities of the diffracted electron
beams as a function of incident electron energy (I-V profiles). The experimentally measured profiles must then be compared to results obtained from fully-
converged dynamical LEED calculations assuming various structural models for
the geometric arrangement in the outermost monolayers. A measure of the
agreement between the experimental results and the predictions of the model
calculations is provided by the R factor (the lower the R-factor value, the
better the agreement). Such an analysis has been performed for the laser-
annealed Si(111)-(1x1) surface [6]. Results of this analysis suggest that the
first interlayer spacing, d_{12}, is contracted by 25.5±2.5% with respect to the
bulk value and that the second interlayer spacing, d_{23}, is expanded by 3.2±1.5%
with respect to the bulk value. The R factor corresponding to this configuration is 0.115 and this value indicates a very good agreement between
calculated and experimental profiles in a conventional LEED analysis suggesting
that the proposed structural model is highly probable. It should be noted that
only ordered structures could be tested in the LEED analysis and thus questions
related to the presence of surface defects (vacancies, steps, etc.) could not
be addressed. Since the laser irradiated surface is known to undergo an irre-
versible transition from the (1x1) to (7x7) at ∿700 K [7], a study of this
transition should aid in distinguishing between differing models for each surface structure.

In view of the results of the LEED analysis just discussed, an examination
of the electronic structure in the surface region of the laser-annealed Si(111)
surface is of interest. Theoretical one-electron-band-structure calculations
predict that such a surface would be metallic with a half-filled band of
dangling bond states at the Fermi energy E_F. Furthermore, the electronic properties of this surface would be very different from that observed for the
annealed Si(111)-(7x7) surface, as well as for the cleaved Si(111)-(2x1) surface [8]. It is assumed that surface states are band-like in all these
calculations.

Angle-resolved and angle-integrated photoemission studies of both valence
band surface states and surface core-level shifts for the laser-annealed
Si(111)-(1x1) surface, prepared in the same manner as for the LEED study, and
for a Si(111)-(7x7) surface, prepared by thermally annealing the (1x1) surface
have been performed [9]. The measurements were made using the display-type
spectrometer at the synchrotron radiation source, Tantalus I. In Fig. 4,
angle-integrated photoemission spectra are presented for a laser-annealed (1x1)
surface, for a (7x7) surface prepared by thermally annealing the (1x1) surface
to ∿1175 K, and for a hydrogen covered surface. The difference between the
solid curve and the dashed curve within ∿3 eV of E_F represents surface state
emission. The principal difference between the spectra obtained for the (1x1)
and (7x7) surfaces is the weak feature seen at the Fermi level on the (7x7)
surface which is 0.5±0.05 eV above the valance band maximum E_v. The energies
and angular distributions of the other two surface states are identical for
both surfaces. In addition, angle-integrated photoemission spectra of the Si
$2p_{3/2}$ core-level for the (1x1) and (7x7) surfaces show surface core-level
shifts that are similar in energy and intensity.

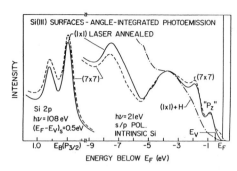

Si(III) SURFACES - ANGLE-INTEGRATED PHOTOEMISSION

(IxI) LASER ANNEALED

INTENSITY

(7x7)

(7x7)

Si 2p
hν=108 eV
(E_F-E_V)_s=0.5eV

hν=21eV
s/p POL.
INTRINSIC Si

(IxI)+H

"P_z"

E_v

1.0   E_B(P_{3/2}) -9   -7   -5   -3   -1   E_F

ENERGY BELOW E_F (eV)

Fig. 4. Angle integrated photoemission spectra for the valence
bands and 2p core levels of laser-
annealed Si(111)-(1x1) and
thermally annealed Si(111)-(7x7)
surfaces. Two prominent surface-
state levels are seen for both,
i.e. "p_z-like" levels at -0.85 eV
and levels at -1.8 eV.

The strong similarity of the valence band surface states and surface core-
level spectra for the laser-annealed (1x1) surface and thermally annealed (7x7)
surface indicate that these surfaces have very similar local bonding geometries
and differ mainly in long-range order involving geometrical rearrangements that
are only a perturbation of the average local bonding geometry. Furthermore,
the absence of any emission in the gap at E_F for the (1x1) surface is not in
accord with predictions of one-electron-band calculations for an unrecon-
structed surface and therefore appears to be inconsistent with the structural
model for this surface determined in the LEED analysis [6]. However, it should
be noted that photoemission can rule out the relaxed ordered (1x1) geometry
only if the surface states are band-like as assumed in one-electron-band
calculations. Correlation effects might be very important for those narrow
surface levels and recent theoretical proposals which include these effects
would make the photoemission data from the Si(111)-(1x1) surface consistent
with the model obtained from the LEED analysis [10,11]. Data obtained in a
recent ion-scattering study of the (7x7) and (1x1) surfaces [12] suggest that,
in agreement with the PES results just presented, the local geometric order in
these two surfaces is very similar. In addition, the values of the interlayer
relaxations obtained in this study are larger than those determined in the LEED
analysis. Results from other PES studies suggest that different laser annealing conditions (depth of melt, regrowth velocity) can result in different surface structures [13]. If this is true, then laser annealing with different
annealing parameters may aid in determining the driving mechanisms for the
reconstruction observed following conventional preparation procedures.

Stepped surfaces. The electronic properties and chemical activity of a surface can be strongly influenced by the presence of defects such as steps.
Although stepped surfaces can be prepared by cleaving or ion-etching, the
control of the step density and ease of reproducibility have proved difficult
using conventional procedures. Thus, it is of interest to determine if well-
annealed surfaces with monatomic steps and uniform terrace widths can be produced using laser annealing techniques. In order to investigate this question
a Si(111) crystal whose surface was cut at 4.3° from a (111) plane in the
[$\bar{1}\bar{1}2$] direction was used [14]. The well-defined (1x1) LEED pattern obtained
from the clean surface and shown in Fig. 5a was observed after irradiating the
surface with five pulses at ∿2.0 J/cm². The pattern indicates the existence of
a stepped surface, and the energies at which a given reflection is split or
nonsplit give specific information on the step height; the angular separation
between split reflections provides information on the terrace widths. With the
use of only a kinematic treatment of single scattering from the top layer and
methods previously discussed [15], it is concluded that the surface consists of
monatomic steps with an average step height of one double layer (3.14 Å) and
terrace widths of ∿45 Å, as illustrated in Fig. 6. The absence of fractional
order reflections, indicative of reconstruction, suggests that the local atomic

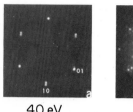 

40 eV   68 eV

Fig. 5. LEED patterns from clean vicinal (111) Si surfaces at primary beam energies of (a) 40 and (b) 68 eV. (a) Laser-annealed, (b) thermally annealed.

Si (111)

LASER ANNEALED — (1x1) WITH SPLIT SPOTS

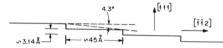

THERMALLY ANNEALED — (7x7)

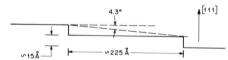

Fig. 6. Schematic illustration of the vicinal surface projected into the (110) plane. Top view is for the laser-annealed surface. Bottom view illustrates a possible configuration obtained with thermal annealing.

arrangement produced by this annealing procedure may be similar to that produced on flat (111) surfaces. The stability of the stepped surface was investigated by subjecting the laser-annealed surface to a series of thermal annealing treatments. As observed for flat (111) surfaces, thermal annealing of the crystal to a temperature greater than ~800 K resulted in a surface from which the (7x7) diffraction pattern shown in Fig. 5b was obtained. All indications of the steps were gone (individual reflections were no longer split), and the (7x7) pattern was obtained over the entire face of the sample. The sharpness of the integral-order reflections is consistent with a surface having domains wider than ~200 Å. In order to maintain the average inclination, multilayer steps must be present as illustrated in Fig. 6. After thermal annealing, the stepped surface arrangement could be regenerated by again irradiating the surface with the laser. These observations indicate that it is possible to produce repeatedly a particular step arrangement by initially cutting the crystal to the desired orientation.

Ion-implanted, laser-annealed silicon. Previous investigations have shown that group III or V implants occupy substitutional sites subsequent to laser annealing, and that, as a consequence of both the high liquid-phase diffusivities and the high values of distribution coefficients, they are able to diffuse into the crystal during the regrowth process after irradiation. In contrast, it has been shown that those implants which do not form covalent bonds exhibit, depending on the implant dose, segregation to the surface as well as the formation of a cell structure subsequent to laser annealing. Although both RBS and secondary-ion mass spectroscopy techniques provide detailed information about the distriution with respect to depth, they provide no information about the concentration in the surface region ($\leq$ 20 Å) and its change with multiple pulse irradiation.

Such information can be obtained by performing experiments using surface sensitive techniques and will be illustrated by examining the case of Si implanted with [75]As. Auger data were obtained from a Si(100) sample implanted with [75]As (100 KeV, $8.3 \times 10^{16}/cm^2$) and subjected to multiple pulse laser irradiation. With these data the ratio of the intensity of the As L_3VV (703 eV) Auger transition to that of the Si KL_3L_3 (1619 eV) is shown in Fig. 7, where it is plotted as a function of the number of pulses. The data show that the relative amount of As in the surface region decreases with an increasing number of pulses, similar to the RBS results obtained for the subsurface region [16]. Although not shown in this figure, little change is observed in the surface concentration after a large number (>15) of pulses where RBS results indicate uniform concentration from the subsurface region down to the liquid-solid interface.

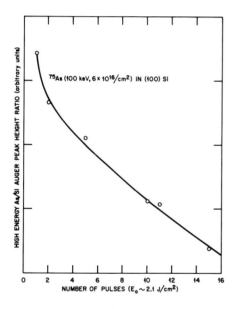

Fig. 7. Plot of the ratio of the As L_3VV to Si KL_3L_3 Auger transition intensities as a function of the number of laser pulses.

To examine the effect of the implanted species on surface order, LEED observations have been made also. Shown in Fig. 8 are LEED patterns obtained from the same [75]As implanted Si(100) crystal. Only a very weak, poorly defined LEED pattern was observed after one pulse. Following two pulses of irradiation, the pattern shown at the top of Fig. 8 was obtained. Integral order beams are observed, as well as streaks between them. With additional pulses the streaks begin to coalesce into half-order reflections, indicating the formation of a (2x1) surface structure. They continue to become sharper and more intense with additional pulses, as shown in Figure 8. However, the pattern observed after ten pulses is not as good as that obtained from a virgin Si(100) crystal, as shown in Fig. 2, and thus indicates the presence of disorder in the surface

326

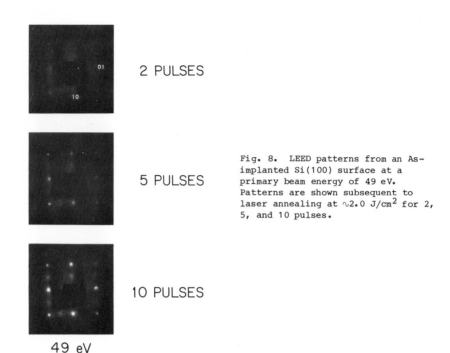

2 PULSES

5 PULSES

Fig. 8. LEED patterns from an As-
implanted Si(100) surface at a
primary beam energy of 49 eV.
Patterns are shown subsequent to
laser annealing at ∿2.0 J/cm^2 for 2,
5, and 10 pulses.

10 PULSES

49 eV

region. It is interesting to note that the (2x1) LEED pattern shows the
existence of the reconstructed surface, similar to that obtained from a virgin
Si(100) crystal. In contrast to this, although AES results obtained from
Si(111) crystals implanted with group III or V dopants are similar to those
obtained for the (100) surface, (1x1) LEED patterns were observed in all cases.
The patterns are of much higher quality after a given number of pulses when
compared to those obtained from the (100) surfaces, and they show no evidence
of ordered lateral reconstruction.

The observation that laser annealing can be combined with ion implantation
to provide semiconductor surface regions containing novel doping concentrations
(superaturated alloys) suggests that these techniques may be used to alter or
tailor the electronic structure in this region. To examine this possibility
photoemission techniques have been used to investigate highly degenerate n-type
Si(111)-(1x1) surfaces as a function of As concentrations up to ∿5x10^{21}/cm^3
(∿10 at.%) and degenerate p-type Si(111)-(1x1) surfaces as a function of B concentrations up to ∿1x10^{21}/cm^3 (∿2 at%).

Angle-integrated photoemission spectra for the valence bands are presented
in Fig. 9 for intrinsic Si(111)-(1x1), degenerate n-type As-doped (4 and 7
at.%) Si(111)-(1x1), and degenerate p-type B-doped (1 at.%) Si(111)-(1x1)
surfaces [17]. The spectra are normalized to constant total emission within
5 eV of E_F and energies are given relative to the valence-band maximum at the
surface (E_v^S). E_F is seen to shift markedly with doping, i.e., from 0.25 eV
above E_v^S for the B-doped sample to the conduction band minimum E_c = 1.1 eV for
the 7% As-doped sample. Relative to intrinsic Si, for highly degenerate (1
at%) B doping, the two states are unaltered and the principal changes are that

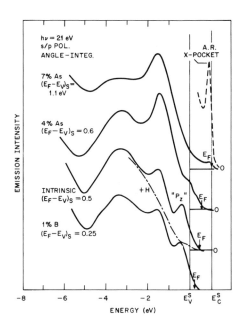

Fig. 9. Photoemission spectra
(PDOS) for the valence bands of
laser annealed (111)-(1x1) surfaces
of intrinsic and highly doped Si.
The levels near -0.4 and -1.3 eV
are due to surface states. $E_v{}^S$,
$E_c{}^S$, and E_F denote the valence-
band maximum, conduction-band
minimum, and Fermi-level positions
at the surface. The dashed-dotted
line shows the effect of an
absorbed monolayer of H on intrinsic
Si.

E_F moves down by 0.25 eV and the surface becomes metallic. More dramatic
effects are seen with As doping. At 4 at.% As doping, the surface states have
become significantly altered, while E_F has increased by 0.1 eV relative to the
intrinsic Si. That is, the upper "sp_z-like" dangling bond state has become
much weaker and shifted upward in energy by ∿0.3 eV; the lower -1.4 eV state
has increased significantly in intensity but it is unshifted in energy, and the
surface has become metallic with new states near E_F. As the doping is further
increased from 4 to 7 at.%, E_F rapidly shifts and becomes pinned at the conduction band minimum E_c. Also, the upper sp_z-like surface-state continues to
diminish in intensity so as to be nearly imperceptible by 7 at.% doping, and
the lower surface state becomes extremely intense. By depositing a thin Au
film on this surface it was possible to show via Si 2p core-level measurements
that E_F remained unchanged (within ∿50 meV). Thus, a "zero-barrier-height"
Schottky barrier was formed, although, for electrical purposes, the Au-Si
interface is undoubtedly shorted because of the extreme degenerate n-type
doping.

SUMMARY

It has been shown that the properties of the surface regions of semiconductor
single crystals can be altered by laser annealing in a UHV environment. The
production of atomically clean surfaces in processing times of << 1 sec and the
restoration of geometric order to the surface region of a damaged crystal indicate that this processing technique provides an alternative approach to conventional surface preparation procedures. When combined with ion implantation,
laser annealing can be used to tailor both the geometric lattice (interatomic
spacings) and electronic structure in the surface region.

These observations indicate that laser annealing of surfaces in UHV has potential as a tool for both surface science and practical application. Although it has been tested extensively only on semiconductors, the results obtained suggest that it will be applicable to a wide range of materials. Additional investigations concerned with the surface properties of materials irradiated with the laser, as well as the changes that occur due to varying the annealing conditions (regrowth velocity, depth of melt, etc.), need to be performed in order to characterize and to understand this processing technique more completely.

REFERENCES

1. D. M. Zehner, C. W. White, and G. W. Ownby, Appl. Phys. Lett. 36, 56 (1980).

2. J. F. M. Westendorp, Z. L. Wang, and F. W. Saris, Laser and Electron Beam Interactions with Solids, B. R. Appleton and G. K. Celler eds., (North-Holland, New York), p. 255.

3. D. M. Zehner, C. W. White, and G. W. Ownby, Appl. Phys. Lett. 37, 456 (1980).

4. D. M. Zehner, C. W. White, B. R. Appleton, and G. W. Ownby, Laser and Electron Beam Interactions with Solids, B. R. Appleton and G. K. Celler eds., (North Holland, New York) p. 683.

5. D. M. Zehner, C. W. White and G. W. Ownby, Laser and Electron Beam Processing of Materials, C. W. White and P. S. Peercy eds. (Academic Press, New York) p. 201.

6. D. M. Zehner, J. R. Noonan, H. L. Davis, and C. W. White, J. Vac. Sci. Technol. 18, 852 (1981).

7. P. A. Bennett and M. B. Webb, Surf. Sci. 104, 74 (1981).

8. D. E. Eastman, J. Vac. Sci. Technol. 17, 492 (1980).

9. D. M. Zehner, C. W. White, P. Heimann, B. Reihl, F. J. Himpsel, and D. E. Eastman, Phys. Rev. B 24, 4875 (1981).

10. C. B. Duke and W. K. Ford, Surf. Sci. 111, L685 (1981)

11. R. Del Sole and D. J. Chadi, Phys. Rev. B 24, 7431 (1981).

12. R. M. Tromp, E. J. van Loenen, M. Iwami, and F. W. Sarris, Solid State Commun. (in press).

13. Y. J. Chabal, J. E. Rowe, and S. B. Christman, J. Vac. Sci. Technol. 20, 763 (1982).

14. D. M. Zehner, C. W. White, and G. W. Ownby, Surf. Sci. Lett. 92, L67 (1980).

15. M. Henzler, Surf. Sci. 19, 159 (1970); 22, 12 (1970).

16. C. W. White and W. H. Christie, Solid State Technol. 23, 109 (1980).

17. D. E. Eastman, P. Heimann, F. J. Himpsel, B. Reihl, D. M. Zehner, and C. W. White, Phys. Rev. B 24, 3647 (1981).

LASER-INDUCED SURFACE DEFECTS

Jean-Marie MOISON and Marcel BENSOUSSAN
Centre National d'Etudes des Télécommunications, Département OMT*,
196 rue de Paris   92220 BAGNEUX - FRANCE

ABSTRACT

The influence of pulsed laser processing (PLP) on the structure and composition of Si, GaAs, InP and Au surfaces has been
investigated as a function of the laser fluence. Above a threshold fluence, all initial structures turn into a (1x1) structure and the V-element content of the compound surfaces is
decreased. The structure change is shown to be related to the
existence of a number of atomic steps on the PLP surface (up
to 10^{14}/cm² equivalent broken bonds). These defects can be
eliminated by further annealings. On the other hand, the
stoechiometry defects, which are less numerous (10^{12}-10^{13}/cm²)
cannot be eliminated. A model for the mechanisms of defect creation and annihilation during PLP is outlined. The incidence of
PLP-induced defects on device technology is evaluated.

INTRODUCTION

The study of laser-induced surface modifications is by now an active field
of research. The main incidence of this topic lies in the use of pulsed-laser
processing (PLP) in the fabrication or the improvement of electronic devices.
On the other band, laser processing has also been demonstrated to be a powerful
surface tool : it can clean very efficiently and very rapidly semiconductor surfaces, and it can induce original surface structures, which cannot be obtained
by other means | 1|. In this work, we have investigated under UHV conditions
the evolution under PLP of the composition and structure of the surfaces of
several materials, Si, Au, and A-B compounds GaAs and InP. We have determined
the threshold fluence at which surface changes occur. Though this threshold is
material-dependent, the nature of the change follows a general trend. It can
be analyzed in terms of 1) stoechiometry defects in the case of compounds, and
2) atomic steps in all cases. The step/superstructure interaction is shown to
play a major part in the surface transition obtained by PLP.

EXPERIMENTAL

The samples are Si (111), GaAs (100), InP (100) and Au (100) single crystals. They are chemically or electrochemically cleaned, then inserted in a UHV
chamber and cleaned by repeated ion bombardments and annealings. In the case of
the A-B compounds, the annealing temperature is kept low enough to prevent the
surface decomposition. When all the contaminant species become undetectable by
AES, the corresponding LEED patterns are the following : (7x7) for Si, c(8x2)
for GaAs, (4x2) for InP, and "(5x20)" for Au. These are the room-temperature
equilibrium (RTE) structures. The surface composition of the A-B compounds is
found to be B-rich by AES measurements.

When this RTE state is reached, the surfaces are irradiated in the UHV chamber. The laser source is an excimer-laser-pumped dye laser. Its wavelength is
5390 Å (2.3 eV), its pulse duration is 3 ns, and its repetition rate is 20 Hz.

Mat. Res. Soc. Symp. Proc. Vol. 13 (1983) ©Elsevier Science Publishing Co., Inc.

The beam is focussed on the sample through a sapphire viewport. Its characteristic features at the impact zone (pulse duration, energy/pulse, energy spatial distribution) are carefully monitored by on-line devices, located at points optically equivalent to the sample. Impact area is about 10^{-3} cm^2 and fluence is varied from 0.01 to 3 J/cm^2 by neutral filters. The sample is then moved in front of the laser beam by a raster-scan mechanism. The scan speed is adjusted so that the whole sample is homogeneously irradiated and each point struck by 25 laser shots. During irradiation, outgoing electron, ion and neutral fluxes are measured by a pulsed charge amplifier, an ion gauge and a mass spectrometer. After the irradiation, AES spectra are taken and LEED is performed.

RESULTS - THE PULSED-LASER-PROCESSED SURFACE

All surfaces behave roughly in the same way : above a threshold fluence, simultaneous ion, electron, and neutral emissions are observed. The initial LEED pattern is changed into a (1x1) pattern with a slightly increased background   . For the A-B compounds, the AES signal of the B element is slightly reduced. All observations for InP, which are described in length elsewhere | 2|, are summarized in figure 1. For the other materials, the results are similar, with changes in the threshold fluence and in the magnitude of the effects. The values of the threshold fluences are listed in table 1. In view of the problems involved in the measurement and even in the definition of a pulsed laser fluence, the absolute values may be off by a large factor (50% ?) ; however, as all experiments were performed under the same conditions, the incertitude on relative values are the ones of table 1. The values for Si are in fair agreement with previous determinations |3|.It must be stressed that the TFs for surface modifications are similar to the fluences used for bulk processing. Hence, all the above-mentioned effects may be unwanted by-products and their features must be analyzed in terms of remnant defects. It is important to note that the results obtained under UHV conditions apply to air-pressure experiments : at the fluences involved, neither initial structure nor initial contamination should matter, and further recontamination which occurs rather slowly acts upon an already cooled sample and should not alter the defect structure.

Near the threshold fluence, the matter removal during irradiation, which is the strongest for InP (Refer. 2), ten times smaller for GaAs and nearly undetectable for Si and Au, remains low in terms of coverage (<0.05 monolayer for InP near the threshold). Hence, its influence on the final laser-processed structure remains minor. The first point is the observation of the (1x1) LEED pattern on the pulsed-laser-processed (PLP) surfaces. We show here that, under our irradiation conditions, a LEED defect pattern is superimposed on the (1x1) pattern, indicating the presence of numerous defects (steps) on the PLP surface. This is evidenced by the observation of (1x1) diffraction spots alternately sharp and diffuse when the primary electron energy is varied. This pattern has been shown to correspond to the existence of steps on an otherwise perfect surface. By plotting the electron wavenumber for either sharp or diffuse spot versus the diffraction index (Henzler's method |4|, we deduce the height of the steps (see table 1). Within the limits of determination they are equal to simple bulk distances : the (100) simple-layer distance for Au, the (111) double-layer distance for Si, and the distance between (100) B-element planes for GaAs and InP. From the angular width of the diffuse diffraction spots, we also deduce the mean value of the randomly-distributed interstep spacing (see table 1). This may in turn be converted into an equivalent defect density. If the steps were ideally distributed along a square net with a spacing of d, the density of defects (one defect per ledge atom) would be ~ 2 a/d, where a is the interatomic spacing on the ledges. The corresponding values are listed in table 1 ; they are only rough estimates which show that the magnitude of the defect density is quite high (> 10^{13} cm^2).

TABLE 1
Surface features of Si, GaAS, and InP laser-processed samples

Surface behavior under laser processing (2.3 eV, 3ns)	Si (111)	GaAs (100)	InP (100)
. Threshold fluence for surface transformation (J/cm²)	0.5±0.05	0.25±0.05	0.17±0.05
. Initial surface structure	(7x7)	c(8x2)	(4x2)
. Final surface structure	(1x1)	(1x1)	(1x1)
. Step formation	YES	YES	YES
. Step height (Å)	3.07±0.10	2.72±0.10	3.05±0.10
. Bulk spacing (Å)	3.14	2.83	2.94
. Interstep mean spacing (Å)	70	30	30
. Annealing necessary for recovering the initial structure	≅400Cx5'	≅250Cx5'	≅150Cx5'
. Initial stoechiometry (AES)		As-rich	P-rich
. Final stoechiometry (AES)		As coverage ↓	P coverage ↓
. Matter ejection near TF	very weak	strong	strong
. Equivalent structural defect density (cm⁻²)	≅10¹³	≅10¹⁴	≅10¹⁴
. Equivalent stoechiometric defect density (cm⁻²)		≅10¹²	≅10¹³
. Complete recover by annealing	YES	NO	NO

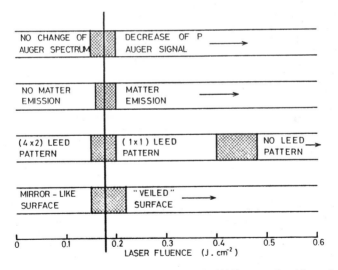

Fig. 1. Summary of various measurements on InP (100) as a function of laser fluence. Photon energy is 2.3 eV ; pulse duration is 3 ns.

Now, in the case of compounds, another source of defects lies in the loss of surface stoechiometry. By assuming that the loss of one V-element atom creates one equivalent defect, the defect density created by this process can be evaluated from the desorption measurements. The values obtained near the TF are listed in table 1. They are smaller than the densities corresponding to step formation. However, an important difference between the two kinds of defects must be noted : while the structural ones may be eliminated by a very mild annealing, the stoechiometric one cannot.

RESULTS - ANNEALING THE PLP SURFACE

We checked the influence of further annealing treatments on the PLP surfaces. In the case of Si and Au, the recover of the initial RTE structure seems perfect, with respect to the intensity and contrast of the LEED pattern. Many PLP/anneal cycles can be achieved without loss of quality. The temperatures and durations required are shown in table 1. It must be noted that they are much lower than the values required to anneal ion-etched samples. LEED observation during the annealing treatment shows that the defect pattern disappears exactly when the RTE reappears.

For A-B compounds, while the results are apparently similar from the point of view of LEED, a basic difference exists : the loss of B-element content of the PLP surface. Annealing then leads to the formation of small metallic islands which gather the excess of A-element, leaving patches with the initial RTE structure. Hence full reversibility can never be obtained, except perhaps near the threshold where the atom loss is very small.

For the InP (100) surface in this fluence range, another extra feature appears : during the anneal, all steps are not eliminated. After several PLP/anneal cycles, the LEED pattern becomes a superposition of the RTE pattern and of the pattern of an ordered step array. The related ledges, which are going alternately up and down, all lie in the <01> direction and their spacing is the four-fold dimension of the RTE unit cell. This rippled structure is stable under all annealing treatment up to the decomposition temperatures.

DISCUSSION - THE PLP STRUCTURE

The (1x1) LEED pattern of the PLP surface has already been observed on Si by many authors and on GaAs |5|. The true meaning of this pattern is still widely discussed. It has been proposed that an ultrathin amorphous layer is formed by PLP and that the (1x1) pattern reveals the underlying bulk configuration |6|. More recently, a more or less bulk-like surface structure with interlayer relaxation has been determined by LEED intensity analysis |7|. Finally, two (2x1)-like |3| and (7x7)-like |8| models have been proposed and discussed. As the irradiation conditions (pulse fluence, wavelength, duration) were most often different in the various papers, a critical comparison of the models remains questionable. The trend towards amorphization as the order range is decreased under increasing laser irradiation, that we indeed observe on InP samples (see figure 1), may possibly lead in Si to a (7x7)-like structure, then to a (2x1)-like one, then to a (1x1)-like one with a thin amorphous layer, and finally to a thick amorphous layer. Consequently, all models may be true, depending on the irradiation parameters and the cooling rate. However, recent investigations of the ruby-laser-processed silicon surface by ARUPS, core-level XPS, and RBS seem to confirm the model of a (7x7)-like disordered PLP surface, up to 2 J/cm² |8|. This should be true in our experiments, at about 0.5 J/cm² for Si. For the other laser-processed surfaces, no information is available by now except the ones mentioned above i.e. their only known features are a (1x1) LEED pattern, on which a defect pattern is superimposed, indicating the presence of constant-height steps distributed at random.

These features of the PLP surfaces have been observed on surfaces prepared otherwise. The RTE structure of Si and Au change reversibly to a (1x1) struc-ture at high temperatures (1140K for Si and 1070K for Au |9|) ; these changes have been attributed to order-disorder transitions. RTE (1x1) surfaces can also be obtained by low-coverage impurity stabilization |10|.

However, in both cases, the structures of the RTE and (1x1) surfaces are not yet clearly determined. Ordered steps have been obtained purposefully on Au and GaAs surfaces |11| by preparation of vicinal samples. The step heights observed are in agreement with the values that we find on PLP surfaces. For silicon sur-faces, steps with the bulk double-layer height are reported on vicinal samples under 800°C, on sputtered samples, and on PLP samples |12|. Hence, the features observed on the PLP surface are classical, and this leaves only the questions of their origin and of their association. The mechanisms of pulsed laser pro-cessing are still under investigation, but whatever the model it is admitted that "melting" of the surface takes place, followed by a fast quenching. During this quenching period, a competition takes place between the elimination of the defects created at the high temperatures involved and the growth of the RTE super-structure. If the defects are eliminated fast enough, the RTE structure is finally obtained. On the opposite, if they are not, they may impede the growth of the superstructure by various mechanisms and appear in the PLP sur-face. Our results are in agreement with the second mechanism involving the inhibition of the growth of the RTE structure by the steps. This is further demonstrated by the results of annealing the PLP structure : as soon as the defects are eliminated, the growth takes place simultaneously and restores the RTE structure. Two mechanisms able to impede the superstructure growth during the quenching period may be suggested : 1) strain fields associated with steps prevent the reconstruction in their immediate vicinity ; if the steps are too close reconstruction does not occur at all, and only a distorted (1x1) lattice remains ; 2) reconstruction does take place, but the existence of the steps prevents the formation of large in-phase domains, and hence the observation of the non-integral LEED spots. The two mechanisms involve a small interstep spacing, and we indeed observe that the PLP interstep distance is only three to five times larger than the unit cell of the RTE superstructure. During the quenching process, the strength of the repulsive step/step interaction is strong enough to prevent the superstructure formation, while during subsequent annealings the superstructure may develop, sweep the steps out of large ter-races, and pile them in incoherent bunches. This process is observed in the case of Si and Cu vicinal surfaces |12|.

A slightly more accurate description can be made in the case of (111) Si. The RTE structure is by now considered to involve an ordered array of steps going alternately up and down |13|. At high temperature, this array gets disor-dered, and PLP quenches this state. The low temperatures required to restore the RTE structure are then easily explained : the random steps need not be annihilated, but merely ordered. In the case of vicinal surfaces, extra steps are purposefully introduced. PLP quenches a situation where the repulsive interaction between them is stronger than the ones involved in the superstruc-ture formation, and where these steps can get into order |12|. An opposite situation is observed after annealing and the extra steps disorder.

In the case of (100) InP, the rippled (stepped) surface we obtain after several PLP / anneal cycles seems to be a configuration more stable than the initial flat surface, and which can be obtained only by PLP (it could pro-bably be obtained by classical annealing if the decomposition of the surface did not occur first).Here again, the PLP-induced random steps need not be eliminated, but merely reordered, and they may act as a necessary precursor state. In both cases, the dynamics of the step creation/annihilation and migration play the leading part in the PLP structure formation.

CONCLUSION

We show here for the first time that the surface modifications induced by PLP on various materials are related to the creation of atomic steps and stoechiometry defects.
A general mechanism involving the blocking of long-order superstructure growth by the PLP-induced defects (steps) is proposed. This model accounts for the surface structure of any PLP sample in a simple and unified way. In the case of monoelement samples,reversibility to the RTE structure is easy, and probably favoured by the very PLP structure. However, stoechiometry defects on the compound surfaces cannot be eliminated. This may prove to be a drawback in electronic device applications, owing to the sheer number of defects, even if they are not all electrically active. On the other hand, PLP may prove to be an excellent surface tool in the preparation of clean original metastable and stable surface structures.(14).

REFERENCES

* Laboratoire associé au C.N.R.S. L.A. 250

|1|D.M. ZEHNER, C.W. WHITE and G. OWNBY, Appl. Phys. Lett. 36, 56 (1980)
   P.L. COWAN and J.A. GOLOVCHENKO, J. Vac. Sci. Technol. 17, 1197 (1980)
|2|J.M. MOISON and M. BENSOUSSAN, J. Vac. Sci. Technol. 21, 315 (1982)
|3|Y.J. CHABAL, J.E. ROWE, and D.A. ZWEMER, Phys. Rev. Lett. 46, 600 (1980)
|4|M. HENZLER, Surf. Sci. 19, 159 (1970)
|5|D.M. ZEHNER, C.W. WHITE, B.R. APPLETON and G.W. OWNBY, Proc. of the
   MRS Annual Meeting - Boston 1981 (North-Holland 1982) p. 603
|6|S.M. BEDAIR and H.P. SMITH, J. Appl. Phys. 40, 4776 (1969)
|7|D.M. ZEHNER, J.R. NOONAN, H.L. DAVIS and C.W. WHITE, J. Vac. Sci.
   Technol. 18, 852 (1981)
|8|D.M. ZEHNER, C.W. WHITE, P. HEIMANN, B. REIHL, F.J. HIMPSEL and D.E. EASTMAN
   Phys. Rev. B24, 4875 (1981)
   F.J. HIMSEL, 16th Int. Conf. on the Physics of Semiconductors, Montpellier
   1982
   R.M. TROMP, E.J. VAN LOENEN, M. IWANI and F.W. SARIS, ibidem (11) and
   Solid State Comm., to be published
|9|P.A. BENNETT and M.W. WEBB, Surf. Sci. 104, 74 (1981)
|10|D.E. EASTMAN, F.J. HIMPSEL and J.F. VAN DER VEEN, Solid State Comm.
   35, 345 (1980)
|11|F. HOTTIER, J.B. THEETEN, A. MASSON and J.L. DOMANGE, Surf. Sci. 65,
   563 (1977)
|12|B.Z. OLSHANETSKY and A.A. SHKLYAEV, Surf. Sci. 82, 445 (1979)
   Y.J. CHABAL, J.E. ROWE and S.B. CHRISTMAN, Phys. Rev. B24, 3303 (1981)
   J.M. MOISON and J.L. DOMANGE, Surf. Sci. 67, 336 (1977)
|13|M.J. CARDILLO, Phys. Rev. B23, 4279 (1981)
|14|J.M. MOISON and M. BENSOUSSAN, Surf.Sci. and Appl.Phys.Lett.,to be published

PART IV
TRANSIENT THERMAL PROCESSING
OF SILICON FOR DEVICES

APPLICATIONS OF LASER ANNEALING IN IC FABRICATION*

L.D. HESS, G. ECKHARDT, S.A. KOKOROWSKI,** and G.L. OLSON
Hughes Research Laboratories, Malibu, California, USA

A. GUPTA, Y.M. CHI and J.B. VALDEZ
Newport Beach Research Center, Hughes Aircraft Company, Newport Beach, California, USA

C.R. ITO and E.M. NAKAJI
Torrance Research Center, Hughes Aircraft Company, Torrance, California, USA

L.F. LOU
Santa Barbara Research Center, Goleta, California, USA

ABSTRACT

Laser annealing is discussed in the context of potential applications in the fabrication of advanced solid state components and integrated circuits. General aspects of the unique temporal and spatial heating distributions that can be obtained with laser heating are presented, and selected examples are given which illustrate the advantage of special time/temperature heating cycles in the processing of specific semiconductor device structures. The performance of silicon and HgCdTe diodes, polysilicon resistors, multiple stacked polysilicon/oxide capacitors, Al/Si ohmic contacts and MOS/SOS transistors fabricated using laser annealing is significantly improved relative to devices fabricated using conventional furnace annealing.

INTRODUCTION

As the efforts in solid state electronics continue along the directions of higher performance, functional packing density, and yield, many challenges arise throughout the entire hierarchy from materials science to packaging technology. Laser and other forms of transient heating provide unique characteristics which can be used to overcome or bypass some of the limitations encountered with conventional processing methods and, additionally, make possible the fabrication of new device and circuit configurations. The former point will be emphasized in this paper and illustrated by examples taken from our current and previous laser annealing studies. The term "laser annealing" originally referred to the use of laser radiation for the removal of ion-implantation damage in semiconductors, whereas it now applies to a much broader class of phenomena in surface science including alloying, surface reactions, annealing, and thin-film crystal growth.

COMPARISON OF HEATING SOURCES

Research studies have established that the first basic step in laser annealing is the rapid conversion of electromagnetic energy from the laser beam to thermal energy within the outer surface region of the semiconductor [1]. As a consequence, the major effects of laser annealing can be understood by

*Supported in part by U.S. Army ERADCOM, Contract No. DAAK20-80-C-0269.
**Current Address: Dikewood, A Division of Kaman Sciences Corporation 2716 Ocean Park Blvd. Santa Monica, CA 90405

considering how this thermal energy is dissipated in time and over spatial dimensions. This aspect of laser annealing is illustrated in Figure 1 for typical laser characteristics and compared with the thermal behavior of samples heated by other methods. With these large differences in time scales and spatial extent of heating it is evident that many new approaches to semi-conductor processing can be devised. The major penalty or compromise in applying this type of materials processing to a production environment is the uncertainty associated with the reproducibility and overall throughput of each particular application that is discovered in research investigations. However, laser and optical technology has become very sophisticated, and it is likely that many of the recent discoveries and applications of laser processing can be adequately "engineered" to meet production standards by the time the semiconductor industry has the need to produce integrated circuits with submicron feature sizes.

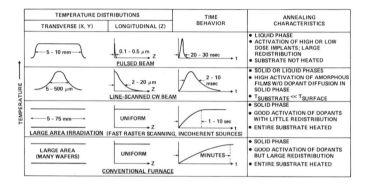

Fig. 1. Thermal profiles and annealing characteristics of different heating sources.

DIODES

Diodes which operate on a combination of impact avalanche and transit time (IMPATT) effects have fabrication requirements which can greatly benefit from laser annealing. The operating frequency and efficiency of these devices is critically dependent on the dimensions of "drift" regions, the quality of p–n junctions, and the resistance of ohmic contacts. As shown in Figure 2, the dimensions of individual diode regions are in the submicron range for D-Band (110–170 GHz) diodes and will be even smaller for Y-Band devices (170–260 GHz). By using laser annealing of the implanted regions, several deleterious diffusions that occur during standard processing can be eliminated. These are (a) in-diffusion of residual contaminant gases, (b) out-diffusion of arsenic from the substrate and (c) redistribution of the implanted profile. Also a high dose shallow implant followed by laser annealing can be used to increase the active carrier concentration at the surface (p^+ region) and thereby reduce the contact resistance. The rf performance for several furnace annealed and laser annealed IMPATT diodes is summarized in Table I along with the furnace and laser annealing parameters used. Microwave power output near the design frequency (140 GHz) for cw diode operation is significantly higher for laser annealed diodes than for those fabricated using furnace annealing. It is

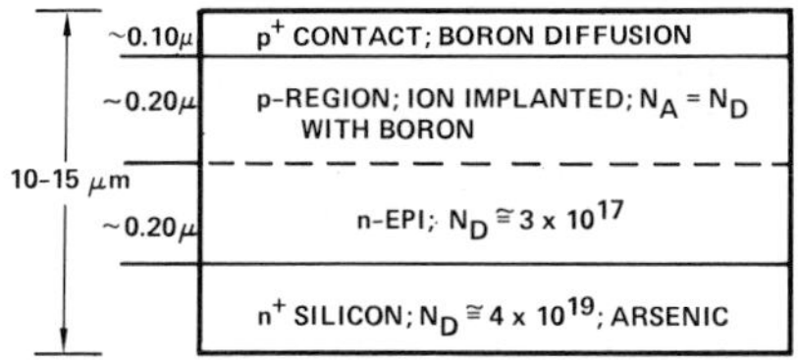

Fig. 2. Nominal dimensions and doping concentrations for 140 GHz IMPATTs.

TABLE I
RF Performance for cw IMPATT Diode Operation

SAMPLE	ANNEALING CONDITIONS	OPERATING CONDITIONS		DIODE OUTPUT	
		I(mA)	V(VOLT)	(GHz)	P(mW)
71A-1	PULSED LASER 6943Å	250	11.7	127	1.9
		275	11.8	135	3.8
71A-3	25 NSEC 1.5 J/CM^2	200	10.9	133	2.7
		230	11.1	136	3.0
69B-1		250	10.0	123	0.16
		264	10.0	124	0.48
69B-2	FURNACE 900°C 20 MIN	300	9.77	121	0.32
		350	9.94	124	0.48
69B-4		365	10.0	131	0.16
72A-1	CW ARGON LASER; T_s = 350°C 3.4 CM/SEC 5.3 W	250	11.9	127	2.1
		280	12.0	136	3.8

apparent that the laser technique has unique advantages for D-band (100-170) GHz IMPATT fabrication and can be expected to be even more important for the processing of Y-band (170-260 GHz) devices which have smaller drift region dimensions. The improved performance of the laser annealed IMPATTs is attributed to an increase in crystalline perfection in the regions damaged by ion implantation, and to improved spatial definition of the diode junction and drift regions.

Although it has been much more successful with silicon than compound materials, laser annealing can also be advantageous for the fabrication of devices in material as delicate as HgCdTe. The structure and I-V characteristics of HgCdTe diodes fabricated using laser annealing are shown in Figure 3. Junctions can be formed in this material by defect doping produced by ion implantation without subsequent annealing, but this technique has several disadvantages. As is apparent from the figure, the laser annealed diodes have two significant improvements, (1) reduced leakage current and (2) higher breakdown voltage. These results were obtained recently and are very encouraging; the work is in its preliminary stages and will be described in greater detail later.

340

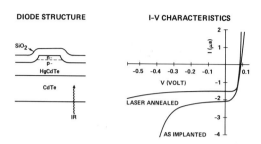

DIODE STRUCTURE   I-V CHARACTERISTICS

Fig. 3. Structure and I-V characteristics of cw Nd:YAG laser annealed HgCdTe diode; B$^+$HgCdTe, 110 keV, 10^{11} cm^{-2}; 40 keV, 5 x 10^{13} cm^{-2}; λ = 1.06 μm, 2.8 W, 120 μm, 2 cm/sec, 10 scans

POLYSILICON COMPONENTS

As another example of the utility of laser processing, we consider the use of polysilicon thin films as resistor elements to replace active load devices in integrated circuits. The reduced power and area consumption of polysilicon resistor load elements compared to depletion load devices represents a significant advantage. However, a major problem with polysilicon resistors is the poor resistivity control that is obtained by conventional processing [2]. By utilizing cw laser annealing in place of furnace annealing we have found an improvement by a factor of five in resistivity control. The comparative data plotted in Figure 4 show that the sensitivity of polysilicon sheet resistance to implant dose is greatly reduced by cw laser annealing. We found that with particular laser conditions a ±10% variation in the doping concentration results in only a factor of two change in resistivity, whereas the furnace annealed material resistivity varies by more than an order of magnitude for the same dopant variation. Analysis showed that this improvement is due to both laser-induced grain size increases and to a significant reduction in the grain boundary trapping density [3].

In order to increase the packing density of VLSI circuits it will also be advantageous to use stacked configurations whenever possible. Examples of solid state devices where simple stacking of oxide and polysilicon layers is commonly used are shown in Figure 5. A basic disadvantage of oxides grown over polysilicon can be traced to asperities on the polysilicon surface which lead to local field enhancement at any applied voltage, and this gives rise to increased leakage currents and premature dielectric breakdown. Application of pulsed laser processing techniques to this problem results in dramatic improvements as shown by the data in Figure 6. Double layer capacitor structures were fabricated with the use of pulsed ruby laser radiation to modify the top polysilicon surface before the second oxide was grown; the number of asperities per unit area, and their size and shape are altered considerably by the laser process. This material modification has the effect of reducing the leakage current by over three orders of magnitude and producing a substantial increase in the breakdown voltage (see Figure 6). It is significant that these beneficial results are produced without inducing any deleterious effects to the underlying materials or interfaces [4]; this particular feature exemplifies the nature and advantage of localized processing. The upper polysilicon can be heated to the melting point by pulsed laser radiation without overheating underlying material, and the entire heating and cooling cycle is accomplished in a fraction of a microsecond [4].

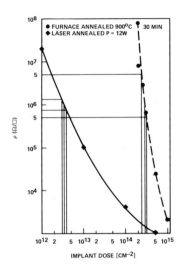

Fig. 4. Sheet resistance of polysilicon as a function of implant dose for cw laser annealed and furnace annealed samples; 0.5 μm polysilicon films were deposited on a 0.1 μm layer of silicon nitride that had been deposited on a (100) silicon substrate.

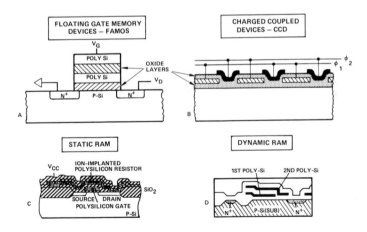

Fig. 5. Examples of microelectronic devices utilizing stacked oxide/ polysilicon structures; (A) floating gate memory device (FAMOS), (B) charge coupled devices (CCD), (C) static RAM, (D) dynamic RAM.

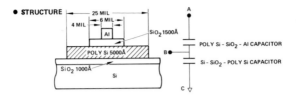

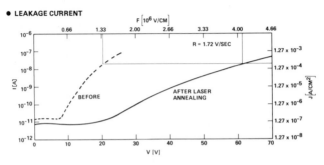

Fig. 6. Schematic diagram of stacked oxide/silicon series capacitor test structure and the upper oxide (1500 Å) leakage current plotted as a function of an applied ramped voltage (1.7 V/sec).

OHMIC CONTACTS

Although relatively simple in structure and components, ohmic contacts play a very important role in high speed integrated circuits. As the trend in miniaturization continues to transistors with submicron dimensions, the importance of minimizing contact resistances and interdiffusion of contact components rapidly increases. This point, and the advantage of laser annealing for contact formation is illustrated in Figure 7. It can be seen that even for very low specific contact resistance, 3.6×10^{-6} Ω-cm^2, the actual resistance of contacts having submicron dimensions is still rather high, and could limit the speed of a VLSI circuit. Because there can be several thousand ohmic contacts in a VLSI circuit it is important to pursue new techniques for lowering contact resistance. The results obtained using laser annealing are promising, both for this purpose and for reducing the spatial migration of the contact materials (see Figure 7a-c). The latter characteristic helps to reduce contact resistance and is also expected to improve ohmic contact yields and lifetimes.

TRANSISTORS

Although transistors with submicron feature sizes are expected to show important speed and packing density advantages in VLSI circuits, the practical realization of minimum size high performance MOSFET and/or bipolar devices will be particularly difficult because common dopants (B, P) diffuse at least 0.25 μm during conventional furnace annealing.

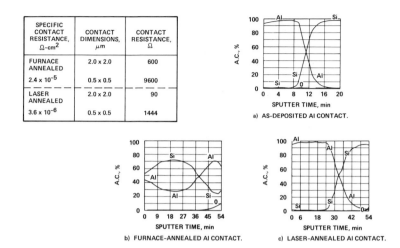

SPECIFIC CONTACT RESISTANCE, Ω-cm^2	CONTACT DIMENSIONS, μm	CONTACT RESISTANCE, Ω
FURNACE ANNEALED	2.0 x 2.0	600
2.4 x 10^{-5}	0.5 x 0.5	9600
LASER ANNEALED	2.0 x 2.0	90
3.6 x 10^{-6}	0.5 x 0.5	1444

a) AS-DEPOSITED Al CONTACT.

b) FURNACE-ANNEALED Al CONTACT.

c) LASER-ANNEALED Al CONTACT.

Fig. 7. Comparison of laser and furnace annealed Al/Si ohmic contacts.

Bipolar transistor fabrication can benefit from laser annealing techniques in a way similar to that described above for IMPATT diodes. A typical bipolar transistor configuration is shown in Figure 8, and the emitter/base, X_{jeb}, and collector/base, X_{jcb}, junction depths are indicated. For X_{jeb} in the range 0.3 to 0.7μm, and X_{jcb} having the values 0.5 to 1.0 μm, satisfactory bipolar devices can be fabricated by furnace processing techniques. However, for devices with higher operating speeds and greater efficiency it is necessary to reduce these dimensions. Side wall capacitance and resistance scale linearly with junction depth and can be reduced significantly by using cw laser annealing to form shallower junctions. Additionally, more abrupt emitter/base junctions increase the emitter efficiency and improve the overall gain of the transistor. Since cw laser annealing techniques provide high electrical activation of ion implanted dopants without altering the dopant spatial profile [5], laser processing combined with ion implantation may provide the key processing technology for

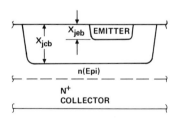

Fig. 8. Expanded cross sectional view of a bipolar transistor illustrating the emitter/base and collector/base junction distance.

producing minimum dimension bipolar transistors with maximum performance characteristics [6]. CW laser annealing of UHV deposited films [7] represents an attractive alternative to ion implantation for the fabrication of bipolar transistors having minimum junction depths and hyper-abrupt junction profiles.

In the case of MOSFET fabrication there are two basic problems that arise as a result of source/drain (S/D) dopant diffusion during annealing. They are illustrated schematically in Figure 9. Lateral diffusion of the S/D dopants under the gate region introduces parasitic gate-source and gate-drain capacitances. These increased parasitic capacitances not only reduce the device operating speed significantly, but additional undesirable effects are also introduced. The short channel length effect decreases the threshold voltage, increases leakage currents and decreases the punch-through voltage. Arsenic has a low diffusivity and can replace phosphorus for submicron devices. However, unless a suitable p-type dopant substitute can be found for boron, it represents the limiting case for CMOS systems. Although it may be possible to reduce the overall thermal cycling time to 30 min at 900°C, this will still cause the source/drain junction distances, X_{jg}, and depths, r_j, to increase by 0.2 µm which would preclude the fabrication of high performance transistors with channel lengths less than 0.8 µm.

Vertical diffusion of S/D dopants leads to a strong dependence of transistor threshold voltage on channel length. Such a steep dependence causes obvious problems in integrated circuit operation and cannot be tolerated for many applications. The effect is shown schematically in Figure 9; it is basically caused by the fact that when the S/D junction depths, r_j, are an appreciable fraction of the channel length, a significant number of the field lines from charges in the depletion volume under the gate can terminate on charges at the edges of the S/D regions rather than on the gate. An approximate analysis

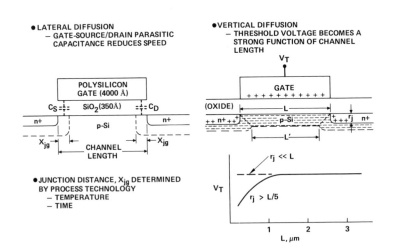

Fig. 9. Effects of S/D dopant diffusion on the performance of short channel (<1.0 µm) MOSFETs.

shows that the effect is greatly reduced when the junction depth is much less than the channel length [8]. Since this could be accomplished with shallow ion implantation and cw laser annealing, this represents another contribution that laser processing can make toward improving the performance of integrated circuit devices.

To illustrate some of the progress achieved with MOSFET fabrication using laser annealing, a process sequence for CMOS/SOS device fabrication is shown in Figure 10, and static electrical characteristics of CMOS/SOS transistors are displayed in Figure 11. In this example, laser annealing was used at step 5 in the fabrication sequence (Figure 10) but not at step 1. The static I_D-V_D characteristics are improved slightly by cw laser annealing but a more significant improvement was found in the dynamic performance of the devices. This is shown in Figure 12 where it is apparent that the transistors with source/drain regions annealed by a scanning cw laser beam have superior performance (greater speed) compared to those fabricated using furnace annealing. These results are very encouraging but only represent a first step in the direction toward the fabrication of high performance MOS transistors with submicron channel lengths.

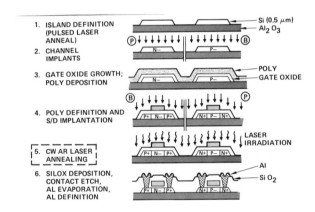

Fig. 10. CMOS/SOS processing sequence; laser annealing can be used in steps 1 and 5 as indicated.

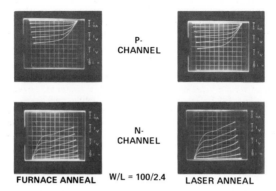

Fig. 11. Comparison of drain current characteristics for MOS/SOS transistors fabricated using either conventional furnace methods or cw laser irradiation for annealing ion-implanted source/drain regions.

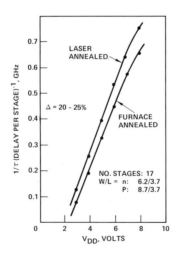

Fig. 12. Ring oscillator speed for oscillators fabricated using either conventional furnace or cw laser annealing methods.

SUMMARY

Laser annealing represents a variety of new and unique heating techniques which can be applied to the solution of device and circuit fabrication problems. Several examples are described above which illustrate the particular advantages of laser processing as it could be used in the fabrication of high speed integrated circuit elements. These and other examples are summarized in Table II where a brief comparison of several types of heating sources is shown; the times and dimensions given in the first two rows refer to the heating periods and depths, respectively. The information given in Tables I and II summarizes many features of laser annealing at its current level of development. It is anticipated that the results obtained from the variety of research and application oriented laser annealing studies that have been conducted will lead to even more advances, and eventually will provide significant contributions to the development of solid state electronics technology.

TABLE II.
Classification of Heating Sources and Semiconductor Processing Applications (preferred operating regimes are shaded)

PULSED	SCANNED CW	LARGE AREA IRRADIATION (FAST RASTER, INCOHERENT)	FURNACE
NANOSECONDS	MILLISECONDS	SECONDS	MINUTES
0.1 - 0.5 μm	2 - 20 μm	UNIFORM	UNIFORM
	SOURCE/DRAIN ANNEALING →		
OHMIC CONTACTS →			
	SILOX REFLOW		
SOS MOBILITY IMPROVEMENT			
	SILICON IMPATT DIODE FABRICATION →		
	POLYSILICON RESISTORS		
STACKED POLYSILICON/OXIDE STRUCTURES →			
	SOI STRUCTURES (LIQUID PHASE)		
	SOI (LEDF) (SOLID PHASE)		
	GATE OXIDE ANNEALING		
STEP COVERAGE			
	HgCdTe DIODES		

HIGH TEMPERATURES	REDUCED TEMPERATURES

REFERENCES

1. Laser-Solid Interaction and Laser Processing-1978, S.D. Ferris, H.J. Leamy, J.M. Poate, eds., AIP Conf. Proc. No. 50 (1979); Laser and Electron Beam Processing of Materials, C.W. White, P.S. Peercy, eds., Academic Press, New York (1980); Laser and Electron Beam Solid Interactions and Materials Processing, J.F. Gibbons, L.D. Hess, T.W. Sigmon, eds., vol. 1, Mater. Res. Soc. Symp. Proc., North Holland, New York (1981); Laser and Electron Beam Interactions with Solids, B.R. Appleton, G.K. Celler, eds., vol. 4, Mater. Res. Soc. Symp. Proc., North Holland, New York (1982).

2. J.M. Andrews, J. Electronic Materials 8, 227 (1979).

3. G. Yaron, L.D. Hess, and G.L. Olson, in Laser and Electron Beam Processing of Materials, C.W. White and P.S. Peercy, Eds. New York, Academic Press, 1980, pp. 626-631.

4. G. Yaron and L.D. Hess, IEEE Trans. Electron Devices ED-27, 573 (1980); G. Yaron, L.D. Hess, and S.A. Kokorowski, Solid-State Electron. 23, 893 (1980).

5. A. Gat, J.F. Gibbons, T.J. Magee, J. Peng, V.R. Deline, P. Williams, and C.A. Evans, Jr., Appl. Phys. Lett. 32, 276 (1978).

6. G.L. Miller, in Laser and Electron Beam Processing of Electronic Materials, C.L. Anderson, G.K. Celler, and G.A. Rozgonyi, eds., Princeton, New Jersey, The Electrochemical Society, Inc., 1980, pp. 83-107.

7. J.A. Roth, G.L. Olson, S.A. Kokorowski, and L.D. Hess, in Laser and Electron-Beam Solid Interactions and Materials Processing, edited by J.F. Gibbons, L.D. Hess, and T.W. Sigmon (Elsevier North Holland, New York, 1981), pp. 413-426.

8. L.D. Yau, Solid-State Electronics 17, 1059 (1974).

DEFECT CONTROL DURING EPITAXIAL REGROWTH BY RAPID THERMAL ANNEALING

H. Baumgart, G. K.Celler, D. J. Lischner*, McD. Robinson, and T. T. Sheng
Bell Laboratories, Murray Hill, New Jersey 07974, USA
*Bell Laboratories, Allentown, Pennsylvania 18103

ABSTRACT

Rapid Thermal Annealing (RTA) with tungsten halogen lamps provides excellent regrowth of silicon layers damaged by ion implantation. In addition to minimizing dopant redistribution, the inherent advantage of this technique is good control of temperature gradients. The latter is instrumental in reducing the density of extended defects in the annealed samples. In contrast, solid phase laser annealing, which involves steep temperature gradients, always leaves interstitial dislocation loops and point defect clusters. We present a comparative study of crystal quality following laser processing and incoherent light annealing as well as furnace annealing of As, P and B ion implanted Si wafers.

INTRODUCTION

During the past five years the search for rapid semiconductor heat treatment techniques has been focused on lasers and electron beams. One of the more recent developments has been the introduction of graphite resistive heaters and high intensity halogen lamps for rapid isothermal annealing. Rapid Thermal Annealing (RTA) with incoherent light from tungsten-halogen lamps allows fast and precise temperature-time cycles in a clean and controlled environment. Compared to conventional furnace annealing the new technique retains the temporal advantage of laser annealing, reconstructing amorphized layers within seconds. Early experiments with continuous lamps for annealing of sheet implants of As and B indicated complete electrical activation of the dopant [1,2,3,4]. The uniform heating of full-size wafers and the potential of high throughput using sequential wafer handling make RTA attractive for semiconductor processing.

In this paper we present a comparative study of crystalline quality and perfection resulting from incoherent light annealing, conventional furnace annealing, and from scanning cw laser annealing. Its purpose is to assess the relative advantages of the three heat treatment techniques for practical applications in annealing of whole wafers.

EXPERIMENTAL METHODS

Ion implanted silicon wafers were annealed in a few seconds by the intense radiation from an air-cooled array of tungsten halogen lamps. The samples are positioned below the lamps in a wafer chamber with a quartz window, that allows annealing in an inert atmosphere. Each wafer is placed on three quartz pins for thermal insulation. Details of the RTA light furnace as well as the spatial and temporal temperature profiles have been published in Ref. 5. In our experiments the samples were $<100>$ and $<111>$ 3" Si wafers with a resistivity of 20-50 Ω cm. They were implanted with 150 keV As, 50 keV B, or 100 keV P, the implant doses ranging from 10^{14} to 10^{16} cm^{-2}.

Solid phase epitaxial recrystallization of the ion implantation amorphized surface layers was accomplished in 10 or 30 sec at a preset temperature. The furnace was ramped up to this temperature in 10 sec and the power reduction also took 10 sec. Wafer temperature was monitored by optical pyrometry. Two sets of samples were

Mat. Res. Soc. Symp. Proc. Vol. 13 (1983) ©Elsevier Science Publishing Co., Inc.

Fig. 1. Transmission X-ray topograph showing the region at the edge of the halogen lamp annealed wafer [1×10^{16} B cm^{-2} implanted at 50 keV].

Fig. 2. TEM cross-section of B implanted Si sample with high density of residual defects after 30 sec 1100°C lamp annealing.

lamp annealed at 900°C, 1000°C, 1100°C and 1200°C. The first set was held at the annealing temperature for 30 sec and the second set only for 10 sec. For comparison, we subjected the third set of wafers to a conventional 30 min furnace treatment at the same temperatures. Finally, we also annealed some samples with a scanning cw laser, in the solid phase regime. The structural perfection and defect annihilation were stu-

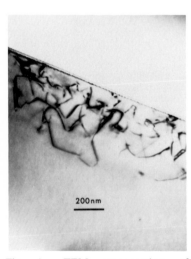

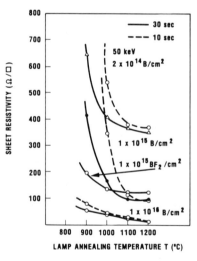

Fig. 3. TEM cross-section of B implanted Si wafer lamp recrystallized in 30 sec at 1200°C [1×10^{16} B cm^{-2} at 50 keV].

Fig. 4. Sheet resistivity of B and BF$_2$ implanted Si samples as a function of lamp annealing temperature.

died by transmission electron microscopy and X-ray topography. Sheet resistivity was determined by four-point probe measurements and the electrically active dopant concentration profiles were measured by spreading resistance.

RESULTS AND DISCUSSION

The aim of RTA is: a) to restore lattice order by epitaxial regrowth of the amorphized implanted region and annealing the damaged underlying substrate region, b) to electrically activate the implanted ions, and c) to minimize dopant redistribution. We have previously demonstrated [4] that RTA processing results in full activation of dopants and little diffusion. Here we focus primarily on the microstructural quality of the regrown layers.

To characterize the degree of perfection and the level of strain and slip we have employed transmission X-ray topography. The topograph in Fig. 1 was recorded with Cu Kα radiation and shows a B implanted wafer [1 × 10^{16} B cm^{-2} at 50 keV] which was lamp annealed at 1200°C for 10 sec. An orthogonal array of slip dislocations is present along the wafer edge, caused by increased radiative cooling at the wafer perimeter [5].

X-ray topography is extremely sensitive to lattice strain in the crystal. The uniform X-ray topographical contrast across the wafer demonstrates that the lamp annealed wafer is bow-free, confirming the uniformity of heating in the RTA light furnace [5]. Since the first two millimeters at the wafer perimeter are not used during device fabrication, slip at the edges is not too critical.

Because of the limited spatial resolution of the X-ray method, details of residual defects are best examined in the electron microscope. Cross sectional TEM micrographs of 1 × 10^{16} cm^{-2} B implanted samples at 50 keV show that a 900°C lamp anneal cycle for 30 sec reconstructs the implanted layer to a very defective single crystal epitaxial layer. The same procedure at 1000°C shows some improvement in defect reduction but even an 1100°C anneal cycle for 30 sec produces a regrown layer with an unacceptably high density of irregularly shaped dislocation loops, as demonstrated in Fig. 2. However lamp annealing at 1200°C for 30 sec yields a significant

200 nm

 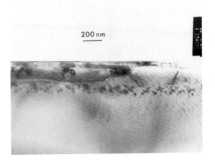

Fig. 5. TEM cross-section of a defect-free, recrystallized As implanted Si wafer. [1×10^{15} As cm^{-2} at 50 keV; 10 sec lamp annealed at 1050°C].

Fig. 6. TEM cross-section of 150 keV As implanted sample [1×10^{16} As cm^{-2}] showing residual defects after lamp annealing at 1100°C for 30 sec.

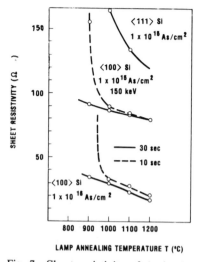

Fig. 7. Sheet resistivity of As implanted Si wafers with <100> and <111> surface orientation.

Fig. 8. SEM-EBIC micrograph of scanning cw laser annealed Si wafer free of slip dislocations.

defect reduction in the regrown layer, as shown in Fig. 3. At this elevated temperature essentially all of the small dislocation loops found in Fig. 2 have been annihilated and only a small fraction of them have grown into very large dislocation loops penetrating into the underlying substrate.

Electrical activation data obtained with a four point probe correlate well with the TEM results. Figure 4 shows the measured sheet resistivity versus lamp annealing temperature. We have included data for boron implantation doses from 2×10^{14} cm^{-2} to 1×10^{16} cm^{-2} as well as for BF$_2$ implants. The data show a steep decrease of the sheet resistivity with temperature to a constant value, that corresponds to nearly complete electrical activation. Since the sheet resistivity also correlates with the defect density in the crystal, these plots confirm the TEM results showing a significant defect reduction in the temperature range up to 1200°C.

When the implanted B is substitutional on Si lattice sites, as evidenced by complete electrical activation, it will cause shrinkage of the lattice [6]. This tetragonal lattice contraction adds a misfit stress to the regrown epitaxial layer and is an obstacle to complete removal of all residual dislocations in high B dose implanted Si, as shown in Fig. 3.

This problem does not exist for As implants since the ionic radius of As is close to that of Si. Indeed, our best results with the RTA lamp furnace were for low energy arsenic implants. Figure 5 shows a perfect solid state epitaxially regrown surface layer without any slip or residual dislocation loops. The sample was amorphized by a 1×10^{15} As cm^{-2} implantation at 50 keV and then subjected to a 10 sec lamp anneal at 1050°C. Complete recrystallization without any defects was also reported for As implantations with doses equal to or lower than 1×10^{15} cm^{-2} at 100 keV or lower energy [7]. However, the same crystalline perfection cannot be maintained for higher energy As implantations. Figure 6 clearly displays residual dislocation loops in the p-n junction region of a p-Si wafer implanted with 10^{16} As cm^{-2} at 150 keV and lamp annealed at 1100°C. It is evident in Fig. 6 that the originally amorphized layer has been perfectly regrown, and that all of the dislocation loops are located at and below the p-n junction where the crystal lattice was damaged but not amorphized. In this

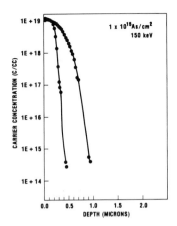

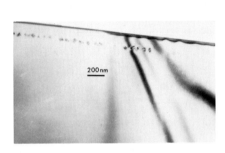

200 nm

Fig. 9. TEM cross-section of sample annealed in the furnace at 1100°C for 30 min, showing dislocation loops in the p-n junction region. [1×10¹⁵ As cm⁻² at 150 keV].

Fig. 10. Electrically active dopant concentration profiles measured by spreading resistance for RTA and furnace annealed Si wafers.

region there is a high density of interstitials and vacancies in the lattice which subsequently condense and form the residual dislocation loops.

The defect-free regrown surface layers for low energy arsenic implantations are attributed to a lower point defect density in a shallow p-n junction region. High energy As implantations produce a broad damage region with a much higher density of point defects. In the latter case longer annealing times and higher temperatures are needed to annihilate the remaining dislocation loops, at the risk of increased dopant redistribution. It is interesting to plot the sheet resistivities as a function of lamp anneal temperature and compare the annealing behavior of <100> and <111> surfaces. Figure 7 shows that the sheet resistivity in <111> As implanted wafers does not drop to the same value as in <100> Si wafers under identical lamp annealing conditions. This suggests that we cannot reach complete dopant activation in <111> Si. Electron microscopy revealed heavily defective and micro-twinned regrown layers in the <111> samples.

Phosphorus ion implanted and lamp annealed samples yield results similar to high energy As implants, but with a slightly higher density of dislocation loops than is shown in Fig. 6.

In the case of laser processing, large area annealing of sheet implants of As, B or P has to be performed in the scanning mode. Compared to RTA by halogen lamps, solid phase laser processing achieved faster local recrystallization, with typical laser dwell times of 10⁻³ sec. However, scanning a focussed cw laser beam across a Si wafer introduces very large temperature gradients. Our data indicate that high densities of dislocation loops are formed on the flanks of each line scan where the highest temperature gradient occurs [8,9]. The EBIC micrograph in Fig. 8 displays a nonuniform residual defect distribution which acts to decrease minority carrier lifetime in the overlapping regions of the parallel line scans. Therefore, for the best epitaxial crystal quality a uniform intensity over the entire wafer is preferable to a scanning hot spot.

This criterion is always satisfied by conventional 30 min annealing in a furnace. The resulting crystal quality is good and the residual defect density is equal to or slightly worse than that obtained after lamp annealing. The TEM cross-section in Fig. 9 shows dislocation loops typically seen after an 1100°C half hour furnace anneal

354

of As implanted Si wafers. Occasionally we observed twins and stacking faults, but the defects are uniformly distributed over an entire wafer and no non-uniform EBIC contrast can be detected. The main drawback of conventional furnace annealing is not the crystal quality, which is only marginally worse compared to halogen lamp annealing, but the very long annealing times that cause a considerable dopant redistribution. The electrically active dopant concentration profiles were measured by spreading resistance on angle lapped samples, to compare lamp annealing with conventional furnace annealing. Carrier concentration data vs. depth are presented in Fig. 10. Within the experimental resolution there is negligible dopant diffusion following RTA with halogen lamps. In contrast, considerable dopant redistribution is detected after a half hour furnace anneal. For direct comparison we have listed in Table 1 the measured p-n junction depth in As, BF_2 and P implanted Si wafers following 30 min furnace annealing and 30 sec halogen lamp annealing.

Table 1: Measured p-n Junction Depth

	Furnace Annealed 30 min 900°C	Lamp Annealed 30 sec 900°C
As 150 keV) $1\times10^{15}cm^{-2}$)	0.91 μm	0.44 μm
BF_2 (50 keV $1\times10^{15}cm^{-2}$)	0.54 μm	0.19 μm
P (100 keV, $1\times10^{15}cm^{-2}$)	0.94 μm	0.57 μm

SUMMARY

RTA is a practical process for large area annealing, retaining the advantage of very short annealing times while avoiding the spatial non-uniformities in the residual defect density which is inherent to scanning cw laser processing. The main advantage of RTA over the laser is the far better temperature uniformity which is instrumental for an effective defect control during epitaxial regrowth. As a consequence of this, RTA with halogen lamps achieves superior crystal quality in the reconstructed layer. Furthermore, X-ray topography reveals practically bow- and slip-free wafers with occasional slip dislocations confined to the wafer edges. The experiments demonstrated that the structural perfection is equal to or better than in conventional furnace annealed samples. There was full electrical activation of the implanted dopants with a minimum of dopant diffusion. These advantages, combined with the potential of high wafer throughput, offer an improvement over other thermal annealing techniques.

REFERENCES

1. K. Nishiyama, M. Arai and N. Watanabe. Jap. J. of Appl. Phys. 19(10)L 563 (1980).
2. R. A. Powell, T. O. Yep and R. T. Fulks. Appl. Phys. Lett. 39, 150 (1981).
3. A. Gat. IEEE Electron Device Lett. EDL-2, 85 (1981).
4. J. L. Benton, G. K. Celler, D. C. Jacobson, L. C. Kimerling, D. J. Lischner, G. L. Miller and McD. Robinson in "Laser and Electron Beam Interactions with Solids", B. R. Appleton and G. K. Celler, eds. (North Holland, New York) p. 765 (1982).
5. D. J. Lischner and G. K. Celler in "Laser and Electron Beam Interactions with Solids", B. R. Appleton and G. K. Celler, eds. (North Holland, New York) p. 759 (1982).
6. B. C. Larson, C. W. White and B. R. Appleton. Appl. Phys. Lett. 32 (12) 801 (1978).
7. M. Haond, D. P. Vu and C. Adès. Electronic Letters, Vol. 18, No. 16, p. 694 (1982).
8. H. Baumgart, O. Hildebrand, F. Philipp and G. A. Rozgonyi. Inst. of Physics (London) Conference Ser. No. 60: Section 2, p. 127 (1981).
9. H. Baumgart, F. Phillipp and H. J. Leamy, in "Laser and Electron Beam Interactions with Solids", B. R. Appleton and G. K. Celler, eds. (North Holland, New York) p. 355 (1982).

RAPID THERMAL ANNEALING OF SILICON USING AN ULTRAHIGH POWER ARC LAMP

R. T. HODGSON, J. E. E. BAGLIN and A. E. MICHEL
IBM Thomas J. Watson Research Center, Yorktown Heights, New York 10598

S. MADER
IBM Corporation, Hopewell Junction, New York 12533

J. C. GELPEY
Eaton Ion Implantation Systems, 16 Tozer Road, Beverly, MA 01915

ABSTRACT

We have used an ultrahigh power, 100kW vortex cooled arc lamp to anneal As^+ implant damage in <100> silicon wafers. When the wafer temperature was held above 1100°C for ~1sec, TEM analysis indicated that the material was free of extended defects. The dopant diffused much more rapidly than would be expected from the usual models. However, preliminary results indicate that defect free material can be produced with dopant movement limited to ~100Å.

INTRODUCTION

In the past few years, experiments on annealing of ion implanted semiconductor wafers have extended the time ranges of interest from milliseconds to picoseconds with the aid of various lasers.[1-4] Even more recently, incandescent and arc sources have been proposed[5-9] as sources of radiant energy which would be more feasible in manufacturing than lasers. The time scales for annealing, measured in seconds, using such sources lie between the extensively investigated regions of laser annealing and oven annealing.

Nanosecond pulsed lasers produce material free of extended defects if sufficient energy to melt below the damage produced by the implanted ions is used. In this case, however, the dopant ions diffuse rapidly in the molten material, and the final dopant distribution is quite different than the "as implanted" distribution. "cw" laser annealing, on the other hand, does not melt the silicon nor move the dopant, but leaves many defects. The question remains - is there a time temperature region which will remove implant damage and not move dopant ions?

Mat. Res. Soc. Symp. Proc. Vol. 13 (1983) © Elsevier Science Publishing Co., Inc.

EXPERIMENT

We have investigated annealing the damage caused by arsenic implants in <100> silicon using a single argon arc lamp. The arc, which is struck through a vortex in water which cools the interior walls of the surrounding quartz tube, can be run continuously with 100 kilowatts of input power. About half of this power appears as visible and near IR light which, when collected and concentrated uniformly on a four inch silicon wafer, can raise the wafer temperature to the melting point in a few seconds.

The temperature of the wafer is measured by scanning the side of the wafer opposite to the lamp with calibrated infrared optical pyrometer (10msec response time). Arsenic implanted wafers were annealed with various time-temperature cycles. Sheet resistivity, He ion backscattering, channeling, and TEM studies were carried out to test the material properties and the distribution of dopant.

RESULTS AND DISCUSSION

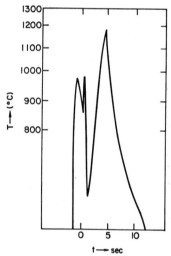

Figure 1 Wafer temperature
vs. time.

5sec 300amp.

A typical trace of the wafer temperature is given in Fig. 1 for a 5 second, 300 amp treatment of a 3″ wafer (Sample 17). The broad peak at the beginning of the trace is pick-up from the lamp starter. The second sharp spike is due to the infrared lamp light which initially passes through the silicon wafer. After the wafer temperature reaches a few hundred degrees, it is quite opaque to the frequencies that the pyrometer measures.

A number of samples implanted with various doses of 40 and 50keV arsenic ions were annealed with various lamp powers and times listed in Table I. TEM analysis showed no extended defects in samples with sufficiently high temperature-time history. As the maximum temperature was lowered, a few dislocation loops could be found (Fig. 2a) and at still lower temperatures many 100-500Å loops were evident (Fig. 2b).

Some of the samples with no extended defects were characterized by He ion backscattering and channeling. Figure 3 shows an example of a random and channeled He ion backscattering profile for Sample 13 annealed with the 5sec pulse depicted in Fig. 1. The 4.4% χ_{min} shows that the silicon is good technical quality. The arsenic trace shows that the deeper arsenic is more completely on lattice sites than that near the surface.

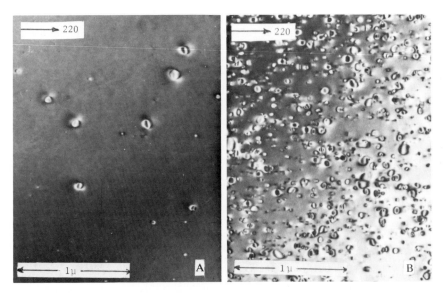

Figure 2 Interstitial dislocation loops in Si implanted with 5E15 As$^+$/cm^2 at 50keV.
(a) Sample 4, annealed 1sec 410amp + 9sec 175amp max temp. 1140°C
(b) Sample 3, annealed 2sec 410 amp max temp. 1100°C

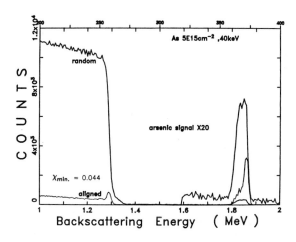

Figure 3   Random and
channeled He ion back-
scattering   spectra   for
5E15cm^{-2},   40keV   As$^+$
implanted sample annealed
with temperature history of
Fig. 1.

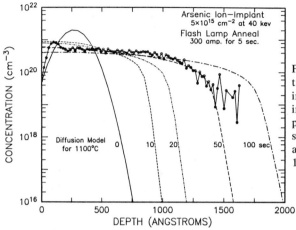

Figure 4 Arsenic concentration measured by grazing incidence backscattering from Sample 13 compared with standard diffusion model calculations for a step temperature rise to 1100°C.

Figure 4 shows a graph of arsenic concentration measured from a high resolution backscattering trace of the same sample as Fig. 3. The concentrations calculated for a standard diffusion model for various times after a step rise in temperature to 1100°C are also shown. Clearly, the arsenic diffuses much faster than calculated since, from Fig. 1, the temperature is above 1100°C for one second. The segregation to the surface is particular marked. The sheet resistivity calculated assuming 100% activation of the dopant profile shown in Fig. 4 was $38\Omega/\square$ which can be compared to the measured value of $47\Omega/\square \pm 0.2\%$.

We have some evidence for fewer dislocation loops using higher power, shorter time annealing cycles. Figure 5 shows a high resolution backscattering picture of a dislocation free sample heated for 3.0sec to a maximum temperature of 1165°C. Very little dopant movement has occurred, although there is some evidence for movement to the surface.

Figure 5 Grazing incidence backscattering spectrum of Sample 19 after 3sec lamp anneal to a maximum temperature of 1165°C. No dislocation loops were found in this sample.

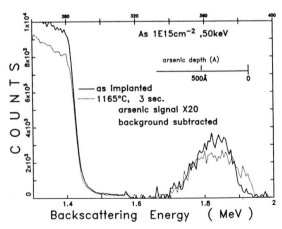

TABLE I
Table of Implant Energy, As$^+$ Ion Dose, Sheet Resistivity, Annealing Conditions,
Maximum Temp. Reached, Dopant Movement, and TEM Count of Extended Defects

Sample	keV	Dose (cm^{-2})	$\Omega/\square$	Annealing Conditions	Max Temp °C	Dopant Moves Å	Defects
1	40	5E15	66.5	300Amp 2.5sec	925	0	Many
2	40	5E15	57.6	350Amp 2.8sec	1085	50	Many
3	40	5E15	59.4	410Amp 2sec	1100	50	Many
4	40	5E15	53.7	410Amp 1 sec, +175Amp 9sec	1140	400	Few
5	50	1E15	105.5	370 Amp 2.5sec	1125		Few
6	50	1E15	104.4	205Amp 10sec	980	0	Many
7	50	5E14	181	380Amp 3sec	1240		None
8	50	5E14	180	211Amp 10sec	990	0	Many
9	50	1E14	579	380Amp 2.7sec	1170		None
10	50	1E14	583	200Amp 10sec	960	0	Many
11	40	1E14	156	300Amp 5sec	1170		None
12	40	1E15	119	300Amp 5sec	1170		None
13	40	5E15	47.3	300Amp 5sec	1180	600	None
14	40	5E14	196	300Amp 5sec	1170	400	None
15	50	1E15		350Amp 1.8sec	865	0	Many
16	50	1E15		350Amp 2.4sec	1020	0	Many
17	50	1E15		350Amp 2.6sec	1070	0	Few
18	50	1E15		350Amp 2.8sec	1100	0	Few
19	50	1E15		350Amp 3.0sec	1165	50	None
20	50	1E15		350Amp 3.2sec	1240	800	None

CONCLUSIONS

Dislocation free wafers can be produced where the ion implanted arsenic moves only a short distance (≤ 50Å) from its as implanted location. The arsenic diffusion coefficient for ion implanted silicon samples irradiated with high power light is much higher than that measured using conventional furnaces.

ACKNOWLEDGEMENTS

We would like to thank Philip Saunders for the backscattering and channeling measurements.

REFERENCES

1. Laser and Electron Beam Processing of Materials, C. W. White, P. S. Peercy, eds., Academic Press (1980).

2. Laser and Electron Beam Solid Interactions, J. F. Gibbons, L. D. Hess, T. W. Sigmond, eds., North Holland (1981).

3. Laser and Electron Beam Interactions with Solids, B. K. Appleton and G. K. Celler, eds., North Holland (1982).

4. G. A. Rozgonyi, H. Baumgart, R. Uebbing and H. Oppolzer, pg. 177 of Ref. 3.

5. Kazuo Nishiyama, Michio Arai and Naozo Watanabe, J.J.A.P. **19**, L563 (1980).

6. R. T. Fulks, C. J. Russo, P. R. Hanley and T. I. Kamins, Appl. Phys. Lett. **39**, 604 (1981).

7. J. L. Benton, G. K. Celler, D. C. Jacobson, L. C. Kimerling, D. J. Lischner, G. L. Miller and M. D. Robinson, pg. 765 of Ref. 3.

8. D. J. Lischner and G. K. Celler, pg. 759 of Ref. 3.

9. Arnon Gat, IEEE **EDL-2**, 85 (1981).

FLAME ANNEALING OF ION IMPLANTED SILICON*

J. NARAYAN AND R. T. YOUNG
Solid State Division, Oak Ridge National Laboratory, Oak Ridge, TN 37830

ABSTRACT

We have investigated flame annealing of ion implantation damage (consisting of amorphous layers and dislocation loops) in (100) and (111) silicon substrates. The temperature of a hydrogen flame was varied from 1050 to 1200°C and the interaction time from 5 to 10 seconds. Detailed TEM results showed that a "defect-free" annealing of amorphous layers by solid-phase-epitaxial growth could be achieved up to a certain concentration. However, dislocation loops in the region below the amorphous layer exhibited coarsening, i.e., the average loop size increased while the number density of loops decreased. Above a critical loop density, which was found to be a function of ion implantation variables and substrate temperature, formations of 90° dislocations (a cross-grid of dislocation in (100) and a triangular grid in (111) specimens) were observed. Electrical (Van der Pauw) measurements indicated nearly a complete electrical activation of dopants with mobility comparable to pulsed laser annealed specimens. The characteristics of p-n junction diodes showed a good diode perfection factor of 1.20-1.25 and low reverse bias currents.

INTRODUCTION

Annealing of ion implantation damage and electrical activation of dopants in semiconductors has been achieved with high-power lasers [1,2] and electron beams [3]. Two useful regimes of crystal growth for semiconductor processing have been clearly identified. First, the liquid-phase-epitaxial growth (LPE) that is involved during high-power pulsed laser irradiation, and second, the solid-phase-epitaxial (SPE) growth involved in scanned continuous wave (CW) laser irradiation. The rates of crystal growth in the former case are of the order of several meters per second, while in the latter the maximum range is around a few tenths of a centimeter per second [4]. Recently, various other transient thermal sources such as high intensity arc lamps [5] and resistively heated graphite sources [6] have been employed to anneal displacement damage involving crystal growth in SPE regime. In this investigation, we have used a hydrogen flame to anneal displacement damage in arsenic and boron implanted silicon specimens. It was found that, below a certain dose of ion implantation, a complete annealing of displacement damage accompanied by a complete electrical activation of dopants was achieved. Above this dose, cross-grid or triangular network of edge dislocations was observed.

*Research sponsored by the Division of Materials Sciences, U. S. Department of Energy under contract W-7405-eng-26 with Union Carbide Corporation.

Mat. Res. Soc. Symp. Proc. Vol. 13 (1983) Published by Elsevier Science Publishing Co., Inc.

362

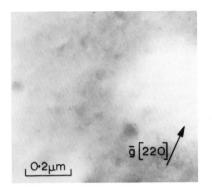

Fig. 1. Bright-field (B-F) electron micrograph showing a "complete" annealing of displacement damage in $^{75}As^+$ implanted (100 keV, 2.0 x 10^{15} cm^{-2}), (001) silicon specimens, annealed by 1150°C flame (interaction time 10 s). Some dark patches in the micrograph are artifacts on the surface due to chemical polishing.

EXPERIMENTAL

Silicon specimens having <100> and <111> orientations (Czochralski grown, 2-6 Ω-cm resistivity, 500 μm thick) were implanted with 100 keV $^{75}As^+$ and 35 keV $^{11}B^+$ in the dose range 1.0 to 10.0 x 10^{15} ions cm^{-2}. The implanted specimens were treated with a hydrogen flame in the temperature range 1050 to 1200°C with total interaction time of about 5 to 10 seconds. The calibration of specimen temperature was done to within ±20°C using an optical pyrometer while heating with the flame the companion silicon specimens. A detailed TEM analysis on these specimens was performed using plan-view and cross-section electron microscopy in a Philips (EM-400) analytical microscope. The Van der Pauw measurements were made to determine carrier concentration (N_s), carrier mobility (μ_h), and sheet resistivity (ρ_h) in the flame-processed specimens. These results from flame annealed (FA) specimens were compared with those from thermally annealed (TA) and laser annealed (LA) specimens. The latter were obtained by using one pulse of a ruby laser (λ = 0.694 μm, pulse energy density = 1.5 J/cm^2, duration = 15 x 10^{-9} s, beam diameter 1.5 cm), operated in the TEM_{00} mode.

RESULTS AND DISCUSSION

Figure 1 shows a plan-view electron micrograph after flame annealing of an arsenic implanted (001) specimens (implanted dose = 2.0 x 10^{15} cm^{-2}). No residual damage, in the form of dislocation loops or point defect clusters, is observed in these specimens. The as-implanted specimens contained an amorphous layer ∿1500 Å thick followed by a band of dislocation loops. Some of the flame annealed specimens were subsequently annealed at 800°C for 30 min to study the coarsening of dislocation loops that may be below the resolution limit of the microscope (∿5-10 Å). However, no such defects were detected after subsequent thermal annealing treatment, indicating a "complete" annealing of displacement damage by flame annealing. The results from (001) specimens implanted with 1.0 x 10^{16} cm^{-2} As$^+$ ions are shown in Fig. 2. The micrograph shows one set of dislocations with Burgers vector a/2[110] for diffraction vector ($\bar{g}$) = [22$\bar{0}$]. The other set of dislocations lying at 90° to this set was in contrast for g = [2$\bar{2}$0]. The nature of these dislocations was determined to be edge type lying in [110] and [1$\bar{1}$0] directions, respectively; both sets of dislocations lie in the (001) plane. From the image shift under $\bar{g}$ = ±[220], it was determined that extra half planes associated with dislocations lie between the dislocation lines and the free surface. The specimens contained a few dislocation loops with average size 200 Å and number density 1.0 x 10^{12} cm^{-3}. Figure 3 shows results

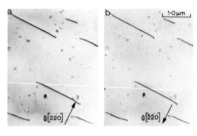

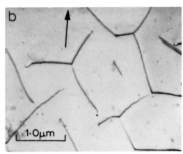

Fig. 2. The electron micrographs (under ± g = [220] diffraction conditions) from ion implanted (100 keV, $^{75}As^+$, 1.0 x 10^{16} cm^{-2}) specimens of (001) Si after annealing with 1200°C flame (interaction time 5 s). One set of the cross grid of dislocations with a/2[110] Burgers vector is in contrast. Some contamination due to chemical polishing is also visible near the surface.

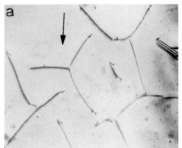

Fig. 3. The electron micrographs from (111) Si specimens annealed under similar conditions as those used in Fig. 2. A triangular network of dislocations with ±g = [220] diffraction conditions is shown.

from (111) Si specimens implanted with 1.0 x 10^{16} cm^{-2} As$^+$ ions. After flame annealing, a triangular network of dislocations is observed, in addition to the presence of a few dislocation loops with average size 130 Å and number density 1.0 x 10^{13} cm^{-3}. The nature of the dislocations was determined to be edge type lying in (111) plane and in <112> directions. From the contrast behavior under g = ±[220], it was determined that extra half planes associated with dislocations lie between the dislocation lines and the free surface.

In the following, we discuss the mechanism of flame annealing and the formation of grids of dislocations. The specimens shown in Figs. 1-3, before flame annealing, contained an amorphous layer (~1500 Å wide) followed by a band of dislocation loops, typically 100-200 Å wide. After annealing in a furnace (at 600°C for 10 min.) the (001) Si specimens implanted with 2.0 x 10^{15} cm^{-2} As$^+$ ions, the entire amorphous layer grew leaving no residual damage in the solid-phase-epitaxial (SPE) grown region, as shown by the cross-section micrograph in Fig. 4(a). The similar heat treatment in the specimens implanted with 10.0 x 10^{15} cm^{-2} As$^+$ ions left some residual damage in the form of defect clusters in the SPE grown layer [Fig. 4(b)]. The annealing occurred by the movement of original crystalline-amorphous (c-a) interface toward the surface leaving behind the band of dislocation loops. From the measurements of thickness of regrown layer versus time, the kinetics of interface motion could be described approximately by the following equation V = 8.0 x 10^{14} $e^{-2.5}$ eV/kT (A/sec) in (001) orientation. The pre-exponential factor was about 1/25 of this value in (111) orientation. From this, it is expected that times less than one millisecond at 1200°C would be required for a complete recrystallization of a 1500 Å thick amorphous layer in (001) orientation. Longer times, of the order of a few seconds, are helpful in removing the residual damage in SPE grown layers.

364

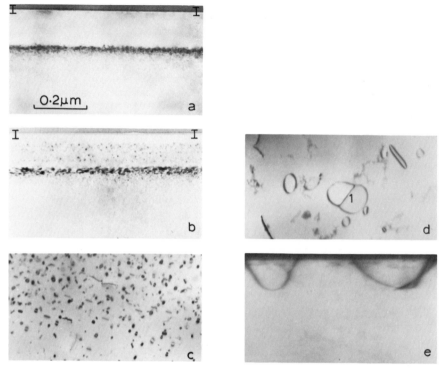

Fig. 4. The electron micrographs illustrating the changes in defect microstructures in ion implanted, (001) Si specimens after different annealing treatments in a furnace: (a) cross-section micrograph from 2.0×10^{15} As$^+$ cm^{-2} specimens after after 600°C/10 min; (b) from 10.0×10^{15} As$^+$ cm^{-2} specimens after 600°C/10 min; (c) plan-view micrograph after annealing at 800°C/6 min; (d) plan-view micrograph after 900°C/20 min annealing; (e) cross-section micrograph after 1000°C/20 min annealing. The dislocation reaction is shown at 1 and I-I denotes the original interface.

The size distribution of dislocation loops in the underlying dislocation band exhibited a relatively small change after annealing at 600°C. As the time and temperature of annealing was increased loop coarsening occurred (average loop size increased while number density decreased) as shown in Fig. 4(c) after annealing at 800°C for 6 minutes. The coarsening of loops occurs by loop coalescence which primarily involves conservative climb and glide of dislocation loops. During conservative climb, the loops move toward each other as a result of material transport within the loop due to pipe diffusion along the dislocations. The nature of these loops was determined to be interstitial type having a/2 ⟨101⟩ Burgers vectors. At higher temperatures of annealing, loop coarsening continues and eventually the dislocation reactions of the type (a/2[101] + a/2[01$\bar{1}$] ⟶ a/2[110]) lead to the formation of a/2⟨110⟩ dislocations that are edge type. One example of this dislocation reaction is shown at 1 in Fig. 4(d). These dislocations, which lie in (001) or in (111) planes, are frequently referred to as 90° dislocations. The above mechanism of dislocation formation

TABLE I.
Electrical Properties of Hydrogen-Torch Annealed Silicon Specimens

Ion Energy Species/Dose	Carrier Concentration N_s (cm^{-2})	Resistivity ρ_h (Ω/cm^2)	Mobility (cm^2/V·S) μ_h
100 keV, As$^+$, 5.5 x 10^{15} cm^{-2} CTA 900°C/30' 100 keV, As$^+$ 3.0 x 10^{15} cm^{-2}[a]	2.4 x 10^{15} cm^{-2}	47	47
FA (1200°C/10 s	3.5 x 10^{15} cm^{-2}	37	47
Laser annealing	3.5 x 10^{15} cm^{-2}	37	47
35 keV, As$^+$ 2.0 x 10^{15} cm^{-2}[a]			
FA 1050°C/10 s	1.6 x 10^{15}	77	51
1100°C/10 s	1.7 x 10^{15}	70	52
1200°C/10 s	2.1 x 10^{15}	50	60
35 keV, B$^+$ 2.0 x 10^{15} cm^{-2}[a]			
FA 1200°C/10 s	3.0 x 10^{15} cm^{-2}	50	42
Laser annealing	3.0 x 10^{15} cm^{-2}	50	42

[a]Implanted nominal dose

is consistent with the location of extra half planes associated with the dislocations. Figure 4(e) shows a (011) cross-section micrograph from a (100) specimen (after annealing at 1100°C for 20 min.) in which dislocations constituting the cross-grid have fully developed. The loop coalescence and the formation of dislocation grids require a relatively high number density of loops in the underlying dislocation band of as-implanted specimens. The effect of conservative climb is significant only if the distance between the loops is within the diameter of the larger loop. The number density and size distribution of the dislocation loops are determined by the total dose and rate, mass of implanted ions, and substrate temperature. The number density of loops decreases with decreasing dose and substrate temperature and with increasing mass of the incident ions [7,8]. In the case of low-dose (2.0 x 10^{15} cm^{-2}) arsenic implanted specimens, the dislocation loop density was sufficiently low that loop coalescence could not occur. Therefore, large loops did not form and small loops annealed by lattice diffusion. Therefore, to form p-n junctions that are free from dislocations, it is necessary to manipulate ion implantation conditions in such a way that the number density of loops in the underlying dislocation band does not exceed a critical value. It should be mentioned that cross and triangular grids of dislocations represent very stable networks of dislocations. These networks cannot be annealed by further thermal annealing. Since these networks are located close to the p-n junctions, they represent a major source of current leakage and poor junction performance. These dislocations, however, can be removed by pulsed laser annealing where depth of melting exceeds the depth of the grid of dislocations [9].

The results of Van der Pauw measurements on arsenic and boron implanted and flame annealed specimens are summarized in Table I. It should be mentioned that the boron implanted specimens contain only dislocation loops in the as-implanted layer. A complete electrical activation of dopants is obtained in both arsenic

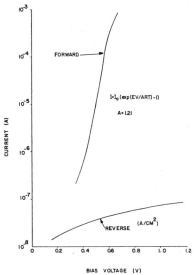

Fig. 5. Mesa diode characteristics from arsenic implanted and flame annealed (1150°/10 s) specimens.

and boron specimens after flame annealing at 1200°C for 10 seconds. As shown in Table I, the hall mobility of flame annealed specimens was comparable to ion implanted, laser annealed specimens. The electrical properties of flame or laser annealed specimens were found to be much better than those processed using conventional thermal annealing (CTA) (for example after 90′°C/30 min. annealing of arsenic implanted specimens) as shown in Table I. We have recently fabricated mesa diodes from flame annealed specimens with diode perfection factor A = 1.20-1.25 and reverse bias current $<10^{-7}$ Acm^{-2}, both indicating very good diodes. The substitutional and total concentrations of arsenic as a function of depth were determined using Rutherford backscattering and ion channeling techniques before and after flame annealing treatments. The as-implanted Gaussian profile showed a minimum broadening <100 Å after annealing at 1200°C for 10 seconds. It was interesting to note that the specimens implanted with a dose of 1.0×10^{16} As^+ ions cm^{-2} (100 keV) attained substitutional concentration of 1.6 $\times 10^{21}$ cm^{-3} near the mean projected range of the ions. This concentration is very close to the retrograde solubility limit of arsenic in silicon. The retrograde solubility limit represents a maximum substitutional concentration under equilibrium conditions at any temperature. However, under SPE growth conditions, it is possible to exceed retrograde solubility limits in ion implanted specimens [7].

In order to investigate the feasibilitiy of flame annealing for device fabrication, mesa diodes were formed on the annealed samples. The I-V characteristics of a typical mesa diode are shown in Fig. 5. The forward dark I-V characteristics could be fitted to the diode equation $I=I_0 [exp (EV/ART)^{-1}]$, and the value of diode perfection factor was extracted to be 1.21. This value of diode perfection factor and the reverse currents $< 10^{-7}$ A/cm^2 are indicative of a superior junction.

CONCLUSIONS

In conclusion, a "complete" annealing of displacement damage can be achieved by flame annealing up to a certain ion implantation dose by the process of solid-phase-epitaxial growth. Above the critical doses, the dislocation loops in the underlying damage band coarsen by loop coalescence, which can lead to the formation of a cross-grid of edge dislocations in (001) specimens and a triangular grid of edge dislocations in (111) specimens. Electrical (Van der Pauw) measurements have shown 100% electrical activation of dopants with Hall mobility comparable to pulsed laser annealed specimens. Superior p-n junction diode characteristics with A = 1.20, have been obtained.

REFERENCES

1. J. Narayan, R. T. Young and C. W. White, J. Appl. Phys. 49, 3912 (1978).
2. C. W. White, J. Narayan and R. T. Young, Science 204, 461 (1979).
3. S. R. Wilson, B. R. Appleton, C. W. White and J. Narayan, p. 481 in Laser-Solid Interactions and Laser Processing-1978, ed. by S. D. Ferris, H. J. Leamy and J. M. Poate, American Institute of Physics, Vol. 50, 1979.
4. W. L. Brown, p. 20 in Laser and Electron Beam Processing of Materials, ed. by C. W. White and P. S. Peercy, Academic Press, New York, 1980.
5. D. J. Lischner, p. 759 in Laser and Electron Beam Interactions with Solids, ed. by B. R. Appleton and G. K. Celler, North Holland, New York, 1982. R. Klabes, J. Mathai, M. Voelskow, G. A. Kachurin, E. V. Nidaev and H. Bartsch, Phys. Stat. Sol (a) 66, 261 (1981).
6. B.-Y. Tsaur, J. P. Donnelly, J.C.C. Fan and M. W. Geis, Appl. Phys. Lett. 39, 93 (1981).
7. J. Narayan and O. W. Holland, Phys. Stat. Sol. (a) 73, 225 (1982).
8. J. Narayan, O. W. Holland and B. R. Appleton, J. Appl. Phys. (in press).
9. J. Narayan and C. W. White, Phil. Mag. 43, 1515 (1981).

ISOTHERMAL ANNEALING OF ION IMPLANTED SILICON WITH A
GRAPHITE RADIATION SOURCE

S. R. WILSON, R. B. GREGORY AND W. M. PAULSON
SRDL, Motorola, Inc., 5005 E. McDowell Road, Phoenix, Arizona, USA

A. H. HAMDI AND F. D. McDANIEL
Dept. of Physics, North Texas State University, Denton, Texas, USA

ABSTRACT

Both (100) and (111) Si wafers were implanted with As, P or
B to doses from 10^{13} to 10^{16}/cm^2 and annealed with a Varian IA-
200 isothermal annealer. The anneal occurs in a vacuum using
infrared radiation for exposure times of 5 to 30 sec. Sheet
resistance (R_s), Hall effect, RBS and SIMS were used to analyze
the wafers. For each dopant a decreasing R_s occurs with increasing exposure time until a minimum value is reached. Longer
anneals produce increased dopant diffusion, and the R_s for As
and P implanted wafers increased unless the wafer was capped
with 0.05 μm of SiO$_2$ which prevents a loss of dopant. The results for (100) wafers are better than for (111) with As doses
$\leq 10^{15}$/cm^2, however at doses $>10^{15}$/cm^2 the (100) and (111) results are comparable. The As implanted, isothermally annealed
layers were thermally stable for As concentrations $\leq$2E 20/cm^3.

INTRODUCTION

For several years ion implantation has been used by the semiconductor industry for introducing impurities into silicon. The proper choice of implantation
energy and dose provides for better control of impurity profiles than can be
achieved by diffusion processes. However, to achieve complete activation of
the dopant and to remove the implantation damage, a high temperature anneal
must be performed. At these high temperatures the common dopants (As, P and
B) will diffuse. However, the need to minimize dopant diffusion during the
anneal has become important as device dimensions become smaller.

Recently the concept of rapid isothermal annealing, where the entire wafer
is heated to 900°C or greater for times on the order of 10 sec. has been investigated as a means to anneal implants with minimal dopant diffusion [1].
Large area raster scanned electron beams [2,3], high intensity arc lamps [4]
and graphite strip heaters [5] have all been used. Varian has developed a system which uses infrared radiation from a resistively heated sheet of graphite
to anneal ion implanted silicon [6-9]. The anneal occurs in a vacuum for exposure time of 5-45 sec. Due to the long absorption length of infrared radiation in silicon, heat is uniformly generated throughout the bulk of the wafer
and thus thermal gradients are minimal. The short anneal times help to reduce
dopant diffusion. Since the radiation is incoherent, and has a distribution
of wavelengths, there are no thin film interference effects, which have limited
the use of lasers for annealing devices and circuits. We have analyzed Si wafers that have been implanted with B, P or As and annealed with a Varian IA-200
isothermal annealer. The effects of different dopants, doses and substrate orientation are discussed. Results on the thermal stability of As implanted layers are presented.

Mat. Res. Soc. Symp. Proc. Vol. 13 (1983) © Elsevier Science Publishing Co., Inc.

SAMPLE PREPARATION AND EXPERIMENTAL PROCEDURE

The substrates used in this set of experiments were (100) and (111) oriented (CZ) Si wafers. . As (100 keV) and . P (50 keV) were implanted into p-type Si to doses from 1.0E 14/cm^2 to 1.0E 16/cm^2. ^{11}B (50 keV) was implanted into n-type Si to doses of 1.0E 13/cm^2 to 5.0E 15/cm^2. The As and P wafers were capped with .05 μm of deposited SiO$_2$ to prevent the loss of the dopant during the anneal. The wafers were annealed in a Varian IA-200 isothermal annealer. This system has been described previously by Downey et al. [7] and is similar to the prototype system used by Fulks et al. [6]. The infrared radiation is generated in a vacuum system using a resistively heated sheet of graphite whose temperature is determined with a thermocouple. The exposure time is controlled with a molybdenum shutter. The wafer is located 1 cm from the graphite during the exposure. All energy transfer between the heater and the Si wafer is by radiation. Both the heater and wafer are surrounded by reflectors to minimize energy loss to the surrounding. The heater temperatures used ranged from 1100°C to 1350°C and exposure times varied from 5 to 30 sec. (Slip lines were observed on wafers which were annealed for times greater than 10 sec. at 1300°C.) The wafers were analyzed subsequent to isothermal annealing using sheet resistance (R_s) and Hall effect techniques to evaluate the electrical properties. The boron profiles were measured by SIMS. The arsenic profiles and the residual damage were measured by RBS analysis.

RESULTS AND DISCUSSION

The R_s versus exposure time for wafers implanted with P or B are plotted in Fig. 1a and 1b respectively. The heater temperature was 1200°C for both sets of data. On the P implanted wafers one half was capped with SiO$_2$ prior to anneal. In both figures the R_s decreases with time up to 15 sec. However, for the P-implanted wafers the R_s was lower on the SiO$_2$ encapsulated portion than on the uncapped portion. In addition, the R_s increases rapidly after 15 sec. for the uncapped half relative to the capped half. This is similar to the results on As implanted Si annealed with and without an SiO$_2$ cap [8,9] where a substantial amount of As was lost unless the wafer was capped. We believe that the same effect is occurring with P. On the other hand, B is not lost at the surface, but merely diffuses into the wafer with increasing exposure time [9]. This explains why the R_s reaches a minimum and then remains essentially constant with increased exposure time. Since at room temperature Si only weakly absorbs infrared radiation, some amount of time is required for the wafer to heat up. The heat up time is a significant portion of the 15 sec. exposure time required to achieve a minimum R_s. For shorter times, the Si does not reach a sufficient temperature long enough for complete annealing [8]. The exposure time necessary to achieve the minimum R_s decreased with increasing heater temperature. The minimum R_s values of 70 ohm/sq for P (1.0E 15/cm^2) and 90 ohm/sq for B (1.0E 15/cm^2) can be compared to 80 and 95 ohm/sq, respectively, obtained by furnace annealing at 950°C, 20 min. in N$_2$.

We have shown previously that Si implanted with As to doses between 1.0E 14/cm^2 and 2.5E 15/cm^2 can be annealed with this technique [8,9]. However, the results for (111) silicon depend on the As dose as shown in Fig. 2. This figure presents R_s versus exposure time for (100) and (111) Si annealed with a heater temperature of 1200°C. In all four cases the wafers were capped with .05 μm of SiO$_2$ to prevent the loss of As during the anneal. For the 1.0E 15/cm^2 dose, the R_s for the (111) wafers are always higher than the (100) wafers. The two results differ by 40% after a 15 sec. exposure. The (111) data are higher even after a 25 sec. exposure. However, the R_s for the 1.0E 16/cm^2 implants agree to better than 10% for each exposure time.

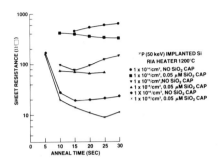

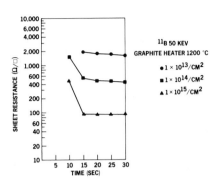

Fig. 1a. R_s versus exposure time for P implanted, isothermally annealed (1200°C) Si.

Fig. 1b. R_s versus exposure time for B implanted, isothermally annealed Si (1200°C) Si.

The RBS data for the (111) Si implanted with As to 1.0E 15/cm^2 and isothermally annealed are presented in Fig 3. The channeling data were taken with a 2.0 MeV He$^+$ beam aligned parallel to a <110> axis. The scattering from As in the random spectrum indicates minimal diffusion of the As and no loss of As has occurred. However, the yield from Si, with the beam in the aligned direction is very high ($\chi_{min_{Si}}$ = 42.6%) indicating considerable damage is still present in the sample. The yield from As with the beam aligned is only slightly lower than the yield in the random direction ($\chi_{min_{As}}$ = 62.6%). The $\chi_{min_{Si}}$ decreases to 13.3% after a 30 sec. exposure. In comparison, the channeling data obtained on the (100) wafer was quite good ($\chi_{min_{Si}}$ = 3.5% and $\chi_{min_{As}}$ = 6.8%) after a 15 sec. exposure.

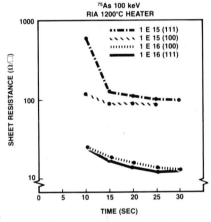

Fig. 2. R_s versus exposure time for As implanted (100) and (111) Si isothermally annealed with 1200°C heater.

Figure 4 presents the channeling data for the (111) Si implanted with As to 1.0E 16/cm^2 and annealed. The scattering yield from the As indicates the dopant has diffused to a depth in excess of 0.35 µm for the 15 sec. anneal. This rapid diffusion of As is due to a concentration enhanced diffusion of As [10,11]. The channeling yield for both the Si ($\chi_{min_{Si}}$ = 3.3%) and As ($\chi_{min_{As}}$ = 7.3%) indicates complete removal of the damage and excellent substitutionality of the As. These results are in good agreement with the results obtained from the (100) wafer implanted to 1.0E 16/cm^2 and annealed for 15 sec. ($\chi_{min_{Si}}$ = 2.6% and $\chi_{min_{As}}$ = 4.5%)

The channeling results presented in Figs. 3 and 4 explain the R_s results of Fig. 2. For (100) Si excellent anneals are obtained regardless of the

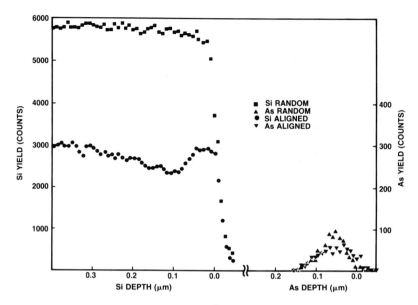

Fig. 3. RBS spectra for As (1.0E 15/cm^2) implanted (111) Si isothermally annealed for 15 sec. with a 1200°C heater.

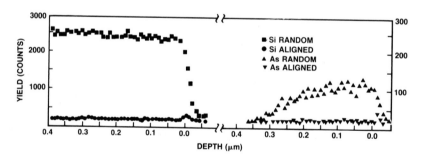

Fig. 4. RBS spectra for As (1.0E 16/cm^2) implanted (111) Si isothermally annealed for 15 sec. with a 1200°C heater.

dose. However, for (111) Si the anneal quality and thus the R_s improves with
increasing dose. Results similar to the 1.0E 16/cm^2 data were achieved for
a dose of 5.0E 15/cm^2 and data for wafers implanted to 5.0E 14/cm^2 were similar
to the 1.0E 15/cm^2 results. Chu et al. reported [12] better anneals for (111)
wafers implanted with As to doses of 1.0E 16/cm^2 than doses of 1.0E 15/cm^2 after a conventional furnace anneal at 1100°C in N$_2$. The poor (111) anneals for
1.0E 15/cm^2 wafers are probably due to the low growth velocity of (111) Si relative to (100) Si [13]. This growth velocity can be enhanced at high dopant
concentrations by the impurity enhanced self-diffusivity of silicon, and is
greatly affected by high concentrations of electrically active impurities [14].

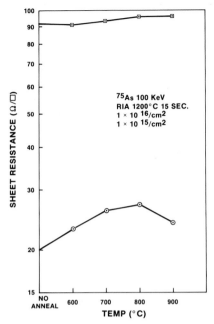

Fig. 5. R_s versus anneal temperature for Si wafers implanted with [75]As (100 keV) to doses of 1.0E 15/cm^2 or 1.0E 16/cm^2 and isothermal annealed for 15 sec. with a 1200°C heater.

In order for these implanted and annealed layers to be useful in device processing, they must be stable under subsequent thermal processing. Si wafers (100) orientation were implanted with As to doses of 1.0E 15/cm^2 or 1.0E 16/cm^2 and annealed with a 1200°C heater and a 15 sec. exposure. Then, pieces from each wafer were subjected to furnace annealing for 30 min. in N_2 at temperatures ranging from 600 to 900°C. The R_s results are presented in Fig. 5. The R_s for the 1.0E 15/cm^2 wafers changes by less than 5% over the temperature range studied. However, the R_s for the 1.0E 16/cm^2 wafer increases by 35% after an 800°C, 30 min. anneal. The channeling data on these samples shows a slight increase in the As χ_{min} (5.4% to 8.7%), indicating As is coming out of solution. An increase was also observed in the Si scattering yield. The RBS shows the As concentration in this case is 4.0E 20/cm^3. The concentration for the 1.0E 15/cm^2 implant was less than 1.0E 20/cm^3 after isothermal annealing and was stable. These results are in good agreement with our previous results from laser annealing where As concentration $\leq$ 2E 20/cm^3 were thermally stable and higher concentrations were not [19].

SUMMARY

In summary we have shown that As, P and B implanted silicon can be annealed quite well with a Varian IA-200 isothermal annealer. It was necessary to cap the As and P implants with 0.05 μm of SiO_2 to prevent a loss of the dopant during the anneal. This was not necessary on the B implants as no loss of dopant occurred. Good anneals were achieved on (100) wafers regardless of dose. However, the anneal quality on (111) wafers improved dramatically for As doses of 1.0E 16/cm^2 compared to 1.0E 15/cm^2. Finally, it was shown that the ion implanted, rapid isothermally annealed layers were thermally stable in the temperature range 600 to 900°C if the As concentration was less than 2.0E 20/cm^3.

REFERENCES

1. C. Hill, Laser and Electron Beam Solid Interactions and Materials Processing, J. F. Gibbons, L. D. Hess and T. W. Sigmon, eds., North Holland, 1981, p. 361.

2. N. J. Shah, R. A. McMahon, J. G. S. Williams and H. Ahmed, Laser and Electron Beam Solid Interactions and Materials Processing, J. F. Gibbons, L. D. Hess and T. W. Sigmon, eds., North Holland, 1981, p. 202.

3. T. O. Yep, R. T. Fulks and R. A. Powell, Appl. Phys. Lett. 38, 162 (1981).

4. R. A. Powell, T. O. Yep, and R. T. Fulks, Appl. Phys., Lett. 39, 150 (1981).

5. B.-Y Tsaur, J. P. Donnelly, John C. C. Fan, and M. W. Geis, Appl. Phys. Lett. 39, 93 (1981).

6. R. T. Fulks, C. J. Russo, P. R. Hanley and T. I. Kamins, 39, 604 (1981).

7. D. F. Downey, C. J. Russo and J. T. White, Solid State Technology 25, 87 (1982).

8. S. R. Wilson, R. B. Gregory, W. M. Paulson, A. H. Hamdi, and F. D. McDaniel, Appl. Phys. Lett. 41, (1982).

9. S. R. Wilson, R. B. Gregory, W. M. Paulson, H. T. Diehl, A. H. Hamdi, and F. D. McDaniel, IEEE Transactions on Nuclear Science, NS-30 (1983) in press.

10. M. Y. Tsai, F. F. Morehead, J. E. E. Baglin and A. E. Michael, J. Appl. Phys. 51, 3230 (1980).

11. Richard B. Fair and Joseph C. C. Tsui, J. Electrochem. Soc. 122, 1689 (1975).

12. W. K. Chu, H. Muller, J. W. Mayer and T. W. Sigmon, Ion Implantation in Semiconductors, S. Namba ed., (Plenum, N.Y., 1975) p. 177.

13. L. Csepregi, J. W. Mayer and T. W. Sigmon, Appl. Phys. Lett. 29, 92 (1976).

14. J. M. Fairfield and B. J. Masters, J. Appl. Phys. 38, 3148 (1967).

15. S. R. Wilson, W. M. Paulson, G. Tam, R. B. Gregory, C. W. White and B. R. Appleton, Laser and Electron Beam Interactions with Solids, B. R. Appleton and G. K. Celler, eds., (North Holland, 1982) 287.

GETTERING OF IMPURITIES BY INCOHERENT LIGHT ANNEALED POROUS SILICON

V. E. BORISENKO, A. M. DOROFEEV
Minsk Radioengineering Institute, P.Browka 6, Minsk, USSR

ABSTRACT

The application of electrochemically formed porous silicon to getter Cu and Au atoms in silicon crystals during incoherent light annealing for 4 to 20 sec is reported. Exposure light power density of 30 W/cm^2 was used to heat the samples up to 950 °C. Improvement of lifetime related to Au and Cu gettering with porous silicon has been observed. Excess impurity interstitials and their enhanced diffusion are supposed to be responsible for the gettering effect.

INTRODUCTION

Heavy-metal impurities introduce energy levels within the bandgap of silicon, which act as a recombination centers, thus decreasing the minority carrier lifetime and increasing the leakage currents of p-n junctions [1]. Gettering appears to be the only manner in which an appropriate yield of high quality devices can be achieved.

Various techniques involving formation of damaged lattice regions or layers with enlarged intrinsic surfaces acting as a condensation sites for metallic contaminations and stacking fault nuclei have been presented to getter unwanted impurities in silicon [2]. Highly activated silicon surface can be provided with porous silicon fabricated by anodic treatment of silicon crystal in HF [3,4].

In order for gettering to occur, long-time high temperature heating is necessary for migration of impurities and their accumulation in a getter. Annealing temperatures ranged from 800 to 1200 °C and annealing cycles of 30 min and longer are inevitably employed [5-7]. Meanwhile, there has been a growing interest in the application of short-time heat processing by intense incoherent light which has been succesfully used to anneal ion-implanted silicon [8-13].

In the present study electrochemically formed porous silicon layer has been used to getter heavy-metal impurities extracted out of a silicon crystal during rapid incoherent light annealing.

EXPERIMENTAL

The experimental arrangement is schematically shown in Fig.1. Structure of the sample, initial metal impurity profile and theoretically estimated impurity redistribution due to their gettering by the porous silicon layer are presented.

Mat. Res. Soc. Symp. Proc. Vol. 13 (1983) ©Elsevier Science Publishing Co., Inc.

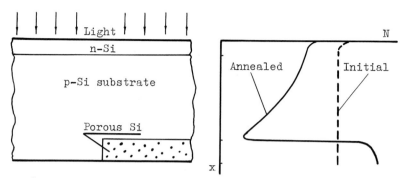

Cross section of the sample   Metal impurity distribution

Fig.1. Schematic view of the experimental arrangement.

Substrates were 0,1 Ohm·cm p-type (111) silicon wafers of 380 μm-thick and 76 mm in diameter. As a first step in a sample preparation, 2,0 μm-thick n-epitaxial layer was formed at the front side of wafers for lifetime measurements. The samples were uniformly saturated with Au or Cu up to the concentrations of $10^{16} - 10^{17}$ atom/cm^3 and chemically etched to remove impurity enreached surface layers.

Porous silicon layer of 8-10 μm-thick was formed in a half of the silicon wafer back side by anodic dissolution in HF solution. The material was characterized with a porosity of 30-40%. It contained 30-100 Å pores and provided significantly enlarged surface area.

The wafers were annealed in the rapid heating apparatus by radiation of halogen lamps [13] for 4 to 20 sec. The exposure light power density was about 30 W/cm^2 which ensured heating of a sample up to the steady-state temperature of 950 °C (as controlled with a chromel-alumel thermocouple bonded to the wafer back surface).

To estimate gettering effect, lifetime measurements (capacitance-time method) and neutron activation analysis have been carried out.

RESULTS AND DISCUSSION

Although, a significantly increased surface has been established to getter impurities, porous silicon has not been applied for this purpose in short-time heating cycles yet. Results illustrating variation of generation lifetime over Cu-doped wafers subjected to incoherent light annealing are shown in Fig.2. The initial average lifetime level is also plotted. Generation lifetime in the ungettered (control) area remains practically unchanged while that in the gettered area is improved by about an order and tends to the level of about 0,5 μsec. The plots in the ungettered region, however, are spread between 0,02 and 0,1 μsec. Such small values are connected with a high concentration of Cu and Au which are known as good agents for reducing gene-

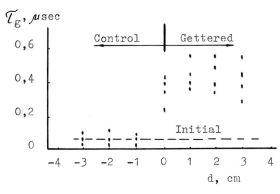

τ_g, μsec

Fig.2. Distribution of generation lifetime in the gettered and un-gettered areas of the 8 sec incoherent light annealed wafer.

ration lifetime.

Effective increasing of the lifetime depends on the regime of light annealing. Fig.3 presents variation of lifetime in the samples doped with Cu under the light pulse duration. The data obtained in the reference furnace annealing samples are displayed in the right part of the plot. Rapid rise of the lifetime occurs with the increase of annealing time from 4 to 8 sec. Maximum value of 0,5 μsec is achieved at 8-10 sec. The subsequent increase of annealing time leads to the permanent decrease of lifetime to the values, which are observed in the furnace annealing experiments.

Variation of the generation lifetime in the Au-doped samples has occured to have the same character.

To investigate the reason of the lifetime improvement, neutron activation analysis has been performed. Contamination of Cu and Au atoms in the porous silicon layer was analysed. The metal impurity concentration was found to be increased by about two orders over the initial level.

Cu concentration of $7 \cdot 10^{18}$ atom/cm^3 and Au concentration of $6 \cdot 10^{17}$ atom/cm^3 were detected in the porous silicon layers after incoherent light annealing for 8 sec.

Dependence of the impurity concentration accumulated within the porous silicon layer upon the annealing time closely corre-

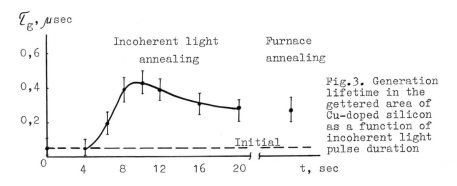

Fig.3. Generation lifetime in the gettered area of Cu-doped silicon as a function of incoherent light pulse duration

lates with the lifetime measurements. This fact allows to conclude that improvement of generation lifetime in the areas of
epi-layer above the porous silicon layer is related to gettering
of Cu and Au atoms by porous silicon.

Analysing a possible mechanism of Au and Cu atoms migration
effective diffusivity of the impurities has been calculated.
Amount of the impurities accumulated in the getter (Q) can be
expressed as follows [14]:

$$Q = N_0 l \left\{ 1 - \frac{8}{\pi^2} \sum_{k=0}^{\infty} \frac{1}{(2k + 1)^2} \exp\left[- \frac{(2k + 1)^2 \pi^2}{l^2} Dt \right] \right\}$$

where N_0 is the initial concentration of the impurity (N_0^{Cu} =
$2 \cdot 10^{17}$ atom/cm^3, N^{Au} = 10^{16} atom/cm^3), l is the sample thickness, D is the impurity diffusivity, and t is the annealing
time.

The diffusivity of Cu and Au atoms derived by the numerical
iteration procedure from the equation is about $3 \cdot 10^{-4}$ cm^2/sec
for incoherent light annealing. That is considerably higher
than both the diffusivities obtained in the reference furnace
annealed samples and the equilibrium diffusivities of Cu
($\sim 2 \cdot 10^{-6}$ cm^2/sec) and Au ($\sim 5 \cdot 10^{-7}$ cm^2/sec) in silicon at 950 °C
[15]. The results show that intense incoherent light enhances
diffusion of Cu and Au atoms during short-time heat processing.

Diffusion of Cu and Au in silicon is known to occur via
dissociative mechanism [14,15]. According to that mechanism,
interstitial impurity atoms are responsible for the fast impurity
migration in the crystal. Hence, concentration of the impurity
atoms in interstitial sites controls the impurity diffusivity.
During incoherent light annealing excess impurity interstitials
may be generated in the surface layer where light power is
absorbed. There are at least two reasons for that. The first one
is intense electron exitations provided lowering the energy of
interstitial formation. Local increase of temperature is the
second possible reason. Moreover, the interstitial migration
into the crystal is promoted with the heat flow direct from
the irradiated surface to the wafer back surface.

CONCLUSION

The experimental results presented demonstrate a possibility
to utilize electrochemically formed porous silicon layers for
gettering fast diffusing impurities in silicon crystal during
incoherent light annealing. Heating time of 8-10 sec has been
found to be sufficient to produce the gettering effect. Incoherent light heating enhances diffusion of Cu and Au atoms. It is
interpreted in terms of excess impurity interstitials generated
at the irradiated surface and their rapid migration supported
with a powerful heat flow across the sample. Meanwhile, the
phenomenon observed is complicated and further investigations
are in progress.

ACKNOWLEDGMENT

The authors would like to thank professor V.A.Labunov for his continuous support of the study and helpful discussion.

REFERENCES

1. A. G. Milnes, Deep Impurities in Semiconductors, (John Wiley & Sons, New York 1973).

2. J. R. Monkowski, Solid State Technol. 24, No.7, 44 (1981).

3. T. M. Buck, J. M. Poate, K. A. Pickar, Surf. Sci. 35, 362 (1973).

4. K. Murase and H. Harada, J. Appl. Phys. 48, 4404 (1977).

5. A. G. Cullis, T. E. Seidel and R. L. Meek, J. Appl. Phys. 49, 5188 (1978).

6. P. Revesz et al., J. Appl. Phys. 49, 5199 (1978).

7. M. J. T. Lo, J. G. Skalnik and P. F. Ordung, J. Electrochem. Soc. 128, 1569 (1981).

8. K. Nishiyama, M. Arai and N. Watanabe, Jap. J. Appl. Phys. 19, L563 (1980).

9. C. Drowley and S. Hu, Appl. Phys. Lett. 38, 876 (1981).

10. R. A. Powell, T. O. Jep and R. T. Fulks, Appl. Phys. Lett. 39, 150 (1981).

11. R. T. Fulks et al., Appl. Phys. Lett. 39, 604 (1981).

12. J. L. Benton et al., in: Laser and Electron Beam Interaction with Solids, ed. by B. R. Appleton and G. K. Celler (North Holland, New York 1982) p. 765.

13. V. E. Borisenko and V. A. Labunov, Phys. Stat. Sol.(a) 72, (1982) (to be published).

14. S. Crank, Mathematics of Diffusion, (Oxford, London 1965).

15. S. M. Hu in: Atomic Diffusion in Semiconductors, ed. by D. Show (Plenum-Press, New York 1973) p. 217.

THE CONTRIBUTION OF BEAM PROCESSING TO PRESENT AND FUTURE
INTEGRATED CIRCUIT TECHNOLOGIES

C HILL
Plessey Research (Caswell) Ltd., Allen Clark Research Centre, Caswell,
Towcester, Northants. NN12 8EQ, England

ABSTRACT

A brief history and present state of beam processing
techniques and applications to silicon integrated circuit
technology are given. The viability of incorporating pulse-
laser controlled doping profiles into the emitter-base
structure of an advanced bipolar transistor is discussed.
The areas of present I.C. technology which will constrain
future device development are identified, and the contribution that beam processing can make in removing these constraints is discussed. The beam processing techniques most
likely to be found in future I.C. technologies are described.

INTRODUCTION

The field of beam processing, which began as laser annealing in 1977, has
now reached a degree of maturity in which applications of the new capabilities
discovered are being actively looked for and, indeed, expected. This paper
attempts to make a realistic assessment of the usefulness of beam processing
in silicon integrated circuit technology over the next few years, and is
necessarily somewhat speculative. However, it is considered that the general
conclusions are sound and that the important factors have been considered. In
order to indicate the scope of the paper, beam processing will be defined as:
the use of energetic beams to induce a heat cycle in, and so change the properties of, the materials that make up a final device structure. Such a definition
clearly includes such operations as annealing, recrystallization, dopant redistribution, topography shaping, inter-reaction of layers, introduction of controlled damage for contaminant gettering: it equally clearly excludes lithography, photolytic deposition, etching, cutting and drilling operations by
beams, which are all well-established fields in their own right. In what
follows, the history and present position of techniques and applications of
beam processing is briefly reviewed: then one particular technique and its
application (adiabatic redistribution of dopants to fabricate novel emitter-
base structures for bipolar transistors) is assessed by the stringent criteria
required to be met by any technique to be used in present and future integrated
circuit technologies: lastly, the areas of application and techniques considered
most promising are described.

HISTORY OF BEAM PROCESSING

The development of the field is accurately mirrored in the papers making up the
five volumes of MRS meeting proceedings from 1978-1982 {1-3} and the 1978 and
1979 ECS Symposia {4,5}. Initially, the basic physics of laser annealing of ion
implantation damage was the dominant interest, but the use of other beams soon
began to be explored: electron beams {6,7}, ion beams {8}, incoherent wide-band
photon radiation {9,10,11}. At the same time, a wide range of materials changes
achievable by beam processing was discovered, including those listed in the

Mat. Res. Soc. Symp. Proc. Vol. 13 (1983) © Elsevier Science Publishing Co., Inc.

382

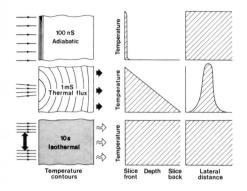

Fig.1 Schematic diagram illustrating the characteristics of the three modes of beam processing. From left to right the characteristics shown are beam configuration, heating pulse duration, temperature contours in the depth of the sample, heat flow, temperature profiles perpendicular and parallel to the plane of the sample surface. From Ref.3 1.

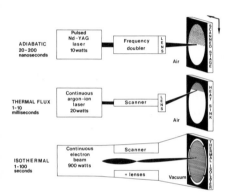

Fig.2 Schematic diagram showing typical equipment used for each of the three modes of beam processing. In each case, a small area beam is scanned to cover the relatively large area of a standard silicon slice. Typical scan rates for a 100 micron spot are adiabatic 200cms/sec: thermal flux 10cms/sec: isothermal 2Km/sec.
From Ref.13.

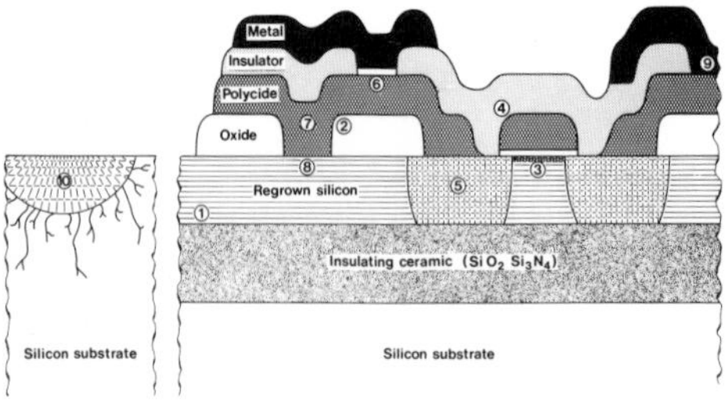

Fig.3 Composite sectional diagram illustrating the applications of beam processing to integrated circuit structures (1) regrown single crystal on amorphous dielectric (2) contoured oxide edge (3) annealed implantation damage with no dopant redistribution (4) contoured step in polysilicon/polycide (5) supersaturated solid solution of dopant redistributed (6) smoothed polysilicon surface to give high quality oxide for capacitors (7) reacted polysilicon & metal to form silicide (8) cleaned surface for Schottky contact (9) ohmic contact (10) beam-induced damage to form gettering sites for impurities. From Ref.13.

MECHANISM AND MODE / MATERIAL CHANGE EFFECTED	LIQUID PHASE PROCESS		SOLID PHASE PROCESS	
	Adiabatic Mode	Thermal Flux Mode	Thermal Flux Mode	Isothermal Mode
RE-ORDERING OF MATRIX MATERIAL	Anneal of Implant Damage; Anneal of Diffusion Damage; Increase in Substitutionality of Dopants		Anneal of Implant Damage	Anneal of Implant Damage
DISORDERING OF MATRIX MATERIAL	Introduction of Point and Line Defects and Associated Electronic Levels; Amorphisation		Introduction of Point and Line Defects and Associated Electronic Levels	
RE-ORDERING OF DEPOSITED MATERIAL	Epitaxial Re-growth of Deposited Si on Crystalline Substrates	Crystalline Regrowth of Deposited Si on Amorphous Substrates	Epitaxial Re-growth of Deposited Si on Crystalline Substrates	
REDISTRIBUTION OF DOPANTS	Novel Controlled Redistribution of Dopants in Silicon	Uniform Distribution of Dopants in Deposited Layers	Minimal Redistribution of Dopants in Silicon	Small Controlled Redistribution of Dopants in Silicon
REACTION OF SURFACE LAYERS	Formation of Silicides; Ohmic Contacts; Synthesis of Semiconductor Films; Surface Cleaning; Surface Oxidation		Formation of Silicides; Surface Oxidation	Formation of Silicides
REDISTRIBUTION OF SURFACE MATERIAL	Shaping Steps in Silicon; Silicon Oxide Smoothing Poly Si; Formation of Waves and Ripples in Si; Topographic Changes in Oxide-Coated Si	Long-Range Redistribution of Deposited Si on Amorphous Substrates		

Table 1 Categories of materials changes that can be effected by beam processing, the techniques and mechanisms by which they can be effected, and the existing applications to semiconductor device fabrication in each category. Note: In heat-sinked applications of the thermal flux mode, as illustrated in Fig.1, the characteristic heating time is 1-10 milliseconds. In freely-radiating applications, e.g. zone-melting recrystallization with high substrate temperatures, the small thermal gradients result in characteristic times of 0.1 - 1 second.

Introduction. As the mechanism of laser annealing became better understood, and successful models for the regrowth process based on extrapolation of existing crystal growth theory were developed, it became evident that the mechanism of material modification did not depend on the particular beam used but rather on the heating cycle induced in the material by the beam. This brought a great simplification into the field and it was possible to classify all the very diverse beam processing experiments into three broad categories based on the beam pulse duration {12}. These modes of processing are summarised schematically in Fig.1. The pulse durations for the three modes correspond roughly to the time taken for the induced heat to flow over three important dimensions of a standard silicon slice viz. the beam absorption depth (2×10^{-4} cms, adiabatic mode); the slice thickness (4×10^{-2} cms, thermal flux mode); and the slice radius (4cms, isothermal). The characteristics of these three modes of material processing are described elsewhere {12}. Historically, they were explored using typical systems illustrated in Fig.2. More recently, wide-band incoherent radiation sources have been favoured for thermal flux and isothermal processing, as the papers on recrystallization and annealing in this volume show. Adiabatic processing remains still the province of the Q-switched laser, apart from some specialised pulsed electron-beam work. A large number of different types of material changes induced by beam processing have been explored over the past four years: those specific to silicon integrated circuit applications are summarised in diagrammatic form in Fig.3. The applications shown mostly require a particular mode of beam processing, and a particular mechanism (solid or liquid phase), though in some cases more than one approach is possible. The available combinations and their classification into types of change have been described {13} and are listed in Table 1. Those indicated with a heavy outline are of particular interest and will be dealt with in more detail.

ADIABATIC BEAM PROCESSING IN BIPOLAR TRANSISTOR FABRICATION

Though all the applications of beam processing in Table 1 have been shown to be perfectly feasible, very stringent criteria have to be met if these applications are to find a place in I.C. technologies. These criteria have been previously outlined {12} and are (i) is control of the materials change adequate? (ii) are secondary effects induced and, if deleterious, can they be minimised? (iii) is the beam processing compatible with the rest of the I.C. technology? (iv) can sufficiently large areas of silicon be heat-treated with adequate throughput? (v) is there a sufficient need for the new technique to justify changing a mature existing technology? These criteria will be applied to what appears to be a promising application, the fabrication of very square and independently controlled dopant distribution profiles by adiabatic multiple-pulse melt and regrowth of the emitter (and/or base) regions of a bipolar transistor {14}. The parameter control required (criterion (1)) is illustrated in Fig.4; the emitter junction depth must be redistributed to an accuracy of ±4% to achieve a specified tolerance of ±15% on transistor gain, for the typical device structure shown. This implies a definite control required on the beam parameters, which has been calculated for each of the three processing modes {13} and is shown diagrammatically in Fig.5. The novel square dopant profiles are only obtained with the adiabatic mode, but conventional solid state redistribution to the same depth by the other two modes is included for comparison. The control required for adiabatic processing in this case (±2%) is just matched by the reproducibility and uniformity of the best Q-switched laser/homogeniser systems. Criterion (ii) is met, since the adiabatic melt and regrowth process also yields perfect single crystal with some point defects which can be annealed out at low temperature {15}. The main stumbling block in meeting criterion (iii) is the omnipresence of dielectric layers in I.C. processing, which are not compatible with pulsed laser heating. These deleterious effects have been summarised {12} and are illustrated in Fig.6. The solution adopted by several

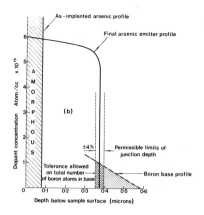

Fig.4 Typical dopant concentration profiles through the emitter-base region of a high performance bipolar transistor structure fabricated by pulsed laser adiabatic redistribution of the emitter dopant. The permissible variation in emitter junction depth consistent with ±15% tolerance in transistor gain is shown. After Ref.13.

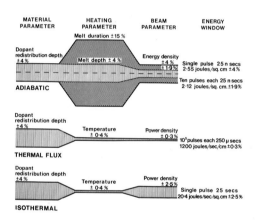

Fig.5 Schematic diagram showing the factors determining the control of beam parameters required for fabrication of an arsenic emitter structure to the tolerances given in Fig.4. The alternative diagrams for adiabatic processing refer to fabrication using one pulse, or ten pulses from a 25 nanosecond ruby laser beam. From Ref.13.

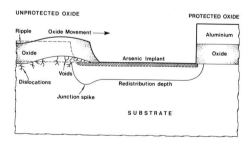

Fig.6 Schematic section through an oxide-coated silicon slice with photoengraved window and arsenic implant showing effects occurring during adiabatic laser processing. The left and right sides illustrate the effects associated with unprotected and protected oxides respectively. From Ref.3 1.

workers is to use a reflecting layer to protect the oxide areas and this has been quite successful {16,17}. The four possible approaches to bipolar transistor fabrication envisaged by the author are shown in Fig.7. The simple technique shown in (a) is technologically unsatisfactory, since it involves vertical etching of the dielectric down to the active device area using a metal mask, and contamination of the active device area occurs. Method (b), using an integral polysilicon mask and emitter region showed great promise in two dimensional test structures (Figs.8a, 8b) especially where an oxide overlayer was used to restrain the molten polysilicon flow and so retain the original edge geometry (Fig.8b). Unfortunately, in small three-dimensional device structures, new problems appear: without capping oxide, unacceptable flow of the polysilicon mask occurs (8c): with capping oxide, the confined polysilicon damages dielectric and cap by molten protrusions (8d). Similar protrusions are reported in silicon-on-insulator structures recrystallized by zone-melting a polycrystalline silicon film sandwiched between two oxide films {18}, and a solution will almost certainly be found to avoid their formation. This particular problem is circumvented in the alternative structures 7c and 7d. In 7c, the metal protective layer is separated from the active dielectric layer by thin nitride spacer. In addition, a portion of the emitter only is redistributed by adiabatic melt and regrowth. By these means, standard wet-etching techniques can be used to self-align the inner emitter window to the protective mask, and the critical emitter-base junction under the dielectric is not melted. This is not, however, a minimum geometry structure because of the need for an inner and outer emitter region. Structure 7d is the most different from existing technology in that the emitter-base planar structure is fabricated first, and isolation and passivation dielectric layers later. Such a technology is entirely compatible with laser processing, and is also a possible route to achieve minimum geometry structures by conventional processing.

The criterion (v) (area and throughput) is quite achievable both from technical and cost points of view, as discussed elsewhere {12}. Thus, all of the first five criteria are met for this novel method of emitter-base fabrication. The last criterion is, however, the most important since it not only decides whether a particular process finds real application, but also provides the driving force to solve the problems in developing the process to that stage. Is there, in fact, a sufficient need for this new process to justify changing from existing techniques?

THE NEED FOR NEW TECHNIQUES IN I.C. TECHNOLOGY

Integrated circuits based on the bipolar transistor are generally chosen because of their high operating speeds. The switching speed is, therefore, a key parameter, and new techniques have previously been accepted if a significant improvement in this one parameter resulted without degradation in transistor gain. The previously universal phosphorus-doped emitter has given way entirely to the arsenic-doped emitter, because the squarer profile of the latter allowed basewidth and charge storage to be reduced, both increasing switching speed. The emitter-base structure typical of today's advanced bipolar technologies is shown in Fig.9a; also shown are three possible ways of obtaining a structure with even higher switching speeds, one of which a future technology must be able to fabricate (9b,c,d). Structure 9b is the one discussed in the previous section, and offers possibilities of profile optimisation not available in 9a, because the emitter-profile shape can be chosen independently of concentration and the base profile is independent of the emitter-processing. Structure (c) has a polycrystalline silicon, arsenic-doped emitter with a thin oxygen-doped layer at the emitter-base junction. The absence of emitter-sidewalls and the asymmetry in electron and hole currents caused by the "oxide" barrier

At laser anneal stage **Final structure**

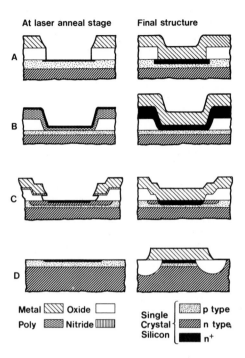

A

B

C

D

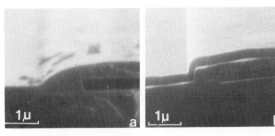

Metal, Oxide, Poly, Nitride, Single Crystal Silicon [p type, n type, n⁺]

Fig.7 Schematic sections
through alternative structures for fabrication of the
emitter region of a bipolar
transistor by pulsed laser
adiabatic processing after
implantation of emitter
dopant. Structures a b & c
incorporate a reflecting
layer to protect the dielectric from the laser pulse.
(a) Self-aligned aluminium
mask, dielectric window and
implant (b) Self-aligned
polysilicon emitter and dielectric window (c) Nested
adiabatically-redistributed
emitter inside conventional
emitter (d) Emitter-base
structure with post-anneal
dielectric formation.

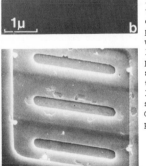

Fig.8 Scanning electron
micrographs of photoengraved
structures based on the emitter fabrication technique
shown in Fig.7b, after pulsed
laser processing. (a) Section
through window edge, no oxide
overlayer (9×1.85 joules/cm^2
pulses) (b) As for (a), but
with 0.4 micron oxide overlayer present during beam
processing (c) Bipolar transistor, emitter and p+ contact windows, no oxide overlayer (5×1.4 joules/cm^2 pulses) (d) As for (c) but with
0.4 micron oxide overlayer
present during processing.

enhances the emitter efficiency by a factor of ten {19}, and is such a step forward from structure 9a that it is the natural successor. Looking to the far future, much higher switching speeds are offered by hot electron transistors {20}, operating on a different principle from conventional bipolar devices. The need for structure 9b is therefore being bypassed. The new technologies required by the new structures are probably isothermal processing (to minimise dopant redistribution during anneal of implants) and molecular beam epitaxy (to fabricate the very thin, highly doped layers required by structure 9d). Thus the only beam processing technique likely to play a part in future bipolar technology is isothermal processing.

MOS technology is being pushed to higher and higher complexity and hence component densities. This requires lateral shrinkage of devices, and also a change in technology as illustrated in Fig.10. A hypothetical future 1 micron process is compared with an existing advanced 2½ micron MOS process. The biggest technology change is the use of silicon-on-insulator material: beam processing may be the technology adopted, though here it is competing with the alternative techniques of oxygen implantation and porous silicon formation. The scaling down of operating voltages and contact areas will require achievement of much lower contact resistances and more perfect anneal of implants if present switching speed/power products are to be maintained. Isothermal processing, both of implants and of metal-silicon contacts, may well be required. From the foregoing, it follows that anneal of implantation damage by new techniques may well be a feature of both MOS and bipolar future technologies and this will be discussed in more detail.

ANNEAL OF IMPLANTATION DAMAGE

The use of ion implantation for controlled introduction of dopants is now an essential feature of all integrated circuit fabrication, and will remain so because of its accuracy, consistency, high throughput and transferability from one factory to another. The need to remove the concomitant lattice damage will also be a feature of present and future I.C. technologies. The heat treatment required to anneal the damage depends very much on the dopant and its dose. In particular, damage in the form of an amorphous layer (high doses of heavy elements) is much easier to fully anneal than finely divided damage in a still essentially crystalline silicon matrix (low doses of light elements). In present production processes, the necessary anneals are all adequately given by the heat-treatments required to distribute the dopants to the necessary depths. In processes now at development stage, described in the previous section, the device dimensions are sufficiently smaller that the heat treatments required by dopant redistribution are not sufficient to anneal all the implants in the structure. Engineers have found ingenious solutions to this problem (e.g. control of defect growth in partially annealed boron base implants of bipolar transistors, anneal of all low dose implants in MOS structures before the critical source/drain redistribution) but in future smaller structures a radically new solution will be needed.

The basic problem is illustrated in Fig.1, in which is compared the times required to anneal two different types of implant with the time to redistribute a high concentration arsenic implant over various distances. For structures which allow 1 micron redistribution (typical production process) anneal of all implants is achieved for all heat-treatments above 950°C and furnace processing is quite satisfactory. However, where distribution must be confined to 0.1 micron (typical development process), the situation is quite different. Fig.1 shows that only at times less than 100 secs and temperatures over 1000°C can full anneal and the required control of redistribution be achieved. This time range is the domain of the isothermal transient annealing techniques, which have

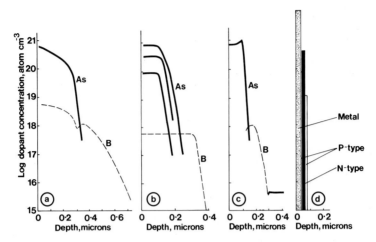

Fig.9 Present and future bipolar transistor structures. Vertical dopant concentration profiles through (a) present production transistor (b) laser-processed structure (c) SIPOS polysilicon emitter structure (d) hot-electron transistor.

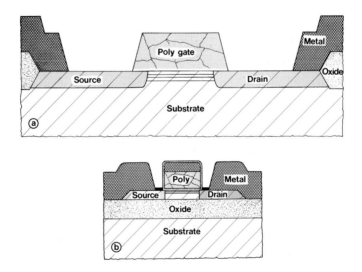

Fig.10 Present and future MOS transistor structures. Vertical section through (a) typical transistor fabricated on 2½ micron process (b) potential transistor fabricated on a 1 micron silicon-on-insulator process.

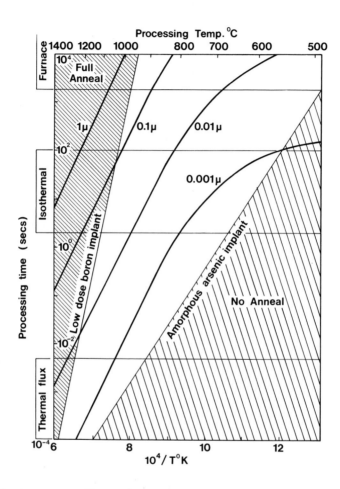

Fig.11 Time-temperature diagram for anneal and redistribution of dopants by solid-state mechanisms. The straight lines represent the minimum time-temperature schedules for full anneal of (a) a 40KeV low dose boron implant and (b) a 40KeV amorphising arsenic implant based on published experimental values (26,27,28). The family of curves gives the loci of time-temperature schedules for redistribution $(3\sqrt{Dt})$ of a 10^{16}ions/cm 40KeV arsenic implant by 1,0.1,0.01 and 0.001 micron.

been shown to be capable of fabricating MOS {21} and bipolar {22} transistors
of at least equal quality to those conventionally fabricated. The criteria for
application are all satisfactorily met, and a wide range of beam sources have
been shown to be satisfactory and equivalent (electron, halogen lamp, arc lamp,
hot carbon strip), and many are described in this volume. Isothermal annealing
is thus assured of a place in future I.C. technology and once adopted may also
find applications in forming good low resistance metal-silicon or metal-metal
contacts with minimum interpenetration.

In the far future, it might be imagined that device dimensions might reduce
to the point where 0.01 micron redistribution only could be permitted: Fig.1
indicates that anneal times of less than 10 millisecs and temperatures of over
1200°C will be required for full anneal. Successful annealing has been carried
out in this time range by thermal flux techniques, notably by a scanned focussed
beam from a CW argon-ion laser {23}. There are problems with laser-induced damage and low throughput but despite this MOS devices and circuits have been
successfully fabricated {24}. However, there are other technology problems to
be solved before such small structures can be reproducibly fabricated, since
the distribution of the original implant can vary by ±0.05 microns due to statistical channelling and variations in mask edge profile.

CONCLUSIONS

The needs of future integrated circuit technology for a process-compatible
technique for full anneal of implants with minimal redistribution of dopants
assures isothermal processing of a place in that technology. The need for a
new silicon-on-insulator starting material for MOS devices establishes zone-
melting recrystallization as one of three competing technologies. Adiabatic
processing is not likely to meet any of the needs of future I.C. technology
with the possible exception of laser-induced damage for gettering {25}.

ACKNOWLEDGEMENTS

I thank my colleague, Alan Butler, for permission to use some of our un-
published material and Plessey Research (Caswell) for permission to publish.
This work has been carried out with the support of the Procurement Executive
MOD, sponsored by D.C.V.D.

REFERENCES

1. Laser-Solid Interactions and Laser-Processing (S D Ferris, H J Leamy,
    J M Poate eds) A.I.P. New York (1979)

2. Laser and Electron-Beam Processing of Materials (C W White, P S Peercy eds)
    Academic Press New York (1980)

3. MRS Symposia Proceedings Volumes 1 (1981), 4 (1982), _, (1983) North-
    Holland New York

4. Semiconductor Characterisation Techniques (P A Barnes, G A Rosgonyi eds)
    466-526 E.C.S. Princeton N.J. (1978)

5. Laser and Electron Beam Processing of Electronic Materials (C L Anderson,
    G K Celler, G A Rosgonyi eds) ECS Pennington N.J. (1980)

6. A R Kirkpatrick, J A Minucci, A C Greenwald IEEE Trans. Elec. Devices
    ED-24 429 (1977)

7. R A McMahon and H Ahmed Electronics Letts. $\underline{15}$ No.2 45 (1979)

8. R T Hodgson, J E E Baglin, R Pal, J M Neri, D Hammer Appl. Phys. Letts. $\underline{37}$ 187 (1980)

9. R L Cohen, J S Williams, L C Feldman, K W Wost Appl. Phys. Letts. $\underline{33}$ 751 (1978)

10. K Nishiyama, M Arai, N Watanade Jap. J. Appl. Phys. $\underline{19}$ L563 (1980)

11. R T Fulks, C J Russo, P R Hanley, T I Kamins Appl. Phys. Lett. $\underline{39}$ 609 (1981)

12. C Hill ref. 3 $\underline{1}$ 361 (1981)

13. C Hill in Laser Annealing of Semiconductors (J M Poate, James W Mayer eds) 479-558 Academic Press New York (1982)

14. C Hill, A L Butler, J A Daly in ref. 3 $\underline{4}$ 579-584 (1982)

15. J L Benton, C J Doherty, S D Ferris, L C Kimerling, H J Leamy, G K Celler, Ref. 2 430 (1980)

16. M Miyao, M Koyanagi, H Tamura, N Hashimoto, T Tokuyama, Jap. J. Appl. Phys. $\underline{19}$-1 129 (1980)

17. C Hill in ref. 5 26 (1980)

18. M W Geis, H I Smith, B Y Tsaur, J C C Fan, D J Silversmith, R W Mountain, R L Chapman this volume (1983)

19. J Graul, A Glasl, H Murrman, $\underline{450}$ IEEE J. Sol. State Circuits $\underline{SC-11}$ 491 (1976)

20. J M Shannon In Nuclear Instruments and Methods Section 5 North Holland New York (1980)

21. R A McMahon, H Ahmed, J D Speight, R M Dobson, ref.5 130 (1980)

22. J D Speight, A E Glaccum, D Machin, R A McMahon, H Ahmed, ref. 3 $\underline{1}$ 383 (1981)

23. A Gat and J F Gibbons Appl. Phys. Letts. $\underline{32}$ 142 (1978)

24. L D Hess, S A Kokorowski, G L Olson, Y M Chi, A Gupta, J B Valdez, Ref.3 $\underline{4}$ 633 (1982)

25. G E J Eggermont, D F Allison, S A Gee, K N Ritz, R J Falster, J F Gibbons, Ref.3 $\underline{4}$ 615 (1982)

26. A Gat, J F Gibbons, T J Magee, J Peng, P Williams, V Deline, C A Evans Jr. Appl. Phys. Lett. $\underline{33}$ 389 (1978)

27. C J Pollard, A E Glaccum, J D Speight Ref.3 $\underline{4}$ 789 (1982)

28. J L Benton, G K Celler, D C Jacobson, L C Kimerling, D J Lischner, G L Miller, McD Robinson Ref.3 $\underline{4}$ 765 (1982)

APPLICATIONS OF A CONTINUOUS WAVE INCOHERENT LIGHT SOURCE (CWILS) TO SEMI-
CONDUCTOR PROCESSING

H.B. HARRISON, S.T. JOHNSON, B. CORNISH, F.M. ADAMS, K.T. SHORT and
J.S. WILLIAMS.
J.M.R.C. Faculty of Engineering, Royal Melbourne Institute of Technology,
Melbourne, 3000, Australia

ABSTRACT

We present results which highlight applications
of a continuous wave incoherent light source in the
processing of semiconductor devices. In particular,
damage removal and activation of ion implanted gallium
arsenide is demonstrated for both capless and capped
annealing of low dose implants at temperatures of
> 800°C for times < 10 s. For gallium arsenide FET
applications, we demonstrate that it is possible to
simultaneously carry out activation, contacting
and interconnection steps by utilizing thermomigration
processes which are not available with conventional
furnace processing. In silicon we demonstrate that
shallow multi layer bipolar structures can be
successfully fabricated with anneal cycles that lead
to supersaturation effects and negligible diffusion.

INTRODUCTION

Rapid bulk heating of semiconductor structures using a continuous wave
incoherent light source (CWILS) offers an attractive alternative to both
conventional furnace annealing and other more rapid and local forms of transient
processing using pulsed lasers, CW lasers and electron beams. For example,
the technique potentially offers high throughput for processing of wafers,
compatability with other device processing steps and controllable alloying of
metals to semiconductors for contacting applications [1]. For the annealing of
ion implanted layers in semiconductors, we have already demonstrated that it is
a decided advantage to be able to achieve dopant activation at high temperatures
(typically 800-1100°C) in a time regime (≃ 3-10s) that can prohibit dopant
redistribution and often provide supersaturated solid solutions [1].

In this paper, we further explore the applications of CWILS in semiconductor
device processing. In particular, we concentrate our attention on i) the
thermal processing of ion implanted GaAs and Si, and ii) intriguing
thermomigration processes in GaAs which result from a finite temperature
gradient across the wafer.

ANNEALING DETAILS

Details of the CWILS apparatus employed in the present study have been
described in detail elsewhere [1]. Basically, radiant light from a 500 W
quartz-iodide lamp is focussed to a 2 cm diameter spot onto the front surface
of a semiconductor sample by means of an aluminium bowl reflector. The
apparatus is mounted in a bell jar to permit annealing in vacuum or under a
desired gaseous environment.

Mat. Res. Soc. Symp. Proc. Vol. 13 (1983) ©Elsevier Science Publishing Co., Inc.

Our particular CWILS annealing situation is depicted in Figure 1 and operates in an isothermal mode. Radiant light is absorbed by the front surface of the sample and emitted from the back. The sample is thermally isolated from its surrounds and thus attains thermal equilibrium rapidly, in times of the order of 1-10 s depending on the incident power and the thermal mass of the sample [1-4]. In this arrangement heat loss is by radiation, in contrast with the alternate heat sunk mode [5] where heat is lost mainly by conduction.

For the isothermal mode depicted in Fig. 1, a small but finite temperature gradient ΔT is established between the front and back surfaces. This temperature gradient provides the driving force for thermomigration of selected materials deposited on the backside of the semiconductor sample [6].

When our particular annealing apparatus is operating at full power, the substrate can attain equilibrium temperatures in excess of $1000^{\circ}C$, depending of substrate type and size. Normally, the lamp is switched off before equilibrium is reached: typical calculated thermal cycles (using the Stefan-Boltzman equation [7]) are shown in Figure 2. As described previously [1], the average sample temperature has been measured by a variety of methods (e.g. thermocouple, optical pyrometer, and observation of melting of metallic eutectics) and the agreement with calculated temperatures is reasonable.

CWILS APPLIED TO GaAS

CWILS annealing has been carried out on variable dose implants of S, Se, Sn and Te into semi-insulating (100) GaAs at energies of 80-120 keV. Capless anneals have been undertaken at temperatures in excess of $800^{\circ}C$ for varying times Δt (as illustrated by the thermal cycles in Figure 2). Typical results are given in Table 1 for 80 keV S^{+} implants to a dose of 3×10^{13} ions cm^{-2}. The electrical measurements indicate only partial activation and low mobilities for annealing times of less than 3s. Rutherford backscattering and channeling measurements typically showed that the implantation damage was removed for such annealing times, with some dissociation of the GaAs surface [8]. However, annealing for times exceeding about 3s produced highest activity (~50%) and mobilities $>$ 300 cm^2/Vs despite the fact that the surface was badly dissociated, as measured by RBS and channeling methods [8]. Times longer than about 8 s produced a marked deterioration in electrical properties, consistent with gross surface dissociation.

Table 1   80 keV $S^{+} \rightarrow$ SI GaAs (3×10^{13} ions cm^{-2})

$\Delta t (> 800^{\circ}C)$	R ($\Omega/\square$)	$\mu (cm^2/Vs)$	Surface Dissociation
1 s.	200 kΩ	–	Slight dissociation
2 s.	50 kΩ	~ 200	Ga+As loss of > 10^{15} atoms cm^{-2}
3-8 s.	~1 kΩ	$>$ 300	Ga+As loss of > 10^{16} atoms cm^{-2}

In a further series of experiments, Sn-implanted samples were capped with SiO_2 and annealed in the temperature range $800-1000^{\circ}C$ for times up to 100 s. In general, electrical properties were superior to uncapped samples as has been previously shown from other studies [3,4,9]. Furthermore, RBS/channeling spectra shown in Figure 3 illustrate complete damage removal and a stoichiometric surface composition (following removal of the SiO_2 cap) for 8 s. annealing at $800^{\circ}C$ of a 1×10^{14} Sn implant. In addition, preliminary high dose studies, to be reported fully elsewhere, indicate that such transient

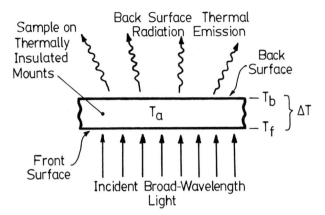

Fig. 1.   Isothermal process configuration for CWILS annealing.

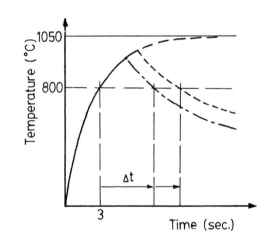

Fig. 2.   Typical thermal cycles for CWILS annealing of GaAs. Δt refers to the time the sample is above 800°C; and 1050°C is the typical full-power equilibrium temperature.

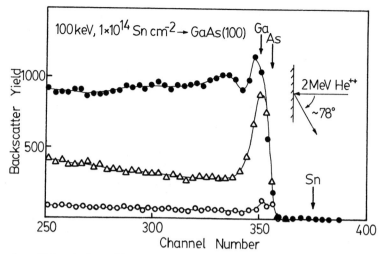

Fig. 3.   RBS and channeling spectra of an SiO$_2$ capped, Sn implant, CWILS annealed above 800°C for $\Delta t = 8s$; random (●), as-implanted <100> (Δ), and annealed <100> (O).

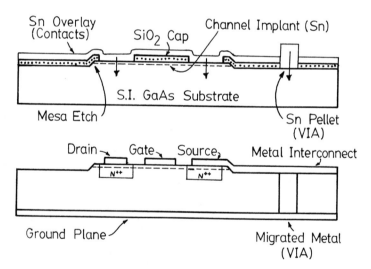

Fig. 4.   Schematic of a GaAs I.C. structure showing simultaneous thermomigration and implant steps annealed by CWILS (above). The final structure is shown below.

annealing methods can lead to the formation of supersaturated solid solutions
of Te in GaAs. This suggests that optimised transient annealing may result in
higher electrical activity than can be obtained by conventional furnace
annealing.

Transient annealing of capped GaAs can be combined with simultaneous
processing of contacts and interconnections, as illustrated schematically in
Figure 4. We highlight three important technological problems relating to
GaAs integrated circuit fabrication. These are, production of high quality
low carrier concentration active layers in S.I. substrates, low ohmicity
contacts to these layers, and interconnects from front surface components to
back surface ground planes. The isothermal CWILS offers potentially
attractive solutions to each of these problems. As we have mentioned above, the
technique can be used to activate low dose implants with the addition of an
SiO_2 cap. During this annealing procedure, with appropriate cuts in the SiO_2
and a limited-source overlay of Sn, thermomigration of Sn into the implanted
layer can be achieved to provide a heavily-doped degenerate layer. This is
illustrated in Figure 5 where we show RBS/channeling spectra which clearly
indicate Sn migration and high solubility (> $10^{20} cm^{-3}$) following CWILS
annealing. In this case, the Sn migrates about 1 μm into the GaAs to provide
a concentration up to about 4 at % over this depth. The experimental details
needed to achieve this migration and the mechanisms involved will be the subject
of a subsequent report. Such layers have provided an excellent base for metal
deposition and consequent formation of low ohmicity, low noise contacts. For
example, direct Al metal deposition, without alloying, typically provides planar
contact resistivities < $10^{-5} \Omega cm^2$ using transmission line measurements [10].
This is commensurate with values from conventional Au/Ge alloyed contacts [11].
Our results show that this CWILS thermomigration step offers considerable
promise for simultaneous fabrication of source, drain and gate contacts in FET
structures. We have also demonstrated that highly conductive front-to-back-
surface interconnects (vias), as depicted in Figure 5, can be formed by more
conventional temperature gradient zone migration [6] of Sn from the back to the
front surface using CWILS annealing. The complete characterisation of these
CWILS annealing processes is the subject of continued investigation.

CWILS APPLIED TO SILICON

As has been previously demonstrated, CWILS techniques can be successfully
employed to anneal ion implanted silicon [1,12]. CWILS techniques also have
application in other areas of silicon processing, for example, high temperature
oxidation, silicide formation, annealing of polycrystalline silicon, and for
controllable alloying of metals. In this study, we have further investigated
CWILS annealing of ion implanted structures, in particular, annealing of multilayer implanted layers.

For this study various doses of In in the range 5 x 10^{12} to 2 x 10^{14} cm^{-2}
have been implanted into <100> n-type silicon at 100 keV. These samples were
CWILS annealed for up to 10 s. at temperatures exceeding $900^{\circ}C$. Typical
behaviour is illustrated in Figure 6. A minimum is observed in the sheet
resistivity at a dose of about 5 x $10^{13} cm^{-2}$. RBS/channeling measurements
(see lower insert) indicate that the loss of activation at higher doses may be
interpreted in terms of precipitation of In, since the peak concentration,
even at 5 x $10^{13} cm^{-2}$, is more than an order of magnitude above the solubility
limit [13] of In in Si ($\approx$9 x $10^{17} cm^{-3}$). It is interesting to note that no
redistribution of the In is obtained during CWILS processing at > $900^{\circ}C$. This
is in contrast to conventional furnace processing where considerable redistribution of In is observed for implants into both <100> and <111> Si.
Previous furnace data [14] for In in <111> Si are also shown in Figure 6,
together with an RBS/channeling insert (upper portion) which illustrates the

398

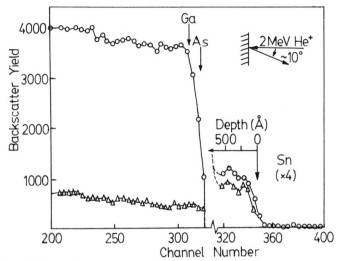

Fig. 5. RBS/channeling spectra showing Sn migration in GaAs; random (O) and
<100> aligned (Δ) spectra.

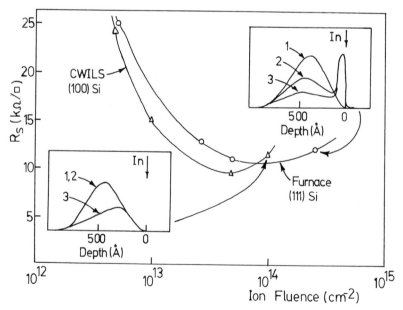

Fig. 6. Comparison of CWILS and furnace annealed In implanted Si. Inserts
show typical random as implanted (1) and annealed (2) In profiles,with <100>
aligned profiles (3).

typical redistribution of In in such anneals.

The lack of dopant redistribution during rapid CWILS processing has considerable implications for dual implants to enable fabrication of shallow p-n junction devices. For example, we have carried out dual implants of In (100 keV, as above) and Sb (60 keV, shallow angle) into <100> silicon and annealed the composite structures by CWILS techniques to successfully fabricate shallow n-p-n (bipolar) structures under appropriate conditions. Bipolar action in such shallow structures is only achieved (for processing at > 900°C) if both Sb and In profiles are highly active (In above equilibrium solubility limit) and no dopant redistribution occurs during annealing. These conditions are indeed satisfied by CWILS annealing.

CONCLUSION

We have demonstrated that isothermal CWILS annealing methods have considerable advantage in the processing of silicon and gallium arsenide structures. In particular, it is possible to induce thermomigration processes and to simultaneously carry out activation, contact and interconnect annealing steps which cannot be achieved by conventional furnace processing.

ACKNOWLEDGEMENTS

We wish to acknowledge financial support under the Commonwealth Special Research Centres Scheme. This project was supported in part by the ARGC, the RRB and AINSE.

REFERENCES

1. H. B. Harrison, M. Grigg, K. T. Short, J. S. Williams and Z. Zylewicz, in "Laser and Electron Beam Processing of Materials", (B. R. Appleton and G. K. Cellar, eds.). Elsevier-North Holland, New York (1982), p. 771.

2. K. Nishiyama, M. Arai, N. Watanabe, Jap. J. Appl. Physics, 19, 10 (1980).

3. N. J. Shah, H. Ahmed and P. A. Leigh, Appl. Phys. Lett. 39, 322 (1981).

4. D. E. Davies, P. J. McNally, J. P. Lorenzo and M. Julian, Electron. Device Lett. EDL3, 102 (1982).

5. A. Lietoila, R.B. Gold, J. F. Gibbons, J. Appl. Phys. 53, 2 (1982).

6. H. E. Cline, T.R. Anthony, J. Appl. Phys. 49, 4 (1978).

7. P. D. Parry, J. Vac. Sci. Tech. 13, No. 2, 1976.

8. A. Rose, J. T. A. Pollock, M. D. Scott, F. M. Adams, J. S. Williams, and E. M. Lawson, these proceedings.

9. B. J. Sealy, Microelectronics J. 13, 21 (1982).

10. G. K. Reeves, H. B. Harrison, IEEE Electron Device Letters, May 1982.

11. A. Christou, Solid-State Electronics Vol. 23, 1979.

12. J. L. Benton, G. K. Cellar, D. C. Jacobson, L. C. Kimerling, D. J. Lischner, G. L. Miller and Mc. D. Robinson, in "Laser and Electron Beam Interactions with Solids", (Eds. B. R. Appleton and G. K. Cellar). (Elsevier-North Holland, New York, 1982) p. 765.

13. F. Trumbore, Bell Syst. Tech J. 39, 206 (1960).

14. R. G. Elliman, H. B. Harrison, D. G. Beanland, Nuclear Instruments and Methods 182/183 (1981).

EFFECT OF PULSE DURATION ON THE ANNEALING OF ION IMPLANTED SILICON WITH A XeCl
EXCIMER LASER AND SOLAR CELLS*

R. T. YOUNG
Helionetics, Inc., San Diego, CA 92123, and Oak Ridge National Laboratory,**
Oak Ridge, TN 37830
J. NARAYAN AND W. H. CHRISTIE[+]
Solid State Division, Oak Ridge National Laboratory, Oak Ridge, TN 37830
G. A. van der LEEDEN,[++] D. E. ROTHE, AND R. L. SANDSTROM
Helionetics, Inc., San Diego, CA 92123

ABSTRACT

The advantages of pulsed excimer lasers for semiconductor
processing are reviewed. Studies of XeCl excimer laser
annealing with pulses of 25 and 70 nsec duration and energy
densities in the range from 0.5-3.0 J/cm^2 are discussed. The
annealing characteristics are described in terms of the
results of melt depth, dopant profile spreading, and electrical properties (sheet resistivity, diode characteristics)
measurements. Solar cells with efficiencies as high as 16.7%
AM1 have been fabricated using glow discharge implantation
and XeCl laser annealing.

INTRODUCTION

The advantages of using pulsed excimer lasers for semiconductor processing
have been demonstrated recently. [1-3] It has been shown that, regardless of
the large differences in the optical properties of Si at the wavelengths of
UV excimer lasers (e.g., 308 nm for XeCl lasers) and ruby (694 nm) or frequency
doubled Nd:YAG (530 nm) lasers, [4] the quality of annealing, assessed in terms
of damage removal in the implanted layer, dopant profile behavior, and junction
properties, is very similar. However, because of the good optical quality of
the beam and perhaps because of the trapezoidal (rather than Gaussian) temporal
shape of the excimer laser pulses, excellent annealing can be obtained without
the use of beam homogenizers and substrate heating [3]. This provides important
simplifications for application of laser annealing to semiconductor processing.
Furthermore, other features of excimer lasers, such as good energy conversion
efficiency, excellent pulse-to-pulse reproducibility, square or rectangular
shapes of the laser beam, and high average power [5] may make laser techniques
commercially viable for large volume material processing.
    Advances in excimer laser technology indicate that, in addition to the capa-
bility for scaling up both the energy per pulse and the pulse repetition rate,
the pulse duration time τ_ℓ can be easily adjusted over a range from 10 to several hundred nsec. Variation of τ_ℓ over this range with solid state lasers is
very difficult, if not impossible. Melting model calculations [6] show that,

---

* Research sponsored jointly by the Solar Energy Research Institute under
  contract BS-0-9078-1 and by Helionetics, Inc.
**Operated by Union Carbide Corporation under contract W-7405-eng-26 with the
  U.S. Department of Energy.
[+]Analytical Chemistry Division
[++]Present address: Guest scientist, Solid State Division, Oak Ridge National
  Laboratory, Oak Ridge, TN 37830

for the same energy density E_ℓ, the depth of melting increases with decreasing pulse duration; there is also a dependence of the regrowth velocity on pulse duration, especially for energy densities E_ℓ near the threshold for surface melting. The influence of the regrowth velocity on the quality of laser regrown layers and the optimum τ_ℓ for energy-efficient semiconductor processing have not yet been investigated in detail.

In the work reported here, we studied the effect of τ_ℓ on the annealing of ion implanted Si by comparing the melting depth, crystal perfection, dopant profiles, and electrical properties of samples annealed with a XeCl laser with E_ℓ in the range of 0.5-3.0 J/cm^2 and for pulse durations of 25 and 70 nsec. We will also report on our recent progress in the development of excimer laser processed silicon solar cells. Probably because of the excellent beam quality and pulse-to-pulse stability, we have been able to fabricate shallow junction solar cells with AM1 efficiencies as high as 16.7%.

EXPERIMENT

An x-ray preionized, discharge-pumped XeCl laser designed and constructed by Helionetics, Inc., was used in this work. The laser was designed so that τ_ℓ can be varied between 20-90 nsec simply by changing the ratio of the gas mixtures. Polished (100) Si wafers were implanted with $^{11}B^+$ at 100 kV and at multiple energies of 25, 50, 100, and 200 kV to create a uniformly damaged layer up to ~6800 Å thick. Samples were annealed by single laser pulses with E_ℓ in the range of 0.5-3.0 J/cm^2 and τ_ℓ of 25 and 70 nsec. The depth of annealing or melting and crystal perfection of the laser regrown layers were determined by transmission electron microscopy (TEM). The dopant redistribution after laser annealing was examined by secondary ion mass spectroscopy (SIMS). The electrical properties of the laser regrown layer were obtained from van der Pauw measurements and I-V characteristics. Solar cells were fabricated from shallow (1 kV) BF_3 gas discharge implanted (100) Si wafers annealed with 55 nsec pulses at E_ℓ = 1.4 J/cm^2.

RESULTS

Effect of τ_ℓ on the annealing of B implanted layers

Melt depth. The depth of melting as a function of laser energy density for 25 and 70 nsec laser pulses was investigated by TEM. Implantation of boron at multiple energies up to 200 kV produced dislocation loops distributed throughout a 6800 Å thick layer. The melting threshold for 25 and 70 nsec laser pulses was found to be 0.5 and 1.0 J/cm^2, respectively. Figure 1 shows cross-section micrographs of samples laser irradiated with E_ℓ = 1.0 and 1.5 J/cm^2, τ_ℓ = 70 nsec and E_ℓ = 1.0 and 2.0 J/cm^2, τ_ℓ = 25 nsec pulses. The complete removal of dislocation loops in the annealed regions and the sharp transition of the loop number densities between the annealed and unannealed regions are consistent with the melting of the annealed regions. Table I summarizes the melting depth data for 70 and 25 nsec pulses. It is interesting to note that considerably deeper melting is achieved with 25 nsec pulses than with 70 nsec pulses, which is in good qualitative agreement with calculations [6]. A complete annealing of displacement damage or dislocation loops is achieved with 2.5 J/cm^2, 25 nsec pulses, but greater than 3.0 J/cm^2 is required with 70 nsec pulses for complete annealing of the 6800 Å thick damaged layer.

Dopant profiles. The effect of τ_ℓ on dopant profile redistribution of 100 kV boron-implanted Si was studied by SIMS. Figure 2 shows dopant profiles typical of ion-implanted, pulsed laser annealed samples. It is interesting to see that an abrupt dopant profile (▲ curve) was obtained on the sample that was annealed just above the threshold for complete annealing (2.5 J/cm^2, 70 nsec) as determined by electrical measurements, and confirmed by TEM. Similar

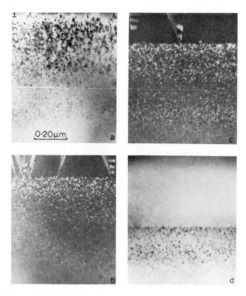

Fig. 1. Cross-section TEM micrographs after laser irradiation with
a) E_ℓ = 1.0 J/cm^2 70 nsec
b) E_ℓ = 1.5 J/cm^2 70 nsec
c) E_ℓ = 1.0 J/cm^2 25 nsec
d) E_ℓ = 2.0 J/cm^2 25 nsec

Table I.
A comparison of melt depth (Å) as a function of laser energy density for 25 and 70 nsec laser pulses.

Pulse duration (nsec) \ Energy density (J/cm^2)	1.0	1.5	2.0	2.5	3.0
25	1700	2900	4300	6800	---
70	----	1350	2850	4580	6160

results were also observed in arsenic implanted Si [1]. This phenomenon may represent the characteristic of the annealing with a trapezoidal shape laser pulse. At the same energy densities, 25 nsec pulses consistently provide greater dopant redistribution than do 70 nsec pulses. A more abrupt junction to a depth of ~7000 Å was obtained without surface damage when multiple (up to five) 25 nsec pulses of 3.0 J/cm^2 were used. This result indicates that excimer laser annealing should be suitable for deep junction device fabrication.

Electrical properties. A comparison of the recovery of the sheet resistivity from van der Pauw measurements) of multiply B-implanted Si samples annealed with E_ℓ in the range from 1.0-3.0 J/cm^2 for the two different laser pulse durations is shown in Fig. 3. It is clearly seen from the figure that 25 nsec pulses provide lower sheet resistivity than do 70 nsec pulses at the energy densities below the threshold for complete annealing of the implanted layer. We should point out that the van der Pauw technique measures the electrical properties of the laser regrown layer only, since any underlying heavily damaged region that remains at a given E_ℓ can be considered as an insulator. Lower sheet resistivities correlate with thicker laser regrown layers, as expected. These results are in good agreement with TEM observations. Diode

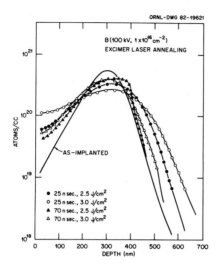

ORNL-DWG 82-19621

B (100 kV, 1 x 10^{16} cm^{-2})
EXCIMER LASER ANNEALING

AS-IMPLANTED

ATOMS/CC

DEPTH (nm)

- ● 25 n sec., 2.5 J/cm^2
- ○ 25 n sec., 3.0 J/cm^2
- ▲ 70 n sec., 2.5 J/cm^2
- △ 70 n sec., 3.0 J/cm^2

Fig. 2. A comparison of concentration profiles of B in Si after laser annealing at 2.5 and 3.0 J/cm^2 with 25 and 70 nsec pulses.

Fig. 3. The recovery of sheet resistivity as a function of laser energy density for the pulses of 25 and 70 nsec.

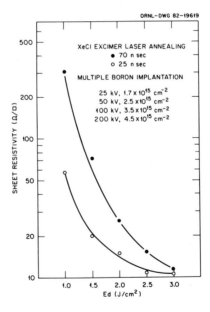

ORNL-DWG 82-19619

SHEET RESISTIVITY (Ω/□)

XeCl EXCIMER LASER ANNEALING
- ● 70 n sec
- ○ 25 n sec

MULTIPLE BORON IMPLANTATION

25 kV, 1.7 x 10^{15} cm^{-2}
50 kV, 2.5 x 10^{15} cm^{-2}
100 kV, 3.5 x 10^{15} cm^{-2}
200 kV, 4.5 x 10^{15} cm^{-2}

Ed (J/cm^2)

junction characteristics were measured on mesa diodes annealed above the
annealing thresholds for the pulses of 25 and 70 nsec. In both cases, the
forward dark I-V characteristics could be fit to the diode equation
$I = I_0[\exp(qV/AKT)-1]$ with A in the range of 1.20-1.40 for applied bias
voltages of 0.35 to 0.55 V, which indicates that satisfactory junctions can be
formed with both pulse durations.
  The results from TEM, sheet resistivity, and diode characteristics are not
sensitive enough to reveal the presence of residual point defects in the laser
regrown layers. Whether or not the short duration laser pulses quenched in
more point defects in the laser regrown layer than do the longer pulses is
still not clear. Deep Level Transient Spectroscopy studies have been started
on samples implanted with Si (40 kV, 5×10^{15} cm^{-2}) and annealed with laser
pulses at the two different τ_ℓ. Preliminary results indicate that DLTS signals
are seen in both cases, but quantitative analysis of the data is not available
at the present time.

Excimer laser processed Si solar cells
  We have stated previously that XeCl excimer lasers have many of the characteristics needed for laser processing of semiconductor devices. One of the
examples that clearly demonstrates this is the processing of single crystal Si
solar cells, which we now discuss.
  Polished 1-3 Ω-cm n-type FZ Si was used in this study. Low energy (1 kV) dc
glow discharge BF_3 implantation was used to form shallow p^+n junction. After
implantation, each wafer was cut into six 1x2 cm samples. Half of them were
annealed with a ruby laser and the other half with the excimer laser. The back
surfaces were made degenerate by laser-induced diffusion of Sb [7,8]. Under
optimum annealing conditions, all of the cells had open circuit voltages V_{oc}
(measured at 28°C) in the range 595-605 mV and fill factors FF in the range
0.77-0.80. However, there were noticeable differences in short circuit
currents J_{sc}. In general, the excimer laser annealed cells showed better J_{sc}
than did the ruby laser annealed cells. This difference can be understood from
internal quantum efficiency measurements on the cells before the application of
antireflection (AR) coatings. The results shown in Fig. 4 demonstrate that the
excimer laser annealed cells have superior response in the blue region of the
solar spectrum. The best discharge-implanted, excimer laser annealed cell
obtained to date, had application of a double layer AR coating (Ta_2O_5 and
MgF_2), had V_{oc} = 610 mV, J_{sc} = 34.7 ma/cm^2, FF = .79 and η = 16.7%. This efficiency is comparable to the highest efficiency Si solar cells made by the most
sophisticated conventional methods.

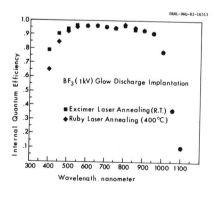

Fig. 4. Comparison of internal
quantum efficiency between
cells fabricated by excimer
laser and ruby laser annealing.

SUMMARY

We have demonstrated experimentally that 25 nsec pulses are more energy efficient in annealing ion implanted damage than are 70 nsec pulses. The quality of annealing, as measured by the crystalline perfection of the regrown layer and by p-n junction characteristics, is very similar for the two pulse durations. However, quantitative information about quenched-in point defects in the layers regrown with the laser pulses of different τ_ℓ is not yet complete, and research in this area is continuing. We also demonstrated in this work that high efficiency solar cells, with enhanced blue response, can be made with glow discharge implantation and excimer laser annealing. We anticipate that this simplified beam processing, after further development, will have significant advantages for the fabrication of high efficiency terrestrial solar cells, as well as for thin space cells.

ACKNOWLEDGMENT

The authors would like to thank R. F. Wood for his careful reading of the manuscript and helpful suggestions.

REFERENCES

1. R. T. Young, G. A. van der Leeden, J. Narayan, W. H. Christie, R. F. Wood, D. E. Rothe, and J. I. Levatter, IEEE Elec. Dev. Lett. EDL-3, 280 (1982).

2. D. H. Lowndes, J. W. Cleland, W. H. Christie, R. E. Eby, G. E. Jellison, Jr., J. Narayan, R. D. Westbrook, R. F. Wood, J. A. Nilson, and S. C. Dass, Appl. Phys. Lett. 41, 938 (1982).

3. R. T. Young, G. A. van der Leeden, R. F. Wood, and R. D. Westbrook, Proceedings of the 16th IEEE Photovoltaic Specialists Conference (September 27-30, 1982, San Diego, CA) (to be published).

4. G. E. Jellison, Jr., and F. A. Modine, Appl. Phys. Lett. 41, 180 (1982).

5. R. T. Young, G. A. van der Leeden, R. D. Westbrook, R. L. Sandstrom, and J. I. Levatter, Proceedings of the 16th IEEE Photovoltaic Specialists Conference (September 27-30, 1982, San Diego, CA) (to be published).

6. R. F. Wood and G. E. Giles, Phys. Rev. B 23, 2923 (1981).

7. J. Narayan, R. T. Young, R. F. Wood, and W. H. Christie, Appl. Phys. Lett. 33, 338 (1978).

8. R. T. Young, R. F. Wood, and W. H. Christie, J. Appl. Phys. 53, 1178 (1982).

PULSED EXCIMER LASER (308 nm) ANNEALING OF ION IMPLANTED
SILICON AND SOLAR CELL FABRICATION*

D. H. LOWNDES, J. W. CLELAND, W. H. CHRISTIE, R. E. EBY, G. E. JELLISON, JR.,
J. NARAYAN, R. D. WESTBROOK, AND R. F. WOOD
Solid State Division, Oak Ridge National Laboratory, Oak Ridge, TN 37830, and
J. A. NILSON AND S. C. DASS
Lumonics, Inc., Kanata (Ottawa), Ontario, K2K 1Y3, Canada

ABSTRACT

   A pulsed ultraviolet excimer laser (XeCl, 308 nm wavelength, 40
nsec FWHM pulse duration) has been successfully used for laser
annealing of both boron- and arsenic-implanted silicon. TEM,
SIMS, and sheet electrical measurements are used to characterize
specimens. C-V and I-V measurements demonstrate that near-ideal
p-n junctions are formed (diode perfection factor A = 1.2).
Electrical activation of implanted ions by single laser pulses is
essentially complete for energy densities $E_\ell \geq 1.4$ J/cm^2, far
below the threshold for substantial surface damage ~4.5 J/cm^2.
Melting model calculations are in good agreement with observed
thresholds for dopant redistribution and for epitaxial regrowth.
Changes in annealing behavior resulting from multiple (1,2,5)
laser pulses are also reported. Finally, we demonstrate the use
of scanned overlapping excimer laser pulses for fabrication of
large area (2 cm^2) solar cells with good performance characteristics. In contrast to pulsed ruby laser annealing, high open circuit voltages can be obtained without the use of substrate heating.

INTRODUCTION

   Pulsed solid state lasers (both ruby and Nd:YAG) are useful for removal of
ion implantation-induced lattice damage [1] and for p-n junction formation [2]
in silicon. However, commercial application of these techniques has been
limited by several disadvantages inherent to pulsed solid state lasers:
(1) Spatial beam nonuniformity which has required using diffuser plates [3] or
diffusing light pipes [4] to uniformly anneal even ~1 cm^2 areas; (2) low repetition rate (~1 pulse per min) for high energy ($\gtrsim$5 J) pulses; and, (3) high
spatial coherence, which causes both troublesome interference patterns
("speckle") and actual damage on a microscopic scale due to diffraction
effects. Solid state laser rods and coatings are also susceptible to damage
at high pulse energies, resulting in low duty-cycle operation. This has led to
the general perception that the necessary scaling up of the single-pulse
annealing method to achieve commercially viable areas (~20-80 cm^2) and pulse
energies (~40-160 J.), could become prohibitively expensive and/or difficult.
   In contrast, the recent development of excimer lasers is of great interest
for potential commercial applications of pulsed laser annealing, such as large
area solar cells. Excimer lasers combine a rectangular annealing beam area,
with highly uniform (±5%) pulsed energy density and only moderate spatial
coherence. Thus, both the need for diffuser plates or light pipes and the most
troublesome spatial interference effects are simultaneously eliminated.
Excimer lasers also offer high (currently 1-100 Hz) pulse repetition rates,
thus making feasible annealing large areas in a scanned, repetitively pulsed

*Research sponsored by the Division of Materials Sciences, U.S. Department of
Energy, under contract W-7405-eng-26 with the Union Carbide Corporation.

mode of operation. Some control of pulse shape and duration (useful for controlling melt depth and duration) is also possible. Finally, fundamental materials-related limits do not appear to be a factor in scaling up present day excimer lasers to obtain commercially attractive pulse areas and energies, while maintaining a high duty cycle.

Because previous studies focused almost exclusively on the use of visible wavelengths, important questions exist regarding the use of pulsed ultraviolet (UV) wavelengths for laser processing of silicon. These questions include knowledge of the threshold energy density, E_ℓ, and the optimum E_ℓ for annealing of ion implantation lattice damage; the quality of annealing that is possible; electrical activation of dopant ions and dopant ion redistribution; the threshold E_ℓ for laser damage; the role of laser pulse shape and pulse duration in controlling both melt depth and crystal regrowth velocity; comparison of the electrical quality of the resulting p-n junctions with those obtained using solid state pulsed lasers; and, the effect of multiple or overlapping pulses on the quality of annealing. In this paper we summarize the results of experiments and model calculations covering these areas.

EXCIMER LASER ANNEALING OF ION IMPLANTED SILICON

High dose (2 and 6×10^{15} ions/cm^2) boron and arsenic implantations were carried out at 27°C into (100) Czochralski silicon wafers of opposite conductivity type to that produced by the implanted ions. Because a thin emitter region is desirable for photovoltaic applications, B (As) was implanted at 5 and 20 keV (20 and 110 keV), resulting in projected ranges of only ~16 nm and ~65 nm. The Lumonics model TE-292X XeCl (308 nm) excimer laser used for annealing produced a 1.5 J (maximum) pulse energy, uniform to ±5% over a 3x3 cm area. Variable E_ℓ (to >5 J/cm^2) was obtained by focusing the excimer beam. The shape of the pulse varied slightly, depending on the age of the gas mixture, but with the FWHM pulse duration in the range ~39±4 nsec.

As is shown in Fig. 1(a), a (001) silicon specimen implanted with 110 keV ^{75}As$^+$ to a dose of 2×10^{15}/cm^2 contained a 112 nm thick amorphous layer above a 40 nm thick layer containing dislocation loops and tangles. Complete annealing of displacement damage was accomplished by irradiation with one 1.4 J/cm^2 laser pulse, as is shown by TEM in Fig. 1(b). This type of annealing was also obtained in previous pulsed laser experiments when a melt front penetrated beyond the damaged layer, permitting high velocity epitaxial regrowth from the defect-free c-Si substrate beneath [6]. The regularity and symmetry of the diffraction spots in Fig. 1(c) confirms the "defect-free" assessment.

Figure 2 shows results of SIMS measurements of boron and arsenic dopant ion redistribution following pulsed excimer laser annealing (PELA). A detectable departure from the as-implanted dopant ion profile was first observed at E_ℓ = 1.15 J/cm^2 for boron and at E_ℓ = 0.98 J/cm^2 for arsenic. This slight difference in threshold for the onset of dopant ion redistribution is probably a result of more complete amorphization and a lower resultant melting threshold produced by the arsenic implant.

Model calculations of the depth and duration of pulsed excimer laser melting, as a function of E_ℓ, were carried out using the thermal melting model [7,8]. For the present case, the calculations assumed a laser pulse with 41 nsec FWHM duration, a trapezoidal shape closely approximating that shown in the inset to Fig. 2(a), an absorption coefficient [5] at 308 nm of 10^6 cm^{-1}, a reflectivity at 308 nm varying from 0.60 to 0.70 between 27°C and the 1412°C melting point of silicon, and temperature dependent values for the heat capacity and thermal conductivity of silicon. These model calculations predict a surface melting threshold of 0.9 J/cm^2 for c-Si, and $E_\ell \sim 1.03$ J/cm^2 for melting to a depth of 50 nm in c-Si. The latter result is in good agreement with our observation (Fig. 2) of boron redistribution to a depth of 50 nm at

409

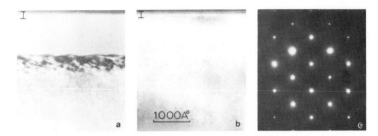

Fig. 1. Annealing of ion implantation damage by a pulsed XeCl laser [308 nm wavelength; pulse shape as shown in Fig. 2(a)]. (a) TEM (110) cross-section micrograph showing the as-implanted amorphous layer and dislocation-type damage beneath. The original surface is denoted by I. (b) Cross-section micrograph after one 1.4 J/cm² laser pulse, showing complete annealing of ion implantation damage. (c) (110) selected area diffraction pattern of the excimer laser-recrystallized silicon shown in (b).

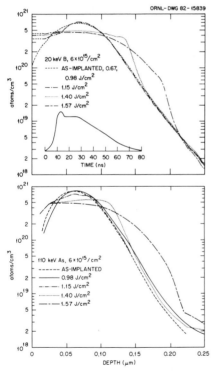

Fig. 2. SIMS measurements of dopant redistribution following single-pulse excimer laser annealing of (top) boron- and (bottom) arsenic-implanted silicon (see text). Inset: Temporal shape of the 308 nm excimer laser pulse.

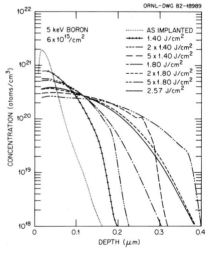

Fig. 3. SIMS dopant redistribution profiles for 5 keV, 6x10¹⁵/cm² boron ion implanted samples, following multiple pulse excimer laser annealing.

E_ℓ = 1.15 J/cm^2, given the slight uncertainty in laser pulse width and the estimated E_ℓ uncertainty of $\pm 10\%$.
Extensive surface damage (easily visible frozen-in ripples and vaporization craters) was found <u>not</u> to occur until $E_\ell \sim 4.5$ J/cm^2, indicating that a wide range of energy densities is available for annealing. Model calculations show that melt depths $\gtrsim 1$ μm can be obtained for E_ℓ = 3.0-3.5 J/cm^2, for 20-60 nsec FWHM pulse durations, with the melt depth greater for the shorter duration pulses. Thus, PELA's combination of inherent beam uniformity and broad range of annealing E_ℓ allows considerable control over the depth of p-n junction formation and of doping profiles.

SHEET ELECTRICAL PROPERTIES AND p-n JUNCTION CHARACTERISTICS

Hall effect and sheet resistivity measurements were used to evaluate sheet electrical properties of the shallow p$^+$ and n$^+$ layers formed by PELA. Electrical activation of As implants by single laser pulses was complete at E_ℓ = 1.4 J/cm^2 and at slightly higher E_ℓ (~1.5 J/cm^2) for B implants. A combination of SIMS and carrier concentration measurements showed that B loss was <5% for $E_\ell \leq$ 1.7 J/cm^2, with As loss <20% over the same E_ℓ range, consistent with the higher As vapor pressure. No significant change ($\pm 10\%$) of sheet carrier concentration was then observed with increasing E_ℓ for $E_\ell \leq$ 2.6 J/cm^2, for either B or As. Sheet resistivities of 24-34 (60-70) $\Omega/\square$ were obtained for the high (low) dose implants for $E_\ell >$ 1.4 J/cm^2. These observations of electrical activation are consistent with our TEM measurements, which show defect-free epitaxial recrystallization for E_ℓ = 1.4 J/cm^2, and with melting model calculations, which predict that E_ℓ = 1.32 J/cm^2 is required to melt crystalline silicon to a depth of 160 nm, the deepest lattice damage in our samples (see Fig. 1).
Mesa diodes containing p-n junctions formed by ion implantation and PELA were evaluated using C-V and I-V measurements. The linearity of plots of C^{-2} vs V demonstrated that the junctions formed are abrupt relative to the zero bias depletion width; I-V measurements on these same mesa diodes revealed perfection factors A = 1.2-1.3 (using I = I_0exp(qV/AkT)). Similar plots of I_{sc} vs V_{oc} for 2 cm^2 solar cells, obtained by varying the illumination level, also yield A=1.2 and confirm the high quality of the large area p-n junction that is formed in excimer laser-recrystallized silicon [12].

MULTIPLE PULSE ANNEALING

The high pulse repetition rate of excimer lasers suggests a scanned, repetitively pulsed mode of operation for potential device applications of PELA. However, an important question for such applications is the effect of multiple annealing pulses, or of overlapping pulses, on dopant ion redistribution since junction depth might be deeper in the overlap region between two pulses than in the adjacent single pulse-annealed regions. Figure 3 summarizes SIMS measurements of the effect of 1,2 and 5 pulses at 1.4, 1.8 and 2.57 J/cm^2 (the last for a single pulse only) on boron redistribution profiles. The flattening of the sharply peaked as-implanted profile and the pronounced "shoulder" followed by a steep drop in dopant concentration are characteristics of multiple pulse annealing [9]. As seen in Figure 3, two pulses at the annealing threshold ~1.4 J/cm^2 result in a dopant profile about 300 Å deeper than for a single pulse, which provides an estimate of the change of p-n junction depth in the overlap region, in scanned, repetitively pulsed annealing.

SUBSTRATE HEATING DURING PELA

Studies of the effect of substrate temperature (T_s) during PELA were carried out at T_s = 27°C, 200°C and 400°C, since an earlier study showed that V_{oc} could be increased by ~30 mV if pulsed ruby laser annealing was carried out at 400°C [3]. The increase in V_{oc} was attributed to a reduction in the emitter region contribution to the minority carrier recombination current. It was speculated that this emitter recombination current reduction was due to a reduction in the concentration of quenched-in <u>point</u> defects, resulting from the reduced epitaxial regrowth velocity which melting model calculations predict should occur with increasing T_s.

In contrast, the present studies, using PELA, showed only a much smaller (<5 mV, and possibly not significant) increase of V_{oc} when T_s was increased from 27°C to 400°C, using both E_ℓ = 1.2 and 1.4 J/cm^2. A similar result, also using PELA, was recently reported by Young et al. [10]. There are at least two possible reasons for this difference between pulsed ruby and excimer laser annealing. First, the epitaxial regrowth velocity may be somewhat slower using the long-tailed trapezoidal pulse shape shown in Fig. 2(a) than with the nearly triangular and somewhat shorter (~25-30 nsec FWHM) ruby laser pulse. One preliminary model calculation suggests that specially "tailoring" the excimer laser pulse shape to consist of an initial intense peak followed by a long, low intensity tail results in substantially extending the surface melt duration and correspondingly reducing the epitaxial regrowth velocity in the near-surface region. This type of laser pulse shaping would provide a method for defect reduction, if the concentration of point defects is proportional to regrowth velocity. However, a calculation for the pulse actually used in our PELA experiments suggests that regrowth velocities are probably in the ~3 m/sec range, not substantially different from velocities expected with the ruby laser. The second possibility is that substrate heating, and the attendant reduction in regrowth velocity, simply compensates <u>in the ruby laser case</u> for residual point defects introduced via use of a diffuser plate (due to "microfocusing" effects [11]). Thus, it may simply be the high spatial homogeneity and relatively low coherence of excimer lasers that makes it unnecessary to use <u>either</u> diffuser plates or substrate heating to form high quality emitter regions at growth velocities ~3 m/sec. This conclusion is supported by the observation (see below) that PELA without substrate heating results in solar cells with V_{oc} = 585 mV, approximately as high as is obtained by pulsed ruby laser annealing <u>with</u> substrate heating to 400°C (and no back surface field). However, a definitive separation of these two effects will have to await further experiments and model calculations.

PELA SOLAR CELLS

B-implanted solar cells were found to be superior to cells using As implants, as was reported also by Young <u>et al.</u> [3]. We obtained the best cell performance from the shallowest B implant (5 keV, $2 \times 10^{15}/cm^2$ into ~4 Ω-cm, Czochralski-grown (100) silicon); results for these cells are reported here. PELA was performed in air at 27°C with the 1x2 cm samples mounted on a pair of stepping motor-driven translators and scanned under the laser beam in synchronization with the laser's 1 Hz repetition rate. The annealing pulse was focused to a 4x4 mm square, with E_ℓ=1.4 J/cm^2. Successive laser pulses were overlapped by slightly more than 50%; thus, all parts of the sample received at least two annealing pulses, and spatial variations of p-n junction depth should be $\leq$300 Å (see discussion above). However, the slightly greater junction depth obtained with two annealing pulses (Fig. 3) should slightly reduce the short wavelength spectral response of these cells, relative to single pulse-annealed cells with a shallower p-n junction.

Contacts to the emitter region were made by evaporating Ti-Pd-Ag layers through a mask that shadowed ~4 1/2% of the cell area. To aid in making a good back surface contact, Sb was deposited on the n-type back surface and a shallow n^+ layer was formed by melting with a single 1.5 J/cm^2 ruby laser pulse. (We

have also carried out laser-induced diffusion of Sb films using the excimer
laser.) Ti-Pd-Ag layers were then evaporated and the front and back contacts
were sintered together at 500°C for 2 min. Solar cells were evaluated at AM1
and 28°C.

The best result obtained from a batch of 3 cells (using the ~4 Ω-cm, ~380 μm
thick silicon substrate material) with (without) a 2-layer Ta_2O_5/MgF_2 AR
coating was V_{OC} = 585 (575) mV, I_{SC} = 34.3 (24.0) mA/cm^2, efficiency η = 15.0
(10.5)% and fill factor FF = 0.74 (0.75); the remaining cells had characteristics very close to these. These results are clearly more representative of the
small number of cells fabricated than of the ultimate capabilities of excimer
lasers. The low FF is believed to be a temporary problem originating in the
metallization/sintering steps with the new 4 1/2% shadow fraction mask; we
routinely achieve FF = 0.78 with an older mask [12]. Similarly, the η increase
resulting from the AR coating is low; a 48-50% increase is normal for 2-layer
AR coatings. Thus, we expect that using existing excimer laser technology, and
simply optimizing the present solar cell processing, $\eta > 16\%$ should be readily
achievable, using even ~4 Ω-cm material and no back surface field.

It should be noted, though, that these performance figures refer to particularly simple cell design and processing. Still higher efficiency (>17% AM1)
p^+nn^+ PELA solar cells can be expected to result from a combination of (1) surface passivation, to reduce surface recombination at the top of the highly-
doped emitter; (2) improvement of the minority carrier diffusion length in the
silicon substrate (through use of float zone-grown silicon and gettering procedures); (3) incorporation of a back surface field (perhaps using PELA at high
E_ℓ to melt to depths ≥ 1 μm); (4) optimization of cell thickness to take advantage of (2) and (3); and, (5) optimization of boron implantation depth and
dose.

ACKNOWLEDGMENTS

We thank P.H. Fleming for assistance in sample preparation and measurements.

REFERENCES

1. See B. R. Appleton and G. K. Celler, eds., Laser and Electron Beam
    Interactions with Solids (North-Holland, New York 1982).
2. R. T. Young, C. W. White, G. J. Clark, J. Narayan, and W. H. Christie in:
    Photovoltaic Solar Energy Conference (Proc. of the Inter. Conf.,
    Luxembourg, Sept. 27-30, 1977; Reidel, Boston 1978) p. 861.
3. R.T. Young, R.F. Wood, and W. H. Christie, J. Appl. Phys. 53, 1178 (1982).
4. A. G. Cullis, H. C. Webber, and P. Bailey, J. Phys. E 12, 688 (1979).
5. G. E. Jellison, Jr., and F. A. Modine, J. Appl. Phys. 53, 3745 (1982);
    G. E. Jellison, Jr., and F. A. Modine, Appl. Phys. Lett., 41, 180 (1982).
6. J. Narayan, J. Fletcher, C. W. White, and W. H. Christie, J. Appl. Phys.
    52, 7121 (1981).
7. R. F. Wood and G. E. Giles, Phys. Rev. B, 23, 5555 (1981).
8. R. F. Wood, G. E. Jellison, Jr., D. H. Lowndes, and G. E. Giles, submitted
    to Phys. Rev. B.
9. C. W. White, S. R. Wilson, B. R. Appleton, and F. W. Young, Jr., J. Appl.
    Phys. 51, 738 (1980).
10. R. T. Young, G. A. van der Leeden, R. F. Wood, and R. D. Westbrook in:
    Proc. of 16th IEEE Photovoltaic Specialists Conf. (to be published).
11. D. H. Lowndes, R. F. Wood, and R. T. Young in: ORNL Technical Report
    TM-7442 (July, 1980) pp. 15-20.
12. D. H. Lowndes, J. W. Cleland, W. H. Christie, R. E. Eby, G. E. Jellison,
    Jr., J. Narayan, R. D. Westbrook, and R. F. Wood, Appl. Phys. Lett., Nov.
    15, 1982.

SCANNING ELECTRON BEAM ANNEALING OF OXYGEN DONORS IN CZOCHRALSKI SILICON

C J POLLARD, J D SPEIGHT
British Telecom Research Labs, Martlesham Heath, Suffolk, England
K G BARRACLOUGH
Royal Signals and Radar Establishment, St Andrews Road, Great Malvern,
Worcestershire, England

ABSTRACT

The destruction of the oxygen donor complex in
Czochralski silicon has been studied using scanning
electron beam annealing in the range $550^{\circ}C$ to $1050^{\circ}C$
for 5s to 1000s. A two stage annealing schedule ensured
a rapid rise to the target temperature and overscanning
provided a uniform, non-distorting heating field. Four
point probe and spreading resistance measurements showed
a very rapid donor destruction rate above $650^{\circ}C$; between
$550^{\circ}C$ and $650^{\circ}C$ the lower donor destruction rate allowed
a study of the annealing behaviour.

INTRODUCTION

The presence of oxygen in silicon has a significant effect on its
mechanical and electrical properties during i.c. processing. Such effects
are seen in Czochralski silicon (Cz-Si) wafers which contain interstitial
oxygen at supersaturated levels up to approx 10^{18} at cm^{-3}, as a result of
the incorporation of oxygen during Cz-Si crystal pulling when the molten
silicon reacts with the silica crucible. Whereas oxygen impurities may
enhance the yield and performance of devices fabricated in Cz-Si wafers by
pinning slip dislocations (1) and gettering impurities away from the active
device regions (2), a disadvantage of oxygen in silicon is the formation of
donors at elevated temperatures. The rate of the oxygen donor formation is a
maximum at $450^{\circ}C$, and is dependent on the fourth power of the oxygen
concentration (3). The exact nature of the oxygen donor (sometimes referred
to as the thermal donor) is not known. From the formation kinetics, Kaiser
et al (4) proposed a model based on SiO_4 complexes built up by successive
unimolecular steps. Infra-red spectroscopy (5), photoluminescence (6,7) and
EPR (8) data indicate that the donor is a composite of various defect
species. Since oxygen donor concentrations as high as 10^{16} at cm^{-3} can
be formed during the cooling of high oxygen Czochralski silicon, it is
important that they are destroyed at the wafer manufacturing stage, and not
reformed during subsequent processing. Thermal donor destruction is normally
accomplished by annealing at temperatures above $500^{\circ}C$ followed by rapid
cooling through $450^{\circ}C$ to room temperature (9).
    This paper describes a preliminary investigation of the annealing of
oxygen donors in Cz-Si in the temperature range $550^{\circ}C$ to $1050^{\circ}C$ using a
scanning electron beam annealing system which provides rapid, clean and
accurately controlled whole-wafer heating conditions.

Mat. Res. Soc. Symp. Proc. Vol. 13 (1983) Published by Elsevier Science Publishing Co., Inc.

414

EXPERIMENTAL

Two batches of 50mm diameter <100> p-type (boron doped) polished Cz-Si
wafers have been examined.

Batch A wafers were prepared from seed- and tail-end sections of a
dislocation free crystal grown at 1.5 mm min^{-1} in a low pressure puller.
The target resistivity was 25 - 35 ohm cm. Interstitial oxygen concentration
at the slice centre was typically 1.85 x 10^{18} cm^{-3} at the seed end and
1.6 x 10^{18} cm^{-3} at the tail end (established using the infra-red ASTM
test method F121-76); substitutional carbon concentrations were typically 6
x 10^{15} cm^{-3} at the seed end and 9 x 10^{15} cm^{-3} at the tail end
(infra-red ASTM test method F123-74).

Batch B wafers were obtained from a commercial source with central
resistivities between 20 and 30 ohm cm. Oxygen concentrations for this batch
were typically 1 x 10^{18} cm^{-3} at the slice centre, and carbon
concentrations were 5 x 10^{16} cm^{-3}.

Batch A wafers were examined in the as-grown state. Batch B wafers had
presumably undergone a stabilisation heat treatment during manufacture and
were therefore given a 450°C anneal in nitrogen for 66 hours to generate
oxygen donors prior to e-beam annealing.

Investigation of the annealing behaviour of the oxygen donors was carried
out in a Lintech scanning electron beam annealing system. This apparatus has
been described previously (10): in these experiments, the sample wafer was
mounted in thermal isolation (but electrically grounded) in the target
chamber of the vacuum system and scanned by a high energy (30keV), medium
power (70W to 450W) electron beam (spot diameter approximately 250um) at
high frequency (25kHz x 100Hz). A square scan field was used which was
slightly larger than the wafer diameter (overscan) to ensure uniform,
distortion-free heating.

The relatively low peak temperature regime investigated in detail
(550°C to 650°C) required correspondingly low scan power densities

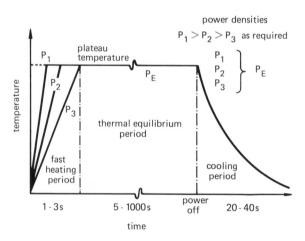

Fig 1;   Thermal Cycle

(2.6W cm^{-2} to 4.1W cm^{-2}): this would have resulted in excessively long
times to reach the target temperature. For example, for a peak temperature
of 600°C (power density 3.3W cm^{-2}), the time to reach 570°C (95% of
the peak temperature) would be approximately 20s. To avoid this, a two-stage
heating schedule was devised, in which a high initial power density (7 - 10W
cm^{-2}) was employed to bring the sample rapidly (ie, in <5s) to the
equilibrium temperature, following which the power density was reduced to
the level needed to maintain the sample in equilibrium at that temperature
(fig.1).

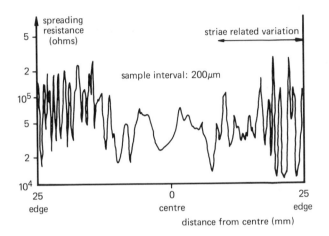

Fig. 2a; Spreading resistance trace vs radius before E-beam annealing

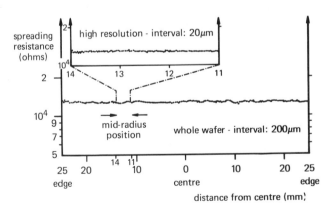

Fig. 2b; Spreading resistance trace vs radius for wafer E-beam annealed at
650°C for 10 seconds

Following the preset exposure time, the power input to the wafer was switched off by means of deflecting the beam onto a water cooled copper plate, allowing the wafer to cool radiatively in vacuo. The initial rate of cooling was approximately $50°C$ s^{-1}, and the time spent between $500°C$ and $400°C$ (the oxygen donor regeneration temperature regime) was less than 4s.

The electrical resistivity of both batch A and batch B wafers was measured by four point probe at 5 mm intervals along two orthogonal diameters before and after electron beam annealing. Oxygen donor concentration changes were calculated from the resistivity changes in the 1 cm diameter central core region.

Selected wafers were also examined before and after e-beam annealing by spreading resistance measurements using a Mazur 210C instrument. Complete diameter scans were made at 200 μm intervals; higher resolution measurements were also made at 20 μm interval around the peripheral, mid-radius and central regions of the wafer.

RESULTS AND DISCUSSION

For temperatures greater than $650°C$ the oxygen donors were rapidly destroyed in both batch A and batch B wafers. For example, 4s at $820°C$, 10s at $720°C$ and 10s at $650°C$ caused complete recovery to the target resistivity, which was radially uniform to within 4%. This treatment therefore gives the same result as a standard stabilisation furnace heat treatment (typically 1-3h at $650°C$). Fig. 2a shows a spreading resistance trace as a function of diameter for an unannealed wafer: this shows the oxygen donor striae characteristic of Cz-Si and an overcompensated n-type central region. After annealing at $650°C$ for 10s, no resistivity variation due to oxygen donors remained (fig 2b), and an extremely flat p-type resistivity profile was observed. Examination by x-ray topography showed the wafers to be slip- and dislocation-free, although the characteristic

Fig. 3   X-ray topograph of wafer e-beam annealed at $850°C$ for 10s.
(Cu K_α radiation: double crystal method)

striation features of the wafer remained (fig 3). This confirms that although the electrical character of the oxygen donors has been destroyed, the localised variation in lattice parameter associated with oxygen and carbon impurities remains.

Due to the very rapid annealing character of the donor complex, it was not possible to study systematically the time dependence of donor destruction for temperatures greater than 650°C; it has also been shown (9) that for temperatures less than 500°C, generation of the donor complex occurs. Consequently, the temperature range 550°C to 650°C was investigated in detail for times ranging from 5s to 1000s in batch A seed end wafers.

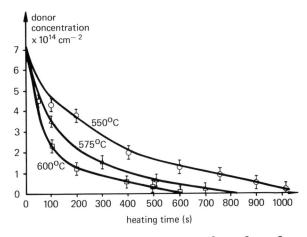

Fig. 4;   Donor concentration vs heating time at 550°C, 575°C, 600°C

Fig 4 shows the degree of donor destruction as a function of time at 550°C, 575°C and 600°C. A rapid initial decrease in the donor density was observed with the destruction rate increasing with temperature. Beyond this initial period, the donor destruction rate dropped systematically with time at all temperatures. This non-linear annealing behaviour suggests that the oxygen donor complex destruction proceeds via a series of intermediate states. This behaviour is consistent with photoluminescence, infra-red spectroscopic and EPR measurements (5,6,7,8) which identify a number of different energy levels associated with the donor complex. It has been shown (11) that pre-annealing at 450°C (i.e., in the thermal donor generation regime) increases the subsequent precipitation of oxygen at elevated temperatures, thus indicating that the donor formation is associated with nucleation sites for subsequent SiO_2 precipitates, with the electrically active nucleation sites coherent with the matrix and the stable, electrically inactive SiO_2 incoherent with the matrix. Thus, the annealing behaviour of oxygen donors (fig. 4) will depend strongly on the thermal history of the sample and how the donors are generated (i.e. cooling rates in the puller, deliberate heat treatments etc.). Oxygen donor formation and destruction heat treatments are important steps in a thermal processing cycle which can dramatically affect yield and performance of VLSI devices subsequently fabricated in Cz-Si wafers.

418

SUMMARY

1. Donor destruction takes place at an extremely rapid rate above 650°C ($>10^{14}$ s^{-1}).
2. Between 550°C and 650°C, the donor destruction rate is sufficiently low to allow a study of partially annealed specimens. This shows that a number of intermediate stages are involved in the donor destruction mechanism.
3. Electron beam annealing is a clean, economical and efficient technique for returning as-grown Czochralski silicon to its target resistivity prior to VLSI processing.

ACKNOWLEDGEMENTS

The authors are grateful to the Director of Research, British Telecom Research Labs for permission to publish this work, and to Dr R W Series, Mrs D Kemp and Mr I M Young for assistance with sample characterisation.

REFERENCES

1. S M Hu, Appl. Phys. Lett. 31, 53 (1977)

2. K W Tan, E E Gardner, W K Tice, Appl. Phys. Lett. 30, 175 (1977)

3. W Kaiser, Phys. Rev. 105, 1751 (1957)

4. W Kaiser, H L Frisch, H Reiss, Phys. Rev. 112, 1546 (1958)

5. D Wruck, P Gaworzewski, Phys. Stat. Sol. 56, 557 (1979)

6. H Nakayama, J Katsura, T Nishino, Y Hamakawa, Jap. J. Appl. Phys. 19, L547 (1980)

7. M Tajima, S Kishino, M Kanamori, T Iizuka, J. Appl. Phys. 51, 2247 (1980)

8. G Feher, Phys. Rev. 114, 1219 (1959)

9. C S Fuller, R A Logan, J. Appl. Phys. 28/12, 1427 (1957)

10. C J Pollard, A E Glaccum, J D Speight, Proc MRS Annual Meeting, Boston 1981, p789 (Eds. B R Appleton, G K Celler: Elsevier Science Publishing Inc)

11. H F Schaake, S C Baber, R F Pinizzotto, Proceedings of Fourth International Symposium on Silicon Materials Science and Technology, Semiconductor Silicon 1981, p273 (Ed. H R Huff, R J Kriegler, Y Takeishi) Electrochemical Society)

ARSENIC IMPLANT ACTIVATION AND REDISTRIBUTION IN P-TYPE SILICON INDUCED
BY PULSED ELECTRON BEAM ANNEALING

D. BARBIER[+], M. BAGHDADI[+], A. LAUGIER[+] and A. CACHARD[++]

+ : Laboratoire de Physique de la Matière, Institut National des Sciences Appliquées
de Lyon, 20 Avenue Albert Einstein 69621 VILLEURBANNE CEDEX (France)

++ : Département de Physique des Matériaux, Université Claude Bernard-Lyon I
43, Boulevard du 11 Novembre 1918, 69622 VILLEURBANNE CEDEX (France)

ABSTRACT

In this work Pulsed Electron Beam Annealing has been used
to activate As implanted in (100) and (111) silicon (140 keV-
$10^{15} cm^{-2}$). With a selected electron beam energy deposition
profile excellent regrowth layer quality and As activation has
been obtained in the 1.2-1.4 J/cm^2 fluence range. As redistribution is consistent with the melting model assuming a diffusivity
of $10^{-4} cm^2/s$ in liquid silicon. As losses might slightly reduce
the carrier concentration near the surface in the case of (100)
silicon. However a shallow and highly active N^+ layer have been
achieved with optimized PEBA conditions.

1. INTRODUCTION

Ion implantation followed by a single submicrosecond high energy pulse is now
considered as among the most attractive doping process for shallow junction formation in
silicon. Implant activation is obtained by liquid phase epitaxial regrowth of the amorphous
surface layer starting from the undamaged substrate. In past years pulsed laser
annealing has been widely used to activate most of the classical dopants in silicon with
various implantation parameters (1). A few works have been devoted to pulsed electron
beam annealing (PEBA) up to now, but it has been recently demonstrated that this
technique can be successfully used to achieve shallow implanted junctions in silicon solar
cells (2)(3). However, as well as with any other pulsed beam technique, annealing kinetics
must be optimized as a function of implantation parameters. Desired doping profiles can
only be obtained by controlling thermal effects of the pulse (molten layer thickness, melt
front velocity, temperature profiles).

This is made easier in the case of electrons than in the case of lasers because the
heat generation function in the material can be largely modified by variation of
acceleration voltages and beam currents.

In this work an optimized electron beam pulse has been used with various fluences to
recrystallize p-type (100) and (111) silicon implanted with arsenic ($10^{15}/cm^2$-140 keV).
RBS and SIMS were used to study annealing effects. Dopant activation has been
investigated by mean of incremental sheet resistance measurements.

2. EXPERIMENT

The machine used for annealing (SPIRE-300) has been described in ref.3. It allows for
adjustment of electron beam parameters (energies,currents, annealing diameter). These
have been selected to yield an energy deposition profile convenient for melting the silicon
surface over a few tenths of microns with a pulse duration of about 50 ns. The mean
electron energy has been kept around 15 KeV and the samples have been processed in a

vacuum chamber (5 x 10^{-6}Torr) with electron energy densities (fluences) ranging from 1 to 1.6 J/cm^2. Thermal effects of the pulse have been studied by solving the one-dimensional heat flow equation starting from time-resolved measurement of the electron beam parameters.

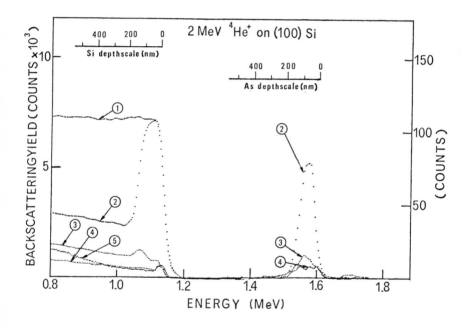

Fig. 1 - (1) RBS spectra of (100) Si implanted with 10^{15} As/cm^2 at 140 KeV in the random direction ;
        (2) aligned not annealed
        (3) aligned after PEBA with 1 J/cm^2 ;
        (4) aligned after PEBA with 1.2 to 1.4 J/cm^2 ;
        (5) aligned after PEBA with 1.6 J/cm^2

3. RESULTS

RBS measurements in the random and aligned position on as-implanted and annealed samples have been performed in order to study crystal regrowth as a function of fluence (Fig.1). The as-implanted samples exhibits an about 2000 Å thick amorphous layer and the experimentally determined threshold for lattice recovery is 1 J/cm^2 for both (111) and (100) silicon. This result is in agreement with the computer calculation of the molten depth which is of the order of the amorphous layer thickness for this fluence threshold. The best annealing results are obtained in the 1.2-1.4 J/cm^2 fluence range for which the Si residual disorder is below 5 % for both orientations and the fraction of substitutional As reaches 97 % for (111) and 93 % for (100) Si. The slightly lower value obtained with (100) Si could be related to differences either in crystalline planes rearrangement or in possible point defects trapping during fast regrowth as a function of orientation(4)(5).

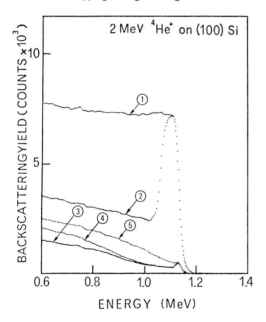

Fig. 2 - RBS spectra of (100) Si implanted with 10^{15} As/cm^2 at 140 KeV in the random direction (1) ; aligned not annealed (2) and after PEBA at 1.6 J/cm^2 : one shot (3), two shots (4), four shots (5).

Moreover a degradation of the crystal beneath the amorphous layer is clearly observed when the fluence is raised to 1.6 J/cm^2 (Fig.2). Defect creation beyond the molten layer is evidenced by a further increase of the dechanneling in the RBS spectrum when the same sample undergoes 2 and 4 overlapping shots at 1.6 J/cm^2. We believe that too high thermal gradients ($\simeq$ 600°C/μm) induced in the solid beneath the molten zone could be responsible for extended defects generation at high fluence (5). Dopant redistribution has been studied by SIMS. Fig.3 shows the spreading of the As profile after a 1.2 J/cm^2 PEBA shot on (100) silicon. A simple diffusion model has first been employed to fit experimental results assuming no segregation effects. The tail of As profiles were well fitted assuming a constant diffusivity of $10^{-4} cm^2/s$ in the molten layer. This value although lower than the one obtained by several authors from laser annealing experiment (6) is of the order of classical dopant diffusivities in liquid silicon. However in the surface region (500 Å) the simple diffusion model did not succeed in fitting the experimental data which exhibit a drop in As concentration compared to the computed points. A good fit of the near surface part of the profiles (solid line on Fig.3) was only possible by assuming As evaporation in the vacuum process chamber during the melting stage. In order to support this assumption other experiments were carried out on phosphorus implanted (100) silicon. It turns out that the fitting of PEBA induced phosphorus profiles obtained by SIMS should account for a higher percentage of dopant losses than in the case of As implants. This result was expected because the phosphorus equilibrium vapor pressure over liquid silicon is higher than the one of arsenic. However SIMS profiles in (111) silicon did not exhibit significant As losses after PEBA as in the case of (100) silicon. Moreover no detectable orientation effect has been observed in As redistribution for the tail of the profiles between (100) and (111) silicon.

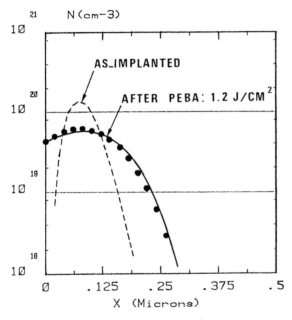

Fig. 3 - As redistribution after a 1.2 J/cm^2 PEBA shot

As electrical activity in the regrowth layers of (100) silicon has been studied by means of incremental sheet resistance measurements with a depth resolution of 150 Å. Fig.4 shows the carrier profiles induced in (100) silicon by PEBA at 1.4 J/cm^2 and 1.6 J/cm^2. Analysis of these profiles gives information consistent with the above described results. As is more than 90 % activated in the 1.2-1.4 J/cm^2 fluence range which is consistent with the low residual defect density and the peak value of the As substitutional fraction in the same annealing conditions. As mentioned above, the As electrical activity between 1.2 and 1.4 J/cm^2 could be limited by a point defect trapping phenomenon depending on the annealing kinetics (4). Moreover annealing with a fluence of 1.6 J/cm^2 results in low electron concentrations in the tail of the profile which causes the As electrical activity to drop to 60 %. This must be related to the poor quality of the regrowth layer for this fluence. Finally the surface part of the carrier profiles exhibits a drop in the electron concentration over the first 500 Å which is in agreement with the shape of the atom profiles. However, even with the rather low implantation dose electron concentrations higher than 4×10^{19} cm^{-3} are obtained over about 2000 Å with the best PEBA conditions (fluence between 1.2 and 1.4 J/cm^2).

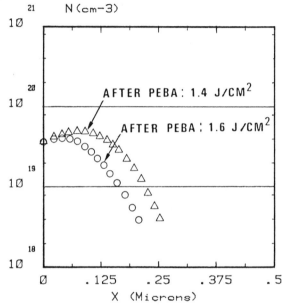

Fig. 4 - Carrier profiles induced by a 1.4 J/cm^2 and a 1.6 J/cm^2 PEBA shots

4. CONCLUSION

This work has shown that PEBA can be successfully used to activate a shallow and low dose As implant in (100) and (111) silicon. The electron energy deposition profile has been adjusted to induce liquid phase epitaxial regrowth of an about 2000 Å thick molten layer in the 1-1.6 J/cm^2 fluence range. Excellent As activation has been obtained between 1.2 and 1.4 J/cm^2. In these conditions, the PEBA induced molten layer extends beyond the amorphous layer and the underlying crystal is moderately stressed by thermal gradients. As redistribution during PEBA can be fitted assuming a diffusivity of 10^{-4} cm^2/s which is consistent with liquid phase epitaxy as annealing mechanism. In the case of (100) silicon, As losses during the melting process might slightly reduce the dopant concentration near the surface.

REFERENCES

1. C.W. White, J. Narayan, R.T. Young, MRS Proceed. AIP 50 New York, 275 (1979)

2. A.C. Greenwald, A.R. Kirkpatrick, R.G. Little, J.A. Minnucci, J. Appl. Phys. 50, 2, 783 (1979)

3. A. Laugier, D. Barbier, G. Chemisky, Proceed. of the 4th E.C. Photovoltaic Solar Energy Conference. Ed. D. Reidel Publishing Company, p. 1007 (1982)

4. L.C. Kimerling, J.L. Benton, Laser and Electron Beam Processing of Materials Academic Press, p. 385 (1980)

5. See "PEBA Induced Deep Level Defects in Virgin Silicon". This symposium

6. C.W. White, S.R. Wilson, B.R. Appleton, F.W. Young Jr, J. Appl. Phys. 51, 738 (1980)

-§-

TRANSIENT ANNEALING OF ION-IMPLANTED SILICON USING A SCANNING IR LINE SOURCE

Y.S. LIU, H.E. CLINE, G.E. POSSIN, H.G. PARKS AND W. KATZ
General Electric Research and Development Center, Schenectady, New York 12345, USA

ABSTRACT

Recent interest in finding an efficient method for transient annealing of ion-implanted silicon has led to studies of various rapid annealing schemes such as graphite heaters and high intensity incoherent light sources as alternative methods to laser annealing. In this paper, we describe a recent study of transient annealing of ion-implanted silicon using a scanning IR line source created by a single tungsten filament enclosed in a quartz envelope. Various dopants (B^+, P^+ and As^+) with fluences of 10^{14} to 10^{16} ions/cm^2 were implanted and annealed under both transient and steady-state thermal conditions. Dopant depth distributions were analyzed using the SIMS technique. Sheet resistance measurements indicated that almost 100% activations of the implanted dopants were achieved. Sensitivities of dopant activation to transient annealing conditions were studied as a function of dopant concentrations, and high-dose As- and B-implanted samples were found to be sensitive to transient thermal cycle, particularly to the peak temperature. Recrystallization was studied with Rutherford backscattering spectroscopy using 2 Mev He$^+$ ions.

INTRODUCTION

Recent studies in laser and electron beam interactions with semiconductors have improved our understanding of transient annealing of ion-implanted semiconductors. Transient annealing provides an alternate method to steady-state furnace annealing for the fabrication of high resolution devices in VLSI-related processes. More recently, the effort to find a more efficient method for transient annealing has led to studies of other rapid annealing methods such as graphite heaters [1] and high intensity light sources [2].

In this paper, we describe results of a transient annealing study of ion-implanted silicon using a scanning IR line source employing a 2 kW quartz lamp. This focused IR line source provided a maximum radiation flux on the order of 300 W/cm^2 over an area of 2 mm by 25.4 cm, resulting in a surface temperature of approximately 1200 °C on the samples.

EXPERIMENTS

The transient annealing apparatus which was used is schematically illustrated in Fig. 1. The IR emitter is a tungsten filament enclosed in a quartz envelope operated at approximately 3000 °C and has an emission spectrum peaked at 0.9 micron, which is below the absorption wavelength edge of Si at one micron. The IR radiation was focused with an elliptically shaped lamp reflector onto a stripe about 2 mm wide by 25.4 cm long. A maximum radiation flux of about 300 W/cm^2 could be reached at the focal plane.

The samples were mounted on the top of a substrate heater operated near 600 °C to reduce temperature increments required during a transient annealing cycle and to reduce the stress induced by the temperature gradient. Biasing the samples at around 600 °C over a time period of minutes had no significant effects on annealing in the present experiments. The substrate heater was mounted on a translating stage such that the samples could be scanned across the heated zone over the 2 mm width defined by the IR lamp reflector.

Mat. Res. Soc. Symp. Proc. Vol. 13 (1983) ©Elsevier Science Publishing Co., Inc.

426

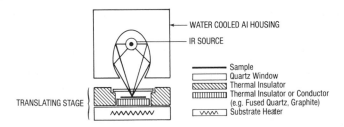

WATER COOLED Al HOUSING

IR SOURCE

Sample
Quartz Window
Thermal Insulator
Thermal Insulator or Conductor
(e.g. Fused Quartz, Graphite)
Substrate Heater

TRANSLATING STAGE

Fig. 1. Transient annealing apparatus using a scanning IR line source.

The steady-state thermal distribution of a scanning line source on a semi-infinite material has been previously analyzed by Cline and Anthony [3], and recently by LeCoz et al. [4]. The steady-state temperature distribution has been shown to depend on the power absorbed at the material surface, the scanning speed, and the thermal diffusivity of the material. Fig. 2(a) shows the temperature distributions at the surface of a semi-infinite substrate heated by a Gaussian-shape heat source. The maximum temperature increases with time until an asymmetric steady-state distribution is reached. The plots were calculated using a parameter $V/2D$ equal to unity, where V is the scanning speed and D is the thermal diffusivity of the material. The steady-state temperature distribution in a slab produced by a scanning line source is shown in Fig. 2(b). This shows that while the thermal distribution is radial near the line source, at a distance equal to $2D/V$ from the line source the distribution becomes asymmetrical. From a thermal analysis, it can be shown that the maximum wafer dimension that can be transient-annealed using the line source method is determined by stress-induced deformation across the wafer [5].

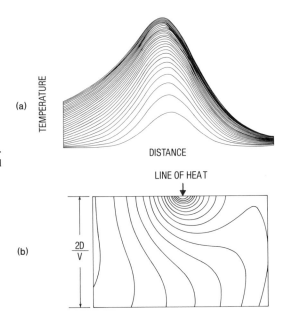

(a)

TEMPERATURE

DISTANCE

Fig. 2. Surface and bulk temperature distributions of silicon heated with a scanning IR line source.

LINE OF HEAT

(b)

$\frac{2D}{V}$

The temperature-time profile of a typical transient annealing thermal cycle is shown in Fig. 3. The sample was placed on a substrate heater biased at 600 °C and was scanned across the heated zone defined by the IR lamp and the lamp reflector. The temperature-time profile was measured by placing a thermocouple on the surface of the sample during scanning. A maximum temperature of 1200 °C was measured.

Silicon wafers of (100) were implanted with B^+, P^+, and As^+ with fluences varying from 10^{14} to 10^{16} ions/cm^2, and were transient-annealed with different peak temperatures. Dopant depth profiles were analyzed using the SIMS technique and were compared with steady-state thermal-annealed samples. Rutherford backscattering spectra using 2 Mev He^+ were studied and compared for transient and steady-state thermal annealing. Sheet resistances were compared.

During transient annealing, samples were placed on a heated substrate biased at a temperature around 600 °C and were scanned at a constant speed of 0.08 cm/sec. The samples were transient annealed at peak temperatures varying from 950 ° to 1200 °C in either atmosphere or argon ambients.

RESULTS AND DISCUSSION

Fig. 4 shows the sheet resistances measured for B^+ samples implanted at dose levels in the 10^{15} to 10^{16} range and transient annealed with various peak temperatures. For high-dose implanted samples, sheet resistance was very sensitive to the variation of peak temperature in the transient annealing cycle; for example, factor of four changes of sheet resistance are shown in Fig. 4 for implant doses of 10^{16} ions/cm^2. Comparisons with sheet resistance data measured for samples annealed with steady-state thermal conditions at 950 °C for 15 minutes, and those calculated using SUPREM show full activation of the implanted dopants.

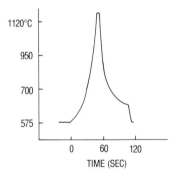

Fig. 3. Temperature-time profile of a typical transient annealing thermal cycle.

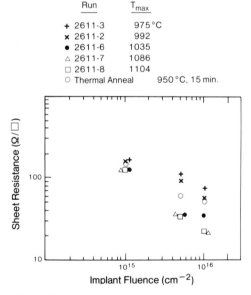

Fig. 4. Sheet resistance measurements and maximum temperature during transient annealing cycles (B^+, 50 keV).

428

Boron depth profiles were analyzed using the SIMS technique. Fig. 5 shows the results of samples B^+ implanted at 50 keV, 5×10^{15} ions/cm^2 and transient annealed at peak temperatures of 1035 °C and 1104 °C. These are compared with SIMS data for the sample under steady-state thermal annealing at 950 °C for 15 minutes. The results show that transient annealing significantly reduced dopant redistribution under the condition that full activation was achieved.

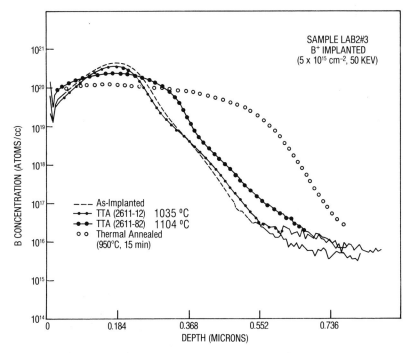

Fig. 5. SIMS boron depth profile.

The Rutherford backscattering spectra for As$^+$ implanted (10^{16}/cm^2, 100 keV) samples transient annealed at peak temperature about 1085 °C are shown in Fig. 6. A dechanneled yield less than 4% is observed, indicating good recrystallization. Sheet resistance measured for these samples was about 100 ohm/□. To determine whether multiple transient annealing cycles would further improve the dechannel yield, some samples were repeatedly transient-annealed up to three thermal cycles at peak temperature of 1085 °C. No significant improvements were observed within the resolution limit of about 1%. Arsenic distributions for samples transient-annealed up to three thermal cycles were determined from random backscattering spectra using 2 Mev He$^+$ ions. These data are shown in Fig. 7 and are compared with the spectra for an as-implanted sample and a sample steady-state thermal-annealed at 1000 °C for 30 minutes. Channeled and random Si backscattering spectra of transient-annealed samples are shown in Fig. 8. Good recrystallization was observed from the channeled spectra shown. A slight improvement was observed for high dose As$^+$-implanted silicon samples transient-annealed twice at the peak temperature of 1085 °C. Good recrystallization was also observed for P-implanted samples annealed under transient thermal conditions.

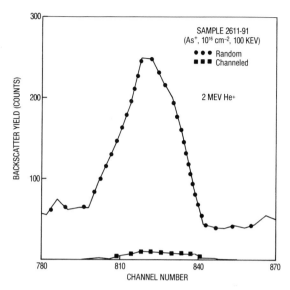

Fig. 6. Channeled and random backscattering spectra of transient annealed As$^+$-implanted silicon.

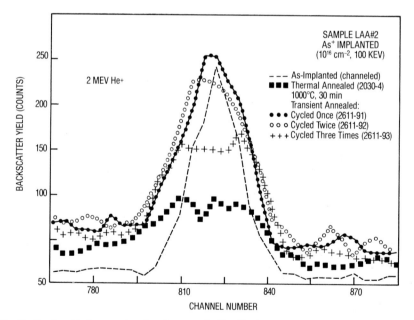

Fig. 7. Backscattering spectra showing arsenic distributions.

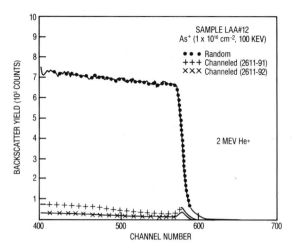

Fig. 8. Channeled and random backscattering spectra of transient annealed As⁺-implanted silicon.

In conclusion, the present study indicates that a scanning IR line source provides a simple method for studying transient thermal annealing of ion-implanted silicon. Present results show that transient annealing using the scanning IR line source resulted in almost 100% electrical activation in B^+, As^+, and P^+-implanted silicon with significantly reduced dopant redistribution. The application of transient thermal annealing to VLSI device processing provides a very attractive alternative to the conventional furnace annealing method and warrants further study.

ACKNOWLEDGMENTS

The authors wish to thank Professor H. Bakhru of the State University of New York at Albany for providing the RBS facility for the current study. We also wish to thank R. Guida for his excellent technical assistance, and G. Smith for carrying out the SIMS measurements.

REFERENCES

1. J.C.C. Fan, T-Y Tsaur, and M.W. Geis, *Laser and Electron Beam Interactions with Solids,* ed. by B.R. Appleton and G.K. Celler, (North-Holland, New York 1982), p. 751.

2. D.J. Lischner and G.K. Celler, Ibid. p. 759; H. B. Harrison, M. Grigg, K.T. Short, J.S. Williams, and A. Zylewicz, Ibid. p. 771.

3. H.E. Cline and T.R. Anthony, *J. Appl. Phys., 48,* 3895 (1977).

4. H.J. Le Coz, E.W. Maby, and D.A. Antoniadis, Abstract, Spring Meeting of the Electrochemical Society, Montreal (1982).

5. H.E. Cline (to be published).

CONDUCTION IN POLYCRYSTALLINE SILICON: GENERALIZED THERMIONIC
EMISSION-DIFFUSION THEORY AND EXTENDED STATE MOBILITY MODEL

A.N. KHONDKER,* D.M. KIM,* AND R.R. SHAH**
*Electrical Engineering Department, Rice University, Houston, Texas 77251
**Texas Instruments, P.O. Box 225012, Dallas, Texas 75265

ABSTRACT

We present a general theory of conduction in polysilicon. The theoretical framework reconciles two
apparently divergent approaches for modeling conduction
processes in polysilicon and provides a physical basis
to correctly interpret and to point out the deficiencies
of previously reported thermionic and thermionic field
emission theory. This model is based on an extended
state mobility in the disordered grain boundary and the
thermionic emission-diffusion theory for conduction of
current. The attractive features of our theory are
(a) it can explain the experimental data without the
use of an artificial factor, f, (b) the conduction process is characterized explicitly by the inherent
material properties of the grain and the grain boundary.
Our model is particularly suited for describing the
electrical properties of laser restructured polysilicon,
where because of large grain size the diffusion process
is expected to be dominant.

INTRODUCTION

Recent advances in laser and electron beam processing of polysilicon films
hold out an attractive promise for 3-d integration in integrated circuit
technology. The ultimate goal of localized transient radiation processing
is to generate device quality material from polysilicon deposited at appropriate locations on insulating or conducting layers during the course of
device fabrication. However, due to the complex nature of recrystallization
phenomena and their dependence on a variety of factors, the generation of
single crystal regions may not always be possible without the use of special
techniques such as seeding, graphoepitaxy, encapsulation, etc. Even so, the
recrystallized, near single crystal regions may contain defects and low angle
grain boundaries. Extensive effort has been made in recent years to fabricate basic unipolar and bipolar active device elements in radiation processed
polysilicon of different grain size. It is, therefore, important to have a
quantitative understanding of conduction mechanisms in polysilicon.
Until recently there has been only one predominant school of thought to
describe conduction in polysilicon: (1) charge carriers generated by the
dopant atoms are trapped by the dangling bonds in the grain boundaries,
(2) the resulting space-charge barrier potential is modeled as symmetric
back-to-back Schottky diodes, and (3) conduction across the barrier is described by invoking thermionic and field emission [1-4]. This approach,
although reasonably successful in computationally modelling electrical properties of polysilicon, can not adequately provide a physical basis for the
conduction mechanism. This approach neglects the physics of conduction
within the grain boundary -- an amorphous material, and, then mathematically

Mat. Res. Soc. Symp. Proc. Vol. 13 (1983) ©Elsevier Science Publishing Co., Inc.

defines an effective mobility for the composite grain and grain boundary system, a physically unclear procedure. Furthermore, this model has to resort to artificial scaling factors, f, to explain the data [Eq. (36), Ref. 2]:

$$\rho = \frac{1}{f} \frac{1}{2Wq^2 p(-L/2)} (2\pi m^* kT)^{\frac{1}{2}} \exp[qV_B/kT] \qquad (1)$$

Here, $p(-L/2)$ denotes the hole concentration at the grain center, W the depletion depth, and qV_B the barrier height. The scale factor used in Ref. [2] was f = 0.06 and f = 0.12 in Ref. [1]. Attempts have since been made to either eliminate [4] or explain [5] this artificial factor. In an attempt to provide a physical basis to conduction, we recently reported an alternative description using both drift and diffusion contributions to current in the crystalline region, and introducing the extended state carrier mobility [6] to reinterpret the conduction mechanism within the disordered grain boundary.

The purpose of this paper is to present a general theoretical framework which reconciles these two apparently divergent schools of thought and, most importantly, which provides a physical basis to correctly interpret and to point out the deficiencies of the previously reported thermionic and field emission theory. This model is based on an extended state and/or hopping mobility in the disordered grain boundary and the thermionic emission-diffusion theory [7] for conduction of current.

THEORY

Consider one dimensional conduction for boron doped (p-type) polysilicon where the dopant segregation is negligible. The schematic energy level diagram for a unit cell of length L polysilicon consisting of a crystallite grain and the boundary is shown in Fig. 1. The grain boundary modeled here as an amorphous semiconductor is characterized by trap states, band tailing (Δ) and mobility shoulder (Δ'). Because of the holes trapped in the boundary, the (electronic) valence band edge, $-qV(x)$ or $E_v(x)$ in the depletion region of the crystallite is lowered by an amount $(q^2 N_A/2\epsilon)(|x| - W)^2$, where N_A is the dopant concentration and W the depletion depth. The corresponding equilibrium profile of mobile holes is symmetric around the grain boundary and is shown in Fig. 2. When an external voltage V_a is applied to the unit cell, a part of it drops across the grain and the remainder, V_{gb}, drops across the grain boundary. Under this condition, the valence band edge is bent accordingly, a quasi-Fermi level is introduced and the local hole concentration departs from the symmetric equilibrium value. The steady state carrier concentration departs from the symmetric equilibrium value. The steady state carrier concentration at the left (p_{mL}) and (p_{mR}) of grain boundary edges can be given in terms of the equilibrium value (p_0) as

$$p_{mL} = p_0 \, e^{[(E_{f0} - E_{fL}) - qV_{gb}/2]/kT} \qquad (2)$$

$$p_{mR} = p_0 \, e^{-[(E_{fR} - E_{f0}) - qV_{gb}/2]/kT} \qquad (3)$$

with p_0 given by

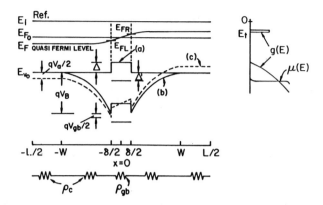

Fig. 1. Schematic energy level diagram of a unit cell for polysilicon under (a) undoped, (b) doped, and (c) doped and biased conditions. Also shown are the density states factor, $g(E)$ and the mobility, $\mu(E)$ for undoped polysilicon. V_a and V_{gb} are the applied voltages across the unit cell and the grain boundary, respectively. Δ is the amount of band tailing, and Δ' represents the mobility shoulder.

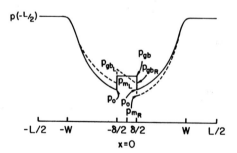

Fig. 2. Schematic hole concentration profile in grain and grain boundary. Solid lines and dashed lines represent the thermal equilibrium and biased conditions, respectively.

$$P_0 = N_v e^{-[E_{f0} - (E_{v0} - qV_B)]/kT}$$
(4)

where E_{fL} and E_{fR} are the values of quasi-Fermi level at the respective grain boundary edges.

The current density in the grain is given by $J = \mu_c p(x)(\partial E_f/\partial x)$, where μ_c is the crystalline mobility and the concentration $p(x)$ can be written as $N_v \exp[-(E_f - E_v)/KT]$. Here N_v, k, T have their usual meanings. Therefore one may write

$$\frac{J}{\mu_c} \int e^{-\frac{E_v}{kT}} dx = N_v \int e^{-\frac{E_f}{kT}} dE_f$$
(5)

Upon integrating the left hand side of Eq. (5) w.r.t. x over $(-L/2, -\delta/2)$ and $(\delta/2, L/2)$, carrying out the corresponding integration w.r.t. E_f on the right and multiplying both sides by $\exp[(E_{v0} - qV_B)/KT]$, one finds after lengthy but straightforward algebra,

$$\frac{JL}{\mu_c kT} Fe^{-\frac{qV_B}{kT}} = 2p_0 \sinh(\frac{qV_a}{kT}) - (p_{mL}e^{\frac{qV_{gb}}{2kT}} - p_{mR}e^{-\frac{qV_{gb}}{2kT}})$$
(6)

Here the term

$$F = 1 - \frac{2W}{L} + \frac{2W - \delta}{L} [D(\sqrt{qV_B/KT})/\sqrt{qV_B/KT}]e^{\frac{qV_B}{kT}}$$
(7)

is expressed in terms of Dawson integral defined as

$$D(a) = e^{-a^2} \int_0^a e^{t^2} dt$$
(8)

We now discuss the charge transport within the grain boundary. J consists, in general, of the drift component and the diffusion component:

$$J = q\mu_{gb}p_{gb}(V_{gb}/\delta) + qD_{gb}(p_{gbL} - p_{gbR})/\delta$$
(9)

where μ_{gb} and D_{gb} denote the extended state hole mobility and diffusion constant respecitvely, whereas p_{gb} denotes the average mobile holes, $(p_{gbL} + p_{gbR})/2$ beyond the mobility gap. Because the grain boundary width, δ, is small and also $qV_{gb} \ll KT$, we have taken external field there-in to be uniform, qV_{gb}/δ, and have linearized the concentration gradient. We have previously derived [6] p_{gb} as

$$p_{gb} = \gamma p_0$$
(10a)

with

$$\gamma = e^{\frac{\Delta}{KT}} [erfc\sqrt{\frac{\Delta + \Delta'}{kT}} + \frac{2}{\sqrt{\pi}}\sqrt{\frac{\Delta + \Delta'}{kT}} e^{-\frac{\Delta + \Delta'}{kT}}]$$
(10b)

Upon inserting Eq. (10) into Eq. (9) there results

$$J = q\mu_{gb}(\gamma p_0)V_{gb}/\delta + q(D_{gb}\gamma/\delta)(p_{mL} - p_{mR}) \tag{11}$$

A major contribution of this work consists of introducing the diffusion component of the current and attaching a novel interpretation to it. In interpreting this term, one may conveniently recast it as $q(V_{gb}\gamma/\delta)(p_{mL} - p_0) - q(V_{gb}\gamma/\delta)(p_{mR} - p_0)$. With an applied bias, p_{mL} rises above p_0 and p_{mR} drops below p_0. Current should, therefore, flow to restore the equilibrium value p_0. This situation is closely analogous to the Schottky barrier diode case, where excess charge carriers are consumed at a rate, v_R, across the metal-semiconductor interface. In this context, we can formally identify the constant $D_{gb}\gamma/\delta$ as an effective recombination velocity, v_R, and the two terms $qv_R(p_{mL} - p_0)$ and $qv_R(p_{mR} - p_0)$ can be attributed to the two opposite currents of back-to-back Schottky barrier diode. This second term of Eq. (11), in fact, establishes contact of our theory with the termionic emission theory as applied to polysilicon. With this interpretation one may rewrite Eq. (11) as

$$\frac{J}{qv_R} = p_0\frac{qV_{gb}}{kT} + (p_{mL} - p_{mR}) \; ; \; v_R = \frac{D_{gb}\gamma}{\delta} \tag{12}$$

For the case where $qV_{gb} \ll kT$, the second term of Eq. (6) can be readily seen to be identical to the right hand side of Eq. (12). Therefore, upon inserting Eq. (12) in Eq. (6) one finds,

$$J = \frac{2qp(-L/2)\sinh(\frac{qV_a}{2kT})}{\dfrac{e^{(qV_B/kT)}}{v_R} + \dfrac{1}{v_D}} \tag{13a}$$

with

$$1/v_D = qLF/\mu kT \tag{13b}$$

and

$$p(-L/2) = p_0 e^{qV_B/kT} \tag{13c}$$

and v_D is an effective diffusion velocity associated with the transport of holes from the grain to the edge of grain boundary.

This expression for current density can be shown to be essentially identical to that obtained by us previously [6], hence it is capable of explaining experimental data on resistivity, mobility, and activation energy without the use of artificial factor, f.

It can be clearly observed from Eq. (13) that J depends on two types of velocities, namely v_D and v_R. For the grain size ranging from 300 Å to 3000 Å, $v_D \gg v_R \exp(-qV_B/kT)$, and the expression for J reduces to

$$J = 2qp(-L/2)v_R e^{-\frac{qV_B}{kT}}\sinh(\frac{qV_B}{kT}) \tag{14}$$

Eq. (14) formally resembles the current density expression, which has been derived using thermionic emission theory [Eq. 10, Ref. 2]. The only difference between the two expressions lies in the value and interpretation of v_R.

In thermionic theory, the charge carriers are regarded as free particles and the corresponding effective recombination velocity, v_R, is specified by the thermal velocity [7], i.e. $v_R = A^*T^2/qN_V$, A^* being the Richardson constant. The typical value of v_R in this case is 4.3×10^6 cm/sec. at room temperature. In our theory v_R is characterized by the inherent boundary properties, e.g. the extended state hole diffusion constant, boundary width, band tailing (Δ) and position of mobility shoulder (Δ'). For the parameter values used in Ref. [6], $v_R \sim 3.25 \times 10^5$ cm/sec. for the data in Ref. [2] and $v_R \sim 8.2 \times 10^5$ for the data of Ref. [1]. The difference in values of v_R between our theory and the previous theories stems from the fact that holes are not considered in our model as free charge carriers crossing the boundary. Furthermore, the need for the artificial factor in previous models can be easily understood if one considers the ratios of v_R values in connection with Eq. (14). The ratio of the two velocities are .075 and 0.18 for the data of Ref. [2] and Ref. [1], respectively, which can be correlated to the corresponding f values used, viz f = 0.06 and f = 0.12.

CONCLUSION

In conclusion, we have presented a generalized model for conduction in polysilicon. The attractive features of our theory are: (a) it can explain the experimental data without the use of f-factor, (b) the conduction process is characterized explicitly by the inherent material properties of grain and the boundary. We feel that our model is particularly suited for describing the electrical properties of laser restructured polysilicon, where because of large grain size the diffusion process is expected to be dominant.

ACKNOWLEDGMENT

This work is supported by the National Science Foundation Grant No. ECS-8009119.

REFERENCES

[1] J.Y.W. Seto, J. Appl. Phys., vol. 46, pp. 5247-5244 (1975).

[2] N.C.C. Lu, L. Gerzberg, C.Y. Lu, and J.D. Meindl, IEEE Trans. Electron Devices, ED-28, pp. 818-830 (1981).

[3] M.M. Mandurah, K.C. Saraswat, and T.I. Kamins, IEEE Trans. Electron Devices, ED-28, pp. 1163-1170 (1981).

[4] N.C.C. Lu, L. Gerzberg, C.Y. Lu, and J.D. Meindl, Electron Device Lett., EDL-2, pp. 95-98 (1981).

[5] C.M. Wu and E.S. Yang, Appl. Phys. Lett., vol. 40, pp. 49-51 (1982).

[6] D.M. Kim, A.N. Khondker, R.R. Shah, and D.L. Crosthwait, IEEE Electron. Device Lett., EDL-3, No. 5, pp. 141-143 (1982).

[7] S.M. Sze, Physics of Semiconductor Devices (New York: John Wiley, 1981).

PROCESSING OF SHALLOW (Rp<150Å) IMPLANTED LAYERS WITH ELECTRON BEAMS

G.B. McMILLAN, J.M. SHANNON* AND H. AHMED
Engineering Department, University of Cambridge, Cambridge CB2 1PZ U.K.

*Philips Research Laboratories, Redhill, Surrey, RH1 5HA, U.K.

ABSTRACT

The multiple-scan method of electron beam annealing has been used to activate shallow (Rp<150Å), highly doped silicon layers produced by ion implantation of arsenic at 10keV. Beam conditions have been optimised ($600Wcm^{-2}$ for 100ms) to produce essentially undiffused layers, as determined by high resolution SIMS, containing high concentrations of electrically active arsenic impurities. Computer modelling of diffusion effects in such layers has been used to identify optimum beam conditions and the calculations have been compared with experimental results. Hot electron device structures, which depend on negligible diffusion and high electrical activity, have been fabricated using the multiple-scan method with a peak annealing temperature of 900°C.

INTRODUCTION

The ability to fabricate shallow (Rp<150Å) highly doped electrically active layers will be an important part of future electronic circuits when scaling down of device dimensions will require the formation of shallow layers for contacting and conductivity control. Such layers also have applications in hot electron devices where band bending is required over interatomic distances (1). The non-equilibrium process of ion implantation can produce electrically active impurity concentrations well in excess of the solid solubility limit following solid phase regrowth at temperatures just above ∿550°C in silicon (2,3). As the annealing temperature is increased, however, precipitation of the excess impurities occurs and diffusion leads to broadening of the dopant profile.

This paper is concerned with the formation of very shallow (Rp<150Å), highly doped, silicon layers by ion implantation of arsenic at an energy of 10keV and processing by rapid isothermal annealing using the multiple-scan electron beam annealing method (4). In producing concentrations of electrically active dopants the aim was to limit profile broadening to a few tens of angstroms (at Cmax/10^3) whereas with conventional annealing of ion implanted material profile broadening of ≈500Å would generally be considered acceptable.

EXPERIMENTAL METHOD AND ANALYSIS TECHNIQUES

Ion implantation at 10keV, with a dose range of 10^{14} to $10^{16}cm^{-2}$, has been used to introduce arsenic impurities into (100) 4-6 Ωcm CZ grown single crystal silicon. (Rp ≈120Å and ΔRp ≈40Å) The material was annealed using the multiple-scan method of electron beam processing (4) which operates in the isothermal regime. 5 x 5 mm samples were processed using the large range of conditions of time and temperature which are available with the multiple-scan method (up to 1.2kW beam power and exposure times from 100 ms to 15 min have been investigated).

Throughout this work layers shallower than 500Å have been investigated and in analysis the resolution limit of several of the diagnostic techniques is being approached. Sheet resistivity measurements using a four-point probe have pro-

vided rapid assessment of the electrical layers after annealing. As punch
through to the substrate can occur when probing thin layers a probe loading of
∿10g has been used and found to give reproducible results. The total dopant
profile both before and after annealing was obtained by using high resolution
Secondary Ion Mass Spectroscopy (SIMS). A DIDA ion microprobe has been used
with a low energy oxygen primary beam to give an enhanced yield of the secondary ions and reduce bombardment induced profile broadening (5). Electronic
gating of the detection system to remove edge effects at the sputter crater
perimeter has been employed and the limitations due to neutrals in the primary
ion beam have been removed. The sample has been biased with respect to the
secondary ion detection optics, thus using energy discrimination, to distinguish
between the $^{75}As^+$ ion and the molecular species $^{29}Si^{30}Si^{16}O^+$. The calibration
of the instrument has been carried out by using known reference samples and
paying attention to the positioning of the sample and secondary ion optics.
The depth scale was derived from Talystep measurements made on a crater formed
after 10 hours erosion (∿2600Å). The depth scale has been corrected for changes
in sputter yield during the initial stages of sputtering (0→39Å) using the relationship proposed by Wittmaack (6). Normal incidence Rutherford backscattering
spectrometry (RBS) has been employed to investigate the lattice regrowth and
dopant position both before and after annealing. The resolution of such RBS is
∿200Å but use has been made of glancing incidence RBS giving a depth resolution
of ∿30Å.

DIFFUSION MODELLING

Computer modelling has been undertaken to consider the effects of diffusion
on the as-implanted profiles in these shallow, highly doped layers. The purpose of the modelling has been two-fold, (a) to identify the optimum beam
conditions to produce high concentrations of electrically active dopant with
essentially no diffusion; and (b) to compare diffusion theory with experimental
results under a variety of annealing conditions of time and temperatures that
have not previously been studied in any detail.

The as-implanted profile has been taken as the starting point for the modelling. (see Fig.1) The diffusion equation was then applied, for the duration of
the thermal cycle produced by the electron beam processing, using the technique
of finite differences.

In calculating the diffusion coefficient, high temperature and high concentration effects must be included in the present work. The total arsenic diffusion coefficient may be written as $D_{AS} = D_iD_eD_vD_c$ where D_i is the temperature
dependent intrinsic arsenic diffusion coefficient, D_e is the built in electric
field enhancement factor, and D_v is the excess vacancy enhancement factor all
of which have been calculated using accepted theory. D_c, the cluster retardation factor, has been calculated according to Hu (7). The diffusion coefficient is thus a function of temperature and impurity concentration. Profile
broadening will therefore be a function of not only the annealing conditions
but the implantation dose. The modelling predicts essentially no diffusion
for annealing times of 100 ms and 1s when the peak temperature is below ≈950°C.
When the peak temperature reaches ≈1000°C and above, significant (>50Å at
$C_{max}/10^3$) diffusion is predicted.

EXPERIMENTAL RESULTS AND DISCUSSION

Initially a direct comparison was made between electron beam annealing and
furnace annealing at temperatures between 550 and 750°C. The annealing time
was 15 minutes for both systems and general agreement was found between the
properties of the annealed material. The investigation of electron beam annealing was then extended to shorter times (15s, 1s and 100ms) and higher temper-

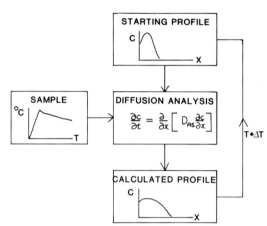

Fig. 1. Block diagram of the diffusion modelling analysis procedure.

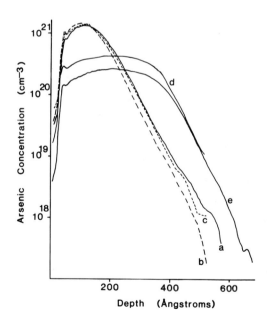

Fig. 2. High resolution SIMS depth profiling of $2 \times 10^{15} cm^{-2}$ As implanted at 10keV into (100) Si and electron beam annealed

a) unannealed

b) 37.5Wcm^{-2} 1s Tmax=610°C

c) 600Wcm^{-2} 100ms Tmax=945°C

d) 67.5Wcm^{-2} 1s Tmax=995°C

e) 75Wcm^{-2} 1s Tmax=1080°C

atures. Representative results are shown in Fig.3 (5×10^{14} and $2 \times 10^{15} cm^{-2}$ As implants) and Fig.4 ($5 \times 10^{15} cm^{-2}$ As implant). These graphs show the sheet resistivity against the peak temperature reached during annealing. They have as a parameter the beam processing time and indicate results for 15 min, 15s, 1s and 100ms duration. Three regimes may be identified : low temperature ($\lesssim 600^{\circ}C$) where only partial recrystallization has occurred during 1s and 15s anneals; an intermediate region where the sheet resistivity is found to be relatively insensitive to the anneal time; and temperatures $\gtrsim 950^{\circ}C$ where diffusion and dopant loss are becoming significant. Similar results have been found for implant doses of 10^{14}, 10^{15} and $10^{16} cm^{-2}$.

Fig.2 shows the results of high resolution SIMS depth profiling of 2×10^{15} cm^{-2} As implanted at 10keV. The implanted, unannealed sample profile curve (a) is essentially the same as that for samples which have been electron beam annealed with $37.5 Wcm^{-2}$ 1s Tmax=$610^{\circ}C$ curve (b) and $600 Wcm^{-2}$ 100ms Tmax=$945^{\circ}C$ curve (c). Therefore, no significant diffusion or dopant loss has occurred under these conditions. The diffusion modelling program has confirmed that these experimental results correspond to theoretical predictions. However, as the temperature is increased diffusion and dopant loss become significant. Curve (d) $67.5 Wcm^{-2}$ 1s Tmax=$995^{\circ}C$ indicates a profile broadening of $\approx 125Å$ at Cmax/10^3 while the SIMS analysis suggests a dopant loss of $\approx 45\%$ or a surface segregation (6). Curve (e) $75 Wcm^{-2}$ 1s Tmax=$1080^{\circ}C$ indicates a dopant loss of $\approx 60\%$ and a profile broadening of 125Å which compares with 75Å predicted using the diffusion modelling program. The dopant loss in the above cases has been confirmed by RBS. Fig.5 shows the channelled and random profiles before and after annealing at $600 Wcm^{-2}$ 100ms and indicates regrowth of the amorphous surface layer while the inset indicates that 75% of the arsenic is on lattice sites (or near lattice sites) after annealing. The same regrowth characteristics have been obtained from a large variety of annealing conditions eg. $750^{\circ}C$ 30s, 5 min and 15 min, $67.5 Wcm^{-2}$ 1s Tmax=$995^{\circ}C$ and $52.5 Wcm^{-2}$ 1s Tmax=$810^{\circ}C$. Glancing incidence RBS has shown no significant diffusion for a $10^{15} cm^{-2}$ As sample annealed with $50 Wcm^{-2}$ 1s Tmax=$780^{\circ}C$.

The minimum sheet resistivities obtained using electron beam annealing were no lower than those obtained following furnace annealing at between 600 and $700^{\circ}C$. Diffusion in 15 min furnace annealing ($4 \times 10^{14} cm^{-2}$ As at 12keV) becomes significant at $750^{\circ}C$ and a profile broadening of 100Å at Cmax/10^3 has been found. The accurate control over time and temperature available with the multiple-scan method has enabled peak temperatures of $950^{\circ}C$ to be reached without diffusion occurring. However, the rate of temperature drop is determined by radiation cooling ($\approx 200 Ks^{-1}$ from $1000^{\circ}C$) and this slow rate has probably allowed some precipitation to occur.

DEVICE APPLICATIONS

Hot electron device structures have been produced by using thin layers with high fields leading to abrupt changes in potential over interatomic distances. In particular, Schottky diodes with barrier height control (8) have been fabricated by implanting arsenic at dose levels between 10^{12} and $5 \times 10^{14} cm^{-2}$ with an energy of 10keV into an epitaxial 5 Ωcm n on n^+ (100) silicon substrate. The diodes were fabricated by implanting the arsenic into windows etched in silicon dioxide. Three different diameters were used to separate bulk effects from those related to the periphery. After implantation and electron beam annealing ($60 Wcm^{-2}$ 1s Tmax=$900^{\circ}C$) nickel was deposited to give a field plate structure. The reverse saturation currents were found to scale with diode area even at the lowest doses which indicated that peripheral leakage was negligible following electron bombardment and annealing of the passivating oxide, at least in barriers with $\Phi \lesssim 0.6eV$. At the highest doses used, the contact was essentially ohmic (see Fig.6), with resistance which is in agreement with that expec-

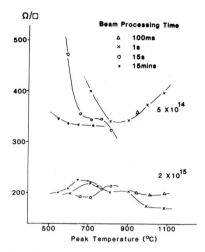

Fig. 3. Sheet resistivity versus peak electron beam annealing temperature, with annealing time as parameter, for 5 x 10^{14} and 2 x 10^{15} cm^{-2} As implanted at 10keV.

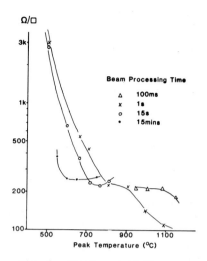

Fig. 4. Sheet resistivity versus peak electron beam annealing temperature, with annealing time as parameter, for 5 x 10^{15}cm^{-2} As implanted at 10keV.

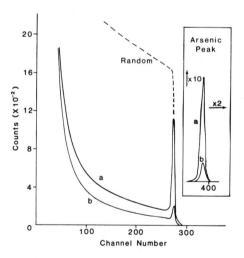

Fig. 5. RBS measurements showing channelled and random spectra for 2 x 10^{15}cm^{-2} As implanted at 10keV a) unannealed b) electron beam annealed 600Wcm^{-2} 100ms Tmax=945°C.

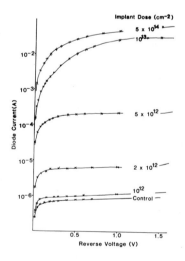

Fig. 6. Reverse current characteristics of Schottky diodes (100µm diameter) with 10keV implanted surface layers and electron beam annealed 60Wcm^{-2} 1s Tmax=900°C.

ted from the epitaxial layer. Following an implant of 5 x 10^{12}cm^{-2} the magnitude of the current and the value of the reverse voltage required to reach saturation (0.35V) both suggested that all the implanted arsenic was electrically active and that negligible diffusion had occurred during electron beam annealing. A furnace anneal at 750°C for 15 min, however, while giving similar values for the reverse current, required a greater reverse voltage (0.55V) before saturation occurred showing that some diffusion of the impurity profile had taken place (8). These results are in qualitative agreement with results discussed earlier.

CONCLUSIONS

It has been shown that multiple-scan electron beam annealing can activate arsenic impurities in very shallow (Rp<150Å) layers with insignificant diffusion at temperatures as high as 950°C. A minimum resistivity of ≈2 x 10^{-4} Ωcm was obtained. At a temperature of 1000°C dopant loss and diffusion became significant. Properties of electron beam annealed and furnace annealed samples at processing temperatures between 550 and 700°C are in agreement. Under these conditions there was negligible diffusion and precipitation. No excess leakage currents were detected from Schottky junction barriers <0.6eV processed using electron beams. A wide range of effective barrier heights should be obtainable by using multiple-scan electron beam annealing at temperatures around 950°C.

ACKNOWLEDGEMENTS

It is a pleasure to acknowledge the help given by A. Gill and B.F. Martin of Philips Research Laboratories, Redhill. The authors are also grateful to J.B. Clegg for his help and advice with the high resolution SIMS. The RBS measurements were made at the University of Surrey.

REFERENCES

1) J.M. Shannon, Jr. Nuclear Inst. and Methods, 182/183, 545 (1981)

2) J.S. Williams and R.G. Elliman, Jr. Nuclear Inst. and Methods, 182/183, 389 (1981)

3) G. Dearnaley, J.H. Freeman, R.S. Nelson and J. Stephen, Ion Implantation, North Holland (1973)

4) R.A. McMahon and H. Ahmed, Electron. Lett. 15, 45 (1979)

5) J.B. Clegg and D.J. O'Connor, Appl. Phys. Lett. 39, 997 (1981)

6) K. Wittmaack and W. Wach, Jr. Nuclear Inst. and Methods, 191, 327 (1981)

7) S.M. Hu in Atomic Diffusion in Semiconductors Ed. by D. Shaw, Plenum Press, New York (1973)

8) J.M. Shannon, Solid-St Electron., 19, 537 (1976)

SELECTIVE LASER ANNEALING FOR DEVICE PROCESSING

I.D. CALDER, A.A. NAEM and H.H. NAGUIB
Bell-Northern Research, Ottawa, Canada K1Y 4H7

ABSTRACT

Selective laser crystallization of undoped polysilicon films has been achieved through the use of a patterned Si_3N_4 anti-reflection (AR) coating. The recrystallized poly-Si beneath the AR cap exhibits an etch rate 50-90% lower than the surrounding uncapped material, allowing anisotropic etching of poly-Si for the fabrication of MOSFET gates. Undercut is reduced by at least a factor of two from unannealed material. Annealed edge profiles are uniform within ± 0.03μm for plasma etching (± 0.05 for wet etching) compared to ± 0.1μm (± 0.25μm for wet etching) for unannealed regions. The sheet resistivity of 0.5μm films doped by phosphorus diffusion was reduced from an initial value of 82±5 $\Omega/\square$ to 40±8 $\Omega/\square$ when the dopant was diffused into recrystallized poly-Si and to a final value of 10.2 ± 0.2 $\Omega/\square$ after a further laser activation step. Potential applications in VLSIC processing are discussed.

INTRODUCTION

Research into cw laser crystallization of polysilicon films has been directed towards the formation of large grain or single crystal silicon films for silicon-on-insulator (SOI) devices [1] or three dimensional integration [2]. The resistivity of heavily doped poly-Si, as used in MOS gates and inter- connects, can also be reduced by a factor of two [3], enabling faster devices to be fabricated. When LA is applied to real devices, the variations in optical and thermal properties across complex structures will cause the attainable temperature to be strongly dependent upon the sample structure. In particular, conditions that will melt one region of a structure may not melt another region, as desired, while raising the temperature of a third region to a destructively high value.

Our objective in this work has been to characterize a technique for selective laser annealing (SLA) of IC structures [2,4], so that only designated regions of poly-Si are melted and crystallized. We present results on the effects of SLA on the grain morphology, etching characteristics and sheet resistivity of poly-Si films. Some potential applications of this technique in the fabrication of conventional and novel MOS devices are outlined.

EXPERIMENTS

Figure 1 illustrates the reflectivity R [5] of solid and liquid Si for Ar laser light as a function of the thickness of a Si_3N_4 antireflection (AR) coating on silicon. These calculations take into account the spectral content of Ar laser light and the dependence of the optical constants on wavelength and temperature [6,7]. They indicate that since we intend to melt the silicon, the selected thickness of the Si_3N_4 AR film should be ~ 53nm, i.e. intermediate between the optimum values for solid and liquid Si. In practice, we use a Si_3N_4 thickness of 60 ± 5nm to improve the uniformity and the quality of the film as an etch mask. Then from the absorptivities (1-R) we can see how much power is required to reach the melt threshold, when the temperature T reaches melting (T_M), and the crystallization threshold, when silicon melts to a significant

Mat. Res. Soc. Symp. Proc. Vol. 13 (1983) ©Elsevier Science Publishing Co., Inc.

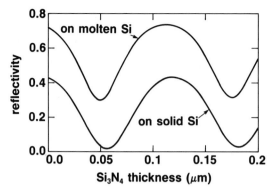

Figure 1. Optical reflectivity for light from an argon laser in the multiline mode. The material is silicon at melting temperature with a coating of Si_3N_4. The optimum AR coating thickness is 56nm on solid Si and 50nm on molten Si.

depth (R = reflectivity for molten Si). If we normalize the absorbed power to P = 1 at the melt threshold for complete absorption, then P = 1.67 is required to reach the melt threshold in uncoated areas while only P = 1.5 is needed to reach the crystallization threshold under the AR coating. In fact the uncoated material does not achieve a deep melt until P = 3.5. Therefore we can selectively crystallize those areas of a silicon film covered by a patterned Si_3N_4 AR cap.

Based on the preceeding analysis, the samples were structured as shown schematically in Fig. 2. A 50nm layer of SiO_2 was thermally grown in dry O_2 at 1000°C on a <100> Si wafer. This was followed by the low pressure chemical vapour deposition of a 0.5µm layer of undoped poly-Si at 650°C and a 60nm film of Si_3N_4 at 800°C. The nitride film was then patterned by standard photolithographic and etching techniques into rectangular islands with dimensions varying from 2µm x 100µm up to 5mm x 5mm.

Laser annealing was carried out in a Coherent System 5000 laser annealer by raster scanning of a focussed argon laser beam in the multiline mode across the surface of the sample. With the use of a heated vacuum chuck, the substrate was maintained at a temperature of 500°C. The 50µm laser spot was scanned with 80% overlap at an incident power of 3.25 - 5.25 watts and a speed of 50 cm/s.

After laser processing the batch was divided into four groups for analysis. The Si_3N_4 was removed from the first group so that the poly-Si could be Secco etched [8] to decorate the grain boundaries and observe the crystallite structure by Nomarski interference microscopy and the SEM. The second group was wet etched to study edge profiles and etch uniformity in the SEM. A solution of $CO_3COOH:HNO_3:HF$ was used since it preferentially removes poly-Si over Si_3N_4 with

Figure 2.

The sample structure.

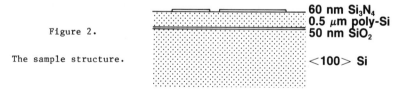

60 nm Si_3N_4
0.5 µm poly-Si
50 nm SiO_2

<100> Si

a ratio of several hundred to one. A similar procedure was carried out for the third group, except that dry etching in a plasma of CF_4 + 5% O_2 was used. In this case a photoresist etch mask was required, since the plasma etched both poly-Si and Si_3N_4. The Si_3N_4 was removed from the fourth group by etching in HF prior to performing a $POCl_3$ deposition for 45 minutes at 900°C, followed by drive-in at 1000°C for 30 minutes. Sheet resistivity was then measured using a four point probe. Then a laser activation step was carried out at a speed of 50 cm/s with a spot size of 40μm and a laser power of 5 watts (below the melt threshold), followed by a repetition of the sheet resistivity measurements.

RESULTS

From the Secco etched group it was observed that samples laser annealed at 4.5 watts showed partial crystallization of AR coated islands, indicating that this laser power is just above the crystallization threshold. At a laser power of 5 watts, a deep melt occurred and the coated poly-Si islands crystallized into large elongated grains. Grains as large as 5μm x 50μm were observed in the 5mm x 5mm Si_3N_4 coated areas. However, these large grains exhibited protrusions or pits similar to those reported by Kamins [9] and Palkuti et al. [10]. No such defects were observed in laser annealed poly-Si films coated with patterned Si_3N_4 islands of size ≤ 20μm x 100μm.

Figure 3 shows the etch rate of AR coated and uncoated poly-Si films, both subjected to identical laser annealing conditions, as a function of etchant and laser power. For wet etching there is a fall in the etch rate from an initial value for unannealed poly-Si followed by an increase in the etch rate with increasing laser power and ultimately a decline in the etch rate at a highest level of incident power. These results indicate that the maximum etch rate ratio between uncoated and coated laser annealed poly-Si films is 2:1 and occurs at a laser power of 4 watts. Etch rates in the plasma were measured only for unannealed and fully crystallized, uncoated poly-Si films, giving a ratio of 11:1. Since no change in the edge profile of plasma etched islands was observed

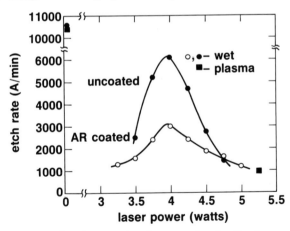

Figure 3. Etching rates for AR coated and uncoated poly-Si, laser annealed under the same conditions. There is a peak of 2:1 in the wet etch rate ratio at a laser power of 4 watts. The plasma etching rate remains essentially constant at the indicated zero power value, until the crystallization threshold is passed, falling then to less than 10%.

446

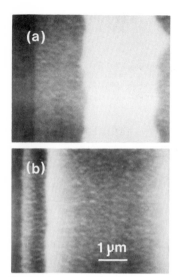

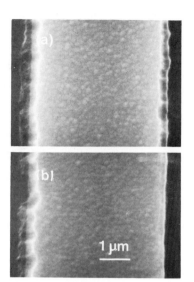

Figure 4. Etch profile of SLA islands after wet etching with the Si_3N_4 cap still present (top view). Here the samples were deliberately over-etched to emphasize edge non-uniformities. The sample (a) was not annealed while for (b) the laser power was 4.25 watts.

Figure 5. Etch profile of SLA islands after plasma etching (top view). The sample (a) was not annealed while for (b) the laser power was 5.25 watts. Because the photoresist mask was slightly misaligned (to the right) with respect to the Si_3N_4 mask, only the right edge profile is meaningful.

until the laser power exceeded the crystallization threshold, this ratio likely held true for selectively annealed, coated material.

SEM micrographs of etched samples before and after SLA are shown in Figs. 4 and 5. As can be inferred from these figures, undercut is reduced by a factor of more than two after SLA. Even more striking is the improvement in edge uniformity. For wet etching the edge uniformity is improved from ± 0.25µm to ± 0.05µm, while for dry etching the change is from ± 0.1µm to ± 0.03µm. Such improvement in edge profiles is highly desirable for VLSI structures.

Doping of poly-Si with phosphorus resulted in a sheet resistivity of 82 ± 5 $\Omega/\square$ for unannealed areas and somewhat lower resistivities (30-45 $\Omega/\square$) for regions that had been previously recrystallized. (Fig. 6) However after the solid phase laser activation step, sheet resistivity was reduced to 1.2 $\Omega/\square$ ±0.2 $\Omega/\square$ from wafer to wafer and among various initial laser recrystallization treatments while within one wafer variations were ± 0.1 $\Omega/\square$.

DISCUSSION AND APPLICATIONS

The experimental results described in the previous section indicate that melting and crystallization of AR coated poly-Si islands occur at a certain threshold laser power. Under the same laser annealing conditions, uncoated poly-Si areas anneal slightly without melting and with only a minor change in grain morphology. Since etch rates are proportional to available surface area, an

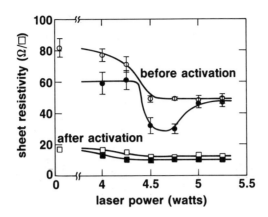

Figure 6. Sheet resistivity of 0.5μm phosphorus doped poly-Si. The filled symbols represent regions left uncoated during the laser crystallization step, while the open symbols indicate samples which had an AR coating. Results are shown before (●,○) and after (■,□) the laser activation step.

increase in grain size will reduce etch rates because fewer grain boundaries exist to enhance the effective surface area. This hypothesis is confirmed since the progression from unannealed poly-Si to LA poly-Si generally corresponds to increasing grain size and decreasing etch rate. there is a maximum change in etch rate of 10-11 times, although for wet etching differences between coated and uncoated regions that have both been similarly laser treated are less (Fig. 3). The lack of grain boundaries also removes most paths for rapid etching into the island edges, so that uniformity is greatly improved.

The absence of pitting of the poly-Si means that careful treatment of the Si_3N_4 or non-standard deposition processes are not required as in the case of blanket Si_3N_4 caps [10]. Since these defects are known to arise from interfacial stresses [10], the absence of pitting under small Si_3N_4 islands can be attributed to stress relief at the island edges.

Etch rates are affected not only by the distribution of grain boundaries, but also by other defects present in the material. The creation and/or absence of defects (in the presence or absence of a Si_3N_4 cap) below the melt or crystallization thresholds may account for the complex behaviour of the wet etch rate as a function of laser power (Fig. 3). This phenomenon apparently does not occur for plasma etching. From the resistivity measurements we can differentiate between the effects of crystal size and degree of activation. There is a reduction by a factor of two when the phosphorus is diffused into large grain poly-Si, although the reproducibility is not good. Since the diffusion length during drive-in is only 65nm, there will be a high concentration of phosphorus near the surface of the large grains, with a consequent reduced electrical activity. Subsequent activation has a much greater effect (4 times reduction in resistivity) and results in a high degree of sample-to-sample uniformity.

From all these results it can be appreciated that the structure shown in Fig. 2, in conjunction with SLA and preferential etching, can be used to produce well-defined small-geometry gates and interconnects with low electrical resistivity for VLSI MOSFETS. Reduced undercut and much improved edge uniformity

448

will allow smaller structures to be patterned reproducibly. When this process is
followed by diffusion of the polysilicon gates while the source/drain regions
are protected, high conductivity small-geometry gates and interconnects can be
obtained. The same techniques can be applied to the fabrication of 3-D devices
[2] so that a film region can be melted and recrystallized without affecting
neighbouring areas.

CONCLUSIONS

A new process has been described, in which a Si_3N_4 anti-reflection coating
is used to achieve selective laser annealing and preferential etching of laser
annealed polysilicon. Compared with standard etching, undercut has been reduced
by a factor >two and edge uniformity has been reduced to ± 0.03μm, without
pitting of the poly-Si, so that precise, reproducible small-geometry features
can be found. In addition, resistivity is reduced by 8 times from unannealed
material, but only after a second (solid phase) laser anneal.

ACKNOWLEDGEMENTS

We wish to express our appreciation to T.W. MacElwee for his work on the
laser annealer, to G.M. Smith for his help with the SEM work, and to J.J. White
and J.P. Ellul for sample preparation and many useful discussions. We
particularly wish to thank Prof. A.R. Boothroyd for his active cooperation and
many helpful suggestions. This work was supported in part by the National
Research Council of Canada through its Industrial Research Assistance Programme
(IRAP Plasma 1100).

REFERENCES

1. T.I. Kamins, K.F. Lee, J.F. Gibbons and K.C. Saraswat, IEEE Trans. Electron
Dev. ED-27, 290 (1980).

2. J.-P. Colinge, E. Demoulin and M. Lobet, IEEE Trans. Electron Dev. ED-29,
585 (1982).

3. T. Shibata, K.F. Lee, J.F. Gibbons, T.J. Magee, J. Peng and J.D. Hong, J.
Appl. Phys. 52, 3625 (1981).

4. J. Sukarai, S. Kawamura, M. Nakano and M. Tagai, Appl. Phys. Lett. 41, 64
(1982).

5. M. Born and E. Wolf, "Principles of Optics" (Pergamon, New York 1959).

6. G.E. Jellison, Jr., and F.A. Modine, Appl. Phys. Lett. 41, 160 (1982).

7. K.M. Shvarev, B.A. Baum and P.V. Gel'd, Sov. Phys. Solid State 16, 2111
(1975).

8. F. Secco d'Aragona, J. Electrochem. Soc. 119, 948 (1972).

9. T.I. Kamins, J. Electrochem. Soc. 128, 1824 (1981).

10. L.J. Palkuti, C.S. Pang, T.J. Magee, R. Ormond, J.B. Price, E. Reed, J.R.
Yeargain and G. Davis, Electrochem. Soc. Ext. Abst. 82-1, 236 (1982).

PULSED ELECTRON BEAM ANNEALING INDUCED DEEP LEVEL DEFECTS IN VIRGIN SILICON

D. BARBIER[+], M. KECHOUANE[++], A. CHANTRE[++], A. LAUGIER[+]

+ : INSA de Lyon, Laboratoire de Physique de la Matière, Bât. 502, 20 Avenue A. Einstein 69621 VILLEURBANNE CEDEX (France)
++ : CNET, B.P. 42, 38240 MEYLAN (France)

ABSTRACT

DLTS has been used to investigate deep level defects induced by Pulsed Electron Beam Annealing (PEBA) in virgin (100) boron doped silicon. Various PEBA conditions were selected resulting in different molten layer thicknesses, melt front velocities and thermal gradient distributions. Discrete hole traps distributed in the regrowth layer were observed in all the annealed samples. The activation energies and thermal signatures of these levels do not correspond to already known defects except for one level which has been assigned to the carbon interstitial substitutional pair. Carbon contamination during irradiation is the most probable explanation for the creation of this defect. Other discrete hole trap levels are likely to be generated by quenching of the molten layer as far as their profiles do not extend beyond the regrowth layer. Moreover, a broad band of levels, characteristic of extended defects, has been observed only on the samples which have suffererd the highest thermal stresses. This band of levels might be related to the generation of dislocation networks as recently observed by means of T.E.M. on the same PEBA processed samples.

I. INTRODUCTION

In recent years it has been demonstrated that Pulsed Electron Beam Annealing (PEBA) can be successfully used to remove implantation damage in silicon by liquid phase epitaxial regrowth for shallow junction formation (1). However the electrical quality of the devices achieved by PEBA as well as with any other transient thermal process can be limited by possible defect introduction in the material depending on the annealing kinetics produced by the pulse.

In this work, Deep Level Transient Spectroscopy has been used to investigate possible electrically active defect generation by PEBA in virgin boron doped (100) CZ silicon as a function of different PEBA kinetics.

II. EXPERIMENT

The apparatus used for annealing is a SPI-300 pulsed electron beam processor which delivers polykinetic electron pulses of 50 ns in duration with variable electron parameters (energy-current). The electron pulses are obtained by discharge of a coaxial capacitor in a plasma field emission diode with a high density graphite cathode and a tungsten mesh as anode. Diode geometry can be modified to yield different pulse parameters. A monitoring system allows one to record current and voltage waveforms during each shot for electron beam time-resolved spectroscopy. A more detailed description of this apparatus is given in ref.2. The electron energy deposition profiles in silicon have been determined by Monte-Carlo simulation and the temperature profiles were calculated by solving the

Mat. Res. Soc. Symp. Proc. Vol. 13 (1983) ©Elsevier Science Publishing Co., Inc.

450

one-dimensional heat flow equation. During irradiation the depth-temperature profile in the specimen is roughly proportional to the time-integrated energy deposition profile which thus controls the thermal gradients distribution in the solid beneath the molten zone. Two different electron beam pulses were used in this work and the resulting normalized depth-dose profiles in silicon are plotted on Fig. 1. For a fixed electron energy density the maximum thermal gradient induced in the solid by pulse n° 1 is about twice the one induced by pulse n° 2 as shown on Fig. 2 which gives the two calculated depth temperature profiles immediately after pulsing at 1.6 J/cm^2.

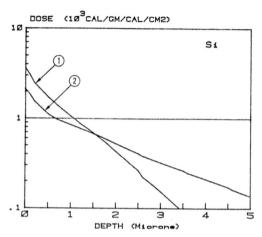

Fig.1 - Monte-Carlo calculation of the normalized electron energy deposition profiles in silicon for pulses n° 1 and n° 2.

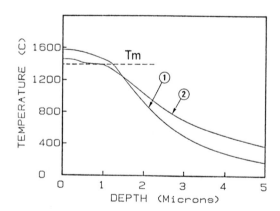

Fig.2 - Computed depth-temperature profiles at the end of a 1.6 J/cm^2 PEBA shot for pulses n° 1 and n° 2.

Moreover, the annealing kinetics strongly depend on the beam fluence as can be seen in table I where are given the main annealing parameters corresponding to the 4 PEBA conditions used in this work. In particular, the melt front velocity decreases quasi linearly in the 0.9-1.6 J/cm^2 fluence range for pulse n°1 while it is about 1.8 m/s at 1.6 J/cm^2 for pulse n°2. After annealing, aluminium Schottky barriers were evaporated at room temperature directly onto the PEBA processed area and electrically active defects were investigated using DLTS (3) with a sensitivity $\delta C/C \sim 10^{-6}$. The experimental set-up has been described in ref.4. With the Schottky barrier structure used only the majority carrier traps could be studied. Before investigating the PEBA processed samples, DLTS spectra on as-grown samples were made for comparison. They did not exhibit any detectable structure.

III. RESULTS

The sample processed with pulse n°1 at 0.9 J/cm^2 and 1.2 J/cm^2 both exhibit two discrete hole trap levels H_1 (0.19 eV) and H_2 (0.55 eV) as shown on spectrum n°1 of Fig.3 for which the biasing conditions correspond to an analyzed surface layer thickness of about 0.4 μ. The concentrations of each trap is the same for both fluences (0.9 and 1.2 J/cm^2).

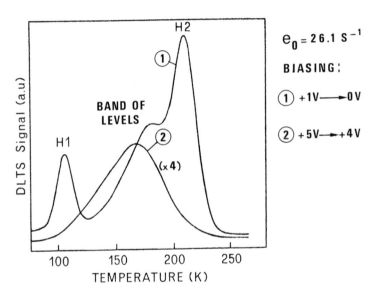

Fig.3 - DLTS spectra for samples annealed with pulse n°1 : (1) at 0.9 and 1.2 J/cm^2, superficial contribution (x < 0.4 μ) ; (2) at 0.9 J/cm^2 inner contribution (x > 0,4 μ)

In addition a broad structure partially superimposed on the peak H_2 is also observed which is likely to be due to a band of levels. The capture cross section of H_1 and H_2 are respectively 7×10^{-16} cm^2 and 3×10^{-12} cm^2. The value obtained for H_2 is very high but the accuracy of the measurements was poor because of the presence of the broad structure. When the biasing conditions are changed to investigate the levels deeper than 0.4 μ in the 0.9 J/cm^2 processed sample the discrete hole traps H_1 and H_2 disappear and the broad band of levels is the only structure left (spectrum n°2 on Fig.3) with only a shift of its maximum. When pulse n°1 is applied at 1.6 J/cm^2 the DLTS spectrum (N°1 of Fig.4) exhibits again two discrete hole trap levels. The first one H$_0$ (0.19 eV) is different from H$_1$ because the capture cross sections and the thermal signatures are not identical and the second one is identical to the level H$_2$ observed in the sample processed at 0.9 and 1.2 J/cm^2. When the surface contribution of the DLTS spectrum is gradually reduced the amplitudes of the peak H$_0$ and H$_2$ are lowered but they do not disappear as can be seen on spectrum n°2 of Fig.4. With the biasing conditions used this indicates that the hole trap levels H$_0$ and H$_1$ are distributed in a layer about 1 μm thick.

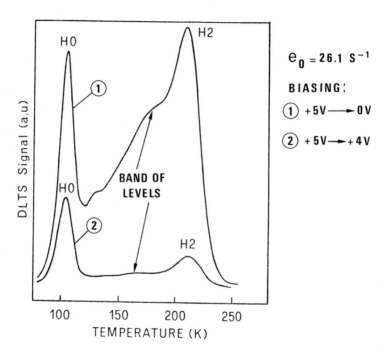

Fig. 4 - DLTS spectra for samples annealed with pulse n°1 at 1.6 J/cm^2. (1) total contribution of the analyzed layer ; (2) inner contribution.

Moreover, the broad band of levels is again observed on the samples processed at 1.6 J/cm^2 with pulse N°1 whatever the biasing conditions (see Fig.2).

On the sample processed with pulse N°2 at 1.6 J/cm^2, the band structure observed on the samples processed with pulse N°1 is not clearly detectable. Moreover, two discrete hole traps H$'_1$ (0.36 eV) and H$'_2$ (0.505 eV),different from those previously described,are observed with a sharp profile confined in a 0.5 μm thick surface layer. The capture cross sections are respectively 1.4×10^{-15}cm^2 for H$'_1$ and 7.6×10^{-14}cm^2 for H$'_2$.

IV. DISCUSSION

The DLTS spectra described above yield the first experimental evidence for the dependence between annealing kinetics and electrically active defects generation by

PEBA in virgin boron doped silicon. However most of the hole trap levels observed in this work cannot be assigned to already known defects. Only the thermal signature of H'_1 is very close to the one of the carbon interstitial-substitutional pair defect (C_i-C_s) which is present in low concentration in as-grown silicon (5). DLTS observations of this defect in N^+P PEBA solar cells have been recently reported by Schott (6). Carbon contamination from the plasma field emission diode during the melting process is the most probable mechanism for generation of C_i-C_s pairs. The shallow distribution of H'_1 supports this hypothesis. One must note that H'_1 is only detected in samples processed with pulse n°2. However, the maximum of H'_1 is located in the same temperature range as the broad band of levels observed in samples processed with pulse n°1. This explains why it cannot be detected in these samples even if the C_i-C_s pair defect is generated as well as in the samples processed with pulse n°2. Analysis of the depth distribution of the defects compared with the PEBA process parameters allows one to make assumptions about the defects creation mechanism and their origin. First, it appears that the discrete hole trap levels are all located in a zone which has been melted by the pulse whatever its thermal effects. Indeed, the DLTS spectra of Fig.3 clearly indicates that for a 0.9 J/cm² PEBA shot with pulse n°1, the profiles of H_1 and H_2 do not extend over more than about 0.4 µ from the surface which is precisely the calculated maximum molten depth given in table I for these PEBA conditions. In addition, if the fluence is raised up to 1.6 J/cm² the peak H_0 and H_2 are still detectable up to about 1 µ from the surface. This result is consistent with an increase of the maximum molten depth up to 1.2 µ. The difference in the thermal signature of H_0 and H_1 may be related to modification in the annealing kinetics between 1.2 to 1.6 J/cm² (melt front velocity, quenching rate). Moreover H'_1 and H'_2 are distributed in a layer of about 0.5 µ which is less than the calculated molten layer thickness for a 1.6 J/cm² PEBA shot with pulse n°2. So we believe that all the discrete hole traps observed are generated by quenching of the molten layer. We suggest point defect trapping at the liquid solid interface as a possible mechanism which has been already proposed by Kimerling for laser annealing (7). However, other results such as precise determination of the hole traps depth-profiles will be necessary to validate this hypothesis.

On the other hand, the broad band of levels observed only in samples processed with pulse n°1 is likely to be due to extended defects generated by thermal stress in the solid beneath the molten zone. Because of the deep profile of the broad hole traps band we suggest that dislocations might be generated in the highly stressed zone of the solid material beneath the melt front. Although the temperature gradients are reduced during resolidification the thermal stress induced in the solid might be sufficiently high to generate dislocations up to the surface during the regrowth mechanism. Moreover, the maximum temperature gradient induced by pulse n°1 is higher even at 0.9 J/cm² than the one induced by pulse n°2 at 1.6 J/cm² (see table I). So the resulting thermal stress could be sufficiently high to produce crystal deformation in the case of pulse n°1 but not in the case of pulse n°2. T.E.M. observations on samples processed with pulse n°1 have revealed dislocation networks initiated beneath the molten layer and extending up to the surface (8). Experiment will be shortly carried out in order to establish the correlation between dislocations and the observed band of levels in PEBA processed samples.

Electron spectrum	Fluence (J/cm²)	Molten layer thickness (µm)	Melt front velocity (cm/s)	Maximum liquid phase duration (ns)	Maximum thermal gradient during irradiation (°K/µm)
1	0.9	0.4	300	130	600
1	1.2	0.8	220	350	650
1	1.6	1.2	120	800	700
2	1.6	1.0	180	650	400

TABLE I

454

V. CONCLUSION

DLTS has been successfully used to investigate electrically active defects induced by PEBA in virgin (100) boron doped C.Z. silicon. Point defect levels originating from the molten zone have been observed whatever the PEBA conditions. However annealing kinetics play a significant role in the defect nature and their spatial distribution. The presence of the carbon interstitial-substitutional pair defect is not suprising because of probable carbon contamination of the specimen during the PEBA melting process. Moreover a broad band of levels characteristic of extended defects are observed in the sample highly stressed by thermal gradients.

ACKNOWLEDGEMENT

We thank D. Bois and G.Chemisky for valuable and exciting discussions.

REFERENCES

1.   A. C. Greenwald, A. R. Kirkpatrick, R. G. Little and J. A. Minnucci
     J. Appl. Phys. 50, 2, 783 (1979)

2.   D. Barbier, A. Laugier and A. Cachard, Proceedings of the International Meeting on the relationship between epitaxial growth conditions and the properties of semiconductor epitaxial layers. Perpignan. J. de Phys. Appl. (to be published) (1983)

3.   D. V. Lang, J. Appl. Phys. 45, 3023 (1974)

4.   A. Chantre, M. Kechouane and D. Bois, Laser and Electron Beam Processing of Materials, MRS Proceedings, Academic Press, p. 385(1981)

5.   J. R. Troxell, Ph.D. Thesis, Lehigh University (1979)

6.   J. T. Schott, H. M. Deangelis and P. J. Drevinsky, J. Electr. Mat. 9, 2, 419 (1980)

7.   L. C. Kimerling, J. L. Benton, Laser and Electron Beam Processing of Materials, MRS Proceedings, Academic Press, p. 385 (1980)

8.   M. Tholommier, D. Barbier, M. Pitaval, M. Ambri and A. Laugier, J. Appl. Phys. (to be published) (1982)

CRYSTALLIZATION OF AMORPHOUS SILICON FILMS BY PULSED ION BEAM ANNEALING

J. GYULAI,* R. FASTOW, K. KAVANAGH, M. O. THOMPSON, C. J. PALMSTROM, C. A. HEWETT,** AND J. W. MAYER
Department of Materials Science, Cornell University, Ithaca, NY 14853
*Permanent Address: Central Res. Inst. for Physics, H-1525 Budapest
**Permanent Address: Dept. of Elec. Engineering, Univ. of California, San Diego, La Jolla, CA 92093

ABSTRACT

Regrowth by pulsed proton beam was studied for evaporated amorphous Si layers, for layers converted to polycrystalline by annealing (both with and without Ge markers) and for implantation-amorphized SOS films. Silicon-on-sapphire showed the lowest threshold for regrowth. Amorphous silicon melted at about 0.2 J/cm^2 lower fluences of protons of 380 keV energy than crystalline Si. Implanted Sb into SOS occupies lattice positions exceeding the solid solubility.

INTRODUCTION

Non-equilibrium annealing techniques, e.g. lasers or electron beams have received attention in the past few years [1]. Recently, pulsed ion sources producing protons or heavier particles were used to form metastable layer structures [2] using the equipment in Plasmaphysics Laboratory at Cornell University [3].
Annealing properties of ion beams are analogous to that of the electron beams. The situation closely resembles that of splat cooling, where a molten layer is in good contact with a heat sink.
In this study, one approach is described in which markers were used to detect regrowth properties of amorphous and crystalline silicon layers. Parallel to this, the measurement of real-time electrical conductivity of the molten layers was also performed [4].

EXPERIMENTAL

In one set of experiments, Si-Si layered structures were produced by vacuum evaporation. Thin layers ($\cong$50 Å) of germanium were deposited as markers between layers of Si. In some cases, a Ge layer was first deposited, in other instances it was sandwiched between Si layers. All samples were subjected to a 350°C heat treatment for 2 hours to increase stability of the layers. A part of the samples was converted to polycrystalline silicon by an additional anneal at 650°C, 2 hours.
A second set of experiments was based on SOS structures, where the amorphous state was reached by room temperature implantation of Xe or Sb at doses of about 10^{15} ions/cm^2.
Pulsed ion beam annealing was performed on a $\leq$450 keV accelerator in Cornell Plasmaphysics Laboratory. The incident energy of protons was chosen between 280 and 400 keV. The total energy flux was tailored by changing the distance between diode and the sample. The incident fluence was measured by an equivalent temperature rise of a free-hanging Ni-platelet spotwelded onto a thermo-

Mat. Res. Soc. Symp. Proc. Vol. 13 (1983) © Elsevier Science Publishing Co., Inc.

couple and placed adjacent to the sample. Pulses were of 2-400 nsec duration. The beam contained a high percentage of shallow-penetrating particles, too. This was shown by masking the Ni-platelet with a 5000 Å thick, self-supporting Si window. Energy readings in this case dropped to one third of their value. Because of the two beam component, comparisons were made between samples of nearly identical layer thicknesses.

Analysis of the samples was mostly made by backscattering spectrometry using MeV He ions of a 1.1 MV General Ionex tandem accelerator. We considered that "melting" occurred, when the Ge markers broadened after the pulse to widths corresponding to a diffusion coefficient of 10^{-4}-10^{-5} cm^2/sec. Some samples were then subjected to TEM studies.

In some cases, for intermediate energies and high fluences, the samples had a structured surface. To avoid possible arguments on evaporation, comparative shots were made through 5000 Å thick Si single crystal windows as cap layers.

RESULTS AND DISCUSSION

Germanium between evaporated layers. Fig. 1 shows results or irradiation of a sandwich with different ion doses. An irradiation with 0.7 J/cm^2 makes the "upper" marker somewhat spread out and causes no alteration of the deeper one. At this energy, polycrystal formation was found and was accompanied by a loss of smoothness of the sample. Polycrystal formation was quasi-oriented. At higher energies, the marker spreads, epitaxial growth takes place and the Ge atoms are on substitutional sites.

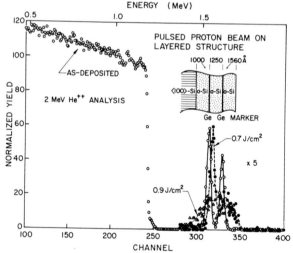

Fig. 1. Random backscattering spectra of Ge markers sandwiched between evaporated Si layers

Pulsed ion beam annealing seems promising to prevent polycrystal formation of silicon before melting temperatures are reached for high-energy pulses. Thus, it was the desire to demonstrate differences in energies of melting between amorphous and crystalline Si. To this end, part of the wafers was converted into polycrystalline layers. After this, both, otherwise identical, samples were irradiated simultaneously. On Fig. 2 only the Ge markers are shown, demonstrating a difference between amorphous and polycrystalline cases.

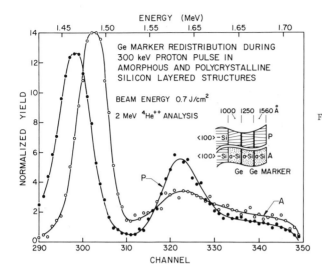

Fig. 2. Comparison of spreading of Ge markers in amorphous and crystalline Si after identical treatment with ion pulse

In a set of experiments, the Ge marker was deposited first and the asymmetric character of the spreading was monitored. This character was considered as a sign of melting, and not an artifact of possible roughening of the surface, except for cases where visible cracks are formed at very high fluences. Fig. 3 shows the spreading of the marker for two different proton fluences. For 0.81 J/cm^2, melting of the amorphous layer was found without melting of the crystal. For samples, irradiated at 0.9 J/cm^2, the crystalline silicon also melts. To separate effects of the shallow and deep penetrating part of the beam, shots were made through 5000 Å Si single crystal windows as cap layers.

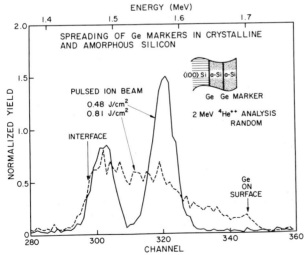

Fig. 3. Asymmetric motion of Ge marker on interface between crystalline and amorphous Si. Another marker was embedded in amorphous Si

This thickness of silicon stops ≤ 50 keV protons and results in a similar decrease of the ion energy. For "through-window" shots, a small tail from the upper Ge marker appeared in the amorphous Si for fluences ≥ 0.4 J/cm^2.

Defects in regrowth. The amorphous layers regrew in a rather perfect form. For polycrystalline samples, a dislocation network is formed which causes extra dechanneling at a depth corresponding to the deeper Ge marker. The dechanneling can be attributed to inadequate annealing of polycrystalline Si as compared to amorphous Si.

TEM studies on samples, where the Ge markers were sandwiched between the amorphous Si layers confirmed the findings with backscattering spectrometry. Fig. 4 shows the quality of regrowth on a polycrystalline sample (1.6 J/cm^2) The regrown layers are single crystals. On the picture the heavy imperfections in the layer are visible. Another example is presented on Fig. 5 for a sample irradiated with a fluence of 1.4 J/cm^2. This shows the structured surface, but otherwise single crystalline nature of the layer.

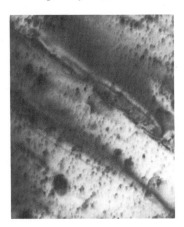 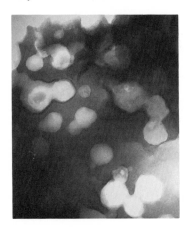

Fig. 4,5. TEM pictures of sandwich structures. Fluences: 1.6 and 1.4 J/cm^2

Implanted SOS samples. Xenon implantation (300 keV, 10^{15} atoms/cm^2) amorphized the silicon layer to a depth of 1500 Å. Successive pulse annealing resulted in regrowth. The crystal perfection was crucially dependent on the energy density. At low energy densities, the amorphous layer does not reorder, at intermediate energy densities good crystal perfection is found and at high energies, polycrystalline formation occurs (Fig. 6). These findings are in accord with unpublished results of F. Stafford and D. Ast (Cornell University) using TEM on similarly treated samples.

At highest energies, the silicon simply "balled up" and/or evaporated. For fluences on Fig. 6 no appreciable spreading of the Xe marker was found.

Antimony implantation was chosen to check differences between inert and "active" dopants. Fig. 7 shows that at proton fluences equivalent to ≥ 0.4 J/cm^2 produce a good quality regrowth. At lower fluences some defects around R$_p$ remain. Curves for 0.42 J/cm^2 show that a fast redistribution of the Sb occurs and, apart from a fraction piling up at the surface, the majority of Sb occupies lattice sites. The concentration of substitutional fraction is 3×10^{20}/cm^3, close to an order of magnitude higher than solid solubility limit.

Melt threshold for SOS is found to be lower by about a factor two. This can be explained by the heating effect of the shallow penetrating component of the beam. (No through-window shots were made on SOS to date.)

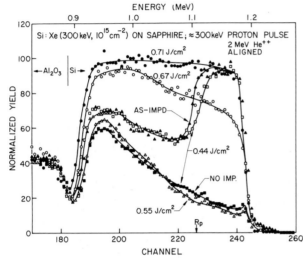

Fig. 6. Aligned spectra of Xe-implanted silicon (on sapphire) for different fluences of the ion pulse

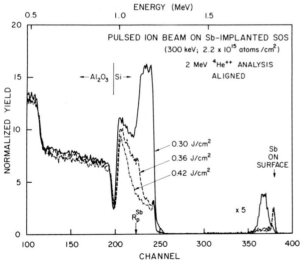

Fig. 7. Aligned spectra of Sb-implanted silicon (on sapphire) for different fluences of the ion pulse

CONCLUSIONS

Regrowth experiments on evaporated amorphous Si layers show polycrystal formation at fluences equivalent to $0.7 - 0.8$ J/cm^2. At higher fluences, good quality regrowth was found with some surface morphology (cracks). Spread of the Ge markers starts at 0.7 J/cm^2, about 0.2 J/cm^2 lower value, than the one leading to melting of the crystalline (or polycrystalline) silicon. Some caution should be in interpretation of this value as a difference in melting energy, because of the low-penetrating component of the beam. However, the identical geometries of the two sets of samples enhance confidence in this value. Masking the samples with 5000 Å self-supporting Si films suggests that the high penetrating part of the beam also causes some rapid spread of Ge in the silicon above 0.4 J/cm^2 fluences. No movement of the Ge into crystalline Si occurs at this fluence.

Regrowth of implantation amorphous layers of Si on sapphire showed an optimum at $0.4 - 0.5$ J/cm^2 as crystal quality is concerned. This fluence of protons was enough to spread a chemically active marker of Sb, while the noble gas marker (Xe) was immobile at this fluence. Majority of the Sb was put into substitutional positions in above-solid solubility concentration.

ACKNOWLEDGMENTS

We wish to acknowledge D. Hammer and the Cornell University Plasma Physics Lab for the use of the pulsed ion beam equipment. Implantations are greatly acknowledged to L-S. Hung (Cornell University). Some of our Si windows were supplied by L. Csepregi (Inst. Festkorpertechnologie, Munich) and I. Adeshida (Cornell University). We highly appreciate their help. One of us (J. G.) is indebted to the Bohmische Physical Society for a grant. The work was supported in part by NSF (L. Toth).

REFERENCES

1. E. G. Laser and Electron-Beam Interactions with Solids (Eds.: B. R. Appleton, G. K. Celler) North Holland, 1982.

2. L. J. Chen, L. S. Hung, J. W. Mayer, J. E. E. Baglin, J. M. Neri, D. A. Hammer, Appl. Phys. Lett. <u>40</u>, 595 (1982).

3. J. M. Neri, D. A. Hammer, G. Ginet, R. N. Sudan, Appl. Phys. Lett. <u>37</u>, 101 (1980).

4. R. Fastow, J. Gyulai, J. W. Mayer. This conference.

PART V
CRYSTALLIZATION
OF SILICON ON INSULATORS–I

BEAM PROCESSING OF SILICON WITH A SCANNING CW Hg LAMP

TIM STULTZ*, JIM STURM AND JAMES GIBBONS
Stanford Electronics Laboratory, Stanford, CA 94305

ABSTRACT

A scanning arc lamp annealing system has been built using a 3" long mercury arc lamp with an elliptical reflector. The reflector focuses the light into a high intensity narrow line source. Silicon wafers implanted with 100 KeV $^{75}As^+$ to $1x10^{15}$ cm^{-2} have been uniformly annealed with a single scan, resulting in complete activation and negligible redistribution of the implanted species. Using a scan rate of 1cm/s, entire 3" wafers have been annealed in less than 10 seconds with this system. The system has also been used to recrystallize thin films of polysilicon deposited on thermally grown silicon dioxide. The recrystallized films contain grains that are typically 0.5-1 mm in width and several centimeters long. Surface texture measurements show the crystallites to be almost entirely (100) in the plane of the film with the orthogonal <100> direction closely paralleling the scan direction. MOSFETs were fabricated in these films with surface mobilities 66% of ones fabricated in single crystal silicon. An epitaxial layer with the same crystallographic features as the recrystallized film was grown on the film itself.

INTRODUCTION

Over the past few years a considerable amount of research has been directed towards developing alternative heat treatment techniques for application to semiconductor processing. The most thoroughly investigated of these has been the use of pulsed and cw laser and electron beams [1]. Work performed using these sources has provided new insights into the mechanisms of thermally induced processes such as solid phase and liquid phase epitaxial growth and thin film recrystallization. In particular, some of the earliest work demonstrated that ion implantation damaged silicon could be annealed with a scanned cw laser such that there was complete electrical activation of the implanted species and virtually no impurity redistribution [2]. In addition, laser recrystallization of fine grain or amorphous films on insulating substrates was found to significantly improve the electronic properties of the film itself [3]. Once these processes were observed and understood, work began to identify alternative beam sources for producing these effects. In this paper we describe the results of using a high pressure Hg arc lamp system for large area annealing of ion implanted crystalline silicon and zone recrystallization of silicon thin films on insulating substrates. We also present data obtained from MOSFETs fabricated in the zone recrystallized film as well as describe the crystallographic features of an epitaxial layer grown on the film itself.

---

*Also Lockheed Palo Alto Research Laboratory

Mat. Res. Soc. Symp. Proc. Vol. 13 (1983) © Elsevier Science Publishing Co., Inc.

ANNEALING SYSTEM

We have built an arc lamp annealing system which consists of a 3 inch long high pressure mercury lamp, an elliptically shaped reflector and a variable speed translation hot stage. In previous work xenon and krypton lamps were used. We chose to use a mercury lamp primarily for the following two reasons. First, the spectral distribution of this lamp is most heavily weighted in the uv range. Consequently, the mean absorption depth for this source is much shallower than that for the other lamps.

The second reason the mercury lamp was chosen over the other available lamps is due to the narrow capillary used for containing the discharge. For example, a typical 3" krypton lamp has a 5 mm capillary whereas the mercury lamp uses a 2 mm capillary. The tighter confinement results in a narrower imaged beam, and thus yields a higher intensity at the sample surface for a given lamp power.

The light from the lamp is focused into a narrow (< 5mm wide) ribbon by a 4" long reflector with an elliptical cross section. By using this shape of reflector to collect and focus the light, we can create an intense linear heat source and still maintain a reasonable working distance between the sample and the lamp. The reflector in our system has major and minor axes of 10cm and 8.6 cm respectively. This configuration gives a magnification factor of about 3.2 and a working distance of about 26mm between the edge of the reflector and the focal plane.

The samples are placed on a heated sample stage, over which the lamp can be translated at speeds up to 10cm/s. A schematic representation of the arc lamp annealing system is shown in Fig. 1.

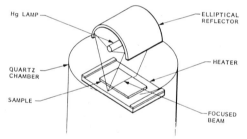

Fig. 1.   Schematic representation of scanning cw arc lamp annealing system.

ARC LAMP ANNEALING OF ION IMPLANTED CRYSTALLINE SILICON

Arc lamp annealing of ion implanted silicon has several distinct advantages over other techniques, in particular laser annealing. First, because the light is incoherent, interference effects which result from the interaction of mono-chromatic light and the dielectric layers on the semiconductor material are eliminated. These effects can lead to nonuniform heating of the irradiated material if not eliminated. Second, the effective processing area for an arc lamp is orders of magnitude greater than that for a laser. For example, an entire silicon wafer can be arc lamp annealed in about 10 seconds, whereas a cw laser takes several tens of minutes to process the same amount of material. This single scan-large area processing also eliminates the periodic nonuniformities observed in raster scanned laser annealed materials. Finally, an arc lamp is significantly less expensive than a laser and is much more efficient to operate.

In the following section, we describe the results of using an arc lamp to anneal ion implanted silicon. We demonstrate that while it has the above advantages over laser annealing, it also provides the benefits of beam annealing over conventional furnace annealing. Specifically it results in high quality, solid phase epitaxial regrowth with complete activation of the implant species and no dopant redistribution.

Annealing experiments were carried out using 2 and 3 inch 1-8 Ω-cm p-type (100) silicon wafers which were implanted with $^{75}As^+$ at 100keV to $1 \times 10^{15}cm^{-2}$. It was found that several combinations of arc lamp power, scan rate and substrate temperature could be used to achieve a wafer temperature sufficient for good annealing, $\sim$ 1000°C. In general, scan rates from 5mm/s to 1cm/s, substrate temperatures of 400°C to 600°C and arc lamp input power of 1kW/inch to 1.5kW/inch gave good results.

Because of the length of the arc lamp, an entire wafer can be annealed in a single scan. Figure 2 is a photograph of a 3" wafer in which the scan was stopped at the center of the wafer. Because of the difference in reflectivities between single crystal and amorphous silicon, the demarcation between the annealed and unannealed regions is readily seen. Transmission electron microscope (TEM) diffraction patterns made from samples taken from the unannealed and annealed regions clearly indicate the amorphous and completely recrystallized nature of the regions, respectively. (The picture is somewhat distorted due to the angle at which the photo was taken.)

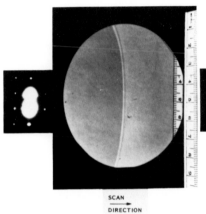

Fig. 2. Photograph of partially annealed silicon wafer, and TEM diffraction patterns from the annealed and unannealed regions.

SCAN DIRECTION

The recrystallized layer was also analyzed by Rutherford Backscattering Spectroscopy (RBS) and directly compared to a sample which was thermally annealed at 850°C for 30 minutes followed by a 10 second 1000°C anneal. Figure 3 shows the backscattering spectra from an as-implanted, arc lamp annealed and thermally annealed sample. As shown, there is no detectable difference between the quality of the regrown material annealed by the scanned arc lamp and the thermally processed sample.

Carrier concentration and mobility of the arc lamp annealed material as a function of depth were determined by means of differential sheet resistivity and Hall effect together with an anodic oxidation stripping technique. These

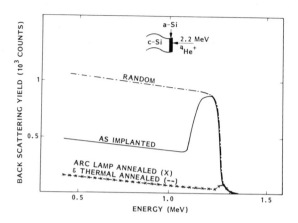

Fig. 3. Rutherford backscattering spectra from as-implanted, arc lamp annealed and thermally annealed samples.

results are shown in Fig. 4, along with the as-implanted profile calculated using LSS theory and published bulk silicon mobilities for the respective impurity concentrations. As shown, no measurable dopant redistribution occurred during the annealing process and the free carrier mobility in the annealed region is as good as found in bulk silicon.

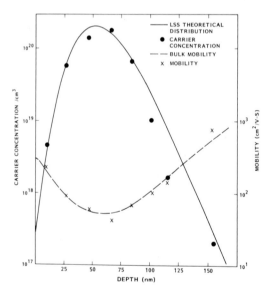

Fig. 4. Carrier concentration and mobility versus depth from an arc lamp annealed sample.

To assess the uniformity of the anneal, four point probe sheet resistivity measurements were made on a wafer which was completely annealed in a single scan. The variation across the wafer from side to side and top to bottom (with respect to the scanning beam) showed no significant variation in either direction, indicating a very uniform annealing of the implanted wafer. These data are shown in Fig. 5.

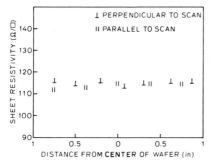

Fig. 5. Sheet resistivity profiles across an arc lamp annealed wafer.

High resolution transmission electron microscopy (TEM) was used to compare the residual dislocation density in arc lamp annealed material to material which was furnace annealed at 1000°C for 1 hour in N_2.

In Fig. 6 we compare the results of a sample which was annealed using 1 kW/inch lamp power, a scan rate of 7 mm/s and a substrate temperature of 450°C, to one which was thermally annealed. As shown the residual dislocation density is approximately the same in both cases.

Deep level transient spectroscopy has also been performed on these samples and compared to furnace annealed as well as other beam (e.g., laser and electron) annealed samples [4]. These results indicate that the arc lamp annealed material has approximately an order of magnitude lower defect density and does not contain the principle defect species identified in the other beam processed material.

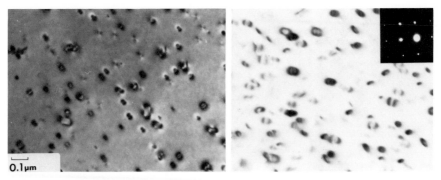

Fig. 6. Bright field TEM micrographs of (a) furnace annealed (1000°C, 1 hr.) and (b) arc lamp annealed samples.

In summary, arc lamp processing of ion implanted silicon provides fast, high quality annealing of the damaged material, complete dopant activation and no redistribution of the dopant species.

RECRYSTALLIZATION OF THIN POLYSILICON FILMS ON SiO_2

Fabrication of device worthy silicon thin films on insulating substrates has been an active area of research for many years. One of the most successful approaches for obtaining this material is by recrystallizing as-deposited fine grain or amorphous films into large grain or single crystal material.

Over 30 years ago zone melting was shown to be an effective grain enlarging recrystallization technique for thin films [5]. In that work, a platinum wire was used to zone melt and recrystallize films of luminescent material on insulating substrates. By 1966 several researchers had grown large grain germanium thin films using an electron beam for the zone melting process [6-9]. More recently, lasers [10,11] and graphite strip heaters have been used for zone melting of silicon films [12-14].

In this section we show that a scanned, high pressure mercury lamp can be used for zone melting and recrystallization of silicon thin films on insulating substrates.

The samples used for these experiments were thermally oxidized single crystal silicon wafers upon which polycrystalline silicon films had been deposited using low pressure chemical vapor deposition (LPCVD). The thermal oxide was 1 μm thick and the deposited silicon film was 600 nm thick. Finally, the films were encapsulated with 1 μm-2 μm of CVD SiO_2.

Three parameters controlling the zone melting process were investigated; arc lamp power, starting substrate temperature and zone scan rate. Although a variety of combinations of these parameters could be used to achieve a molten zone in the deposited film, the following general observations were made.

If the scan rate was too fast and/or the lamp power-substrate temperature combination was too low, partial melting of the film and/or severe agglomeration would occur. If the starting substrate temperature was too low, the substrate would be highly damaged after the recrystallization process, exhibiting a large number of slip lines. If the combination of substrate temperature and lamp power was too high, melting of the underlying single crystal substrate would occur. Finally, within the operating region of a stable molten zone, an increase in scan rate was found to increase the spacing between the subgrain boundaries.

Our best results to date were obtained using a lamp power of 1 kW/inch, a substrate temperature of 1150° C and a scan rate of 3 mm/s. The following results were obtained under these experimental conditions.

The zone recrystallized films using our system exhibited features similar to those found in zone recrystallized films using other techniques. The film consisted of very large grains, typically 0.5 mm to 1 mm in width and up to several centimeters long. These large grains were obtained without using any intentional seeding techniques. Such techniques have been shown to eliminate large angle grain boundaries and yield a film composed entirely of low angle grain boundaries [12]. The individual grains were composed of many subgrains, with subgrain boundaries typically from 50 μm to 100 μm apart. Figure 7 is an optical photomicrograph of a section of recrystallized film which has been Secco [15] etched to delineate the subgrain boundaries. In Fig. 8 we show a scanning electron micrograph which was taken under channeling contrast conditions. This figure clearly indicates that the surface normals of the regions separated by the subgrain boundaries have slightly different orientations.

Because of the large grain size in the recrystallized films, conventional crystallographic techniques for evaluating the structure of the films are very time consuming and complex. This is because the incident beam, i.e., electron

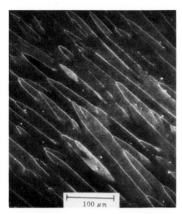

Fig. 7. Photomicrograph of zone re-
recrystallized film which has been
Secco etched to delineate the subgrain boundaries.

Fig. 8. Scanning electron micrograph
of zone recrystallized film taken
under channeling contrast conditions.

or X-ray, is smaller than or on the order of the average grain size of the
material. Consequently the data obtained from a single diffraction pattern is
representative of at most a few grains. Thus in order to obtain any statistically meaningful information about the crystallographic composition of the
film, many patterns over a large area of the film must be taken and analyzed.

Geis et al. have described a technique for observing the orientation and
size of large grains in zone recrystallized films by using an anisotropic
etch in combination with a square array of etch pits [16]. We have used this
approach to evaluate our zone recrystallized film. Using conventional photolithographic techniques, round holes on 50 μm centers were fabricated in
the SiO_2 encapsulant after the underlying material was zone recrystallized.
The samples were then immersed in a 50% by weight solution of KOH in DI H_2O
at 80° C for 6 - 8 minutes. This solution preferentially etches in the <110>
direction [22]. Consequently the symmetry and alignment of the etch pits can
be related to the crystallographic nature and in-plane orientation of the
grains being etched.

Figure 9 is a photomicrograph of a recrystallized film which has been preferentially etched in the manner described above. The first feature to note
is that the etch pits are square, indicating a <100> surface texture of the
film. We have also observed hexagonally shaped etch pits in some of our films
which indicates a <111> texture. However, this texture was found only in
some very localized areas and on just a few samples. All the material we
have recrystallized has exhibited predominantly a <100> texture.

The second feature to notice is that the etch pit diagonals closely parallel
the zone scan direction, indicating a <100> texture in the growth direction.
Also shown in Fig. 9 is a grain boundary as evidenced by the misalignment of
the etch pits on either side.

DEFECTS IN ZONE RECRYSTALLIZED FILM

The principal defects we have observed in the zone recrystallized film and/or
substrates are: (1) slip and warpage, (2) substrate melt through, (3) protrusions, (4) voids and (5) agglomeration or dewetting.

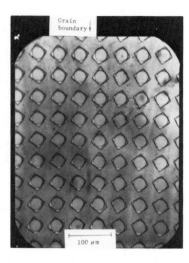

Fig. 9. Photomicrograph of zone recrystallized film illustrating misalignment of etch pits due to the
presence of a grain boundary.

Slip and warpage damage to the single crystal silicon substrate are a result
of the large thermal stresses the substrate sees during the processing. In
general, higher substrate temperatures and very uniform substrate heating reduce
or eliminate these problems.

When the molten zone (or an area of it) becomes too hot, melting of the
crystalline substrate beneath it can occur. Under these conditions the thermal
oxide can rupture resulting in a large 'melt through' pit, where the surface
film and substrate have fused together. This is shown in Fig. 10(d).

If the film to be recrystallized contains impurities, they can segregate to
the sub-boundaries during the solidification process. This segregation will
depress the melting tmeperature of the film in that region. Consequently, it
will solidify after the zone front has passed that region and the neighboring
film has solidified. Since silicon has a positive expansion coefficient upon
solidification, these small molten areas which are constrained by the surrounding solid silicon must expand upward, resulting in small protrusions at the
surface of the film. Figure 10(b) illustrates this case. Note how the protrusions all lie along the low angle grain boundaries. In general, clean material,
reduced scan speeds and steep thermal gradients will prevent this defect.

We have also observed small areas in the zone recrystallized film where the
film has thinned out and exposed the underlying thermal oxide. These 'voids' are
believed to be a result of some local contamination at the interface between the
oxide and the deposited film. Figure 10(c) is a photomicrograph of this defect.

The last defect, agglomeration and dewetting is a gross feature which generally renders the film useless. This is the case when the film balls up or
gathers into separate unattached regions, as shown in Fig. 10(a). The driving
force for this reaction is the minimizing of free energy by reducing the surface
to volume ratio of the film. The encapsulation layer is intended to minimize
or eliminate this effect. Some investigators have found that two layer encapsulants (18) are more effective than the single layer SiO_2. We have observed,
however, that some deposited films are less prone to agglomeration than others
and can be relatively insensitive to the encapsulant. In either case, this
problem is perhaps the most demanding and the solution must be understood and
universally reproducible if this process is to become a useful technique.

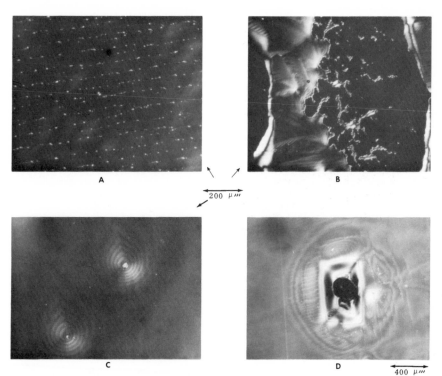

A   B

200 μ

C   D

400 μ

Fig. 10. Defects in zone recrystallized films (a) agglomeration, (b) protrusions, (c) voids, and (d) substrate melt through.

MOSFETS IN ZONE RECRYSTALLIZED FILM

In order to investigate the electronic properties of the zone recrystallized films, n-channel metal-oxide-semiconductor field-effect-transistors were fabricated using an aluminum gate process. The transistors had channel lengths of 10 μm, 20 μm and 50 μm with all channel widths being 200 μm. The transistors were fabricated such that there were channels running both parallel and perpendicular to the low angle grain boundaries. In this way the effect of the low angle boundary orientation on the channel mobility could be assessed. Channel implants of 1.5×10^{11} cm^{-2} boron and phosphorus were used for the enhancement and depletion mode devices, respectively. The implants were thermally annealed under such conditions as to insure a uniform dopant distribution through the film thickness, yielding a net impurity concentration of 2×10^{15} cm^{-3}. The gate oxide thickness was 1000 Å and all devices were isolated by an island etch.

A circular bulk resistor pattern was also included to evaluate bulk properties of the film. Resistivity data obtained from these devices indicate that there was complete activation of the phosphorus implant and single crystal bulk mobilities in the film. Surface mobility data were obtained by analyzing transistor IV characteristics from a curve tracer. In Table I we summarize the surface mobility data obtained from the MOSFETs. As shown, the average

472

TABLE I
Summary of N-channel MOSFET surface mobilities ($cm^2/V-S$)

| Channel Length + Direction | Arc Lamp Recrystallized | | Single Crystal Control |
	Enhancement	Depletion	Enhancement
50 μm parallel	436	334	537
50 μm perpendicular	418	399	
20 μm parallel	--	432	573
20 μm perpendicular	382	415	
10 μm parallel	345	332	591
10 μm perpendicular	335	357	

Conclusions: no sub-boundary orientation effects on surface mobility.

surface mobilities for the enhancement and depletion mode devices in the zone recrystallized film ranged from 335 $cm^2/V-s$ to 436 $cm^2/V-s$. No systematic variation due to low angle grain boundary orientation or channel length was observed. Included in Table I are the surface mobilities of enhancement mode devices simultaneously fabricated in single crystal material. The average mobility for these devices is 567 $cm^2/V-s$. Thus the transistors fabricated in the recrystallized film had surface mobilities approximately 66% of those fabricated in single crystal material.

EPITAXIAL LAYERS ON ZONE RECRYSTALLIZED FILM

Although 0.5 μm silicon films can be used for MOSFET devices, thicker films are desirable for bipolar applications. In light of this need, a 14 μm epitaxial layer was grown on a 0.5 μm zone recrystallized film. The sample preparation and zone recrystallization were performed as described in the previous section. After processing, the 2 μm SiO_2 encapsulation layer was removed from the sample. The sample was then cleaned and placed in a standard conventional epitaxial reactor.

It was found that the epitaxial layer replicated and enhanced the surface features of the zone recrystallized film. In particular, the low angle grain boundaries were clearly evident in the topography of the epitaxial layer. Figure 11 is a Nomarski photomicrograph of the surface of the film illustrating these features. Diffraction patterns of epitaxial layers grown on as-deposited as well as zone recrystallized material were taken using a Read camera. These patterns, shown in Fig. 12, clearly indicate that while the layer grown on the as-deposited film is fine grained with no preferred orientation, the layer grown on the recrystallized film is large grained and highly oriented. In order to evaluate the macroscopic crystallographic features of the epitaxial layer, the preferential etch pit method previously described was used. These results are presented in Fig. 13. As shown, there are no facets in the etch pits formed in the layer grown on the as-deposited film. However the etch pit pattern in the layer grown on the recrystallized film clearly indicates a (100) film orientation. The photomicrograph shown in the lower portion of Fig. 13 was taken using light scattered off the sample surface. All the etch pits within a single grain will have their facets aligned. This alignment however will vary from grain to grain. Thus this technique can be used to

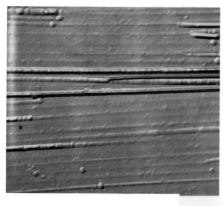

Fig. 11. Nomarski photomicrograph of surface of epitaxial layer grown on zone recrystallized film.

200 μm

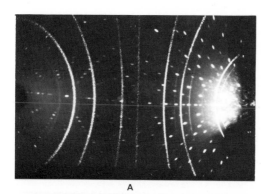

A

B

Fig. 12. READ camera diffraction patterns from (a) epi on as-deposited film, (b) epi on zone recrystallized film.

474

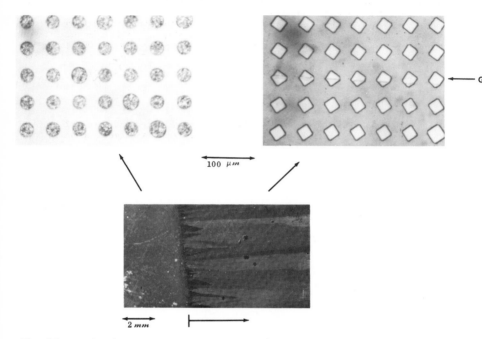

GB

100 μm

2 mm

Fig. 13. Etch pits and scattered light photo of epi on zone recrystallized film.

delineate and observe the grains in the film. As shown, the grains in the epitaxial layer tend to be 0.5 mm to 1 mm in width, as was observed in the zone recrystallized films.

In summary, an epitaxial layer which replicates the surface and crystallographic features of a zone recrystallized thin film substrate have been grown. Work in progress includes evaluating the electronic properties of these epitaxial layers.

SUMMARY

We have built a scanning Hg arc lamp system which can be used for thermal processing of semiconductor materials. The Hg lamp was chosen over other available high intensity sources due principally to its large near UV spectral content which results in better optical coupling with the silicon being processed. The scanning Hg arc lamp system was used to anneal ion implanted silicon and to zone recrystallize silicon thin films on insulating substrates. The annealing experiments have demonstrated that Hg lamp irradiation can produce a high quality diffusionless anneal in less than 15 seconds. The recrystallization experiments have demonstrated that large grain, highly oriented thin films of silicon on amorphous substrates can be produced using this system.

The recrystallized film had a (100) in plane orientation and exhibited a <100> texture in the direction of the zone scan. Enhancement and depletion mode MOSFETs were fabricated in the zone recrystallized film. The electron surface mobilities were greater than 60% of those measured in devices fabricated in single crystal silicon, demonstrating that the electronic properties of the film are suitable for device applications. Finally, a 14 μm epitaxial layer was grown on a zone recrystallized film. This layer contained the same crystallographic features as the zone recrystallized film used as a substrate and has potential for bipolar device applications.

ACKNOWLEDGMENTS

The authors wish to thank D. Reynolds for his continued interest and support of this work. This work has been supported by Lockheed Missiles and Space Co., Inc. and DARPA Contract No. MDA 903-81-C-0294.

REFERENCES

1. See for example, Laser and Electron Beam Solid Interactions and Materials Processing, Gibbons, Hess & Sigmon eds. (North-Holland, N.Y., 1981)

2. A. Gat, J. F. Gibbons, T. J. Magee, J. Peng, V. R. Deline, P. Williams, and C. A. Evans, Jr., Appl. Phys. Lett. 32, 276 (1978)

3. R. A. Laff, and G. L. Hutchins, IEEE Trans. Elect. Dev. ED-21, 743 (1979).

4. N. Johnson, T. Stultz, J. Gibbons (to be published).

5. E. Leitz, Brit. Pat. 691, 335 (1950).

6. G. G. Gilbert, T. O. Poehler, and C. F. Miller, J. Appl. Phys. 32, 1597 (1961).

7. J. Maserjian, Sol. St. Elect. 6, 477 (1963).

8. T. O. Poehler and G. B. Gilbert, Single Crystal Film, Francombe, Sato, eds. (Pergamon Press, 1964) pp. 129-135.

9. S. Namba, J. Appl. Phys. 37, 1929 (1966).

10. A. Gat, L. Gerzberg, J. F. Gibbons, T. J. Magee, J. Peng, J. D. Hong, Appl. Phys. Lett. 33, 775 (1978).

11. T. J. Stultz and J. F. Gibbons, Appl. Phys. Lett. 39, 498 (1981).

12. J. C. C. Fan, M. W. Geis, and B. Y. Tsaur, Appl. Phys. Lett. 38 365 (1981).

13. R. F. Pinizzotto, H. W. Lam, B. L. Vaandrager, Appl. Phys. Lett. 40, 388 (1982).

14. H. J. Leamy, Laser and Electron Beam Interactions with Solids, Appleton, Celler, eds., (North-Holland, N.Y. 1982) p. 467.

476

15. F. Secco d'Aragona, J. Elec. Chem. Soc. <u>119</u>, 948 (1972).

16. M. W. Geis, H. I. Smith, B.-Y. Tsaur, J. C. C. Fan, E. W. Maby and D. A. Antoniadis, Appl. Phys. Lett. <u>40</u>, 158 (1982).

17. D. L. Kendall, <u>Ann. Rev. Mater. Sci.</u>, Vol. 9, Huggins, Bube, Vermilyea, eds., 373 (1979).

18. E. W. Maby, M. W. Geis, Y. L. LeCoz, D. J. Silversmith, R. W. Mountain, D. A. Antoniadis, IEEE Elect. Dev. Lett. <u>EDL-2</u>, 241 (1981).

ZONE-MELTING RECRYSTALLIZATION OF SEMICONDUCTOR FILMS

M. W. GEIS, HENRY I. SMITH, B-Y. TSAUR, JOHN C. C. FAN, D. J. SILVERSMITH,
R. W. MOUNTAIN AND R. L. CHAPMAN
Lincoln Laboratory, Massachusetts Institute of Technology, Lexington,
Massachusetts 02173

ABSTRACT

The use of zone melting recrystallization (ZMR) to prepare large-grain (and in some cases single-crystal) semiconductor films is reviewed, with emphasis on recent work on Si on SiO_2. Encapsulants are generally required to minimize contamination and decomposition, induce a crystalline texture, improve surface morphology and prevent agglomeration. In the case of Si, the solid-liquid interface is faceted, which gives rise to subboundaries. These can be entrained by laterally modulating the temperature through the use of an optical absorber on top of the encapsulant. Control of thermal gradients and in-plane crystallographic orientation are important for reliable entrainment.

Preparation of high-quality semiconductor films on insulating substrates has been the subject of numerous investigations. One of the more promising techniques for producing such films is zone-melting recrystallization (ZMR) which is accomplished by scanning a molten zone through the semiconductor film. The resulting film usually consists of large grains with a specific crystallographic texture, and has electrical properties approaching those of bulk single crystal material. In this article we review briefly the development of ZMR and discuss its application to Si on SiO_2.

The concept of ZMR was first discussed by Leitz[1] as a possible method for obtaining single-crystal films of ZnS. Mitchell[2] et al. used a zone melting technique to produce single-crystal films of AgCl and AgBr. Later, ZMR was used to produce crystalline films of Ge, InSb and, in recent years, Si on SiO_2. Various methods have been used to produce the molten zone. For relatively low-melting-temperature materials, such as AgCl, AgBr and certain organics, a small hot plate, which was moved with respect to the substrate, was sufficient. Materials with higher melting temperatures, such as Si and Ge, require higher power densities, which can be obtained with electron or laser beams. If the substrate is heated to near the melting temperature of the semiconductor by some auxiliary means, a resistively heated wire or carbon rod, or a focused lamp, can be used to supply the additional heat necessary to melt a narrow zone in the semiconductor.[15-18]

Most investigators have found it necessary to use an encapsulant layer on top of the film to be recrystallized by zone melting. These encapsulants serve the obvious purposes of minimizing contamination and decomposition of the film during ZMR. However, they also induce a crystalline texture, improve the surface morphology of the film, and reduce or prevent agglomeration. For example, an SiO_2 encapsulant was found necessary to obtain a (100) texture in ZMR of Si films on SiO_2.[9] As shown in Fig. 1, the surface morphology of a Si film improves with the SiO_2 encapsulant thickness.[19] In the case of ZMR of InSb, the natural oxide, In_2O_3, was used as an encapsulant.[6]

Often, a molten semiconductor will not wet a substrate and will "bead up" or agglomerate during ZMR. Maserjian[4] found that agglomeration could be avoided by keeping the molten zone very narrow, a few 100 μm wide. However, such narrow molten zones usually require highly focused energy sources such as electron or laser beams. When a hot wire or carbon rod is used, the molten zone is considerably wider, ~1-2 mm, and an encapsulant layer is needed to stabilize the melt against agglomeration. Spivak et al.[20] discovered that an encapsulant of native In_2O_3 would stabilize an InSb molten zone. Unfortunately, in the case of Si on SiO_2 an SiO_2 encapsulant alone is not sufficient except when the substrate is patterned with a surface relief structure, or the Si film is patterned into islands or stripes.[9,19] However, the addition of Si_3N_4 to an SiO_2 encapsulant will stabilize molten Si during ZMR, as shown in Fig. 2. Various combinations of SiO_2 and Si_3N_4 have been reported to be effective encapsulants. The role of the Si_3N_4 and the mechanism by which encapsulants induce texture and stabilize a molten zone are still not understood.[19] Table 1 summarizes ZMR methods and results for a number of materials.

Zone-melting recrystallization of Si on an SiO_2 substrate generally produces films composed of a few large grains ~1 mm wide, extending the full length of the region scanned and having (100) texture. The <100> direction is generally within ~15° of the scan direction. Films of a single orientation have been produced by seeding from a single crystal,[21,22] as indicated in Fig. 3, by zone-melting through planar constrictions,[23,24] as shown in Fig. 4, and by so-called cross-seeding.[4,11] In all cases, the films contain subboundaries which are arrays of dislocations or, equivalently, low-angle grain boundaries. (We use the term subboundary to refer to those low-angle boundaries which form during solidification from a single seed, and the term grain boundary to refer to the region where crystals grown from two distinct seeds meet.)

Subboundaries are the result of faceting of the liquid-solid interface[19] as shown in Fig. 5. They are not eliminated by seeding.[25] The result shown in Fig. 5 was obtained by quenching the molten zone during ZMR, thereby revealing the general morphology of the liquid-solid interface. The true liquid-solid interface is bounded by (111) planes and, since the Si film has a (100) texture, the (111) facets form 90° angles where they intersect at the film plane. The subboundaries originate at the interior corners of the faceted interface; thus if the lateral locations of the facets can be fixed, the subboundaries can be arranged to lie along lines parallel to the zone motion direction.[26,27] Such an entrainment of the subboundaries can be accomplished by spatially modulating the temperature in the film as shown in Fig. 6. A carbonized photoresist grating is used for this purpose. The grating lines on top of the encapsulant absorb more radiant energy than the open areas and, as a result, the regions under the lines are the last to freeze. The interior corners tend to locate in these regions. Figure 7 shows an example of ZMR of a Si film using a carbonized photoresist grating to position or entrain the subboundaries. Although subboundaries can be entrained, new subboundaries sometimes form between the entrained subboundaries giving the film a feathering appearance as shown in Fig. 8.

A simplified model of the liquid-solid interface[27] which can explain the formation of new subboundaries and the characteristic wishbone patterns is shown in Fig. 9. Three consecutive times during ZMR are depicted. The semiconductor film is assumed to have a (100) texture, with <100> direction parallel to the direction of zone motion. Growth occurs by the addition of atoms at ledges or steps which sweep rapidly across the facets. The distance a chevron-shaped pair of facets extends into the liquid region, which determines the spacing between the subboundaries originating at the interior corners, is limited by two temperatures, T_1 and T_2, shown as dotted isothermal contours. T_2 is the temperature at which the generation of new ledges occurs at such a rate that the forward advance of the (111) facets just matches the imposed zone-scanning velocity. At temperatures close to T_2 the speed with which ledges sweep across a (111) facet is much greater than the rate of forward advance of the solidification front; hence the rate of forward advance is determined by the rate of generation of new ledges. The velocity of ledge motion decreases as a ledge encounters higher and higher temperatures. Eventually, when T_1 is reached, the forward velocity of the ledge just matches the zone scanning velocity. T_1 is probably very close to the melting temperature of Si, T_m. Subboundaries tend to be parallel to the direction of zone motion, but small perturbations can cause them to move laterally. Often two subboundaries coalesce in a wishbone pattern and form a single subboundary, as in Fig. 9(b). When the distance between adjacent subboundaries becomes so large that the chevron formed by a pair of facets would extend beyond T_1, the front of the chevron becomes flattened, as indicated in Fig. 9(b). This flattened face is not stable and soon develops an interior corner, Fig. 9(c), which becomes the source of a new subboundary when it falls back to T_2, as depicted in Fig. 9(c). This simplified model is discussed in more detail in reference 27.

Figure 10 provides experimental evidence for the model showing both the formation of new subboundaries by the flattening-dimpling mechanism and the coalescence of subboundaries. The micrograph shows a recrystallized Si film where the subboundaries have been delineated by etching and appear as dark vertical lines. Dopant variations in the film appear as zig-zag lines running left to right. The dopant variations were produced by pulsing the upper heater strip at 120 Hz to demark the instantaneous liquid-solid interface.

The feathering appearance in Fig. 8 and the formation of new subboundaries can be minimized by controlling the thermal contours and the inplane crystallographic orientation during ZMR. Figures 11(a) and 11(b) depict two different temperature contours. In Fig. 11(a) the depth of modulation is small and subboundaries form between entrained subboundaries. In Fig. 11(b) the depth of modulation is sufficient to eliminate feathering. Although entrainment of subboundaries by temperature contour modulation has been demonstrated, full reproducibility has not been achieved. For example, we have not been able to eliminate feathering over an entire 2 inch diameter sample. However, by using a vacuum chuck, as shown in Fig. 12, the conduction of heat vertically is improved and more consistent entrainment is obtained. Figure 13 illustrates a type of feathering that can arise when the Si <100> direction is not parallel to the direction of zone motion. The grid of etch pits[28] shows that two crystal grains are present and that the grain boundary is entrained. On the lower side of the grain boundary the <100> direction is parallel to the direction of zone motion and the axis of the entrainment grating. On the upper side the <100> direction is at an angle of 40° to the entrainment lines. As a result of this misorientation, the pattern of subboundaries has an asymetric, "feathered" appearance, while there is minimal feathering for <100> parallel to the grating axis.

Several techniques have been developed to eliminate subboundaries from limited areas of Si films recrystallized by zone melting. Cellers et al.[30] have developed a technique to eliminate subboundaries from areas 50 x 500 μm long, as shown in Fig. 14. This technique uses an oscillatory motion of the liquid-solid interface during ZMR. During the remelting portion of the cycle, any newly formed subboundaries are preferentially melted and do not immediately reform during the solidification portion of the ZMR cycle. Subboundaries have also been controlled by patterning the Si film prior to ZMR.[24,31] The patterning tends to position the faceted liquid-solid interface so as to terminate subboundaries at the pattern edges.

In addition to subboundaries, recrystallized Si films contain a variety of topographic defects. Small pools of molten Si are often trapped by the advancing solidification front. When these molten pools solidify, the Si expands forming protrusions, as shown in Fig. 15.[25] Why these molten pools form in the Si film is not understood. They might be caused by impurties or reflect the presence of local hot spots.[19,32,33] When a Si film is recrystallized by zone-melting the substrate often warps causing height variations of the order of 25 μm.[34] Such height variations can present difficulties in lithography or other stages of wafer processing.

Despite subboundaries, protrusions, and substrate warpage, Si films produced by ZMR have good electrical quality[35]. The electron and hole mobilities are superior to those obtained with silicon on sapphire, SOS,[36] and devices have been obtained in ZMR Si films.[37] If the present difficulties can be overcome, ZMR of Si may be of commercial value.

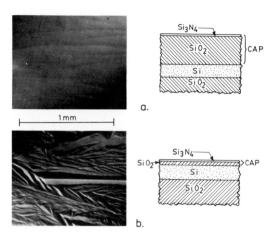

a.

b.

Figure 1. Optical micrographs of recrystallized Si films for two thicknesses of the SiO_2 component of the SiO_2/Si_3N_4 encapsulation layer: (a) SiO_2 is 2 μm thick and the Si film is smooth (b) SiO_2 is 0.2 μm thick and the Si film is highly faceted. In both cases the Si_3N_4 was 30 nm thick.

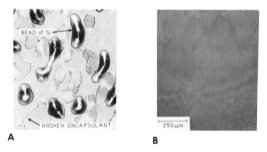

A   B

Figure 2. (a) Micrograph of a ZMR Si film with only a SiO_2 encapsulant. Note that the film agglomerates or "beads up". (b) Micrograph of a similar ZMR Si film encapsulated with a SiO_2 and Si_3N_4 composite. The Si film is continuous.

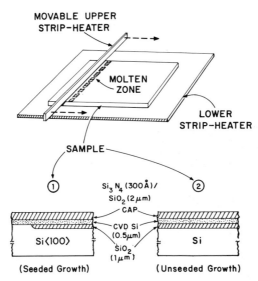

Figure 3.   Schematic illustration of ZMR of Si using a strip-heater oven.
The lower left illustration shows an example of a substrate in
which the bulk single-crystal Si is used to seed the film.
The lower right shows an example of a substrate in which the
Si film self seeds.

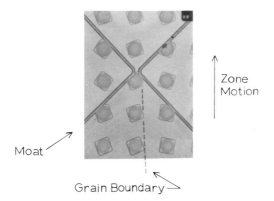

Figure 4.   Selection of a single orientation from two initial grains by
ZMR through a planar constriction in the form of an
hourglass.

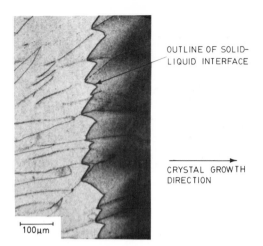

OUTLINE OF SOLID-
LIQUID INTERFACE

CRYSTAL GROWTH
DIRECTION

100μm

Figure 5. Optical micrograph of a recrystallized Si film that was rapidly
cooled by a He jet during the course of zone motion. We believe
the sudden cooling produced on approximate demarcation of the
solid-liquid interface. The actual facets are close to (111)
planes. The micrograph illustrates that subboundaries originate
at the interior corners of the faceted solid-liquid interface.

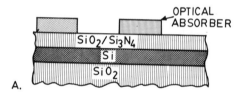

OPTICAL
ABSORBER

SiO₂/Si₃N₄

Si

SiO₂

A.

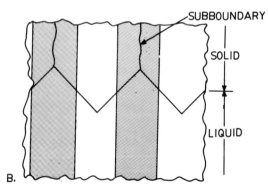

SUBBOUNDARY

SOLID

LIQUID

B.

Figure 6.

(a) Schematic illustration
of the use of a grating of
optical absorber stripes on
top of an encapsulation
layer to modulate the temperature profile and entrain
subboundaries. (b) Illustration of the entrainment of
subboundaries under the
middle of the optical
absorber stripes.

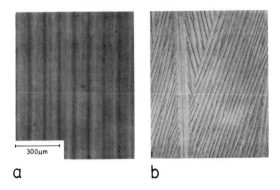

Figure 7. (a) Entrainment of subboundaries in a 1-μm-thick Si film using a 100 μm period, 50 μm linewidth, grating of carbonized photoresist. (b) Typical pattern of subboundaries in an identical sample recrystallized without an entrainment pattern.

Figure 8. Micrograph showing entrained subboundaries about 100 μm apart. Between the entrained subboundaries new subboundaries form and move to the left coalescing with the entrained subboundaries.

484

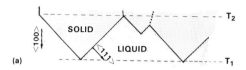

(a)

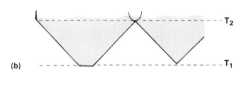

(b)

Figure 9.

Simplified model of the solid-
liquid interface, shown in a top
view, at three consecutive times.
The isothermal contours, T_1 and
T_2, are defined in the text. Sub-
boundaries are shown originating
at interior corners. The characteristic wishbone pattern is due to
the coalescence of two interior
corners.

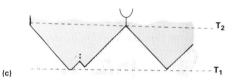

(c)

Figure 10.

Micrograph of a Si film recrystallized by zone-melting from the top
down showing experimental evidence
in support of the model presented
in Fig. 9. The instantaneous
solid-liquid interface was demarked
by pulsing the upper strip-heater
at 120 Hz. Approximately every
twelfth demarcation line is traced
in ink to clarify the time
evolution of the solidification
front.

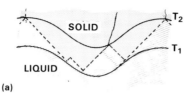

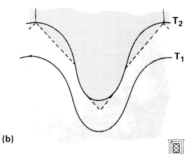

Figure 11. Schematic illustration of a model for the solid-liquid interface when the isothermal contours are sinusiodally modulated. (a) Depth of modulation is insufficient to achieve full entrainment. (b) Depth of modulation is sufficient to induce entrainment.

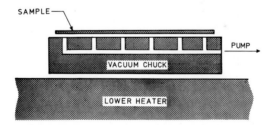

Figure 12. Schematic diagram of a vacuum chuck which is placed between the sample and the lower heater to obtain better thermal contact and vertical thermal conduction.

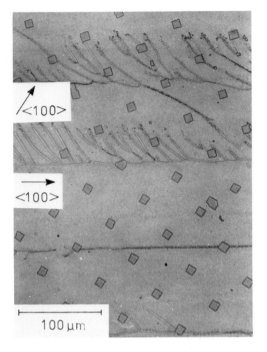

Figure 13.

Optical micrograph showing the dependence of entrainment on azimuthal crystallographic orientation. The diagonals of the etch pits are $\langle 100 \rangle$ directions.

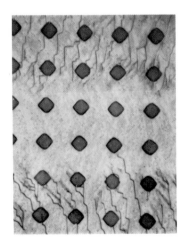

Figure 14.

Optical micrograph of a continuous Si film on SiO_2 recrystallized by a laser beam moved in an oscillatory path. The trace of oscillation is shown at right. The sample was moved under the laser beam. (Courtesy of G. K. Celler).

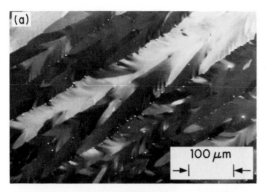

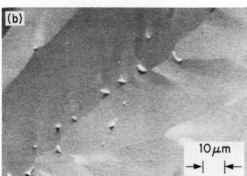

Figure 15.

(a) Micrograph of recrystallized Si film taken with a scanning electron microscope operated in the backscatterning mode. Protrusions appear along the subboundaries. (b) An enlarged micrograph of (a).

TABLE I
Summary of ZMR of semiconductors

Material	Encapsulant	ZMR Technique	Texture	References
AgCl, AgBr	Glass Plate	Hot Plate	(100)	2,3
Ge	None	E-Beam, Laser	None, (110)	4,5
InSb	In_2O_3, SiO	E-Beam Heated Wire	(111), (110)	6
Si	SiO_2, Si_3N_4	E-Beam, Laser, Heated Rod, Lamp	(100), (111)	7-19

The authors are grateful to C. V. Thompson for helpful discussions, C. L. Doherty, K. E. Krohn, J. M. Lawless, G. H. Foley for technical assistance, and M. Finn for Auger Analysis.

This work was sponsored by the Defense Advanced Research Projects Agency, the Department of the Air Force, and the Department of Energy.

488

H. I. Smith, one of the authors, is with the Department of Electrical Engineering and Computer Science, Massachusetts Institute of Technology, Cambridge, MA, and is a consultant at Lincoln Laboratory.

REFERENCES

1. British Patent 691, 335, E. Leitz, published May 13, 1953 "Improvements in or Relating to Methods of Making Luminescent Screens".

2. J. W. Mitchell, Proc. R. Soc. London A 371, 149 (1980).

3. J. M. Hedges and J. W. Mitchell, Philos. Mag. 44, 223 (1953).

4. J. Maserjian, Solid-State Electron. 6, 477 (1963).

5. J. C. C. Fan, H. J. Zeiger, R. P. Gale, and R. L. Chapman, Proc. 14th IEEE Photovoltaic Specialists Conference, San Diego, CA, 1980.

6. A. R. Billings, J. Vac. Sci. Technol. 6, 757 (1969).

7. R. A. Laff and G. L. Hutchings, IEEE Trans. Electron Devices ED-21, 743 (1974).

8. D. K. Biegelsen, N. M. Johnson, and M. D. Mayer, Laser and Electron Beam Solid Interactions and Material Processing, Edited J. F. Gibbons, L D. Hess, and T. W. Sigmon (Elsevier, New York, 1981) p. 487.

9. M. W. Geis, D. A. Antoniadis, D. J. Silversmith, R. W. Mountain, and H. I. Smith, Appl. Phys. Lett. 37, 454 (1980).

10. T. I. Kamins, J. Electrochem. Soc. 128, 1824 (1981).

11. M. W. Geis, H. I. Smith, B-Y. Tsaur, J. C. C. Fan, E. W. Maby, and D. A. Antoniadis, Appl. Phys. Lett. 40, 158 (1982), and E. W. Maby, M. W. Geis, Y. L. LeCoz, D. J. Silversmith, R. W. Mountain, and D. A. Antoniadis, Electron Device Lett. ED-2, 241 (1981).

12. M. W. Geis, D. A. Antoniadis, D. J. Silversmith, R. W. Mountain, and H. I. Smith, J. Vac. Sci. Technol. 18, 229 (1981).

13. J. A. Knapp and J. T. Picraux, these proceedings.

14. B-Y. Tsaur, J. C. C. Fan, M. W. Geis, D. J. Silversmith, and R. W. Mountain, Appl. Phys. Lett. 39, 561 (1981).

15. R. F. Pinizzotto, H. W. Lam, and B. L. Vaandrager, Appl. Phys. Lett. 40, 388 (1982).

16. R. A. Lemons, these proceedings.

17. T. J. Stultz, J. Sturm, and J. F. Gibbons, these proceedings.

18. A. Kamgar and G. A. Rozgonzi, these proceedings.

19.   M. W. Geis, H. I. Smith, B-Y. Tsaur, J. C. C. Fan, D. J. Silversmith, and R. W. Mountain, J. Electrochem. Soc. 129, 2813 (1982).

20.   J. F. Spivak and J. A. Carrel, J. Appl. Phys. 36, 2321 (1965).

21.   J. C. Fan, M. W. Geis, and B-Y. Tsaur, Appl. Phys. Lett. 38, 365 (1981).

22.   M. Tamura, H. Tamura, and T. To Kuyama, Jpn. J. Appl. Phys. 19, 123 (1980).

23.   H. I. Smith, H. A. Atwater, and M. W. Geis, Extended Abstract, 161 Meeting of Electrochemical Society, Montreal, Canada, May 9-14, 1982, session on Growth of Single Crystals on Amorphous Substrates.

24.   H. A. Atwater, H. I. Smith, and M. W. Geis, Appl. Phys. Lett. 41, 747 (1982).

25.   J. C. C. Fan, B-Y. Tsaur, R. L. Chapman, and M. W. Geis, Appl. Phys. Lett. 41, 186 (1982).

26.   J. D. Colinge, G. Auvert, and J. M. Temerson, Extended Abstract, 161 Meeting of Electrochemical Society, Montreal, Canada, May 9014, 1982, session on Growth of Single Crystals on Amorphous Substrates.

27.   M. W. Geis, H. I. Smith, C. V. Thompson, D. J. Silversmith, and R. W. Mountain, to be published in J. Electrochem. Soc., Also M. W. Geis et al., Electronic Materials Conference (Fort Collins, Colorado, June 1982).

28.   K. A. Bezjian, H. I. Smith, J. M. Carter, and M. W. Geis, J. Electrochem. Soc. 129, 1848 (1982).

30.   G. K. Celler, L. E. Trimble, K. K. Ng, H. J. Leamy, and H. Baumgart, Appl. Phys. Lett. 40, 1043 (1982).

31.   W. G. Hawkins, J. G. Black, and C. H. Griffiths, Appl. Phys. Lett. 40, 319 (1982).

32.   R. A. Lemons, M. A. Bosch, J. Cheng, and H. J. Leamy, Electronic Materials Conference, (Fort Collins, Colorado, June, 1982).

33.   H. J. Leamy, C. C. Chang, H. Baumgart, R. A. Lemons, and J. Cheng, Mater. Lett. 1, 33 (1982).

34.   R. F. Pinizzotto, Electronic Materials Conference, (Fort Collins, Colorado, June 1982).

35.   B-Y. Tsaur, J. C. Fan, M. W. Geis, R. L. Chapman, S. R. J. Brueck, D. J. Silversmith, and R. W. Mountain, these proceedings.

36.   B-Y. Tsaur, J. C. C. Fan, R. L. Chapman, and M. W. Geis, Appl. Phys. Lett. 40, 322 (1982).

37.   B-Y. Tsaur, J. C. C. Fan, R. L. Chapman, M. W. Geis, D. J. Silversmith, and R. W. Mountain, to be published in Electron Dev. Lett.

DETECTION OF ELECTRONIC DEFECTS IN STRIP-HEATER CRYSTALLIZED SILICON THIN FILMS

N. M. Johnson, M. D. Moyer, and L. E. Fennell
Xerox Palo Alto Research Centers, Palo Alto, CA   94304

E. W. Maby and H. Atwater
Massachusetts Institute of Technology, Cambridge, MA   02139

ABSTRACT

Electronic defects in strip–heater crystallized silicon thin films have been investigated with capacitance–voltage (C–V), deep–level spectroscopic, and scanning–electron microscopic techniques.   For electrical characterization the crystallized silicon films were used to fabricate inverted metal–oxide–silicon capacitors in which degenerately doped bulk silicon substrates provided the gate electrode.   High–frequency C–V characteristics yield effective fixed–charge densities in the oxide of $\leq 2\times10^{11}$ cm^{-2}.   Trap–emission spectra, recorded with deep–level transient spectroscopy on both p–type and n–type capacitors, indicate a continuous distribution of deep levels throughout the silicon bandgap.   The Si–SiO$_2$ interface is considered to be the principal source of this deep – level continuum, since the films are essentially single crystal with a low density of subgrain boundaries; the effective interface – state density is $\leq 2.5\times10^{10}$ eV^{-1} cm^{-2}.   A discrete energy level, detectable above the background continuum, appears in the upper half of the silicon bandgap; it may identify a point defect in the bulk of the silicon film with a spatially uniform density of approximately 1×10^{13} cm^{-3}.   On lateral p–n junction diodes, electron–beam – induced – current images reveal enhanced diffusion of arsenic along structural defects intersecting the junction.

INTRODUCTION

The rapid advancements and potential technological impact of beam–crystallized silicon thin films, as evidenced by the numerous papers on the subject in this symposium, have stimulated the need for comprehensive characterization of electronic defects and their correlation with materials processing and device performance. In a recent study [1], deep–level transient spectroscopy (DLTS) was used to measure electronic defect levels in silicon films which had been crystallized with a scanning cw Ar–ion laser.   The measurement revealed a continuous distribution of deep levels in the lower half of the silicon bandgap, with the Si–SiO$_2$ interface considered to be a major source of these levels. In a separate study [2], lateral p – n junction diodes where fabricated in similarly crystallized silicon thin films for analysis of the metallurgical junction with electron – beam – induced currents (EBIC).   This investigation revealed that large–angle incoherent grain boundaries provide efficient paths for dopant diffusion during device processing.

Mat. Res. Soc. Symp. Proc. Vol. 13 (1983) ©Elsevier Science Publishing Co., Inc.

In the present study DLTS and capacitance–voltage techniques were applied to detect electronic defect levels and EBIC was used to examine lateral p–n junction profiles in strip–heater crystallized silicon thin films. This crystallization technique has been shown to produce large grains (of the order of millimeters wide) permeated by a low density of subgrain boundaries [3,4].

DEVICE FABRICATION

In preparation for strip–heater crystallization, a 0.5 μm thick film of polycrystalline silicon was deposited by low–pressure chemical vapor deposition (LPCVD) onto a 0.52 μm thick thermal oxide over single–crystal silicon substrates. For electrical measurements the silicon substrate was degenerately doped for n^+ conductivity. The polycrystalline silicon film was encapsulated first with a 2 μm layer of silicon dioxide and then with 30 nm of sputtered silicon nitride [3].

The composite thin film structure was crystallized with a graphite strip heater as has been described previously [5]. The substrate was heated to 950°C, and the temperature of the graphite strip was estimated to be approximately 1580°C. The distance between the substrate and strip was 1.5 mm, and the scan speed was 1 mm/sec. The crystallization was performed in a flowing argon ambient. It was verified by transmission electron microscopy that the strip–heater crystallized silicon films consisted of large grains, with grain dimensions of the order of millimeters, which were permeated with subgrain boundaries spaced $10{\rightarrow}50$ μm apart.

For electrical evaluation of strip–heater crystallized silicon film, MOS capacitors were fabricated with an inverted electrode configuration. A schematic of the test device is shown in Fig. 1. The processing steps will be described specifically for p–type devices. After chemically removing the dielectric encapsulating layers, the crystallized silicon film was implanted with boron at 100 keV to a dose of 1×10^{12} cm^{-2}. The dopant was redistributed and activated with a furnace anneal at 1000°C for 2 hrs. Next, boron was shallowly implanted at 10 keV to a dose of 5×10^{14} cm^{-2} and activated, without significant redistribution, at 900°C for 20 min. to form a shallow p^+ layer for ohmic contact to the "backside" of the silicon film. The MOS capacitors were completed by vacuum depositing an aluminum–silicon alloy (1% Si) over the silicon film, photolithographically patterning the metal layer to define back electrodes, plasma etching the exposed silicon film to form the mesa structure depicted in Fig. 1, and finally sintering the devices at 450°C in forming gas for 20 min. The high–dose boron implant was found to be essential for maintaining an ohmic contact over the range of temperatures required for DLTS, and the Al/Si alloy was used to minimize contact alloying during sintering.

Lateral p–n junction diodes were fabricated for EBIC characterization of metallurgical junctions in strip–heater crystallized silicon films. The p region of the diode was doped by ion implanting boron at 80 keV to a dose of 2×10^{12} cm^{-2}, and the n region was implanted with arsenic at 150 keV to a dose of 1×10^{15} cm^{-2}. The dopants were redistributed and activated with a furnace anneal at 1000°C for 4 hrs. An oxide layer grown during the furnace anneal was chemically removed from the

Fig. 1. Schematic cross section of an inverted MOS capacitor consisting of a beam–crystallized silicon thin film on thermally grown silicon dioxide over single–crystal silicon. The silicon film contains residual grain and/or subgrain boundaries (GB) and is shown doped for p–type conductivity with a p^+ layer for ohmic back contact.

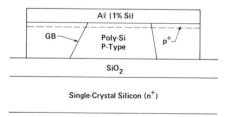

Aℓ (1% Si)

GB — Poly-Si P-Type — p^+

SiO₂

Single-Crystal Silicon (n^+)

entire diode area, then an Al/Si film was deposited and patterned for device contacts, and finally the devices were sintered for ohmic contact formation. The diodes are 24 μm wide, which is comparable to the separation between adjacent subgrain boundaries. A finished diode is shown in Fig. 6 (a).

INVERTED MOS CAPACITORS

Inverted MOS capacitors were fabricated with both p–type and n–type conductivities. This was particularly useful for DLTS measurements on MOS capacitors since deep levels in only the majority–carrier half of the silicon bandgap are detectable [6]. Capacitance–voltage (C–V) measurements provided supplemental defect characterization and set – up information for DLTS.

The C–V characteristic for a p–type MOS capacitor is shown in Fig. 2. Specifically, the high–frequency (1 MHz) equivalent–parallel capacitance is plotted versus dc gate bias (V_G). Also shown in Fig. 2 is the gate–voltage dependence of the energy, relative to the silicon valence – band maximum at the surface ($E_{V,Surf}$), at which the Fermi level (E_F) intersects the Si–SiO$_2$ interface; this was computed from the high–frequency C–V curve with the assumption of a uniform doping density [7]. For voltages less than -4.5 V, the capacitor is in accumulation. The range from -4.5 V to approximately 12 V is the depletion region, and for $V_G > 12$ V the capacitor is in inversion. For uniform doping densities the demarcation between accumulation and depletion is termed the flatband voltage (V_{FB}) and provides an estimate of the effective density of fixed charge in the oxide [7]. The measured flatband voltage of -4.5 V corresponds to a fixed charge density of 1.5×10^{11} cm^{-2}.

In Fig. 3 is shown a DLTS spectrum, obtained in the constant–capacitance (CC) mode [6], for a p–type capacitor. The spectrum was obtained by biasing the capacitor in depletion (i.e., $V_G = 8$ V) and periodically pulsing the gate voltage into

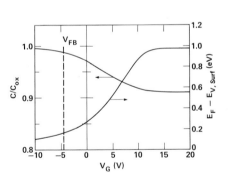

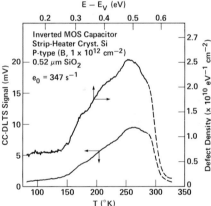

Fig. 2. High–frequency capacitance–voltage characteristic for a p – type (boron–doped) MOS capacitor in a strip–heater crystallized silicon film. Also shown is the gate–voltage dependence of the energy at which the Fermi level intersects the Si–SiO$_2$ interface.

Fig. 3. CC–DLTS spectrum and estimated defect distribution obtained for an MOS capacitor in p–type strip–heater crystallized silicon. The arrows identify the appropriate scales for each curve.

accumulation in order to populate deep levels throughout the film with majority carriers (in this case, holes). After a trap filling pulse, the interface states and deep levels in the depletion layer possess a non–equilibrium distribution of trapped charge which relaxes by thermal emission of holes to the valence band. In Fig. 3 the emission signal is non–zero from the lowest investigated temperature to above room temperature. Such a spectrum is representative of hole emission from deep levels which are continuously distributed in energy. Continuous deep – level distributions have been demonstrated for grain boundaries in bulk polycrystalline silicon [8,9] and are also characteristic of the Si–SiO$_2$ interface on bulk single–crystal silicon [6,10].

In order to numerically estimate the defect density, it may be assumed that interface states are the sole contributor to the emission signal; this is supported by the negligible density of grain boundaries and the low density of subgrain boundaries in strip – heater crystallized silicon films. Then the spectrum in Fig. 3 can be analyzed to obtain the defect distribution shown in the same figure. The energy scale was calculated for a hole capture cross–section of 1×10^{-14} cm^2, which is based on similar measurements of interface states in MOS capacitors on p–type bulk single–crystal silicon [11]. The defect density varies slowly with energy reaching a maximum value near midgap of approximately 2.5×10^{10} eV^{-1} cm^{-2}. If the spectrum in Fig. 3 were due to hole emission from unresolved bulk deep levels, the spatially uniform defect density would be of the order of 10^{13} cm^{-3}.

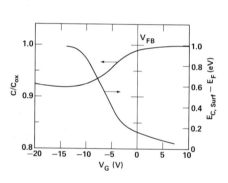

Fig. 4. High–frequency capacitance–voltage characteristic for an n–type (phosphorus–doped) MOS capacitor in strip–heater crystallized silicon. Also shown is the gate–voltage dependence of the energy at which the Fermi level intersects the Si–SiO$_2$ interface.

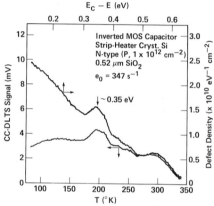

Fig. 5. CC–DLTS spectrum and estimated defect distribution for an MOS capacitor in n–type strip–heater crystallized silicon. The arrows identify the appropriate scales for each curve.

The high–frequency C–V curve for an n–type capacitor is shown in Fig. 4. The n–type conductivity was obtained by implanting the crystallized silicon film with phosphorous to doses of 1×10^{12} cm^{-2} for the bulk doping and 5×10^{14} cm^{-2} for the shallow n$^+$ layer; otherwise device fabrication was as described in the previous section. Also shown in Fig. 4 is the energy in the silicon bandgap at which the Fermi level intersects the Si–SiO$_2$ interface as a function of gate voltage; the energy is plotted relative to the silicon conduction – band minimum at the surface ($E_{C,Surf}$).

A significant feature of the C–V curve is the gentle increase of the capacitance for gate voltages less than -14 V. This is the inversion region for the n–type capacitor, in which the high–frequency capacitance ideally remains constant and at its minimum value [7]. The same phenomenon, with a more pronounced increase of the capacitance, has also been observed in n–type MOS capacitors fabricated on Ar–ion laser crystallized silicon [12]. In Fig. 4, the absence of a well–defined minimum capacitance, which is used to determine the flatband voltage [7], permits only an estimate of the fixed oxide charge. A flatband voltage of zero (as indicated in the figure) yields a positive charge density of roughly 1×10^{10} cm^{-2}, after accounting for the contribution to V_{FB} from the metal – semiconductor work function difference. This must be considered a lower limit for the fixed charge density in the n – type capacitors.

The effect which has been observed in the inversion region of n – type capacitors on beam – crystallized silicon films can occur when minority carriers (holes) are supplied to the interface in response to the small–amplitude ac signal of the capacitance measurement. This may arise from band bending in the vicinity of grain or subgrain boundaries, perhaps with Fermi – level pinning at the boundary, which leads to the formation of an inversion layer. The layer would be localized in the vicinity of the boundary and extend along the boundary through the film. This would provide a reservoir of holes which would diffuse laterally along the silicon surface when the capacitor is biased in inversion. The more pronounced effect in Ar – ion laser crystallized films could reflect the high density of large – angle grain boundaries in that material, as compared to the low density of subgrain boundaries in the strip – heater crystallized films.

A CC–DLTS spectrum for electron emission in an n–type MOS capacitor is shown in Fig. 5. The signal arises from electron emission from deep levels in the upper half of the silicon bandgap and is non–zero over the entire investigated temperature range, which again suggest a continuous defect distribution. However, at approximately 200 K there is a resolvable peak superimposed on the continuum. Also shown in Fig. 5 is the effective interface – state density for the continuum, computed with an assumed value of 1×10^{-15} cm^2 for the electron capture cross – section. The density decreases gradually with increasing energy from the conduction band toward midgap. The emission peak may identify the discrete energy level of a point defect in the bulk of the silicon film, with an activation energy for thermal emission of electrons of approximately 0.35 eV and a spatially uniform density of 1.2×10^{13} cm^{-3}. However, the microscopic origin of this level has not been determined.

LATERAL P–N JUNCTION DIODES

Large–angle incoherent grain boundaries intersecting metallurgical p–n junctions in laser–crystallized silicon have been shown to provide efficient paths for arsenic diffusion during device processing [2]. It has also been shown that coherent microtwins do not serve as arsenic diffusion paths [13,14]. In the present study scanning electron microscopy was used to assess the metallurgical junction in strip–heater crystallized silicon films. Results are shown in Fig. 6. The secondary–electron image in Fig. 6 (a) displays the device configuration. The ion–implantation–defined p–n junction is situated midway between the metal electrodes which are visible at the top and bottom of the figure. The EBIC image is shown in Fig. 6 (b). The n$^+$–region is in the lower half of the silicon island, and the p–region is in the upper half. The EBIC image displays high electron (minority–carrier) collection efficiency in the p–region and low hole collection in the n$^+$ – region, the latter largely being due to the high doping concentration which reduces the free – carrier diffusion length. The key item of interest in the present

(a) Secondary Electron Image   (b) EBIC

Fig. 6. Lateral p–n junction diode fabricated in strip–heater crystallized silicon.

study is the spike of low charge – collection efficiency extending from the junction into the p–region (near the left end of the junction). This feature can be ascribed to the enhanced diffusion of arsenic along a structural defect which intersects the metallurgical junction [2]. In enhancement – mode field – effect transistors, it has been shown that such perturbations of a metallurgical junction can introduce current shunting paths between source and drain contacts [15,16]. In strip – heater crystallized silicon films, subgrain boundaries may also be providing efficient dopant diffusion paths during device processing; however, this remains to be determined by transmission electron microscopy.

CONCLUDING REMARKS

Electronic defects in strip–heater crystallized silicon thin films were detected with capacitance–voltage, deep–level spectroscopic, and scanning–electron microscopic techniques. Several observations can be drawn from the results presented here. First, the deep–level distribution appears to be dominated by a continuum in both the upper and lower halves of the silicon band gap. The continuous defect distribution is most likely associated with the $Si–SiO_2$ interface, due to the low density of predominantly subgrain boundaries in these films. The effective interface–state density of $\leq 2.5 \times 10^{10}$ eV^{-1} cm^{-2} is closely approaching the state of the art in thermally oxidized bulk single – crystal silicon. An electron emission peak with an activation energy of approximately 0.35 eV may be the signature of a bulk defect with a spatially uniform density of $\sim 1 \times 10^{13}$ cm^{-3}, although the defect has not been chemically or microscopically identified. A contaminant which has previously been detected in graphite strip – heater crystallized silicon is carbon [17]. This impurity when incorporated in the silicon lattice as an interstitial contributes a hole – emission deep level at 0.27 eV and as a carbon interstitial – carbon substitutional complex displays a hole – emission level at 0.36 eV [18]. The absence of any clearly defined peaks in the DLTS hole – emission spectrum (Fig. 3) argues against the presence of high concentrations of electrically active carbon in the films examined here. Finally, both capacitance–voltage data and EBIC images strongly suggest that subgrain boundaries can affect device operation. In particular, the EBIC image suggests that subgrain boundaries may serve as paths for enhanced diffusion of dopants (e.g., arsenic) during device processing. This would preclude attaining high yields of low – leakage short–channel transistors in strip–heater crystallized silicon films which are permeated with subgrain boundaries. However, further study of the above issues is required.

ACKNOWLEDGMENTS

The work at MIT was supported by the Defense Advanced Research Projects Agency, Contract # N0014 – C – 80 – 0622, and by the Department of Energy, Contract # DE – AC02 – 80 – ER13019.

REFERENCES

1. N. M. Johnson, M. D. Moyer, and L. E. Fennell, Appl. Phys. Lett. 41, 560 (1982).
2. N. M. Johnson, D. K. Biegelsen, and M. D. Moyer, Appl. Phys. Lett. 38, 900 (1981).
3. E. W. Maby, M. W. Geis, Y. L. LeCoz, D. J. Silversmith, R. W. Mountain, and D. A. Antoniadis, IEEE Electron Device Lett. EDL – 2, 241 (1981).
4. M. W. Geis, H. I. Smith, B – Y. Tsaur, J. C. C. Fan, E. W. Maby, and D. A. Antoniadis, Appl. Phys. Lett. 40, 158 (1981).
5. J. C. C. Fan, M. W. Geis, and B – Y. Tsaur, Appl. Phys. Lett. 38, 365 (1981).
6. N. M. Johnson, J. Vac. Sci. Technol. 21, 303 (1982).
7. E. H. Nicollian and J. R. Brews, MOS Physics and Technology (Wiley, New York, 1982).
8. C. H. Seager, G. E. Pike, and D. S. Ginley, Phys. Rev. Lett. 43, 532 (1979).
9. J. Werner, W. Jantsch, K. H. Froehner, and H. J. Queisser, Grain Boundaries in Semiconductors (Elsevier, New York, 1982), eds. H. J. Leamy, G. E. Pike, and C. H. Seager, pp. 99 – 104.
10. N. M. Johnson, D. K. Biegelsen, and M. D. Moyer, Physics of MOS Insulators (Pergamon, New York, 1980), eds. G. Lucovsky, S. T. Pantelides, and F. L. Galeener, pp. 311 – 315.
11. N. M. Johnson, D. K. Biegelsen, and M. D. Moyer, J. Vac. Sci. Technol. 19, 390 (1981).
12. N. M. Johnson, unpublished.
13. D. K. Biegelsen, N. M. Johnson, R. J. Nemanich, M. D. Moyer, and L. Fennell, Laser and Electron Beam Interactions with Solids (Elsevier, New York, 1982), eds. B. R. Appleton and G. K. Celler, pp. 331 – 336.
14. N. M. Johnson, D. K. Biegelsen, and M. D. Moyer, Grain Boundaries in Semiconductors (Elsevier, New York, 1982), eds. H. J. Leamy, G. E. Pike, and C. H. Seager, pp. 287 – 296.
15. T. I. Kamins and B. P. Von Hersen, IEEE Electron Device Lett. EDL – 2, 313 (1982).
16. K. K. Ng, G. K. Celler, E. I. Povilonis, R. C. Frye, H. J. Leamy, and S. M. Sze, IEEE Electron Device Lett. EDL – 2, 316 (1981).
17. R. F. Pinizzotto, H. W. Lam, and B. L. Vaandrager, Appl. Phys. Lett. 40, 388 (1982).
18. L. C. Kimerling, Radiation Effects in Semiconductors 1979 (Inst. Phys. Conf. Ser. 31, Bristol, 1977), p. 221.

CHARACTERIZATION AND APPLICATION OF LASER INDUCED SEEDED-LATERAL EPITAXIAL Si LAYERS ON SiO_2

M.MIYAO, M.OHKURA, T.WARABISAKO and T.TOKUYAMA
Central Research Laboratory, Hitachi Ltd.,
Kokubunji, Tokyo 185, Japan

ABSTRACT

Electrical and crystal properties of seeded lateral epitaxial Si are evaluated as a function of distance from seeding area with the aid of a micro-probe RHEED and MOSFET fabrication. The results indicate that the quality of a grown layer is as good as that of bulk Si crystal for most of the epitaxial layer. However, at the SiO_2 edge, electrical properties are somewhat poor due to the existence of dislocation and residual stress. Element devices useful for SOI structures are fabricated. Electrical properties of MOSFET's with double active areas indicate that surface and bottom regions of the epitaxial layer are all of device worthy quality. Insulated control gate bipolar type transistors are proposed and some preliminary results are shown.

INTRODUCTION

Silicon on an insulating substrate (SOI) [1,2,3] is a very attractive material structure for future VLSI. Moreover, it is raising expectations of achieving three dimensional devices and circuits.

The various techniques which realize SOI structures reported to date are summarized in table 1. Techniques shown in categories (A) and (B) can be realized using a conventional oven or strip heater as a heating tool. These techniques can provide large SOI crystals, especially good crystal quality was achieved in technique (B). However, the high temperature heating (1000 C or more) required in these processes would degrade any circuits that might be included in the substrate, when so called three dimensional devices are considered.

On the other hand, localized heating techniques using a laser or electron beam, which are of recent interest, can complete the process at less than 500 C substrate temperature. However, the techniques shown in category (C) only resulted in large grain poly.Si films. The crystal is far from LSI application quality, though grain growth on the transparent substrate showed some potential for flat panel display driver circuits.

Consequently, a new, low temperature SOI process with good crystal quality, is needed for three dimensional integrated device fabrication. The key factor is obviously to control the crystal orientation of the grown layer. This is because, many process parameters, i.e. etching speed, oxidation rate, impurity diffusion coefficient are strongly dependent on crystal orientation. Moreover, variation in such electronic properties as carrier mobility and MOS interface properties that originate in the different orientations of the crystal grains in which each elementary device is located, degrade total LSI characteristics.

However, large area SOI crystals will not always be necessary for circuits with high packing density. This is because the active area of elemental devices has become smaller in recent years. The essential thing is how to fabricate SOI crystals of excellent quality at desired positions.

These requirements prompted our group to develop bridging epitaxy (seeded lateral epitaxy) using pulsed ruby laser irradiation [4,5].

Mat. Res. Soc. Symp. Proc. Vol. 13 (1983) ©Elsevier Science Publishing Co., Inc.

	high temperature process (oven heating)	low temperature process (localized beam heating)
without orientation control	(A) CVD poly Si	(C) poly Si island on or in insulator
with orientation control	(B) Si on sapphire CVD epi. lateral overgrowth deep oxygen impl. strip heater melting	(D) seeded lateral epitaxy

TABLE 1. SOI techniques reported to date

However, limited control of the solidification process due to pulse laser irradiation resulted in a small lateral growth (a few microns) beyond the SiO_2 edge. This work influenced many researchers [5-9] who used a scanning Ar laser (or electron beam as well), which is recognized to have many adjustable parameters, in order to obtain large oriented crystals. Today, oriented crystal growth has already been extended to 500 μm over SiO_2.

In line with this background, this paper describes recent progress in laser induced seeded lateral epitaxy. First, topics are focused on the necessity of seeding upon recrystallization. Second, electronic properties of the grown layer are discussed in connection with crystal quality. Finally, some novel device structures will be discussed.

CRYSTAL GROWTH OF SEEDED LATERAL EPITAXY

Seeding Effect Upon Recrystallization

The basic idea of obtaining single crystal Si with a unique, certain orientation is called "lateral seeding". It comes from micro-Bridgeman growth [10]. The principle behind this process is schematically shown in Fig.1. Irradiation by laser or electron beam of sufficient energy first melts the deposited Si layer. With the scanning of the beam, the melted zone moves laterally. If the cooling process occurs mainly from the Si substrate through an insulator window area, seeding from the Si single crystal substrate surface is possible.

The linearly graded temperature profile in the scanning direction (higher temperature at the melting front) is the force producing lateral growth. In order to realize crystal growth with a speed in the cm/sec range, i.e. approximately equal to the scanning speed, a temperature gradient of 10^3 K/cm is necessary in the lateral direction.

Such a condition can be obtained by utilizing an insulator with the thermal conductivity (K) that is less than that of Si (0.2 cal/cm.sec.deg). To utilize materials having such K values, but which are close to that of Si, film thickness must be increased. With larger film thickness and/or lower thermal conductivity, the temperature gradient becomes dominant. This leads to higher crystal growth speed, so that higher annealing through-put can be expected.

Temperature contour lines for the above conditions are schematically shown in Fig.1. Crystal growth direction and average cooling rate are also shown. Since the growth front propagates perpendicular to the thermal

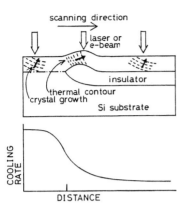

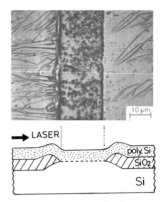

FIGURE 1. Schematic of beam-induced seeded lateral epitaxial process

FIGURE 2. Optical micrograph and schematic cross section of laser annealed and Secco-etched poly Si on either Si or SiO_2

contour lines, liquid phase epitaxial growth first occurs vertically from the Si substrate and then propagates laterally over the SiO_2, as shown in the figure.

However, any large change in thermal flow due to the thermal conductivity difference between Si and the insulator (SiO_2; 0.0035 cal/cm.sec.deg, Si_3N_4; 0.0041 cal/cm.sec.deg) causes a large change in cooling rate and crystal growth direction in the vicinity of the SiO_2 edge. In addition, stress due to thermal expansion difference (Si; $3.2x10^{-6}K^{-1}$, SiO_2; $5x10^{-7}K^{-1}$, Si_3N_4; $2.8x10^{-6}K^{-1}$) causes crystal defects [8]. Consequently, growth features in this area should be investigated with great care.

Evaluation Of Crystal Orientation

One example of lateral epitaxial growth is shown in Fig.2 (poly.Si and SiO_2 thickness are both 350 nm). After the scanning of the Ar laser beam (P=8W, v=50cm/sec) from left to right (a-b-c), crystallinity was examined by Secco etching. Different features of the grown layer are summarized as follows; poly.Si with large grain size (region(a)), single crystal Si with high defect densities (region(b)), and single crystal Si growth about 10um from the SiO_2 edge (regions(c) and (d)).

Crystal orientation of the grown layer is precisely investigated with the aid of a newly developed micro-probe RHEED [13]. In the system [14], a field emission electron source with a probe beam focus of 0.1 μm was adopted. The probe beam was, first, scanned over the whole sample to obtain a scanning electron microscope image on the display. Then the beam was directed to the desired location on the sample by referring to the image. Thus, a diffraction pattern for a very small area (0.1 μm) was observed.

Results from various regions of the grown layer at the optimum condition for recrystallization (P=8W, v=50cm/s) are shown in Fig.3. Single crystal

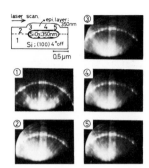

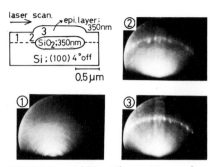

FIGURE 3. *Diffraction patterns from an SOI sample (laser irradiation at optimum condition) obtained by micro-probe RHEED.*

FIGURE 4. *Diffraction patterns from an SOI sample (laser irradiation at low energy) obtained by micro-probe RHEED.*

diffraction patterns with Kikuchi lines were obtained all over the recrystallized layers. In addition, the diffraction patterns were all identical to the pattern obtained from the seeding area and substrate ((100) 4 degree off the axis). The preferential crystal orientation in a random crystal regrowth over the SiO_2 is reported to be just (100) [12]. Thus, it is hard to believe that crystal growth of this orientation ((100) 4 degree off the axis) was unexpectedly accomplished.

Results obtained at a condition of lower laser energy (P=7W, v=40cm/sec) irradiation are shown in Fig.4. Single crystal diffraction patterns with Kikuchi lines were obtained from the recrystallized area on the SiO_2 substrate, although the seeding area was still polycrystalline. However, the crystal orientation was also (100) 4 degree off the axis. In order to solve such a contradiction, precise investigation was carried out in the vicinity of the SiO_2 edge. As a result, single crystal growth in the seeding area was found only within 0.5μm of the SiO_2 edge (Fig.4-2). This result can be phenomenologically explained as follows; in the irradiation at low laser energy, complete melting of the deposited poly.Si layer occurred on the SiO_2. In the seeding area, only the surface portion was melted. However, prevention of heat flow by the SiO_2 walls enhanced the melting depth at the SiO_2 edge. Consequently, vertical liquid phase epitaxial growth initiated at the SiO_2 edge and then propagated over the SiO_2 substrate.

These results confirmed that seeded vertical epitaxial growth from the substrate is essential to obtaining single crystal Si on SiO_2. This is contrary to the recent findings of Magee et al [5]. Moreover, it was found that a very small seeding area (0.5 μm) was enough for lateral epitaxial growth.

Refinements Of Seeded Lateral Epitaxial Technique

Control of temperature profiles parallel and/or perpendicular to the growth direction is the most important factor in crystal growth. A linearly graded temperature profile is essential in the growth direction, as described in the previous section. On the other hand, a convex temperature profile (lower temperature at the center) is necessary perpendicular to the growth direction in order to prevent random nucleation from the borders. These conditions can be achieved by utilizing a beam shaping technique (crescent

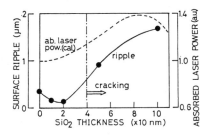

FIGURE 5. *Influence of encapsulation upon recrystallization ; open circles indicate surface morphology and broken line is normalized absorbed laser power calculated by interference effect.*

[15], donut [16] or tilted elliptical beam [17]) and/or special sample structure (Si island recessed into SiO_2 tubs [18] or antireflecting stripes [19]). In these latter methods, location of the single crystal Si is controlled by highly reproducible photolithographic means, which offers a big advantage for device fabrication.

Despite such efforts to obtain an ideal temperature profile, single crystal lateral growth was limited to only 30 μm. The limiting factor for crystal growth is considered to be defect formation due to stress, which is caused by the different thermal expansion coefficients (K) for Si and SiO_2 [Lam's model [8]], and/or zone refining of impurities contained in the deposited film [Leamy's model [11]]. Consequently, a clean process (surface cleaning and deposition of high purity Si) should be developed. In addition, development of new materials which have the same thermal expansion coefficient as that of Si becomes more important.

If seeding zones are lined along the beam scanning path, the seeding process can be repeated during crystal growth, which enlarges the SOI crystal area. This idea leads to the conclusion that utilization of transverse seeding is better than longitudinal seeding [8]. Celler et al. [18] and Ishiwara et al. [20] used a scanning laser or electron beam in two orthogonal directions, and produced SOI crystals as far as 500 μm from the seed.

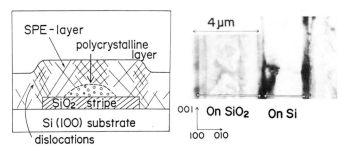

FIGURE 6. *(A) Schematic cross section view of the sample after the first-step annealing. (B) TEM photograph of the SPE layer on SiO_2. (from Y. Kunii et al. [23])*

From the stand-point of device application, changes in surface morphology become a serious problem. Surface ripple due to mass flow during melting is a frequently observed phenomenon. Vapor phase epitaxial growth after smoothing of the rippled surface by etching is one practical way to obtain a smooth, and thick epitaxial layer, although, this requires high temperature processing.

Encapsulation effects by SiO_2 during laser annealing are shown in Fig.5. These results show some improvement in surface morphology. In addition, the descrepancy between experiments and calculation (effective laser power absorption by poly.Si) based on interference effect indicated that such improvement did not originate from the anti-reflection effect but from the physical capping effect. However, utilization of a thick cap resulted in cracking of the deposited Si film.

Multi-layer capping with different softening temperature (T_s) layers might improve these phenomena. For example, utilization of a thin, top capping film (high T_s) might prevent mass flow and a thick, bottom film (low T_s) might act as a buffer layer for stress reduction. However, recently reported data [1-3] about capping effects are scattered. This does not seem to support our simple speculation.

Thermal detachment (stripping off) of molten Si from the SiO_2 is another problem that appeared during seeded lateral epitaxy. This phenomenon is caused by the large difference in temperature rise between the layers on the Si and SiO_2. Very recently, Sakurai et al.[21] succeeded in optimizing absorbed laser power in poly.Si, on both Si and SiO_2 using locally varied anti-reflection film.

Lately, investigation of another SOI technique has been reported by Ohmura et al [22]. They proposed the possibility of solid phase seeded lateral epitaxial growth. After that, Kunii et al. [23] developed this method using special surface cleaning (H_2 anneal at 1100 C plus Si surface etching with HCl diluted with H_2 and Ar) and two-step furnace annealing (550 C, 9 days and 1150 C 4 hrs).

One example of the results is shown in Fig.6 which indicates that crystal growth was limited to only 4 μm from the seeding zone. The key factor which limits the propagation of epitaxial growth is nucleation of the poly.Si randomly occuring during furnace annealing. Heating only the small desired part and scanning adjusted to the regrowth speed are effective in preventing such nucleations. Thus, the beam annealing process will become more important.

ELECTRICAL PROPERTIES OF THE EPITAXIAL LAYER

Electrical properties of a grown layer [24] should be evaluated as a function of distance from the seeding area, because a large change in crystal growth features occurs in the vicinity of SiO_2 edge. In line with this, n-channel MOSFET's were fabricated in various regions of the epitaxial layer [25,26].

Plan and cross-section views of FET's (channel length: 8 μm, channel width: 15 μm) are shown in Fig.7, where the gate regions for the FET's indicated as (a),(b),(c) and (d) are located in the corresponding epitaxial regions shown in Fig.2. In the fabrication, boron implantation ($6x10^{11}$ cm^{-2}, 100 keV) and furnace annealing were first used to reduce resistivities of seeded lateral epitaxial layers. Source and drain regions were formed by self-aligned arsenic implantation ($1x10^{16}$ cm^{-2}, 80 keV) to the poly.Si gate (350 nm) and furnace annealing (1000 C, 40 min).

Drain current and voltage characteristics of typical samples are shown in Fig.8. Channel mobilities derived from the I-V characteristics are

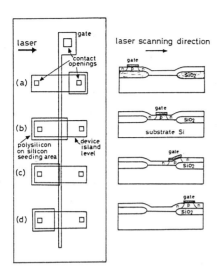

FIGURE 7. Plan and cross section views of FET's.

summarized in Fig.9 as a function of device location from the seeding area.

In region (a), a dramatic increase in poly.Si grain sizes due to laser annealing improved electron mobility by a factor of 100. However, grain boundaries randomly distributed in this region (Fig.2) cause scattering of electron mobilities in a wide range (100 - 500 cm^2/V.s).

Recently, the effects of grain boundaries upon electrical properties have been investigated [27]. Results indicate that grain boundaries perpendicular to the current path decreased electron mobility. On the other hand, enhanced impurity diffusion along grain boundaries cause short circuiting between sources and drains, so that grain boundaries parallel to the current path increase leakage current. These results clearly suggest that crystal quality in region (a) is still far from the LSI application level.

In region (b), mobility values of 300 cm^2/V.s were obtained for samples without laser annealing, which suggests grain growth during furnace annealing. After laser annealing, high mobility (500 cm^2/V.s) and low threshold voltage (0.5 V) were obtained, although crystal defects still remained, as shown in Fig.2. Recent observation by cross-sectional TEM [7] have indicated that dislocations are mainly localized near the interface between the deposited film and substrate. These are considered to originate from contamination of the substrate surface before poly.Si deposition.

In region (c), mobility value was 30% lower and threshold voltage was 0.5 V higher than for the other regions. TEM measurements indicated dislocations in this region also. Large differences in temperature rise and cooling rate during and after laser irradiation produce residual stress discontinuities and could enhance dislocation formation. Such crystal degradation is the main reason for the low electron mobility. In addition, concentrations of boron atoms implanted and annealed in the channel regions could be gathered around this region. They could cause a threshold voltage shift in the positive direction.

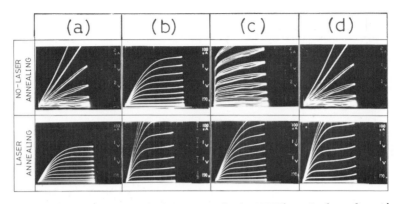

FIGURE 8. *I-V characteristics for typical MOSFET's. Scales of vertical axis, horizontal axis and gate step are (2μA,1V,2V) for samples (a),(c) and (d) without laser annealing, (100μA,1V,1V) for (b) without laser annealing and (a),(b),(c) and (d) with laser annealing, respectively.*

In region (d), a representative mobility value identical to that for bulk crystal was obtained. Scattering of this mobility value was very small, to the contrary of that in region (a).

Electronic properties in the subthreshold region indicate a low leakage current (0.7 - 1.0 pA/micron channel width at V_G=-5V, V_D=5V) and a low tailing factor (90mV/decade). The leakage current level is comparable to that for bulk Si, and is much smaller than that of published SOI data [28], i.e., seeded lateral epitaxial Si on a Si_3N_4 substrate. These results suggested that negative charge density at the bottom interface, which caused a serious problem when using a Si_3N_4 substrate, is very small when using a SiO_2 substrate. Experiments with the back gate MOSFET's [26] indicated that negative charge density at the grown layer and the oxide are in the low to mid 10^{11} cm^{-2} range.

Such interface structures caused by molten Si and insulator have not appeared in conventional Si processing technology, and physical models of the structures have not been developed yet.

APPLICATION OF SEEDED EPITAXIAL LAYER FOR NOVEL DEVICES

The first idea for development of three-dimensional devices was a stacked device fabricated in sequentially grown layers, which was proposed by Geis et al. [29]. In order to realize such sophisticated devices, the influence of the recrystallization process upon underlying device characteristics must first be clarified. In the case of electron beam annealing, severe damage is induced in the Si substrate. Recently, Saitoh et al. [30] investigated this phenomena using the device structure shown in Fig.10. They found out that such damage was easily removed by low temperature (400 - 500 C) annealing. Moreover, they determined that the main reason for damage generation was not melting and/or subsequent rapid cooling, but the high charge density (10^{-7} C/cm^2) of penetrating electrons. In other words, their experiments suggested that an SOI process, useful for stacked devices, can be developed by using laser beam annealing instead of electron beam annealing.

A more realizable idea for three dimensional devices is the utilization

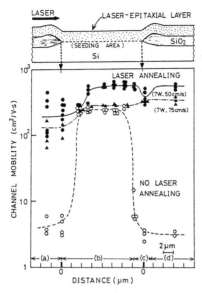

FIGURE 9. *Values of channel mobility summarized as a function of distance from SiO_2 edge.*

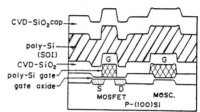

FIGURE 10. *Cross section view of a sample used in e-beam annealing. (from S.Saitoh et al [30])*

of one SOI layer for two electrically active regions, which was proposed by Gibbons et al. [31]. They produced a shared gate CMOS inverter by fabricating a second MOS channel on top of the gate electrode of a bulk device.

In the case of seeded lateral epitaxiy, one SOI layer provides three regions; the top and bottom regions of the epitaxial layer and seeding region. These regions should be effectively utilized for novel devices.

The electrical characteristics of these regions are shown in Fig.11. The p-n junction characteristics in the seeding area (Fig.11-B) indicate that the reverse current level is lower than 10^{-7} A/cm² at 0.1 V bias. The lowest value obtained is 6 x 10^{-8} A/cm². The ideality factor in forward operation is in the range 1.2 - 1.4.

These relatively poor characteristics are considered to be the result of residual defects at the interface between Si substrate and the recrystallized layer. Special cleaning of the Si substrate, which was described in the previous section (Fig.6), should improve the interface characteristics.

A conventional front-gate MOSFET (Fig.11-C) indicates that electrical properties of the top region in the epitaxial layer are good, as already discussed in the previous section. The characteristics of a back-gate MOSFET, in which the poly.Si gate was shorted to the source and the Si substrate was used as the gate, is shown in Fig.11-D. In this operation, a p-n diode is connected between the back-gate and drain. Therefore, the forward biased I-V characteristics of the p-n diode are superimposed onto the original I-V characteristics, as shown in the figure.

Apparent electron mobilities of back-gate MOSFET's are summarized in Fig.12-A as a function of laser scanning speed. "Apparent mobility" means the value expressed by $\mu \cdot (L_{mask}/L_{channel})$, where μ, L_{mask} and L_{chann} are

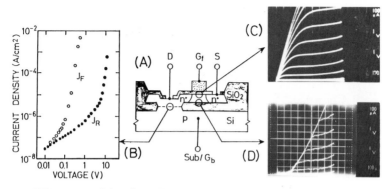

FIGURE 11. (A) Schematic cross section of a seeded SOI device.
(B),(C) and (D) are I-V characteristics of a diode in the seeding
area, a front gate MOSFET, and a back gate MOSFET, respectively.

electron mobility, gate length , and channel length, respectively. An
extremely high apparent mobility (2000 cm^2/V.s) was obtained at a special
scanning speed (30 cm/s). The failure rate of MOSFET's, i.e., the
probability of source and drain being short circuited, is shown in Fig.12-B.
Results indicate that both irradiation conditions, in which electron mobility
and failure rate showed maximum values, coincided well. This suggests that
side diffusion of arsenic atoms during furnace annealing becomes dominant at
this annealing condition.

Thus, abnormally high apparent electron mobility is concluded to be the
result of the large difference between L_{mask} and $L_{channel}$.

When scanning speed is decreased, the failure rate decreases and
apparent electron mobility reaches a normal value. Moreover, it was found
that this scanning speed (20 cm/s) corresponds to the threshold annealing
condition needed to obtain single crystal growth over SiO$_2$.

These results lead to the important suggestion that insufficient melting
during laser annealing causes stress or defects at the back interface, which
enhances arsenic atom diffusion.

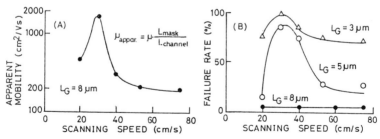

FIGURE 12. Apparent electron mobility (A), and failure rate (B) of
back gate MOSFET's as a function of laser scanning speed.

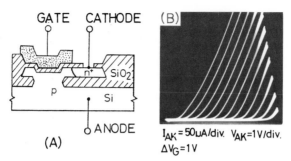

GATE   CATHODE   (B)

I_{AK} = 50uA/div. V_{AK} = 1 V/div.
ΔV_G = 1 V

(A)

*FIGURE 13. Schematic cross section (A), and I-V characteristics (B)
of a MOS-gate-controled-bipolar type Trs.*

A novel device that utilizes this seeded lateral epitaxial structure
has been fabricated. The schematic cross section and I-V characteristics for
this device are shown in Fig.13-(A),(B). This device is essentially a
lateral p-n diode [32], in which the junction barrier can be controlled by an
external MOS gate. This is because the SOI layer thickness (0.35µm) is
thinner than the Debye length in the p-type region under the MOS gate.
Therefore, under a certain gate bias, the forward p-n junction current can be
cut off even at a higher forward bias.

Anode to cathode I-V characteristics were measured as a function of gate
bias. They show a triode-like behavior. This device has the inherent high
input impedance of a MOS gate and a low output impedance due to the bipolar
action of the device.

CONCLUSION

The detailed electrical and crystal properties of a seeded lateral
epitaxial layer have been described based upon our recent work. The results
show that this SOI technique is a powerful means to fabricate unique device
structures, in particular small geometry, more sophisticated devices.
However, there remain some problems mainly concerned with crystal growth by
the melting scheme. Calculations of temperature profile and cooling rate are
necessary in order to understand the crystal growth features of this
multi-layered structure, i.e. cap-insulator/ deposited Si film/ insulated
substrate on crystal Si. Investigation of the physical structure and
properties of the back interface is also important from both fundamental
interest and device application view points.

ACKNOWLEDGEMENT

The authors would like to thank Drs. I.Takemoto, N.Hashimoto and
H.Sunami for their help in device fabrication. Drs. M.Tamura, M.Ichikawa
and N.Natsuaki are also acknowledged for the TEM and micro-probe RHEED
observations and for their useful discussions.

REFERENCES

1) Laser and Electron Beam Processing of Materials edts: C.W.White and P.S.Peercy (Academic Press, New York 1980).
2) Laser and Electron Beam-Solid Interactions and Material Processing edts: J.F.Gibbons, L.D.Hess and T.W.Sigmon (North-Holland, New York 1981).
3) Laser and Electron-Beam Interactions with Solids edts: B.R.Appleton and G.K.Celler (North-Holland, New York 1982)
4) M.Tamura, H.Tamura and T.Tokuyama: Jpn.J.Appl.Phys. 19 L23 (1980).
5) M.Tamura, H.Tamura, M.Miyao and T.Tokuyama: Jpn.J.Appl.Phys. 20 (1981) Suppl.20-1 P.43.
6) T.J.Magee, L.J.Palkuti, R.Ormond, C.Leung and S.Graham: Appl. Phys.Lett. 38 248 (1981).
7) T.I.Kamins, T.R.Cass and C.J.Dell'Oca, K.F.Lee, R.F.W.Pease and J.F.Gibbons: J.Elec.Chem.Soc. 128 1151 (1981).
8) H.W.Lam, R.F.Pinizzotto, A.F.Tasch Jr: J.Elec.Chem. Soc 128 1981 (1981).
9) M.Tamura, M.Ohkura and T.Tokuyama: Jpn.J.Appl.Phys. 21 (1982) Suppl.21-1, P.193.
10) D.K.Biegelsen, N.M.Johnson, D.J.Bartelink and M.D.Moyer:in ref (2).
11) H.J.Leamy:in ref(3).
12) M.W.Geis, H.I.Smith,B.Y.Tsaur, J.C.C.Fan, E.W.Maby and D.A.Antoniadis: Appl.Phys.Lett. 40 158 (1982).
13) M.Ohkura, M.Ichikawa, M.Miyao and T.Tokuyama; to be appeared in Appl.Phys.Lett. (Dec. 1982).
14) M.Ichikawa and K.Hayakawa; Jpn.J.Appl.Phys. 21 145 (1982).
15) T.J.Stultz and J.F.Gibbons; Appl.Phys.Lett. 39 498 (1981).
16) S.Kawamura, J.Sakurai, M.Nakano and M.Takagi; Appl.Phys.Lett. 40 394 (1982).
17) L.E.Trimble, G.K.Celler, K.K.Ng, H.Baumgart and H.J.Leamy; in ref(3).
18) G.K.Celler, L.E.Trimble, K.K.Ng, H.J.Leamy and H.Baumgart: Appl.Phys.Lett. 40 1043 (1982).
19) J.P.Colinge, E.Demoulin, D.Bensahel and G.Auvert: Appl.Phys. Lett. 41 346 (1982).
20) H.Ishiwara, M.Nakano, H.Yamamoto and S.Furukawa; 1982 International Conf. on Solid State Device, Tokyo.
21) J.Sakurai, S.Kawamura, M.Nakano and M.Takagi: Appl.Phys.Lett. 41 64 (1982).
22) Y.Ohmura, Y.Matsushita and K.Kashiwagi; Jpn.J.Appl.Phys. 21 1152 (1982).
23) Y.Kunii, M.Tabe and K.Kajiyama; 1982 International Conf. on Solid State Device, Tokyo.
24) H.W.Lam, Z.P.Sobczak, R.F.Pinizzotto and A.F.Tasch Jr.; IEEE Electron Device 29 389 (1982).
25) M.Miyao, M.Ohkura, I.Takemoto, M.Tamura and T.Tokuyama; Appl.Phys.Lett. 41 59 (1982).
26) M.Miyao, M.Ohkura, I.Takemoto, M.Ichikawa, M.Tamura and T.Tokuyama; Digest of Tech. Paper p177 (1982 International Conf. on Solid State Device, Tokyo).
27) K.K.Ng, G.K.Cell, E.I.Povilonis, R.C.Frye, H.J.Leamy and S.M.Sze; IEEE Electron Device Lett. 2 316 (1981).
28) H.W.Lam; in ref (3)
29) M.W.Geis, D.A.Antoniadis, D.C.Flanders and H.I.Smith; Inter.Electron Device Meeting Technical Digest P.210, Dec. 1979.
30) S.Saitoh, K.Higuchi and H.Okabayashi; Digest of Tech. Paper P.171, (1982 International Conf. on Solid State Device, Tokyo).
31) J.F.Gibbons and K.F.Lee; IEEE Electron Device Lett. 1 117 (1980).
32) Y.Ohmura; Appl.Phys. Lett. 40 528 (1982).

OXYGEN AND NITROGEN INCORPORATION DURING CW LASER RECRYSTALLIZATION OF
POLYSILICON

C.I. DROWLEY AND T.I. KAMINS
Hewlett-Packard Laboratories, 3500 Deer Creek Road, Palo Alto, CA   94304

ABSTRACT

The incorporation of nitrogen and oxygen in polysilicon
has been examined by SIMS. The analysis, combined with C-V
measurements and ion implantation, has been used to correlate the incorporation of the two species with the fixed-
charge density at the back polysilicon/SiO_2 interface.
Laser recrystallization with a silicon-nitride encapsulation
layer results in the inclusion of 2-4 x 10^{17} cm^3 nitrogen
atoms in the polysilicon; if an oxide capping layer is used,
the nitrogen level observed is at the background of the SIMS
system ($\sim 10^{15}$ cm^{-3}). Either type of capping layer results
in 3-4 x 10^{18} cm^{-3} oxygen atoms being incorporated into the
polysilicon. Implantation of nitrogen into the polysilicon
before recrystallization increases the fixed-charge density
$N_{f,b}$) at the back interface, while implanted oxygen decreases $N_{f,b}$. The high $N_{f,b}$ found with a nitride capping
layer is attributed to deposition of nitrogen or SiN_x at
the back interface.

INTRODUCTION

Recent investigation [1,2] has shown that the fixed-charge density ($N_{f,b}$) at
the back surface of a laser-recrystallized polysilicon film on silicon dioxide
is affected by the choice of encapsulation layer on the front surface during
recrystallization. Silicon-nitride encapsulation layers lead to values of $N_{f,b}$
up to the high-$10^{11}$to low-10^{12} cm^{-2} range, while silicon-dioxide layers (either
deposited or grown) can reduce the value of $N_{f,b}$ to the low 10^{11} cm^{-2} range.

The cause of the different $N_{f,b}$ resulting from the two different encapsulation layers has not previously been reported and is the subject of this paper.
We have examined chemical differences in the films and interfaces arising from
reactions between the SiO_2 or Si_3N_4 and the molten silicon during recrystallization. Molten silicon is known to attack both Si_3N_4 and SiO_2 [3], so that chemical attack is a reasonable possibility. The solubilities of nitrogen and
oxygen in liquid silicon are 6 x 10^{18} cm^{-3} and 2.2 x 10^{18} cm^{-3}, respectively [4].
In order to reach these limits, less than one monolayer of oxide or nitride
would be dissolved in a 0.55µm thick polysilicon film.

Two possible models may explain the effect of a chemical reaction on the back
interface properties. First, if a layer of the SiO_2 at the back interface is
dissolved during laser melting and the oxygen is incorporated into the polysilicon film, the number of dangling bonds at the back interface may increase,
degrading the electrical characteristics of the interface. The reaction between the underlying SiO_2 and the molten silicon may be less with an oxide
encapsulation layer, since the equilibrium concentration of oxygen in the
silicon could be at least partially satisfied by oxygen from the top layer. A
nitride capping layer would not be expected to provide significant oxygen, so
that the dissolution of the underlying SiO_2 would be greater (and the resulting

Mat. Res. Soc. Symp. Proc. Vol. 13 (1983) ©Elsevier Science Publishing Co., Inc.

electrical properties would be worse) than with an oxide capping layer. We call this model the "dissolution model."

The second model involving chemical attack assumes that the encapsulation layers are partially dissolved during recrystallization and that some of the oxygen or nitrogen is incorporated into the film. After subsequent processing some of the included oxygen or nitrogen moves to the back interface (and other interfaces; e.g., grain boundaries) forming SiO_x and SiN_x. The SiN_x at the back interface (formed with the nitride cap only) would cause poorer electrical properties than would occur at an ideal Si/SiO_2 interface, thus explaining the observed difference in behavior with the two encapsulants. This second model is called the "deposition" model.

EXPERIMENTAL RESULTS AND DISCUSSION

Chemical effects have been correlated with the observed difference in $N_{f,b}$ with the different encapsulation layers by electrical characterization (C-V), SIMS, and controlled introduction of dopants via ion implantation.

Results of C-V measurements on wafers implanted with oxygen or nitrogen prior to laser recrystallization are shown in Fig. 1. The samples were recrystallized using a Gaussian, CW Ar-ion laser beam scanned at 25 cm/sec. The beam power was 9-10W; the beam radius was 40-50μm; the melted track width was 40μm; and the scan overlap was 50% of the melted track width. The substrate was heated to 500°C. An inverted C-V structure [5] was used with a 0.01 ohm-cm, n-type bulk wafer as the gate and a 200 nm-thick, thermally grown SiO_2 "gate" dielectric between the bulk wafer and the polysilicon. The polysilicon was 0.55μm thick and was doped with boron prior to recrystallization so that about half the film thickness was depleted at inversion.

For the unimplanted wafers (center of Fig. 1), $N_{f,b}$ is much higher (1.9 × 10^{11} cm^{-2}) for the Si_3N_4-capped sample than for the SiO_2-capped sample (1.0 × 10^{11} cm^{-2}), consistent with previous results [1,2]. The effect of oxygen implantation is also shown; oxygen implantation through a nitride cap has little effect until ~1× 10^{16} cm^{-2} oxygen is implanted. At this concentration $N_{f,b}$ decreases dramatically. This effect has been reproduced several times. A 10^{16} cm^{-2} oxygen implantation through the oxide cap reduces $N_{f,b}$ to the mid-10^{10} cm^{-2} range. The O^{16} and N^{14} were implanted at 115 kV and 100kV, respectively.

Nitrogen implantation has the opposite effect of oxygen implantation; nitrogen implanted through an oxide cap noticeably increases $N_{f,b}$. This increase is not likely to result from implantation damage; the energies for the oxygen and nitrogen implants were chosen so that the peak of the implant was at the center of the polysilicon layer for both species. Since O^{16} and N^{14} have approximately the same mass and atomic number, the difference in induced damage between the two species should be small. Consequently, the effect on $N_{f,b}$ is most likely a chemical effect.

SIMS profiles on samples similar to those used for the electrical measurements were obtained using a Cameca IMS 3a system with a C^{s+} beam [6]. Unless otherwise noted, the concentrations were calibrated using implanted standards of oxygen and nitrogen in silicon; the impurity yields were normalized by the matrix yield. All of the capping layers were chemically removed prior to profiling.

First, in order to demonstrate dissolution of the capping layer, profiles were obtained on bulk FZ wafers which were coated with either SiO_2 or Si_3N_4 and

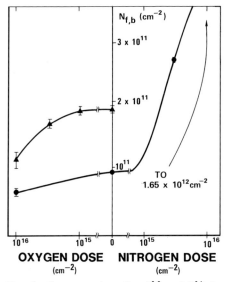

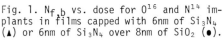

OXYGEN DOSE NITROGEN DOSE
(cm^{-2})   (cm^{-2})

Fig. 1. $N_{f,b}$ vs. dose for O^{16} and N^{14} implants in films capped with 6nm of Si_3N_4 (▲) or 6nm of Si_3N_4 over 8nm of SiO_2 (●).

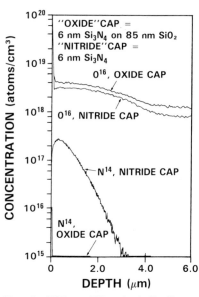

Fig. 2. SIMS profiles in bulk float zone silicon wafers.

then laser melted. The laser melting conditions were identical to those used for the C-V wafers, except that the beam radius was 60μm, and the powers used were 15.4W for the nitride cap and 11W for the (anti-reflection) oxide-capping layer. The estimated depth of melting is 4-5μm. The SIMS profiles are shown in Fig. 2. The nitride-capped sample contained significant levels (3 x 10^{17} cm^{-3}) of nitrogen, peaked near the surface and decreasing into the bulk; in the oxide-capped sample the nitrogen signal was at the background level of the system (10^{15} cm^{-3}). Both samples had 3 x 10^{18} cm^{-3} oxygen at the surface, decreasing to the SIMS system background of about 10^{18} cm^{-3} in the bulk. These measurements demonstrate dissolution of the encapsulation layers. Oxygen introduction from the nitride cap is consistent with a thin native oxide on the surface; it is also possible that the thin (6 nm) nitride is discontinuous, allowing air to be in contact with bare molten Si. The penetration depth of the impurities is consistent with liquid-phase diffusion.

Similar concentrations are seen in thin polycrystalline-silicon films (Figs. 3 and 4) deposited on 1.4μm of SiO$_2$. These samples used the same recrystallization conditions as the bulk FZ wafers, except that the power used was 10-11W. Figure 3 shows the oxygen and nitrogen profiles for a nitride-capped, laser-recrystallized film before and after a heat treatment to simulate device processing. Prior to heat treatment, the oxygen level is approximately constant at 3 x 10^{18} cm^{-3}, while the nitrogen level increases from 10^{17} cm^{-3} at the front of the film to 2 x 10^{17} cm^{-3} at the back. After the heat treatment, the oxygen level has decreased to the high 10^{17} cm^{-3} range, while the nitrogen level has decreased sharply to the low 10^{15} cm^{-3} range (comparable to the 4.5 x 10^{15} cm^{-3} solid solubility of nitrogen in silicon). The nitrogen also appears to have sharp peaks at both interfaces, although these may be instrument artifacts. The oxygen in oxide-capped samples behaves similarly to that in nitride-capped

514

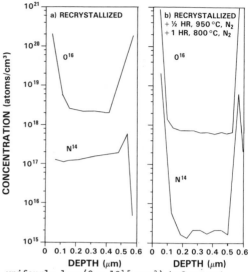

Fig. 3. SIMS profiles of O^{16} and N^{14} in a laser-recrystallized, 0.55μm polysilicon film on 1.4μm of SiO_2. The film was capped with 6 nm of LPCVD Si_3N_4.
a) After recrystallization,
b) After a post-recrystallization heat treatment to simulate transistor processing.

samples, but the nitrogen level is uniformly low (2 x 10^{15} cm^{-3}) before and after heat treatment.

Nitride-capped polysilicon layers implanted with 10^{16} cm^{-2} O^{18} were also profiled using SIMS (O^{18} was chosen because of the higher sensitivity of SIMS to this isotope). The as-implanted, recrystallized, and heat-treated profiles are shown in Fig. 4. The recrystallized samples show a significant redistribution of oxygen compared to the as-implanted sample; little difference is seen between the oxygen profiles in the recrystallized sample and in the heat-treated sample. The as-implanted sample has a low nitrogen concentration, as expected. The profiles in the recrystallized sample and in the heat-treated sample are remarkably similar, especially when compared with the profiles shown in Fig. 3. The heat-treated sample has a slightly lower nitrogen concentration toward the front surface and a more pronounced peak at the back interface than does the recrystallized sample.

The striking difference between the behavior of the nitrogen profiles in the unimplanted and implanted samples may be related to the high oxygen concentration of 10^{20} cm^{-3} in the implanted sample. This concentration, which is about 1.5-2 orders of magnitude greater than the solid solubility, provides a strong driving force for the formation of precipitates (SiO_x) or gas bubbles. These, in turn, could getter the nitrogen, thus pinning the profile. Supporting evidence for this hypothesis is seen in SIMS isotope images and TEM images (Fig. 5).

The SIMS images of O^{18} and N^{14} in the O^{18}-implanted, laser-recrystallized (but not heat treated) sample are shown in Figs. 5(a) and (b). Both images show very localized concentrations of O^{18} and N^{14}. In contrast, an unimplanted, nitride-capped, recrystallized sample [Fig. 5(c)] shows a relatively uniform concentration of N^{14}, with occasional depleted regions (dark bands). The TEM image obtained from the O^{18}-implanted, laser-recrystallized (but not heat-treated) sample [Fig. 5(d)] shows 20 nm-diameter bubbles or voids (often seen with defect clusters). The estimated volume fraction of these bubbles is about 1%, comparable to the atomic percentage of oxygen in the film. These bubbles

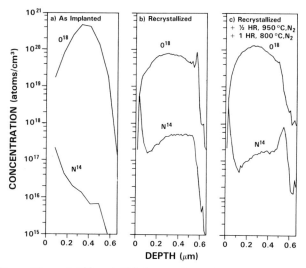

Fig. 4. SIMS profiles of O^{18} and N^{14} in a 0.55μm thick polysilicon film. The film was capped with 6 nm of LPCVD Si_3N_4 and implanted with 1 x 10^{16} cm^{-2} O^{18} at 115 kV. a) As-implanted, b) After recrystallization, c) After heat treatment.

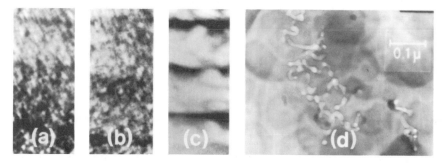

Fig. 5. (a,b) SIMS images of O^{18} (a) and N^{14} (b) for film of Fig. 4(b). (c) N^{14} SIMS image for unimplanted, nitride-capped, recrystallized polysilicon film. (d) TEM image of film of Fig. 4(b), showing bubbles in the film. The areas shown in (a), (b) and (c) are ~30μm wide.

have not been observed in unimplanted, laser-recrystallized samples; either the bubbles or the associated defects could be gettering sites for the nitrogen, explaining the "pinned" nitrogen concentration profile.

In addition to the chemical models presented above, other possible causes of the different backside fixed-charge densities with the two different encapsulation layers may be contamination or strain effects. Contamination seems

516

unlikely, since the differences in $N_{f,b}$ between the two types of encapsulation have been consistently observed over several runs in our laboratory [2]; they have also been observed at other laboratories using different deposition equipment [1]. Strain effects also seem highly unlikely, since the $N_{f,b}$ resulting from SiO_2 layers varies little with formation conditions; it is also independent of the presence of an Si_3N_4 layer above the SiO_2.

CONCLUSIONS

Several conclusions can be drawn from the results presented above:

1) Significant dissolution of the capping layer can occur during laser recrystallization. Nitride caps lead to incorporation of both nitrogen and oxygen in the melt, while oxide caps allow incorporation of oxygen, but not nitrogen.

2) $N_{f,b}$ increases when nitrogen is implanted and decreases when oxygen is implanted, suggesting that the variation in $N_{f,b}$ is a <u>chemical effect</u>.

3) The presence of oxygen in both the nitride- and oxide-capped samples reduces the likelihood of the dissolution model as the cause of the increased $N_{f,b}$, since the oxygen concentration in the film is the same with the two capping layers.

4) High levels of nitrogen (well above solid solubility) are present after recrystallization with a nitride cap. The nitrogen moves readily during subsequent heat treatments, lending support to the deposition model (with the high $N_{f,b}$ caused by nitrogen or SiN_x at the back interface). This possibility is further supported by the reduction in $N_{f,b}$ when high-dose oxygen implants are used to "pin" the nitrogen.

ACKNOWLEDGEMENTS

The authors would like to acknowledge Jeff Hiller for his assistance with SIMS analysis, Tom Cass for TEM examination, Curt Heimberg for his experimental assistance, and Dragan Ilic and Fred Schwettmann for their continued support and encouragement.

REFERENCES

1. H.P. Le and H.W. Lam, Electron Dev. Lett. EDL-3, 161 (1982).

2. D.J. Bartelink and T.I. Kamins, Paper Q-2, Electronic Materials Conference, Fort Collins, Colorado, June, 1982.

3. R.E. Chaney and C.J. Varker, J. Electrochem. Soc. 123, 846 (1976).

4. Y. Yatsurugi, N. Akiyama, Y. Endo, and T. Nozaki, J. Electrochem. Soc. 120, 975 (1973).

5. T.I. Kamins, K.F. Lee, and J.F. Gibbons, Electron Dev. Lett. EDL-1, 5 (1980).

6. SIMS profiling was performed at C.A. Evans, Assoc., San Mateo, CA.

CW LASER ANNEALING OF ION IMPLANTED OXIDIZED SILICON LAYERS ON SAPPHIRE

G. ALESTIG, G. HOLMÉN AND S. PETERSTRÖM
Department of Physics, Chalmers University of Technology, S-412 96 Göteborg, Sweden

ABSTRACT

CW laser annealing has been performed on silicon on sapphire (SOS) implanted with boron or phosphorus ions to a dose of 10^{15} ions/cm^2. The laser irradiation was done both with and without an oxide layer on top of the silicon and from both the silicon and the sapphire side. Sheet resistivity and Hall effect measurements were used for the analysis of the samples. Good annealing and high activation of the dopants were obtained for both oxidized and unoxidized SOS. For samples irradiated from the silicon side, the needed laser power changed depending on the thickness of the oxide. For samples irradiated from the sapphire side, the needed laser power was independent of oxide thickness.

INTRODUCTION

Silicon on sapphire (SOS) is an interesting material for the fabrication of electronic devices. If ion implantation is used as a doping method, the crystal structure must be restored and the dopants activated by a post-implantation annealing process. In recent years, the use of laser or electron beams for annealing of semiconductors has attracted much attention [1]. However, relatively few investigations of SOS annealed in this way have been made [2-4].
In this work we report on scanned CW (continuous wave) laser annealing of boron or phosphorus implanted SOS. Since most device processing techniques incorporate the use of SiO$_2$ on Si, we have studied annealing of samples with an oxide overcoat as well as samples without oxide.
A unique feature of silicon on sapphire is the possibility of laser annealing from the sapphire side. This would make it possible to anneal areas otherwise masked by contacts etc. In the work presented here, we have performed both back side and front side laser annealing. Sheet resistivity measurements were used for the analysis of the samples. On some samples Hall effect measurements were made.

EXPERIMENTAL PROCEDURE

The material used in the investigations was intrinsic (> 50 Ωcm), 0.6 μm (100) silicon on sapphire. The thickness of the sapphire was 0.33 mm. Since some of the samples were to be irradiated from the sapphire side, the SOS wafers were purchased with a back side optical polish. Prior to the ion implantation, the silicon on some of the wafers was oxidized in dry oxygen at 1100°C. Three different oxide thicknesses were grown; 490 Å, 830 Å, and 1190 Å. The samples were then implanted using the Chalmers 400 kV ion accelerator with either 10^{15} ^{11}B ions/cm^2 at 150 keV or 10^{15} ^{31}P ions/cm^2 at 250 keV. The implantations were performed at 10° off the surface normal and at room temperature.
After the implantations, the samples were cut into pieces measuring 7 x 7 mm^2 and laser annealed with a CW Ar ion laser, operating in the TEM$_{00}$ mode on the 5145 Å line. Focusing was obtained with a 100 mm focal length lens, giving a calculated beam spot diameter of 34 μm (at $1/e^2$ intensity points) [5]. The

laser beam was scanned in the vertical direction at a speed of 1 cm/s by a movable mirror. Between each vertical scan, the sample was translated horizontally in steps of 9 μm.

After the laser annealing, the oxide of the samples was removed in buffered HF. The sheet resistivity of the silicon layer was then measured with a four-point probe [6].

Hall effect measurements were made on some unoxidized samples so that additional information on the electrical properties of the laser annealed SOS could be obtained. These samples measured 9 x 14 mm^2 and the van der Pauw technique [7] was used for the measurements.

RESULTS AND DISCUSSION

An immediately observable effect of the laser irradiation was a characteristic colour change of the implanted silicon layer. The colour of the boron implanted samples changed gradually back to the colour of unimplanted SOS as the laser power density was increased. The colour of the phosphorus implanted samples changed more abruptly under the same treatment. This difference is probably caused by the fact that at the doses and energies used, the phosphorus implanted layer is almost amorphous, while the boron implanted samples still are essentially crystalline. The phosphorus implantations are expected to remain amorphous until the laser power density is high enough for recrystallization to start. The annealing of defects in the boron implanted samples can on the other hand occur at lower power densities.

Observation of the samples under an optical microscope showed that for low laser power densities the surface was smooth and free from any structure. For high power densities, however, the laser beam produced visible surface damage. The threshold powers for surface damage are given in table I. The damage is probably due to surface melting [8].

TABLE I
Threshold powers for the onset of surface damage

Oxide thickness	Front side irradiated		Back side irradiated	
	B	P	B	P
No oxide	1.7 W	1.7 W	1.1 W	1.1 W
490 Å	1.4 W	1.4 W	1.0 W	1.0 W
830 Å	0.9 W	1.0 W	1.0 W	1.0 W
1190 Å	1.0 W	1.0 W	1.0 W	1.0 W

When a focused beam is used for annealing, it is important that the overlap is sufficient between the lines in the scanning pattern. This was checked by making a single scan of the laser beam on some unoxidized samples and then measuring the widths of the annealed lines. For 1.5 W incident power on the silicon side of the SOS wafer, the apparent linewidths were 38 μm and 17 μm for the boron and phosphorus implanted samples, respectively. The difference in apparent linewidth is probably due to the same mechanisms as those causing the difference in colour-change behaviour, as discussed above. The measured linewidths show that the chosen distance between scanning lines, 9 μm, is quite sufficient to produce good overlapping.

The results of the sheet resistivity measurements are shown in figure 1. As expected, the required power density for good annealing depends strongly on the oxide thickness for the front side irradiated samples. This is caused by the ability of the oxide to act as an antireflection layer for the incident light

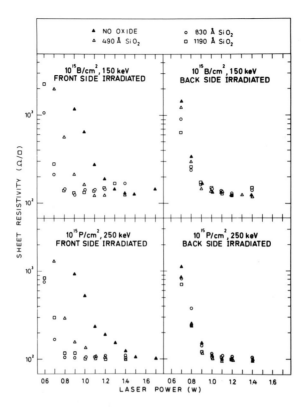

Fig. 1. Sheet resistivity of laser annealed SOS as a function of laser power.

[9,10]. The amount of light absorbed in the silicon layer versus oxide thickness can be calculated using standard multilayer theory [11]. Since the optical data for the implanted silicon probably lie somewhere between the data for single crystalline and amorphous silicon, calculations were done for both. The optical data used were: $n_{sapphire}$ = 1.77 [12], n_{SiO_2} = 1.46, n_{Si} (single crystal) = 4.05 + 0.028 i [13], n_{Si} (amorphous) = 4.80 + 0.77 i [14].

Figure 2 shows that for front side irradiation, a strong variation in the laser light absorption depending on the oxide thickness is expected. This variation correlates well with the measured sheet resistivities. The lowest absorption is expected for unoxidized SOS, and the measurements showed that these samples required the highest laser powers. The samples with an oxide thickness of 830 Å or 1190 Å needed the lowest laser powers to obtain good annealing. These oxide thicknesses are close to the thickness where the calculations predict the maximum laser light absorption.

It can also be seen from figure 2 that for back side irradiation, there is no variation in the laser light absorption for amorphous SOS and only a small variation for single crystalline SOS. This is in agreement with the curves in figure 1 which show no oxide thickness dependence for back side irradiation.

The sheet resistivity curves show that laser irradiation from the back side produces as good annealing as front side irradiation does. This might prove

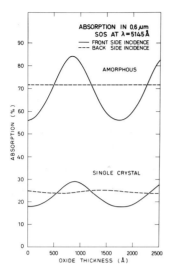

Fig. 2. Calculated absorption versus oxide thickness for amorphous and single crystalline 0.6 μm silicon on sapphire at λ = 5145 Å. The solid curves represent front side irradiation and the dashed ones back side irradiation.

very valuable in some device processing applications. The curves also show that the power levels needed for annealing of SOS are much lower than the corresponding levels for bulk silicon. In our measurements, about 1.5 W was sufficient to anneal unoxidized SOS, whereas for bulk silicon annealed under conditions similar to ours, about 9 W is needed [15]. The difference is explained by the much lower thermal conductivity of sapphire as compared to silicon. The sheet resistivities obtained for laser powers above the threshold for good annealing are for the boron implantations about 130 $\Omega/\square$ and for the phosphorus implantations about 105 $\Omega/\square$. For comparison, we calculated sheet resistivity values by numerical integration of LSS range statistics [16], assuming 100% activation and using bulk silicon mobilities as determined by Irvin [17]. The results were 112 $\Omega/\square$ and 70 $\Omega/\square$ for boron and phosphorus, respectively. The measured sheet resistivities are rather close to these values. This indicates that by a laser annealing process it is possible to restore the crystal structure and to move the dopants to lattice sites where they are electrically active. The differences between calculated and measured sheet resistivities are probably due to the lower mobility in SOS as compared to bulk silicon. This assumption is confirmed by the Hall effect measurements discussed below.

When the laser power density is increased above the level for surface damage (table I), it can be seen from figure 1 that no immediate change in sheet resistivity occurs. However, the rough surface probably renders this annealing regime unusable for device applications. If the power density if further increased, the sheet resistivities start to increase. This can be seen for the boron curves in figure 1. In the phosphorus case, the increase occurs at power densities slightly higher than the ones included in figure 1.

From the Hall effect measurements we calculated the effective number of carriers per cm^2, $(N_s)_{eff}$, by

$$(N_s)_{eff} = \frac{r}{e\,R_s}$$

where R_s is the sheet Hall coefficient and r, which was set to unity, is the

ratio between the Hall mobility and the conductivity mobility. We also calculated the effective mobility, μ_{eff}, by

$$\mu_{eff} = \frac{R_s}{\rho_s}$$

where ρ_s is the sheet resistivity.
    Table II presents the results of the Hall effect measurements. As shown,
100% activation of the dopant is achieved for the boron implanted samples, while
the activation of the phosphorus implanted samples seems to be somewhat lower.
However, it must be remembered, that the value of r might deviate from unity.
For electrons it tends to be slightly more than one and for holes slightly less
than one [18]. If this is taken into consideration in the calculations, it
would lead to a higher activation of the phosporus and a lower activation of
the boron than indicated in table II.

TABLE II
Effective mobility, μ_{eff}, and effective sheet carrier concentration, $(N_s)_{eff}$, as
determined by Hall effect measurements.

Ion	Laser	$\mu_{eff}\left[cm^2/Vs\right]$	$(N_s)_{eff}\left[cm^{-2}\right]$
B	1.4 W front side	44	1.0×10^{15}
	1.0 W back side	41	9.6×10^{14}
P	1.5 W front side	67	7.6×10^{14}
	1.0 W back side	63	7.8×10^{14}

    The mobility values in table II show that the mobility is higher for the
phosphorus implantations. This explains why these samples have lower sheet
resistivities than the boron implanted ones. It is also worth noting, that no
significant difference between front side and back side irradiation can be found
in the Hall effect measurements.

CONCLUSIONS

    We have shown that CW laser annealing of boron or phosphorus implanted SOS
produces low sheet resistivities and high activation of the dopants. The mobility values attained are, however, somewhat lower than the corresponding values
of bulk silicon. Good annealing is possible both with and without an oxide
layer on top of the SOS. The ability of the oxide to act as an antireflection
layer must be taken into consideration when the power density for annealing is
chosen. We have also shown that laser annealing is possible by irradiation from
the sapphire side. In this case the power density needed is independent of the
oxide thickness.
    In the work presented here we have used rather low laser powers, maximum 1.7
W. This shows that laser annealing of SOS is feasible, even with fairly small
lasers. On the other hand, by using higher laser powers and consequently larger
spot radii, it would be possible to reduce the time needed to anneal an SOS wafer.

ACKNOWLEDGEMENTS

    The authors are indepted to Prof O. Almén for his interest in this work.
Special thanks are also due to the other members of the ion physics group.
M. Soondra made the excellent figures.
    Financial support was recieved from the Swedish Boards for Technical Develop-

522

ment and the Swedish Natural Science Research Council.

REFERENCES

1. See, for example, Laser and Electron-Beam Solid Interactions and Materials Processing, J.F. Gibbons, L.D. Hess, and T.W. Sigmon, eds. (North Holland, New York, 1981).

2. M.E. Roulet, P. Schwob, K. Affolter, W. Lüthy, M. von Allmen, M. Fallavier, J.M. Mackowski, M.A. Nicolet, and J.P. Thomas, J. Appl. Phys. 50, 5536 (1979).

3. W. Lüthy, K. Affolter, H.P. Weber, M.E. Roulet, M. Fallavier, J.P. Thomas, and J. Mackowski, Appl. Phys. Lett. 35, 673 (1979).

4. I. Golecki, G. Kinoshita, and B.M. Paine, Nucl. Instr. and Meth. 182/183, 675 (1981).

5. Lasers in Industry, S.S. Charschan, ed. (Van Nostrand, New York 1972).

6. F.M. Smits, Bell System Tech. J. 37, 711 (1958).

7. L.J. van der Pauw, Philips Res. Repts. 13, 1 (1958).

8. J.S. Williams, W.L. Brown, H.J. Leamy, J.M. Poate, J.W. Rodgers, D. Rousseau, G.A. Rozgonyi, J.A. Schelnutt, and T.T. Sheng, Appl. Phys. Lett. 33, 542 (1978).

9. H. Okabayashi, M. Yoshida, K. Ishida, and T. Yamane, Appl. Phys. Lett. 36, 202 (1980).

10. T.C. Teng, Y. Shiau, Y.S. Chen. C. Skinner, J.D. Peng, and L.J. Palkuti, in Laser and Electron-Beam Solid Interactions and Materials Processing, p. 391. J.F. Gibbons, L.D. Hess, and T.W. Sigmon, eds. (North Holland, New York, 1981).

11. G.R. Fowles, Introduction to Modern Optics (Holt, Rinehart and Winston, New York, 1968).

12. Handbook of Optics, W.G. Discroll and W. Vaughan, eds. (McGraw-Hill, New York, 1978).

13. T. Smith and A.J. Carlan, J. Appl. Phys. 43, 2455 (1972).

14. T. Motooka and K. Watanabe, J. Appl. Phys. 51, 4125 (1980).

15. D.H. Auston, J.A. Golovchenko, P.R. Smith, C.M. Surko, and T.N.C. Venkatesan, Appl. Phys. Lett. 33, 539 (1978).

16. J.F. Gibbons, W.S. Johnson, S.W. Mylroie, Projected Range Statistics (Halsted Press, New York, 1975).

17. J.C. Irvin, Bell System Tech. J. 41, 387 (1962).

18. N.G.E. Johansson, J.W. Mayer, and O.J. Marsh, Solid State Electronics, 13, 317 (1970).

BEAM SHAPING FOR CW LASER RECRYSTALLIZATION OF POLYSILICON FILMS

P. ZORABEDIAN, C.I. DROWLEY, T.I. KAMINS, AND T.R. CASS
Hewlett-Packard Laboratories, 3500 Deer Creek Road, Palo Alto, CA 94304

ABSTRACT

A shaped laser beam has been used for laterally seeded recrystallization of polysilicon films over oxide. Direct maps of the shaped-beam intensity distribution in the wafer plane are correlated with the grain structure of the recrystallized polysilicon. Using 60% overlapping of shaped-beam scans along <100> directions, we have obtained seeded areas one mm wide and 50 to 500μm long. These consist of 40μm-wide adjacent single-crystal strips regularly separated by low-angle grain boundaries extending laterally away from the seed openings. The spacing between grain boundaries is equal to the scan spacing, providing a means for controlling the location of grain boundaries in otherwise defect-free, single-crystal films.

INTRODUCTION

Single-crystal silicon films which are free of all grain boundaries are desirable for the application of recrystallized silicon-on-insulator structures to VLSI. Energy-beam shaping can be used to prevent heterogeneous nucleation and promote growth of long individual grains. The cw Ar-ion laser offers the potential for good control over the recrystallization process because of the ability to produce a narrow molten zone and to shape the beam with a variety of optical techniques. Recent work with argon lasers has concentrated on shaping the laser spot either by inserting masks in the TEM$_{00}$ (Gaussian) mode [1] or by forcing the laser to run in the TEM$_{01}$* mode [2]. In those experiments the shape of the beam at the wafer was inferred, but was not directly observed. In our experiment the beam was shaped by using a 90° metal wedge to mask the trailing edge of a TEM$_{01}$* beam, and the intensity distribution at the wafer surface was directly mapped with a computer-controlled photodiode system [3]. This correlation was useful because the heat diffusion time is much less than the laser dwell time for the spot sizes and scan speeds typically used for recrystallization [4]. Therefore, the local temperature responds to the local laser intensity as the beam is swept across the polysilicon, and the details of the intensity distribution in the laser spot strongly affect the recrystallization process.

In the first part of this study, the beam-shaping technique was characterized by correlating the grain structure of isolated scans with the experimental shaped-beam intensity distributions. Subsequently, the shaped beam was used for laterally seeded recrystallization.

MATERIALS and APPARATUS

A Coherent System 5000 laser annealer was used, with modifications for mapping and masking the beam. In this system, the beam size at the wafer surface is adjusted by varying the working distance between the lens and the wafer. The samples studied were three-inch wafers containing a 0.55μm LPCVD polysilicon layer over a 1.4μm SiO$_2$ film on a (100)-oriented silicon substrate.

Mat. Res. Soc. Symp. Proc. Vol. 13 (1983) ©Elsevier Science Publishing Co., Inc.

The stabilization layer was 6 nm of LPCVD Si_3N_4. On some wafers, the oxide under the polysilicon was patterned by a LOCOS process to define seed openings in which the polysilicon was directly in contact with the single-crystal silicon. These seed openings were oriented to lie either along the <110> direction or along the <100> direction. On all wafers with seed openings, the beam was rastered at 10 cm/sec perpendicular to the edges of the openings. The substrate was preheated to 500°C. The laser output mode was varied by changing the radius of curvature of the output mirror. With a 15-meter radius-of-curvature mirror, the mode was TEM_{00} (Gaussian). Installation of a 5-meter radius-of-curvature mirror changed the mode to predominantly $TEM_{01}*$ with the intensity at the center approximately 50% of the maximum intensity. This so-called "donut" beam was not azimuthally uniform but exhibited intensity variations or bright spots around the ring. These nonuniformities were sensitive to very slight changes in laser alignment, but were quite stable for a fixed set of mirror alignments after the laser stabilized for one hour.

LONG-GRAIN STRUCTURE

To evaluate the effect of beam shaping on recrystallization, samples were recrystallized with nonoverlapping, unseeded scans, Wright etched [5] to delineate grain boundaries, and then examined with Nomarski interference-contrast microscopy. An unmasked $TEM_{01}*$ mode was used for the initial experiments, and the resulting grain structure was found to be very sensitive to the detailed intensity profile of the beam. At low power, recrystallization occurred near both lateral edges of the beam, forming a double recrystallized track with unmelted polysilicon in the center. As the beam power was increased slightly above the 6 W threshold for recrystallization, long-grain structure was observed in a very narrow power range (<0.5W) (Fig. 1—top); at higher power, the usual chevron grain structure was formed (Fig. 1—bottom). The beam map revealed a bright spot at the leading edge (Fig. 2). Because the laser melting was in the time-independent regime, the long grains were formed when the intensity of the leading edge was sufficient for recrystallization but that of the less-intense trailing edge was not, so that a re-entrant beam shape was effectively formed. Donut beams in which the leading-edge peak was absent did not produce long grains.

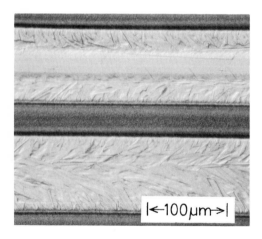

Fig. 1. Nomarski micrograph of recrystallized tracks in polysilicon over SiO_2 (after etching of grain boundaries). Top: Single crystal formed by a re-entrantly shaped beam scanned left to right. Bottom: Chevron grain structure formed by a convex beam scanned right to left.

|←100μm→|

In subsequent experiments, a 90° metal wedge was introduced approximately in
the center of the donut beam to mask the trailing edge, causing about 25% insertion loss. Double tracks were again formed at low power; above the threshold
for recrystallization at the beam center, long grains were reproducibly formed
over a wider power range than without masking. Outside of the confocal region
of the beam waist, the shaped beam was characterized by a re-entrant trailing
edge and laterally displaced intensity peaks (Figs. 3 and 4). Grains up to 30μm
wide and 500μm long were obtained when the beam was scanned with a concave
trailing edge to control the solidification front; when the scan direction was
reversed and the trailing edge was convex, chevron grains were formed.

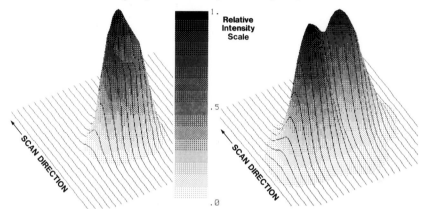

Fig. 2.   Intensity map of a TEM$_{01}$* donut
beam with a bright spot at the
leading edge.

Fig. 3.   Intensity map of shaped
beam formed by masking the
trailing edge of a donut
beam.

NORMALIZED SHAPED BEAM INTENSITY CONTOURS

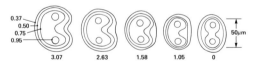

DISTANCE FROM FOCAL PLANE (mm)

Fig. 4.   Normalized intensity contours of a shaped beam at several displacements
from the focal plane.

Transmission electron microscopy verified the microstructure produced by the
shaped beam. Figure 5 shows a transmission electron micrograph of a segment of
a long grain. Transmission electron diffraction showed that most of the unseeded long grains had a (111) plane perpendicular to the surface, but no preferred in-plane orientation was found along the scan direction.

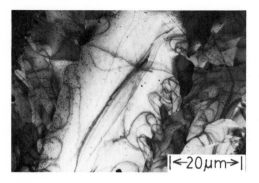

Fig. 5. TEM micrograph of a long grain formed by a shaped beam.

|←20μm→|

Divergence of the laser beam caused its size to increase away from the lens focal plane. Because of diffraction, the shape of the intensity contours also varied with wafer position, becoming more re-entrant with increasing distance from the focal plane, especially at lower relative intensities (Fig. 4). The power range for long-grain formation also increased with distance (Fig. 6). In general, grains recrystallized at lower power were more uniform and contained fewer defects. The existence of the long-grain power window and its variation with beam size and shape can be explained qualitatively as follows: assume that the molten-zone boundary is defined by a certain intensity contour in the input beam (this neglects heat diffusion in the underlying substrate). For a low-power beam, the melt boundary will follow an intensity contour near the peak of the beam. As the beam power increases, the melt boundary occurs at lower relative intensity contours. Because of diffraction, the contours rapidly become convex for beams near the focal plane (Fig. 4 — right), destroying the condition for long-grain formation. For the larger beams farther from the focal plane (Fig. 4 — left), the contours stay concave at the trailing edge down to lower relative intensities, allowing long grains to form over a wider power range.

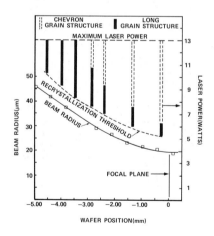

Fig. 6. Power ranges for long-grain formation by a shaped beam for several displacements from the focal plane. "Recrystallization threshold" refers to recrystallization at the center of the beam.

LATERALLY SEEDED RECRYSTALLIZATION

The shaped beam was applied to overlapped scanning with lateral seeding to control the film orientation. The effects of beam power, amount of overlap, and crystallographic scan direction were evaluated. The transverse beam width was about 80μm, producing an isolated recrystallized track about 100μm wide in poly-silicon over oxide at 12 W. A range of beam powers was used for scan spacings from 20 to 80μm. For <100>-directed scans with approximately 60% overlap (30-50μm scan spacing), single-crystal strips were regularly linked together without intervening polycrystalline areas. Long, low-angle (1-2°) grain boundaries were formed parallel to the scan direction between adjacent single-crystal strips. These were separated by the scan spacing, and had the appearance of "stitching" the single-crystal strips together (Fig. 7). The areas between these "stitching" grain boundaries were essentially free of defects, indicating that, under proper conditions, overlapped scanning combined with lateral seeding could remove most of the defects that appeared in the isolated long grains.

The overall width of the "stitched," single-crystal areas was 1 mm, equal to the width of the seed opening. The strips extended away from the seeding edges for varying distances, from a minimum of approximately 50μm to 500μm. The single-crystal strips abruptly terminated in polycrystalline regions with heavy twinning and different crystal orientations than the seed. The length of lateral seeding was greatest when the minimum power necessary for inducing vertical epitaxy over the seed openings was used. The mechanism for nucleation of the polycrystalline areas is not presently understood, but this observation implies that the shaped solidification front produces lateral seeding more readily when

Fig. 7. Nomarski micrograph of laterally seeded Si film formed by 60% overlapped shaped-beam scans along the <100> direction (after etching).

Fig. 8. Nomarski micrograph of laterally seeded film formed by 60% over-lapped shaped-beams scans along the <110> direction (after etching).

528

the melt temperature is lower.   With larger amounts of overlap (20µm scan spacing) the results were similar to those obtained with a Gaussian beam or an unmasked donut beam and were characterized by a film free of grain boundaries for 10-20µm from the edge of the seed. Smaller amounts of overlap (60-80µm scan spacing) produced adjacent, apparently independent recrystallized tracks with irregular areas of single-crystal and polycrystalline material.

For <110> scanning, the recrystallized films were also free of high-angle grain boundaries near the seed, but they contained a "herring-bone" pattern of low-angle grain boundaries which encroached on the single-crystal areas (Fig. 8). This result correlates with our earlier lateral-seeding experiments without beam shaping, which indicated that low-angle boundaries in laterally seeded films tend to lie along <100> directions.

REFINEMENT OF BEAM-SHAPING OPTICS

Two additional features have recently been incorporated into the system. A cylindrical-lens telescope has been inserted behind the mask to shift the effective focal plane for one axis of the shaped beam, leaving it unchanged for the orthogonal axis. This allows the shape to be elongated transverse to the scan to produce a wider recrystallized track, or parallel to the scan to accentuate the re-entrant shape. As an alternative beam-shaping technique, phase-diffraction gratings have been designed using a one-dimensional Fast Fourier Transform and have been fabricated in quartz disks by etching in an oxygen-freon plasma. Beams at the wafer surface with laterally displaced intensity peaks have been obtained using these gratings.

CONCLUSIONS

A beam-shaping technique for laterally seeded recrystallization consisting of masking a $TEM_{01}*$ argon laser beam has been developed. The importance of accurate measurement and control of the intensity distribution at the wafer surface has been demonstrated by correlating the beam shape and the power range for long-grain formation. Laterally seeded films have been produced, with large single-crystal areas separated by low-angle grain boundaries confined to the lines of overlap between adjacent scans in a controlled and reproducible manner.

ACKNOWLEDGEMENTS

We wish to thank Curt Heimberg for TEM sample preparation and assistance with the optical system, and Drs. Dragan Ilic and Fred Schwettmann for encouragement during the course of the experiment.

REFERENCES

1. T.J. Stultz and J.F. Gibbons, Appl. Phys. Lett. 39, 498 (1981).

2. S. Kawamura, J. Sakurai, M. Nakano, and M. Takagi, Appl. Phys. Lett. 40, 394 (1982).

3. B.P. Von Herzen, T.I. Kamins, and C.I. Drowley, Journal Electrochem. Soc., 128, 2695 (1981).

4. C.I. Drowley, C. Hu, and T.I. Kamins, Electrochem. Soc. Meeting (Montreal, May 1982), abs. 145.

5. M. Wright Jenkins, Journal Electrochem. Soc., 124, 757 (1977).

A COMPARISON OF CW LASER AND ELECTRON-BEAM RECRYSTALLIZATION OF POLYSILICON IN MULTILAYER STRUCTURES

C.I. DROWLEY* AND C. HU[+]
Hewlett-Packard Laboratories, 3500 Deer Creek Road, Palo Alto, CA 94304*
Electronics Research Laboratory, University of California, Berkeley, CA 94720[+]

ABSTRACT

A thermal model for beam–induced melting of polysilicon in polysilicon/insulator/silicon structures has been used to compare CW Ar[+] laser and electron-beam melting processes. The laser melting process exhibits a decreasing recrystallization threshold power and an increasing substrate melting threshold power with increasing insulator thickness. The decrease in recrystallization threshold with increasing insulator thickness is generally smaller for electron beams because of their deep penetration. The substrate melting threshold is approximately independent of insulator thickness for e-beams since no change in the power absorption occurs on melting of the polysilicon. As a result, the process window between the onset of recrystallization and melting of the substrate is narrower for e-beams than for laser beams. Also, essentially no window will exist between the onset of melting over a seed and substrate melting under an oxide if lateral epitaxy is performed with an electron beam.

INTRODUCTION

Both CW laser and electron beams have been used to melt and recrystallize polysilicon in polysilicon/insulator/silicon structures [1,2]. Active devices have been fabricated in the recrystallized films, with similar electrical characteristics [2,3]. The two techniques are qualitatively similar; however, no comparison of the two methods has previously been made to reveal relative advantages and drawbacks of the two techniques.

The comparison of laser and electron beam melting of polysilicon is based on a model for laser melting in multilayer structures [4]. The model is used to calculate the temperature and reflectivity profiles in a polysilicon/insulator/silicon structure and has successfully predicted melt linewidths, melt thresholds, and recrystallization thresholds for the laser melting technique [4,5].

LASER MELTING

The model calculates the temperature of the polysilicon layer in the multilayer structure. For laser heating, all of the power is assumed to be absorbed in the polysilicon, and the temperature drop across the polysilicon is neglected. One-dimensional heat flow from the polysilicon through the insulator to the substrate is assumed; the substrate heat flow is given a three-dimensional treatment similar to that of Lax [6]. The beam is assumed to be stationary; this approximation is valid over a wide range of scan speeds and beam sizes [5]. An axially symmetric, Gaussian beam is assumed.

Using these assumptions, the linearized temperature at the surface of the

Mat. Res. Soc. Symp. Proc. Vol. 13 (1983) ©Elsevier Science Publishing Co., Inc.

substrate is:

$$\theta_{sub,surf}(r) = \frac{1}{K_{si}(T_0)} \int_0^\infty u \, F(u) \, (1-R(u))G(r,u)du \qquad (1)$$

where F is the Gaussian input flux, R is the reflectivity, K_{si} is the silicon thermal conductivity at the backside temperature T_0, r is the radius, and $G(r,u)$ is given by

$$G(r,u) = \frac{2}{\pi u} \, K((r/u)^2) \qquad r<u \qquad (2)$$

$$= \frac{2}{\pi r} \, K((u/r)^2) \qquad r>u$$

$$= 1 \qquad r=u$$

where K is the elliptic integral of the first kind.

The linearized temperature is related to the true temperature by the Kirchoff transform [6]:

$$\theta(r,z) = \frac{1}{K_{si}(T_0)} \int_{T_0}^T K_{si}(u)du \qquad (3)$$

We use $K_{si}(T) = 7.177 \sqrt{T-493.8}$ (T in °K), which fits the data within 5% in the range 673°K<T<1685°K [5]. The true temperature is

$$T_{sub,surf}(r) = 493.8 + [\frac{K_{si}(T_0)\theta}{14.354} + \sqrt{T_0 - 493.8} \,]^2 \qquad (4)$$

Assuming one-dimensional heat flow through the insulator, the temperature at the polysilicon surface is

$$T_{poly}(r) = T_{sub,surf}(r) + F(r)(1-R(r)) \, t_{ins}/K_{ins} \qquad (5)$$

where the second term on the right side is the temperature drop across the insulator. K_{ins} is the insulator thermal conductivity (assumed independent of temperature) and t_{ins} is the insulator thickness.

The reflectivity will be a function of radius once melting occurs. The reflectivity and temperature are thus both unknown; they may be found using a relaxation algorithm previously presented [4].

The model may be used to determine the melting threshold of the polysilicon, the recrystallization threshold (defined below), and the substrate melting threshold (the power at which $T_{sub,surf}(0) = T_{melt}$). A typical plot of the different thresholds vs. underlying oxide thickness is given in Fig. 1. Several interesting features of the curves may be noted. The surface melting threshold decreases with increasing oxide thickness as a result of the increased thermal insulation. Once the surface has started to melt, the reflectivity increases from the solid value to that of liquid silicon as the power is increased; the surface temperature remains at the melting point. Proportionally less power is absorbed because of the increased reflectivity (i.e., negative feedback). When the reflectivity has increased to the liquid value, the feedback is no longer effective and the surface temperature can rise beyond the melting point, allowing the melt to penetrate to the back interface. This behavior defines the recrystallization threshold. Finally, the substrate melting threshold is even higher than the recrystallization threshold, and increases with increasing oxide thickness to a limiting value of 2.2X the surface melting threshold with no oxide present. The increase is a consequence of the lateral dimension of the polysilicon melt increasing as the oxide thickness increases. Since the molten region (with high reflectivity) is larger, the fraction of power absorbed will be smaller for thicker oxides, increasing the substrate melting threshold. This shielding of the substrate by the reflectivity of the molten silicon is one of the primary differences between laser and electron-beam recrystallization. The limiting power is determined by the condition that the entire surface has the

Figure 1. Surface melting threshold. recrystallization threshold, and substrate melting threshold as a function of underlying oxide thickness for laser melting. The solid lines are the results of the model; the symbols denote experimental data. The curves assume $R_{solid} = 0.40$ and $R_{liquid} = 0.71$.

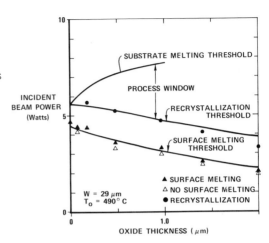

reflectivity of molten silicon.

The curves given in Fig. 1 may be used for other beam sizes and insulators if "normalized" axes of power/radius (P/w) and "normalized insulator thickness" $(t_{ins}/K_{ins}*w)$ are used. The curves are only valid for bare polysilicon ($R_{solid}= 0.40$, $R_{liquid}=0.71$) or for very thin encapsulation layers.

ELECTRON BEAM MELTING OF POLYSILICON

The heating of the multilayer structure is different for e-beams than for laser beams. First, the depth of penetration of the e-beam can be quite large (5μm for a 25 kV e-beam in Si or SiO_2), so that the power is not completely absorbed in the polysilicon layer. Second, melting has no significant effect on the e-beam power absorption, so that the "feedback" present with laser melting is negligible for e-beams.

The power loss of an e-beam as it penetrates light elements (Z=10 to 14) is modeled as [7]

$$\frac{dP(r)}{dz} = \frac{fI(r)V_b}{R_g} \left[0.60 + 6.21 \left(\frac{Z}{R_g}\right) - 12.40 \left(\frac{Z}{R_g}\right)^2 + 5.69 \left(\frac{Z}{R_g}\right)^3 \right] \quad (6)$$

where V_b is the beam voltage, f is the fraction of incident electrons absorbed, and R_g is the Gruen range (an estimate of the maximum penetration depth) given by R_g (μm) = $0.0181(V_b(kV))^{1.75}$. The beam flux is assumed to be

$$F(r) = I(r)V_b = \frac{I_t V_b}{\pi w^2} \exp(-(r/w)^2) \quad (7)$$

where I_t is the total current and w is the 1/e radius of the power. The lateral spread of the beam as it penetrates the solid is neglected. For beam sizes in the range used in recrystallization (w=20-50μm) and for beam energies <25 kV, the lateral spread should be negligible [8].

The energy dissipation depends only weakly on atomic number [7]; it is much more dependent on the density of the material being traversed. Dissipation in the $Si/SiO_2/Si$ structure can be analyzed by changing the depth scale in the Si to the equivalent thickness of SiO_2. The density of Si is 2.33 g/cm^3 while that of SiO_2 is 2.20 g/cm^3. The "normalized" depth in Si is thus 1.06X the true depth. In this manner, the universal dissipation curve can be applied to the composite structure and the power absorbed in the polysilicon, the oxide, and the substrate can be determined. Figure 2 illustrates the variation of power dissipation with depth and beam voltage for electrons incident on a "normalized" $Si/SiO_2/Si$ structure. The 5 kV beam is entirely absorbed in the polysilicon while the 15 and 25 kV beams penetrate more deeply. In all cases, the penetration depth is still small compared to typical beam sizes used in recrystallization.

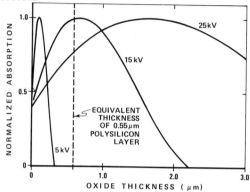

Figure 2. Power absorption as a function of depth in a "normalized" multilayer structure for 5 kV, 15 kV, and 25 kV electron beams. The maxima have been normalized to 1.0.

The temperature drop across the oxide no longer has the form shown in Eq. 5 because of the power absorption in the oxide. Rather, it has the form

$$\Delta T_{ox}(r) = f\, F(r)\, [g_1 \frac{t_{ox}}{K_{ox}} + g_2 + g_3 \frac{t_{ox} + t_{poly,eff} - R_g}{K_{ox}}] \tag{8}$$

The first term (g_1) arises from power absorption in the polysilicon:

$$g_1 = 0.60\, t_p + 3.105\, t_p^2 - 4.133\, t_p^3 + 1.423\, t_p^4 \tag{9}$$

where $t_p = \frac{t_{poly,eff}}{R_g} = t_{poly} \frac{\rho_{si}}{\rho_{ox} R_g}$. If $t_p > 1$, $g_1 = 1$ and $g_2 = g_3 = 0$. The second term (g_2) is the temperature drop resulting from absorption in the oxide, and is given by

$$g_2 = \frac{R_g}{K_{ox}} [h_1(t_p + t_{ox}/R_g) - h_1(t_p)] \tag{10}$$

where

$$h_1 = -g_1 u + 0.30 u^2 + 1.035 u^3 - 1.033 u^4 + 0.285 u^5 \tag{11}$$

If $t_p + t_{ox}/R_g > 1$, the parameter in the first term on the right side of Eq. 10 should be replaced by 1. The third term (g_3) is non-zero only if $t_{poly,eff} + t_{ox} > R_g$; in this case $g_3 = 1$.

Since the penetration depth of the electrons into the substrate is small compared to the beam size, little error is introduced if we assume that all power flowing to the substrate is absorbed at the substrate surface. The linearized temperature as a function of radius is then given by

$$\theta(r) = \frac{f I_t V_b}{2\sqrt{\pi}\, K_{si}(T_0) w}\, I_0(r^2/2w^2)\, \exp(-(r^2/2w^2)) \tag{12}$$

where I_0 is the modified Bessel function of the first kind. By inverting the Kirchoff transformation and adding the temperature drop across the oxide, we can obtain the total temperature rise in the polysilicon.

$$T_{poly} = 493.8 + \left[\frac{K_{si}(T_0)\theta}{14.354} + \sqrt{T_0 - 493.8}\right]^2 + \Delta T_{ox}(r) \tag{13}$$

Figures 3(a) and (b) show curves of the melting thresholds for electron-beam melting. Several interesting features should be noted. Since the "feedback" effect is absent in electron beam melting, the temperature of the polysilicon rises above the melting point with any increase in power once melting occurs. Deep penetration of the melt then occurs, initiating large-grain recrystallization. Thus, the melting threshold coincides with the recrystallization threshold. In Fig. 3(a), the slope of the melting/recrystallization threshold curves is more gradual for higher beam voltages (greater penetration depths). This behavior arises from the decreased temperature drop across the oxide as the penetration depth increases. The curve for V_b=5 kV is identical to the melting threshold curve for laser melting since all of the power is absorbed in the polysilicon layer. The 15 kV curve decreases more gradually; however, for oxide thicknesses greater than 1.4μm all of the power is absorbed in the polysilicon and oxide layers and this curve approaches the 5 kV curve beyond this thickness. Fig. 3(b) illustrates the effect of varying the beam size. The influence of an increasing insulator thickness is smaller for large beams because the temperature drop across the insulator is proportional to the flux (p/w^2) while the substrate temperature is proportional to power/radius (p/w). The relative influence of the temperature drop across the insulator thus decreases with increasing beam size, and the curves become "flatter."

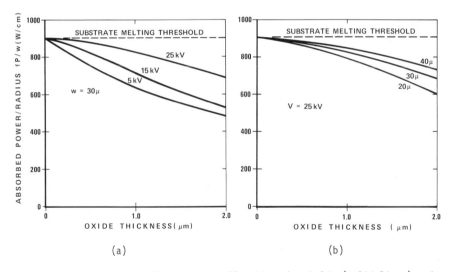

(a)   (b)

Figure 3. The surface melting/recrystallization thresholds (solid lines) and substrate melting threshold (dashed line) for electron-beam melting. a) Different beam voltages, constant size and b) constant beam voltage, different beam sizes.

534

The lack of "feedback" in e-beam melting also implies that the substrate melting threshold is independent of oxide thickness. The functional form of the absorbed power is not changed by melting; thus, the melting threshold of the substrate is the melting threshold with no oxide present. The substrate melting threshold is close to the recrystallization threshold over a wide range of oxide thicknesses for the 25 kV beam resulting in a small process window for this case. The constant substrate melting threshold in turn implies that it will be nearly impossible to avoid substrate melting when performing lateral epitaxial growth, since a power of at least the substrate melting threshold is required to melt the entire thickness of the polysilicon over the seeding window. This problem can conceivably be avoided by making the seeding window much narrower than the beam, in which case the heating of the polysilicon above the window should be dominated by the heating of the oxide. The resulting melting threshold will be smaller than the value over a large opening (calculated above), increasing the process window between recrystallization and substrate melting under the oxide.

CONCLUSIONS

Laser recrystallization is significantly affected by the "negative feedback" provided by the increase in reflectivity on melting. The "feedback" effect shields the substrate so that the substrate melting threshold increases with increasing insulator thickness. This increase in substrate melting threshold, combined with the decrease in recrystallization threshold with increasing insulator thickness, provides a wide process window for both unseeded recrystallization and for lateral epitaxy.

Electron-beam recrystallization lacks any "feedback" effect. Consequently, the substrate melting threshold is independent of insulator thickness. The finite penetration of the electron beam can significantly reduce the decrease in recrystallization threshold with increasing insulator thickness. For these reasons, the process window for unseeded recrystallization tends to be smaller for the e-beam technique than for the laser technique. Additionally, e-beam lateral epitaxy will be difficult to perform without melting the substrate under the oxide-covered regions unless special precautions are taken.

REFERENCES

1. A. Gat, L. Gerzberg, J.F. Gibbons, T.J. Magee, J. Peng, and J.D. Hong, Appl. Phys. Lett. 33, 775 (1978).

2. T.I. Kamins and B.P. von Herzen, Electron Dev. Lett. EDL-2, 313 (1981).

3. K.F. Lee, J.F. Gibbons, K.C. Saraswat, and T.I. Kamins, Appl. Phys. Lett. 35, 173 (1979).

4. C.I. Drowley, C. Hu, and T.I. Kamins, Extended Abstracts of the Electrochemical Society Spring Meeting, Montreal, Canada, 1982 (Electrochemical Society, N.J.), p. 234.

5. C.I. Drowley, Ph.D. Dissertation, University of California, Berkeley, CA (1982) (unpublished).

6. M. Lax. Appl. Phys. Lett. 33, 786 (1978).

7. T.E. Everhart and P.H. Hoff, J. Appl. Phys. 42, 5837 (1971).

8. A. Neukermans and W. Saperstein, J. Vac. Sci. Tech. 16, 1847 (1979).

PART VI
CRYSTALLIZATION
OF SILICON ON INSULATORS–II

LASER-INDUCED CRYSTALLIZATION OF SILICON ON BULK AMORPHOUS SUBSTRATES: AN OVERVIEW

D. K. Biegelsen, N. M. Johnson, W. G. Hawkins,* L. E. Fennell and M. D. Moyer,
Xerox Palo Alto Research Centers, Palo Alto, CA 94304

ABSTRACT

In this paper we review the current understanding of laser-induced silicon thin film crystal growth on bulk amorphous substrates. We propose a model for oriented nucleation and show that the silicon reflectivity jump on melting coupled with radiant heating lead naturally to this autonucleation mechanism. We then survey various techniques for control of lateral epitaxial growth and conclude with the results of some recent electrical device characterization.

INTRODUCTION

Attempts to produce crystalline silicon layers on amorphous substrates have progressed rapidly in the past few years. Most techniques rely on sweeping a radiantly-heated molten zone through the silicon to achieve lateral epitaxial crystallization. As we will discuss below, there are many important characteristics common to all these methods. The most prevalent amorphous substrate has been the dielectric coated crystalline silicon wafer. The dielectric has usually been a thermally grown oxide or a deposited silicon-oxide or nitride layer. The next most studied substrate has been bulk silica glass. SiO_2 has useful features such as a much lower thermal diffusivity, transparency in the visible, high absorptivity in the infrared and compatibility with IC processing restrictions. Investigations using other, "dirtier," but in some ways more desirable non-epitaxial substrates such as encapsulated borosilicate glasses, mullites (alumina/silica composites) and metallurgical grade polysilicon are still in their early stages. Work using incoherent sources (e.g., arc lamps and strip heaters) and crystalline silicon substrates are covered in other papers in this symposium. Therefore, although many aspects of the work to be discussed are quite general, in this paper we will emphasize work in laser-induced crystal growth of silicon on bulk amorphous SiO_2 substrates.

To develop a reliable single crystal thin film technology one must understand and control the growth processes - i.e. autonucleation and continued lateral epitaxy - and subsequent steps of integrated circuit device fabrication. Because most of the transistor processing steps are standard IC art, we will focus here primarily on the crystal growth phenomena.

*Permanent address: Xerox, Webster Research Center, Webster, NY 14580

Mat. Res. Soc. Symp. Proc. Vol. 13 (1983) ©Elsevier Science Publishing Co., Inc.

CONTROL OF NUCLEATION

Consider first some particular optical properties of silicon and silicon dioxide. Silicon strongly absorbs radiation with energy greater than its bandgap. Sub–gap radiation is also strongly absorbed when a sufficient density of free carriers is present, e.g., at high temperatures. SiO_2, by contrast, is transparent above 0.5 eV (2 μm) and strongly absorbing in the infrared with a peak near 9 μm. For slowly scanned cw radiation the heat flow is in steady state (i.e., independent of scan speed). The absorption length in hot silicon at all the commonly used wavelengths (0.4 μm – 11 μm) is then less than 0.5 μm (i.e., light entering the silicon is almost totally absorbed in the thin film). The reflectivity of silicon increases discontinuously on melting. Again, for all common wavelengths, and for the same incident flux, this results in approximately half as much absorption by the liquid phase as for the solid.

The increased light rejection by the melt has several very important consequences. Suppose a thin silicon film is irradiated by a uniform light source. The heat is conducted away into the substrate. As shown schematically in Fig. 1, as the power is increased, the film temperature increases. At P_1 the solid is absorbing sufficient power to maintain the film just below the equilibrium melting temperature, T_m. Now, if the film were entirely melted, incident power $P_2 \simeq 2P_1$ would be required to hold the film at T_m. Between these two powers the film must break into coexisting liquid and solid regions to self–limit the absorbed flux. The size of these lamellae [1] depends on (1) the heat flow and (2) the thermodynamics of nucleation [2]. For example, for some power between P_1 and P_2, if a large region of width D is molten, power is locally insufficient to maintain the liquid at T_m. Depending on the amount of heat flow from the hot solid regions, the temperature at the center of the liquid lamella will undercool by amount ΔT. This is indicated in Fig. 2 by the curve labeled "heat flow." Now, the liquid below T_m is not the equilibrium phase. Therefore, there will constantly be occurring local fluctuations into the solid phase.

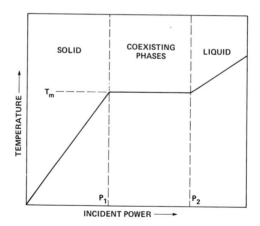

Fig. 1.   Sample temperature versus incident power. P_1 and P_2 denote the onset of melting and total melting.

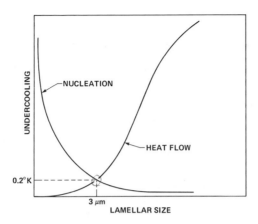

Fig. 2.   Size of solid and liquid lamellae versus peak undercooling at liquid center. Curve marked "heat flow" is derived from thermal and optical properties of sample; curve marked "nucleation" is calculated from thermodynamic data.

However, because there is a large interfacial free energy incurred by creating a liquid/solid boundary, only those nuclei with volume to interface ratios above a certain value [3] will be stable and not melt back. This size depends on ΔT. The relation $D\Delta T$ = constant is labeled "nucleation" in Fig. 2. The intersection of these two countervailing graphs determines the average lamellar size and undercooling. Figure 3 is a video image of black body radiation from an unencapsulated silicon film illuminated with a Gaussian CO_2 laser beam.   The lamellae are ~ $3\mu m$ in diameter and the calculated undercooling is ~ 0.2°K. The same sample heated conductively melts homogeneously. Thus, as shown in Fig. 4, in continuous silicon samples for which the dominant heating is radiant, the molten zone is surrounded by a "slush" zone [4].   During motion of the zone the solid lamellae grow (frequently in a filamentary mode) into the undercooling liquid from the receding solid material. Figure 5 shows an example of this filamentary growth of the solid from the coexistence zone as the laser power is reduced.

The discontinuous reflectivity rise on melting has positive and negative ramifications for crystal growth. The bad news is that there is a strong tendency towards undercooling of the liquid and concomitant interfacial instabilities. Such negative thermal profiles are known to result in twin growth, dendritic growth, cellular growth and liquid inclusion [5]. Indeed, these are the dominant defects observed in "nearly single" thin film silicon crystal growth. The good news is that (1) the radiant melting process is quite insensitive to power fluctuations between P_1 and P_2 ("self–limiting") and (2) the phase coexistence zone allows oriented nuclei to develop fully.   We will explain this point in more detail here.

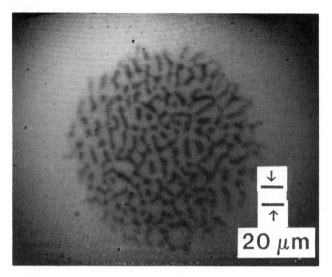

Fig. 3. Video image of black body radiation of inhomogeneous melting. Solid is bright. Solid lamellae are ~ 3 μm in size.

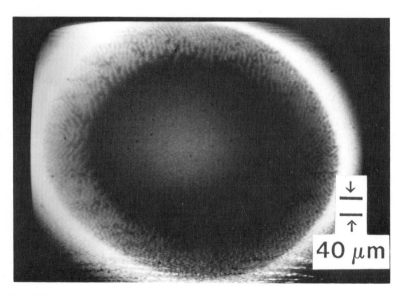

Fig. 4. Video image of blackbody radiation for $P > P_2$ at spot center. Note phase coexistence zone surrounding completely molten region.

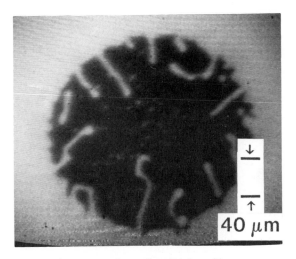

Fig. 5   Video image of spot cooling after total melting.

Let us consider again the energetics of nucleation (for example, a cylindrically shaped solid region in the liquid with height, h, and diameter, D. At a temperature ΔT below the equilibrium melting temperature, T_m, a fluctuation into the solid phase lowers its free energy by approximately $LV\Delta T/T_m$ where L is the latent heat of melting and $V = \pi D^2 h/4$ is the nuclear volume. However, in creating an interface between the liquid and the solid, an extra interfacial free energy of γC is incurred, where $C = \pi Dh$ is the interfacial area of the nucleus. There will also be an energy term related to the difference in interactions of the liquid and solid with the substrate and encapsulating layer (or vacuum). This term is proportional to the area of the nucleus, $A = \pi D^2/4$. Finally, other less important terms such as meniscus energies could be included. When the total free energy difference of the nuclear fluctuation is negative, the nucleus is stable. Otherwise, the nucleus dissolves again into the melt. The "third" term above is here effectively a volume term, so that the total energy change for a nucleus can be written,

$$\Delta E = - L'\pi D^2\, h\Delta T + \gamma \pi Dh.$$

Thus, the nucleus is stable when $D\Delta T \geq$ constant. Now because the solid is crystalline, the interfacial terms are axially anisotropic. For example, we thus expect, and often see, facetting in the coexisting phases. Furthermore, if the nuclei are large (i.e. ΔT is small), then the term proportional to the surface area between the solid and substrate/encapsulant will be relatively important. For SiO_2, it is empirically found that the low bond density silicon {100} planes result in minimum bonding energy. It is at least plausible that the Si {100} planes are energetically preferred in contact with SiO_2. The areal bond density of vitreous SiO_2 and {100} Si are very close [6]. Given the dihedral flexibility of the oxygen bonding, the {100} planes can bond with very little strain being developed in the oxide.

The elegant benefit of the coexistence phase can now be appreciated. The mixture is essentially composed of many nuclei in an undercooled melt. The stable nuclei are restricted from enlarging by the radiant coupling mechanism, i.e. increased size results in a heat imbalance and remelting. The result is that because the nuclei exist for very small values of ΔT and for long times, the relatively small energy differences associated with axial orientation can become fully influential.

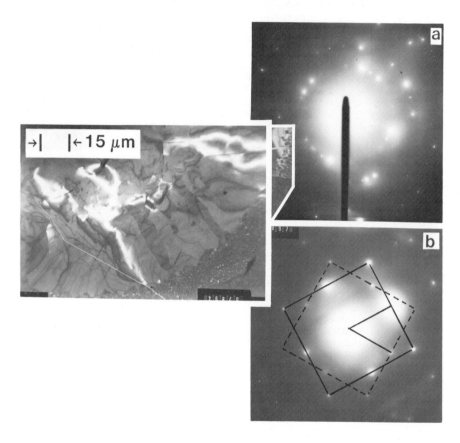

Fig. 6   Dark field transmission electron micrograph of material heated with a static spot similar to Fig. 4. Power was slowly reduced through total solidification. Inset A shows a polycrystalline diffraction pattern for crystallites at the periphery. Inset B is a double exposure selected area diffraction for the bright regions of the image.

In Fig. 6 we show a dark field transmission electron micrograph of material similar to that seen in Fig. 4. The sample was heated to complete melting in the center by a Gaussian spot and cooled slowly. In view is the edge of the recrystallized spot lying within a field of fine grain poly silicon. Several features are evident. First, crystallization proceeds by near-radial growth from the periphery, i.e., parallel to the thermal gradients. Second, the nuclei have a preferential axial orientation normal to the substrate. Inset (A) is a selected area diffraction pattern from a portion of the slush region. The ring structure indicates that the crystallites have {100} texture and random in-plane orientation. The TEM image was made by selectively aperturing only a single {220} diffraction spot. Because of slight sample bending, only two grains had significant Bragg scattering. The large grain/small grain interfaces, which lay approximately perpendicular to the thermal gradients, are the nucleation sites of the two grains, rotated by ~ 55±5°. Inset B is a double exposure of the diffraction patterns for the two grains. The respective {220} directions are rotated by 57°. This is a simple demonstration of the strong tendency we observe for ⟨220⟩ in-plane alignment with the local thermal gradient. The ⟨220⟩ direction is the trace of {111} planes. It is frequently the case that the most rapid silicon growth occurs by extension of {111} planes. It seems evident then that the in-plane orientation is the result of kinetic competition. When a line spot is scanned, many similarly oriented grains with low-angle grain boundaries grow epitaxially behind the zone. We should point out here that strip heater produced material has in-plane crystal orientation with ⟨100⟩ lying within ~ 15° of the thermal gradient. This qualitative difference in the kinetic selection possibly results from the much shallower thermal gradients, associated with substrate thermal biasing to ~ 1200°C, and concomitant tendency towards interface instability. The two different kinetic regimes have been described by Tiller [7].

To our knowledge no literature exists describing the kinetics and energetics of thin film crystal growth when interfacial (texturing) energies are important. This is an exciting area for future study. In summary, it is clear that autonucleation is a natural outgrowth of radiant coupling to silicon.

CONTROL OF GROWTH

Given that a desired seed can be selected, it is necessary then to control the subsequent growth so that lateral epitaxy can be sustained. One must make negligible the probability of spontaneous breakdown of the growth interface and ensure that competitive nucleation does not occur. To this end the solidification interface must be controlled [8]. Several methods have been used to generate desirable interface geometries all of which essentially manipulate the in-plane thermal profiles at the trailing edge of the molten zone.

Continued growth of a crystalline region is primarily driven by the local thermal gradient. It is obvious then that irradiation of uniform material with a Gaussian spot leads to competitive grain growth from the (cooler) sides of the moltem zone. A zone with a concave trailing edge has been shown to lead to stable single crystal growth with the width of the single crystal material extending to the limits of the concave region (Fig. 7) [9]. All finite spots must have convex regions, so trails of polycrystallites will exist. However, by overlapping scans with a zone tilted so that seeding occurs in the preceding region of single crystal material, Stultz and Gibbons [10] have been able to avoid the regions of disordered poly between scans. Concave crystallization fronts are non-optimal for growth stability because anisotropy in the kinetics of crystallization leads to variations along the curved interface. These variations result in twinning and other defect nucleation. To achieve a flat interface then one can use a line source or use inhomogeneities in the sample to transform a radiation profile into the desired thermal profile. On bulk

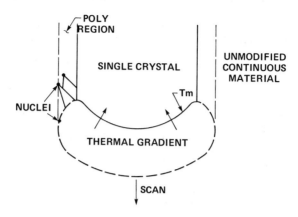

Fig. 7. Sketch of crystallization induced by spot with concave melting isotherm.

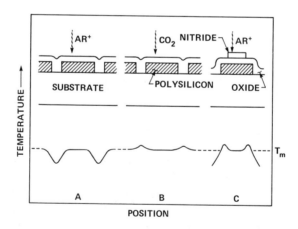

Fig. 8. Above: cross sections of encapsulated silicon stripes on quartz substrates. Below: corresponding temperature plots. (a) Ar^+ laser irradiation; (b) CO_2 laser irradiation; (c) silicon nitride dual dielectric control of visible absorption.

SiO$_2$ substrates, random cracking of continuous silicon films due to differential contraction relative to the substrate has been a problem [11]. This has been circumvented by patterning the silicon prior to the laser melting step. Therefore, a self-aligning thermal profile mechanism has been the predominant technique used with bulk substrates.

In Fig. 8 we illustrate a few examples of methods to control the optical power absorbed. Figure 8A shows a cross sectional view of stripes of silicon on a silica substrate encapsulated in a deposited dielectric (usually SiO$_2$ and/or Si$_3$N$_4$). Visible laser radiation is absorbed in the silicon but not the substrate, so that the temperature varies laterally as shown. This lateral variation results in a convex melting isotherm at the trailing edge of the zone. In Fig. 8B a far infrared source is used instead. The silicon tends to self-limit at T$_m$, but the silica can continue to heat. One thus obtains edge-enhanced heating and concave isotherms. By varying the "moat" width between stripes, the degree of concavity can be controlled. Furthermore, for a CO$_2$ laser with $\lambda \sim 10$ μm and moat widths $\sim 2-3$ μm, diffraction effects lead to much less extra absorption in the substrate than one would calculate from the geometric cross section [12]. Thus, a very nearly flat crystallization front can be easily achieved. Figure 8C represents one of many more complex multilayered thin-film structures for generating lateral thermal profiles. Here the thickness of Si$_3$N$_4$ rails and SiO$_2$ encapsulant are designed to result in an antireflection layer at the stripe edges and enhanced reflection in the center.

Utilizing the technique sketched in Fig. 8B, we have been able to achieve high yields of oriented single crystal stripes. Figure 9 shows a typical example. More work is needed to ensure reliable crystallization processing. The main defects are twinning and low angle grain boundaries – both of which result from growth interface instabilities related to the inherent tendency of radiantly-heated silicon to undercool. Otherwise the material is surprisingly free of dislocations, faults and precipitates. Suffice it to say then that the crystal growth process is rapidly becoming well-controlled on a laboratory scale.

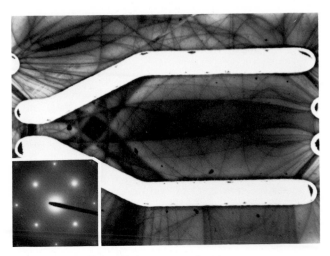

Fig. 9. Single crystal "island" in patterned stripe.

ELECTRICAL PROPERTIES

We turn now to consider the electrical properties of thin film MOS transistor devices made in this material. We first note that it is indeed demonstrably necessary to use oriented single crystal material for device fabrication. Figure 10 shows a plot of correlated measurements of low–field electron channel mobilities and the number of grain boundaries transiting the channel from source to drain. It is readily seen that the presence of grain boundaries would result in large variability in device performance. (Another adverse feature of the presence of incoherent grain boundaries is the enhanced dopant diffusion along the boundaries – which causes electrical short circuits in short channel devices [13].)

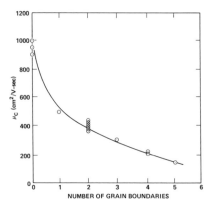

Fig. 10. Plot of measured electron channel mobility and number of grain boundaries transiting from source to drain in same islands.

As was pointed out by the Lincoln Lab group [14], the high electron mobilities observed in single crystal material arises from the large, uniform tensile stress in the silicon due to differential contraction after solidification [15]. The stress splits the degeneracy of the conduction band minima and effectively reduces the average conduction electron mass (thereby increasing the mobility). This stress splitting is anisotropic; it is maximal for {100} texture and zero for {111}. For this reason as well as for uniformity in processing and crystal growth, texturing is important.

Figure 11 shows the dependence of drain current on gate voltage for an n–channel enhancement–mode transistor fabricated in a single–crystal silicon island on fused silica. The leakage current is less than 1 pA per μm of channel width, and the device characteristics yield a low–field electron channel mobility of ~ 1000 cm^2/V–sec. (For further details, see Refs. 16 and 17.) The device results obtained thus far demonstrate that laser crystallization can be used to fabricate high–performance TFTs on bulk amorphous substrates and thereby provide the basis for a silicon–on–insulator technology. The next step is to evaluate circuits fabricated in this material and to establish credible performance and yield statistics.

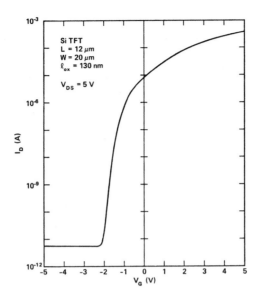

Fig. 11. Drain current I_D versus gate voltage V_G for a single–crystal silicon TFT on bulk glass.

FUTURE

In a review paper covering the field of laser–induced crystallization from seeding to fruition, it seems appropriate to attempt a prognostication of the future methods and areas of application, as well as obstacles. Since such exercises are generally unerringly wrong, we will be brief. First it should be noted that only three years ago the concept of a thin film single crystalline silicon TFT technology was at the dream level. The successes of many research groups have brought us today to a stage of laboratory reality. To convert this to a technology will undoubtedly require an even greater effort. However, the incentives are strong. Unique capabilities will be enabled by such a technology. Applications include large area distributed switching and driving, e.g. matrix addressed I/O tablets or integrated one and two dimensional transducer arrays and drive electronics. It is likely that many of these configurations will not be compatible with heating the entire substrate to temperatures > 1000 C. Laser heating can be an efficient, spatially controllable and non–destructive technique. Whereas little wafer warpage or distortion has been observed, multiple photolithographic operations on large size substrates will certainly encounter problems. Stress arising from use of non–silicon substrates may create obstacles in subsequent processing or fragility. A useful present and future area of study is into inexpensive substrates with thermal expansion coefficients more nearly matching silicon [18]. The immediate technological issue though is how stable and reliable the crystal growth process can be made. The early results are most encouraging.

ACKNOWLEDGMENTS

We would like to thank Drs. Michael Geis, Ross Lemons, Conyers Herring, Robert Street and James Mikkelsen for very useful conversations.

REFERENCES

1. M. A. Bösch and R. A. Lemons, Phys. Rev. Letters $\underline{47}$, 1151 (1981).

2. W. G. Hawkins and D. K. Biegelsen (submitted for publication).

3. J. H. Holloman and D. Turnbull, Prog. in Metal Phys. $\underline{4}$, 333 (1953).

4. J. E. Moody and R. H. Hendel, J. Appl. Phys. $\underline{53}$, 4364 (1982); and Michael Geis, private communication.

5. W. A. Tiller, Art and Science of Growing Crystals, J. J. Gilman, ed. (John Wiley, New York, 1964) p. 276.

6. W. A. Tiller, private communication.

7. W. A. Tiller, Science $\underline{146}$, 871 (1964).

8. D. K. Biegelsen, N. M. Johnson, D. J. Bartelink and M. D. Moyer, Appl. Phys. Letters $\underline{38}$, 150 (1981); S. Kawamura, J. Sakurai, M. Nakano and M. Takagi, Appl. Phys. Letters $\underline{40}$, 394 (1981).

9. D. K. Biegelsen, N. M. Johnson, D. J. Bartelink and M. D. Moyer, in Laser and E–Beam Solid Interactions, J. Gibbons, L. Hess and T. Sigmon, eds. (North Holland, New York, 1980), p. 487.

10. T. J. Stultz and J. F. Gibbons, Appl. Phys. Letters $\underline{39}$, 6 (1981).

11. N. M. Johnson, D. K. Biegelsen and M. D. Moyer, ref. 9, p. 463.

12. J. van Bladel, Electromagnetic Fields, (McGraw–Hill, New York, 1964), p. 410.

13. N. M. Johnson, D. K. Biegelsen and M. D. Moyer, Appl. Phys. Letters $\underline{38}$, 900 (1981).

14. B.-Y. Tsaur, J. C. C. Fan and M. W. Geis, Appl. Phys. Letters $\underline{40}$, 322 (1982).

15. S. A. Lyon, R. J. Nemanich, N. M. Johnson and D. K. Biegelsen, Appl. Phys. Letters $\underline{40}$, 316 (1982).

16. N. M. Johnson, D. K. Biegelsen, H. C. Tuan, M. D. Moyer and L. E. Fennell, Electron Device Letters (in press).

17. N. M. Johnson, D. K. Biegelsen, L. E. Fennell, M. D. Moyer, M. J. Thompson, H. C. Tuan, A. Chiang, these proceedings.

18. For example, R. A. Lemons, M. A. Bösch, A. H. Dayem, J. K. Grogan and P. M. Mankiewich, Appl. Phys. Letters $\underline{40}$, 469 (1982).

THE EFFECTS OF SELECTIVELY ABSORBING DIELECTRIC LAYERS AND BEAM SHAPING ON RECRYSTALLIZATION AND FET CHARACTERISTICS IN LASER RECRYSTALLIZED SILICON ON AMORPHOUS SUBSTRATES

G. E. Possin, H. G. Parks, S. W. Chiang* and Y. S. Liu
General Electric Research and Development Center
Schenectady, New York 12301

ABSTRACT

Selective absorption, using patterned dielectric films, and beam shaping were used as means for improving the recrystallization of LPCVD polysilicon islands on fused quartz. IR imaging of the laser heated region was used to optimize and control the recrystallization. MOSFETs were fabricated in laser-recrystallized silicon islands on amorphous substrates using a standard n-channel poly-gate process. Devices with various channel lengths and widths were fabricated, and the dependence of threshold voltage, channel mobility, and leakage on recrystallization conditions and device dimensions was studied.

INTRODUCTION

This paper will describe an investigation of the use of a CW Argon ion laser to recrystallize polysilicon islands on quartz substrates. Both shaped beams and selective absorption using dielectric layers were used to control the shape of the melt front during recrystallization. Infrared imaging of thermal radiation from the heated region was used to optimize the beam shape and dielectric structures and to improve the control of the recrystallization conditions. The quality of the achieved recrystallization was investigated using grain boundary etching. MOSFETs were fabricated in the recrystallized islands using a standard self-registered polysilicon gate technology. The influences of channel length and width on the device properties were investigated.

IR IMAGING AND TEMPERATURE MONITORING

The laser annealing system consisted of an Ar ion laser and 40 mm microscope objective for beam focusing. The wafers were scanned under the beam using a numerically controlled table with an accuracy of one micron and a maximum scanning speed of 10 cm/sec. Beam power, beam blanking, and table scanning were controlled using a HP85 computer. The beam power was set using a voltage applied to the feedback reference of the plasma supply. In the 1 to 2 watt range used for these experiments, the power was stable to better than 1%. The power was checked between scans and reset automatically as required. A silicon vidicon camera with 1000 mm lens viewed the sample through a dichroic mirror and the same microscope objective. IR transmitting glass filters (Corning 7-69 and 7-57) were used to provide an infrared image of the laser heated region. Because of the 1 micron wavelength cutoff of the silicon vidicon, only a narrow IR band around 1.0 micron was imaged. Quantitative monitoring of the IR vidicon signal during set-up and scanning provides a useful reference for determining the laser power needed for good recrystallization. A beam expander consisting of a converging-diverging lens pair whose separation could be controlled was used to control the beam size. By introducing a small additional beam divergence, large spot sizes (> 50 microns) could be achieved with short focal length objectives (< 40 mm). A significant advantage of

---

* present address, IBM, East Fishkill

Mat. Res. Soc. Symp. Proc. Vol. 13 (1983) ©Elsevier Science Publishing Co., Inc.

this spot size control method, as compared to deliberate defocus of the objective or the use of longer focal length objectives, is that useful magnifications of 500X can be obtained on the TV monitor. This is very useful for set-up and monitoring of recrystallization.

The calibration of this IR signal is illustrated in Figure 1. Unpatterned areas of quartz wafers with 500 nm of polysilicon, 110 nm of LPCVD oxide, and 30 nm of silicon nitride or with just the silicon and nitride layers were scanned with the beam at a velocity of 5 mm/sec. The figure plots the relative camera signal from the center of the heated region as a function of laser power. Neutral density filters were used to prevent saturation of the vidicon. The IR radiation at a wavelength λ per unit wavelength is given by,

$$W = \epsilon C_1 \lambda^{-5} [\exp(C_2/\lambda T) -1]^{-1} \qquad (1)$$

where ϵ is the emissivity, $C_2 = 1.4388$ cm deg , and C_1 is a material independent constant. For $\lambda = 1000$ nm and $T < 4000°$, the -1 in the denominator can be dropped. If it is assumed that $T = aP + T_0$ where T_0 is the ambient temperature, P is the beam power, and a is a constant, then Equation (1) can be written as:

$$\ln (W) = \ln \epsilon + \ln (C_1/\lambda^5) - \frac{C_2}{\lambda a} (1 + T_0/aP)^{-1} p^{-1} \qquad (2)$$

Over a limited power range the factor $(1+T_0/aP)$ is approximately constant. Hence a plot of log (IR signal) versus $1/P$ should be linear over a limited power range. From the slope, the proportionality constant a can be determined. This calibrates the IR signal in terms of temperature for the emissivity of a particular sample. For specimens which are opaque at the monitored IR wavelength, the emissivity is $(1-R)$, where R is the reflection coefficient at the monitored wavelength. The * indicates the calculated melting point based on the slope of the IR signal plot. The arrow indicates the power at which melting of the silicon occurred as determined by Secco etching of the polysilicon. The flat region of IR signal versus power is also consistent with melting as discussed subsequently. Also indicated in the figure as ϵ_{ox} is the calculated change in log (IR signal) due to the change in emissivity of the nitride capped

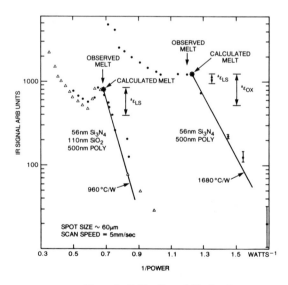

Figure 1. Calibration of IR signal.

layer as compared to the nitride + oxide capped layer. This explains why the observed IR signal at melting is less for the oxide capped layer since it has lower emissivity. This data indicates that the IR signal behavior is consistent with Equation (2). Once it is calibrated for a particular emissivity, it can be used to monitor the recrystallization process. In particular the melting power threshold can be determined at any particular point on a patterned sample independent of spot size and shape and substrate temperature. Because of the 30 ms frame time of the vidicon, absolute calibration is possible only for stationary or slowly scanning beams.

Calculation of Absorption

Figure 2 shows the calculated optical absorption at 500 nm for a nitride capped silicon layer and nitride + oxide capped layer as a function of the nitride capping layer thickness. The solid curve is for optical constants appropriate for molten silicon $n = (2.3 - j4.8)$ [1] and the dotted curve is for for crystalline silicon $n = (4.75 - j1.0)$ [2-4] below the melting point. Room temperature values for the oxide and nitride indices were used because these are not very temperature dependent. Note the large drop in absorbed power after melting. The drop causes an instability since there is a range of powers in which the silicon layer is not stable in either the molten or solid phase. The thermal conductivity of silicon is 60 times larger than quartz so the thin silicon layer can support almost no vertical temperature gradient. Hence the entire silicon layer must be molten or solid. Therefore there should be a range of power above the melting threshold where the layer oscillates between the molten and solid phase. This situation is different from laser annealing on silicon substrates where the high conductivity of the substrate can indirectly stabilize the melt formation because the melt depth changes slowly with increasing laser power [5]. This instability is the cause of the power range in Figure 1 where the IR signal is constant or drops with increasing power. The IR signal can drop with increasing power because the emissivity of liquid silicon is less than that of solid silicon. The amount of drop in IR signal due to the emissivity change is indicated in the figure

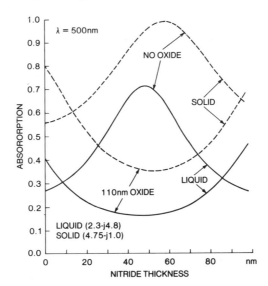

Figure 2. Calculated optical absorption at 500 nm for nitride and nitride + oxide capping layers.

552

as $\Delta \epsilon_{LS}$. Just above the melt threshold the oscillation between liquid and solid predominantly dwells in the solid state. As the power increases, the fractional time spent in the liquid state increases. The IR signal then drops because the emissivity of the liquid state is smaller. Evidence for these oscillations can be seen in the IR signal as variations in the width of the molten region and as patterns of alternate melting and non melting during scanning at powers just above the melting threshold. Studies of the dynamics of this oscillation will be discussed elsewhere.

SELECTIVE ABSORPTION AND LASER RECRYSTALLIZATION

For this work structures such as shown in Figure 3 were used. The use of moated island structures was first discussed in this context by Johnson, Biegelsen, Moyer and Bartelink [6-8]. The laser beam is scanned vertically along a strip of moated islands. The devices are fabricated inside the 18x50 micron island. The sea area outside the island provides power absorption outside the island, which improves the profile of the melt front inside the island. A concave or at least flat melt front is desirable to prevent grain boundaries, which can nucleate at the island edges, from propagating towards the island center [6-8]. All recrystallization was done at a nominal substrate temperature of 400°C. The wafers were covered with a quartz plate separated by a 3 mm air gap to minimize convective cooling of the wafer surface.

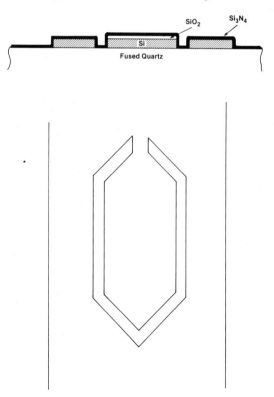

Figure 3. Moated island structure.

Polysilicon 0.5 microns thick was deposited by low-pressure chemical vapor deposition (LPCVD) on 0.015 in. thick, high-purity fused quartz substrates. The silicon was capped with 0.1 microns of LPCVD silicon dioxide, which was removed except above the islands (Figure 3). The wafers were then capped with 20, 30 or 40 nm of LPCVD silicon nitride. The absorbed power in the 50 micron wide sea region is increased relative to the island by the addition of a 100 nm thick LPCVD oxide layer under the nitride cap. For nitride thicknesses of 20, 30, and 40 nm the absorbed power ratio between the two areas should be 1.8, 2.9, and 3.9 respectively, above the silicon melting point. The ratios are smaller below the melting point. For a beam size larger than the 50 microns wide sea this increased power absorption in the sea should produce a concave melt front even though the trailing beam edge is convex. Figure 4 shows IR images of such island structures heated with a ≈ 50 micron diameter laser spot and a laser power of ≈ 1 watt. These images were taken below the melting threshold where the absorbed power ratios are less. Hence the temperature distribution for the 40 nm case is similiar to what is observed well above the melting threshold for the 20 nm nitride cap.

Figure 5 shows some typical recrystallized islands, after removing the oxide/nitride cap by wet etching and Secco etching to delineate grain boundaries. This sample had a 30 nm nitride cap and was recrystallized using a 2.0 watt, 40 micron diameter beam scanned at 2.5 cm/sec. Note that recrystallization is very good in the island area as compared to the sea. Such good results cannot be consistently reproduced. The results shown in the top two micrographs of

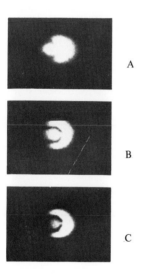

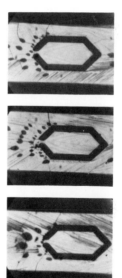

Figure 4. IR imagers of laser heated regions of the moated island structures with 0.1 microns of oxide in the island region and capped with nitride of thickness (a) 0.02 microns, (b) 0.03 microns and (c) 0.04 microns. Temperature below the melting point of silicon in all cases.

Figure 5. Secco etched islands after recrystallization.

554

Figure 5 are typical of most of the islands, but about 10 to 30% of the islands have some grain boundaries revealed by Secco etching as shown in the bottom micrograph of Figure 5. Many of these appear to be low-angle grain boundaries, which commonly occur in silicon films laterally regrown on amorphous substrates [7,9-10]. All three islands in Figure 5 are from the same area of the wafer and the same laser scan.

PHASE PLATE BEAM SHAPING

An alternate way to achieve a concave melt front is to shape the incident beam.[7] A patterned dielectric film on an optical flat inserted in the beam path before the objective lens can produce a large variety of beam shapes. If the dielectric thickness is one half wave thick standard Fourier optical theory [11] predicts that the field intensity at the focal plane will be

$$V\ (x_f,y_f) = \int\int dxdy\ I(x,y)\ T(x,y)\ \exp\ [-j2\frac{\pi}{\lambda}f(xx_f + yy_f)] \qquad (3)$$

where $T(x,y)$ is a function, in the plane of the phase plate, which is -1 inside the dielectric regions and +1 outside. The incident field intensity $I(x,y)$ is assumed to be Gaussian for purposes of simulation. The power density is given by V^2. Standard Fourier transform programs can be used to calculate the integral. Phase plates were prepared using sputtered indium tin oxide (ITO) patterned using standard semiconductor lithography techniques and an HCl-HNO$_3$ etch. Figure 6 compares the results for one of the more complicated patterns investigated. The figure shows the calculated power distribution, the phase plate dielectric pattern, and photographs of the beam intensity incident on the phase plate and the IR image of the focused beam incident on a capped polysilicon film on a quartz substrate. Less exotic patterns such as stripes or edges can produce double lobed beams. Significant flexibility in the beam shape obtained from a single pattern can be realized by changing the position of the plate in the beam, varying the divergence of the beam using the telescope, or varying the focus of the objective. Refocusing of the camera lens can compensate for defocusing of the objective and maintain the focus of the IR image. A significant advantage of this type of beam shaping is its simplicity and power efficiency. Unlike aperture methods of shaping, this type of shaping loses no power, and a relatively inexpensive phase plate which is easily fabricated can provide the same beam shape in any laser system. A significant disadvantage of shaped beams over

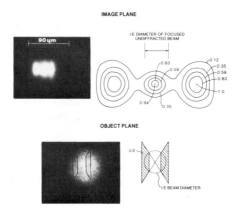

Figure 6. Comparison of calculated and observed beam shape using phase plate beam shaping.

selective dielectric absorption is that more critical beam alignment relative to the substrate is required. Using a double lobed beam shape on these samples requires a beam alignment better than 2 microns along the entire scan or the absorbed power distribution will change significantly along the scan. By comparison the selective absorption is self registering and the alignment is less critical.

FET RESULTS

Devices were fabricated using a standard, poly-gate, all-oxide, ion implant MOSFET process. All oxidations and annealing were done at 950°C or lower. N-channel enhancement devices were fabricated using As source drain implants and a B threshold control implant of 8E11 at 50keV and on one half of each wafer a deep back channel implant of 8E11 at 180 keV. Devices were fabricated with channel widths between 1 and 18 microns and channel lengths between 1 and 12 microns. All the recrystallized islands were the nominal 18 micron width as shown in Figure 3. The channel widths were defined after laser recrystallization by dry etching. Some islands were not etched back.

For the larger channel width and length devices, where short channel and narrow width effects are small, the average channel mobility in the linear region was between 400 and 500 cm²/volts-sec. The extrapolated threshold voltage determined in the linear region was 1.2± 0.3 volts. The calculated threshold voltage using SUPREME was 0.6 volts. Subthreshold leakage currents at -1 volts gate bias and 2.5 volts source-drain bias were in the 1 pA range for most of the devices with or without a back channel implant. Even devices with 3 micron channel lengths had similiar low leakages. The leakage current at 0 volts gate bias was typically 0.1 nA. The 0 volts leakage currents were 10 to 100 larger without a back channel implant. On some wafers islands which were etched back had a larger and broader range of leakage currents than the unetched islands.

Under the best laser processing conditions device yields in excess of 90% were observed. The dominant failure mode appears to be cracks in the islands or substrates due to the thermal expansion mismatch. The use of smaller island sizes or a substrate with better thermal match could eliminate this problem.

ACKNOWLEDGMENTS

The authors gratefully acknowledge the technical assistance of Renato Guida.

REFERENCES

[1] K.M. Shvarev, B.A. Baum, and P.V. Gel'd, "Optical Properties of Liquid Silicon," *Sov. Phys. Solid State, 16,* #11, pp. 2111-12 (1975).

[2] G.E. Jellison, Jr. and D.H. Lowndes, "Optical Absorption Coefficient of Silicon at 1.152 microns at Elevated Temperatures," *Appl. Phys. Lett., 41(7),* pp. 594-6, (1982).

[3] M.A. Hopper, R.A. Clarke, and L. Young, "Thermal Oxidation of Silicon," *J. Electrochem. Soc., 122,* pp. 1216-22, (1975).

[4] T. Sato, "Spectral Emissivity of Silicon," *Jap. J. Appl. Phys., 6,* pp. 339-347 (1967).

[5] S.A. Kokorowski, G.L. Olsen, and L.D. Hess, "Thermal Analysis of CW Laser Annealing Beyond the Melt Temperature," *Electron-Beam Solid Interactions and Materials Processing,* Gibbons, Hess, and Sigmon ed., (Elsevier North Holland, N.Y., 1981), pp. 139-46.

[6] N.M. Johnson, D.K. Biegelsen, and M.D. Moyer, "Processing and Properties of CW Laser-Recrystallized Silicon Films on Amorphous Substrates," in *Laser and Electron-Beam Solid Interactions and Materials Processing*, Gibbons Hess and Sigmon ed., (Elsevier North Holland, N.Y. 1981), pp. 463-70.

[7] D.K. Biegelsen, N.M. Johnson, D.J. Bartelink, and M.D. Moyer, "Laser Induced Crystal Growth of Silicon Islands on Amorphous Substrates," *Laser and Electron-Beam Solid Interactions and Materials Processing*, Gibbons Hess and Sigmon ed.,(Elsevier North Holland, N.Y. 1981), pp. 487-94.

[8] D.K. Biegelsen, N.M. Johnson, D.J. Bartelink, and M.D. Moyer, "Laser Induced Crystal Growth of Silicon Islands on Amorphous Substrates Multilayer Structures," *Appl. Phys. Lett., 38*, (3), pp. 150-2 (1981).

[9] H.J. Leamy, C.C. Chang, H. Baumgart, R.A. Lemons, and J. Cheng, "Cellular Growth in Micro-Zone Melted Silicon," *Materials Letters*, pp. 33-36 (1982).

[10] D.K. Biegelsen, N.M. Johnson, R.J. Neumanich, M.D. Moyer and L.E. Fennell, "Correlated Electrical and Microstructural Studies of Recrystallized Silicon Thin Films on Bulk Glass Substrates," in *Laser and Electron-Beam Solid Interactions and Materials Processing*, Appleton and Celler ed., (Elsevier North Holland, N.Y., 1982), pp. 585-90.

[11] Joseph W. Goodman, *Introduction to Fourier Optics*, (McGraw Hill, N.Y., 1968), pp. 77ff.

GROWTH OF SILICON-ON-INSULATOR FILMS USING A LINE-SOURCE ELECTRON BEAM[*]

J. A. Knapp and S. T. Picraux,
Sandia National Laboratories, Albuquerque, New Mexico 87185

ABSTRACT

A swept line-source electron beam has been used to study unseeded Si-on-insulator crystallization at beam scan speeds of 150-1500 cm/s. For a particular sample configuration a maximum linear crystallization velocity of ~ 350 cm/s was observed. At higher sweep speeds, competing nucleation occurred at intervals across the film. Both the limit in crystallization velocity and the intervals between nucleation are tentatively explained by a simple model.

INTRODUCTION

A variety of approaches to the fabrication of Silicon-on-Insulator (SOI) films are being explored [1-3]. Our studies have centered on the use of a line-source electron beam [4,5], both exploring the unique capabilities of such an energy source and examining questions of relevance to all approaches. The results reported here deal with the dependence of the quality of SOI formation on the speed of the sweeping e-beam. This applies to the practical question of just how fast an SOI layer can be successfully formed, and addresses such issues as minimizing the time for contamination and the substrate temperature. These questions are also important to other approaches to SOI formation. Our results indicate that, for the particular sample and beam configuration used here, the advancing recrystallization front breaks into multiple fronts above a sweep speed of ~ 350 cm/s. This phenomenom can be understood by the use of a simple model of the heat flow in the layered structure.

EXPERIMENTAL

A schematic of the line-source electron beam system is shown in Fig. 1. The beam sweeping and beam diagnostics have been reconfigured since our earlier SOI results were reported [4,5]. The system uses a sheet beam focussed to a line measuring 2 cm x ~ 1 mm at the sample table, with a peak current density of up to ~ 9 A/cm^2 and energy of 10-50 keV. Samples are swept under the beam on a rotating table at up to 5000 cm/sec. The table also includes a slit used for determining the beam profile and power density.

The sample configuration used in this study was similar to that used at other laboratories in SOI studies [3]. Specifically, <100> Si wafers were covered with 1 μm of SiO_2, 0.5 μm of poly Si, and a capping layer of 1 μm SiO_2 and 500 Å of Si_3N_4. No seeding was provided for the SOI growth. This configuration is not expected to be ideal for our system, but was used to provide a reference point to previous work. After treatment, samples were examined with Nomarski optical microscopy, both before and after etching the cap off, and after delineating the grain structure with a Secco etch [6]. The orientation of the recrystallized films was determined by electron channeling patterns formed with secondary electrons in a scanning electron microscope.

RESULTS

Figure 2 shows the relationship between beam power density and sweep speed

---

[*]This work performed at Sandia National Laboratories supported by the U. S. Department of Energy under contract number DE-ACO4-76DP00789.

Mat. Res. Soc. Symp. Proc. Vol. 13 (1983) ©Elsevier Science Publishing Co., Inc.

LINE-SOURCE ELECTRON BEAM SYSTEM

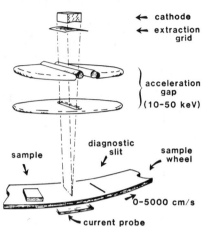

← cathode

← extraction grid

$\Big\}$ acceleration gap (10-50 keV)

Figure 1. Schematic of the line-source electron-beam system.

diagnostic slit

sample

sample wheel

0-5000 cm/s

current probe

(effectively, the beam dwell time) observed for melting and recrystallizing the isolated Si layer. Also shown as an inset is a schematic of the sample structure. Significant damage, such as cracking, resulted at high power, while low power was insufficient to completely melt the Si layer. All samples were held on a substrate at 500°C during treatment. The dashed lines in Fig. 2 represent, the approximate boundaries of the experimentally observed window in power density/sweep speed for reasonable results. Just as we had observed for samples using much thinner isolating dielectric layers under the poly Si, uniform, damage-free regrowth correlates well with: 1) complete melting of the poly Si layer and 2) no melting of the Si substrate. For poly Si layers over 0.25 μm of Si_3N_4 the window was observed to be ~ 1% of the power density at 1200 cm/s. The present sample configuration uses SiO_2 instead of Si_3N_4, for lower thermal conductivity, and a 4x thicker layer. The result is an ~ 10% window for

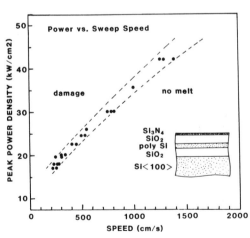

Figure 2. Power density vs. beam sweep speed for uniform, damage-free crystallization. Insert is a schematic of the sample configuration.

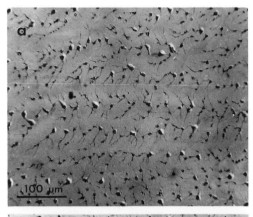

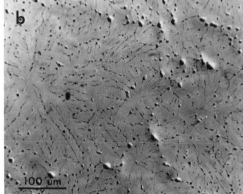

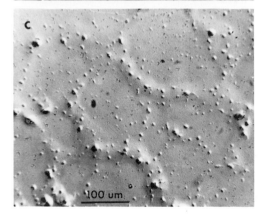

Figure 3. Nomarski optical micrographs of SOI films treated with a beam swept from left to right at speeds (a) 280 cm/s, (b) 500 cm/s, and (c) 890 cm/s. (a) and (b) have been etched to delineate grain boundaries.

success, which compares favorably both with our own previous results and the reported window using a graphite strip heater as the heat source [7].

The overall results can be divided into three regimes by sweep speed. In the first regime, below ~350 cm/sec, the films were characterized by grain growth which followed the beam in a roughly linear fashion, broken up into numerous subgrain boundaries with typical dimensions of 10-50 μm width, but extending in places several hundred μm in length. Figure 3(a) is a Nomarski optical micrograph of such a sample after Secco etching. The dark spots along sub-grain boundaries are defects which often result in small protrusions into the capping layer. This topology suggests that small pockets of trapped liquid Si may solidify after the surrounding material. Electron channeling patterns show that in this regime the films regrow with an approximate <111> orientation perpendicular to the film.

The second regime of film growth was found for sweep speeds 350-500 cm/s. Parts of the grain growth in these films follow the beam, but at intervals new nucleation and grain growth appears, resulting in crystallization back toward the original front as well as a new front following the beam. Figure 3(b) shows a micrograph of such a sample. Several nucleation sites are visible in the micrograph, with more growth in each case along the direction of the beam sweep. These results imply that the crystallization speed in the film could not keep up with the beam sweep speed. As might be expected, electron channeling shows no preferred orientation of these grains, but rather a variety of orientations in each film. The defects along sub-grain boundaries are still present, with larger protrusions occurring along ridges where competing crystallization fronts meet. Another feature that begins to appear at higher sweep speed is an occasional small hole in the recrystallized films. The cause of these holes is not understood at present.

The density of holes in the Si film increases at higher sweep speed. This is a common feature in the third regime, corresponding to sweep speeds above ~ 500 cm/s. As the speed is increased, the films quickly become dominated by nucleations at intervals of ~ 100-200 μm, with little evidence of preferred growth along the sweep direction. Clearly, the sweep velocity is now much faster than the crystallization velocities along the film. Such a sample is shown in Fig. 3(c), in this case without a Secco etch because of the presence of holes in the Si layer.

The most notable feature of the films which depends on sweep speed is the distance between nucleations of new growth fronts. Below about 350 cm/s, the crystallization proceeds as far as the beam sweeps, with no new nucleations occurring. Above ~ 350 cm/s, nucleations appear at intervals which decrease

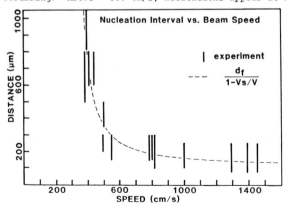

Figure 4. Nucleation intervals observed in SOI films treated at various sweep speeds. The dashed line is a prediction from Eq. 1, using d_f = 100 μm and V_s = 350 cm/s

with increasing sweep velocity. Figure 4 plots this dependence with data from several different samples. Each bar represents the observed range of nucleation spacings on one sample, swept at a particular velocity.

The dependence shown in Fig. 4 may be understood on the basis of a model shown schematically in Fig. 5. In part (a) the temperature profile of the Si layer is shown for a beam velocity less than 350 cm/s. The molten zone in the Si layer crystallizes at a point determined by the undercooling in the liquid, moving at a velocity V_s equal to the beam velocity V. The latent heat of fusion given up by the front has sufficient time to diffuse away in this case. As V is increased and V_s tries to keep up, however, the heat of fusion will start to limit the undercooling ahead of the front, and hence V_s. The heat of fusion will tend to collect ahead of the front because the thermal conductivity of liquid Si is ~6x that of solid Si and ~100x higher than that of SiO_2 [8]. (This is analogous to the situation in explosive crystallization of an amorphous film, where the heat of fusion from the crystallizing film moves ahead and melts the amorphous layer [9]). The magnitude of V_s as limited by the heat of fusion in a sample configuration such as this was roughly estimated at 100–500 cm/s using the ratios of thermal conductivity and finite-element heat flow calculations for bulk Si. This is the same order of magnitude as that indicated by the experiment (V_s = 350 cm/s). Although heat-flow calculations indicate V_s may be larger at larger V, we shall take the experimentally observed value for the following model.

As the crystallization front falls behind for V > V_s, the undercooling far ahead of the front will eventually reach a point sufficient for nucleation of a new crystallization front. This situation is shown in Fig. 5(b). The distance d_f ahead of the old front will be determined by the details of heat flow in the layer, including the heat of fusion from the old front. An analytical calculation of a moving heat source in liquid Si was used to estimate d_f ~100 μm. Given d_f and V_s, it is straightforward to show that the distance d_n between which nucleations occur is given by

$$d_n = \frac{d_f}{1 - V_s/V} \qquad (1)$$

The dashed line in Fig. 4 is a plot of eq. 1), using d_f = 100 μm and a constant V_s = 350 cm/s (which represents the experimentally observed value). The match of the curve shape to the data gives encouragement that the model is at least qualitatively correct. A more detailed numerical simulation is in preparation. An in-depth understanding of the crystallization of SOI films

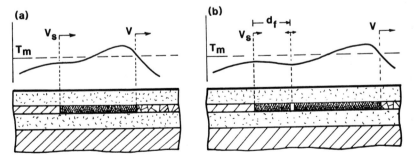

Figure 5. Schematics of the temperature distribution in a SOI film being treated by the scanning e-beam, for two cases: a) V_s = V, and b) V_s < V.

will help optimize sample configurations and beam parameters for uniform, reproducible growth.

CONCLUSIONS

With the present sample and beam configuration for unseeded SOI growth, we observe a maximum velocity of beam-driven crystallization of ~ 350 cm/s. A qualitative model has tentatively identified why this velocity is limited and predicted the spacing of nucleation in films swept faster. A more deed understanding of this process will be valuable in optimizing beam sweep speed and sample configurations for the fabrication of SOI films.

ACKNOWLEDGEMENTS

We are grateful to R. E. Asbill for technical assistance and to W. R. Anderson and D. M. Follstaedt for assistance with the electron microscope analysis. We are also grateful to W. R. Dawes for supplying the samples.

REFERENCES

1. Laser and Electron-Beam Solid Interactions and Materials Processing, J. F. Gibbons, L. D. Hess and T. W. Sigmon, eds. (North Holland, NY 1981).

2. Laser and Electron-Beam Interactions with Solids, B. R. Appleton and G.K. Celler, eds. (North Holland, NY 1982).

3. M. W. Geis, Henry I. Smith, B-Y. Tsaur, John C. C. Fan, E. W. Maby and D. A. Antoniadis, Appl. Phys. Lett. 40, 158 (1982).

4. J. A. Knapp and S. T. Picraux, J. Appl. Phys. 53, 1492 (1982).

5. J.A.Knapp, et al., p. 511 in ref.2.

6. F. Secco d'Aragona, J. Electrochem. Soc. 119, 948 (1972).

7. "Electronics", Vol. 55, p. 45, June 2, 1982.

8. J. Phys. Chem. Ref. Data, Vol. 3, Suppl. 1, 1974.

9. H. J. Leamy, W. L. Brown, G. K. Celler, G. Foti, G. H. Gilmer, and J. C. C. Fan, Appl. Phys. Lett. 38, 137 (1981).

CHARACTERISTICS OF RECRYSTALLISED POLYSILICON ON SiO_2 PRODUCED BY DUAL
ELECTRON BEAM PROCESSING

J.R. DAVIS, R.A. McMAHON AND H. AHMED
Cambridge University Engineering Department, Trumpington Street,
Cambridge, England, CB2 1PZ

ABSTRACT

This paper describes the production of large areas of
precisely oriented, defect-free, single crystal silicon films
on SiO_2 by dual electron beam heating of deposited polysilicon
using lateral epitaxy. Defects which occur in the film far
from the seeding window are characterised, and the dependence
of the area of the defect-free region on the processing
conditions is discussed.

INTRODUCTION

The conversion of polysilicon layers on insulators to single crystal Si
would be an important technological advance (1). In this paper we report the
production of extended areas of precisely oriented, defect-free single crystal
films by the lateral seeding technique (2), and using a dual electron beam
processing system (3,4). This system, with its precisely controllable high
power beams, which have good energy coupling to the silicon, allows us to
investigate a variety of heating techniques. The two most commonly reported
recrystallisation systems employ focussed laser beams or large graphite strip
heaters. In the former system (2), which has been shown to be capable of
producing small islands of defect free silicon, the specimen is heat-sinked at
a relatively low temperature, creating large thermal gradients from the laser
spot to the backside of the wafer. Because the small spot of a low power beam
has to be rastered over the surface of the cool specimen, the laser system has
a poor wafer throughput. The graphite strip heater (5), on the other hand, has
a high throughput and does not produce large temperature gradients, but it does
hold the substrate at high temperatures for long enough to cause changes to any
previously processed substrate structures. The characteristic low angle grain
boundaries in the recrystallised silicon do not prevent the fabrication of MOS
transistors, but the material would be unacceptable in VLSI circuits where
leakage and lifetime are critical.

TECHNIQUES

The specimen is supported in thermal isolation so that both its surfaces are
accessible to electron beams of 20-30keV energy. One beam provides isothermal
heating of the specimen by multiply scanning its back surface. A total power
of 500 W is available in the bottom beam, which can bring the specimen temperature to a radiation-limited thermal equilibrium of, typically, $1100°C$ in about
4 seconds. The upper beam provides the small additional amount of heat,
required to give localised melting of the polysilicon film, by means of a line
beam (3,4). In one method of forming the line-beam, a line source of electrons
is imaged directly onto the specimen. In other methods the emission from a
point source gun may be either focussed into a small circular spot which is
rapidly deflected in one direction by a triangular waveform to time-synthesise a
line, or it may be shaped by electromagnetic optics into a stationary line spot
(4). With the latter two methods the size of the line beam was typically

Mat. Res. Soc. Symp. Proc. Vol. 13 (1983) © Elsevier Science Publishing Co., Inc.

564

3mm x 100μm, and the beam power just to cause melting was 15-30 W. The molten
zone may be swept across the specimen surface either by electromagnetically
scanning the line beam in a direction approximately perpendicular to its long
dimension, or by moving the specimen on a motorised stage through the stationary
beam.

Both (111) and (100) orientation substrates have been used in this work, and
in the latter case the long seeding grooves have been aligned in the <100> and
<110> directions. Nonrecessed, steam-grown isolation oxides from 0.2μm to 2μm
thick were covered by LPCVD polysilicon films of between 0.5 and 1μm thick. No
attempt was made to profile the oxide steps at the edges of the seeding windows,
but the wafers were given a short dip in HF to clean the window areas immediately prior to the polysilicon deposition. After deposition of the undoped CVD
capping oxide the wafers were annealed at 1000°C in nitrogen before being cut
into 10 x 10mm chips for beam processing.

RESULTS AND DISCUSSION

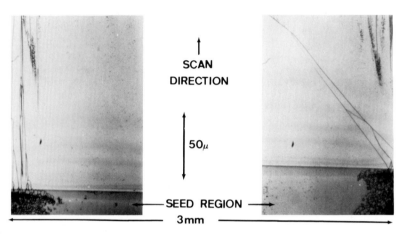

SCAN
DIRECTION

50μ

◄——— SEED REGION ———►
◄——————————— 3mm ———————————►

Figure 1. Nomarski micrograph of part of a Secco etched recrystallised
stripe seeded from a (100) substrate.

Fig. 1 is a Nomarski micrograph of part of a typical recrystallised stripe
after removal of the capping oxide and Secco etching to reveal any defects.
The polysilicon has grown epitaxially on the substrate in the seed window and
the single crystal growth continues up the step and across the isolating oxide.
At the extreme edges of the stripe the line beam power is insufficient to melt
the polysilicon in the seed-window through to the substrate, and so grain boundaries propogate over the oxide, but seeding is generally uninterrupted for the
remainder of the stripe width. At some distance from the seed edge defects are
observed at a gradually increasing density, until the film degenerates into an
array of subgrains. The distance which single crystal propagates before the
onset of defects, and the characteristics of the defects, are functions of the
specimen preparation and the processing conditions. Fig. 2(a) shows a typical
pattern of feather-shaped defects for silicon seeded from (100) substrates; in
this case the seed edges and the long dimensions of the beam are aligned in the
<110> direction. These defects, which are shown in the transmission electron
micrograph of Fig. 2(b) to comprise arrays of dislocations, eventually coalesce
to form low angle grain boundaries running at approximately 45° to the scan

direction, that is along <001>. The misorientation of the sub-grains, as
determined by their selected area diffraction patterns is $\lesssim 3^{\circ}$.

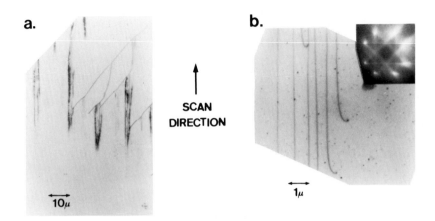

SCAN
DIRECTION

10μ

1μ

Figure 2. Optical (a) and TEM (b) micrographs of typical defects in silicon
recrystallised with a (100) substrate and <110> seed windows.

Another form of defect is occasionally observed with the (100)/<110> seeding
combination. This can be seen in Fig. 3(a) as a line of low contrast (even
after etching) running parallel to the seeding edge, and often extending for
hundreds of microns. Selected area diffraction patterns either side of the
line show it to be caused by a small tilt of the <100> axis about the perpendicular to the specimen surface. The TEM of a similar region (Fig. 3(b)) shows
that the line initiates a row of stacking faults. The effect occurs independently of the usual 'feather' defects, and is also sometimes observed in subgrains near to the next seed window; it seems likely that it is caused by the
stress induced by mass flow near the oxide edges.

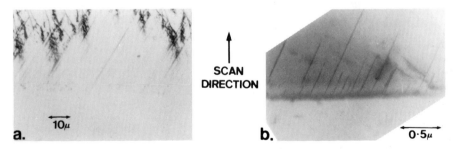

SCAN
DIRECTION

10μ

a.

b.

0·5μ

Figure 3. Optical (a) and TEM (b) micrographs of defects occasionally
observed running parallel to the seeding edge with (100)/<110> substrates.

Recrystallised films seeded from (111) substrates have the characteristic
defects shown in Fig. 4, consisting of a scallop-shaped line of dislocation
tangles parallel to the seed-edge and coalescing into sub-boundaries which run
in the scan direction (along <1$\bar{1}$0>). In general, the density of sub-boundaries
is higher than for equivalent (100) substrates, and they also appear at a smaller distance from the seed edge (typically 20-30µm).

The mechanism responsible for the onset of dislocations at a characteristic
distance from the seed edge is still not fully understood. Lam (2) has suggested that the effect is caused by the stress created in the film as it

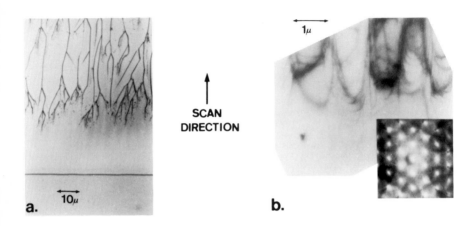

SCAN
DIRECTION

a.   10µ   b.

Figure 4. Optical (a) and TEM (b) micrographs of typical defects in
recrystallised polysilicon seeded from (111) substrates.

expands on solidification. We have investigated this possibility by recrystal-
lising some samples which have had the capping oxide removed, so that the solidifying film is free to expand vertically. The window of usable beam powers for
uncapped specimens is very narrow due to agglomeration of the polysilicon, but
when seeded recrystallisation does occur the width of the defect-free region is
the same as that for similar specimens with the cap left on. This result
suggests that if stress is the cause of the defects then it is that in a horizontal plane at the Si/SiO$_2$ interface due to the difference in linear expansion,
rather than in a vertical plane due to the silicon volume change.

Our best results, in which defect-free regions of 100-150µm width were regularly achieved, were obtained on (100) substrates with 1µm of underlying
(isolating) oxide. The increased thermal impedance of thicker (2µm) oxides
often causes delamination when the beam power is high enough to melt the polysilicon in the seed-windows, and 0.2µm and 0.5µm oxides produced dislocation-
free regions approximately 30 and 50µm wide respectively. This dependence of
the onset of defects on the oxide thickness, if confirmed, would suggest that
the defects occur when the direction of heat loss from the molten zone becomes
predominantly downwards through the underlying oxide, rather than backwards to
the seed region through the just-solidified silicon film. This change creates
a vertical temperature gradient in the film, and allows nucleation to occur at
the Si/SiO$_2$ interface rather than in the bulk of the film.

With graphite strip heaters, the distance between sub-boundaries increases as the speed at which the molten zone is scanned is decreased. For line beam scan speeds in the range 1 to 100 mm/sec, however, we have found no variation in the width of the defect-free region, and hence we prefer to operate at high sweep speeds where interaction of the two beam powers is reduced, and control of the process is easier.

For laser recrystallisation with heat-sinked specimens, it has been reported (1) that the quality of the recrystallised film improves with increasing substrate temperature. In contrast, we have varied the background temperature from 500°C to 1300°C without producing any systematic changes in the width of the defect-free region. The optimum substrate temperature in our system is determined by the production of slip lines at low substrate temperatures (which require correspondingly high line beam powers) and by localised melting of the substrate below the isolating oxide at high substrate temperatures.

Results obtained with the shaped line-spot (4) and the time synthesised line are, in general, very similar. In the time-synthesised method, the frequency of the oscillation has to be high enough to prevent the polysilicon solidifying between passes of the beam without requiring too high a transient temperature under the circular spot. It was found that a 100 KHz triangular wave was sufficient to create a continuous molten line of polysilicon; this relatively low frequency compared to the calculations of Schiller (6) is due to the high background temperature with the concomittant low thermal conductivity of silicon. There are, however, significant fluctuations in local temperature as the beam traces its oscillating path, particularly at the sides of the line. These fluctuations create greater strain in the capping layer, and so increase the tendency for the polysilicon to delaminate by forming small holes, usually along the seed window edge, thus preventing seeded growth on the oxide. It is expected that this problem will be overcome by using higher frequencies and by better control of the deposition conditions.

To be usable as a substrate for VLSI processing, the recrystallised films must have a surface topography comparable to a normal wafer, and no worse than that of a finished integrated circuit. Fig. 5 is a Nomarski micrograph of a 1° bevel through a recrystallised stripe and shows that the surface does indeed remain flat except for a slight ripple at the seed edges. This 'snowplough' effect is caused by the expansion of the film on solidification, and the extra heat sinking which is caused by the exposed substrate. In these specimens the width of the seed region is similar to that of the line beam (100µm). Narrowing the seed region to a fraction of the beam width, which is also desirable

Figure 5. Nomarski micrograph of a 1° bevel through a recrystallised stripe.

568

to conserve chip area, will reduce its heat sinking effect, and so should
result in even flatter surfaces.

CONCLUSIONS

   We have demonstrated that a dual electron beam processing system can produce large areas of defect-free, precisely oriented, single crystal silicon
films without requiring coplanar surfaces. The extent of crystallisation
before dislocations intrude is remarkably insensitive to the beam processing
conditions, but is a poorly characterised function of the wafer preparation
procedure. It is believed that better control of the substrate preparation
conditions will allow fuller exploitation of the versatility of the dual-beam
system to optimise the characteristics of the recrystallised films.

ACKNOWLEDGEMENTS

   We would like to thank J.D. Speight and J.C. Eustace of British Telecom
Research Laboratories for providing the polysilicon specimens. J.R. Davis
and R.A. McMahon acknowledge a Research Scholarship and a Research Fellowship
from British Telecommunications and GEC plc, respectively.

REFERENCES

1. H.J. Leamy in Laser and Electron Beam Interactions with Solids,
    B.R. Appleton and G.K. Celler eds. (Elsevier North Holland, New York,
    1982), 459.

2. H.W. Lam, R.F. Pinizzotto and A.F. Tasch, J. Electrochem Soc. 128,
    1981 (1981).

3. R.A. McMahon, J.R. Davis and H. Ahmed in Laser and Electron Beam
    Interactions with Solids, B.R. Appleton and G.K. Celler eds (Elsevier
    North Holland, New York, 1982) 783.

4. J.R. Davis, R.A. McMahon and H. Ahmed, Electronics Lett., 18, 163 (1982).

5. B-Y Tsaur, J.C. Fan, M.W. Geis, D.J. Silversmith and R.W. Mountain,
    Appl. Phys. Lett. 39, 561 (1981).

6. S. Schiller, S. Panzer and R. Klabes, Thin Solid Films, 73, 221 (1980).

LINEAR ZONE-MELT RECRYSTALLIZED Si FILMS USING INCOHERENT LIGHT

AVID KAMGAR, G. A. ROZGONYI* AND R. KNOELL
Bell Laboratories, Murray Hill, New Jersey

ABSTRACT

Incoherent tungsten radiation has been used to recrystallize polysilicon films deposited on SiO_2, grown on 75 mm diameter wafers. These films have been analyzed by a variety of techniques such as optical microscopy, Auger spectroscopy and TEM cross-sectioning. These films are large (> 50 mm dia.) single crystals, and contain fewer defects and impurities than similar films recrystallized in graphite strip heater ovens. They do, however, contain the subgrain boundaries found previously, even when high purity MBE polysilicon is used. In vertical TEM cross-sectioning, in addition to the subgrain boundaries, we have seen crystallized particles of 50-200 Å size in the Si film at both the upper and lower SiO_2 boundaries. Although these particles could be related to formation of the subgrain boundaries, no definite correlation has been made.

INTRODUCTION

Potential advantages of silicon films on insulating substrates (SOI) as an alternative to silicon on sapphire (SOS) for fabricating high speed, radiation hard circuits have created much interest in the SOI technology. Many laboratories are investigating the recrystallization of Si films using various techniques such as lasers [1], electron beams [2] graphite strip heaters [3-5] and incoherent light [6-8]. Recently we reported successful recrystallization of large area single crystal films on oxidized three inch wafers using tungsten lamps [6]. Here we will present additional structural and chemical analysis of these films using optical microscopy, Auger spectroscopy, and horizontal and vertical cross-sectional transmission electron microscopy (TEM) in a Philips model 400 equipped with EDAX (energy dispersion analysis by x-ray).

EXPERIMENT

Figure 1(a) illustrates the three main parts of the system. The wafer is housed in a rectangular quartz tubing which is flushed by argon. A high intensity panel light, employing six tungsten lamps, heats the wafer from the bottom to 1000-1300°C. A line-heater provides the additional heat, from the top, for melting the polysilicon. The line-heater consists of a tungsten lamp and two elliptical reflectors which focus the light into a narrow strip. In this configuration the line heater is drawn by a motor, at the desired

---

*   North Carolina State University, Raleigh, NC.

Mat. Res. Soc. Symp. Proc. Vol. 13 (1983) ©Elsevier Science Publishing Co., Inc.

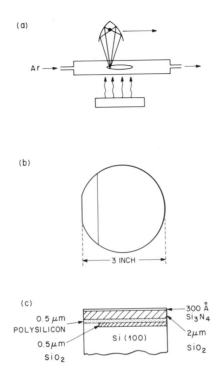

(a)

Ar →

(b)

⟵ 3 INCH ⟶

(c)

0.5 μm
POLYSILICON

0.5 μm
SiO₂

Si (100)

300 Å
Si₃N₄

2μm

SiO₂

Fig. 1   Schematic view of the recrystallization system (a), and the multi-layered wafer
(b) and (c).

speed, to drag the molten zone from the single crystal edge of the wafer to the end. A detailed description of the system can be found in Ref. [6].

The multi-layered structure was fabricated on 75 mm diameter (100) Si wafers as shown schematically in Figs. 1(b) and (c). The planar structure was obtained by using Si_3N_4 as an edge mask while ~1 μm of thermal oxide was grown, and thinning the oxide to 0.5 μm. Stripping the Si_3N_4, depositing 0.5 μm of LPCVD polysilicon, and then encapsulating [9] with 2 μm of CVD SiO_2 and 300Å of LPCVD Si_3N_4 completed the sequence.

In the recrystallization process an elongated molten zone (~ 4mm width at its center) was initially created in the seed area of the wafer. The molten zone then traversed at a speed of around 0.3 mm/sec and recrystallized a large area (>50 mm dia.) of the wafer. A small temperature nonuniformity across the wafer prevented melting the whole polycrystalline film. This nonuniformity is in part due to a more efficient radiative cooling around the perimeter of the wafer, and partly due to the artifact that the width of the bottom lamp unit was less than 75 mm.

RESULTS

After removal of the dielectric encapsulating layers, the films were analyzed using Nomarski contrast optical microscopy. The surface appeared mirror smooth, without any protrusions, unlike those reported to exist on films recrystallized using a graphite strip heater [5] and [10]. A Schimmel etch [11], however, revealed the existence of subgrain boundaries much like those observed in previously reported SOI films [3-7]. A micrograph of the wafer near the seed area is shown in Fig. 2. The seed area can be seen at the bottom of the figure, above which there is a transition region containing interference fringes due to a thickness change at the seed boundary. The subgrain boundaries with their familiar wish-bone structure start at about 50 μm from the seed boundary. These boundaries run generally in the scan direction, and are on the average about 20 μm apart. Note that the scan direction is from bottom to top in this figure.

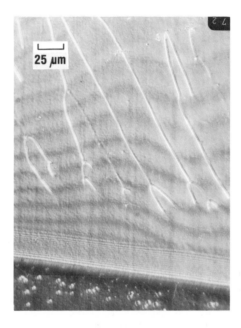

Fig. 2   Nomarski optical micrograph of a wafer near the seed area.

We have studied the films both in horizontal and vertical TEM cross-sections. The horizontal TEM micrographs revealed the subgrain boundaries having a structure much like those reported in Ref. 4. Regions between the subgrain boundaries were defect free. The vertical cross-sectioning, however, revealed not only the subgrain boundaries, an example of which is shown at the center of Fig. 3, but also small precipitates at each oxide-Si film interface. Note that the boundary of the silicon substrate and the thermally grown oxide appears perfectly smooth, while at the boundaries of the regrown Si film with its two neighboring SiO_2 layers, a high density of precipitates is observed. Figure 4 shows an enlarged view of another subgrain boundary, wider than the previous one, with its characteristic dislocation network. The crystallographic misorientation between the two sides of the subgrain boundary in Fig. 4, as indicated by the larger variation in the electron channeling contrast of the two sides, is greater than that of the two sides of the boundary in Fig. 3. This is in agreement with the wider dislocation network of the boundary in Fig. 4 as compared with that of Fig. 3.

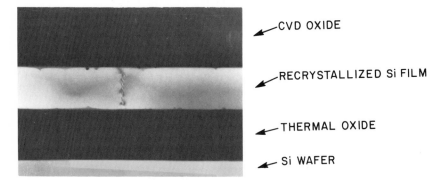

CVD OXIDE

RECRYSTALLIZED Si FILM

THERMAL OXIDE

Si WAFER

Fig. 3   TEM vertical cross-sectional view of the multi-layered structure after the recrystallization.

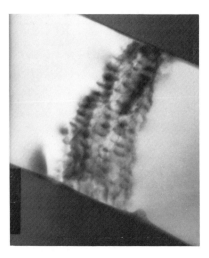

Fig. 4   TEM micrograph illustrating the dislocation network in a subgrain boundary in the vertical cross-sectioning.

Figure 5 is an enlarged TEM micrograph of one of the pocket-like structures seen in Fig. 3. The spots at the center are two crystallized particles with 50-200Å dimension. These particles seem to be embedded in a pocket of amorphous SiO_2 which extends into the Si film. The pockets at the film-SiO_2 boundary all have similar structures, but they contain different numbers of particles varying from one to three or more. Selected area electron diffraction patterns of these particles show that they are single crystals, but due to their small size we have not yet been able to identify their nature. We suspect that these particles may be crystallized SiO_2. However, more detailed converging beam electron diffraction analysis is needed for a definite identification of them.

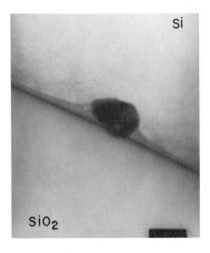

Fig. 5   A pocket-like extension of amorphous SiO_2 into the recrystallized film containing two crystallized particles.

Auger spectroscopy of the subgrain boundaries and the nodes at which these boundaries meet did not reveal any detectable impurities. This is in contrast to the results of the strip heater recrystallized films of Ref. [5] where a variety of impurities were found, or the results reported in Ref. [4] concerning carbon impurities. Energy dispersive X-ray analysis of the particles in cross-section, and horizontal TEM of the nodes did not identify any elements above mass 23. The fact that we were unable to identify impurities in our recrystallized films is consistent with our use of a clean lamp source and a flowing argon ambient.

We performed another experiment in order to rule out the impurity segregation as a cause for the formation of subgrain boundaries [5]. This was to replace the usual LPCVD polysilicon by molecular beam epitaxy, MBE, deposited polysilicon. This film was free of contaminant to .01 ppm, yet, after recrystallization, it also contained the subgrain boundaries as well as the particles.

SUMMARY AND CONCLUSIONS

We have made detailed structural and chemical analysis of the silicon films recrystallized in a linear zone-melt radiation oven. We found that these films had mirror smooth surfaces without any protrusions. They had no large angle grain boundaries. They were free of impurities, however, they did contain the low angle grain boundaries. The films also contained small (50–200Å) particles at the two boundaries with the SiO_2 layers. The highly pure recrystallized films deposited by MBE also contained both the subgrain boundaries and the particles. Hence, we conclude that impurity segregation is not likely to be responsible for the formation of low angle grain boundaries. We speculate that structural inhomogeneities at the two Si film-SiO_2 boundaries are the more likely causes for such structural defects as the subgrain boundaries or the particles in the Si film. This speculation is substantiated by the absence of particles at the boundary of the thermally grown SiO_2 and the Si bulk, see Fig. 3.

ACKNOWLEDGEMENTS

We wish to thank S. M. Sze for his continued support and interest, E. Labate for his technical assistance, C. C. Chang for the Auger Spectroscopy, E. Povilonis for the preparation of the wafers and J. C. Bean for the MBE deposited polysilicon films.

REFERENCES

[1]   See for example: "Laser and Electron-Beam Interactions with Solids" Editors B. R. Appleton and G. K. Celler (North-Holland, NY 1982).

[2]   T. I. Kamins and B. P. Von Herzen, IEEE Electron Device Lett. *EDL-2*, 313 (1981).

[3]   See: M. W. Geis, H. I. Smith, B-Y. Tsaur and J. C. C. Fan, Appl. Phys. Lett. *40*, 158 (1982), and references therein.

[4]   R. F. Pinizzotto, H. W. Lam, and B. L. Vaandrager, Appl. Phys. Lett. *40*, 388 (1982).

[5]   H. J. Leamy, R. A. Lemons, C. C. Chang, H. Baumgart, J. Cheng and E. Lane, Mat. Lett. *1*, 33 (1982).

[6]   A. Kamgar and E. Labate, Mat. Lett. Nov. 1982 (in press).

[7]   D. P. Vu, M. Haond and D. Bensahel, Jour. Appl. Phys. Dec. 1982 (in press).

[8]   G. K. Celler, M. D. Robinson and D. J. Lischner, Appl. Phys. Lett., Jan. 1983 (in press).

[9]   E. W. Maby, M. W. Geis, T. L. LeCoz, D. J. Silversmith, R. W. Mountain, and D. A. Antoniadis, IEEE Elec. Der. Lett. *EDL-2*, 241 (1981).

[10] J. C. C. Fan, B-Y. Tsaur, R. L. Chapman and M. W. Geis, Appl. Phys. Lett. *41*, 186 (1982).

[11] D. G. Schimmel, Jour. Electrochem. Soc. *126*, 479 (1979).

LATERAL EPITAXIAL GROWTH OF THICK POLYSILICON FILMS ON OXIDIZED 3-INCH WAFERS

G. K. CELLER, McD. ROBINSON, D. J. LISCHNER*, and T. T. SHENG
Bell Laboratories, Murray Hill, New Jersey, 07974, USA
*Bell Laboratories, Allentown, Pennsylvania 18103, USA

ABSTRACT

Recently we reported the first successful crystallization from the melt of 15μm thick Si films on SiO$_2$. In this technique of Lateral Epitaxial Growth over Oxide (LEGO), crystallization is initiated at seeding vias in SiO$_2$ and propagates over the amorphous insulator, resulting in high structural quality of the grown film. Uniform and bow-free crystallization of complete 3-inch wafers occurs in <60 sec in a special furnace, with wafers placed between a bank of high intensity tungsten halogen lamps and a water cooled base.

In this paper we provide further details of crystallization procedure and new structural characterization data. The mechanism of LEGO crystallization and its limitations are also discussed.

INTRODUCTION

Crystallization of *thin* Si films on insulating substrates has been investigated by several groups [1,2]. Lasers, electron beams, and strip heaters have been successfully employed to obtain crystalline Si films, marred only by the presence of low angle grain boundaries (sub-boundaries). We recently demonstrated the first successful seeded crystallization of *thick* Si films over SiO$_2$ [3]. The films were free of sub-boundaries and contained only a low density of dislocations. In this paper we provide a more extensive description of the crystallization process, review structural characterization data, and discuss the physical limitations of the epitaxial overgrowth.

The method of Lateral Epitaxial Growth over Oxide (LEGO) is based on uniform melting of silicon layers deposited over amorphous insulators, such as SiO$_2$, with an extended radiative heat source. Periodic openings in the insulator link deposited Si with the crystalline silicon substrate. These openings or vias are critical to the process; they serve as seeding regions for epitaxial crystallization, but also impose lateral temperature gradients necessary for propagation of the solid-liquid interface. Typically, Si over the entire 3-inch wafer is melted simultaneously, in a matter of seconds, and then allowed to recrystallize over 10 - 100 sec period. The resultant films are single crystalline, free of *any* grain boundaries, and contain only moderate densities of dislocations and stacking faults. Lack of low-angle grain boundaries in LEGO experiments is in a distinct contrast to strip heater results [2] where lineage is invariably observed, and attempts to eliminate it have been so far futile.

EXPERIMENTAL DETAILS

The sample furnace has been described before [3,4]. Briefly, it consists of two rectangular chambers, with lateral dimensions 10 x 12.5 inches, positioned one above the other. The upper chamber contains a bank of air-cooled tungsten halogen lamps suspended below a gold plated reflector. Two quartz windows separate the lamps from the wafer chamber. Wafers are placed on quartz pins ~0.5 in. above the water cooled aluminum base. Uniform heating of the top surface with intense photon flux and equally uniform radiative cooling of the back surface cause a vertical temperature gradient of ~100°C/cm. This allows controlled melting of the films deposited on Si wafers. An approximately planar melt front penetrates into the wafer to a depth determined by the radiative flux and the length of exposure. Cooling is predominantly radiative, with convection significant only at T < 700°C. The lamps are ramped up in power linearly over 10 sec, held at a constant output for 10 to 100 sec and ramped down with a predetermined slope. Power densities of 60 to 90W/cm² were found appropriate for melting. Processing is usually done in oxygen atmosphere.

Fig. 1. Pattern defined in oxide before silicon deposition.

Sample preparation involved oxidation, photolithography and wet etching, one- or two-step Si deposition and encapsulation with a second oxide film. First, $2\mu m$ of steam oxide was grown on float zone (100) Si wafers and patterned into rectangular islands, as shown in Fig. 1. In the initial experiments, Si was deposited on the wafers in two steps. $1.4\mu m$ of epitaxial Si was deposited selectively in the openings in the SiO_2 by reaction of $SiCl_4$ and H_2 at 1150°C [5], after which temperature was reduced to 1050°C and flow of $SiCl_4$ was replaced with SiH_4. The desired thickness of silicon was grown in these conditions over the entire wafer, at a rate of $\sim 0.5\mu m/min$. In further experiments selective filling of the moats with crystalline Si was found unnecessary and was eliminated. In all cases, because of the high deposition temperature, Si grew epitaxially over the exposed parts of wafer, in contrast to LPCVD deposition, where polycrystalline layers are usually formed. The surface of the polycrystalline regions was microscopically rough, whereas the the epitaxially grown regions were mirror-like. To complete the structure, a $2\mu m$ layer of LPCVD oxide was deposited over the whole wafer. The microstructural features of an as-deposited wafer are shown in a cross-sectional optical micrograph in Fig. 2(a). The sample has been Schimmel etched [6] for 10 sec to delineate the grain boundaries and other extended defects.

RESULTS AND DISCUSSION

At 78W/cm² the entire deposited film and some underlying substrate melted. The substrate solidifies first, as soon as the energy influx is reduced below the radiative loss from the wafer. Crystallization of deposited silicon starts at the openings in oxide and then propagates laterally over the SiO_2. In Fig. 2(b), a cross-section of a recrystallized sample is shown, prepared in exactly the same way as that of Fig.2(a). This Nomarski contrast optical micrograph reveals the excellent quality of the recrystallized Si; no grain boundaries of any kind are present and only a few dislocations. From the same Figure it can be deduced that the melt penetrated $\sim 10\text{-}15\mu m$ into the substrate, as delineated by a slight discoloration edge. The boundary marks a discontinuity in doping density between the recrystallized silicon and the unmelted part of the substrate.

A detailed examination of recrystallized cross-sections with a transmission electron microscope (TEM) confirmed the absence of any grain boundaries in melt-processed films. In Fig. 3, a cross-sectional TEM micrograph includes Si on both sides of the buried SiO_2. No defects are detected in this segment, located $\sim 50\mu m$ away from the nearest seeding moat. In an electron diffraction pattern, in the inset, features from all three regions are superimposed. It contains a halo caused by amorphous SiO_2 and *one* set of diffraction spots, in evidence of the epitaxial nature of the regrown Si layer.

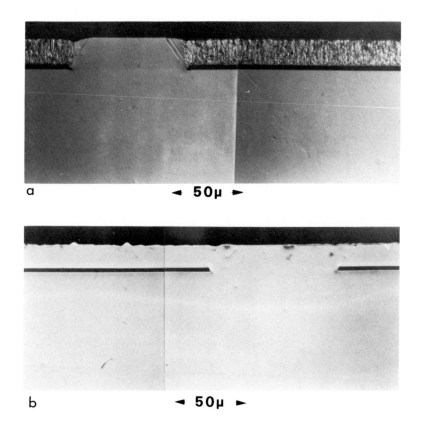

Fig. 2. Optical micrographs of two wafer cross-sections after 10 sec Schimmel etching; (a) as deposited sample, (b) after lamp melting and recrystallization.

In Fig. 4, another cross-sectional TEM micrograph is shown, representing the area at and near the opening in SiO_2. Again it is evident from the extinction contours that Si is single crystalline on both sides of the oxide. No extended defects are detectable. One unexpected feature is the presence of small voids right under the edge of the oxide window, on the substrate side.

Typical bow in 3-inch wafers after recrystallization was $<20\mu m$. This slight distortion is caused by a temperature drop at the edges, where radiative cooling is more efficient. For the same reason a 1-2mm zone along the wafer perimeter is often incompletely recrystallized.

Epitaxial overgrowth requires lateral temperature gradients locally, to drive the solidification front from the seeds and across the sample. In many recent papers [1,2] suitable temperature profiles were achieved with scanned heat sources, such as laser beams and wire-like heating strips. In our approach the heat source is stationary and

578

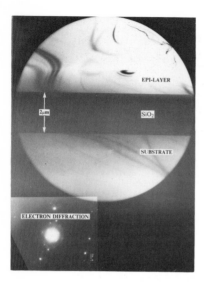

Fig. 3. TEM cross-section of recrystallized Si film. An electron diffraction pattern, in the inset, was integrated over the three regions shown in the micrograph.

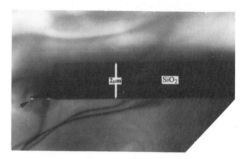

Fig. 4. TEM cross-section of recrystallized Si near a via. An arrow points to the voids.

uniform, and the silicon-insulator structure provides the necessary lateral temperature gradients. During melting the lamp power output is kept at a constant peak value for several seconds, after which it is slowly ramped down to assure that cooling is dominated by radiation from the back surface. Heat loss from the molten zone is higher through the vias than through the oxide islands, causing the solidification front to move upward through the vias, then laterally over the oxide islands. When it reaches the top surface in the vias, the resulting increase in emissivity further enhances the lateral temperature gradients.

In our experiments, films as thin as $7.5\mu m$ and as thick as $150\mu m$ have been transformed into single crystalline. We were also able to grow epitaxially more silicon on top of recrystallized films, while preserving its structural quality. Contrary to our initial expectations, the extent of lateral crystal growth is not limited by competitive random nucleation of polycrystals. We have never observed partial lateral epitaxy, i.e., seeded growth over some distance with polysilicon covering the rest of the oxide island. Instead, the main process limitation is redistribution of Si during solidification, away from the seeding regions, that causes thinning of the film at the edges of oxide pads and in extreme cases exposes SiO_2. The excess Si forms an elevation near the center of the island. It is interesting to note that the oxide cap accommodates these macroscopic changes of topography without breaking. On the other hand, the microscopic roughness of the as-deposited Si is replicated in the oxide cap and is preserved through the process. Clearly, at the melting temperature the SiO_2 is softened sufficiently to allow for macroscopic changes in topography, but not enough to affect its micro-texture.

An angle lapped section through the centers of several islands is shown in Fig. 5, where thickness variations are enhanced 10 times by the lapping angle. This allows resolving a slight distortion of the buried oxide, in addition to the silicon redistribution. An angle lapped section through the center of one $700\mu m$ island is shown in Fig. 6. The surface was Schimmel etched to show an array of dislocations right under the peak, where solidification

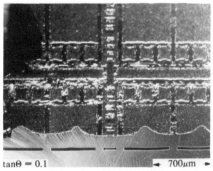

Fig. 5. Angle-lapped sample after recrystallization at 78W/cm².

Fig. 6. Angle-lapped and Schimmel etched center of a recrystallized island.

fronts approaching from four sides, meet. Inferior material quality at the intersection of colliding lateral growth fronts has been reported previously in laser [7] and strip heater [8] crystallization experiments. These zones with a high density of dislocations probably cannot be entirely eliminated, but their location is easily controlled and can be limited to a small fraction of the wafer area.

A top view of much larger recrystallized islands, in Fig. 7, also illustrates the effect of redistribution. The pattern of islands in this wafer resembled a checkerboard, with $3.5\mu m$ thick square pads of oxide. A $30\mu m$ layer of deposited Si was recrystallized over these insulating pads, and a silicon spike is seen in the center of each square.

Results in Figs 2-5 were for samples crystallized at 78W/cm². This was about 10% above the threshold power, causing the melt front to penetrate into the substrate below the patterned oxide. Fig. 8 shows the microstructure obtained at 67W/cm², just below the threshold for epitaxial recrystallization. The polysilicon was transformed into very large grains penetrating the entire film, indicating complete melting of Si over the oxide. The epitaxial Si over the via, however, shows no signs of having been molten, judging from the

Fig. 7. Top view of a recrystallized checkerboard sample.

Fig. 8. Cross-section of an incompletely recrystallized sample, after 60 sec etching.

580

stacking faults and surface topography characteristic of the CVD epi/poly interface. It appears that a narrow zone of polysilicon adjacent to the via also remained solid through the heating cycle, preventing any seeded regrowth.

At a higher power setting of 74W/cm^2, complete crystallization occurred with only partial melting of the epitaxial regions over the vias and without any melting under the SiO_2. In this case there was little Si redistribution. The power window of epitaxial regrowth without melting into the substrate is, however, very narrow and easily shifted by wafer-to-wafer variations of surface emissivity. Therefore, for consistent crystallization higher powers were used and some excessive melting was tolerated.

CONCLUSIONS

Single crystalline Si films were formed on the SiO_2, free of any grain boundaries, coherent or incoherent, and with a low density of other extended defects. The crystallization process requires melting of the entire deposited polysilicon film with a stationary extended radiative heat source. Heating must be uniform laterally but allow for a vertical temperature gradient. The lateral gradients necessary for seeded crystallization are imposed by the pattern defined in the buried oxide. The method is most suitable for relatively thick films, in the 10-200μm range. Thinner films also recrystallize as single crystalline, but their continuity can be disrupted by redistribution of silicon away from the seeding areas.

ACKNOWLEDGMENTS

Technical assistance of R. F. Benjamin, R. A. Cirelli, R. L. MaSaitis, and L. E. Trimble is gratefully acknowledged.

REFERENCES

[1] *Laser and Electron Beam Interactions with Solids,* B. R. Appleton and G. K. Celler, eds. (North Holland, New York 1982).

[2] *Laser-Solid Interactions and Transient Thermal Processing of Materials,* J. Narayan, W. L. Brown, and R. A. Lemons, eds., (North Holland, New York 1983).

[3] G. K. Celler, McD. Robinson, and D. J. Lischner, Appl. Phys. Lett. **42,** (Jan. 1, 1983).

[4] D. J. Lischner and G. K. Celler, in *Laser and Electron Beam Interactions with Solids,* B. R. Appleton and G. K. Celler, eds. (North Holland, New York 1982), pp. 759-764.

[5] P. Rai-Choudhury and D. K. Schroder, J. Electrochem. Soc. **118,** 107 (1971).

[6] D. G. Schimmel, J. Electrochem. Soc. **126,** 479 (1979).

[7] M. Tamura, H. Tamura, M. Miyao, and T. Tokuyama, Jap. J. Appl. Phys. **20,** Suppl. **20-1,** pp. 43-48 (1981).

[8] J. C. C. Fan, B.-Y. Tsaur, and M. W. Geis, Appl. Phys. Lett. **39,** 308 (1981).

CRYSTALLIZATION OF SILICON FILMS ON GLASS: A COMPARISON OF METHODS*

ROSS A. LEMONS*, MARTIN A. BOSCH**, AND DIETER HERBST**
*Los Alamos National Laboratory, MS D429, Los Alamos, NM 87545
**Bell Laboratories, Holmdel, NJ 07733

ABSTRACT

The lure of flat panel displays has stimulated much research on the crystallization of silicon films deposited on large-area transparent substrates. In most respects, fused quartz is ideal. It has high purity, thermal shock resistance, and a softening point above the silicon melting temperature. Unfortunately, fused quartz has such a small thermal expansion that the silicon film cracks as it cools. This problem has been attacked by patterning with islands or moats before and after crystallization, by capping, and by using silicate glass substrates that match the thermal expansion of silicon. The relative merits of these methods are compared. Melting of the silicon film to achieve high mobility has been accomplished by a variety of methods including lasers, electron beams, and strip heaters. For low melting temperature glasses, surface heating with a laser or electron beam is essential. Larger grains are obtained with the high bias temperature, strip heater techniques. The low-angle grain boundaries characteristic of these films may be caused by constitutional undercooling. A model is developed to predict the boundary spacing as a function of scan rate and temperature gradient.

INTRODUCTION

Two primary applications for crystallized silicon on amorphous substrates are dielectrically isolated devices and matrix displays. For dielectric isolation, most of the work has been done on thermally oxidized silicon wafers. This substrate is fully compatible with integrated circuit processing and eliminates thermal expansion problems. However, for the display application, silicon wafers are too small, too opaque, and too expensive. For a display medium such as a liquid crystal twist cell, a large area, transparent, inexpensive substrate is needed. Transistors fabricated on this substrate could provide threshold switching for each display element and decoding, multiplexing, and line driver circuitry.

Thin film transistors have been studied for over 20 years. Various semiconductors, such as CdS,[1] CdSe,[2] amorphous hydrogenated silicon,[3-6] and polycrystalline silicon,[7-9] have been investigated. For display element switching, the mobility of these materials is adequate, but for peripheral circuitry, much higher mobility is needed. The necessary mobility can be obtained by melting a

---

*Research performed by the authors was done at Bell Laboratories, Holmdel, NJ.

Mat. Res. Soc. Symp. Proc. Vol. 13 (1983) ©Elsevier Science Publishing Co., Inc.

deposited silicon film to produce large crystal grains. With this material, both switching and control circuitry potentially could be fabricated on a single substrate.

The work on this problem has concentrated on transparent fused quartz substrates. In most respects, fused quartz is ideal. It is available with high purity and in arbitrarily large areas. It can be heated to the silicon melting point without deformation, and it is resistant to thermal shock. Unfortunately, the thermal expansion coefficient of fused quartz ($\alpha \sim 0.5 \times 10^{-6}/^{\circ}C$) is small compared to that of silicon ($\alpha \sim 3.5 \times 10^{-6}/^{\circ}C$). As the silicon cools from the melt, its contraction relative to the fused quartz can produce severe crazing of the film. This problem is illustrated in Fig. 1. A 1.0-μm-thick silicon film was chemically vapor deposited on a 0.5-mm-thick fused quartz plate. A 2-W argon ion laser beam focused to a $\sim$250-μm-diam spot was scanned across the film. This produced a pool of molten silicon (Fig. 1A) that solidified at the trailing edge with single crystals up to 200 μm x 20 μm in size. A network of cracks (Fig. 1B) typically appears a few seconds after the film solidifies.

Recently, various radiant energy sources have been substituted for the laser. Geis et al. demonstrated uniform melting with a resistively heated graphite strip in close proximity to the silicon film.[10] Similar results are obtained by focusing radiation from a tungsten filament[11,12] or an arc lamp[13] on the film. This increases the working distance and reduces potential contamination. To achieve a narrow melt zone across the full width of the substrate, the sample must be heated to 1000–1300°C. In the graphite strip heater case, this bias temperature is provided by placing the sample directly on a resistively heated graphite plate. Other systems use a bank of incandescent lamps.[12] The distinguishing features of the zone melting techniques are 1) high bias temperature, 2) slow scan speed (0.1–1.0 mm/s), and 3) much wider melt perpendicular to the scan than along it.

To prevent the molten silicon from agglomerating, the film is encapsulated with SiO_2 (Fig. 2A). By making the encapsulation 1–2 μm thick, the surface remains flat, and the resulting crystal texture is uniform.[10] Films crystallized by this method typically have very large grains (10 mm x 1 mm) with (100) crystal planes parallel to the substrate. As shown in Fig. 2B, a characteristic feature of silicon films that have been microzone melted at high bias temperature is a network of low-angle grain boundaries.[14] These

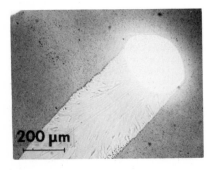

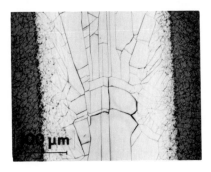

Fig. 1A. Reflection micrograph of a Si film on fused quartz during laser melting, showing large grains trailing the high reflectivity molten pool.

Fig. 1B. Reflection micrograph of the crack network that forms shortly after cooling.

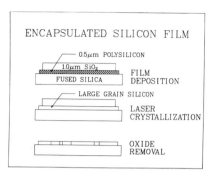

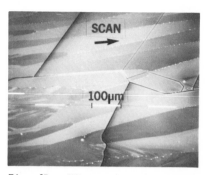

Fig. 2A. Schematic showing temporary crack suppression with an encapsulating film.

Fig. 2B. Electron channeling micrograph of a strip heater melted
Si film on fused quartz. Cracks
form when the encapsulation is removed.

boundaries separate crystallites with a 1-2° difference in the [100] direction out of the plane. A model to predict the spacing of these boundaries
is developed in the last section.

The uniformity of crystal orientation in these films substantially improves
electron mobility. Indeed, on fused quartz substrates where the film is under
tensile stress, mobilities better than in bulk silicon have been measured.[15]
Unfortunately, this tensile stress also produces cracks (Fig. 2B). These
cracks are less dense than in laser melted films, but they still make circuit
fabrication impractical.

CRACK SUPPRESSION TECHNIQUES

An encapsulating SiO_2 film >100 nm thick suppresses crack formation.
However, when the SiO_2 is completely removed, a network of cracks inevitably
forms during subsequent processing. Much better performance is obtained by
only removing the SiO_2 cap from the small islands where devices are to be
built. As shown in Fig. 3, this substantially reduces the number of cracks.

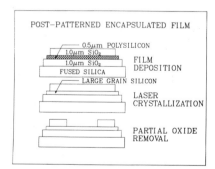

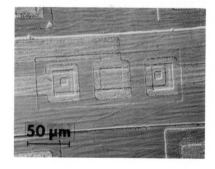

Fig. 3A. Schematic showing partial cap removal to suppress
cracks.

Fig. 3B. PMOS transistor fabricated in a crack-free region.

584

A large fraction of the devices are functional.[16] This film was melted with a ∿30 μm x ∿1 cm cylindrically focused argon ion laser beam in combination with an 1100°C bias temperature. This hybrid technique allows large-area zone melting with steeper temperature gradients than a strip heater provides, thus giving better melt stability and more closely spaced low-angle grain boundaries, as discussed in the last section.[17]

An alternative method is to define islands in the crystallized silicon before removing the SiO_2 cap. The idea is that sufficiently small islands can withstand the stress without cracking. There are two variations to this approach. In the first, a photoresist layer is patterned into islands and then baked to protect the underlying SiO_2. The exposed SiO_2 is removed with a HF solution. The underlying Si and the photoresist are then removed with a KOH solution. Finally, the SiO_2 over the islands is removed. Unfortunately cracks that form between the encapsulated islands often penetrate far enough under the cap to nucleate cracks within the islands when the overlying cap is removed.

In a variation of this technique, moats are patterned in the photoresist to define islands. Both the SiO_2 and the Si are sequentially removed from the moats before the SiO_2 over the rest of the area is removed. These steps are shown schematically in

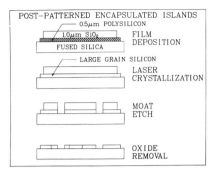

Fig. 4A. Schematic showing patterning with moated islands after crystallization.

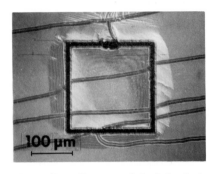

Fig. 4B. Unsuccessful island in which cracks propagate through the fused quartz.

Fig. 4A. The idea is that cracks are suppressed by the SiO_2 while the moats are fabricated. The moats would then prevent cracks from propagating into the islands when the SiO_2 is removed. As shown in Fig. 4B, surface cracks in the fused quartz often allow cracks to cross the moats into the islands.

A better solution is to pattern the silicon film prior to crystallization, as shown schematically in Fig. 5A. This technique was used by Kamins and Pianetta in their early work on silicon crystallization on fused quartz.[18] The idea is to define islands small enough to withstand the stress. Isolated islands, however, have two problems. The first is mass transport. When silicon melts, the volume is reduced ∿12%, which causes the silicon island to pull away from its original boundary, often leaving voids along the edge. Conversely, when the silicon solidifies, it expands ∿12% to form a sharp pinnacle with the last material to freeze. The remedy for this problem is to pattern the silicon into long, narrow strips, rather than islands. The molten zone is then passed along the length of the strip. In this way, the contraction and expansion problems are relegated to the ends, and the center is uniform and flat (Fig. 5B). The second problem is edge cooling. If the silicon is melted with radiation that is weakly absorbed in the surrounding fused quartz, the

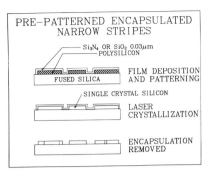

PRE-PATTERNED ENCAPSULATED NARROW STRIPES

Si₃N₄ OR SiO₂ 0.03μm
POLYSILICON

FILM DEPOSITION AND PATTERNING
FUSED SILICA

SINGLE CRYSTAL SILICON
LASER CRYSTALLIZATION

ENCAPSULATION REMOVED

Fig. 5A. Schematic showing patterning into moated islands prior to crystallization.

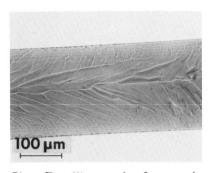

Fig. 5B. Micrograph of a crack-free, pre-patterned strip showing nucleation due to edge cooling.

edges of the strips remain cooler than the center. This causes crystals of random orientation to nucleate along the edge (Fig. 5B), resulting in reduced device performance.

The edge cooling can be remedied by depositing additional absorbing films, such as polysilicon, either above or below the isolated strips.[19] A simpler approach is to define the strips with a narrow moat.[18,19] In this way, radiation absorbed in the surrounding polysilicon suppresses edge cooling. One problem with the moat technique is that cracks can propagate through the surface of the fused quartz. The polysilicon surrounding the moats should also be patterned to reduce cracks.

Another solution to the edge cooling problem is to shape the laser beam to preferentially heat the edges.[19] Various masking and interference techniques have been used for this purpose. Another elegant approach is to melt the silicon with radiation that is absorbed more strongly in the fused quartz than in the silicon. This is easily done with the $\lambda = 10.6$ μm line of a CO_2 laser and naturally provides the desired temperature profile.[20] With a hotter edge, a crystal nucleus forming in the center can spread to make a single crystal island.

Long, narrow strips with internal moats and CO_2 laser heating have been combined very successfully at the Xerox Research Center.[21] As shown schematically in Fig. 5A, the central moated region usually solidifies as a single crystal, typically 100 μm x 25 μm. When the encapsulation is removed, the outer silicon may crack but the inner islands are substantially free of cracks.

THERMAL EXPANSION MATCHED SUBSTRATES

To varying degrees the patterning techniques described above reduce the cracking problem but do not eliminate it. The obvious alternative is to select a substrate whose thermal expansion closely matches that of silicon. In itself, this is not difficult. A number of silicate glass compositions meet this criterion. They have good transparency and surface finish, and they are substantially cheaper than fused quartz. What they lack is the purity, thermal shock resistance, and high softening temperature of fused quartz.

We have tested three types of silicate glass: Corning 1723, an alumina-silicate glass; Corning 7059, a low soda barium-borosilicate glass; and Corning 7740, a borosilicate glass better known as Pyrex.[22] These three types were selected for high annealing point (710°C, 639°C, and 560°C, respectively), resistance to thermal shock, and low alkali metal content.

The annealing point is defined as the temperature at which the glass viscosity is 4×10^{13} poise. Above this temperature, nonuniform stress can easily distort the plate. Empirically, we found that if these glasses were supported on a flat substrate they could be heated to 620°C for up to 1 hour. This is the temperature used for silicon deposition by pyrolytic decomposition of SiH_4.

The low softening temperature of the silicate glasses requires the silicon to be melted by surface heating. This is easily done with a focused argon ion laser beam. As shown in Fig. 6, the resulting crystal structure is comparable to that obtained on fused quartz. The zone melting techniques can not be used with silicate substrates because the high bias temperature would melt the glass.

The thermal expansion of these glasses is plotted in Fig. 7 as a function of temperature for comparison with the expansion of silicon and fused quartz. These glasses follow the thermal expansion of silicon up to ∿600°C, but they expand substantially more than silicon at higher temperatures. This applies a tensile stress to the silicon during heating that is relieved when the silicon melts. As the silicon cools from the melt, it is under a compression that effectively suppresses cracking.

The key problem with the silicate glass substrate is that impurities in the glass can diffuse into the silicon, ruining the electronic characteristics. As shown in Fig. 8, alkali metals are a particular problem. This sample of type 7740 glass was coated with 1.0 μm of SiO_2 before the 1.0-μm Si layer was deposited. The secondary ion mass spectroscopy depth profile shows that sodium from the glass diffused through the silicon before the deposition was complete. The boron in the glass was effectively blocked by the SiO_2 layer even after laser melting.

To isolate the glass impurities from the silicon, composite buffer layers are required. Our best results were obtained on type 7059 glass using a 1.0-μm layer of SiO_2 for adherence, a 200-nm layer of Si_3N_4 to block alkali metal diffusion, and a second 1.0-μm SiO_2 layer to provide a better electrical interface for the 1.0-μm Si layer.

As yet, devices have not been fabricated in these films. Novel processing techniques must replace the high-temperature steps in conventional integrated circuit processing. In particular, the gate oxide, which is usually grown at 1150°C, must now be grown at 650°C or below. Two possibilities are plasma oxidation and high-pressure oxidation. The risk of contaminating existing silicon processing facilities with impurities from the glass has also precluded device fabrication.

LOW-ANGLE GRAIN BOUNDARY FORMATION

As mentioned in the introduction, a prominent feature of silicon films that have been microzone melted at high bias temperature is a network of low-angle grain boundaries. These boundaries have a characteristic wishbone shape and separate crystallites with a 1-2° difference in orientation. The origin of these low-angle grain boundaries is an instability of the melt front, which leads to faceting. The low-angle boundaries form where adjacent facets meet at the trailing edge of the melt.

These facets have been studied both by rapid quenching of the melt[23] and by direct microscopic observation of the melt front,[24] as shown in Fig. 9C. Both techniques show that the facets are most commonly bounded by (111) crystal planes, which are the usual limiting planes for silicon crystal growth.

Melt front instability and faceting are commonly produced in bulk crystal growth by constitutional undercooling.[25,26] In this section, we develop a constitutional undercooling model for the melt front instability in microzone crystallization of thin silicon films.

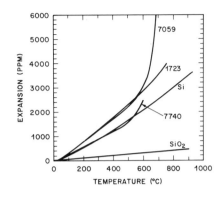

Fig. 6. Electron channeling micrograph of large grains produced by
laser melting a Si film deposited on
a silicate glass substrate with a
SiO₂ buffer layer.

Fig. 7. Comparison of the thermal
expansion of Si, fused quartz, and
glass types 1723, 7059, and 7740 as
a function of temperature.

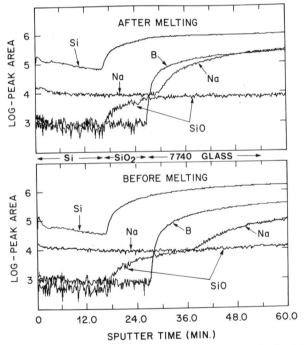

Fig. 8. Secondary ion mass spectroscopy analysis of
the composition in depth of a 1.0-μm Si film deposited on a type 7740 glass substrate with a 1.0-μm SiO₂
buffer layer. The composition after melting (top) is
compared with the same sample before melting (bottom).

To explore the effects of constitutional undercooling, consider the hypothetical phase diagram (Fig. 9A). This phase diagram has a character similar to the published silicon-oxygen binary system[27] but has been simplified for illustration. For very low impurity concentration, we assume that both the liquidus and solidus are straight lines and that the segregation coefficient is $k = 0.1$, as shown in Fig. 9B. Thus, if the impurity concentration in the solid is $C_o = 1 \times 10^{18}/cm^3$, the impurity concentration in the adjacent liquid is $C_o/k = 1 \times 10^{19}/cm^3$. As the melt zone is scanned, impurity atoms are rejected from the solid, building up a concentration gradient in the liquid ahead of the interface, as shown in Fig. 9D. In the simple diffusion model,[28] this concentration gradient is given by the expression:

$$C(x) = C_o \left[1 + \frac{(1 - k)}{k} \exp \left(- \frac{R}{D} x \right) \right] \tag{1}$$

where R is the scan rate and D is the diffusion coefficient of the impurity in the liquid. When the concentration gradient is referred to the phase diagram, it implies that the freezing temperature of the liquid (T_L) also has a spatial dependence given by

$$T_L(x) = T_o - mC_o \left[1 + \frac{(1 - k)}{k} \exp \left(- \frac{R}{D} x \right) \right] \tag{2}$$

where T_o is the freezing temperature of pure silicon and m is the slope of the liquidus. This curve is plotted in Fig. 9E using convenient parameters.

In the simple theory of constitutional undercooling, a linear temperature gradient of slope G imposed on the melt is assumed. If

$$G < \frac{mC_o R}{D} \frac{(1 - k)}{k} \tag{3}$$

there is a region (Fig. 9E) where the actual temperature is below the equilibrium freezing temperature. This region is constitutionally undercooled. Therefore, a perturbation of the interface that enters this region will tend to grow.

Let us assume that the projection grows until it no longer has undercooled liquid ahead of it. The projection length would be the distance from the average interface to the intercept of the temperature curve and the equilibrium freezing temperature curve, as marked with the dashed line in Fig. 9E.

If the projection is constrained by (111) growth facets, it will have the characteristic triangular shape. Thus, the spacing of adjacent low-angle boundaries will be comparable to the length of the undercooled region (Fig. 9F). This allows us to calculate the dependence of grain boundary spacing on both temperature gradient and scan speed. These relations are most easily expressed as a function of the length of the undercooled region, x_o. We obtain

$$G = \frac{mC_o}{x_o} \frac{(1 - k)}{k} \left[1 - \exp \left(- \frac{R}{D} x_o \right) \right] \tag{4}$$

which is plotted as $x_o(G)$ in Fig. 10, and

$$R = -D/x_o \cdot \ln \left[1 - \left[k/(1 - k) \cdot Gx_o/mC_o \right] \right] \tag{5}$$

which is plotted as $x_o(R)$ in Fig. 11. These curves indicate that the undercooled region and the boundary separation will go to zero for a sufficiently high temperature gradient or for a sufficiently slow scan speed.

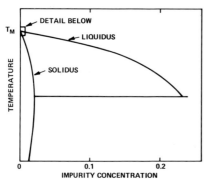

Fig. 9A. Hypothetical phase diagram for impure Si.

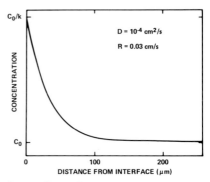

Fig. 9D. Impurity concentration profile in the liquid.

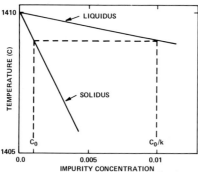

Fig. 9B. Linear approximation for the phase diagram at low-impurity concentrations.

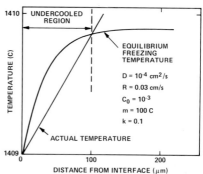

Fig. 9E. Equilibrium freezing temperature profile and the imposed temperature gradient produce a constitutionally undercooled region.

Fig. 9C. Reflection micrograph of melt front faceting during strip heater crystallization.

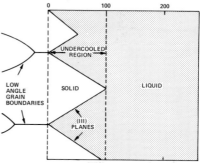

Fig. 9F. Faceted protrusions span the undercooled region and form low-angle grain boundaries where they meet.

The general character of these curves corresponds with observation. Increasing the temperature gradient reduces the boundary separation. For example, the boundaries in Fig. 2B that have a 50-100-μm spacing were produced with a 1-mm-wide graphite strip. Whereas the 5-10-μm spacing in Fig. 3B were produced with a ∿30-μm-wide laser beam that provides a much steeper temperature gradient. As yet, we have not been able to obtain a sufficiently steep gradient to eliminate the boundaries.

It has also been observed that the boundary spacing is reduced by lowering the scan speed. Geis has reported a $x_o \propto R^{1/2}$ dependence,[23] which does not show the sharp knee of Fig. 11. Our measurements show a small increase in spacing with scan speed, in keeping with the flat portion of Fig. 11, but we have not observed the abrupt decrease in spacing at slow speeds.

This simple model does not take into account lateral impurity segregation. In a metal that does not facet, lateral segregation causes an increase in boundary separation as the scan speed is reduced because there is more time for the impurity to diffuse.[28] However, in a material such as silicon where the facets are bounded by (111) planes, the separation of adjacent projections is constrained. This should reduce the effect of lateral segregation. Nevertheless, lateral segregation may make the change in boundary spacing with scan speed less abrupt.

In considering which impurities might cause constitutional undercooling, oxygen and nitrogen are potential candidates. These elements are commonly found in the substrate or capping layer or in the atmosphere in which the melting is performed. Distinguishing such impurities from the capping layer may be difficult because they may segregate to the surface. In addition, only small concentrations may be needed. The examples used a concentration of $C_o = 10^{-3}$. However, as seen in Eqs. (4) and (5), the key factor is k/mC_o. If k were smaller or m were larger, a much smaller concentration would produce the same result.

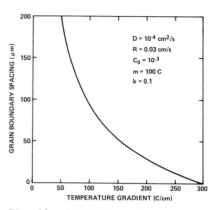

Fig. 10. Predicted grain boundary spacing as a function of the thermal gradient.

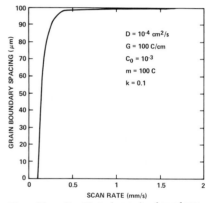

Fig. 11. Predicted grain boundary spacing as a function of scan rate.

CONCLUSION

The crystallization of silicon films on fused quartz is an attractive technology for control circuitry on flat panel matrix displays. Thermal stress cracking, however, remains a problem. Although patterning techniques have substantially suppressed cracking, the device yields needed for displays require further improvement.

Silicate glass substrates exchange the thermal stress problem for contamination and processing problems. The development of an expansion matched, pure, high melting temperature substrate could make an important impact on this field.

REFERENCES

1. P.K. Weimer, Proc. IRE 49, 1462 (1962).

2. T.B. Brody, F.C. Luo, Z.P. Szepetti, and D.H. Davies, IEEE Trans. Electron Devices ED-22, 739 (1975).

3. P.G. LeComber, W.E. Spear, and A. Ghaith, Electron. Lett. 15, 179 (1979).

4. H. Hayama and M. Matsumura, Appl. Phys. Lett. 36, 754 (1980).

5. M. Matsumura and N. Yasuo, J. Appl. Phys. 51, 6443 (1980).

6. M.J. Thompson, N.M. Johnson, M.D. Moyer, and R. Lujan, Program 39th Annual Device Research Conf. VIA-8 (1981).

7. M. Matsui, Y. Shiraki, Y. Katayama, K.L.I. Kobayashi, A. Shintani, and E. Maruyama, Appl. Phys. Lett. 37, 936 (1980).

8. S.W. Depp, A. Juliana, and B.G. Huth, Proc. 1980 International Electron Devices Meeting, IEEE Cat. CH1616-2/80, 703 (1980).

9. G. Auvert, D. Bensahel, A. Georges, V.T. Nguyen, P. Henoc, F. Morin, and P. Coissard, Appl. Phys. Lett. 38, 613 (1981).

10. M.W. Geis, D.A. Antoniadis, D.J. Silversmith, R.W. Mountain, and H.I. Smith, Appl. Phys. Lett. 37, 454 (1980).

11. G.K. Celler and Mc.D. Robinson, Laser-Solid Interactions and Transient Thermal Processing of Materials, J. Narayan, W.L. Brown, and R.A. Lemons, Eds., (Elsevier North Holland, New York, 1983), this volume.

12. A. Kamgar and G.A. Rozgonyi, ibid.

13. T.J. Stultz, J. Sturm, and J.F. Gibbons, ibid.

14. E.W. Maby, M.W. Geis, Y.L. LeCoz, D.J. Silversmith, R.W. Mountain, and D.A. Antoniadis, Elect. Device Lett. EDL-2, 241 (1981).

15. B.Y. Tsaur, J.C.C. Fan, and M.W. Geis, Appl. Phys. Lett. 40, 322 (1982).

16. D. Herbst, M.A. Bosch, R.A. Lemons, S.K. Tewksbury, and T.R. Harrison, "PMOS Transistors Fabricated in Large Area Laser Crystallized Si on Silica," Elect. Lett., in press (1983).

17. M.A. Bosch and R.A. Lemons, Electronic Materials Conference (Fort Collins, Colorado, 1982).

18. T.I. Kamins and P.A. Pianetta, IEEE Elect. Devices Lett. EDL-1, 214 (1980).

19. D.K. Biegelsen, N.M. Johnson, D.J. Bartelink, and M.D. Moyer, Laser and Electron-Beam Solid Interactions and Materials Processing, J.F. Gibbons, L.D. Hess, and T.W. Sigmon, Eds. (Elsevier North Holland, 1982), 487.

20. W.G. Hawkins, J.B. Black, and C.H. Griffiths, Appl. Phys. Lett. 40, 319 (1980).

21. N.M. Johnson, H.C. Tuan, M.D. Moyer, M.J. Thompson, D.K. Biegelsen, L.E. Fennell, and A. Chiang, Laser-Solid Interactions and Transient Thermal Processing of Materials, J. Narayan, W.L. Brown, and R.A. Lemons, Eds., (Elsevier North Holland, 1983), this volume.

22. R.A. Lemons and M.A. Bosch, Appl. Phys. Lett. 40, 703 (1982).

23. M.W. Geis, H.I. Smith, B.Y. Tsaur, J.C.C. Fan, D.J. Silversmith, and R.W. Mountain, J. Electrochem. Soc. 129, 2812 (1982).

24. R.A. Lemons, M.A. Bosch, H.J. Leamy, and J. Cheng, Electronic Materials Conference (Fort Collins, Colorado, 1982).

25. W. Bardsley, J.S. Boulton, and D.T.J. Hurle, Solid-State Electron. 5, 395 (1962).

26. D.E. Holmes and H.C. Gatos, J. Appl. Phys. 52, 2971 (1981).

27. H.F. Wolf, Semiconductor Data, (Pergamon, Oxford, 1969).

28. W.C. Winegard, An Introduction to the Solidification of Metals, (The Institute of Metals, London, 1964).

ELECTRICAL CHARACTERISTICS AND DEVICE APPLICATIONS OF
ZONE-MELTING-RECRYSTALLIZED Si FILMS ON SiO_2

B-Y. Tsaur, John C. C. Fan, M. W. Geis, R. L. Chapman, S. R. J. Brueck,
D. J. Silversmith, and R. W. Mountain
Lincoln Laboratory, Massachusetts Institute of Technology,
Lexington, Massachusetts 02173

ABSTRACT

Device-quality Si films have been prepared by using
graphite strip heaters for zone melting poly-Si films deposited on SiO_2-coated substrates. The electrical characteristics of these films have been studied by the fabrication and
evaluation of thin-film resistors, MOSFETs and MOS capacitors. High yields of functional transistor arrays and ring
oscillators with promising speed performance have been
obtained for CMOS test circuit chips fabricated in recrystallized Si films on 2-inch-diameter Si wafers. Dual-
gate MOSFETs with a three-dimensional structure have been
fabricated by using the zone-melting recrystallization technique.

INTRODUCTION

The utilization of dielectric isolation to achieve improved circuit performance for VLSI has stimulated great interest in the development of both Si-
on-insulator (SOI) materials and device isolation techniques. The materials
and techniques that have been investigated include Si-on-sapphire (SOS), Si
films prepared by graphoepitaxy [1], energy-beam-recrystallized Si films [2],
Si films grown on SiO_2 by vapor-phase lateral overgrowth [3,4], isolation by
oxygen implantation [5], and isolation by oxidation of porous Si [6]. We have
recently developed [7,8] a recrystallization process that uses a movable
graphite strip heater for zone melting poly-Si films, encapsulated with a composite SiO_2/Si_3N_4 layer, that have been deposited on SiO_2-coated Si substrates. Well oriented Si films up to 3 inches in diameter have been
obtained by this technique [9]. The structural properties of zone-
melting-recrystallized Si films have been reported elsewhere [8]. Briefly,
unseeded films consist of elongated grains with (100) texture that are typically a few millimeters by a few centimeters on a side, with their long axis
aligned approximately parallel to the direction of molten-zone motion. Within
each grain there are many fine-line defects, typically spaced 20-30 μm apart,
that also run roughly parallel to the direction of molten-zone motion. TEM
analysis indicates that these defects (which we call sub-boundaries) are dislocation arrays that are associated with angular deviations in orientation of
the order of one degree or less.

In this paper, the electrical characteristics of zone-melting
recrystallized Si films will be described in detail. The majority-carrier
transport properties and minority-carrier generation lifetime have been studied by the fabrication and evaluation of thin-film resistors, MOSFETs and MOS
capacitors. Mobility enhancement occurs in Si films recrystallized on SiO_2-
coated fused silica and sapphire substrates because of the large thermal
stress in these films. Test CMOS circuit chips have been fabricated on 2-
inch-diameter SOI films to evaluate the material uniformity and speed
performance of SOI/CMOS devices for integrated-circuit application. Dual-gate

Mat. Res. Soc. Symp. Proc. Vol. 13 (1983) ©Elsevier Science Publishing Co., Inc.

MOSFETs with a three-dimensional structure have also been fabricated by utilizing the zone-melting recrystallization technique.

MAJORITY-CARRIER TRANSPORT

The predominant defects in zone-melting-recrystallized Si films are sub-boundaries. We have studied the effects of these defects on carrier transport by experiments on thin-film resistors and n-channel MOSFETs [10]. Thin-film resistors with dimensions of 18 x 180 μm were fabricated in n-type recrystallized films with carrier concentrations of 1 or 2 x 10^{17} cm^{-3} obtained by implantation of P$^+$ ions. The resistor bars were oriented with their long axis either parallel or perpendicular to the sub-boundaries. Resistance measurements were carried out over the temperature range between -20 and 80°C. For resistors parallel to the sub-boundaries, the resistance is proportional to $T^{3/2}$, the same temperature dependence as for bulk Si. For resistors perpendicular to the sub-boundaries, the resistance also increases monotonically with temperature but is slightly higher and has a more complex temperature dependence. The difference in resistance between the perpendicular and parallel devices is the resistance contribution due to the sub-boundaries. This contribution has been analyzed [10] by using the charge-trapping model for potential barrier formation. The trapping state density determined for the sub-boundaries is ~ 7 x 10^{11} cm^{-2}, substantially lower than the values of 2-4 x 10^{12} cm^{-2} observed for CVD poly-Si films.

The effect of sub-boundaries on carrier transport has also been studied by measuring the surface electron mobility in two sets of n-channel MOSFETs with configurations such that electron transport is parallel to the sub-boundaries in one set and perpendicular to the sub-boundaries in the other. The devices have a gate length and width of 45 μm and channel doping concentration of 5 x 10^{15} cm^{-3}. For each set of transistors, the distribution of mobility measured for 92 randomly selected devices is shown in Fig. 1. The devices with electron transport parallel to the sub-boundaries exhibit an average mobility of 647 cm^2/V-s with a standard deviation of 18 cm^2/V-s, while the corresponding values for the other set are 635 and 22 cm^2/V-s. The difference in mobility is small, consistent with the sub-boundary parameters determined from the data for thin-film resistors. Since the electron concentration in the strongly inverted surface layer of MOSFETs is quite high (> 10^{18} cm^{-3}), the sub-boundary resistance in this layer is quite small. Consequently the mobilities observed in these

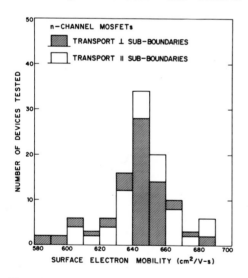

Fig. 1. Distribution of surface electron mobility for n-channel MOSFETs fabricated in zone-melting-recrystallized Si films with electron transport either parallel or perpendicular to the sub-boundaries.

devices are comparable to those in single-crystal Si devices with the same doping concentration.

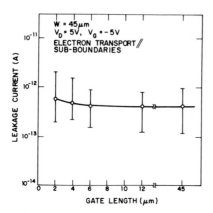

Fig. 2. Leakage current as a function of gate length for n-channel MOSFETs fabricated in zone-melting-recrystallized Si films.

Although sub-boundaries do not have a significant effect on majority-carrier transport, MOSFET performance might be degraded by enhanced dopant diffusion occurring along the sub-boundaries during device fabrication. For narrow-channel MOSFETs fabricated in laser-recrystallized Si films, dopant diffusion along grain boundaries was found to result in shorting between source and drain or in high source-to-drain leakage currents [11,12]. We have fabricated n-channel MOSFETs with gate lengths ranging from 2 to 45 µm in zone-melting-recrystallized films. Source and drain dopant activation was accomplished by annealing at 900° for 30 min. Figure 2 shows the source-to-drain leakage current as a function of gate length. The leakage current is extremely low and nearly independent of gate length. This result indicates that dopant diffusion is not greatly enhanced by the sub-boundaries.

MINORITY-CARRIER GENERATION LIFETIME

Planar MOS capacitors have been fabricated [13] in zone-melting-recrystallized Si films and in epitaxial Si layers grown by CVD on top of the recrystallized films. Figure 3 shows a photomicrograph and schematic cross-

(a)

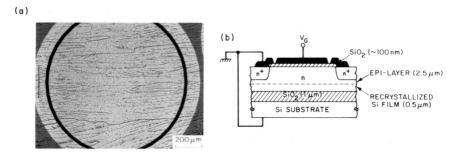

Fig. 3. (a) Optical micrograph and (b) schematic cross-section diagram of planar MOS capacitor fabricated in Si epilayer grown on recrystallized film.

section diagram of a capacitor fabricated in an epilayer. The pulsed MOS capacitor method was used to measure the minority-carrier generation lifetimes

[13]. For n-type recrystallized films with a carrier concentration of 1 x 10^{17} cm^{-3}, the lifetimes are 0.2-0.5 µs. For n-type epilayers with a carrier concentration of 2 x 10^{15} cm^{-3}, the lifetimes are 0.8-1.3 µs, compared to values of ~ 2 µs for similarly doped layers grown on single-crystal Si<100> wafers in a parallel run. The sub-boundaries in the recrystallized Si films and epilayers do not seriously degrade the lifetime, although devices with higher sub-boundary densities appear to have smaller lifetimes. The relatively high lifetime values indicate that zone-melting recrystallization does not introduce significant amounts of impurities into the Si films despite its high processing temperature. The lifetime values also suggest the possiblility of bipolar device applications utilizing SOI structures prepared by zone-melting recrystallization.

STRESS-ENHANCED MOBILITY

The experiments discussed so far all involved zone-melting recrystallization of poly-Si films deposited on SiO$_2$-coated Si substrates. We have also utilized the movable strip-heater technique to recrystallize poly-Si films deposited on SiO$_2$-coated fused silica and sapphire substrates [14]. The crystallographic properties of the recrystallized films are similar to those of the films on Si substrates. However, since the thermal expansion coefficient of Si is 5-10 times that of fused silica but only 40-50% that of sapphire, the Si films recrystallized on fused silica substrates are under a large tensile stress while those on sapphire substrates are under a large compressive stress. The thermal stress σ_{xy} parallel to the surface, as determined from lattice constant measurements made by x-ray diffraction analysis on several films of each type, averages +9.6 kbar for films on fused silica and -10.2 kbar for those on sapphire, where the "+" sign denotes tensile stress. The films recrystallized on SiO$_2$-coated Si substrates are nearly stress free, as expected.

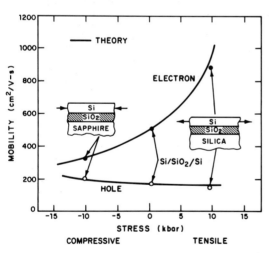

Fig. 4. Mobility vs stress for n- and p-type recrystallized Si films on SiO$_2$-coated Si, fused silica and sapphire substrates.

The Hall mobilities of majority carriers have been measured by the van der Pauw technique on n- and p-type films that were doped by implantation of 5 x 10^{12} cm^{-2} P$^+$ and B$^+$ ions, respectively, followed by furnace annealing at 950°C for 30 min. Implantation and annealing did not cause a detectable change in either surface morphology or stress in the films. Figure 4 shows the Hall mobility as a function of thermal stress. For the films on fused silica, the electron mobility is 74% higher and the hole mobility 14% lower than the corresponding values in unstressed Si films. For the films on sapphire, the changes are opposite in sign, with electron mobility 35% lower and hole mobility 10%

597

higher than those of unstressed films. Because of the large stress enhancement, the electron mobility in the films on fused silica, 870 cm^2/V-s, is even higher than the value of ~ 750 cm^2/V-s measured for bulk single-crystal Si with the same carrier concentration.

Since the recrystallized Si films have (100) texture and a high degree of in-plane orientation, their electrical properties can be expected to approximate those of <100>Si single-crystal films that are under isotropic two-dimensional stress. For such a single-crystal film the relative change in mobility is given by:

$$\Delta\mu/\mu_0 \simeq (\pi_{11} + \pi_{12})\sigma_{xy} / [1 + (\pi_{11} + \pi_{12})\sigma_{xy}] \quad , \tag{1}$$

where $\Delta\mu$ is the absolute change in mobility, μ_0 is the mobility of an unstressed film, and π_{11} and π_{12} are piezoresistance coefficients [14]. We have used Eq. (1), together with reported values of π_{11} and π_{12} for Si determined by measurements employing uniaxial stress, to calculate $\Delta\mu/\mu$ as a function of σ_{xy}. The calculated values of $\mu(\sigma_{xy}) = \mu_0 + \Delta\mu$ are plotted in Fig. 4. These values (solid curves) are in satisfactory agreement with the experimental results, in view of the polycrystallinity of the recrystallized films.

The large mobility enhancement for electrons in recrystallized films on fused silica substrates has been confirmed by the characteristics of n-channel MOSFETs. Figures 5(a) and 5(b) show the drain I-V characteristics for devices

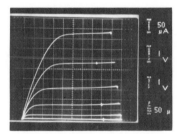 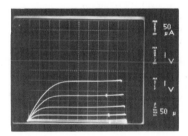

Fig. 5. Drain I-V characteristics of n-channel MOSFETs fabricated in Si films recrystallized on (a) fused silica and (b) Si substrates.

fabricated on fused silica and Si substrates, respectively. The surface electron mobilities deduced from the linear regions of the I-V curves are 1050 and 620 cm^2/V-s, respectively. The latter value is comparable to that for single-crystal Si devices. The mobility for the devices on fused silica is the highest value obtained for Si MOSFETs. Enhancement of electron mobility has also been observed by Johnson, et al. [15] in single-crystal Si islands produced by CO$_2$ laser recrystallization on fused silica substrates.

SOI/CMOS TEST CIRCUITS

The most important application of SOI materials is in high-density and high-speed integrated circuits. Higher packing density can be achieved be-

598

cause of the simplicity of device isolation, and higher operating speeds result from the reduction in parasitic capacitance. As previously reported
[16,17], both n- and p-channel MOSFETs fabricated in the recrystallized Si
films exhibit electrical characteristics comparable to those of single-crystal
Si devices. In order to carry out a more critical evaluation of zone-
melting-recrystallized SOI films for IC applications, we have designed a CMOS
test circuit chip for fabrication in these films. The test chip, which is
based on a 5 μm design rule, contains n- and p-channel transistor arrays, ring
oscillators, inverter chains, and various test devices for process control.
The objectives of utilizing this design are to assess the uniformity of the
SOI films and to determine the speed of SOI/CMOS circuits. Film uniformity is
required for obtaining a high yield of functional circuits, while speed is a
key parameter in deciding between alternative material technologies for VLSI.
The CMOS test chips were fabricated on SOI structures consisting of a 0.5-
μm-thick recrystallized Si film, a 1-μm-thick SiO$_2$ layer, and a 1 Ω-cm p-type
Si<100> wafer 2 inches in diameter. The fabrication process used involves a
total of six photomask steps with poly-Si gate and self-aligned ion-implanted
source and drain. Figure 6 is a photomicrograph of a finished SOI/CMOS chip,

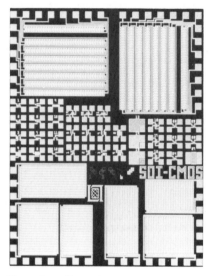

which measures about 3 x 4 mm. About
100 such chips were fabricated on
each of three wafers. A 31-stage
ring oscillator with fan in and fan
out of one and a 231-stage inverter
chain are located at the upper left.
A similar circuit, rotated by 90°, is
placed at the upper right. The purpose of the rotation is to examine
the effects of sub-boundaries on circuit performance. Carrier transport
is approximately parallel to these
boundaries in one circuit and approximately perpendicular to them in the
other. Test transistors, single-
stage inverters, gated diodes, capacitors and test patterns are located
in the middle portion of the chip.
Two n-channel transistor arrays consisting of 360 or 533 parallel devices and a p-channel array consisting of 460 parallel transistors are
located at the lower left. The individual transistors, which have a 5 μm
gate length and 20 μm gate width, are
spaced 5 μm apart. Three similar arrays, rotated by 90°, are located at
the lower right.

Fig. 6. Photomicrograph of SOI/CMOS
test chip fabricated in zone-melting-
recrystallized Si film.

To evaluate the uniformity of the SOI films, we have investigated the
performance of all the transistor arrays on a wafer with 98 test chips. Since
the transistors in each array are connected in parallel, failure of a single
device results in failure of the entire array. Of the 588 arrays, 490 were
functional while 98 failed because of source-to-drain short or open circuits.
For 62 of the inoperable devices, localized metallization defects such as incomplete etching of the Al or poor contacts were found by microscopic inspection. Thus the overall yield of functional arrays exceeds 90% when the obvious fabrication defects are discounted. The total number of transistors in
all the arrays is 2.65 x 10^5. If it is assumed that each of the failed arrays

contains one defective device, the transistor failure rate is 3.7 x 10⁻⁴, or only 1.4 x 10⁻⁴ if the known fabrication defects are taken into account.

The operating characteristics of the functional transistor arrays are quite uniform from chip to chip. For the three types of arrays in which carrier transport is approximately parallel to the sub-boundaries, the average values and standard deviations for the transconductance are as follows: p-channel (measured at $V_D = V_G = -5$ V), 52 ± 2 mS; 360-transistor and 533-transistor n-channel (measured at $V_D = V_G = +5$ V), 78 ± 2 and 115 ± 3 mS, respectively. The arrays in which carrier transport is perpendicular to the boundaries have similar characteristics, except that the transconductance values are ~ 5% lower.

Measurements have also been made on all of the 31-stage ring oscillators on the 98-chip wafer. For the two orientations differing by 90°, 82 of the 98 oscillators in each set are functional. Again, most of the failures can be attributed to obvious metallization defects. The output waveform and operating characteristics of a typical functional oscillator are shown in Figs. 7(a) and (b), respectively. The circuit starts to oscillate at a supply

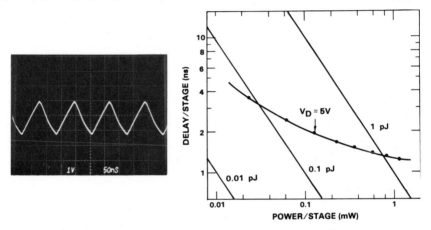

Fig. 7. (a) Output waveform and (b) switching delay time of 31-stage SOI/CMOS ring oscillator as a function of power dissipated per stage.

voltage V_D of 1.5 V. At $V_D = 5$ V, the switching delay time and dissipated power are respectively 2 ns and 0.13 mW per stage, for a power-delay product of 0.26 pJ. The operating speed can be attributed to the high carrier mobilities in the recrystallized Si films and the reduced parasitic capacitance of the SOI structure. Figure 8 shows the distributions of switching delay time per stage for the two sets of oscillators. Both distributions peak at about 2 ns with a standard deviation of ~ 0.1 ns. This similarity indicates that the sub-boundaries, despite their large numbers, do not have a significant effect on circuit performance. For the oscillators with delay times exceeding 3 ns, the metallization appears to be imperfect. The slow speed of these circuits may therefore be due to an increase in contact resistance.

The yield of inverter chains on the test chips is comparable to the yield of ring oscillators. The performance of the functional chains has been tested

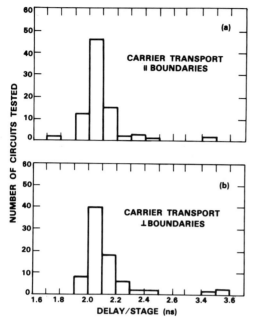

Fig. 8. Distribution of switching delay times for 31-stage ring oscillators with carrier transport (a) parallel and (b) perpendicular to the sub-grain boundaries.

by two methods: (1) an input signal is supplied, and the propagated signal is monitored at different output stages, and (2) the input and output are connected together so that the circuit functions as a ring oscillator. The chains exhibit normal inverter characteristics and switching delay times similar to those measured for the ring oscillator circuits.

For each of the other 2-inch-diameter SOI wafers on which CMOS chips were fabricated in this study, we have tested the transistor arrays, ring oscillators, and inverter chains on about 10 chips distributed over the area of the wafer. All the functional circuits measured are comparable in performance to those on the wafer for which all 98 chips were tested.

DUAL-GATE MOSFETs

The usefulness of zone-melting-recrystallized SOI structures in the fabrication of three-dimensional (3-D) integrated devices is limited by the high temperature required for the recrystallization process. However, some kinds of 3-D devices can be fabricated by using the recrystallization technique. To demonstrate this capability, we have designed and fabricated one such device, a dual-gate MOSFET. The fabrication process is shown schematically in Fig. 9. A CVD poly-Si film was deposited on a SiO_2-coated Si substrate and patterned by the local-oxidation-of-Si (LOCOS) process, which also defined a buried poly-Si gate. After growth of gate oxide on the gate, a second CVD film was deposited and encapsulated with the standard SiO_2/Si_3N_4 cap (not shown in the diagram). The sample was processed in the strip-heater system to recrystallize both the second CVD film and the buried gate. A conventional poly-Si gate process was then used to fabricate MOSFETs in the top recrystallized Si

DUAL-GATE MOSFETs

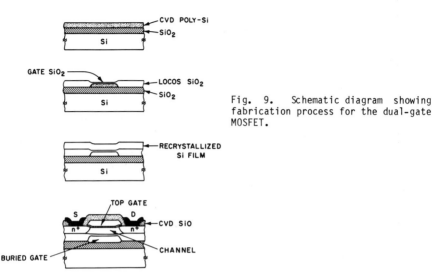

Fig. 9. Schematic diagram showing fabrication process for the dual-gate MOSFET.

film. Figure 10 shows a cross-section SEM micrograph of a finished n-channel dual-gate MOSFET. The gate oxides for the top and buried gates are 800 and 2000 μm thick, respectively.

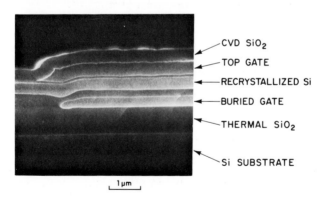

Fig. 10. Cross-section SEM micrograph of a finished dual-gate MOSFET.

Since the two gates of the dual-gate MOSFET can independently modulate the channel, this device can be used as a two-input pass or NOR gate. Alternatively, if the two gates are connected together the device should have a high

602

transconductance because both the top and bottom inversion channels contribute carrier conduction. In addition, the device should have a low leakage current because back-channel conduction can be eliminated by using the buried gate to control the bottom channel. Typical I-V characteristics for top-gate, buried-gate and dual-gate (both gates connected together) operation are shown in Fig. 11. The transconductance for dual-gate operation is equal to the sum of the transconductances for top- and buried-gate operation.

TOP GATE

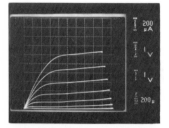

BOTTOM GATE

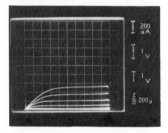

DUAL GATE

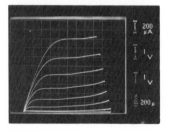

Fig. 11. Drain I-V characteristics of a dual-gate MOSFET for (a) top-gate, (b) buried-gate, and (c) dual-gate (both gates connected together) operation.

CONCLUSION

Studies of thin-film resistors and n-channel MOSFETs have shown that the majority-carrier transport properties of zone-melting-recrystallized Si films are comparable to those of single-crystal Si. The sub-boundaries, which are the predominant defects in these films, have negligible effects on carrier transport and MOSFET characteristics. As a result of thermal stress, mobility enhancement up to ~ 75% for electrons and ~ 10% for holes has been observed in Si films recrystallized on SiO_2-coated fused silica and sapphire substrates, respectively. Pulsed MOS capacitor measurements have shown that the minority-carrier generation lifetimes are in the microsecond range for both recrystallized Si films and epitaxial layers grown on these films. High yields of functional transistor arrays and ring oscillators with promising speed performance have been obtained for CMOS test circuit chips fabricated on 2-inch-diameter SOI wafers. Dual-gate MOSFETs using SOI structures prepared by zone-melting recrystallization have been fabricated to demonstrate the applicability of this technique to 3-D devices.

The results reported here show that zone-melting recrystallization yields SOI structures of sufficient quality to be seriously considered for utilization in VLSI circuits. It remains to be demonstrated that wafers suitable for conventional IC processing can be prepared by this technique. The material problems associated with large-area wafers, such as film uniformity and wafer flatness, should be addressed. Elimination of sub-boundaries in the films also remains an important task, even though these defects do not seriously degrade MOSFET performance. We anticipate that the zone-melting recrystallization process can be developed sufficiently to yield SOI materials for VLSI applications.

603

ACKNOWLEDGEMENTS

We acknowledge A. J. Strauss for fruitful discussions, G. W. Turner for electrical measurements, and P. M. Nitishin for SEM characterization. This work was sponsored by the Department of the Air Force and the Defense Advanced Research Projects Agency.

REFERENCES

1. M. W. Geis, D. C. Flanders, and H. I. Smith, Appl. Phys. Lett. <u>35</u>, 71 (1979).

2. See, for example, <u>Laser and Electron-Beam Interactions with Solids</u>, eds. B. R. Appleton and G. K. Celler (North-Holland, New York, 1982).

3. D. D. Rathman, D. J. Silversmith, J. A. Burns, and C. O. Bozler, J. Electrochem. Soc. <u>127</u>, 501C (1980).

4. L. Jastrzebski, J. F. Corboy, and R. Pagliaro, Jr., Electrochem. Soc. Meeting, Montreal, Canada, 1982, Extended Abs. No. 149.

5. K. Izumi, M. Doken, and H. Ariyoshi, Electron. Lett. <u>14</u>, 593 (1978).

6. K. Imai, Solid-State Electron. <u>24</u>, 159 (1981).

7. E. W. Maby, M. W. Geis, L. Y. LeCoz, D. J. Silversmith, R. W. Mountain, and D. A. Antoniadis, IEEE Electron Dev. Lett. <u>EDL-2</u>, 241 (1981).

8. M. W. Geis, H. I. Smith, B-Y. Tsaur, J. C. C. Fan, E. W. Maby, and D. A. Antoniadis, Appl. Phys. Lett. <u>40</u>, 158 (1982).

9. J. C. C. Fan, B-Y. Tsaur, R. L. Chapman, and M. W. Geis, Appl. Phys. Lett. <u>41</u>, 186 (1982).

10. B-Y. Tsaur, J. C. C. Fan, M. W. Geis, D. J. Silversmith, and R. W. Mountain, IEEE Electron Dev. Lett. <u>EDL-3</u>, 79 (1982).

11. N. M. Johnson, D. K. Biegelsen, and M. D. Moyer, Appl. Phys. Lett. <u>38</u>, 900 (1981).

12. K. K. Ng, G. K. Celler, E. I. Povilonis, R. C. Frye, H. J. Leamy, and S. M. Sze, IEEE Electron Dev. Lett. <u>EDL-2</u>, 316 (1981).

13. B-Y. Tsaur, J. C. C. Fan, and M. W. Geis, Appl. Phys. Lett. <u>41</u>, 83 (1982).

14. B-Y. Tsaur, J. C. C. Fan, and M. W. Geis, Appl. Phys. Lett. <u>40</u>, 322 (1982).

15. N. M. Johnson, D. K. Biegelsen, H. C. Tuan, M. D. Moyer, and L. E. Fennell, IEEE Electron Dev. Lett. (to be published).

16. B-Y. Tsaur, M. W. Geis, J. C. C. Fan, D. J. Silversmith, and R. W. Mountain, Appl. Phys. Lett. <u>39</u>, 909 (1981).

17. B-Y. Tsaur, M. W. Geis, J. C. C. Fan, D. J. Silversmith, and R. W. Mountain, <u>Laser and Electron-Beam Interactions with Solids</u>, eds. B. R. Appleton and G. K. Celler (North-Holland, New York, 1982), p. 585.

THIN-FILM TRANSISTORS IN CO_2-LASER CRYSTALLIZED SILICON FILMS ON FUSED SILICA

N. M. Johnson, H. C. Tuan, M. D. Moyer, M. J. Thompson, D. K. Biegelsen, L. E. Fennell, and A. Chiang
Xerox Palo Alto Research Centers, Palo Alto, CA   94304

ABSTRACT

Thin-film transistors (TFT) have been fabricated in scanned CO_2 laser-crystallized silicon films on bulk fused silica. In n-channel enhancement-mode transistors, it is demonstrated that an excessively large leakage current can be electric-field modulated with a gate electrode located beneath the silicon layer. This dual-gate configuration provides direct verification on bulk glass substrates of back-channel leakage as has recently been demonstrated for beam-crystallized silicon films on thermal oxides over silicon wafers. With the application of deep-channel ion implantation to suppress back-channel leakage, high-peformance TFTs have been fabricated in single-crystal silicon films on fused silica. The results demonstrate that scanned CO_2 laser processing of silicon films on bulk glass can provide the basis for a silicon-on-insulator technology.

INTRODUCTION

Beam-crystallized silicon layers on bulk glass substrates have been used to fabricate thin-film transistors (TFTs) [1,2]. While the semiconducting layer has generally possessed large-angle incoherent grain boundaries which limit device performance [3,4], high-performance transistors have recently been reported for films containing predominantly subgrain boundaries and crystallized with a swept graphite strip-heater [5] and for single-crystal silicon films crystallized with a scanned CO_2 laser [6]. An additional mechanism which has been found to limit performance in n-channel metal-oxide-semiconductor (MOS) TFTs is back-channel current leakage, due apparently to a high density of residual positive space charge in the dielectric substrate [7]. This mechanism has been successfully suppressed by using deep-channel ion implantation to prevent the formation of a conducting inversion layer along the back interface of the silicon layer [5,7]. In this paper are presented results from a study of transistors fabricated in CO_2 laser-crystallized silicon films on bulk fused silica. In n-channel enhancement mode devices, it is shown with dual-gate test devices that an excessively large leakage current can be electric-field modulated with a gate electrode located beneath the silicon layer. This provides direct verification on bulk glass substrates of back-channel leakage. High-performance TFTs were fabricated on bulk fused silica by using deep-channel ion implantation to suppress back-channel leakage and laser-crystallized

Mat. Res. Soc. Symp. Proc. Vol. 13 (1983) ©Elsevier Science Publishing Co., Inc.

single–crystal silicon films to eliminate the detrimental effects of residual grain boundaries on device operation.

DEVICE FABRICATION

Conventional microelectronic processing techniques were combined with laser crystallization to fabricate thin–film silicon transistors. In this section is summarized the processing steps used in the fabrication of the low – leakage high – performance TFTs which are discussed in the section entitled "Single – Crystal Silicon TFTs"; variations in processing for evaluating back – channel leakage and for fabricating dual – gate transistors on fused silica are noted in the next section. A layer of silicon, 0.57 μm thick, was deposited by low–pressure chemical deposition (LPCVD) on polished fused silica substrates and patterned to promote the nucleation and growth of single crystals in predefined areas during scanning laser crystallization. The patterned silicon was then encapsulated with 200 nm of silicon dioxide in preparation for crystallization. The silicon film was crystallized by raster scanning a CO_2 laser beam. Details are presented elsewhere on silicon patterning [8 – 10] and CO_2 laser crystallization [6,11 – 13]. The test devices were n–channel, enhancement–mode transistors with a channel length of 12 μm and width of 20 μm. The source and drain regions were doped by implanting As^+ at 180 keV to a dose of 1×10^{15} cm^{-2}. The silicon islands were then thermally oxidized in dry O_2 at 1000°C to produce a 130–nm thick oxide, with a post oxidation anneal in N_2 (20 min.) to reduce the density of fixed oxide charge and interface states. The channel was then doped by ion implanting boron in two steps: (1) a shallow implant at 60 keV to 2.8×10^{11} cm^{-2} and (2) a deep implant at 200 keV to 4.7×10^{12} cm^{-2}. The channel dopant was thermally activated (without significant redistribution) and ion–implantation damage in the gate oxide was removed with an additional furnace anneal after channel implantation. After selectively removing the oxide layer from the ends of the islands, an Al–Si alloy (1% Si) was vacuum deposited and patterned for device electrodes. Transistor fabrication was completed with a sinter at 450°C for 30 min. in forming gas.

DUAL–GATE TFTs

It has recently been demonstrated that beam–crystallized n–channel silicon transistors on thermal oxides over bulk silicon display a back–channel leakage current [7]. This is apparently due to residual positive space charge in the silicon–dioxide isolation layer which induces an n – type inversion layer along the lower interface of the silicon film. In the present study it was found that transistors fabricated on fused silica with a single low–dose implantation of boron (e.g., 1×10^{11} cm^{-2}) into the channel, which was redistributed during gate oxidation, displayed large leakage currents. To verify that this was a back–channel leakage, after device fabrication the bulk fused silica substrate was thinned by mechanical polishing to a thickness of approximately 18 μm. Then a metal electrode was deposited over the back surface of the substrate and used as a lower gate electrode. In Fig. 1 is shown the dependence of the drain current on source – drain voltage for two biases applied to the back electrode. Figure 1(a) shows the current – voltage curves with the back gate grounded and for top – gate biases of 0 and 2 V, and Fig. 1(b) displays the curves with the back gate biased to – 400 V. In this otherwise conventional single – gate "upright" TFT on fused silica, the additional substrate electrode demonstrates that the leakage current is readily modulated with an electric

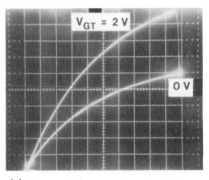 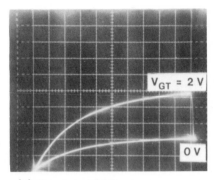

(a) V_GB = 0 V **(b) V_GB = – 400 V**

Fig. 1. Electric–field modulation of the back–channel leakage current in a laser–crystallized TFT with a second metal gate electrode deposited over the back surface of the (thinned) fused silica substrate. Voltage applied to the top gate is designated V_{GT}, and the back–gate voltage is V_{GB}. The vertical axis is drain current (20 μA/div), and the horizontal axis is source – drain voltage (1 V/div).

field applied to the lower surface of the silicon channel.

A practical dual–gate TFT configuration, which also displays back – channel leakage, was realized on bulk fused silica by depositing and patterning a degenerately doped polycrystalline silicon layer on the substrate to provide the lower gate electrode. This polysilicon gate electrode was coated first with 100 nm of LPCVD silicon nitride (to serve as a dopant diffusion barrier during subsequent processing) and then with 400 nm of LPCVD silicon dioxide. Channel doping of the semiconducting silicon layer was as described in the preceding paragraph. No

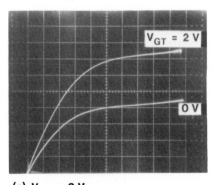

(a) V_GB = 0 V **(b) V_GB = – 12 V**

Fig. 2. Electric–field modulation of the back–channel leakage current in a laser–crystallized dual – gate TFT on fused silica. Voltage applied to the top gate is designated V_{GT}, and the back–gate voltage is V_{GB}. The vertical axis is drain current (100 μA/div), and the horizontal axis is source – drain voltage (1 V/div).

attempt was made to optimize the laser crystallization conditions for growth of single – crystal silicon, and transmission electron microscopy (TEM) revealed that the silicon islands consisted of large – grain polycrystalline silicon, comparable to previously reported devices [2,3,10]. Results from curve – tracer characterization are shown in Fig. 2. As in the previous example, a negative bias applied to the lower gate electrode significantly reduces the measured channel current. The negative bias suppresses the formation of the fixed – charge – induced inversion layer and thereby pinches off the source – to – drain leakage path along the lower surface of the silicon layer.

SINGLE–CRYSTAL SILICON TFTs

With deep–channel ion implantation to suppress back–channel leakage, high–performance TFTs were fabricated in islands of single–crystal silicon on fused silica. The single crystallinity of the transistors was established with transmission electron microscopy after device evaluation. An example is shown in Fig. 3 with a TEM micrograph of a 20-μm wide silicon island; the metal electrodes and oxide layer were chemically removed in TEM specimen preparation. The island is free of gross structural defects such as large–angle incoherent grain boundaries and subgrain boundaries; a small fraction of the islands contains twins. It was also determined that the island possesses (100) crystal orientation, as has been previously reported for CO_2 laser crystallization of patterned silicon films on fused silica [6,11,12].

The current–voltage characteristics for single–crystal silicon TFTs reveal high performance. The dependence of drain current (I_D) on gate voltage (V_G) for a source–drain voltage (V_{DS}) of 5 V is shown in Fig. 4. The leakage current dominates for gate voltages of less than – 1 V and is below 10 pA. This low level of leakage is a consequence of the deep–channel boron implant which suppresses back – channel leakage [5,7].

Fig. 3. Bright–field transmission electron micrograph of a single–crystal silicon island after TFT fabrication. The island is 20 μm wide.

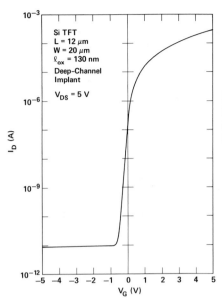

Fig. 4. Dependence of the drain current I_D on gate voltage V_G for a single–crystal silicon TFT on fused silica. The source–drain voltage V_{DS} is 5 V.

Transistor I–V characteristics were used to determine the threshold voltage and electron channel mobility. Figure 5 is the I_D vs. V_G characteristic in the linear regime of transistor operation. The drain current is observed to vary slightly sublinearly with gate voltage above threshold. Similar performance is observed in bulk single–crystal silicon MOSFETs [14]. The threshold voltage was estimated, as on bulk silicon devices [15], by linearly extrapolating from the point of maximum slope on the nominally linear I–V curve to zero drain current which yields 0.6 V. The gradient of the I–V curve in Fig. 5 defines the transconductance g_m of the transistor from which the channel mobility can be computed [14]. In Fig. 6 is plotted both the

Fig. 5. Drain current I_D versus gate voltage V_G in the linear region of transistor operation (V_{DS} = 0.1 V) for a single–crystal silicon TFT on fused silica.

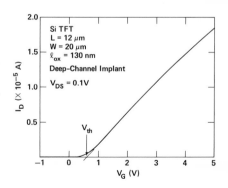

610

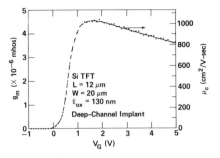

Fig. 6. Dependence of transconductance g_m and computed electron–channel mobility μ_c on gate voltage for a single–crystal silicon TFT on fused silica.

transconductance and channel mobility. Above threshold, the transconductance rapidly increases to a maximum value of ~ 4.5 × 10^{-6} mho and then decreases slowly with gate voltage. The solid line identifies the range of transconductances over which the channel mobility can be computed from the linear characteristics. The mobility decreases slowly with gate voltage from a maximum of greater than 1000 cm^2/V sec. A similar dependence of channel mobility on the transverse electric field in the inversion layer has previously been reported [14] and may be ascribed to the increasing density and spatial confinement of electrons in the inversion layer with increasing gate voltage, which contribute to increased electron scattering with a concomitant reduction in the mobility.

DISCUSSION AND CONCLUSIONS

The results presented above demonstrate that scanned CO_2 laser crystallization can be combined with conventional microelectronic processing to fabricate high–performance silicon thin–film transistors on bulk fused silica. The back–channel leakage which degrades n–channel device operation is readily suppressed with a high–dose deep–channel boron implant. It has been previously shown that a single dose of boron (e.g., 1 × 10^{12} cm^{-2}) implanted into the channel and redistributed during oxidation is also effective in suppressing back channel leakage, provided the dose exceeds the areal density of positive space charge in the dielectric substrate [6]. The generally recognized advantage of deep–channel implantation combined with a low–dose shallow implant for front–channel conduction is the largely independent control of back–channel leakage and threshold voltage, which is a critical requirement in the design of logic circuitry.

Grain–boundary degradation of TFT operation motivated the attainment of single–crystal silicon thin films on bulk fused silica. An important consequence of this achievement has been the realization of high electron channel mobilities as shown in Fig. 6. Such high mobilities have recently been reported for transistors fabricated in strip–heater–crystallized silicon films on fused silica and ascribed to a stress–induced enhancement of the electron mobility which is substantial for (100)–oriented silicon films under tensile stress [16]. The laser crystallized silicon films on fused silica have also been shown to be under tensile stress [4,17] and display a preferred (100) crystal orientation [6,11,12]; our results are thus consistent with the proposed stress–enhancement mechanism. The results presented in this paper demonstrate that scanned CO_2 laser crystallization of silicon films on bulk glass can provide the basis for a silicon–on–insulator technology.

REFERENCES

1. T. I. Kamins and P. A. Pianetta, Electron Device Lett. EDL-1, 214 (1980).
2. N. M. Johnson, D. K. Biegelsen, and M. D. Moyer, Insulating Films on Semiconductors (Springer-Verlag, New York, 1981) eds. M. Schulz and G. Pensl, pp. 234-237.
3. N. M. Johnson, D. K. Biegelsen, and M. D. Moyer, Appl. Phys. Lett. 38, 900 (1981).
4. D. K. Biegelsen, N. M. Johnson, R. J. Nemanich, M. D. Moyer, and L. Fennell, Laser and Electron Beam Interactions with Solids (Elsevier, New York, 1982), eds. B. R. Appleton and G. K. Celler, pp. 331-336.
5. B-Y. Tsaur, M. W. Geis, J. C. C. Fan, D. J. Silversmith, and R. W. Mountain, Laser and Electron-Beam Interactions with Solids (Elsevier, New York, 1982), eds. B. R. Appleton and G. K. Celler, pp. 585-590.
6. N. M. Johnson, D. K. Biegelsen, H. C. Tuan, M. D. Moyer, and L. E. Fennell, Electron Device Letters (in press).
7. H. W. Lam, Z. P. Sobczak, R. F. Pinizzotto, and A. F. Tasch, Jr., IEEE Trans. Electron Devices ED-29, 389 (1982).
8. D. K. Biegelsen, N. M. Johnson, D. J. Bartelink, and M. D. Moyer, Appl. Phys. Lett. 38, 150 (1981).
9. D. K. Biegelsen, N. M. Johnson, D. J. Bartelink, and M. D. Moyer, Laser and Electron-Beam Solid Interactions and Materials Processing (Elsevier, New York, 1981), eds. J. F. Gibbons, L. D. Hess, and T. W. Sigmon, pp. 487-494.
10. N. M. Johnson, D. K. Biegelsen, and M. D. Moyer, Laser and Electron-Beam Solid Interactions and Materials Processing (Elsevier, New York, 1981), eds. J. F. Gibbons, L. D. Hess, and T. W. Sigmon, pp. 463-470.
11. W. G. Hawkins, J. G. Black, and C. H. Griffiths, Appl. Phys. Lett. 40, 319 (1982).
12. W. G. Hawkins, J. G. Black, and C. H. Griffiths, Laser and Electron Beam Interactions with Solids (Elsevier, New York, 1982), eds. B. R. Appleton and G. K. Celler, pp. 529-534.
13. D. K. Biegelsen, N. M. Johnson, W. G. Hawkins, L. E. Fennell, and M. D. Moyer, these proceedings.
14. S. M. Sze, Physics of Semiconductor Devices (Wiley, New York, 1981), 2nd ed., ch. 8.
15. R. Tremain, private communication.
16. B-Y. Tsaur, J. C. C. Fan, and M. W. Geis, Appl. Phys. Lett. 40, 322 (1982).
17. S. A. Lyon, R. J. Nemanich, N. M. Johnson, and D. K. Biegelsen, Appl. Phys. Lett. 40, 316 (1982).

TEMPERATURE PROFILES INDUCED BY STRIP HEATERS

M. L. Burgener and R. E. Reedy
Naval Ocean Systems Center, San Diego, California, 92152

ABSTRACT

We present a model based on analytic solutions to the
heat equation that calculates the temperature profiles
induced by line shaped heat sources. The sensitivity of
isotherms to various heating conditions including the effect
of tailoring power profiles in both curved and straight
sources is discussed. Conditions are presented that result
in "dogboned" shaped isotherms. Also shown is the effect
that inhomogeneities (e.g., power density fluctuations or
defects in the material) have on the ideal temperature
profiles.

INTRODUCTION

Recent work in the localized melting of silicon on insulator (SOI)
structures has shown the desirability of shaping these heat sources to obtain
large regrown crystallites. By shaping a circular laser such that flat isotherms
were produced at the melt—solid interface Stultz and Gibbons increased
crystallite size from 5X15 micrometers to 25X65 micrometers[1]. Bosh and Lemons
produced a line source from an argon laser beam and also reported large grain
crystals regrown along the flat temperature front[2]. Evaluation of the thermal
profiles that led to improved crystallization was the motivation for the work
reported here.

In this paper we report a model which predicts isotherms for a curved or
straight line source, with any linear power density along the source. The model
is a solution to the linear heat equation, hence for the case of a slow moving
heat source the true temperature can be represented by a sum of linear
temperatures and the temperature induced by any shaped, slow moving source can be
represented by an appropriate sum of solutions of line sources[3]. In this work,
we present the basic mathematical derivation of the model, and its application to
specific examples of line sources. Also, the effect of hot and cold spots along
the line source will be shown, thus giving a qualitative feel for the effect that

Mat. Res. Soc. Symp. Proc. Vol. 13 (1983) Published by Elsevier Science Publishing Co., Inc.

source or materials inhomogeneities have on the temperature profiles.

THEORY

The three dimensional linear heat equation is given by

$$\nabla^2 \theta = \frac{-Q}{K_o} \tag{1}$$

where θ is the linear temperature, K_o is the thermal conductivity at room temperature and Q is the heat source[4]. The solutions to this equation for a line source in homogeneous material is[5]

$$\theta = c \int \frac{Q(r') \, d^3 r'}{|\vec{r} - \vec{r}'|} \tag{2}$$

where the integration is along the line source. For the case of a straight line source of constant power density, equation 2 becomes

$$\theta = c \, LOG_e \left[\frac{\sqrt{x^2 + (y-a)^2} + (y-a)}{\sqrt{x^2 + (y-b)^2} + (y-b)} \right] \tag{3}$$

where b–a is the length of the line, y is measured along the line from the center and x is measured perpendicular to the line. Appropriate changes in the integration are necessary to handle curved line sources or sources with nonconstant power density.

RESULTS

Figures 1 and 2 show results of numerical integration of equation 2 followed by algorithms to find and plot isotherms. Figure 1 shows isotherms for various line heat profiles absorbed in a silicon substrate. Figure 1(a) is for the case of a straight line source with uniform power density along its length. As can be seen, there is a curvature toward the center of the source. This is similar to the graphite strip heaters used to melt SOI structures. Figure 1(b) shows the effect of curving the line heat source: there are two regions with reverse curvature from each other separated by an approximately straight isotherm. For melting of SOI films, the power and curvature should be such that the melt zone would be in the inner region.

Figure l(c) is for the case of two regions of constant linear power density, with the ends having a higher linear power density than the center. The addition of hot ends to the otherwise uniform line source straightens out and eventually reverses the curvature of the isotherms. Similar effects are found for the case shown in figure l(d), in which the power density increases linearly from the center to a maximum at the ends. After investigating numerous power density profiles, we have drawn the conclusion that addition of power to the ends of the line sources results in the dogbone shape temperature profile that may reduce competitive recrystallization during SOI processing. Considering the ease with which the profile of figure l(c) can be implemented, it seems to be the preferred shape of those we have evaluated.

Figure 2 shows the temperature profiles generated by a uniform line source 1 cm long in which an inhomogeneity exists half way from the center to the end of the source. The temperature profiles are 30 micrometers from the source, and are very flat for the unperturbed case. Figure 2 shows the effect of hot spots of different linear dimension. The 10 micrometer spot has a linear power density twice that of the rest of the line source, while the smaller ones have the same total power as the 10 micrometer spot. The most striking feature of these curves is that relatively small perturbations cause significant changes in the temperature profiles. For example the 1 micrometer wide inhomogeneity causes a bulge in the temperature profile almost 50 micrometers wide. The case of a 10 micrometer cold spot with half the surrounding power density causes a cusp in the temperature profile approximately 40 micrometer wide. We conclude from these figures that perturbations in thermal profiles can extend significantly beyond the dimension of the inhomogeneity that induced the perturbation, and that this could lead to thermally induced defects in regrown crystals.

We have modeled inhomogeneities as perturbations in the power profile, but they can represent variations in the materials as well. For example, a defect or step in the silicon dioxide layer of an SOI structure can cause higher or lower heat conduction which can be viewed qualitatively as a region of lower or higher incident power in the silicon film. Our goal was to obtain an estimate of the effect that small perturbations in the heat flow process have on the temperature profiles.

SUMMARY

By using an analytic solution to the heat equation we have shown temperature profiles created by line sources of various shapes and power density profiles. We have made the observation that a line source with extra power on the ends causes a dogbone shape temperature profile and that the specific shape and location of the extra power density does not have a major effect. We have also shown the effect that small inhomogeneities have on the temperature profiles created by line sources and have concluded that small inhomogeneities can have a relatively large effect on resultant temperature profiles.

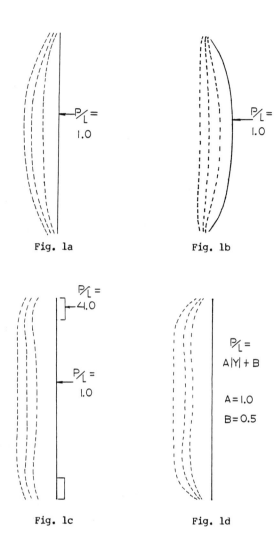

Figure 1. Isotherms for various line heat sources.

a) Straight line source with uniform power density.
b) Curved line source with uniform power density.
c) Uniform line source with additional power on ends.
d) Straight line source with power density increasing linearly from the center.

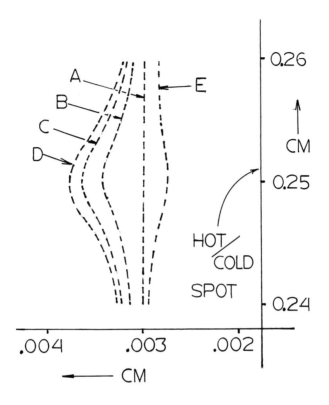

Figure 2. Effect of inhomogeneities on shape of isotherms.

Isotherm a is for a straight line source with
uniform power density. Isotherms b-d are for an
inhomogeneity with the power=0.1% of the total incident
power. The inhomogeneity dimensions are b: 10 microns
c: 5 microns d: 1 micron. Curve e: a 10 micron cool
spot with half the surrounding incident power density.

618

REFERENCES

[1] T.J. Stultz, J.F. Gibbons, "Structural and Electrical Characterization of Shaped Beam Laser Recrystallized Polysilicon On Amorphous Substrates" Laser and Electron Beam Interactions With Solids, Elsevier Science Publishing Company, N.Y. 499(1982).

[2] M.A. Bosh and R.A. Lemons, "Large Area Laser Crystallization of Si Films," Electronic Materials Conference, Ft. Collins, CO, June 23-25, 1982.

[3] M.L. Burgener and R.E. Reedy, "Temperatures Distributions Produced In a Two-Layer Structure by a Scanning CW Laser or Electron Beam," J. Appl. Phys. 53(6), June 1982.

[4] H.S. Carslaw and J.C. Jaeger, Conduction of Heat in Solids, Oxford University Press, London, 89(1959).

[5] J.D. Jackson, Classical Electrodynamics, John Wiley & Sons, Inc., N.Y. 34(1975).

PART VII
COMPOUND SEMICONDUCTORS

TRANSIENT ANNEALING OF ION IMPLANTED GALLIUM ARSENIDE

J.S. WILLIAMS
JMRC, Faculty of Engineering, Royal Melbourne Institute of Technology,
Melbourne   3000, Australia.

ABSTRACT

This paper provides a brief overview of the application
of transient annealing to the removal of ion implantation
damage and dopant activation in GaAs. It is shown that both
the liquid phase and solid phase annealing processes are more
complex in GaAs than those observed in Si. Particular
attention is given to observations of damage removal, surface
dissociation, dopant redistribution, solubility and the
electrical properties of GaAs. The various annealing
mechanisms are discussed and areas in need of further
investigation are identified.

INTRODUCTION

Over the past few years, considerable attention has been given to
transient annealing of ion implanted semiconductors. Most of the
important results of this research effort have been reported or reviewed
at the previous symposia in this series and appear in the corresponding
proceedings [1 - 4]. Annealing has been carried out in time regimes
which span 18 orders of magnitude, using,for example, picosecond laser
pulses to rapidly melt and refreeze near-surface layers at one end of
the time spectrum, to the utilization of solid phase annealing via more
conventional furnace processing at the other end of the time scale.
Although considerably more effort has been directed towards Si, transient
annealing has much intuitive appeal for thermal processing of GaAs. For
example, major difficulties are experienced using furnace processing;
namely, in removing implantation damage, low activation of n-type dopants
and the necessity of protecting the surface against dissociation and As
loss. However, the initial hope that transient processing might overcome
such difficulties has, to a large extent, not been realised in practice.
As outlined in previous review articles [5 - 7], the results of transient
annealing of GaAs attest to the difficulties of completely removing damage
and activating dopants, regardless of the annealing regime (solid or
liquid phase crystallisation) or of the particular time regime employed.
This brief review summarises the available results which have been
obtained using transient thermal processing of ion implanted GaAs. It
assesses the present understanding of the various annealing processes,
both in the liquid phase and solid phase regimes, and identifies some key
areas which are open for further investigation.

ANNEALING IN THE LIQUID PHASE REGIME

Since existing reviews [7,8] provide an accurate assessment of the
current status of rapid liquid phase annealing of GaAs, only a brief
summary of the important points is given here.

Mat. Res. Soc. Symp. Proc. Vol. 13 (1983) ©Elsevier Science Publishing Co., Inc.

Damage removal and dissociation

Figure 1 summarises the typical behaviour which results when nanosecond energy pulses of increasing energy density are incident upon ion implanted GaAs. As the energy density within a single pulse is increased a thin near-surface region will melt at a particular threshold energy density. The absolute value of this threshold depends upon several parameters, including the nature of the incident radiation (e.g. ruby or Nd:YAG laser light, electron beams), the pulse length, and the level of near-surface damage in the implanted GaAs. The melt threshold (of between 0.15 to 0.25 J cm^{-2}) illustrated in Fig.1 is typical of pulsed ruby laser irradiation using a pulse length of around 25 nsec [9-13] .

Following melting, the quality of the rapidly resolidified crystalline layer depends largely upon the melt penetration depth and the nature of the damage in the original implanted layer. The inset in the upper left of Fig.1 illustrates a possible as-implanted structure, containing an amorphous surface layer (of, say, 1000-1500 Å in thickness) and a deeper (transition) band of crystalline damage. For this situation, a polycrystalline layer results upon resolidification if the melt depth does not penetrate into the underlying single crystal. This behaviour has been clearly observed by Campisano et al [9] and by Cullis [14]. As the incident energy density is increased further, the melt depth eventually penetrates to the underlying crystal (upper right inset in Fig.1) and liquid phase epitaxy results upon resolidification. As illustrated in Fig.1, for an amorphous layer of about 1000Å this may occur at an energy density for a pulsed ruby laser of around 0.4 J cm^{-2} [8].

Under some implantation conditions the damaged layer may not be completely amorphous but,rather, consist of a highly defective crystalline layer [13,15]. In this situation, single crystal material will invariably result upon refreezing regardless of the melt penetration depth. However, if the melt depth does not penetrate beyond the extent of the implant-damaged layer, poor quality epitaxial growth will result in which defects at the melt boundary are propagated back to the surface during liquid phase epitaxy. This behaviour has been observed by Cullis [14] and Fletcher et al [13] and is illustrated in Fig.2.

Even with the situation of an amorphous GaAs layer, defect-free crystallisation will only result when the melt punches through the band of underlying defects into the perfect bulk crystal. Although near-perfect liquid phase crystallisation has been observed under such conditions [13,14], several studies [10,12-14, 16-19] have shown that melt-phase annealing can lead to significant surface damage, consisting of excess surface Ga and defects extending from the surface into the bulk. This behaviour results from surface evaporation and preferential loss of the more volatile As component [16,20] during the melt duration. The severity of this effect increases with the deposited energy density and is apparent even at power levels significantly below 1 J cm^{-2} [13,17,20]. Thus, the occurrence of surface damage sets an upper limit to the energy density for optimum melt-phase recrystallistion of GaAs. The energy density window for optimum near-perfect annealing (Fig.1) is, consequently, quite narrow. In cases of thick implant-damaged layers, it may not be possible to obtain defect-free regrowth (i.e. melt penetration into the underlying perfect crystal) without excessive surface damage and As loss [21].

Dopant redistribution and solubility.

The dopant redistribution and supersatured solid solutions which can result from liquid phase annealing of GaAs have recently been reviewed in detail [8]. The observations are essentially similar to those found during melt-phase annealing of Si, namely i) redistribution of implanted impurities within the molten surface layer and ii) segregation and solute

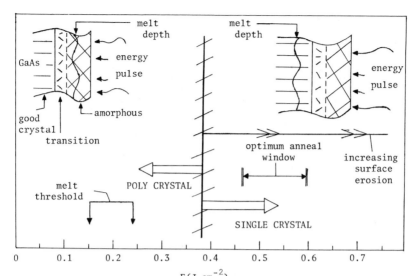

Fig.1. A summary of the important damage removal regimes as a function of
        energy density for pulse-annealed GaAs (ion implanted). See text
        for details.

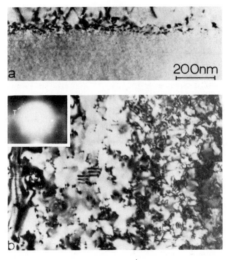

Fig.2. Pulsed ruby laser annealing of Se[+] implanted GaAs at an energy
        density of 0.25 J cm[-2]. a) Cross-section electron micrograph showing
        twins propagated back to the surface; b) Plan-view electron micro-
        graph. Inset; electron diffraction pattern showing twin spots (T).
        From Ref. 13.

trapping at the rapidly moving liquid-solid interface during refreezing.
Evidence for redistribution has been obtained from RBS [22], SIMS [11,12,23]
and electrical profiling techniques [24,25] for several n and p-type
dopants. Segregation of impurities to the surface has also been observed
[12,22,26]. Furthermore, channeling data have clearly indicated the
presence of supersaturated solid solutions (e.g. Te in GaAs at
$> 5 \times 10^{20} cm^{-3}$ [22]). Less direct evidence for supersaturation is obtained
from electrical measurements which show that the activities are higher
than those which can be achieved by furnace annealing [22-27].

Despite the above observations, few quantitative data exist in GaAs
on i) the amount of supersaturation and ii) the influence of the liquid-
solid interface velocity on segregation and solute trapping of impurities.
In addition, the lack of basic thermal conductivity and diffusivity data
in both the solid and liquid phases of GaAs has hampered attempts to
model dopant redistribution and segregation processes [11,12].

Electrical properties and stability
    The important electrical characteristics obtained from pulse-annealed
GaAs are summarised below:
i)   High doses of both n-and p-type dopants can be activated to
concentrations well above $10^{19} cm^{-3}$ [21-27] and, thus, well above doping
concentrations which can be achieved by conventional furnace annealing.
However, the measured active fraction, particularly for n-type dopants,is
typically ≤30%, although substitutional fractions of close to 100% are
measured [22,24].
ii)   Low doses ($<10^{14} cm^{-2}$) of n-type dopants are not active following
pulsed annealing [7,8].
iii)   Mobilities of n-type layers are generally poor (<800 cm^2/Vs) [7,8],
although some higher values (~1400 cm^2/Vs) have been reported [23].
iv)   Layers activated by pulsed annealing have poor thermal stability [28];
substantial reductions in the electron concentrations have been observed
at temperatures as low as 300°C. Interestingly, the results of
Pianetta et al [28] show that the substitutional location of (Te) dopants
does not alter whilst the active concentration drops dramatically during
subsequent thermal annealing.

More detailed reviews of the above observations have recently been
given by Anderson [7], Williams [8] and Eisen [29]. Anderson [7] has also
provided an excellent overview of models to account for the poor electrical
properties resulting from pulsed annealing of GaAs. Basically two
types of proposals have been put forward to explain both the low activity
of n-type dopants and the poor mobilities. These are a) quenched-in
point defect complexes [30-34] and b) the pinning of the Fermi level below
mid-gap by "internal surfaces" [7,35]. As is breifly indicated below,
neither proposal has yet been unequivocally verified. Several studies
have indicated that the lack of activation for low dose n-type dopants
following pulsed annealing is a general result regardless of the starting
material (Cr doped, undoped, etc. [36]), the nature of the initial
implantation damage [17], and the degree of surface dissociation [33].
These studies have attributed the inactivity to quenched-in point defect
complexes. However, although DLTS [30,31] and optical techniques [34,37]
have indicated the presence of deep levels following pulsed annealing, the
measured concentrations (~$10^{15} cm^{-3}$) do not appear to be high enough to
give the necessary level of compensation. Alternatively, Anderson [7]
has suggested that the presence of extended defects following pulsed
annealing may be responsible for Fermi level pinning as proposed by
Fan et al [35]. He argues that this explanation could account for the
measured higher p-type than n-type activity and low mobilities. He further

suggests that most electrical measurements may have been made on pulsed-annealed samples containing extended defects, since the annealing window for defect-free annealing is narrow. However, in order to better understand the poor electrical properties, more detailed studies must be undertaken to characterise and quantify point defect complexes and to monitor the influence of extended defects on the electrical behaviour.

ANNEALING IN THE SOLID PHASE REGIME

Time scales and annealing methods
Annealing in the solid phase can be carried out in time scales greater than about 10^{-3} sec using a number of methods as indicated in Fig.3. For times significantly less than 1 sec, scanning continuous wave annealing (SCWA) using lasers or electron beams has been employed [38-41] In this regime, a focused energy beam heats a local region on the wafer to high temperatures. The scan rate effectively determines the dwell or annealing time, and the incident power determines the local near-surface temperature. It will be useful later to distinquish between slow scan (ss) and fast scan (fs) regimes, corresponding to anneal times of ~1 sec and ~10^{-2} sec, respectively.

Rapid bulk annealing (RBA) can be performed at times exceeding about 1 sec using a variety of thermal pulse annealing (tpa) techniques,mainly employing strip heaters [42-44] and incoherent light sources [45-47] , operating either in vacuum or an inert ambient. Multiply-scanned electron beams (mseb) have been utilized [48] to bulk heat wafers up to high temperatures for times of the order of 1 sec. With the RBA techniques, the rate of temperature rise and hence 'the minimum anneal time is essentially determined by the incident power density and the degree of thermal isolation of the samples. Of course, annealing times greater than a few minutes are readily accommodated by conventional furnace annealing (FA).

The following sections outline the various observations which have been made by processing ion implanted GaAs in the FA,RBA, and SCWA regimes. The curves on the right of Fig.3, to which we return later, provide a summary of the important observations.

Furnace annealing
The typical furnace annealing behaviour is illustrated in Fig.4. Three distinct annealing stages can be identified. A sharp amorphous-to-crystalline transition occurs at temperatures of 150-200°C. This transition can be detected as a reduction in the damage which is measured by channeling [15,49-52] (as shown in Fig.4) or, more directly, by transmission electron microcopy (TEM) [53,54]. However, this basically eptiaxial growth leads to a highly defective crystalline surface layer [53,54] which can consist of a high density of twins, under the particular implant conditions of Fig.4, to more complex polycrystalline structures at considerably higher implantation doses. Under some implantation conditions, involving liquid nitrogen target temperatures and low doses, it is possible to obtain reasonably good epitaxial growth [50-52]. However, as reviewed elsewhere [8] such conditions are not typical; the more general behaviour is for stage I annealing (recrystallisation) to result in highly defective crystalline surface layers.

A second, broad annealing stage in the temperature range ~400 to 800°C is characterised by the removal of the extended crystalline defects. By about 600°C, the twins and large-angle boundaries essentially give way to dislocation loops and other defect clusters [53]. This stage is indicated

626

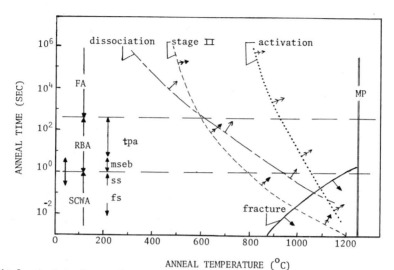

Fig.3. A plot of anneal time vs anneal temperature illustrating the various solid phase annealing regimes and methods. The various curves show the onset of damage removal (stage II), activation (n-type), dissociation ($> 10^{16}$ atoms cm^{-2}) and fracture during the annealing of ion implanted GaAs. See text for a full explanation of symbols.

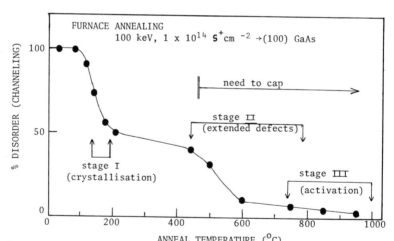

Fig.4. A typical illustration of the various annealing stages during furnace annealing of ion implanted (100) GaAs. The experimental points indicate the removal of S⁺ ion implantation damage as measured by channeling (from Ref. . See text for a full explanation.

in Fig.4 as a marked drop in disorder as measured by channeling. Annealing up to 800°C leads to a further reduction in the dislocation density [53]. Removal of extended defects does not guarantee electrical activation. Particularly for n-type dopants, a third annealing stage (activation) occurs at temperatures of 900-1000°C. It is important to note that the GaAs surface must be protected against dissociation and As loss (by encapulating layers or As overpressures [55,56] for annealing temperatures exceeding about 500°C. Detailed reviews of the electrical properties of ion implanted GaAs have been given by Eisen [55]and Donnelly [56].

Rapid bulk annealing

Several recent studies[42,44-48] have demonstrated that rapid bulk annealing in the time scale 1 to 100 sec can result in good electrical activation and high mobilities for ion implanted n-type GaAs layers. However, optimum electrical properties, at least as good as can be achieved by conventional furnace annealing, are only obtained when the surface is encapsulated during RBA[42,46-48]. More recent channeling studies[47]have indicated some interesting features of RBA of GaAs and these are illustrated in Fig.5. Fig.5a indicates good crystallinity (as measured by channeling of an uncapped 3 x 10^{15}Te cm^{2}implanted GaAs wafer annealed at 850°C for 8 sec using an incoherent light source[47]. However, the hatched area on the annealed spectrum indicates excess surface Ga. Other studies [20] have shown that 3 x 10^{16}As atoms cm^2(~ 30 monolayers) have been lost during this anneal. Despite this somewhat excessive dissociation, the bulk crystallinity is good and high substitutionality (~ 70%) is obtained, giving a peak substitutional concentration of ~10^{21}cm^{-3}.This value is almost certainly above the equilibrium solid solubility limit of Te in GaAs, despite lack of reliable solubility data. In addition, the Te profile following annealing (after normalisation for the 30 monolayers of dissociation loss) is not significantly different from the as-implanted profile. However, despite the apparently good crystal growth and high Te substitutionality, no activity could be measured. This suggests that the annealing conditions have accomplished stage II and not stage III recovery of the implanted GaAs. Fig.5b illustrates typical channeling spectra obtained from capped samples (1 x 10^{14}Sn cm^{-2} implantation, in this case) annealing by RBA at 950°C for 30 sec. The spectra indicate good annealing with little evidence for dissociation, almost total incorporation of Sn onto lattice sites and no detectable migration of Sn. Such annealing conditions give high activity (>30%) and mobility (> 1000cm^2 /Vs) consistent with recent observations of other workers[42,46,48]. Thus, the typical electrical properties of capped samples annealed by RBA for longer times and at high temperatures are at least as good as those obtained by optimum furnace annealing. However RBA offers two major advantages over conventional furnace annealing, as pointed out previously by Sealy [43]:namely, the achievement of good electrical properties without significant dopant diffusion and without the need for extreme care in dopositing encapsulating layers.

Scanning CW annealing

Localised solid phase annealing in times less than 1 sec has not been entirely successful as a means of activating dopants in ion implanted GaAs. The major observations are summarised with reference to Fig.6. For fast, single scanning (Fig.6b), the use of high incident power invariably leads to surface slip and cracking. Even for employment of heated target stages (up to 500°C) to reduce the thermal gradients, the attainable surface temperature is insufficiently high (for ~10^{-2}sec dwell time) to provide good (stage II) annealing without surface cracking [37-41] . However, the use of slow or multiple scanning and heated target stages (Fig.6c), to locally anneal in the time scale of the order of 1 sec, can provide better

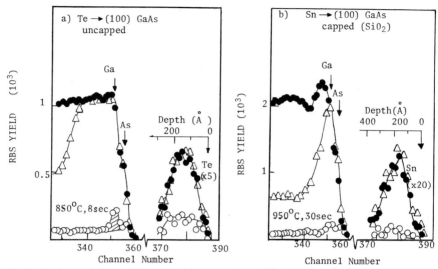

Fig.5. High resolution RBS/channeling spectra indicating typical annealing results obtained with an incoherent light source in the time regime 5 - 100 sec. a) Uncapped annealing of 100 keV 3 x 10^{15} Te^+ implanted GaAs. b) Capped annealing of 100 keV, 1 x 10^{14} Sn^+ implanted GaAs. Random (●) and <100> (O) spectra after annealing and <100> spectra (Δ) before annealing are shown.

Fig.6. Schematic illustration of the various annealing regimes during scanning CW annealing of ion implanted GaAs. See text for a full explanation.

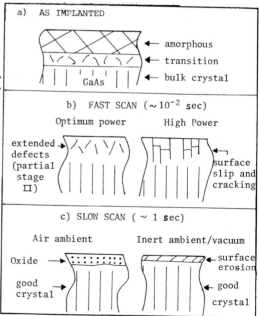

results. However, annealing in air at temperatures above about 950°C
results in the formation of a surface β - Ga$_2$O$_3$ layer [41]. Despite this,
good single crystal with high activation of dopants can be achieved in the
underlying material [41]. Annealing in an inert atmosphere or in vacuum
can similarly result in good activation, but some dissociation of the surface
is inevitable [8]. Thus, SCWA methods operating in the slow scan or multiscan
mode can provide partial activation of dopants, although surface dissociation
limits these applications to deep implanted layers.

Summary and suggested models

The results of solid phase annealing in the various time regimes are
summarised in Fig.3. The region to the right of the curve labelled stage II
indicates the temperature and time regime which typically gives good crystal
recovery and high substitutionality of dopants, but which does not usually
provide high electrical activity. Similarly, the activation curve defines
the regime which can achieve best activation for n-type dopants in GaAs.
However, for acceptable activation of p-type dopants it may be possible to use
shorter times and lower temperatures. As indicated by the dissociation curve,
it is not possible to fully activate n-type layers without capping, regardless
of the time/temperature regime employed. The dissociation curve indicates the
regime where >10^{16} atoms cm^{-2} (ie > 30 monolayers) are removed from the surface
during transient annealing. Surface slip and fracture must also be considered
during rapid local annealing at high temperatures.

Several explanations have been put forward to account for the poor
quality epitaxial growth which takes place during solid phase annealing of
GaAs. The disruption of the local stoichiometry by collisional damage has
been suggested [49,15] as likely to inhibit epitaxial growth. Nissim et al[51]
observed a critical energy deposition density above which eptiaxy was poor,
and Williams et al [52] proposed a non-planar growth process to account for
channeling measurements of recrystallisation in GaAs. Both of these processes
were considered to arise from local stoichiometry imbalance. Indeed, more
recent high resolution TEM studies [54,58] have clearly identified non-planar
growth and subsequent twin formation during annealing of GaAs. Furthermore,
Sadana[58] has observed intriguing surface-crystallisation behaviour. Speriosu
et al [59] have recently shown a transition from elastic to plastic behaviour
during ion implanation of GaAs and suggest that such structural changes may
contribute to the poor epitaxial growth. All of these studies clearly
indicate the complexity of the annealing processes in GaAs as compared with
those observed in Si. A complete explanation of the observed poor epitaxial
recovery must await further,more detailed investigations.

The difficulty in realising good electrical characteristics with ion
implanted GaAs following solid phase annealing has been addressed in some detail
by Eisen [55] and Donnelly [60] . The explanations invoked earlier in this
paper to account for poor activity following pulsed annealing are also
applicable to solid phase annealing. Briefly, point defect complexes[55,56,60]
have long been suggested as possible contributors to low activity.
Bhattacharya et al [61] have recently suggested dopant — Ga vacancy and
dopant — antisite defect complexes to account for compensating effects
in sulphur implanted GaAs. Alternatively, Shahid et al [62]have found a
correlation between dislocation density and electrical activity following
multiply scanned electron beam annealing of Se implanted GaAs. This latter
result lends support to the proposal of Fermi level pinning at internal
surfaces put forward by Donnelly[60]and Fan et al[36] to explain low n-type
activity. More detailed defect characterisation studies are needed to clarify
these proposals.

CONCLUSION

Transient annealing of ion implanted GaAs has not realised the initial
promise of overcoming long standing problems with conventional furnace
annealing for activation of dopants. Both liquid phase and transient solid
phase annealing can result in damage removal and location of dopants onto
substitutional sites but this behaviour does not guarantee good electrical
characteristics. It would appear that residual point defect complexes or low
concentrations of extended defects may control the electrical behaviour.

This review has highlighted several aspects of transient annealing of GaAs
where insufficient data exist to enable a complete understanding of the various
annealing processes. Consequently, the following areas are suggested as
possible avenues for further research.
i) Detailed investigations into the solid phase epitaxial regrowth process
to fully characterise damage removal mechanisms.
ii) Measurements of the solid solubility limits under both liquid and solid
phase annealing conditions to provide data which are not yet available.
iii) Correlate impurity segregation and solute trapping with the liquid-solid
interface velocity to give basic data on rapid melting and solidification
processes in GaAs.
iv) Detailed studies of the type, character and concentration of both point
defect complexes and extended defects to enable correlation with the observed
electrical properties.

ACKNOWLEDGEMENTS

The ARGC, AINSE and the Commonwealth Special Research Centres Scheme are
acknowledged for financial support.

REFERENCES

1. Laser-Solid Interactions and Laser Processing - 1978, S.D. Ferris,
H.J. Leamy and J.M. Poate, eds, AIP Conf. Proc. 50.
(New York, Amer. Inst. Phys., 1979).

2. Laser and Electron Beam Processing of Materials, C.W White and
P.S. Peercy, eds. (New York, Academic, 1980).

3. Laser and Electron Beam-Solid Interactions and Materials
Processing, J.F. Gibbons, L.D. Hess and T.W. Sigmon , eds.(North-Holland,
New York, 1981).

4. Laser and Electron Beam Interactions with Solids, B.R. Appleton and
G.K. Celler, eds. (North-Holland, New York, 1982).

5. F.H. Eisen, in Ref. 2. p. 309.

6. J.S. Williams and H.B. Harrison, in Ref.3, p. 209.

7. C.L. Anderson, in Ref. 4, p. 653.

8. J.S. Williams, Ch.11 in: Laser Processing of Semiconductors,
J.M. Poate and J.W. Mayer eds. (Academic Press, New York 1982), p.383.

9. S.U. Campisano, G. Foti, E. Rinini, F.H. Eisen, W.F. Tseng, M.A. Nicolet
and J.L. Tandon, J. Appl. Phys. 51, 295 (1980).

10. T.N.C. Venkatesan, D.H. Auston, J.A. Golovchenko and C.M. Surko, in
Ref.1, p. 629.

11. D.H. Lowndes, J.W. Cleland W.N. H. Christie,and R.E. Eby, in Ref.3.p.223

12. R.F. Wood, D.H. Lowndes, and W.H. Christie in Ref.3, p. 231.

13. J. Fletcher, J. Narayan, and D.H. Lowndes, in Defects in Semiconductors J. Narayan and T.Y. Tan, eds. (North Holland, New York, 1981),p. 421.

14. A.G. Cullis, these proceedings.

15. J.S. Williams and M.W. Austin, Nucl. Instr. Meth. 168, 307 (1980)

16. R.A. Barnes, H.J. Leamy, J.M. Poate, S.D. Ferris, J.S. Williams and G.K. Celler, Appl. Phys. Lett. 33, 965 (1978).

17. K.Gamo, Y Yuba, A.H. Ornaby, K. Murakami, S. Namba and Y. Kawasaki, in Ref.2,p. 322.

18. M.H. Badawi, B.J. Sealy, S.S. Kular N.J. Barrett, N.Emerson, G.Stephens, G.R. Booker and M. Hockley in Ref. 2, p. 354.

19. T. Inada, K. Tokunaga, and S. Taka, Appl. Phys. Lett. 35, 546 (1979).

20. A. Rose, J.A. Pollock, M.D. Scott, J.S. Williams, F.M. Adams, and E.M. Lawson, these proceedings.

21. J.L. Tandon, I. Golecki, M.A. Nicolet, D.K. Sadana, J. Washburn, Appl. Phys. Lett. 35, 867 (1979).

22. P.A. Barnes, H.J. Leamy, J.M. Poate, S.D. Ferris, J.S. Williams and G.K. Celler, in Ref. 1., p. 647.

23. S.G. Liu, C.P. Wu and C.W. Magee, in Ref.2, p. 341.

24. B.J. Sealy, M.H. Badawi, S.S. Kular, and K.G. Stephens, in Ref.1., p.610.

25. T. Inada, S. Kato, Y. Maeda, K. Tokunaka, J. Appl. Phys. 50, 6000 (1979).

26. H.B. Harrison and J.S. Williams, in Ref.2, p. 481.

27. P.A. Pianetta, C.A. Stolte and J.L. Hanson, Appl. Phys. Lett. 36.697 (1980).

28. P.A. Pianetta, J. Amano, G. Woolhouse, and C.A. Stolte, in Ref.3,p.239.

29. F.H. Eisen, in Ion Implantation and Beam Processing, J.S. Williams and J.M. Poate, eds. (Academic Press, Sydney, 1983).

30. N.G. Emerson and B.J. Sealy, Electron. Lett, 16, 512 (1980).

31. P. Mooney, J.C. Baurgoin, and J. Icole, in Ref.3, p. 255.

32. C.L. Anderson, H.J. Dunlap, L.D. Hess, G.L. Olson, and V.Y. Vaidyanathan, in Ref. 2, p. 334.

33. D.E. Davies, J.P. Lorenzo, E.F. Kennedy, and T.G. Ryan, in GaAs and Related Compounds, Inst. Phys. Con.f Ser, 63, 389 (1982).

34. M. Takai, S. Ono, Y. Ootsuka, K. Gamo and S. Namba, in Ion Implantation Equipment and Techniques, H. Ryssel et al, eds (Springer-Verlag, Munich (1982).

35. J.C.C.Fan, R.L. Chapman, J.P. Donnelly, G.W. Turner, and C.O. Bozler, in Ref.3,p.261.

36. I. Golecki, M.A. Nicolet M. Maenpaa, J.L. Tandon, C. Kirkpatrick, D.K. and J. Washburn, in Ref.2,p. 347.

37. D.H. Lowndes and B.J. Feldman, in Ref.4, p. 689.

38. J.S. Williams, M.W. Austin and H.B. Harrison in Thin Film Interfaces and Interactions, J.E.E. Baglin and J.M. Poate, eds (ECS, Princeton 1980) p. 187.

39. J.C.C. Fan, R.L. Chapman, J.P. Donnelly, G.W. Turner, CO. Bozler, Appl. Phys. Lett. 34, 780 (1979).

40. G.L. Olson, C.L. Anderson H.L. Dunlap, L.D. Ness, R.A. Mc Farlene, and K.V. Vaidyanathan, in Laser and Electron Beam Processing of Electronic Materials, C.L. Anderson, G.K. Celler and G.A. Rozgonyi eds. (ECS, Princton 1980).p. 467.

41. Y.I. Nissim and J.F. Gibbons, in Ref.3, p. 275.

42. B.J. Sealy and R.K. Surridge, IBMM - 78 Conf. Proc. J. Gyulai et al, eds. (Budapest, 1979), p. 487.

43. B.J. Sealy, Microelectronics Journal 13, 21 (1982).

44. R.L. Chapman, J.C.C. Fan, J.P. Donnelly, B-Y Tsaur, Appl. Phys. Lett. 40, 805 (1982).

45. M. Arai, K. Nishiyama, and N. Watanabe, Japan J. Appl. Phys. 20, L124 (1981).

46. D.E. Davies, P.J. Mc Nally, T.G. Ryan, K.J. Soda and J.H. Comer, to be published.

47. H.B. Harrison, F.M. Adams, B. Cornish, S.T. Johnson, K.T. Short and J.S. Williams, these proceedings.

48. N.J. Shah, H. Ahmed, P.A. Leigh, Appl. Phys. Lett. 39, 322 (1981).

49. K. Gamo, T. Inada, J.W. Mayer, F.H. Eisen, and C. G. Rhodes, Radiation Effects 33, 85 (1977).

50. M.G. Grimaldi, B.M. Paine, M.-A. Nicolet, and D.K.Sadana, J. Appl. Phys. 52, 4038 (1981).

51. Y.I. Nissim, L.A. Chrstel, T.W. Sigmon, J.F. Gibbons, T.J. Magee and R. Ormond, J. Appl. Phys. 52, (1981).

52. J.S. Williams, F.M. Adams and K.G. Rossiter, Thin Solid Films 93, 139 (1982).

53. S.S. Kular, B.J. Sealy, K.G. Stephens, D.K. Sadana and G.R. Booker Solid State Electron. 23, 831 (1980).

54. J. Narayan, J. Appl. Phys. (in press).

55. F.H. Eisne, in Ion Beam Modification of Materials. J. Guylai, T. Lohner and E. Paztor eds. (Hung. Acad. Sci. Budapest 1978) p.147.

56. J.P. Donnelly, Inst. Phys. Conf. Ser. 33, 167 (1977).

57. F.M. Adams, J.S. Williams and H.B. Harrison, to be published.

58. D.K. Sadana, private communication

59. V.S. Speriosu, B.M. Paine, M.A. Nicolet and H.L. Glass, Appl. Phys. Lett. 40, 604 (1982).

60. J.P. Donnelly, in Ion Beam Modification of Materials-1980. R.E. Benenson et al eds. (North Holland Amsterdam 1981) p.553.

61. R.S. Bhattacharya, A.K. Rai, Y.K.Yeo, P.P. Pronko and Y.S. Park, to be published.

62. M.A. Shahid, S. Moffatt, N.J. Barrett, B.J. Sealy, K.E. Puttick, B.J. S.T. (to be published).

NEUTRON ACTIVATION MEASUREMENTS OF As AND Ga LOSS DURING
TRANSIENT ANNEALING OF GaAs

A. ROSE, J.T.A. POLLOCK AND M.D. SCOTT
CSIRO, Division of Chemical Physics, L.H.R.L., N.S.W. 2232. Australia

F.M. ADAMS AND J.S. WILLIAMS
Faculty of Engineering, R.M.I.T., Vic. 3000.

E.M. LAWSON
Australian Atomic Energy Commission, L.H.R.L. N.S.W. 2232. Australia

ABSTRACT

    Significant dissociation is normally detected under non-optimised transient
annealing of GaAs. We have utilised neutron activation to measure As and Ga
loss from virgin and implanted material annealed under various transient
conditions. Complementary RBS data are reported. In particular, surface
dissociation has been measured as a function of pulsed ruby laser power and for
several combinations of time and surface temperature using an incoherent light
source and a vitreous carbon strip heater.
    The results indicate that neutron activation analysis offers a powerful
tool to identify the conditions required to minimise GaAs dissociation during
annealing. For example, ruby laser pulses of energy 0.29-1.38 J cm^{-2} caused
As loss of 4 - 90 x 10^{14} cm^{-2}.

INTRODUCTION

    Transient annealing methods for the removal of the radiation damage
accompanying ion implantation of semiconductors have been widely reported
[1,2]. For silicon, the annealing conditions required for the restoration of
single crystallinity and electrical activation of the doped layer are
reasonably well established. The situation for GaAs is less satisfactory [3].
Initial enthusiasm for transient annealing, based on the expectation that
reduced dissociation would accompany such methods and allow uncapped processing,
has been tempered by the understanding that Ga and As loss still occur and that
the optimum conditions for successful annealing need to be much more closely
defined than is the case for silicon. A method for the accurate measurement of
this loss is a desirable prerequisite to the definition of optimum annealing
conditions and subsequent electrical characterisation. In this paper we report
data which suggests that neutron activation analysis can meet this requirement.

EXPERIMENTAL

    Virgin (100) oriented GaAs was primarily used in this work. Some samples
were implanted with Te under conditions as detailed in Tables I and II. As and
Ga loss associated with three transient methods were measured; pulsed ruby
laser, incoherent quartz halogen light and a vitreous carbon strip heater. In
each case, As and Ga lost during the anneal were deposited on quartz catcher
slides located above and close to, but not touching, the GaAs samples [4].
These slides were then used in the neutron activation and Rutherford
backscattering analyses (NAA and RBS) for As and Ga.

Mat. Res. Soc. Symp. Proc. Vol. 13 (1983) ©Elsevier Science Publishing Co., Inc.

TABLE I
Arsenic and gallium loss measured for various laser
energies incident on GaAs samples

Laser energy (joules)	Number As atoms cm^{-2} (x 10^{14})	Number Ga atoms cm^{-2} (x 10^{14})	As:Ga atom ratio
0.29	7.5	0.9	8.2
0.36	4.0	1.6	2.5
0.42	9.5	2.8	3.4
0.57	33.0	17.0	1.9
0.58	8.9	2.7	3.3
0.74	23.0	16.0	1.4
0.83	27.0	13.0	2.1
1.08	43.0	21.0	2.0
1.10	36.0	23.0	1.6
1.38	90.0	86.0	1.0
0.64[a]	5.6	3.1	1.8
0.70[a]	6.6	3.8	1.7

[a] 3 x 10^{15} cm^{-2}, 100 keV Te

TABLE II
Arsenic and gallium loss measured as a function
of time and temperature

Time (mins)	Temperature (°C)	Number As atoms cm^{-2} (x 10^{15})	Number Ga atoms cm^{-2} (x 10^{15})	As:Ga atom ratio
		Strip heater; helium atmosphere		
0.5[a]	1020	160.0	230.0	0.7
0.5[a]	850	9.0	12.0	0.8
0.5[a]	820	2.1	0.5	4.6
1.0[a]	820	5.8	0.5	11.4
2.0[a]	820	5.2	1.3	4.0
5.0[a]	820	4.5	25.0	0.2
		Quartz lamp; vacuum		
0.5[a]	850	48.3	46.9	1.0
0.5[b]	850	107.6	119.4	0.9
0.5[c]	850	254.7	240.5	1.1

[a] Virgin
[b] 5x10^{15} cm^{-2} 100 KeV Te
[c] 3x10^{15} cm^{-2} 100 KeV Te

Annealing Methods

A Q-switched ruby laser operating in multimode with a pulse width of 25
nsec and calibrated against a volume absorbing calorimeter was used for the
laser work. A quartz homogeniser with an exit face diameter of 10 mm was
introduced to produce a pulse of uniform spatial intensity. All laser
annealing was carried out in air with energies in the range 0.29 to 1.38 J
cm^{-2} incident on the sample. This experimental set-up and range of energies
are similar to those used by other workers [5, 6, 7].

Incoherent light annealing was carried out using focussed light from two quartz – halogen lamps to heat GaAs samples to temperatures near 850°C for 30 secs in vacuum. Another series of solid phase anneals were performed in flowing He by placing GaAs samples on a 6.5 cm long x 1.25 cm wide x 0.05 cm thick vitreous carbon strip heater. Dissociation was measured as a function of temperature from 800-1020°C with the heater switched on for 30 s, and as a function of time, 30-300 s, at a constant temperature of 820°C. With these latter methods, sample temperatures were measured by a chromel-alumel thermocouple contacting either the sample or an adjacent control. The accuracy of the vacuum measurement is in doubt and should be used as a guide only.

NAA and RBS Analyses

Neutron activation of the As and Ga on the catcher slides was carried out using the Australian Atomic Energy Commission's Reactor Moata, where a neutron flux of 6.6×10^{11} n cm^{-2}s^{-1} is available. Thermal neutron capture cross sections and decay product half lives for As and Ga are such that both could be determined to better than 10^{13} atoms cm^{-2}.
The relevant capture and decay equations together with the gamma ray counted are;

$$As^{75} + n \longrightarrow As^{76} \xrightarrow{26.3 \text{ h}} Se^{76} \text{ , } 559 \text{ KeV } \gamma\text{-ray}$$
$$\text{and } Ga^{71} + n \longrightarrow Ga^{72} \xrightarrow{14.1 \text{ h}} Ge^{72} \text{ , } 834 \text{ KeV } \gamma\text{-ray}$$

Selected quartz catchers were analysed by RBS as an alternate method for determining As and Ga deposits. In addition, several annealed GaAs wafers were examined using both backscattering and channeling to determine the extent of surface dissociation and damage, and redistribution of implanted species.

RESULTS AND DISCUSSION

Typical spectra obtained by NAA and RBS of the catcher slides are shown in Figures 1 and 2, respectively. Under the particular RBS analysis conditions employed in the present study, sensitivity for combined As and Ga was 10^{13} cm^{-2}. However, overlap of the Ga and As peaks made it difficult to measure the individual Ga and As amounts. For total Ga and As loss greater than 5×10^{15} cm^{-2} it was not possible to adequately separate the Ga and As peaks. This limitation is not experienced with NAA, where sensitivity for Ga and As is conservatively 1×10^{13} cm^{-2} and is not degraded by increasing quantities of deposit. In cases where the combined As and Ga content was measured by both techniques, excellent agreement resulted.

At all but the highest pulse energies, As losses were greater than those for Ga, as measured by NAA (Table I and Figure 3). As shown in Figure 4, for pulse energies >1 J cm^{-2}, the excess Ga can be easily observed by channeling studies on the annealed GaAs surface. Also, laser annealing at pulse energies >0.3 J cm^{-2} resulted in surface melting. For example, as shown by the channeling spectra in Figure 5, for pulse energy of 0.85 J cm^{-2}, removal of lattice damage, redistribution and accommodation of Te on substitutional sites are obtained. Excess Ga may also be detected at the annealed GaAs surface in the spectra obtained with this sample. These results (Figures 4 and 5) are consistent with previously reported studies [6, 7, 8, 9] and imply that good annealing has occurred. Nevertheless, as reviewed by Anderson [10], channeling is not sensitive to low concentrations of extended defects or to point detect clusters, which are the likely cause of the less than satisfactory electrical properties generally measured in pulse annealed GaAs. The corresponding catcher slide for the sample whose spectra are shown in Figure 5, revealed a combined As and Ga loss of >10^{15} cm^{-2}.

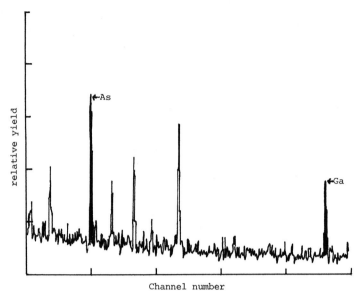

Fig.1 Typical NAA spectra showing 559 KeV
       As peak and 834 KeV Ga peak

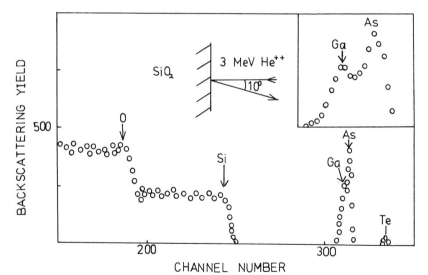

Fig. 2 RBS spectra from a quartz catcher corresponding to incoherent light
       source annealing of GaAs for a $T_{max.}$ of 850°C attained after
       4.5 sec. Insert shows the expanded signal from Ga and As.

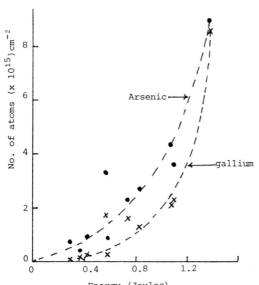

Fig.3 As and Ga losses as a function of
incident laser energy

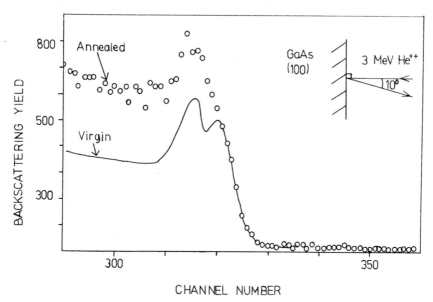

Fig.4 Channeling ⟨100⟩ spectra from virgin and pulsed laser annealed (1.4 J cm^{-2})
GaAs.

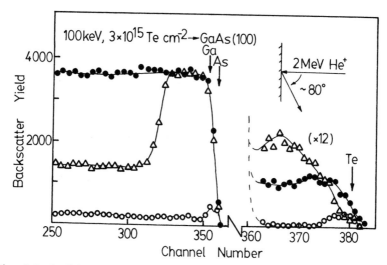

Fig. 5 Random(•) and channeling ⟨100⟩ spectra before (▲) and after (○) pulsed laser annealing at 0.9 J cm^{-2} of 100 KeV 3×10^{15} Te cm^{-2} implanted GaAs.

We have been unable to find any published measurements of Ga and As loss during transient (10^{-2} to 10^{2} sec) solid phase annealing. Our NAA results presented in Table II are preliminary and taken together with RBS data allow the following points to be made. Clearly, annealing between 800 and 1020°C resulted in substantial loss for uncapped GaAs. These estimates are conservative due to heating of the quartz catcher during the anneal. In this regard, it was noted that anomalously low As/Ga ratios were measured for anneals extending to high temperatures or long times. Using implanted samples we have shown that damage removal and incorporation of Te onto substitutional sites, as revealed by channeling data, are obtained despite substantial Ga and As loss. Rough indications are that a combined loss of about 5×10^{16} cm^{-2} takes place. Nevertheless, it is possible that the removal by dissociation of ~ 75 A of GaAs may be tolerated, particularly for higher energy implants. However, recent reports [11, 12, 13] suggest that it is necessary to encapsulate GaAs during transient solid phase annealing to obtain best electrical properties.

CONCLUSIONS

The results demonstrate that NAA may be used to accurately measure the individual Ga and As loss associated with capless transient annealing. For combined Ga and As loss, the NAA results are consistent with RBS data. Good annealing of implanted samples was obtained with laser pulse energies of ≈ 0.9 J cm^{-2} where NAA indicated Ga and As loss of 2 and 4×10^{15} cm^{-2}, respectively. Equivalent loss measured with solid phase annealed samples was an order of magnitude higher.

ACKNOWLEDGEMENT

We thank R. A. Clissold and M. Farrelly for their experimental assistance.

REFERENCES

[1] "Laser and Electron Beam Processing of Electronic Materials", Eds. C. L. Anderson, G. K. Celler and G. A. Rozgoni, ECS, Princeton, N.J. (1980).

[2] "Laser and Electron Beam Solid Interactions and Materials Processing", Eds. J. F. Gibbons, L. D. Hess and T. W. Sigmon, North Holland, N.Y. (1981).

[3] J. S. Williams and H. B. Harrison, ref 2, page 209.

[4] K. Gamo, Y. Yuba, A. H. Oraby, K. Wurakami, S. Namba and K. Kawaski, "Laser and Electron Beam Processing of Materials" Eds. C. W. White and P. S. Peercy, Academic Press, N.Y. 1980.

[5] A. C. Cullis, H. C. Welser and P. Bailey, J. Phys. E. Sci. Inst. 12, 452 (1979).

[6] R. F. Wood, D. H. Lowndes and W. H. Christie, ref. 2, page 231.

[7] D. H. Lowndes, J. W. Cleland, W. H. Christie and R. E. Eby, ref. 2, page 22.

[8] P. A. Barnes, H. J. Leamy, J. W. Poate, S. D. Ferris, J. S. Williams and G. K. Celler, Appl. Phys. Lett. 33, 965 (1978).

[9] J. Fletcher, J. Narayan and D. H. Lowndes, page 421, "Defects in Semiconductors", Eds. J. Narayan and T. Y. Tan, North Holland, N.Y. (1981).

[10] C. L. Anderson, page 653, "Laser and Electron Beam Interaction with Solids", Eds. B. R. Appleton and G. K. Celler, North Holland, N.Y. (1982).

[11] D. E. Davies, P. J. McNally, P. J. Ryan, K. J. Soda and J. J. Comer, Presented at GaAs conference, Japan, Sept. 1982.

[12] M. J. Shah, A. Ahmed, I. R. Sanders and J. F. Singleton, Elect. Letts. 16, 433 (1980).

[13] R. L. Chapman, J. C. C. Fan, J. P. Donnelly and B-Y. Tsaur, Appl. Phys. Lett. 40, 805 (1982).

RAPID ISOTHERMAL ANNEALING OF Si IMPLANTED SEMI-INSULATING GaAs BY MEANS OF HIGH FREQUENCY INDUCTION HEATING.

A.CETRONIO, M.BUJATTI*, P.D'EUSTACCHIO and S.CICERONI
SELENIA S.p.A., Research Laboratories, Via Tiburtina, 00131 ROMA, ITALY.

ABSTRACT.

Experimental results obtained by rapid isothermal annealing of Si implanted semi-insulating GaAs are presented and compared with results obtained by electron-beam and conventional furnace annealing. We shall show that for high Cr-doped material the problem of redistribution cannot be avoided and that for undoped material, this technique yields very good activation (near 100%) and high mobility (approximately 4400 cm^2/V.sec).

INTRODUCTION.

Transient and furnace annealing of low dose ion implanted semi-insulating GaAs for MESFET applications have revealed that:

a) pulsed laser and electron-beam processes ($<$100 nsec duration) produce no, or at best very low activation (1,2);

b) scanning C.W. laser or electron beam processes yield reasonably good activation and that the electrical properties improve with increasing dwell times from approximately 1 msec to several seconds, achieving in the latter case near 100% activation and high electron mobility (3);

c) furnace annealing processes at temperatures $>$ 800°C yield an encapsulant and anneal time dependent activation (typically $<$ 100%), a phenomenon which is more pronounced in undoped Liquid Encapsulated Czhocralsky (LEC) materials.

From these observations and from recently published results with a graphite strip heater (4) it is apparent that a rapid isothermal annealing (RIA) technique, of several second duration, should have all the advantages of the best results obtained with C.W. electron-beam annealing, without the disadvantages of furnace annealing, and with the added advantage that the necessary equipment is much less expensive and simpler to use.

In this paper the results obtained by RIA will be presented and compared with results obtained by conventional furnace and multy-scanned electron-beam annealing.

EXPERIMENTAL.

The system used for this study was a Lepel high frequency (150 KHz) induction heating unit which acting onto a thin-walled (circa 2 mm) graphite sample holder (shown in fig.1a) by skin-effect was capable of attaining temperatures in the range 700 to 1200°C in approximately 5 seconds. The temperature of the sample holder enclosed in a pyrex tube through which is passed a stream of nitrogen gas was measured by both a chromel/alumel thermocouple and an optical pyrometer. The annealing cycle time t_a was taken as the time necessary to reach maximum temperature and as such the profile shown in fig.1b corresponds to a 10

*Present address: Santa Rosa Technology Ctr., Hewlett-Packard, Santa Rosa, CA

Mat. Res. Soc. Symp. Proc. Vol. 13 (1983) ©Elsevier Science Publishing Co., Inc.

second cycle.

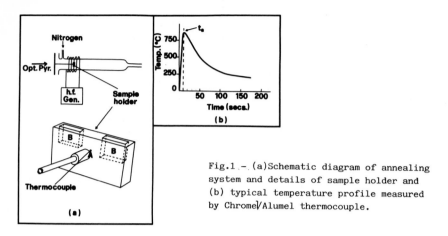

Fig.1 - (a)Schematic diagram of annealing system and details of sample holder and (b) typical temperature profile measured by Chromel/Alumel thermocouple.

High and low Cr-doped and undoped (100) semi-insulating GaAs substrates were implanted 7° off the normal axis with 200 keV Si ions with a dose of $8x10^{12}cm^{-2}$. Following this the samples were encapsulated with a 3000Å thick Si_3N_4 film deposited by sputtering and then activated by RIA.

The sheet carrier concentrations of the activated Si after RIA were obtained from saturation current (I_s) measurements assuming a saturation velocity $V_{s=}$ $=1.2x10^7$ cm/sec. Free electron and mobility depth profiles were also obtained from I_s measurements and the initial slope of the I-V characteristic (5) respectively, by means of a differential stripping technique using a dilute solution of ammonium hydroxide and hydrogen peroxide.

RESULTS.

The experimental results of sheet carrier concentration versus annealing cycle time for a predetermined temperature revealed a strong dependence between the type of material used, (i.e. wether high or low Cr-doped or undoped) and the activation process. Since this phenomenon appears to be related to the quantity of Cr present in the material our presentation has been divided in high Cr and low or no Cr content materials.

High Chrome Content Materials.

This material is typically Bridgman grown with a Cr concentration in the range $5x10^{16}$ to $1.5x10^{17}$ cm^{-3}. As shown in figure 2, in this case the sheet carrier concentration with annealing cycle time t_a at a temperature of approximately 850°C exhibits a complicated activation process. First there is a rapid increase in activation with increasing t_a up to 20 sec, and then followed by a drop in activation and a final increase to a maximum sheet carrier concentration of approximately $4x10^{12}cm^{-2}$ for a 100 sec annealing cycle.

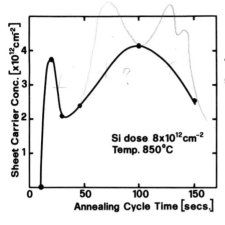

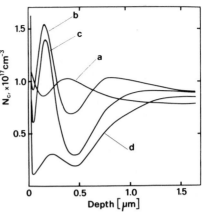

Fig.2 - Sheet carrier concentration versus annealing cycle time for high Cr-doped material implanted with a Si dose of $8\times10^{12}cm^{-2}$ and annealed at 850°C.

Fig.3 - SIMS analysis of Cr distributions for the high Cr doped material after (a) 20 sec, (b) 30 sec, (c) 45 sec and (d) 150 sec annealing at 850°C.

The decrease in sheet carrier concentration for t_a greater than 100 sec is at present unclear, however the drop in carrier concentration for t_a approximately 30 sec has been shown by SIMS analysis (6) to be reasonably well correlated to Cr redistribution. As shown by the chrome depth profiles in fig.3, a 20 sec annealing cycle (a) gives rise to a gradually increasing Cr accumulation towards the surface with a region slightly depleted of Cr at a depth of circa 0.15 um. With increasing annealing cycle time (i.e. 30 and 45 secs) apart from the Cr build up at the GaAs surface (b and c) there is also a pronounced Cr accumulation in the region of the Si implant followed by a corresponding Cr depletion further in depth. For an annealing cycle time of 150 secs (d) the Cr is found to be all

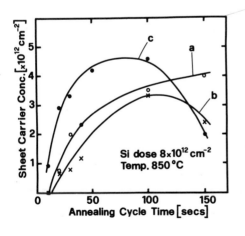

Fig.4 - Sheet carrier concentration versus annealing cycle time for (a) undoped LEC (b) low Cr-doped LEC and (c) low Cr-doped Bridgman grown materials implanted with a Si dose of $8\times10^{12}cm^{-2}$ and annealed at 850°C.

accumulated at the surface and thus creating a trough depleted of Cr (i.e. mean value approximately $2x10^{16}cm^{-3}$) circa 0.5 um deep.

The carrier concentration profiles corresponding to annealing cycle times of 20, 30, 45 and 100 secs respectively in fig.2 are in reasonably good agreement with the Cr depth profiles in fig.3. In fact for t_a = 20 and 100 secs. the profiles are found to be undistorted with the maximum in carrier concentration ($\hat{n}$) at a depth of approximately 0.2 um and with a corresponding mobility of circa 3600 $cm^2/V.sec$. For t_a = 30 and 45 sec, corresponding to Cr accumulation in the Si implanted region, the carrier concentration profiles are found to be smaller in magnitude, distorted and with a much reduced mobility at $\hat{n}$ of approximately 1800 $cm^2/Vsec$.

Low Chrome Content Materials.

This material includes i) Cr-doped Bridgman grown with $N_{Cr}=1x10^{16}cm^{-3}$; ii) Cr-doped LEC grown with $N_{Cr}=5x10^{15}cm^{-3}$; and iii) undoped LEC grown. As shown in figure 4 for these materials the activation process appears to be much simpler, with the sheet carrier concentration increasing with increasing t_a up to 100 sec in all three cases, indicating (unlike the high Cr doped material) the absence of noticeable Cr redistribution. The only feature in common between the characteristics presented in fig.4 and that in fig.2 is the decrease in sheet carrier concentration, with increasing t_a above 100 secs, for the Cr-doped materials. For the undoped LEC material the carrier concentration increases slightly with increasing t_a above 100 secs indicating that this material is the most stable for the annealing conditions investigated.

The carrier concentration and mobility profiles for these materials at t_a = 100 and 150 secs are shown in figure 5. As illustrated for t_a = 100 secs the LEC grown materials have attained the maximum possible activation, and are well behaved (i.e. undistorted) as compared with the theoretical profile (7), with a mobility at $\hat{n}$ of approximately 3700 and 4600 $cm^2/V.sec$ for the low Cr-doped and undoped material respectively. The behaviour of the low Cr-doped Bridgman grown material is found to be unusual, indicating a much broader carrier concentration profile with a reasonable mobility of circa 4100 $cm^{-2}/V.sec$ at $\hat{n}$. The significance of this result is at present not very clear, but tentatively could be attributed to unidentified impurity redistribution (most probably Cr).

As shown by the results for t_a = 150 secs the profiles for the low Cr-doped materials have become distorted (i.e. reduced $\hat{n}$ and shifted) with corresponding decreases in mobility. However the undoped LEC grown material has retained the well behaved carrier concentration profile and mobility (i.e. approximately 4400 $cm^2/V.sec$ at $\hat{n}$) attained with t_a = 100 secs.

DISCUSSION.

For the high Cr-doped material the redistribution with increasing annealing time appears to be in reasonable agreement with the result obtained with multy-scanned electron-beam annealing (3), indicating that two gettering sources exist for the Cr, namely gettering at the surface due to encapsulant induced stresses and gettering at the implantation damage zone. Since prior to annealing the Cr is uniformity distributed in the GaAs material, (N_{Cr} approximately $9\times10^{16}cm^{-3}$ in this case), then for short anneal times (such as 20 secs) gettering at the surface and at the ion implant damage zone are equally possible. As shown by (a) in fig.3 such a situation could result in a slight surface accumulation of Cr with a corresponding depletion at a depth of approximately 0.15 um and then a slight accumulation of Cr further in depth due to initial Cr movement towards the implant damage zone. For slightly longer anneal times (such as 30 to 45 secs) since the surface vicinity has already been depleted of Cr then gettering at the implant damage zone predominates, resulting in a large Cr accumulation in that region and a corresponding depletion at a depth of approximately 0.4 um, as shown by (b) and (c) in fig.3. For long anneal times (such as 150 secs) the thermal mismatch at the cap/GaAs interface predominates resulting in further Cr gettering at the surface and thus the eventual formation of a trough depleted of Cr as shown by (d) in fig.3.

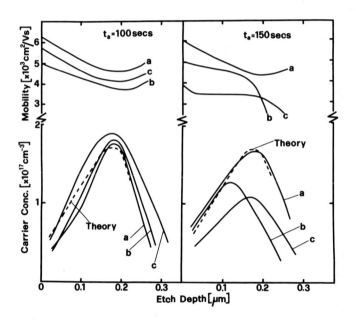

Fig.5 – Carrier concentration and mobility profiles for (a) undoped LEC (b) low Cr-doped LEC and (c) low Cr- doped Bridgman grown materials for t_a=100 and 150 secs.

Unlike conventional furnace annealing, with RIA the Cr redistribution is very rapid, most probably due to stresses induced by fast thermal excursions, and thus problems related to high Cr-doped material cannot be avoided with this technique.

For the low Cr-doped and the undoped LEC material Cr redistribution is not evident, as shown by the characteristics in fig.4. However as indicated by the "broad" carrier concentration profile for t_a=100 secs for the low Cr-doped

Bridgman grown material and as indicated by the drop in sheet carrier concentration for t_a=150 secs for all Cr-doped materials an impurity redistribution is more than likely. Such impurities could be Cr, even though in low concentrations, and most probably oxygen which is known to be present in both low Cr-doped materials.

As expected the undoped LEC grown material results in a sheet carrier concentration which increases with increasing annealing cycle time, up to t_a=100 secs and then remains virtually unchanged with further increase up to t_a=150 secs illustrating the stability of this material. In fact as shown the carrier concentration profile is in good agreement with the theoretical profile and the mobility (i.e. approximately 4600 cm/V.sec)is near to the theoretical maximum expected for this material.

CONCLUSION.

From the experimental results presented we can conclude that:

1) Very good activation with maximum mobility is possible with rapid isothermal annealing.

2) Cr redistribution is accelerated (as compared with furnace annealing) and thus problems related to high Cr content materials cannot be avoided with RIA.

3) LEC material with low or no Cr proves to be the most stable with this annealing technique.

REFERENCES.

1. I.Golecki, M.A.Nicolet, M.Maenpaa, J.L.Tandom, C.G.Kirkpatric, D.K.Sadana and J.Washburn, in Laser and Electron Beam Processing of Materials, Edited by C.W. White and P.S. Peercy, Academic Press, New York (1980) p.347.

2. F.H.Eisen, in Laser and Electron Beam Processing of Materials, Edited by C.W. White and P.S. Peercy, Academic Press, New York (1980), p.309.

3. M.Bujatti, A.Cetronio, R.Nipoti and E.Olzi, Appl. Phys. Lett. 40(4), 334 (1982).

4. R.L.Chapman, John C.C.Fan, J.P.Donnelly and B.Y.Tsaur, Appl. Phys. Lett. 40(9), 805 (1982).

5. R.A.Pucel and C.F.Krumm, Electronics Letters 12, 240 (1976).

6. SIMS analysis by Charles Evans and Associates.

7. M.Bujatti and F.Marcelja, IEEE Proc., 128 Pt., 97 (1981).

ELECTRICAL DEFECT ANALYSIS FOLLOWING PULSED LASER IRRADIATION OF UNIMPLANTED GaAs*

D. PRIBAT, S. DELAGE, D. DIEUMEGARD, M. CROSET,
Thomson-CSF/LCR, Domaine de Corbeville, 91401 Orsay, France.
P.C. SRIVASTAVA, J.C. BOURGOIN,
Groupe de Physique des Solides de l'E.N.S. Université Paris VII,
Tour 23, 2 Place Jussieu, 75221 Paris, France.

ABSTRACT

Current-voltage, capacitance-voltage and defect spectroscopy techniques are used to characterize the electrical properties of GaAs crystals after pulsed laser irradiation with either a Nd-YAG or a Ruby laser. I(V) and C(V) measurements performed in conjunction on Au/GaAs Schottky structures after laser irradiation at low energy density show an important barrier lowering, of the order of 300mV. Carrier compensation up to $6 \times 10^{16}/cm^3$ is observed in a subsurface layer whose thickness increases with deposited laser energy density. D.L.T.S. is used to study the tail of laser induced defects behind the heavily compensated layer. Finally the results are compared to those obtained following conventional thermal treatment.

INTRODUCTION

It is now recognized that pulsed laser irradiation of either ion implanted or virgin GaAs, while providing layers free of extended defects for an adequately chosen energy density |1|, does leave or introduce |2,3,4,5| point defects which affect both electrical and optical |2,6| properties of the processed material. However the lack of electrical activation which has so far been reported for low dose n-type implantations has not been clearly related to laser generated active defects |4,6|. This non activation phenomenon was rather attributed to implantation defects left after laser irradiation |2,3,6|.

The purpose of this paper is to clarify the situation regarding defect introduction and carrier removal in uniformly doped, unimplanted GaAs material, irradiated with single pulses from Ruby or Nd-YAG lasers. For this we studied the electrical characteristics of Au/GaAs Schottky structures by means of Current-Voltage, Capacitance-Voltage and Defect Spectroscopy techniques.

We report an important barrier lowering which is accompanied by bulk defect generation as the deposited energy is raised. The results are compared to those obtained following conventional face to face thermal treatment.

EXPERIMENTAL DETAILS

The study has been performed on n-type doped GaAs wafers with (100) surface orientation. Wafers from different suppliers were used, including Metal Research (CZ grown, Te doped : $4-6 \times 10^{16}/cm^3$, ingots A and B) and Sumitomo (Bridgmann grown, Si doped : $2-3 \times 10^{16}/cm^3$, ingot C). Three groups of samples have been prepared from each ingot, namely, a reference, a thermally annealed, and a laser irradiated group. All the samples were etched about $10 \mu m$ thick prior to any operation.

The thermal treatment consists in baking the samples at 850°C for 30mn in an H_2 atmosphere. The As partial pressure is maintained by capping each sample with a polished As doped silicon wafer.

---

* Work partially supported by D.R.E.T.

Mat. Res. Soc. Symp. Proc. Vol. 13 (1983) © Elsevier Science Publishing Co., Inc.

The lasers used were a Q-switched Ruby laser (pulse duration 25ns, λ = 0.69μm) and a Nd-YAG laser (pulse duration 20ns). The Nd-YAG laser operated either in the green (λ = 0.53μm), in the infra-red (λ = 1.06μm) or in a mixture of the two. The energy in the laser spots was made uniform through the use of either a quartz pipe (Nd-YAG laser) or a diaphragm (Ruby laser). In both cases the beam diameters were of 6mm.

After irradiation, the Schottky devices ($\emptyset$ = 340μm) were fabricated by evaporating a front Au contact through a mask. The backside AuGe contact was sputter deposited and subsequently alloyed at 400°C during 10mn in flowing H_2 prior to Au deposition.

Careful characterizations were first performed on reference samples from each ingot. Carrier concentrations were measured by C(V) techniques using a HP 4275 A LCR Bridge. The ideality factors of the investigated devices ranged from 0.99 to 1.12. The D.L.T.S. spectra recorded on devices from ingot A exhibited practically one single peak corresponding to a 0.83eV ionization energy (E_C - 0.83eV). This trap level is related to a native defect in bulk n-GaAs |7| and its depth concentration was found to be constant around $4\times10^{15}/cm^3$. D.L.T.S. spectra of Sumitomo samples also exhibited one single peak corresponding to a 0.36eV trap level which has also been reported as a native defect in bulk n-GaAs |2,8|. Its concentration was found to be around $6\times10^{15}/cm^3$. No trap levels were detected in the B ingot at a concentration higher than a few $10^{13}/cm^3$.

RESULTS AND DISCUSSION

Laser annealing

Fig. 1 shows the I(V) characteristics of Schottky devices (Sumitomo GaAs) built on (i) reference material, (ii) low energy irradiated material (0.53μm(20%)+1.06μm(80%) /480mJ.cm^{-2}) and (iii) high energy irradiated material (0.53μm + 1.06μm (80%) /1J.cm^{-2}). A first thing to be pointed out is that despite of the use of a quartz pipe to homogenize the beam, there is an important scattering in experimental observations for the diodes irradiated at 1 J.cm^{-2} (curves labelled C-1 and C-2). This is attributed to some speckle effects in laser energy distribution, the grinding of the input face of the quartz pipe probably being too rough.

Apart from this fact, it can be noticed that, for a **definite** voltage, the reference devices present some orders of magnitude lower currents than the ones irradiated at 480mJ.cm^{-2}, while exhibiting the same slope. From the values of the reverse saturation currents, barrier heights of 0.95V and 0.65V are calculated respectively for the reference and the low energy irradiated devices.

For the high energy irradiated samples, no quantitative data regarding barrier height can be extracted from the I(V) plots. The behavior for the latter samples is characteristic of a series resistance effect. Effectively, high resistance values are measured after high energy irradiation, when an ohmic contact is substituted to the rectifying Au contact |9|. This effect is thought to originate from carrier trapping and scattering on defect centers in the laser melted zone. Nevertheless, if not quantitative, these plots indicate that the barrier height for the high energy irradiated samples is more of the order of 0.65V than of the order of 0.95V, a picture which will be confirmed by low frequency C(V) measurements.

Barrier heights have also been evaluated from the intercept of the $C^{-2}(V)$ plots with the voltage axis (see fig. 2). Values of $\simeq$ 1V and $\simeq$ 0.65V are deduced respectively for the reference and 480mJ.cm^{-2} irradiated samples. Another thing to be noticed from the examination of the latter figure, is the little dependance of the capacitance values on the measurement frequency (either 10KHz or 1MHz), indicating that few relatively deep levels are present in the layers irradiated at 480mJ.cm^{-2}. In this case the laser effects are primarily surface effects. This barrier lowering phenomenon explains the decrease in the zero bias depletion width of Al/GaAs Schottky structures reported by Yuba et al. |7| after Ruby laser irradiation.

A completely different behavior is found to occur when the high-energy-irradiated devices are examined. At a measurement frequency of 1MHz, a good straight line results from plotting C^{-2} versus V, but large apparent barrier heights are found, which can be interpreted by considering the existence of an insulating layer beneath the surface |3,10|.

This effect is illustrated on fig. 3 for samples (ingot B) irradiated with a Ruby laser at 600mJ.cm^{-2}.

The measured capacitance C_M is such that $C_M^{-1} = C_I^{-1} + C_{SC}^{-1}$, the capacitance C_I of the insulating layer being in series with the capacitance C_{SC} of the depleted semiconductor. The capacitance C_I can be determined from the $C^{-2}(V)$ plot, since for a voltage equal to the built in potential C_{SC} is infinite and $C_M^{-1} \simeq C_I^{-1}$. According to the I(V) characteristics (fig.1) and to the low frequency C(V) plots — at 10KHz most of the traps are supposed to respond — (fig. 3), the built in potential has to be taken around 0.65V. From the knowledge of C_I, the thickness of the insulating layer can be derived and the exact characteristics C_{SC} (V_{SC}) are obtained by taking out the contribution of C_I in C_M and V. Hence the carrier concentration profile behind the insulating layer can also be evaluated.

This is shown on fig. 4 for a Sumitomo sample irradiated in green at 500mJ.cm^{-2}. The calculated capacitance of the insulating layer for this sample is $\simeq$ 43pF, corresponding to a thickness of $\simeq$ 2400 A. The computer corrected C(V) plot of fig 4-b allows the calculation of the carrier profile behind the laser induced insulating layer as shown in fig. 4-c. The calculated values are distributed around 3×10^{16}/cm^3 (initial doping), indicating few laser effects behind the insulating layer.

This picture is consistent with a melting model, in which one expects an abrupt transition at the liquid-solid interface between the molten layer and the underlying crystal. In the recrystallized melted zone, high concentration of defects are quenched, due to the rapid transition from the liquid to the solid state. As a consequence, the initially present carriers are heavily compensated in that layer whereas, in the unmelted region behind the maximum of the melt-front penetration, few defects are generated.

Fig. 5 shows the increase of the thickness of the insulating layer as a function of the deposited laser energy density. Each value is the average of 10 measurements. The spread in the reported values arises from the inhomogeneities in the laser energy distribution which induce local overheating and enhanced degradations. Nevertheless, within the frame of the melting model |11,12| and keeping in mind that the defects are thought to be generated by the melting process, the increase in the thickness of the insulating layer is consistent with the increase of the melted thickness as the laser energy density is raised.

Fig. 6 (full line) shows the D.L.T.S. spectrum recorded on virgin Sumitomo material. As pointed out in the former section, only the 0.36eV trap level is resolved in the virgin material and its capture cross section is of the order of 10^{-14}cm^2. The investigated thickness, calculated from the capacitance values for the bias voltages used in this D.L.T.S. experiment, lies approximately within 1000A from surface. For the laser irradiated samples, due to the barrier lowering phenomenon and to the presence of the high concentration of defects within the very subsurface layer, important shunt leakage currents are observed as the samples are positively biased (see for instance fig. 1 for the 1 J.cm^{-2} irradiated devices which present a similar behavior). This precludes any exact depth determination from capacitance measurements, for the steady state bias voltage (+ 0.2V). Only the zero bias value (see fig. 4-a) is reliable, indicating a depth of $\simeq$ 3000A. Moreover, in the insulating layer, the Fermi level cannot be raised during the application of the 0.7V -100 μs pulse. Consequently, the D.L.T.S. signal comes entirely from behind the insulating layer and what we observe is the tail of laser induced defects. This explains why defect concentrations, as determined from D.L.T.S. measurements in laser irradiated layers, are quite low.

Fig. 6 (dashed line) also shows the D.L.T.S. spectrum from an irradiated device (green 0.5 J.cm^{-2}). Defect levels are generated in the range 0.15 - 0.36eV from the conduction band. However they cannot be clearly separated from the 0.36eV level which does not anneal during laser irradiation. These results are consistent with the ones reported by Emerson and Sealy |4|, as the D.L.T.S. spectrum they observe after Ruby laser irradiation also corresponds to relatively shallow levels. The trap concentration evaluated from fig. 6 is at most a few times 10^{15}/cm^3 when the spectrum is corrected from the contribution of the insulating layer. No evidence for laser induced defects was found for reverse bias conditions, i.e. at depths greater than 3000A. A last thing to be pointed out is

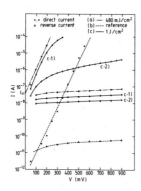

Fig. 1. I(V) characteristics of virgin
and laser irradiated samples.
(a) 480mJ.cm⁻² (b) reference (c) 1J.cm⁻²
laser irradiated (YAG laser, mixture of
green and infra-red).

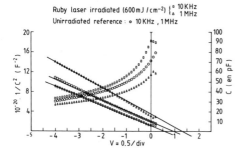

Fig. 3. C(V) characteristics of reference
and 600mJ.cm⁻² irradiated sample (Ruby laser)

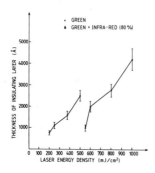

Fig. 5. Thickness of insulating layer
as a function of laser energy density
(YAG laser, green and green + infra-red).

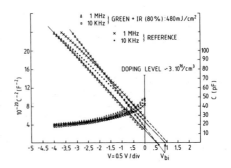

Fig. 2. C(V) characteristics of
reference and 480mJ.cm⁻² irradiated
samples (YAG laser).

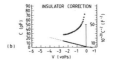
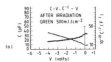
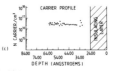

Fig. 4. C(V) characteristics and
carrier profile on 500mJ.cm⁻²
irradiated sample (YAG laser, green)
(a) as irradiated (b) corrected of
insulating layer (c) corrected
carrier profile (see text).

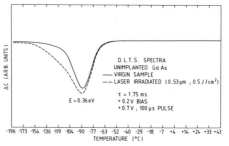

Fig. 6. D.L.T.S. spectra of a virgin
sample and 500mJ.cm⁻² irradiated sample
(YAG laser, green).

that the annealing of the 0.83eV level (when present in the starting material),deep in the material, beyond the melted layer |3,4,7| is not in contradiction with the above interpretation. Effectively, this annealing certainly takes place via a solid phase radiation enhanced mechanism, as evidenced by the elimination of the 0.83eV level after ion implantation only |3|.

Thermal annealing

Fig. 7 shows two D.L.T.S. spectra of thermally annealed ingot B samples. No signal is obtained with a - 2V bias and a 1V pulse amplitude, i.e., in a layer lying within $\simeq$ 1700 and 2500A from the surface. However, when the bias is - 0.4V and the pulse amplitude 0,9V, in order to investigate the very subsurface layer (from $\simeq$ 900A to 1500A), a large peak is observed around 200K for an emission rate of 100s^{-1}. The change in peak position with temperature when the emission rate is varied gives an ionization energy of 0.18eV; this value is not accurate because the shape of the spectrum varies strongly with the bias (fig. 8), i.e. with the electric field (the higher the electric field, the faster the emission rate). This effect is characteristic of phonon assisted tunnel emission |13|. For this reason the signature ln eT^{-2} versus T^{-1} has been determined with an electric field as low as possible. The characteristic filling time of this trap which has been determined by studying the peak amplitude as a function of pulse duration is $\simeq$ 1ms. It corresponds to a rather small capture cross section ($\simeq$ 10^{-20}cm^2) as compared to the usual values obtained for other traps in GaAs. The concentration of this trap is $\simeq$ 3x10^{16}/cm^3. The existence of this trap which is located only near the surface is probably the consequence of a loss of As atoms during the thermal treatment. This is consistent with the results of Chiang and Pearson |14| who concluded to the presence of a high concentration of donor traps, near the surface, associated with As vacancies, after annealing at 850°C for 24 hours. It therefore seems that the way the partial pressure of As is maintained during the face to face annealing is not completely effective. The 0.83eV level which was present in the starting material is not observed; it has annealed out as expected.

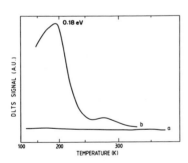

Fig. 7. Typical D.L.T.S. spectra of face to face thermally annealed sample, at the emission rate of 100s^{-1} (a) - 2V bias, 1V pulse amplitude (b) - 0.4V bias, 0.9V pulse amplitude.

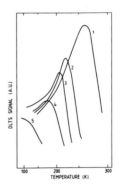

Fig. 8. D.L.T.S. spectra of face to face thermally annealed sample at the emission rate of 100s^{-1} for various bias voltages (1) : 0.1V, (2) : 0.2V, (3) : 0.3V, (4) : 0.4V, (5) : 0.5V and with a constant pulse amplitude of 0.6V.

CONCLUSION

I(V), C(V), and D.L.T.S. data recorded on laser irradiated n-GaAs samples have been interpreted in terms of a barrier lowering phenomenon, together with the formation of a subsurface insulating layer. The insulating layer originates from heavy carrier compensation (at least up to $6x10^{16}/cm^3$), due to the quenching of defects as the laser melted layer recrystallizes. This would tend to indicate that the inability of pulsed laser irradiation to activate low dose n-type implantations is first related to laser generated defects.

Finally, the comparison of the results obtained on the defect density introduced in the GaAs material after laser processing or after face to face thermal processing, is in favor of the latter process. Only in this case a material free of defect (with the exception of a thin surface layer) is obtained.

ACKNOWLEDGMENTS

The authors would like to thank J.F. Morhange for performing the Ruby laser irradiations. J. Perrocheau is also acknowledged for helpfull discussions and for performing some of the Nd-YAG laser irradiations. Thanks are also due to J. Bourquin for her attention and patience in typing the manuscript.

REFERENCES

1. J. Fletcher, J. Narayan, D.H. Lowndes in : Defects in Semiconductors, ed. by J. Narayan and T.Y. Tan (Elsevier, North Holland, 1981) p. 421.

2. S. Nojima, J. Appl. Phys. 53, 5028 (1982).

3. P.M. Mooney, J.C. Bourgoin, J. Icole in : Laser and Electron Beam Solid Interactions and Materials Processing, ed. by J.F. Gibbons, L.D. Hess and T.W. Sigmon (Elsevier North Holland 1981) p. 255.

4. N.C. Emerson, B.J. Sealy, Electron. Lett. 16, 512 (1980).

5. D.E. Davies, J.P. Lorenzo, E.F. Kennedy, T.G. Ryan in : Proceedings of the 1981 International Symposium on GaAs and Related Compounds (Oiso, Japan) p. 255.

6. M. Takai, S. Ono, Y. Ootsuka, K. Gamo, S. Namba in : Proceedings of the 4th International Conference on Ion Implantation Equipment and Techniques Berchtesgaden September 1982 (to be published).

7. Y. Yuba, K. Gamo, K. Murakami, S. Namba, Appl. Phys. Lett. 35, 156 (1979).

8. Y. Yuba, K. Gamo, S. Namba, Ref (5) p. 221.

9. D. Pribat, S. Delage, unpublished data.

10. D. Pribat in : Cohesive Properties of Semiconductors Under Laser Irradiation, ed. by L.D. Laude (Martinus Nijhoff Publishers, The Hague, in press).

11. R. Tsu, J.E. Baglin, G.J. Lasher, J.C. Tsang, Appl. Phys. Lett. 34, 153 (1979).

12. R.F. Wood, D.H. Lowndes, W.H. Christie in Ref. |3| , p. 231.

13. D. Pons, S. Makram-Ebeid, J. Physique 40, 1161 (1979).

14. S.Y. Chiang, G.L. Pearson, J. Appl. Phys. 46, 2986 (1975)

ELECTRON BEAM PROCESSING OF SEMICONDUCTORS

H. AHMED AND R.A. McMAHON
Cambridge University Engineering Department, Trumpington Street,
Cambridge, England, CB2 1PZ

ABSTRACT

Electron beams can transfer energy very efficiently to
semiconductors. Systems have been developed for rapid heating
to temperature around 1000°C under a variety of conditions from
adiabatic to isothermal. Pulsed, focused, line and synthesized shaped beams are used to obtain a wide range of thermal
cycles. The following applications are described: the
annealing of ion-implanted Si, particularly the activation of
As implants and shallow implants (Rp<150Å), the annealing
of Si and Se in GaAs, the e-beam processing of implanted
silicon devices and the improvement of SOS substrate quality.
Localized annealing by a computer controlled e-beam and the
recrystallization of deposited films on insulators are also
considered.

INTRODUCTION

The ease of generation, control and energy transfer to solids associated with
electron beams has led to their incorporation in many systems for thermal processing of materials. Initially, the principal interest was in the annealing of
ion-implantation damage in silicon and other semiconductor materials. More
recently new applications have emerged; the recrystallization of deposited films,
reactions such as silicide formation, and the exploration of fundamental phenomena in materials in previously inaccessible time scales of less than 1s. To
perform these thermal processes, the electron beam must be capable of heating
the material to high temperatures, ranging from 500°C to 1500°C. The heating
may be localized to a region of a few microns, or uniform over the area of a
wafer, the changes in the material properties can take place in the solid or in
the liquid phase and process times range from 10ns to several minutes. Many
different systems have evolved, which may be grouped by their characteristic
process times.

All e-beam processing systems rely on the transfer of energy from the beam to
the solid. When the beam strikes a solid a few electrons are backscattered, but
the bulk of the electrons deposit their kinetic energy as heat, over a volume
dependent on the beam energy and substrate material. The Monte Carlo method
enables the energy distribution in the solid to be computed; Figs. 1a to 1d show
equal energy density contours for a scanning e-beam at 10kV and 50kV, with 1μm
and 10μm diameter spots (1). The charge deposited is 1C/m and the outer contour
is for 10^{14} J/m³. For the 10μm diameter spot the beam approximates to a uniform
surface source, whereas the 1μm result shows the smallest volume in which the
energy can be deposited. The heat generated by the beam diffuses at a rate
determined by the thermal properties of the material. For short heating times,
for example in Si <1μs, the diffusion is small and the process can be called
adiabatic. If the heating continues, a steady temperature gradient is established, and the process is called thermal flux (2). With e-beams, a wider
range of heating patterns are possible, in both space and time, than from other
kinds of beams that have been applied to semiconductor processing.

Mat. Res. Soc. Symp. Proc. Vol. 13 (1983) © Elsevier Science Publishing Co., Inc.

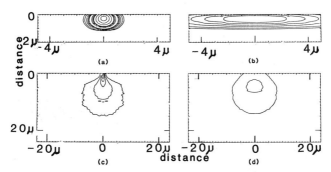

Fig.1 Monte-Carlo simulation of e-beam deposited energy in silicon for various
beam diameters and energies. (a) 1μm, 10keV, (b) 10μm, 10keV, (c) 1μm, 50keV,
(d) 10μm, 50keV.

E-BEAM PROCESSING SYSTEMS

Adiabatic
     Pulsed electron beam systems operate in the adiabatic regime, with character-
istic heating times of 100ns. The best known system is that developed by the
Spire Corporation (3), which was intended for the annealing of ion implanted
layers in the large scale, energy efficient, production of solar cells. The
equipment is essentially a planar diode, comprising a field emission cathode
with a mesh anode, shown schematically in Fig. 2a. The wafer, up to 4" diameter, forms a target following the anode. A pulse of electrons, with an
energy density of around 1J/cm^2 and duration of 75ns, melts the surface layer of
the semiconductor. In the case of ion implants, dopants are incorporated in
electrically active substitutional sites during recrystallization from the
single crystal substrate. Similar heating regimes are possible, in principle,
by switching or blanking continuous beams but the maximum power density,
0.5MW/cm^{-2}, available from cathodes for these beams is inadequate for effective
adiabatic processing. Furthermore the Spire system delivers a nominally uniform
power density over a wafer sized area whereas a continuous beam will have gaussian
current distribution profile and usually small diameter compared to the wafer
size.

Single scan annealing systems
     Several annealing systems have been based on continuous e-beams. The heat
flow from the point of entry of the beam, and the consequent temperature distribution, depends on the power and the shape of the beam, and on the thermal
properties of the material. It also depends on whether the specimen is simultaneously heated or cooled by another source, and on whether the beam is
stationary or scanning. The exact analysis of the heat flow is difficult and no
complete solution has been published. Regolini et al (4) reported a system in
which a 300μm diameter beam was scanned over a heat sinked specimen (Fig. 2b).
Larger areas than the beam size were annealed by overlapping successive scans.
For slow speeds of 10cm/s or less the heat flow is effectively in a steady state.
The temperature rise under the spot is then proportional to the beam power divided by its radius, and depends on the thermal properties of the material.
Krimmel et al (5) did not use a heat sink, but chose beam powers and scan speeds
such that the specimen did not heat excessively. Line beam systems offer
faster annealing of large areas (Fig. 2c). Yu et al (6) used a 10 x 2mm line
beam, through which a heatsinked wafer was moved mechanically. Knapp et al (7)
reported on an electrostatically scanned line source of 20 x 1mm, scanning at

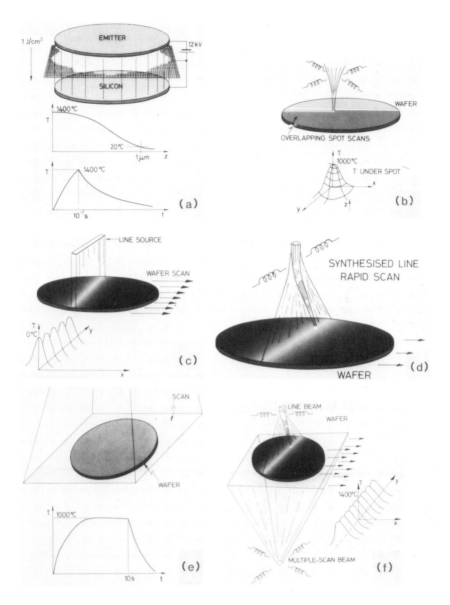

Fig.2 Methods of e-beam thermal processing and associated temperature profiles.
(a) Pulsed. (b) Single-scan (round beam). (c) Single-scan (line beam).
(d) Single-scan (synthesized line-beam). (e) Multiple-scan (round beam).
(f) Dual beam (combining (e) and (d)).

Fig. 3 Localised annealing of
ion implanted silicon. Line
width about 3μm.

speeds up to 10m/s. Temperature calculations are presented in this paper for
the heat flow which is not in a steady state.

Multiple scan annealing systems

In all the single scan systems the aim is to effect the desired material
changes in one pass of the beam but by using multiple passes of the electron
beam different and often more controlled, heat patterns are possible. A widely
used method is the multiple scan method, reported by McMahon and Ahmed (8) who
contrast it with single scan methods. In the multiple scan method, rapid scanning of a continuous e-beam ensures uniform energy deposition, so that the
effect is equivalent to a mean power density equal to the beam power divided by
the scanned area (Fig. 2d). The specimen is in thermal isolation, so heat is
lost by radiation only, and the resulting heating cycles have been described
in detail (9,10). Power densities of $\sim$20W/cm ^{2} over wafer areas, give peak
temperatures of 1050°C in 10s while power densities of several kW/cm ^{2} are
available for experiments on chips heating them to a similar temperature in 0.01
to 0.1s. The technique employs a scanned area slightly greater than the specimen size, to ensure isothermal heating. Multiple scanning systems create beam
shapes that are impossible to obtain directly from guns, but which are smaller
than the specimen size. Bentini et al (11)describe a system in which a large
beam (7mm diameter) is scanned in one direction to form a line 125mm long, while
the specimen, which is clamped and cooled, moves through the beam. Yep et al
(12) report a similar arrangement, in which a 270W beam, 0.25" diameter was
scanned to form a 1.5" x 3" rectangular spot through which 2" diameter wafers
were mechanically moved, taking about 20s to anneal.

Localized e-beam heating systems

In contrast to many systems, where the aim is to anneal large areas, the
heating of semiconductors by a finely focussed beam to influence the electrical
properties of a localised region is a novel microfabrication technique. McMahon
and Ahmed (13) annealed selected areas of boron implanted n-type silicon to form
diodes. The edge resolution was similar to the beam diameter (50μm) and the
dwell time was 60μs, intermediate between adiabatic and steady state heat flow
conditions. Ratnakumar et al (14) annealed micron sized areas using a modified
SEM, with heatsinked specimens. Recent developments (15) have enabled the
annealing of structures with a resolution of 1μm, as illustrated in Fig. 3 which
shows an array of annealed areas formed by a computer controlled electron beam.

Concentration cm^{-3}

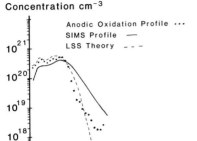

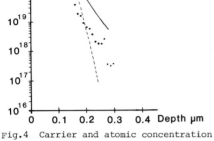

Fig.4 Carrier and atomic concentration profiles for e-beam annealed Si implanted with 5×10^{15}, 160keV and 10^{15}, 40keV arsenic.

Fig.5 Sheet resistance of 10^{16}As implanted Si following long e-beam anneals (a) without capping (b) with 0.5μm vapox cap.

Dual e-beam systems

Although a variety of heat patterns are attainable with single electron beam systems, it is difficult to obtain both a localized effect and a background effect simultaneously. Time sharing with one beam performing both functions is possible, but difficult to achieve in practice. Combining an electron beam with some other heat source such as a hot chuck offers some advantage. A flexible approach is a two beam, or indeed a multiple beam, system (16). Several possibilities are described all with a thermally isolated specimen and isothermal background through the multiple scan method. The second beam can be variously a line, a spot or an inset scan. The line is derived either from a line gun, a deformed spot or sythesized by rapid scanning, the latter is shown schematically in Fig. 2d. Provided that the power deposited by the second beam is small compared to the power flow to the sample from the isothermal heating which is continuous, the second beam will not greatly perturb the mean specimen temperature. The precise temperature distribution under the local heating beam is difficult to calculate, but under appropriate conditions localised line and spot heating patterns are attainable.

Basic mechanism

Ion implantation into silicon results in damage to the lattice which varies from a relatively low density of defects in an essentially intact crystal, to complete amorphization, depending on dopant, dose, energy and implantation conditions. This damage must be removed for devices to function,by annealing, either in liquid phase, where reordering is in nanoseconds once melting occurs, or in the solid phase. In the latter mode, annealing takes place at a temperature dependent rate, and the final structure contains residual defects depending on the implantation conditions. E-beam processing offers advantages over conventional solid phase annealing in a furnace. Pulsed e-beams melt the semiconductor, redistributing impurities, enabling previously impossible doping profiles to be attained. Pulsed electron beam annealing has been used to activate ion implants and devices but the conclusion was that the equipment was not capable of adequate uniformity of annealing over a 3 in diameter wafer (17). Rapid, solid phase annealing can either activate dopants without significant diffusion, or perform precisely controlled diffusions.

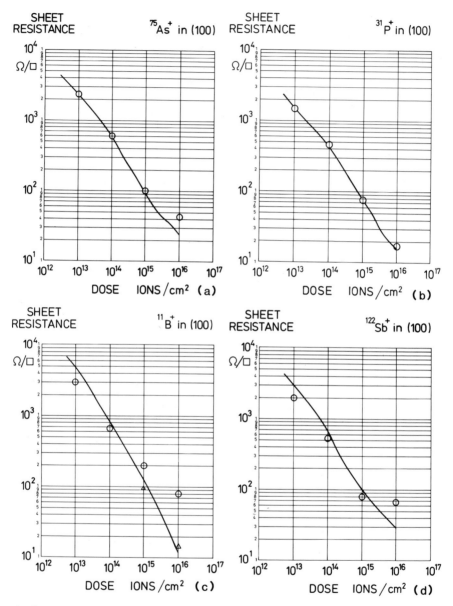

Fig.6 Activation of 100keV implants in (100) Si for a range of doses.
(a) As, (b) P, (c) B (o- standard anneals, Δ - extended anneals)
(d) Sb.

Limited data on the activation of ion implants by single scan line and spot systems has been published (4,5,7). Ion implants can be activated by single scan electron beam methods, but there is difficulty in controlling the temperature adequately within the window between 'no-effect' and damage, in the form of melting or thermally induced slip lines (2). The systems of Yep et all (12) for annealing diodes and Bentini et al (11) for phosphorous implants where large area spots were synthesized by rapid scanning were more successful. A more controllable method is the multiple scan technique (8), in which the electron beam is rapidly scanned over an area just larger than the chip or wafer to be annealed. The next sections discuss several applications of isothermal processing.

Isothermal annealing of ion implants in silicon

Rapid isothermal annealing by the multiple scan method activates implants of arsenic, boron, phosphorous, antimony and gallium, with doses up to the solid solubility limit, and the associated diffusion can be controllable between negligible levels to several times the implant depth (18, 19). Arsenic implants find wide application in VLSI technology, particularly in the source/drain regions of NMOS devices, and in the emitters of bipolar transistors. It is therefore taken as an illustration of the properties of e-beam annealing. Small MOS transistors require a shallow junction, yet need a low sheet resistance, implying high doping levels in the drain and source regions. An arsenic implant of 5.10^{15} ions/cm^2 at 160keV with an additional implant of 10^{15} ions/cm^2 at 40keV to ensure low contact resistance meets this requirement. Above 700°C, activation of the implant becomes rapid, taking a fraction of a second. Fig. 4 shows a carrier concentration profile obtained by anodic oxidation and stripping, an atomic concentration profile from SIMS and an LSS prediction of the impurity profile. These measurements were made following an anneal with a power density of 25W/cm^2 for 5s, reaching a peak temperature of about 1000°C, and the close agreement of the curves indicate close to full electrical activation allied with negligible diffusion.

Bipolar emitter structures require even shallower junctions, perhaps using an implant of 10^{16} As/cm^2 at 40keV. The behaviour of this implant under different e-beam annealing treatments illustrates diffusion, activation and dopant evaporation. Firstly a comparison of sheet resistance of uncapped samples and samples capped with 0.5μm vapox, following anneals at 1280°C shows these effects (Fig. 5). Both types of samples had the same sheet resistance after a 5s anneal, but the sheet resistance of uncapped samples rises with increasing annealing time whereas that of capped samples decreases to 16 Ω/□ after heating for 20s. Junction depth measurements and SIMS profiles show that in both cases diffusion is very large, the junction depth increasing from 0.15μm to 0.5μm. However, with the uncapped material a large amount of dopant is lost, whereas capping prevents the evaporation of arsenic from the surface and the decrease in sheet resistance reflects diffusion of the arsenic.

E-beam processing provides an excellent tool for studying annealing and activation effects in many other dopants, eg phosphorous, boron, gallium, antimony, aluminium, over a range of time and temperature conditions etc. Figures 6a to 6d show, respectively, the activation of arsenic, phosphorous, boron and antimony implants over a dose range from 10^{13} to 10^{16} ions/cm^2 at 100keV by an anneal with a peak temperature of ∿1000°C, ensuring minimal diffusion. Except in the case of phosphorous all the implants at 10^{16}/cm^2 exceed the solid solubility limit of the dopant so do not activate fully. Otherwise the measured sheet resistances are close to those predicted for fully active, undiffused implants.

Processing of shallow implants in Si

Shallow implanted layers (Rp <150Å) have been studied because of their pot-

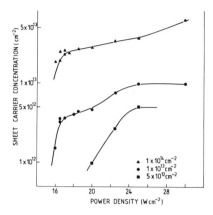

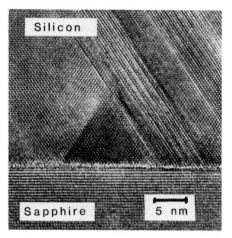

Fig.7 Sheet carrier concentration in Se implanted GaAs after 5s e-beam anneals.

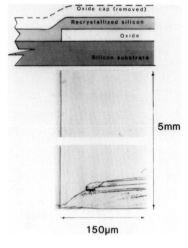

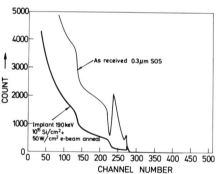

Fig.8 Cross-sectional TEM of silicon and sapphire showing defects in the Si originating from the interface.

RBS spectra of SOS showing improvement in crystalline quality after silicon implantation and e-beam processing.

Fig.9 Recrystallisation of deposited polysilicon by dual e-beams. (a) Silicon-on-insulator configuration. (b) Defect free recrystallized region. (c) Defects at the edge.

ential use in scaling down the dimensions of conventional devices and in the fabrication of hot electron devices. The annealing of implants of As into Si at low energies (10keV) and doses in the range 10^{14} to 10^{16}cm^{-2} has been studied. Sheet resistance measurements after low temperature anneals either in a furnace or by multiple-scan electron beams at temperatures around 700°C for 15 minutes have been shown to correspond to measurements after short time anneals of 100ms at 600W/cm^2 corresponding to a maximum temperature of ~945°C. At higher temperatures, >1000°C, diffusion and dopant loss are significant and the profile is broadened, as demonstrated by high resolution SIMS (20).

Hot electron devices, in particular Schottky diodes with barrier height control, have been fabricated by implanting As at doses ranging from 10^{12} to 5 x 10^{14}cm^{-2} at 10keV energy. The characteristics were measured after a 1s, 60W/cm^2 anneal corresponding to a peak temperature of 900°C in the heating cycle. With optimum e-beam annealing conditions high electrical activity was obtained with an almost undiffused concentration profile for both implant annealing experiments and device fabrication.

Devices

Rapid isothermal annealing with e-beams has been used in processing ion-implanted silicon devices including diodes, bipolar and MOS transistors. Diodes formed by implantation of boron and arsenic through windows in oxide into n and p- type wafers respectively have been annealed. These diodes had low reverse leakage currents and breakdown voltages close to those of identical devices on furnace processed wafers (21). The forward characteristics of arsenic implanted diodes also compared closely with control diodes but the boron diodes had excess forward resistance, because of low surface concentration since the diffusion that normally takes place is relied upon to ensure low contact resistance was absent. The process was extended to MOS transistors whose characteristics indicated insignificant changes in oxide interface (22). SEM examination of cleaved devices confirmed that diffusion was less than 0.1μm in a gate length of 3μm (23). There was no evidence of excessive leakage and the devices compared well with the control devices fabricated with the standard process. Subsequent work on batches of n-channel MOS devices supports earlier results (24). Devices with nominal gate lengths from 0.7μm to 20μm, and with source and drain regions formed by implantations of 2.10^{16}As/cm^2 at 160keV to give a 0.3μm junction depth, and an additional implant of 5.10^{15}/cm^2 at 40keV to ensure low contact resistance were processed. Annealing was at a peak temperature of 1050°C, for a heating time of 10s. Diffusion was limited to 50 nm and there was no damage to gate oxide. Wafers annealed with the beam falling on the front surface showed no difference in electrical characteristics to those annealed from the back surface. E-beam annealed devices in general had slightly lower channel mobilities and substrate leakage currents than similar furnace annealed devices.

Processing of GaAs

The processing of GaAs and other III-V compounds with electron beams presents greater difficulty than silicon because of the narrow window of temperatures between satisfactory activation of ion implants and decomposition of the material. The precise control of temperature and time that is available with e-beam methods has enabled successful experiments to be carried out on silicon and selenium implanted GaAs. Low dose n-type implants ~10^{12}cm^{-2} are needed for the channel of GaAs FETS and high doses are needed for making low noise, ohmic contacts. The annealing of low doses is technologically important and experiments have been conducted on silicon implanted GaAs without encapsulation of the substrate (25). Activation levels in excess of 60% and high values of mobility were measured using 5s anneals at temperatures around 850°C. It was proposed that the short time of the anneals is a means of avoiding thermal conversion and SIMS and other methods were used to investigate compensation effects in Cr doped substrates implanted with silicon (26). Experiments have also been reported on Se implanted GaAs for a wide range of doses (27). The electrical results on silicon nitride capped samples annealed for 5s were very similar to those obtained by furnace annealing with high activation and mobility being obtained consistently (Fig. 7). Anneal times of only 5s were found to be necessary and again it was observed that variation in the content used for compensation are relatively insensitive to this annealing method. A scanning line e-beam method has also been used to process nitrogen implanted GaAsP and results comparable with conventional processing have been obtained (6).

Adiabatic e-beam techniques using 100ns pulses of $\sim 0.5 J/cm^2$ have been applied to annealing ion implanted GaAs layers. High dose implants have been annealed using a narrow window of incident flux so that levels above the solid solubility limit have been reached but the mobilities are low and the concentration decreases on heating to $250^\circ C$ (28). Low dose implants have not been annealed successfully with the pulsed e-beam method probably because of relatively high levels of background impurities or of defects introduced in the nonequilibrium processing.

Application to device fabrication on SOS

Rapid isothermal e-beam annealing advances SOS technology in two ways: improvement of the starting substrates and dopant activation in fabricated devices. As grown silicon-on-sapphire layers often contain extended defects, originating at the interface as shown in the TEM of a silicon sapphire interface region (Fig. 8). Implantation with silicon can amorphize the interface region, breaking up these defects, yet leaving a single crystal layer at the silicon surface capable of seeding regrowth back to the interface. Furnace annealing permits some improvement over as-grown material, but the rapid heating by e-beams to a higher temperature than those possible in a furnace is yielding better material, as determined by initial RBS studies (29). Boron, phosphorous and arsenic gave sheet resistances close to, but always slightly lower, than the same dose in bulk silicon after implantation in SOS and e-beam annealing (10).

Recrystallization of films on insulators

The versatility of electron beam techniques is demonstrated by the application to recrystallization of polysilicon films deposited on insulators into single crystal films. The requirement for precise control of substrate temperature, both spatially and temporally, is conveniently obtained with a dual electron beam system (16) (Fig. 2f). Precisely oriented, defect-free single crystal films have been produced with this system by seeding the recrystallized layer from the substrate through windows in the underlying SiO_2 film. Fig. 9a shows the substrate structure. A CVD oxide layer is used to cap a polysilicon layer which is deposited on top of a thermally grown oxide on a silicon wafer. Both (100) and (111) substrates have been used and seeding windows have been aligned along different crystalline directions to investigate the growth parameters (30). During processing the line beam of the dual system melts the polysilicon in the seeding windows and sweeps a narrow molten zone of polysilicon over the oxide step and across the wafer. The solidified layer is crystalline and defect free for distances up to 150µm from the seeding window edge and across the full length of the line beam. Fig. 9b shows interference contrast micrographs of regions near the centre and Fig. 9c near the edge of the recrystallized film which is about 5mm wide. The micrographs were taken after removing the capping oxide and Secco etching to reveal the defects at the edge. Beyond $\sim$150µm dislocations occur, gradually increasing in density to form low angle grain boundaries.

The reasons for the onset of dislocations is not yet fully understood but it is not believed to be a fundamental limitation of the method. The flatness of the recrystallized surface is generally comparable to that of commercially produced Si wafers and the processing technique is potentially capable of development into a commercially viable method of preparing silicon-on-insulator substrates for device manufacture. For processing 3" diameter wafers a total dual beam power of around 4kW and a processing speed of 100mm/s is required.

Pulsed electron beams have also been used to recrystallize polysilicon films, increasing the grain size to ∿1μm which reduced the sheet resistance very considerably (31). Line beam techniques have encountered problems with beam control which have led to gross damage to the polysilicon layer (32). A circular e-beam spot, by overlapping scans, has produced material with chevron shaped grains. MOS transistors with $\mu=120/cm^2S$ were successfully fabricated in this material.

CONCLUSIONS

A range of e-beam annealing systems have been described, which find application in many semiconductor fabrication processes. E-beams are efficiently generated, easily deflected and are absorbed equally effectively by most semiconductor materials. Single pass methods are difficult to control, and maintaining uniform conditions over specimens is also difficult. Multiple scanning of the beam improves uniformity and mounting substrates in thermal isolation gives reproducible results and high throughput. Rapid isothermal processing (RIA) by the multiple scan method is a precision annealing technique in time and temperature which avoids thermal stresses in wafers. It is successfully applied in silicon and III-V device processing for activating implants of all technologically important dopants. In SOS, RIA offers improvements in the quality of the silicon film. For some applications, local heating is required, for example in sweeping a molten zone across a wafer in polysilicon recrystallization. Here the combination of isothermal background heating and an exceptionally uniform line beam formed by rapid scanning make controlled processing possible. The ease of deflection of an electron beam is the key to localized annealing on a microscale. Computer based pattern generators drive the beam to anneal selected areas only, making a pattern of interconnects which changes from wafer to wafer. E-beam systems are just beginning to impact semiconductor processing. Their versatility and compatibility with other fabrication methods, especially ion beam systems, should ensure them a vital role in the realization of VLSI.

ACKNOWLEDGEMENTS

The contributions from members of the Microcircuits Research Group and D.J. Smith at Cambridge University. D.J. Godfrey and M.G. Pitt (GEC) and J.D. Speight and P. Leigh (BTRL) are gratefully acknowledged.

REFERENCES

1.   D.F. Kyser and K. Murata, Proc. 6th Int. Conf. Electron and Ion Beam Sci. and Technol. Ed. R. Bakish (1974).

2.   C. Hill†

3.   A.R. Kirkpatrick, J.A. Minucci, A.C. Greenwald, IEEE Trans. Electron Devices 429 (1979).

4.   J.L. Regolini, J.F. Gibbons, T.W. Sigmon and R.F.W. Pease, Appl. Phys. Lett. 34, p. 410 (1979).

5. E.F. Krimmel, H. Oppolzer, H. Runge, W. Wondrak, Phys. Stat. Sol(a) 66, 565 (1981).

6. T. Yu, K.J. Soda, B.C. Streetman, J. Appl. Phys. 51, 4399 (1980).

7. J.A. Knapp and S.T. Picraux, Appl. Phys. Lett. 38, 873 (1981).

8. R.A. McMahon and H. Ahmed, Electron Lett. 15, p. 45 (1979).

9. H.J. Smith, E. Ligeon and A. Bontemps, Appl. Phys. Lett. 37, 1036 (1980).

10. N.J. Shah, R.A. McMahon, J.G.S. Williams and H. Ahmed †

11. G.G. Bentini, R. Galloni and R. Nipoti, Appl. Phys. Lett. 36, 661 (1980).

12. T.O. Yep, R.T. Fulks and R.A. Powell, Appl. Phys. Lett. 38, 162 (1981).

13. R.A. McMahon and H. Ahmed, J. Vac. Sci. Technol. 16, p. 1840 (1979).

14. K.N. Ratnakumar, R.F.W. Pease, N.M. Johnson and J.D. Meindl, Appl. Phys. Lett. 35, 463 (1979).

15. H.T. Sun. Unpublished work.

16. R.A. McMahon, J.R. Davis and H. Ahmed.*

17. T.I. Kamins and P.H. Rose, J. Appl. Phys. 50, 1308 (1979).

18. R.A. McMahon and H. Ahmed, IEE Proc. 129, Pt. 1 (1982).

19. D.B. Rensch and J.Y. Chen. To be published by E.C.S. (1982).

20. G.B. McMillan, J.M. Shannon and H. Ahmed. Published in this issue.

21. R.A. McMahon, H. Ahmed, J.D. Speight and R.M. Dobson, Electron Lett. 15, 433 (1979).

22. C.J. Pollard, A.E. Glaccum and J.D. Speight.*

23. J.D. Speight, A.E. Glaccum, D. Machin, R.A. McMahon and H. Ahmed.†

24. D.J. Godfrey. Unpublished work.

25. N.J. Shah, H. Ahmed, I.R. Sander and J.F. Singleton, Electron Lett. 16, 433 (1980).

26. M. Bujatti, A. Cetonio, R. Nipoti and E. Olzi, Appl. Phys. Lett. 40, p. 334 (1982).

27. N.J. Shah, H. Ahmed and P.A. Leigh, Appl. Phys. Lett. 39, 322 (1981).

28. P.A. Pianetta, C.A. Stolte and J.L. Hansen, APL., 36, 597 (1980).

29. M.G. Pitt. Unpublished work.

30. J.R. Davis, R.A. McMahon and H. Ahmed, Electron Lett. 18, 163 (1982).

31. T.I. Kamins and A.C. Greenwald, Appl. Phys. Lett. 35, p. 235 (1979).

32. J.A. Knapp, S.T. Picraux, K. Lee, J.F. Gibbons, T.O. Sedgwick & S.W. Depp.*

33. T.I. Kamins and B.P. Van Hergen, IEEE EDL., 2, 313 (1981).

† in Laser and Electron Beam Solid Interactions, eds. J.F. Gibbons, L.D. Hess and T.W. Sigmon, New York: North Holland (1981).

* in Laser and Electron Beam Interaction with Solids, eds. B.R. Appleton and G.K. Celler, New York: North Holland (1982).

OPTICAL AND ELECTRICAL PROPERTIES OF Cd Te LASER-SYNTHETIZED

L. BAUFAY, A. PIGEOLET, R. ANDREW and L.D. LAUDE
Université de l'Etat, Avenue Maistriau, 23 7000 MONS Belgium.

ABSTRACT

Optical and electrical characterization of CdTe synthe-
tized by laser irradiation of a multilayer film of alternately Cd and Te is achieved. Optical absorption measurements evidence the good quality of these films and show
that they have behaviour comparable to the single crystal.
The influence of the irradiation conditions on the electrical properties of such CdTe films is discussed; they
are compared to single crystal from the point of view of
resistivity. It is shown that it is possible to prepare
by this means samples devoid of impurity states in the
middle of the forbidden gap. Finally, the ohmicity of Au,
Al, Cr, ITO and non irradiated Cd/Te sandwich   contacts
is tested.

INTRODUCTION

Aside from the applications of laser annealing to essentially monoelement systems, as for example the recrystallisation of amorphous or implanted layers in
Si or Ge, there are also a number of reports of what can be termed laser-induced
synthesis of compounds involving two or more elements in stoechiometric ratio.
We may include here the formation of metal silicides [1,2] by irradiation of metal overlayers on silicon wafers, and the formation of semiconducting oxides [3-
7] by irradiation of the corresponding metallic films in oxygen rich ambiances.
We have previously reported the formation of the compounds AlSb, AlAs,
CdSe, CdTe and $CdTe_xSe_{1-x}$ [8-10] by irradiation of metal sandwich structures of
these materials on thermally insulating substrates and have proposed a model [11-
14] for the observed reaction which takes account of the heat of formation involved. Note that whereas laser annealing of compound semiconductors does not in
general lead to very satisfactory results, as compared to the success with silicon, our synthesis technique differs in that the starting point is not the compound but the separate constituents, and we have reason to believe that the liberation of heat of formation during the reaction, leading as it does to high local temperature gradients and a kind of zone refining action, allows us to produce relatively high quality films even when the components before irradiation
may be as much as 1-2% away from the correct stoichiometric ratio. The purpose
of the present article is to examine more closely the electrical and optical properties of CdTe films produced in this way.

FILMS PREPARATION

Metallic films preparation
Cd and Te layers are successively and alternately condensed onto sputter-clea-
ned glass substrates using a multiple crucible e-gun source and a quartz thickness monitor at a base pressure of $10^{-6}-10^{-7}$ torr. Thickness and number of elemental layers are adjusted to give as nearly as possible (uncertainty 2%) equal
numbers of Cd and Te atoms in the overall film. In order to standardize production and have the more volatile component (Cd) buried in the film away from the

outer surfaces, the successive layers are always Te/Cd/Te/Cd/Te for 1000 Å thick
sample. In several cases, we have also used a glass substrate coated with a
1700 Å thick homogeneous ITO layer [15] characterized by an electrical resistivity of about 10 Ωcm.

Laser irradiation

Using all red lines of a CW krypton ion laser beam, such metallic samples are
irradiated in air, at room temperature, in order to produce the semiconducting
compound CdTe. The gaussian beam, characterized by a diameter of 1.6 mm, is
scanned along the x-axis on the sample, in the x-y plane, at speed of 1 cm/s, at
a power beam of about 1-2 watts. Irradiation is performed using either the direct beam or a beam focussed by a lens (cylindrical or spherical lens). The most
homogeneous trace is produced when we set into the beam path a cylindrical lens
and when the elliptical spot thus obtained is scanned perpendicularly to its
great axis. In order to transform a large area, such traces can be overlapped
by displacing the sample along the y-axis at a speed of 30 μm/s, while a 2 cm
long scan is performed along the x-axis.

CHARACTERISATION

Structural

The CdTe films produced have been examined by transmission electron microscopy and found to produce diffraction data exactly corresponding to that in the
literature [16] with a grain size of 100-5000 Å, depending on the parameters of
irradiation conditions.

Optical

Optical transmission measurements are performed between 900 and 400 nm; the
same results being obtained whatever the type of laser irradiation. Fig. 1 shows
the relation between $(\alpha h\nu)^2$ and $h\nu$ (α is the absorption coefficient and $h\nu$ is the
photon energy), it indicates that the fundamental absorption edge, associated
with a direct transition occuring at the center of the Brillouin zone Γ is equal
to $E_0 = 1.45 \pm 0.02$ eV. Moreover, Fig. 2 which shows the optical transmission and its
second derivative as a function of the photon energy indicating a second structure
$E_0 + \Delta_0 = 2.37 \pm 0.03$ eV, associated with the transition occuring between spin-orbit
split off valence band edge states at Γ and conduction band states. Present data agree quite well with published data [17,18] which give $E_0 = 1.47$ eV and $\Delta_0 =$
0.92eV.

Electrical measurements

The structural and optical data show that the material produced is indeed CdTe
but do not give any information on the transport properties, carrier type, etc.,
a close control over which is of course essential for any practical applications.
The most simple electrical measurement is the resistivity of the sample. To perform such a measurement, we simply measure the resistance of a transformed region
between adjoining non irradiated regions (we have observed that the contacts in
this case are ohmic).

For specimens having the same structural and optical characteristics, and in
some cases taken from different regions of the same support slide, we can obtain
values of ρ varying from 10^{-3} to 10^6 Ωcm, but we can distinguish principally
two groups of samples : low (10^{-3}-10^{-1} Ωcm) and high resistivity (~10^6Ωcm) films.
This large variation should be compared to published data on single crystals of
CdTe [19,20] which cover almost exactly the same resistivity range depending on
the precise preparation conditions, and resulting carrier concentration. It is
natural to suppose that non-uniformity of deposition and/or irradiation (since
the spot size in our scanning system varies slightly with position) are the cause of such variations since by very carefully controlling the exact conditions

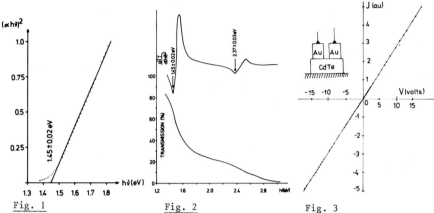

Fig. 1

Fig. 2

Fig. 3

Fig. 1   Optical absorbance characteristic of a CdTe film.

Fig. 2   Optical transmission vs photon energy and its second derivative of a CdTe film. Structures are located.

Fig. 3   I-V characteristic of two gold contacts evaporated on top of the CdTe film. Gold appears to be an ohmic contact.

we can reproduce samples with the same resistivity values and, in particular, with the highest values of 10^6-10^7 Ωcm corresponding to the highest values reported for single crystals and probably corresponding to the purest films.

We then investigated the behaviour of different contact materials by depositing contacts onto the film. For two Au contacts the I-V characteristic is shown in Fig. 3 (the same behaviour has been observed when contacts are formed by the adjoining untransformed material). In both cases, the contacts are ohmic. Fig. 4 and 5 show the I-V characteristic of one Au and one Al or Cr contacts where the second contact is clearly non-ohmic and analysis of the log I=f(V) curve (fig. 4b and 5b) shows a Schottky barrier of height 0.83 and 0.72 V respectively for the Al and Cr contacts [21] (we assume that the Richardson constant is equal to 120 A/cm^2/K^2). We have also experimented with contact formed between the film and ITO covered glass. In this case, the contact appears to be "nearly" ohmic although the results are not easily reproductible since here we are obliged to measure with a deposited gold contact separated from the ITO conducting surface by only the thickness of the film, 1000 Å, so that occasional pinholes in the thin film due to dust or other impurities lead to current shunts (see fig. 6).

Using the ohmic contacts formed by untransformed material, we have measured the resistivity as a function of temperature from 180 K to 600 K (fig. 7). The figure indicates that the width of the mobility gap is 1.62±0.10 eV. It is worth noting that intrinsic conduction manifests itself down to ~400 K; below that temperature, extrinsic conduction is controlled by impurity activation. One single activation energy at 0.16±0.10 eV is detected down to 180 K. These results and a calculation of the position of the Fermi level [22] as a function of the temperature and the impurity concentration indicate that the impurity concentration can be less than $10^{16}-10^{17}$ cm^{-3}. Moreover, it is possible to obtain a material characterized by a forbidden gap nearly devoid of impurity states except close to the bands.

Attempts to measure carrier concentration and mobilities by either Hall effect

668

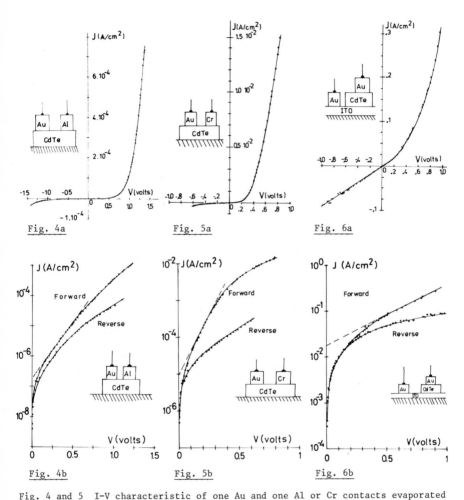

Fig. 4a   Fig. 5a   Fig. 6a

Fig. 4b   Fig. 5b   Fig. 6b

Fig. 4 and 5   I-V characteristic of one Au and one Al or Cr contacts evaporated
                on top of the CdTe film (a : linear scales; b : semilogarithmic
                scales).
Fig. 6   I-V characteristic of one Au and one ITO contacts. These contacts
                and the CdTe film are set in a sandwich configuration.

or thermoelectric power measurements have so far been unsuccessful due to the
thin film form of the specimens and the resulting very high resistances encountered, whereas the preparation of thicker films on which such experiments might
be more successfully performed are currently held up due to lack of sufficiently powerful laser beam.

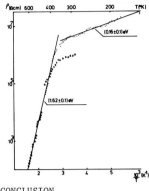

Fig. 7 Logarithm of the resistivity vs temperature inverse
       In this measurement, the contacts are Cd/Te sandwiches which have not been irradiated. The CdTe is synthetized by a single scan of the direct gaussian beam.

CONCLUSION

CdTe synthesis by laser irradiation of multilayered film of Cd and Te has been described. This semiconducting compound has been shown to present the same optical properties as single crystal and as good quality films. Moreover, electrical transport measurements which have been successfully performed on our films indicate that they have comparable behaviour to single crystal; the same range of resistivity and the same classification in low and high resistivity categories. Note that in the single crystal case low resistivity material is generally n-type with higher mobilities than the p-type semiconductor characterized by high electrical resistivity [19] . This difference has been attributed to the presence of defects promoted or not by the thermal history of the crystal. Presumably, the irradiation conditions would influence the thermodynamic processes, which induce the formation of CdTe films, and then govern the sample resistivity as the grain size. However, the possibility of producing material free from impurity states in the middle of the forbidden gap has been clearly shown. A careful control of the preparation and irradiation conditions would lead us to the production of CdTe high purity thin films by using this versatile and low cost technique. Finally, different materials : Au, Al, Cr and ITO have been tested in order to find ohmic contacts and the right element to be used in a Schottky barrier; surprisingly, gold is in this work a good ohmic contact.

ACKNOWLEDGEMENTS

This work is supported by R&D "Energy" Program of the Belgian Ministry for Science Policy.

REFERENCES

1. S.S. Lau, Martti Maenpaa and James W. Mayer, Proceedings of MRS Annual Meeting 1980, vol. 1, P. 547.

2. M. Wittmer and M. von Allmen, Ibid p. 533.

3. Y.S. Liu, S.W. Chiang and F. Bacon, Ibid p. 117

4. I.W. Boyd, J.I.B. Wilson, J.L. West, Thin Solid Films, 83 (1981) L173.

5. S.M. Metev, S.K. Savtchenko, K.V. Stamenov, V.P. Veiko, G.A. Kotov and G.D. Shandibina, IEEE J. Qu. El., QE-17 (1981), 2004.

6. M. Schneider, H. Tews, R. Legros, J. of Cryst. Growth, 59 (1982) p.293-296.

7. R. Andrew, L. Baufay, A. Pigeolet and M. Wautelet, to be published in the Proceedings of MRS, Symposium of Laser Diagnostics and Photochemical Processing for Semiconductor Devices (1982).

8. R. Andrew, M. Ledezma, M. Lovato, M. Wautelet and L.D. Laude, Appl. Phys. Lett., 35 (1979), p. 418-420.

9. R. Andrew, L. Baufay, L.D. Laude, M. Lovato, M. Wautelet, J. de Phys., 41 (1980), C4-71.

10. L. Baufay, D. Dispa, A. Pigeolet, L.D. Laude, J. of Cryst. Growth, 59 (1982) p. 143-147.

11. R. Andrew, L. Baufay, A. Pigeolet, L.D. Laude, Proceedings of MRS Annual Meeting 1981, vol. 4 p. 719.

12. L. Baufay, R. Andrew, A. Pigeolet, L.D. Laude, Thin Solid Films, 90 (1982) p. 69-74.

13. R. Andrew, L. Baufay, A. Pigeolet, L.D. Laude, J. of Appl. Phys. 53 (1982) p 4862-4865.

14. L. Baufay, R. Andrew, A. Pigeolet, L.D. Laude, to be puglished in Physica.

15. Courtesy of "Glaceries Saint-Roch - Auvelais - Belgium".

16. ASTM tables, such CW laser beam induces the cubic structure formation.

17. T.H. Myers, S.W. Edwards, J.F. Schetzina, J. Appl. Phys. 52 (1981) p, 4231.

18. J.O. Dimmock in : Proc. Intern. Conf. on II-VI Semiconducting Compounds, ed. D.G. Thomas (Benjamin, New York, 1967) p. 227.

19. J.B. Mullin, C.A. Jones, B.W. Straughan, A. Royle, J. of Cryst. Growth, 59 (1982), p. 135-142.

20. N.Kh. Abrikosov, V.F. Bankina, L.V. Poretskaya, L.E. Shelimova and E.V. Skudnova in : Semiconducting II-VI, IV-VI and V-VI Compounds (Plenum, New-York, 1969) p. 31.

21. S.M. Sze in : Physics of Semiconductor Devices (Wiley-Interscience, New York 1969), p. 393-403.

22. Ref. 21, p. 32-38.

THERMAL PULSE ANNEALING OF $Hg_{1-x}Cd_xTe$

K. C. DIMIDUK[a], W. G. OPYD[b], M. E. GREINER, J. F. GIBBONS AND T. W. SIGMON
Stanford Electronics Laboratories, Stanford, CA 94305

ABSTRACT

Thermal pulse annealing has been used to modify the near
surface of $Hg_{1-x}Cd_xTe$. Using anneals of approximately 260°C
for seven seconds, the crystal quality of epitaxial HgCdTe
surfaces can be improved as observed by MeV He$^+$ ion channeling. Similar anneals have also been used to repair the
damage resulting from a 250 keV, 10^{15} $^{11}B/cm^2$ implant into
HgCdTe held at LN_2. For higher temperatures and/or longer
anneals, surface Hg loss is observed. Rutherford Backscattering measurements are used to measure this loss. The
resulting loss rate data is described by $N_□ = A \exp (-\Delta E/kT)$
where A and ΔE depend on the material composition with A =
10^{29}, $\Delta E = 1.8$ eV and A = 10^{36}, $\Delta E = 2.6$ eV for x = 0.23 and
0.4, respectively.

INTRODUCTION

$Hg_{1-x}Cd_xTe$, a variable bandgap semiconductor widely used for infrared
detection, is difficult to process due to materials defects and potential Hg
loss. To aid in chosing annealing parameters, that will not cause significant
Hg loss, we have utilized the time-temperature regions allowed by a thermal
pulse anneal to measure the loss rate as a function of temperature. This
extends the work of Takita et al. [1], who measured Hg loss from HgTe, to the
ternary compound $Hg_{1-x}Cd_xTe$. Our results are also consistent with the absence
of Hg loss at room temperature reported by Silberman et al. [2] for this
material. However, we do find a dependence on composition (x).
     Magee [3] has reported significant near surface defects in as grown epitaxial
$Hg_{1-x}Cd_xTe$. We have used the data obtained on Hg loss to develop an anneal
that improves the crystalline quality, as observed by MeV ion channeling without
causing significant Hg loss.

EXPERIMENTAL

Annealing experiments were performed on epitaxial layers of $Hg_{.64}Cd_{.36}Te$
grown on CdTe from a Te rich melt at about 500°C. As shown in ref. [4], the
top layer of the growth plays a critical role in the enhancement of damage
diffusion. A higher dislocation density in this top layer of the growth is
reported [6]. We have performed our experiments for two cases: 1) as grown,
and 2) with 3 μ of that top layer chemically removed [5].
     The experiments to measure Hg loss were done on bulk, single crystal
$Hg_{.6}Cd_{.4}Te$ of unspecified orientation. Previous results, also included, were
done on similar crystals with x values of 0.24 and 0.225.
     Thermal processing of these samples was carried out in a thermal pulse
annealing system [6,7]. This system consists of two 1,000 W quartz tungsten
lamps placed inside conical reflecting surfaces. Sample temperatures up to
900°C for times of 2-90 sec can be obtained with this system. Sequential pulses

Mat. Res. Soc. Symp. Proc. Vol. 13 (1983) ©Elsevier Science Publishing Co., Inc.

are used to obtain longer anneal times. MeV ^{4}He ion backscattering and channeling are used to monitor near surface Hg loss and crystal quality following thermal pulse processing.

RESULTS

Initially we looked at several pieces of $Hg_{.64}Cd_{.36}Te$ using Rutherford backscattering in random and channeled directions. Figure 1 shows the results for "as grown" and "prepared" pieces. Comparing the dechanneling rate in the channeled direction we see a lower value for samples receiving the surface preparation indicating improvement of the crystalline quality. Also shown in Fig. 1 are data for samples annealed at 260°C for seven seconds. It can be seen from this data that a further reduction in dechanneling has resulted, implying further improvement in crystal quality. Also, dechanneling in the annealed as grown sample is lower than that of the unannealed prepared sample.

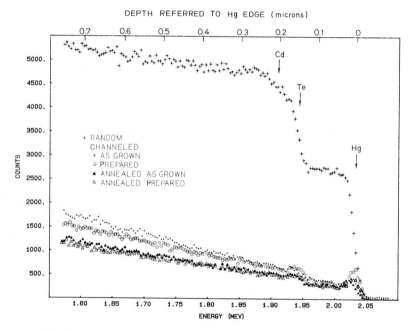

Fig. 1. RBS spectra of epitaxial $Hg_{.64}Cd_{.36}Te$: (+) random, (•) as grown, channeled, (o) prepared, channeled, (▲) annealed, as grown, channeled, (Δ) annealed, prepared, channeled.

In Fig. 2 we show ion channeling data for a HgCdTe sample following ion implantation and annealing. The implantation conditions were 10^{15} $^{11}B/cm^2$ at 250 keV with the sample held near LN_2 temperature during the implant. The anneal consisted of a 260°C eight second thermal pulse. Data before implantation, after implantation and after implantation and annealing is shown. A preanneal was not used on this sample. The anneal is seen to improve the crystalline quality of the material, as measured by ion channeling, to a value better than the starting material.

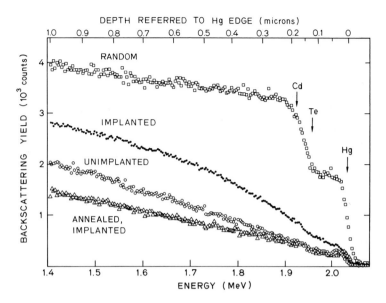

Fig. 2. RBS spectra showing effect of annealing of boron implanted sample, (□) random, (o) unimplanted, channeled, (●) implanted, channeled and (Δ) annealed, implanted channeled.

The Hg loss measurements were made on $Hg_{.6}Cd_{.4}Te$ samples annealed between 316° and 388°C for times ranging from 4.5 seconds to 455 seconds. Rutherford backscattering was used to compare the samples before and after the anneal. Following annealing, loss of Hg in the near surface region is observed in the RBS spectra as movement of the Hg signal to lower energies. By comparsion of this signal to that for an unnanealed sample, the number of Hg atoms lost can be calcuated. This is converted to a loss rate by dividing by the respective anneal times.* This data is plotted vs 1/T in Fig. 3. The error bars correspond to an uncertainty of ± 10°C in temperature and 1-15% in Hg loss/sec. The larger uncertainty in Hg loss corresponds to the higher temperature measurements. This is a result of the shorter anneal times contributing more error to the calculation of the Hg loss rate. Other less significant sources of error include the material composition and statistical fluctuation in the data. A straight line that gives the best least squares fit to the data points results in an activation energy and rate constant of $\Delta E = 2.6$ eV and $A = 10^{36}/cm^2$- sec, respectively. Errors in A and ΔE are taken to be within the range of values that still provide a reasonable fit to the data. Using the above assumptions the error in ΔE is ± .4 eV.

Figure 3 also includes data measured by Takita et al. [1] and our previous results for x = .23 material for comparison. In Table 1 we list the activation energies and rate constants for HgTe, $Hg_{.77}Cd_{.23}Te$, $Hg_{.6}Cd_{.4}Te$ and CdTe. For CdTe, A is assumed to be zero as CdTe contains no Hg to lose. A pattern of increasing ΔE and A with decreasing Hg content is observed.

_____
*(Ref. 7 describes this calculation in detail)

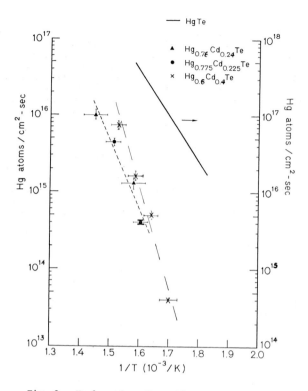

Fig. 3. Hg loss from $Hg_{1-x}Cd_xTe$ as a function of temperature, (▲) x = .24, (●) x = .225, (x) x = .4, — x = 0 (from Takita et. al. [1]).

TABLE I
Constants for Hg Loss Rate Equation $N_\square = A \exp(-\Delta E/kT)$

Material	A (at/cm²-sec)	ΔE(ev.)
CdTe	0	---
$Hg_{.6}Cd_{.4}Te$	10^{36}	2.6
$Hg_{.77}Cd_{.23}Te$	10^{29}	1.8
HgTe	1.9×10^{27}	1.2

DISCUSSION

The use of a thermal pulse anneal to improve the crystalline quality of epitaxial HgCdTe has been demonstrated. Both as grown surfaces and surfaces prepared for low residual defects were improved. The best crystalline quality as measured by MeV ion channeling is for the annealed, low defect surfaces. The thermal pulse annealing system has also been successfuly used to anneal the implantation damage resulting from a 10^{15} cm^{-2} boron implant at 250 keV. For this case, the annealed, implanted sample has better crystalline perfection than before implantation, approaching that of an annealed, unimplanted sample. Further work to measure crystalline perfection of samples first pre-annealed, then implanted and annealed is in progress.

We have utilized the temperature-time control afforded by the thermal pulse anneal to measure the temperature dependence of the loss of near surface Hg in HgCdTe. This data has been fit to an exponential of the form A exp $(-\Delta E/kT)$. The spread in values for A and ΔE are accounted for by the uncertainty in the temperature and Hg loss measurements. This spread should be taken into account when using this equation to estimate Hg loss for a given anneal. Even though the least squares fit is better than the error bars indicate, we do not feel the absolute temperatures are known to better than ± 10°C. The relative temperatures may well be more precise, thereby accounting for the good fit and implying less error in the activation energy.

Experimentally, we have noticed a trend of increasing activation energy, ΔE, and rate constant, A, with decreasing Hg content. Although the exact values are difficult to obtain (see Fig. 3) the range of values for each composition follows this trend.

If we extrapolate our data for x = .23 or x = .4 to room temperature using the equation and parameters of Table 1 we expect to see fewer than 10^5 atoms/ cm^2-day lost. This is consistent with the absence of Hg loss reported by Silberman et al. [2] at room temperature.

SUMMARY

A 260°C seven second thermal pulse anneal has been shown to improve the crystal quality, as observed by MeV channeling of "as grown" and "prepared" epi-taxial $Hg_{.64}Cd_{.36}Te$ layers. At 260°C for eight seconds the damage caused by a 250 keV $^{11}B/cm^2$ implant is annealed.

The thermal pulse annealer has also proven to be a useful experimental tool in accessing a time-temperature regime that has allowed Hg loss measurements from HgCdTe layers to be quantified. We have measured the loss rate for various compositions and temperatures and fit the data to an equation of the form $N_\square$ = A exp $(-\Delta E/kT)$ where A and ΔE depend on the composition parameter x.

ACKNOWLEDGMENTS

We would like to thank the Defense Advanced Research Projects Agency, (MDA-903-80-C-0238), in particular S. Roosild and R. Reynolds for their support of this work, Rockwell International for the epitaxial HgCdTe and Texas Instruments, Honeywell and Santa Barbara Research Center for the bulk HgCdTe. K. Dimiduk would also like to thank the National Science Foundation for fellowship support.

REFERENCES

a. Formerly under maiden name K. L. Conway.

b. Also with Lockheed Palo Alto Research Laboratory, Palo Alto, CA   94304.

1. K. Takita, K. Masuda, H. Kudo and, S. Seki, Appl. Phys. Lett. 37 (5), 460 (1980).

2. J. A. Silberman, P. Morgen, I. Lindau, W. E. Spicer, and J. A. Wilson, J. Vac. Sci. Technol. 21 (1), 154 (1982).

3. T. J. Magee and P. Raccah, Proc. of IRIS Specialty Group on Detectors, San Diego, July 1982.

4. L. O. Bubulac, W. E. Tennant, R. A. Riedel and T. J. Magee, J. Vac. Sci. Technol. 21 (1), 251 (1982).

5. To be published, private communication with L. O. Bubulac, Rockwell International Science Center, Thousand Oaks, CA 91360.

6. K. L. Conway, W. G. Opyd, M. E. Greiner, J. F. Gibbons, T. W. Sigmon, and L. O. Bubulac, Appl. Phys. Lett. 41 (8), 750 (1982).

7. K. L. Conway, W. G. Opyd, J. F. Gibbons, and T. W. Sigmon, Proc. IBMM II, Grenoble, 1982.

PULSED UV LASER IRRADIATION OF ZNS FILMS ON SI AND GAAS

H.S. REEHAL, C.B. THOMAS, J.M. GALLEGO and G. HAWKINS
Department of Applied Physics, University of Bradford, BD7 1DP, U.K.
C.B. EDWARDS
Laser Division, SERC Rutherford Appleton Laboratory, Chilton, Oxon, U.K.

ABSTRACT

Pulsed XeCl (308 nm) and XeF (351 nm) laser annealing of undoped and Mn-implanted single crystal and polycrystalline ZnS films deposited upon either Si or GaAs substrates is described. Annealing energy densities in the range $\sim$ 0.6 - 2.5 J cm^{-2} have been employed, with the specimens held under an Ar environment at $\sim$ 100-200 psig pressure. Film thicknesses in the range $\sim$ 0.2 - 1.0 μm have been investigated, with the substrates cleaned in various ways prior to deposition. In all the cases investigated, laser annealing results in polycrystalline regrowth of the films as shown by RHEED analysis. Furthermore the RHEED patterns obtained are consistent with the presence of both the cubic and the hexagonal phases of ZnS, except for thin ($\sim$ 0.2 μm thick) films annealed above $\sim$ 2 J cm^{-2}. These show a purely cubic structure on both Si and GaAs, possibly due to regrowth from the underlying cubic substrates.

INTRODUCTION

ZnS is a semiconductor with a large, direct band-gap of $\sim$ 3.6 eV and has potential applications for light-emitting devices. Thin films of the material doped with Mn luminescent centres have been studied extensively over the last decade or more [1]. Recently interest has been shown in devices consisting of ZnS:Mn layers deposited upon Si substrates [2] and the present authors have demonstrated laser activation of the Mn centres in these structures [3,4]. Pulsed Xe Cl (308 nm) radiation was utilised and the importance of annealing in an inert atmosphere at several atmospheres pressure was demonstrated. In this paper we present results on the structure of the films for a variety of film/substrate, film deposition and annealing parameters. Additionally, results of SIMS profile measurements on Mn-implanted specimens are presented.

EXPERIMENTAL DETAILS

The ZnS films were deposited onto n-type (100) Si wafers either by evaporation from solid source material in an ultra-high vacuum (UHV) system or by radio-frequency (RF) sputtering from a solid target in an argon atmosphere. Although Si has a good lattice match with the cubic phase of ZnS, the mismatch in the thermal expansion coefficients is not insignificant. For this reason films were also deposited, by RF sputtering only, onto (100) semiconducting n-type GaAs wafers for comparison purposes. This material has a better thermal match with ZnS but the lattice match is poorer. The relevant thermal and lattice data is summarised in Table 1.

For film deposition by evaporation in UHV, the Si substrates were thermally cleaned at $\sim$ 1000°C in a vacuum of $\sim$ 10^{-10} Torr to remove the native oxide layer. The ZnS evaporation was performed at a rate of $\sim$ 1.5 Å sec^{-1} with the substrates held at $\sim$ 280°C. These conditions led to good quality single crystal

Mat. Res. Soc. Symp. Proc. Vol. 13 (1983) © Elsevier Science Publishing Co., Inc.

films as has previously been reported in the literature [6]. In the case of
films sputtered onto Si two types of substrate cleaning procedure were employed.
The first consisted of a chemical clean followed by a 400°C bake at 10⁻⁶ Torr.
Alternatively in-situ argon-ion etching just prior to deposition was utilised.
GaAs substrates were subjected only to the former cleaning procedure. During
sputtering the Ar gas pressure was $\sim 10^{-2}$ Torr and the substrates were
maintained either at room temperature or $\sim 200°C$, with deposition rates lying in
the range 20-60 Å min⁻¹. All the sputtered films were polycrystalline. The
film thicknesses employed ranged between $\sim 0.2 - 1.0$ μm. Selected specimens
were implanted with 300 KeV Mn ions at a dose of $\sim 3-5 \times 10^{15}$ ions cm⁻² at
A.E.R.E. Harwell.

In view of the large band-gap of ZnS, the laser anneals were carried out
with an electron beam-pumped rare-gas halide laser emitting in the UV, as
previously described [3,4]. The 308 nm XeCl line was used predominantly
although some anneals were carried out using the 351 nm radiation from XeF.
Typical pulse widths were in the range 30-40 ns and energy densities of up to
~ 2.5 J cm⁻² could be obtained by focussing the beam. No beam homogenising
apparatus was used; however beam uniformity over the annealed areas was
estimated to be no worse than $\sim \pm 10\%$. During annealing the specimens were in
an Ar environment in a specially constructed cell capable of being pressurised
to ~ 1000 psig. It has been shown that significant vaporisation of ZnS occurs
if the anneals are carried out at atmospheric pressure [3,4]. In the present
studies annealing pressures of $\sim 100-200$ psig were used as no significant
improvements have been noticed at higher pressures.

Reflection high energy electron diffraction (RHEED) was utilised to
determine film structure before and after laser annealing. The data was
obtained in a Philips EM 200 electron microscope operating at 80 kV. Specimen
surfaces were also studied optically using Nomarski interference microscopy.
The SIMS profiles were determined at Imperial College, London.

RESULTS

Figure 1 shows characteristic RHEED patterns from ~ 0.8 μm thick single
crystal films on Si subjected to various treatments. The data for the virgin
film, Figure 1a shows a spot pattern corresponding to cubic ZnS but additional
reflections are observed. These can be accounted for by invoking the presence
of twinning and double diffraction effects but the presence of a small amount of
hexagonal material cannot be ruled out [7]. Results obtained after Mn
implantation and a 1 hour, 500°C thermal anneal of the implanted film are shown
in Figures 1b and 1c respectively. From the re-emergence of the spots in
Figure 1c, thermal annealing appears to be effective in restoring some
structural order. It is also interesting to note that there is no hint of the
additional satellite spots of Figure 1a in Figure 1c, the latter being
characteristic of purely cubic material. The data shown in Figure 1d was

TABLE 1
Thermal expansion coefficients and lattice parameters for Si, GaAs and
ZnS [5]

Semiconductor	Lattice parameter (Å)	Expansion coefficient at 300 K (10⁻⁶K⁻¹)
Si	5.431	2.33
GaAs	5.654	5.8
ZnS cubic	5.409	
hexagonal	a = 3.814, c = 6.258	$\sim$6.2

obtained after the Mn implanted films had been irradiated with a 1.5 J cm^{-2}
XeCl pulse under ∿ 100 psig of Ar; sharp rings characteristic of randomly
orientated crystallites are observed.   In addition to rings corresponding to
the cubic phase of ZnS, extra reflections are present which can be accounted
for by invoking the presence of hexagonal material and/or twinning and subsequent double diffraction [7].   Nomarski micrographs taken on the samples of
Figure 1 are shown in Figures 2a and 2b.   The latter shows the relatively

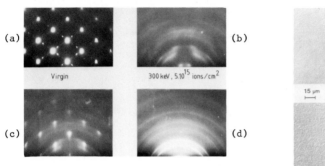

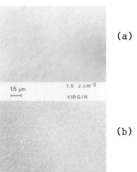

Fig. 1. RHEED patterns from ∿ 0.8 μm thick
single crystal films on Si (a) Virgin
specimen, (b) After Mn implantation
(c) Implanted specimen after thermal
annealing (d) Implanted specimen after a
1.5 J cm^{-2} XeCl pulse under 100 psig Ar.

Fig. 2. Nomarski micrographs
of specimens of Figure 1:
(a) Laser annealed film
(b) Virgin, as-implanted or
thermally annealed films.

TABLE II
Summary of RHEED Analyses on sputtered films annealed under ∿ 150 psig of Ar

Sample Type	Substrate Cleaning	ZnS thickness μm	RHEED RESULTS		
			XeCl		XeF
			∿0.6-1.8 J cm^{-2}	≳ 2 J cm^{-2}	∿1.3-1.8 J cm^{-2}
ZnS:Mn/Si	Chemical only	∿ 0.2	PR	PS	PR
ZnS/Si	Ar-ion etching	∿ 0.2	PR	PS	PR
ZnS/Si	"	∿ 1.0	PT	PR	PR
ZnS/GaAs	Chemical only	∿ 0.2	PR	PS	PR

PR = Sharp polycrystalline rings consistent with cubic + hexagonal phase
       material.
PS = Polycrystalline rings + spots; purely cubic phase material.
PT = As for PR but initial fibre texture of films also present at ≲ 1 J cm^{-2}.

rough surface appearance of the virgin films.   As-implanted and thermally
annealed specimens had an identical appearance.   On the other hand, the
micrograph of the laser annealed specimen shown in Figure 2a is significantly
different, exhibiting a cellular structure.   Behaviour similar to the above
has been seen on all our epitaxial films on Si, which lay in the thickness
range $\sim$ 0.6 - 1.0 μm, up to XeCl annealing energy densities of $\sim$ 2 J cm^{-2}.
Thinner epitaxial specimens have not yet been investigated.

A more extensive study has been conducted on sputtered films using both
XeCl and XeF radiation.   For this two film thicknesses of $\sim$ 0.2 μm and $\sim$
1.0 μm were employed; also both undoped and Mn-implanted films were investigated.   All the as-deposited films were polycrystalline with a cubic structure
and varying amounts of fibre texture.   The latter was most pronounced when Si
which had been argon-ion etched prior to deposition was used as the substrate.
Table II summarises the RHEED results obtained after laser annealing these
specimens.

In all cases polycrystalline regrowth is obtained irrespective of substrate preparation, type or film doping effects.   Any initial fibre texture in the
films is eliminated except for the 1 μm films where it decreases with increasing annealing energy density.   Typical RHEED patterns for the 0.2 μm films on
either Si or GaAs taken below and above $\sim$ 2 J cm^{-2} are shown in Figures 3a and
3b.   The former exhibits rings additional to those of cubic ZnS, like the data

Fig. 3. Typical RHEED patterns
from laser annealed 0.2 μm thick
films on Si or GaAs (a) 1.5 J cm^{-2}
(b) 2.3 J cm^{-2}. XeCl radiation,
150 psig Ar pressure   (a)

(b)

of Figure 1d, which are consistent with the presence of either hexagonal ZnS or,
possibly, double diffraction effects.   Figure 3b on the other hand suggests
the presence of purely cubic phase material; the spots on the rings imply a
larger grain size relative to films annealed at lower energy densities.   For
the 1.0 μm films on Si, RHEED patterns similar to Figure 3a were observed up
to $\sim$ 2.5 J cm^{-2}.   However below $\sim$ 1 J cm^{-2} evidence of initial fibre texture
in the films was also present.

Nomarski micrographs of the 0.2 and 1.0 μm thick films on Si annealed
using XeF or XeCl radiation reveal varying amounts of surface rippling, often
with a cellular structure [4].   On the other hand, the 0.2 μm films on GaAs
substrates exhibit severe cracking after XeCl anneals.   This deleterious
effect is however absent when XeF radiation is used.   These latter observa-

Fig. 4. Nomarski micrographs
of $\sim$ 0.2 μm thick sputtered
films on GaAs annealed at 1.3 J cm^{-2}
under 150 psig of Ar (a) XeCl
radiation (b) XeF radiation

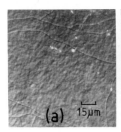
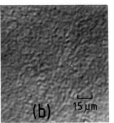

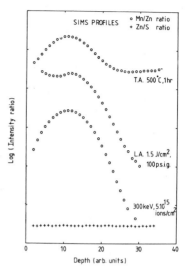

Fig. 5. SIMS profiles on Mn-implanted films of Figure 1 before and after laser and thermal annealing

tions are illustrated by the micrographs of Figure 4.

Finally, Figure 5 shows the results of SIMS profile measurements on Mn implanted films before and after laser and thermal annealing. The data was obtained on the specimens of Figure 1 and shows the flattening of the initially Gaussian Mn profile in the first ∿ 2000 Å of the film after laser annealing. This supports the hypothesis of laser-induced surface melting and dopant redistribution within this ∿ 2000 Å thick molten region.

DISCUSSION AND CONCLUSIONS

Our results show that XeCl and XeF pulsed laser annealing of ZnS films on Si or GaAs results in polycrystalline regrowth under the conditions described above. For the 1.0 μm sputtered specimens, regrowth probably occurs from the underlying polycrystalline film as it is unlikely that melt front penetration to the underlying substrate occurs (see Figure 5). Similarly the Mn-implanted single crystal films probably regrow from implantation damaged layers as Figure 5 suggests. Epitaxial layers implanted at lower energies are being investigated to clarify the situation. In contrast, for 0.2 μm thick specimens, it is likely that melt-front penetration to, and regrowth from, the underlying substrates takes place, particularly above ∿ 2 J cm^{-2}. Consequently these regrown layers exhibit a purely cubic structure characteristic of the substrates. The fact that both the ZnS/Si and ZnS/GaAs film/substrate combinations have either imperfect thermal or lattice matches (see Table I) may well account for the observed polycrystalline nature of these cubic-phase regrown layers; in any case it could contribute, with other possible factors, to the lack of epitaxial regrowth. It should also be noted that film crystallinity in these specimens (and also in the other cases studied) does not appear to depend on interface cleanliness or Mn doping of the films.

Of the other points to note is the observation of a possible hexagonal phase in the films after laser annealing, as described above. ZnS exhibits two stable allotropes, cubic and hexagonal. The latter is the stable form at high temperatures, with the hexagonal to cubic transition being reported to occur at

$\sim 1150^{\circ}$C [8]. Thus competition between the two allotropes may possibly occur during regrowth after melting (the melting point of ZnS being $\sim 1830^{\circ}$C under pressure [4]). For regrowth primarily within the films (as for the thick specimens and for thinner samples annealed at relatively low powers) this may lead to loss of crystallinity and mixed cubic plus hexagonal material as observed. For regrowth from the underlying cubic substrates, however, the hexagonal phase may well be suppressed and purely cubic material obtained as discussed above. However further work is required to confirm this and the identification of the additional structure in the RHEED patterns with a hexagonal phase, rather than with twinning and double diffraction effects.

The room temperature absorption coefficient of ZnS films at 351 nm and 308 nm has values of $\sim 4 \times 10^4$ cm^{-1} and $\sim 1 \times 10^5$ cm^{-1}, respectively [9]. Thus, following Tsu et al[10] one might expect the cooling rates after laser annealing to be greater for the XeCl anneals. However no significant effects due to this possibility have been observed in the crystal structure of the regrown material, although it could perhaps explain why severe film cracking occurs on GaAs substrates when using XeCl radiation but not when the XeF line is used (see Figure 4). Finally the cellular microstructure seen in many of the annealed films (eg Figure 2a) may also be due to microcracking, at least in the thin films. However other explanations are also possible and a fuller discussion, with transmission electron microscopy data, will be published elsewhere.

ACKNOWLEDGEMENTS

We would like to thank G. Jackson for technical assistance and R.S.R.E. Malvern for supplying some of the sputtered specimens. Thanks are also due to Mr A Nicholls of Leeds University for the use of RHEED facilities and Dr J. Kilner of Imperial College for the SIMS analysis. Finally we would like to thank S.E.R.C. and C.V.D. for financial support.

REFERENCES

1. C. Hilsum in: Solid State Devices, J.E. Carroll ed. (Inst. Phys., London 1980) pp. 1-20.

2. J.M. Gallego, H.S. Reehal and C.B. Thomas, S.I.D. 82 Digest, to be published.

3. H.S. Reehal, C.B. Edwards, J.M. Gallego and C.B. Thomas, Proc Vth National Quantum Electronics Conf. Hull, 1981 (to be published by Wiley).

4. H.S. Reehal, J.M. Gallego, C.B. Edwards, Appl. Phys. Lett. 40, 258 (1982).

5. A.G. Milnes and D.L. Fencht, Heterojunctions and Metal - Semiconductor Junctions (Academic Press, New York, 1972).

6. P.L. Jones, C.N. Litting, D.E. Mason and V.A. Williams, J. Phys. D1, 283 (1968).

7. T.G.R. Rawlins, J. Mater. Sci. 5, 881 (1970).

8. A. Addamiano and M. Aven, J. Appl. Phys. 31, 36 (1960).

9. P. Savjani, M. Phil. Thesis, University of Bradford, to be submitted.

10. R. Tsu, R.T. Hodgson, T.Y. Tan and J.E. Baglin, Phys. Rev. Lett., 20, 1356 (1979).

INCOHERENT FOCUSED RADIATION FOR ACTIVATION OF IMPLANTED InP

J.P. Lorenzo, D.E. Davies, K.J. Soda, T.G. Ryan, P.J. McNally*
Rome Air Development Center, Hanscom AFB, MA 01731
*Comsat Laboratories Clarksburg MD 20871

ABSTRACT

InP layers implanted with Si^+ ions are successfully activated through the use of a focused incoherent energy source. High mobility values and excellent carrier concentrations, both directly comparable to values attained using conventional furnace annealing, are achieved with this technique. Under optimized anneal conditions, electrical evaluation also indicates no surface anomalies such as those encountered with pulsed laser or electron beam annealed InP.

The annealing technique described here makes use of the output from two broad spectrum 1000 watt incandescent sources. Through the use of elliptical mirrors, a combined thermal image of these sources is formed as the output at a common focus. The sample is placed at this focus during the anneal cycle. Anneal time is controlled electronically and for these experiments is varied for 1 to 15 sec. Substrate temperature is measured with a low-mass thermocouple recorder readout system placed at the focus in contact with the sample. Annealing temperatures range from 750°C to 950°C. To prevent phosphorus loss and other surface anomalies, dielectric encapsulation and a solid phosphorus proximity source technique are examined. While the proximity technique has proven useful, PSG/SiO_2 double layered dielectric encapsulants appear to provide the most consistent results.

INTRODUCTION

Pulsed laser and electron beams have been used to anneal implanted damage in group III-V semiconductors but have met with limited success. While electrical activation can often exceed the limits of solid solubility and crystalline properties are excellent, mobility for these annealed layers is severly limited. In addition for InP, Rutherford backscattering has shown pronounced phosphorus loss at the surface. Electrical measurements indicate a related highly conducting surface layer both of which inhibit the usefulness of the annealed region unless extreme care is exercised in chosing the anneal conditions.[1]

In view of these limitations a technique for thermal processing using incoherent energy [2] is investigated as an alternative to conventional furnace techniques and to the more recent "beam" annealing systems. Although anneal time is longer than for a typical pulsed laser or E-Beam, there is still a significant reduction in process time over furnace techniques. It will be seen that moderation of the anneal cycle is advantageous in that many anomalies associated with nano second "Beam" annealing are eliminated. Since annealing occurs in the solid phase, frozen in defects appear not to be a problem.

Mat. Res. Soc. Symp. Proc. Vol. 13 (1983) Published by Elsevier Science Publishing Co., Inc.

Activation efficiencies and mobility values of implanted layers comparable to those obtained using furnace annealing are measured.

The experiments to be described deal mainly with activation of n^+ implants in InP using incoherent focused radiation. This system has also been used to activate implanted ions in GaAs using different time/temperature conditions.(2)

EXPERIMENTAL

The annealing system used here consists of two elliptical mirrors combined such that one foci of each coincide along a common axis. A broad spectrum 1000 watt incandescent source is placed at each of the remaining foci and the combined thermal image of these sources is formed as an output at this common focus (axis). For this system the mirrors are ~6" long with a focal distance of ~ 1.5 the major axis ~ 3.0", and the minor axis ~ 2.5". Substrate temperature is measured with a low-mass thermocouple recorder-readout system placed at the common focus in contact with the sample. Anneal time is controlled electronically and for these experiments is varied from 1 to 15 sec. Experimental temperatures range from 750°C to 950°C. Some typical thermal cycles are plotted in Figure 1. As shown, the rise time is approximately 2 sec and temperature is constant to within a few degrees during the anneal. With the power off, the rate of cooling is approximately 56°c/sec and 127°c/sec to 600°C. This represents a significantly extended quench time when compared to pulsed laser and E-beam annealing. This "gentle" quench is assumed to be responsible at least in part for the improvement in measured mobility.

The material used in these experiments is semi insulating (10^7 ohm-cm) iron (Fe) doped InP. Wafers are approximately fifteen mils thick and oriented 3° off the <100>. For doping, Si^+ ions are implanted with a fluence of 4×10^{14} cm^{-2} at an energy of 200 Kev. Both room temperature (amorphous) and 200°C implants are compared. After implantation, individual chips approximately 140 mils square are selected. Sample size is chosen to be convenient for Hall evaluation. Larger samples can be annealed by scanning the sample through the beam. To prevent phosphorous loss some surfaces are coated with a dielectric, others are left bare to be annealed using a proximity source.

Proximity annealing takes place with the implanted InP surface in close contact with a substrate containing phosphorous. The substrates were purchased from Carborundun Co. as a phosphorous diffusion source and produce P_2O_5 from the thermal decomposition of SiP_2O_7. This technique avoids the extra process step of dielectric deposition and possible surface defects due to strain associated annealing using caps such as Si_3N_4. Encouraging results were achieved with SiP_2O_7 (see table 1), however, further work will be required with this technique to improve its reliability.

Dielectric encapsulants are deposited at 330°C in a room pressure reaction chamber by the decomposition of silane. Double layered dielectrics grown with this system and consisting of phosphorous doped SiO_2 at the interface with an SiO_2 overcoat have provided the best and most consistent results.

Next these implanted substrates are placed at the common focus of the lamp system and annealed with varying time and temperature cycles for the different capping techniques and implant conditions(table 1). After annealing the dielectric cap is removed with HF and Au/Ge/Ni contacts evaporated to measure sheet concentration and mobility. Many of the samples were also profiled using differential Hall measurements in conjunction with controlled anodic layer stripping. This technique is used routinely to evaluate electrically

TABLE 1
Comparison of Typical sheet mobility and carrier concentration for
R.T. and 200°C (Si^+, 4 x $10^{14}cm^{-2}$, 200kev) implants annealed under
a variety of experimental conditions. (*200°C Implants)

ANNEAL CAP	TIME (sec)	TEMP (°C)	$Ns(cm^{-2})$ $(x10^{14})$	$\mu_s(cm^2/V\text{-}sec)$
SiP_2O_7	3	760	1.0	507
SiP_2O_7	12	840	1.32	594
SiP_2O_7	14	820	1.5	537
*SiP_2O_7	6	860	1.41	1043
*SiP_2O_7	12	800	1.83	954
*PSG/SiO_2	10	790	2.38	1033
*PSG/SiO_2	10	830	3.2	324
PSG/SiO_2	10	800	2.5	673
PSG/SiO_2	10	810	3.4	661
PSG/SiO_2	10	830	3.2	654
Si_3N_4	10	800	3.6	470
PH_3/Thermal	10 min	775	2.98	730
PH_3/Thermal	10 min	775	2.65	740

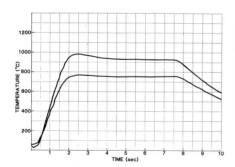

Fig. 1. Typical thermal cycles obtained during annealing at 750°C and 950°C using a low-mass thermocouple with recorder for sensing temperature.

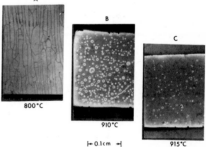

Fig. 2. (A) Pyrolytic SiO_2 on InP. (B) Single-layered PSG on InP heated to 910°C. (C) Double oxide (PSG + SiO_2) results in drastic reduction of pinholes. At 800°C no leaks occur.

active carrier distribution in III-V materials and is described elsewhere.(3)

RESULTS

A comparison of the surface integrity of InP after annealing in this system with various dielectric caps is shown in Fig 2. For Fig. 2a pyrolytic SiO_2 ~ 1100Å thick is deposited on InP and subjected to a 1.2sec 800°C thermal cycle. In a previous study this oxide was used to control the optical energy during CW laser annealing for implanted Ge diodes (4). In that work it was observed that the SiO_2 had good tensile strength and could be cycled to well above the Ge melt point with no ill effects. For the present case, due to differences in expansion coefficients between SiO_2 and InP, this oxide cracks resulting (5) in phosphorous loss, poor surface morphology and degraded electrical characteristics. The same is true for Si_3N_4 capped samples (table 1) Next a 3000Å, 4% phosphorous doped SiO_2 was deposited and heat cycled to ~900°C. The results in Fig. 2b indicate that although cracks are avoided, phosphorous leaks occur. Circular patterns are evident presumably as a result of high phosphorous vapor pressure and pinholes in the oxide. Much improved electrical characteristics are observed, and are most likely due to the isolation of phosphorous loss to small individual islands. The actual dielectric cap used for this work takes advantage of the positive aspects of both types of protective oxides. In figure 2c, the results of annealing at 915°C with the combined the oxides of 2a and 2b are shown. In practice, reducing the phosphorous doped portion of the oxide cap to approximately 1100Å has been found to give the best results. It is shown that at 915°C much reduced phosphorous leakage occurs using this double oxide where the P-doped SiO_2 is in direct contact with InP. It should be pointed out that in order to show the advantage of this double system, the tests in 2b and 2c were performed at 900°C well above the actual temperatures used in these experiments for good activation. When this double oxide is used for annealing InP (t < 15 sec) up to ~ 800°C no apparent phosphorous loss occurs. To attest to the superior quality of this $P-SiO_2/SiO_2$ it is interesting to note that GaAs similarly coated has withstood temperatures up to 1100°C with no obvious surface deterioration.

Elevated InP temperature (200°C) during implantation ($4 \times 10^{14}Si^+cm^{-2}$) has a significant impact on measured sheet mobility after lamp annealing as is the case for conventional thermal processing (6). Mobility values average >1000 cm^2/V-sec over the implanted region with carrier concentration consistently at the mid 10^{18} cm^{-3} level for several samples. In contrast peak values of carrier concentration (Fig 3 & 4) routinely exceed 10^{19} cm^{-3} for room temperature implants annealed under the same conditions with mobility averaging ~ 600 cm^2/V-sec. Sheet concentration for carrier temperature implants is consistently measured to be in the $3.4 \times 10^{14}cm^{-2}$ range (> 80% activation efficiency) when annealed with a double oxide cap. Best results are obtained in both cases for 10 sec, 800°C anneal cycles. A comparison of these and other results measured after annealing under varying experimental conditions is contained in table 1. For 200°C implants annealed at 800°C, carrier concentration and mobility profiles indicate good results for both proximity source (SiP_2O_7) anneals and double oxide capped anneals. The reduced peak carrier concentration and slight broadening evident after annealing high fluence, 200°C implants in this system appears to be consistent with observations of Donnelly et al (6) for furnace annealed implants at 750°C (Fig. 3). In an attempt to further improve peak concentration levels for 200°C implants with this lamp system, an investigation of higher temperature annealing is in progress. Differences in measured mobility of room temperature and 200°C implant anneals may not be entirely attributed to the substrate temperature during implantation. Typical mobility profiles using optimized anneal conditions for 200°C and 25°C implants are shown in

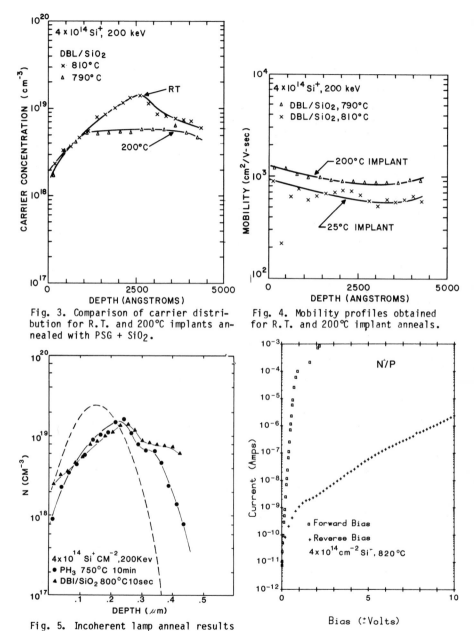

Fig. 3. Comparison of carrier distribution for R.T. and 200°C implants annealed with PSG + SiO₂.

Fig. 4. Mobility profiles obtained for R.T. and 200°C implant anneals.

Fig. 5. Incoherent lamp anneal results compared to conventional furnace anneal and to LSS theory.

Fig. 6. I-V characteristics for N⁺/P lamp fabricated Mesa Diodes.

Figure 4. Extrapolating from the bulk mobility data of Astles et al (7), less than 50% of the improvement observed can be attributed to the lower carrier concentration of the 200°C implant anneal. For Figure 5, a typical lamp annealed profile is compared with both the LSS and a conventional thermal profile. Both experimental peaks are shifted off the theoretical LSS peak but closely coincide with each other. This shift is consistently observed for experimental data obtained after conventional annealing of implanted InP(6). This excess broadening evident in the lamp annealed profile may be the result of annealing at 800°C, 50°C higher than for the conventional furnace process.

P-type $10^{17}cm^{-3}$ Zn doped substrates implanted with Si^+ ions (4 x $10^{14}cm^{-2}$ at 200kev) and annealed as described in the previous text are used in a preliminary study to form n^+ on p InP diodes. After annealing, mesa type diodes are formed by masking with black wax and using a 5% w/o Iodic acid etch. Au/Zn and Ni/Ge/Au are evaporated for p and n contacts respectively. Figure 6 represents a typical I-V characteristic obtained for this process. Yield for these devices was > 80% with high uniformity across the annealed sample. Device diameter (fig 6) is ~ 650µm and the calculated ideality factor from the forward I-V is n = 2. Even though these results are somewhat preliminary, the reverse characteristics compare favorably with other InP diodes (8).

SUMMARY

Deposited oxides containing phosphorous are combined with a focused incoherent lamp system to successfully anneal Si^+ implanted InP. For room temperature implants, greater than 80% electrical activation is achieved resulting in peak carrier concentration > $10^{19}cm^{-3}$. Of equal importance is the high mobility value of > 1000 $cm^2V^{-1}sec^{-1}$ using 200°C implants a significant improvement over values achievable with pulsed laser or electron beam annealing and comparable to the best values obtained with conventional thermal techniques. Mesa diodes are fabricated using this technique and preliminary results indicate low-reverse dark current.

REFERENCES

1. D.E. Davies, J.P. Lorenzo, T.G. Ryan, and J.J. Fitzgerald, Appl. Phys. Lett 35, 631, 1979.

2. D.E. Davies, P.J. McNally, J.P. Lorenzo and M. Julian, IEEE Elec. Device Lett. 102, 1982.

3. J.P. Lorenzo, D.E. Davies and T.G. Ryan, J. Electrochem. Soc. 126, 118, 1979.

4. J.P. Lorenzo, K.J. Soda and D.E. Davies, Laser and Electron-Beam Interactions with Solids B.R. Appleton and G.K. Celler, editors 1982.

5. K.V. Vaidyanathan, H.L. Dunlap, and C.L. Anderson, Int. Symp. GaAs and Related Compounds, Japan, 1981.

6. J.P. Donnelly and G.A. Ferrante, Solid State Elec. Vol 23, 1151, 1980.

7. M.G. Astles, F.G.H. Smith and E.W. Williams, J. Electrochem. Soc. 120, 1750, 1973.

8. T.P. Lee and C.A. Burrus, Appl. Phys. Lett. 36, 587, 1980.

PART VIII
METALLIC SURFACE ALLOYS

LASER-GENERATED GLASSY PHASES

MARTIN VON ALLMEN
Institute of Applied Physics, University of Bern, Sidlerstr. 5,
CH-3012 Bern, Switzerland.

ABSTRACT

Short laser pulses have recently been discovered
to be powerful means for glass formation in binary
alloys as well as in some pure elements. Due to
relevant cooling rates in excess of 10^{10} °K/s, glass
formation is achieved in binary systems over most of
the phase diagram, even in systems not previously
known as glass formers. We first give a review of
recent progress in laser melt quenching, and then
discuss the kinetic conditions of glass formation by
laser irradiation.

INTRODUCTION

Glassy metals have been made by mechanical quenching methods
for almost a quarter of a century. The different quenching
methods are usually compared on the basis of the average cooling
rates achieved; cooling rates ranging up to 10^6 °K/s are expected
to arise with the best mechanical methods. The use of laser
pulses as a means to quench materials into a glassy state has
first been attempted only a few years ago and exploitation of the
method has barely started. In laser quenching heat is produced in
a solid sample by very short irradiation. A surface layer of a
width of typically one micron or less is heated, melted and
subsequently cooled by conduction into the bulk of the specimen.
Owing mainly to the small width of the molten zone the heat flux
during cooling after laser irradiation exceeds that achievable by
mechanical quenching by orders of magnitude. Relevant cooling **rates**
reach 10^{10} °K/s under typical conditions. Not surprisingly, the
faster cooling is found to result in a drastically extended range
of materials that can be obtained and studied in the glassy state.
In what follows we first give a - necessarily sketchy - overview of experimental results obtained to date. In the second part
we analyse the criteria of glass formation from the point of view
of laser quenching.

EXPERIMENTAL RESULTS

Recent experimental work on laser-quenched glassy phases tends
to fall into one of two categories: In the first, short pulses
are used with the goal to extend previous limits of glass formation to new materials and material combinations, so far mainly to
Si-metal systems. In the second, mostly continous laser beams in
a scanning mode are adopted in work typically aimed at achieving
surface hardening of machine alloys [1]. While there is conside-

Mat. Res. Soc. Symp. Proc. Vol. 13 (1983) ©Elsevier Science Publishing Co., Inc.

692

rable technological interest in the latter application, its
results have not been shown to be qualitatively different from
those of splat cooling. We shall therefore consider the first
category only.

Alloys
  Procedure. Alloying of a single deposited surface layer to a
substrate by means of short laser pulses tends to produce complicated microstuctures, due to the nonuniformity of the melt composition prior to solidification [2]. Cellular growth patterns
with cell size down to a few tens of nm have been found as well as
apparently amorphous zones. Quantitative investigations in such
structures are difficult.
  A more controlled technique to investigate the properties of
laser alloyed binary structures was developed in [3] and has
since been applied to a number of material systems. Here several
alternating layers of metal and Si are vapor-deposited onto inert
substrates, e.g. onto sapphire. The thicknesses of the layers
are adjusted such as to yield the desired average film composition. In order to guarantee complete intermixing of the elements
during the lifetime of the melt, individual layers must be no more
that about 15 nm thick if ns pulses are used to melt the structure; total film thickness may be a few 100 nm. The laser pulse
energy required is of the order of 1 J/cm^2 , depending mainly on
the reflectivity of the multilayer.
  The reflectivity of metals, as a rule, decreases with temperature. This has the consequence that the choice of pulse
energy is critical, and that "hot spots" in the laser intensity
distribution tend to be amplified. The problem arises mainly with
laser wavelengths in the infrared, where cold metals have high
reflectivities. In the experiments described here the surface
layer was chosen to be Si, which not only serves the purpose of an
antireflection coating, but also has the property of stabilizing
the light absorption process, as its reflectivity increases upon
melting.
  The described procedure is exemplified in Figure 1, which
shows ^{4}He$^+$ backscattering spectra of a Pt-Si film (20 at.% Si)
before and after irradiation with a 30 ns Nd laser pulse.

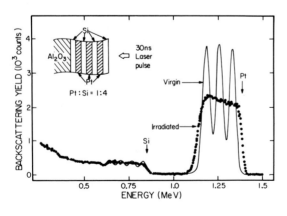

Fig. 1.
Backscattering
spectra of a
Pt-Si sample
before and
after laser
mixing and
quenching.

Such lasers deliver typical pulse energies of about two tenth of a Joule, enough to transform a spot of about 5 mm diameter. More extended areas can be transformed by applying several partially overlapping shots. Experimental results obtained with this method in various systems are summarized in the following.

Au-Si system. Au and Si are well known to form a metallic glass by splat cooling at compositions near the eutectic (17 at.% Si). Although the technological interest in this system is limited, it may serve as a model case for many features of laser quenching. Amorphous films were prepared by the above method with compositions ranging from 9 to 91 at.% Si. Figure 2 shows the Au-Si phase diagram with the composition of investigated films marked by arrows. The amorphous - crystalline transition temperature T_{ac} (determined by monitoring the electrical resistivity while heating the films at 3 °C/min) is indicated by dots. It can be seen that, while T_{ac} increases monotonically with the Si content, its slope is very different in different regions of the phase diagram. Pure amorphous Si is known to crystallize around 650 °C. The presence of only 9 at.% Au decreases the crystallization temperature to about 80 °C. Another dramatic decrease is observed near the Au-rich end. The glasses with 9 at.% Si had to be kept in liquid Nitrogen during and after the irradiation.

The compositional range of the Au-Si glasses reported here greatly exceeds previously established limits for glass formation by melt quenching. Glass-forming ability is thought to be related to melting point depression [4] which is largest at the eutectic point. The most Si-rich glasses shown in Figure 2 have melting point depressions (relative to the weighted average of the melting points of the pure elements) of only about 2%. This may be compared to the 20% depression which is generally found to be necessary for glass formation by splat-cooling [5]. It may be added that neither of the pure constituents can be obtained in a glassy state by irradiation with Q-switch pulses at the present wavelength. Pure semiconducting amorphous Si can be obtained by shortening either the pulse duration or the absorption length (see below); pure amorphous Au is not thought to be realizable.

Careful annealing of the amorphous Au-Si films resulted in nucleation and growth of a metastable silicide of hexagonal structure (c/a ratio = 1.24); its stoichiometry has not yet been worked out. The temperature at which the metastable silicide decays into Au and Si is around 150 °C, as indicated by the triangles in Figure 2 . It may be added that irradiation of the same samples with longer pulses (300 μs instead of 30 ns) directly leads to formation of the same metastable silicide.

Optical studies. The availability of glassy phases covering most of the phase diagram invites comparative studies of glass properties as a function of composition. Figure 3 shows the optical reflection of glassy Au-Si films at three different wavelengths as a function of glass composition; annealing brings about dramatic changes in the reflectivity spectrum for certain compositions. The dc resistivities of the glassy films vary from 10 μΩcm (9 at.% Si) to 10 mΩcm (91 at.% Si). In fact, it turns out that the glasses with more than 80 at.% Si are not metallic, but semiconducting. A simple classical model allowing for bound and free electron contributions to the dielectric function has been developped, which explains the main features of the optical properties as a function of composition [6].

Pd-Si and Pt-Si systems. From a thermodynamical point of
view, the main difference between these systems and Au-Si is the
presence of stable intermediate phases. Pd-Si is known as a glass
former from splat cooling (at the metal-rich eutectic, 18 at.%
Si), whereas Pt-Si is not. Laser quenching yields glassy phases
over most of the phase diagrams in both systems [7]. A conspi-
cious exeption are the congruently melting compounds: These are
found to be crystalline after irradiation for the mentioned set of
parameters. It appears, as a general conclusion from many
different experiments, that glass forming ability at the cooling
rates characteristic of ns laser pulses scales with the compositional "distance" to the nearest congruently melting phase
(including pure elements), rather than with melting point
depression. We shall come back to this point in the second part
of this paper.

How does the availability of equilibrium compounds affect the
stability of the amorphous film? Figure 4 shows the crystallization temperatures T_{ac} as a function of composition for Pt-Si,
indicating the same general trend as in Figure 2 - the stability
increases with the Si content. Even the congruently melting
phases Pt_2Si and $PtSi$ do not significantly deviate from this
general trend (the open dots in Figure 4 were measured with films
amorphized after laser irradiation by Xe implantation; this may
slightly increase the crystallization temperature). Whereas the
trend of T_{ac} is the same as in Au-Si, the absolute numbers are
not: The Pt-Si glasses are stable up to 400 °C and then transform
into metastable silicides among which Pt_2Si_3 or Pt_4Si_9 have been
identified [7].

Band structure studies. Glassy phases offer the possibility
to study the mutual influence of different atomic species in close
contact at compositions outside those of stoichiometric compounds.
Here laser quenching is able to extend the range of investigation
dramatically. The tendency of the metal-silicon glasses to become
semiconducting above a certain Si content, which is also observed
in the Pd-Si and Pt-Si systems, clearly illustrates how strongly
electronic properties can be made to vary within the same binary
system. A convenient method to investigate such effects is
electron spectroscopy of glassy phases of different composition.
As an example, Figure 5 shows UPS spectra of the electron distribution up to the Fermi level, measured at two different photon
energies in laser quenched Pd-Si glasses [8]. The peak position
and other features of the distribution can be seen to shift by
amounts up to several eV as a function of composition. Similar
observations were made for Pd and Si core levels. These results
are preliminary and further work is required in order to draw more
quantitative conclusions. The same study also showed the equivalence of mechanically quenched glasses with laser quenched ones of
the same composition: spectra obtained from near-eutectic Pd-Si
glasses prepared by both methods were indistinguishable.

V-Si and Nb-Si systems. These are pronounced high-temperature systems and very unlikely glass formers from the criteria of
splat cooling, the melting point "depression" being negative over
extended ranges of composition. Nevertheless, a series of glasses
has been obtained by laser quenching in both systems [9]. Some of
the glasses are stable up to 600 °C, in spite of the possible
presence of quenched-in crystalline nuclei. Some of the amorphous
films form yet unidentified metastable crystals upon annealing.

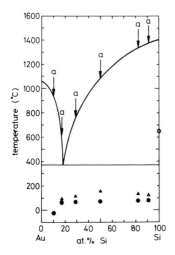

Fig. 2. Au-Si phase diagram.
Dots: amorphous-crystalline
transition temperature; triangles: decay temperature of
metastable silicide.

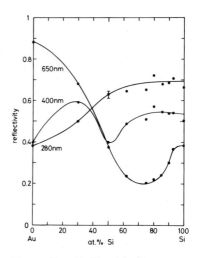

Fig. 3. Reflectivity vs.
composition of glassy Au-Si
films at various wavelengths.

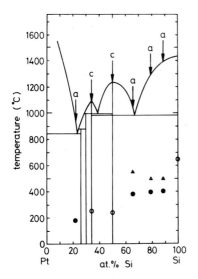

Fig. 4. Pt-Si phase diagram.
"a" and "c" denote films amorphous and crystalline after
irradiation. Symbols have the
same meaning as in Figure 2.

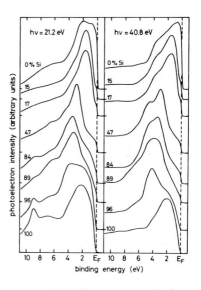

Fig. 5. UPS spectra of glassy
Pd-Si films, as well as pure
Pd and Si, taken at two diffe-
rent photon energies.

Pure elements.
    Except for some nonmetals like B or Se, none of the pure ele-
ments has ever been obtained in a glassy state by splat cooling.
The reason is that crystal nucleation and growth is fastest in
pure melts where compositional constraints at the crystal-melt
interface are absent. Therefore, the cooling rates required to
bypass crystallization in pure elements are larger than in alloy
melts.
    The production of amorphous Si by laser quenching is now well
established [10, 11, 12]. Here bulk specimen, rather than thin
deposited films, were irradiated and amorphous zones formed at the
surface of the crystalline material. The experimental findings
[10, 11] suggest that, not unexpectedly, the "glass forming power"
of laser irradiation increases with decreasing pulse duration and
absorption length. The tendency for glass formation is also found
to depend on the crystallographic orientation of the irradiated Si
wafer; <111> oriented material is easier to amorphize than <100>
oriented one [12].
    Finally, the occurrence of an "essentially amorphous" region
was even reported in pure Al [13] after irradiation by a 15 ns
Ruby laser pulse. The layer was found to contain small grains or
crystalline AlN, originating probably from adsorbed nitrogen at
the surface of the specimen before irradiation.

KINETIC CONDITIONS FOR GLASS FORMATION BY LASER IRRADIATION

Theory
    The objective of melt quenching is to bypass crystal nuclea-
tion and growth. Therefore, the conditions for glass formation
are to be derived from nucleation and growth theory. The growth
velocity of a crystal into its undercooled melt, according to
classical theory, is essentially given by

$$u(T) = u_o \, e^{-Q_u/kT} \, [1 - \exp(-L_m \Delta T/NkT_m T)] \tag{1}$$

    Here u_o is a constant which depends on the detailed structure
of the crystal-melt interface, T is the interface temperature, L_m
is the heat of melting per unit volume, N is the particle density,
$\Delta T = T_m - T$ is the undercooling, and Q_u is the activation energy
for molecular rearrangment at the interface (assumed to be a
simple thermally activated process). Q_u is sometimes estimated
from the activation energy for viscous flow in the undercooled
melt [14]. The growth velocity increases rapidly with increasing
undercooling just below the melting point, reaches a maximmum and
approaches zero at large undercooling. The maximum growth velocity is believed to be between 10 and 20 m/s in the case of pure
Si [12]; for pure metals it is probably much higher.
    The rate of crystal nucleation, too, shows an optimum tempe-
rature range, determined by a compromise between driving force and
particle mobility. In addition, the nucleation rate depends on
time: if a melt is suddenly undercooled, measurable nucleation
sets in only after a "time-lag". The time-lag t_n may be interpreted as the time required to establish an equilibrium population of
clusters corresponding to the new temperature [15]. The rate of
homogenous nucleation may be expressed in a simplified form as

$$I(T,t) = \begin{cases} 0 & \text{for } t < t_n, \\ I_\infty(T) & \text{for } t > t_n \end{cases} \qquad (2)$$

where I_∞ is the steady state nucleation rate given by

$$I_\infty(T) = I_0 \, e^{-Q_I/kT} \, \exp(-\gamma^3 T_m^3/L_m^2 \Delta T^2 kT). \qquad (3)$$

Here γ is the crystal-melt interface energy per unit area and I_0 is a constant. The time-lag t_n should be minimum in the temperature range where I_∞ is at its maximum. A lower limit to the minimum time-lag may be estimated from the ratio of the number of atoms in a critical nucleus to the molecular rearrangment frequency. This yields values of the minimum time-lag of the order of ns for metallic melts. At large undercooling the nucleation rate approaches zero, in a similar way as the growth velocity. At the same time the time-lag, which is expected to scale approximately as the melt viscosity, becomes very large.

What the above well-known equations do not allow for is the crucial fact that the compositions of the crystal and the melt usually differ. Consider a binary system A-B, with a crystal of composition x_s growing in a melt of composition x_ℓ, where x is expressed as atomic fraction of that species, say species B, in which the crystal is richer than the melt. Upon growth of the crystal the melt near the interface is enriched in A and depleted in B. This, in turn reduces the crystal growth velocity, due to the limited supply of B at the interface. If we take the growth velocity to scale with the relative supply of B at the interface

$$u' = u \, (x_\ell/x_s)_{\text{interface}} \qquad (4)$$

where u is of the form of (1), then it can be shown that growth slows down with the grown distance z approximately like

$$u'(z) = uy/[1 + (1-y)uz/D] \qquad (5)$$

where $y = x_0/x_s \lesssim 1$ is the ratio of the bulk melt composition to the crystal composition, and D is the diffusivity of B in the melt. Setting, for illustration, u = 1 m/s, y = 0.8 and D = 10^{-5} cm^2/s, u' is found to decrease to half of its initial value of 0.8 m/s after growth by only 5 nm. Clearly the reduction of growth velocity is the more severe the more y differs from 1, and the larger the "diffusionless" growth velocity u is.

Based on the rates for crystal nucleation and growth as a function of temperature and time, one may formulate a condition for glass formation by requiring that the crystallized volume fraction X, given for small X by

$$X(T,t) \sim I(T,t) \, u^3(T) \, t^4 \qquad (\text{for } X \ll 1) \qquad (6)$$

stays below some arbitrary limit, e.g. 10^{-6}. On the other hand, in the case of irradiation of single-crystalline material crystal nucleation is not required. A criterion for glass formation in this case would be that the l-s interface velocity, as calculated from a solution of the heat flow problem alone, exceeds the maximum growth velocity predicted by (1) and (5).

Equation (6), with empirical values for the material parame-

ters, has been used to predict metallic glass formation by splat
cooling [16]. If applied to laser quenching, a few additional
remarks on the predictions of nucleation and growth theory seem
appropriate. The usual theory assumes that the volume of the
undercooled melt is large compared to that of a critical nucleus,
and that it is uniformly undercooled. However, in the presence of
the large thermal gradients characteristic of laser quenching only
a narrow layer of melt is effectively undercooled at any given
time; its thickness is clearly of the order (for one-dimensional
heat flow)

$$\Delta z \sim \Delta T / (\partial T / \partial z) \qquad (7)$$

where z is the coordinate perpendicular to the interface. If Δz
comes in the order of the critical nucleus size, as may be the
case with short pulses, the nucleation rate should be substantially reduced [14]. The lifetime of a molten phase for ns pulses
is typically of the order of 10^{-7} s; the time during which a given
portion of the melt is undercooled,

$$\Delta t \sim \Delta T / (\partial T / \partial t) \qquad (8)$$

is likely to be an order of magnitude smaller and approaches the
range of the expected minimum nucleation time-lag.
    How can the above ideas be applied to predict whether a parti-
cular set of conditions does or does not result in glass formation? Obviously the criteria of interfacial kinetics must be
combined with a description of the laser induced heat flow and
energy turnover in the sample. This can be done by writing the
heat flow equation in the following generalized form:

$$(\partial H / \partial t) = (\partial / \partial z)(K \partial T / \partial z) + A(z,t) + (\partial L / \partial t) \qquad (9)$$

    Here H(T) is the enthalpy per unit volume, A is the volumetric
rate of heat production by light absorption, and L is the latent
heat content of the material, which varies between 0 and L_m, the
heat of melting per unit volume. (Mixing effects are neglected in
the present discussion). The rate $(\partial L / \partial t)$ is implicitely taken to
be infinite in the usual sort of heat flow calculations [17],
while the possibility of glass formation by melt quenching is, of
course, due to its finiteness. Equation (9) can be solved numerically, provided a workable expression for $(\partial L / \partial t)$ can be found.
In prinicple, $(\partial L / \partial t)$ is given by L_m $(\partial X / \partial t)$, where X has to be
obtained from a generalized version of (6), valid for $0 \leqslant X \leqslant 1$.
    The experimental findings on the compositional range of glass
formation cited above suggest that the difference in composition
between the melt and the nearest congruently melting phase is a
key variable. The crudest way to account for this difference is
to scale X with y, as defined after equation (5). If the nucleation rate is taken to be proportional to y as well, we have $X \propto$
y^4. Such scaling certainly underestimates the impact of a compositional difference, as obvious already from equation (6).
    Further implications of the ideas sketched above will be dis-
cussed elsewhere. Here we will just demonstrate the usefulness of
equation (9) in predicting glass formation by means of a sample
calculation. To this end, a simple Ansatz is used for X and (9)
is solved by the usual finite-difference procedure.

Model calculations

It is assumed that nucleation is required for crystal growth, i.e., that no seed is initially present. The rate of crystallization is taken to be

$$(\partial X/\partial t) = R \cdot f(t) \cdot (1 - T/T_m)^n \exp(-Q/kT) \qquad (10)$$

where R is an adjustable constant. The time-dependent part $f(t)$ allows latent heat to be absorbed or liberated only after a specified time has elapsed since the volume element was first superheated or undercooled, in accordance with (8). Condition (7) is ignored. The variation of (10) with temperature is schematized in Figure 6 for $n = 3$ and $Q = 10 \, kT_m$; here positive values mean liberation, negative ones absorption of latent heat. There is admittedly much arbitrariness in the particular choice of parameters in (10); however, it turns out that the detailed shape of the curve (apart from its width and its height) is of little influence on the results of the calculation. To simplify the presentation of the numerical results, a liquid-solid interface is arbitrarily defined at the position of that volume element which has absorbed or liberated 50% of the latent heat of melting. This interface does not have a physical meaning in all cases, as shown below.

Figure 7 shows the interface movement in the case of a 30 ns laser pulse of 10 MW/cm of absorbed fluence, incident on bulk metal specimen with an absorption length of 10 nm. Typical values for the thermal data were chosen and kept constant during the process, in order to facilitate interpretation of the results. In order to demonstrate the importance of the crystallization rate, 3 curves are shown which differ only by a scale factor in the rate (10). For curve a) the peak rate (assumed to occur at reduced temperature $T/T_m = 0.81$ in all cases) is $3 \cdot 10^9 \, s^{-1}$, for curve b) it is $6 \cdot 10^6 \, s^{-1}$ and for curve c) is is $3 \cdot 10^6 \, s^{-1}$.

The largest rate (curve a) results in crystal growth at a small undercooling ($T/T_m = 0.98$) and a nearly constant velocity of 3.2 m/s, which is limited by heat flow, rather than by the intrinsic growth kinetics. Curves of this kind are well-known from laser annealing studies [17]; only in this case does the l-s interface as defined above, have a clear physical meaning: it is the isothermal plane at which latent heat is liberated. Curve b) presents a case close to the limit of crystal growth: After an initial period of slow growth (limited by $\dot{X}$), the undercooling grows larger and the interface accelerates; eventually it moves at a velocity close to maximum ($T/T_m = 0.85$). During the last stage the whole remaining melt is strongly undercooled. The l-s interface has lost its meaning, since the crystallization process is of a volume nature. Curve c) is just beyond the limit for crystal growth: The undercooling at the interface exceeds the critical value ($T/T_m = 0.81$) after a few tens of ns and the interface stops. The liquid cools down without a phase transition, and the latent heat remains stored in the structure, which may thus be called a glass.

Figure 8 shows the surface temperature as a function of time for the cases of Figure 7. The numbers in the Figure indicate the instantanous value of the cooling rate at the interface (which may slightly differ from that at the surface). In all three cases the cooling rate is of the order of 10^{12} °K/s just after the pulse has ended. This cooling rate is of little relevance since it occurs

700

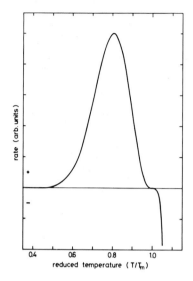

Fig. 6. Temperature-de-
pendent part of the crys-
tallization rate used to
solve equation (9); (+):
crystallization, (-):
melting.

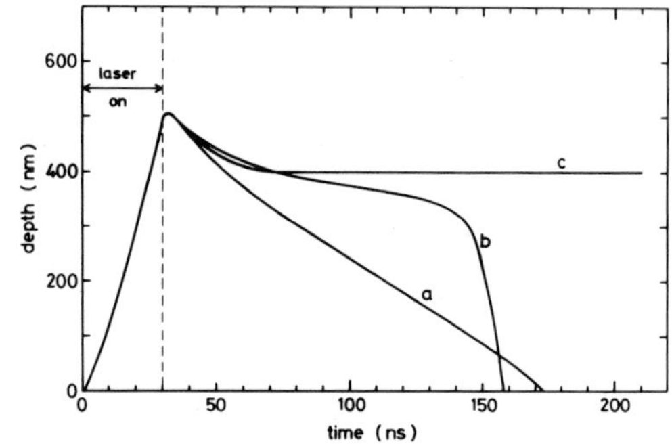

Fig. 7. Interface position in a
bulk metal sample, irradiated by
a 30 ns pulse, calculated with 3
different scale factors for the
rate function in Figure 6.

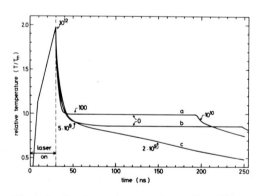

Fig. 8. Surface temperature for the
cases of Figure 7. Inserted numbers
give momentary values of the
cooling rate at the interface.

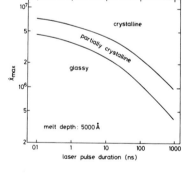

Fig. 9. Variation of the
pulse duration required
for glass formation with
peak crystallization rate
(for constant melt depth).

in the superheated melt. The cooling rate first decreases by heat flow and then, in addition, by liberation of latent heat. It stays close to zero until crystallization is completed in cases a) and b). In case c) liberation of latent heat is suppressed. The cooling rate stabilizes around $2 \cdot 10^9$ °K/s until the temperature has dropped well below the range where the rate (10) is appreciably different from zero: the glass is stable. The calculation shows, however, that even in this case some ten percent of the latent heat has been emitted, which may be interpreted as partial nucleation having taken place. Only with a ten times smaller scale factor for the rate function is nucleation found to be completely suppressed; the cooling rate then remains around 10^{10} °K/s until solidification is completed.

The peak value of $\dot{X}$ largely determines whether glass formation is achieved for a given set of parameters (thermal data, laser pulse duration and energy, etc.). Figure 9 shows the ranges of laser pulse duration and peak crystallization rate $\dot{X}_{max}$ for which the model predicts the material to be glassy, partially crystalline and fully crystalline after cooling. (Partially crystalline means that the residual latent heat L left in the structure after cooling to $T_m/2$ is between 0 and $0.9 \ L_m$). The pulse energy was scaled such as to keep the maximum melt depth contstant at 500 nm. It can be seen that a reduction of the rate by a factor of two increases the critical pulse duration by about one order of magnitude. This could be brought about, in terms of the scaling of $\dot{X}$ with the melt composition discussed above, by going from a melt with y = 1 to one with y = 0.84. (The leveling-off of the glass demarcation line at $X_{max} \approx 5 \cdot 10^6 \ s^{-1}$ is due to the constant melt depth which is somewhat unrealistic for pulses below a few ns.)

Graphs similar to Figure 9 can be constructed for any experimental parameter of interest, such as melt depth, absorption length, etc. Experimental data on glass forming ranges and glass temperatures can in turn be utilized to "calibrate" the crystallization rate. Work along these lines is in progress.

ACKNOWLEDGMENT

This work was supported, in part, by the Swiss Commission for the Encouragment of Scientific Research .

REFERENCES

1. E.M. Breinan, B.H. Kear, C.M. Banas: Physics Today, November 1976, 44-50.

2. See, e.g., M. von Allmen, S. S. Lau, T. T. Sheng, M. Wittmer, in: Laser and Electron Beam Processing of Materials, C. W. White, P.S. Peercy, eds. (Academic Press 1980), pp. 524-529.

3. M. von Allmen, S.S. Lau, M. Mäenpää, B.Y. Tsaur, Appl. Phys. Lett. 36, 205-207 (1980).

4. I. W. Donald, H. A. Davies, J. Non-Cryst. Solids 30, 77-85 (1978).

5. H. J. Vind Nielsen, J. Non-Cryst. Solids 33, 285-289 (1979).

6. E. Huber, M. von Allmen, to be published.

7. M. von Allmen, S. S. Lau, M. Mäenpää, B. Y. Tsaur, Appl. Phys. Lett. 37, 84-86 (1980).

8. P. Oelhafen, R. Lapka, J. Krieg, M. von Allmen, K. Affolter, to be published.

9. K. Affolter, M. von Allmen, H. P. Weber, M. Wittmer, submitted to J. Non-Cryst. Solids.

10. P. L. Liu, R. Yen, N. Bloembergen, R.T. Hodgson, Appl. Phys. Lett. 34, 864-866 (1979).

11. R. Tsu, R. T. Hodgson, T. Y. Tan, J. E. Baglin, Phys. Rev. Lett. 42, 1356-1358 (1979).

12. A. G. Cullis, H. C. Webber, N. G. Chew, in Laser and Electron Beam Interaction with Solids, B. R. Appleton, G. K. Celler, eds. (North Holland 1982), pp.

13. P. Mazzoldi, G. Della Mea, G. Battaglin, A. Miotello, M. Servidori, D. Bacci, E. Jannitti, Phys. Rev. Lett. 44, 88-91 (1980).

14. D. Turnbull, Contemp. Phys. 10, 473-483 (1969).

15. D. Kashiev, Surf. Sci. 14, 209-220 (1969).

16. D. R. Uhlmann, J. Non-Cryst. Solids 7, 337-348 (1972).

17. See, e.g., J. C. Wang, R. F. Wood, P. P. Pronko, Appl. Phys. Lett. 33, 455-458 (1978).

PULSED LASER MELTING OF METALLIC GLASSES

ANIMESH K. JAIN AND D.K. SOOD
Nuclear Physics Division, Bhabha Atomic Research Centre,
Bombay-400 085, India

G. BATTAGLIN, A. CARNERA, G. DELLA MEA, V.N. KULKARNI AND
P. MAZZOLDI
Unita GNSM-CNR, Instituto di Fisica, Universita di Padova,
35100 Padova, Italy

R.V. NANDEDKAR
Materials Science Laboratory, Reactor Research Centre,
Kalpakkam-603 102, India

ABSTRACT

We have studied pulsed ruby laser melting of
metallic glasses - Fe40Ni40P14B6 and Fe40Ni40B20.
Both amorphous and crystallised samples were used,
and some of them were implanted with Ru to serve
as a marker species. SEM measurements on surface
topography and RBS (1.8 MeV He$^+$) depth profiling
of Ru marker redistribution were used to establish
melting above about 1 J/cm^2. Heat flow calculations and liquid phase diffusion analysis of the
Ru marker species under a moving melt front have
been performed. The observed redistribution of Ru
upto 2.5 J/cm^2 is consistent with an effective
$D \sim 5 \times 10^{-5}$ cm^{2}s^{-1}. The flat Ru depth profiles at
higher energy densities are shown to arise from
convection effects. Results on crystallised and
amorphous samples are similar. Preliminary TEM
measurements of quenched regions on a crystallised
sample show an unexpected absence of amorphous
phase.

INTRODUCTION

Pulsed laser treatment (PLT) of materials has emerged as a
viable technique to form metastable surface alloys [1-4]. The rapid
heating with sufficiently intense laser irradiation results in a
molten surface region which resolidifies at a very fast rate due to
conduction of heat to the bulk. Extremely high cooling rates ($\sim 10^9$
Ks^{-1}) and ultrafast solid-liquid interface motion at velocities of
several metres per second [5,6] are factors responsible for producing a metastable phase after laser treatment. The technique thus
has an analogy with conventional splat cooling technique. Numerous
studies have been reported on metastable alloy formation by PLT of
semiconductors [7], as well as metals [8]. However, very few
studies are available on PLT of alloys. In particular, no study has
been reported on PLT of metallic glasses, though formation of a
glassy phase by PLT of deposited films has been studied [9].
In this paper we present our results on pulsed laser melting of

Mat. Res. Soc. Symp. Proc. Vol. 13 (1983) ©Elsevier Science Publishing Co., Inc.

two different metallic glasses - METGLAS 2826 (Fe40Ni40P14B6) and METGLAS E4040 (Fe40Ni40B20). Occurrence of melting is established by SEM analysis and redistribution of a marker species implanted in some samples. Detailed diffusion analysis of a marker profile is presented. Preliminary TEM results are reported.

EXPERIMENTAL

Ribbons (2 and 3 mm wide, 50 microns thick) of METGLAS 2826 and METGLAS E4040 were used in the as received condition for the present study. Irradiations were carried out in air with single pulses ($\sim$ 18 ns FWHM) from a ruby laser, at energy densities upto 5.4 J/cm^2. The spatial profile of the laser beam was homogenised using a ground diffuser plate. The diffused beam was concentrated on the target using a lens. A 3 mm diameter aperture was used almost in contact with the sample to irradiate a well defined area. Some samples were implanted with 120 keV Ru$^+$ ions to a dose of 1×10^{16} ions/cm^2 to serve as a marker species for monitoring surface melting. In addition, some samples (both implanted, as well as unimplanted) were transformed into a crystalline phase by annealing for 1 hr in a vacuum furnace prior to laser treatment. The annealing temperature was chosen to be about 50C above the recrystallisation temperature T_c, and was 468C for METGLAS 2826 (T_c = 414C), and 502C for METGLAS E4040 (T_c = 451C). After annealing, the samples were found to be very brittle. The laser treated surfaces were examined by SEM. The Ru depth profiles were measured by Rutherford backscattering of 1.8 MeV He$^+$ at 160 degrees scattering angle. Preliminary TEM studies have been carried out on some samples.

RESULTS AND DISCUSSION

Fig. 1 shows the SEM micrographs (backscattered electron mode) of a crystallised, as well as amorphous METGLAS E4040 irradiated at various laser fluences. For the crystallised sample, there is a definite smoothening of the surface at 1 J/cm^2, as compared to the untreated surface (0 J/cm^2). Such a smoothening is known to occur at energy densities close to the melt threshold (see e.g. reference 10). A generally smooth surface, with isolated crater like 'pock marks' is seen at higher energy densities upto 2.5 J/cm^2. These are perhaps caused by localised evaporation at surface irregularities. For the as-received (amorphous) samples also, the surface features are essentially similar to those on the crystallised ones (Fig. 1). However, the untreated surface of the amorphous sample is seen to be relatively smoother than that of the crystallised sample. This feature continues even after laser treatment, as can be seen by comparing the two micrographs at 2.5 J/cm^2 in Fig. 1. Laser treatment at 3.5 J/cm^2 and above, produces a very coarse surface characteristic of a turbulent liquid layer frozen instantaneously.

The Ru depth profiles (obtained by RBS) begin to show a noticeable redistribution at 1.5 J/cm^2, which becomes quite prominent at 2.5 J/cm^2 (Fig. 2) signifying sufficiently large melt depth and duration. The redistributed Ru profiles are analysed on the basis of liquid phase diffusion with a time dependent melt depth. The melt kinetics is obtained from one-dimensional heat flow calculations incorporating vapourization [5]. Detailed thermophysical properties of metglasses as a function of temperature are not

CRYSTALLISED METGLAS E4040

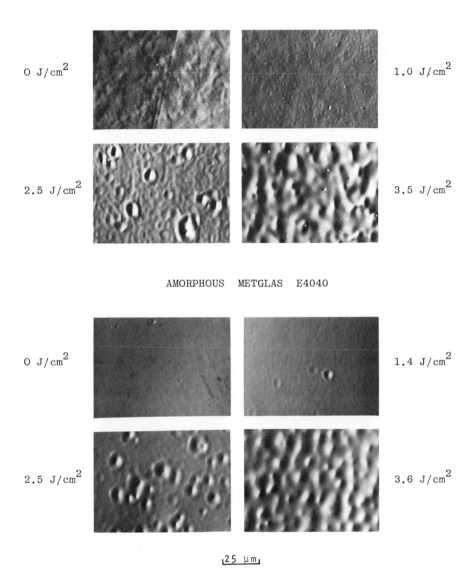

O J/cm^2 1.0 J/cm^2

2.5 J/cm^2 3.5 J/cm^2

AMORPHOUS   METGLAS   E4040

O J/cm^2 1.4 J/cm^2

2.5 J/cm^2 3.6 J/cm^2

⌊25 μm⌋

Fig. 1. Scanning Electron Micrographs of crystallised and amorphous METGLAS E4040 (Fe40Ni40B20) irradiated at indicated laser energy densities.

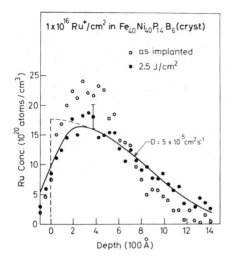

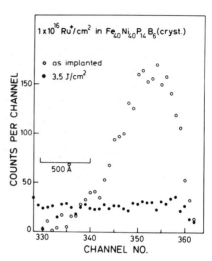

Fig. 2. *The observed and calculated Ru depth profiles for a crystallised METGLAS 2826 sample laser treated at 2.5 J/cm².*

Fig.3. *The Ru depth profiles for a crystallised METGLAS 2826 before and after laser treatment at 3.5 J/cm².*

available in literature. As a first approximation, heating calculations have been performed using thermophysical and optical properties of iron, except that the melting point was taken as 965C to approximate the behaviour of METGLAS 2826. This value of melting point was based on Fe-P and Ni-P phase diagrams [11] and is well within the range of melting points reported for metglasses of Fe40Ni40P14B6 composition [12,13]. At 1.5 J/cm², we obtain a melt duration of 116 ns and a maximum melt depth of 0.47 micron. Corresponding values for 2.5 J/cm² are 176 ns and 0.60 micron respectively. These values are used for the liquid phase diffusion calculations. The calculated diffused profile was folded with a Gaussian detector resolution function of FWHM 300 Å. The folded profile was fitted to the experimental points, with liquid phase diffusivity, D and segregation coefficient, k as adjustable parameters. Results of such a calculation for the profile after 2.5 J/cm² laser shot in METGLAS 2826 (crystallised) are shown in Fig. 2. The calculated profile for $D = 5 \times 10^{-5}$ cm²s⁻¹ (continuous curve) is in good agreement with the experimental data. This value of D is in agreement with typical liquid phase diffusivities. However, in our recent study on PLT of U film on a metallic glass [14], a D value smaller by an order of magnitude has been obtained.

At 3.5 J/cm², there is a remarkable redistribution of Ru into a flat depth profile (Fig. 3), which extends beyond the Ni(Fe) edge. Such a profile can not be interpreted in terms of liquid phase diffusion with reasonable D values. This anomalous redistribution of the marker species, coupled with the SEM observation of a coarse surface topography (Fig. 1), indicates existence of convection in

250 nm

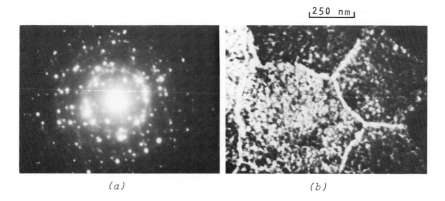

(a)   (b)

Fig. 4. TEM results on crystallised METGLAS E4040 after laser
treatment at 3.5 J/cm² - (a) SAD pattern, (b) bright field image.

the melt [15].

In order to examine the structure of laser quenched phases, TEM
investigations are being carried out and some preliminary results
are given here. The as received samples show a ring pattern characteristic of very fine grain material. Laser treatment at 3.5 J/cm²
does not restore the amorphous phase in both types of the crystallised metglasses, as shown in Fig. 4 for METGLAS E4040. The SAD
pattern (Fig. 4a) is typical to a crystalline material and the
bright field image (Fig. 4b) shows grains $\sim$ 500 nm size. Several
precipitate particles ($\sim$ 25 nm) are seen distributed in the grain
volume, while the grain boundaries are preferentially decorated
with them. This lack of amorphisation is rather surprising, since
the cooling rates during PLT ($\sim 10^9$ Ks^{-1}) are much higher than
those prevalent during splat quenching ($\sim 10^6$ Ks^{-1}) for formation
of these metglas ribbons. A possible reason for this lack of
reamorphisation could be a significant deviation in the melt composition from the glass forming zone. Such an alteration could arise
from one or more of the following factors: 1) substantial evaporation loss of phosphorus from the melt during PLT; 2) inadequate
homogenisation of composition in the melt from the melt produced
from the crystallites in the as-crystallised substrate; or 3) segregation of one or more components at the liquid-solid interface
during solidification. More complete TEM studies are in progress.

ACKNOWLEDGEMENTS

We thank R.W. Cahn for supplying the metglas samples,
G. Dearnaley for Ru implants, M. Sundararaman and G.E. Prasad for
SEM work, E. Jannitti for laser treatment and A. Armigliato for
TEM measurements.

REFERENCES

1.   C.W. White and P.S. Peercy, eds., Laser and Electron Beam
     Processing of Materials (Academic Press, New York 1980).

2. J.F. Gibbons, L.D. Hess and T.W. Sigmon, eds., Laser and Electron Beam Solid Interactions and Material Processing (North Holland, New York 1981).

3. B.R. Appleton and G.K. Cellar, eds., Laser and Electron-Beam Interactions with Solids (Elsevier North Holland, New York 82)

4. Proceedings of International Workshop on Ion Implantation, Laser Treatment and Ion Beam Analysis of Materials, Bombay, February 9-13, 1981, to be published in Radiation Effects (in press).

5. Animesh K. Jain, V.N. Kulkarni and D.K. Sood, Appl. Phys. 25, 127 (1981).

6. P. Baeri, G. Foti, J.M. Poate, S.U. Campisano and A.G. Cullis, Appl. Phys. Lett. 38, 800 (1981).

7. J.M. Poate in : Proc. Microscopy of semiconducting Materials Conf., Oxford, April 6-11, 1981, Inst. Phys. Conf. Ser. No. 60 : Section 2, p. 69.

8. D.K. Sood, in reference 4.

9. C.W. Draper, F.J.A. den Broeder, D.C. Jacobson, E.N. Kaufmann and J.M. Vandenberg, in reference 3.

10. G. Battaglin, A. Carnera, G. Della Mea, P. Mazzoldi, E. Jannitti, A.K. Jain and D.K. Sood, Jour. Appl. Phys. 53, 3224 (1982).

11. M. Hansen, Constitution of Binary Alloys (McGraw Hill, New York 1958), p. 693 and 1027.

12. D.G. Morris, Act. Metall. 29, 1213 (1981).

13. J. Steinberg, S. Tyagi and A.E. Lord, Jr., Act. Metall. 29, 1309 (1981).

14. G. Battaglin, A. Carnera, G. Della Mea, V.N. Kulkarni, P. Mazzoldi, A.K. Jain and D.K. Sood, to be published.

15. Animesh K. Jain, V.N. Kulkarni and D.K. Sood, Nucl. Instrum. Meth. 191, 151 (1981).

PULSED PROTON BEAM ANNEALING OF Co-Si THIN FILM SYSTEMS

L. J. CHEN
Department of Materials Science and Engineering, National Tsing Hua University, Hsinchu, Taiwan, ROC.
L. S. HUNG and J. W. MAYER
Department of Materials Science and Engineering, Cornell University, Ithaca, NY 14853
J. E. E. BAGLIN
IBM T. J. Watson Research Center, Yorktown Heights, NY 10598

ABSTRACT

Cobalt ($\sim$300Å) and $CoSi_2$ ($\sim$1000Å) thin films on Si have been annealed by intense proton beams. RBS and TEM were performed to study ion beam annealing effects.

For ion beam energy densities above about 1 J/cm^2, epitaxial $CoSi_2$ layers were formed for both Co and polycrystalline $CoSi_2$ on Si. At low energy densities, Co_2Si was found to coexist with Co. The results are discussed in terms of eutectic melting processes.

INTRODUCTION

Metal silicides have in recent years received a great deal of attention in the microelectronics industry as Schottky barriers, ohmic contacts, gate electrodes and interconnects. The silicides are generally formed by reacting a thin layer of metal film with substrate silicon at elevated temperatures [1].

Within the past few years, intense energy beams such as laser, electron and ion beams have been used to react metal/Si thin films to form silicides [2]. The techniques offer several advantages over traditional furnace annealing such as shorter reaction times, localized heat treatment and less stringent ambient control. Most of the investigations in this area have been limited to pulsed laser, CW laser and CW electron beam annealing. The principal results obtained with scanned CW electron beams are similar to those produced with scanned CW lasers. Cellular structures were found to result from pulsed laser annealing. The formation of such cellular structures is attributed to the formation of cells by segregation as the melt front moves toward the surface. CW laser or electron beams, on the other hand, produced basically uniform silicide layers similar to those induced by furnace annealing.

High power pulsed ion beams have been used [3] to process nickel thin films on silicon. At high energy densities, cellular structures were observed similar to those produced in laser annealing. However, at lower energy densities, epitaxial silicide layers were formed. This represented the first report of the formation of epitaxial silicide layers by pulsed beam annealing in the melt-quench regime. At lower energy densities, polycrystalline layers containing a mixture of Ni_2Si, $NiSi$ and $NiSi_2$ were formed with an average composition

Mat. Res. Soc. Symp. Proc. Vol. 13 (1983) ©Elsevier Science Publishing Co., Inc.

within the layer near that of NiSi [4]. The structural changes induced by pulsed ion beams were explained in terms of a melting mechanism. In this paper, we report the results of pulsed proton beam annealing of Co-Si thin film systems.

EXPERIMENTAL PROCEDURES

Cobalt thin films, 300Å in thickness, were e-gun deposited on (001) and (111) oriented, 2Ω-cm, n-type Si at room temperature. Some of the samples were annealed in a He atmosphere for 1 hr to form polycrystalline $CoSi_2$ on Si. A magnetically insulated diode, driven by a Marx generator, was used to produce an intense pulsed proton beam. When the Marx bank charging and pulse forming network is fired, a voltage chosen between 150 and 350 keV is applied to that diode for typically 300-400 ns. The central ion current, 50-100 A/cm^2, was monitored with a cylindrical Faraday cup in the sample plane. The details of the equipment and experimental procedures for pulsed ion beam annealing have previously been reported [5]. Each sample was exposed to the ion pulse in a region where the (calibrated) beam intensity falls gradually over the sample length, enabling the study of beam intensity dependence from a single shot.

Rutherford backscattering (RBS) and channeling with MeV He$^+$ ions were used to study the formation and composition of the silicides. For transmission electron microscope (TEM) studies, discs, 3 mm in diameter, were cut ultrasonically from 4 cm long strips of the annealed samples which had been irradiated at a continuous range of ion beam energy densities from zero up to the maximum. Discs were thinned by chemical polishing from the silicon side [6]. The metal side of the sample was protected by an electronic wax during polishing. A JEOL-200CX scanning transmission electron microscope was used to study microstructures induced by ion beam annealing.

RESULTS AND DISCUSSION

With high H$^+$ beam energy density, the fine grain cobalt thin films on Si were found to transform to epitaxial $CoSi_2$ on Si. Figures 1 and 2 show micrographs of as-deposited Co fine grains and epitaxial $CoSi_2$ with curved grain structures, respectively. The RBS spectrum from the epitaxial sample is shown in Fig. 3, together with the corresponding channeling spectrum. The backside depletion of the Co signal in the non-channeled spectrum may be caused by the existence of small islands of silicon located near the $CoSi_2$/Si interface. The presence of such islands is consistent with the dark patches in Fig. 2. The RBS spectrum indicates a surface composition of Co:Si = 1:2. The poor channeling displayed by this layer might be caused partly by the curvature in the $CoSi_2$ grains which can be seen in Fig. 2.

For low H$^+$ energy densities, weak diffraction rings corresponding to Co_2Si in addition to strong Co diffraction rings were detected in the diffraction pattern. An example is shown in Fig. 4 with dark field micrograph showing Co and Co_2Si grains, 100-400Å in size, and the diffraction pattern.

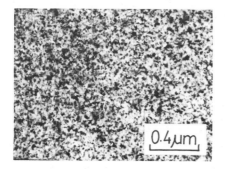

Fig. 1   As-deposited cobalt grains
         on silicon.

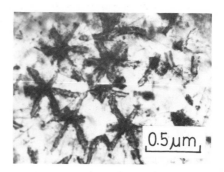

Fig. 2   Epitaxial CoSi$_2$ on Si formed
         by high energy density H$^+$
         bombardment

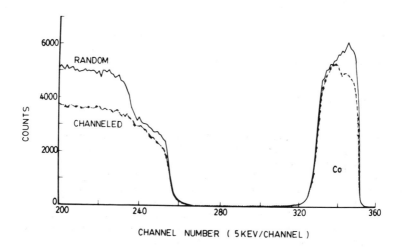

Fig. 3   Random and aligned spectra for 2.3 MeV He ions incident along the [111] axis of
Co/Si after H$^+$ annealing.

712

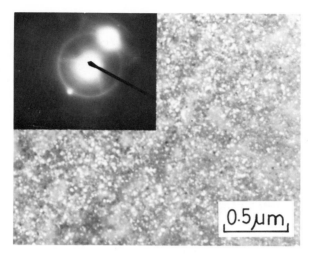

Fig. 4 Dark field micrograph and inset diffraction pattern showing Co and Co_2Si grains, resulting from H+ annealing of Co/Si at low energy density.

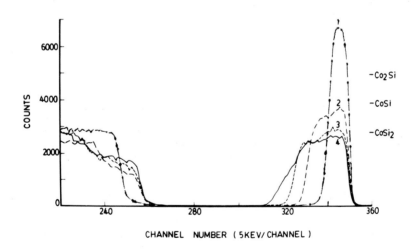

CHANNEL NUMBER (5KEV/ CHANNEL)

Fig. 5 RBS spectrum of a Co/Si sample annealed with H+ at maximum energy density. Curves 1-4 represent RBS for specimens annealed with increasing energy density.

Figure 5 shows backscattering spectra for 2.3 MeV He$^+$ incident on a Co thin film, 300Å in thickness, on (111) Si annealed at maximum energy density. The curves in the figure do not represent equal intervals of energy density; they were selected to highlight the main features of the annealing effects. Each arrow on the figure indicates the height of the Co curve which would be expected if the layer consisted of the composition indicated. Curves 1-4 in the figure represent RBS spectra for samples annealed with increasing energy density.

The effect produced by maximum energy density H$^+$ bombardment of pre-formed CoSi$_2$, about 1μm grain size, on silicon was to transform the polycrystalline layer to epitaxial CoSi$_2$. Examples are shown in Fig. 6. Cracks were seen to develop at the surface, perhaps caused by thermal stresses during cooling. RBS spectra showed that for high energy density H$^+$ beam annealing, the epitaxial layers were slightly Si rich indicating the incorporation of some silicon from the substrate.

(a)   (b)

Fig. 6 Micrographs showing (a) polycrystalline CoSi$_2$ on Si, (b) epitaxial CoSi$_2$ on Si after H$^+$ annealing.

None of the samples displayed cellular structure of the kind shown by Van Gurp et al [7] for laser-annealed Co(250 Å)/Si<100> samples. This may simply indicate that their laser produced a greater melt depth than did the ion beams used in the present work. Van Gurp et al. did not see x-ray diffraction patterns characteristic of CoSi$_2$ in these samples, but they discussed the possibility that signals from epitaxial CoSi$_2$ were masked by the closely similar signals from Si<100>. The present results confirm the ability of pulsed annealing to produce epitaxial CoSi$_2$.

The present data suggest a process in which the temperature rise produced by the beam pulse leads to the growth of a eutectic melt at the Co-Si interface. The composition of the melt at the moment of cooling would depend upon the beam energy density and the thickness of the deposited film. Precipitation of Co$_2$Si and Co from the low-temperature eutectic liquid, or of CoSi$_2$ (epitaxially) from the higher temperature eutectic would be possible, and would be qualitatively consistent with the present observations. Further study will be required in order to develop a clearer picture of the mechanisms for silicide phase growth produced by pulsed ion beam treatment.

ACKNOWLEDGMENTS

The authors acknowledge partial support of the research by VHSIC (AF Wright-Patterson P33615-80-C-1197) and ROC National Science Council through Grant No. NSC-71-0404-E007-04.

REFERENCES

1. J. W. Mayer and K. N. Tu, in: *Thin Films-Interdiffusion and Reactions*, edited by J. M. Poate, K. N. Tu and J. W. Mayer, Wiley, New York (1978).

2. T. W. Sigmon, in: *Laser and Electron-Beam-Solid Interactions and Materials Processing*, edited by J. F. Gibbons, L. D. Hess and T. W. Sigmon, Materials Research Society Proceedings, North Holland, New York (1981) p. 511.

3. L. J. Chen, L. S. Hung, J. W. Mayer, J. E. E. Baglin, J. Neri and D. A. Hammer, Appl. Phys. Lett. 40, 595 (1982).

4. L. J. Chen, L. S. Hung, J. W. Mayer and J. E. E. Baglin, in: *Metastable Materials Formation by Ion Implantation*, edited by S. T. Picraux and W. J. Choyke, Materials Research Society Proceedings, North Holland, New York (1982) p. 319.

5. J. E. E. Baglin, R. T. Hodgson, W. K. Chu, J. M. Neri, D. A. Hammer and L. J. Chen, Nuclear Instruments and Methods 191, 169 (1981).

6. L. J. Chen, L. S. Hung and J. W. Mayer, Appl. of Surface Science, 11, 202 (1982).

7. G. J. van Gurp, G. E. J. Eggermont, Y. Tamminga, W. T. Stacy and J. R. M. Gijsbers, Appl. Phys. Lett. 35, 273 (1979).

PULSED ION BEAM INTERFACE MELTING OF PtCr AND CrTa ALLOYS ON Si STRUCTURES

C. J. PALMSTROM AND R. FASTOW
Department of Materials Science and Engineering, Cornell University, Ithaca,
New York 14853

ABSTRACT

Composition profiles of thermal and pulsed ion beam annealed PtCr/Si and CrTa/Si have been determined using Rutherford backscattering analysis. Thermal annealing resulted in layered phase separation with Pt-silicide in the PtCr/Si case, and $CrSi_2$ in the CrTa/Si case, forming at the Si surface. Pulsed ion beam annealing, 300-380 keV protons at energy densities $\sim 0.75-1.6$ J/cm^2, produced interface melting with no layered phase separation.

INTRODUCTION

For energetic light ions impinging on a material surface, their energy is deposited relatively uniformly in depth up to the ion range (~ 3 μm for 300 keV H^+ in Si). Hence, unlike laser annealing, where all the energy is primarily deposited at the surface, pulsed ion beam annealing may be considered as a rapid thermal heating of the surface region. This technique, with typical $\sim 200-400$ns pulses of 300-380 keV H^+ ions with energy densities $\sim 1-3$ J/cm^2, has been used successfully to anneal ion implanted Si (1,2) and single metals on Si (2-4). Chen et al. (3,4) observed that by tailoring the energy density of the pulsed ion beam, epitaxial $NiSi_2$ could be formed on Si. For pulsed ion beam annealing of Si using 200ns proton pulses with energy density ~ 1.5 J/cm^2 the surface remains molten for a duration of ~ 1μs (5). This time regime requires diffusion coefficients $\geq 10^{-5}$ cm^2/sec in the metal-Si reactions. Baglin et al. (2) suggested that a melt starts at the metal-Si interface and that the amount of melt increases with increasing energy density.

Studies have been made on metal alloy films on Si in the hope of limiting the Si consumption during silicide formation and hence producing a shallow silicide contact (6). The principle behind this idea is that by using an alloy of two metals, one which forms a silicide at relatively low temperature (T_ℓ), and another which forms a silicide at high temperature (T_h), the consumption of Si can be limited during contact silicide formation. Thus, when the metal alloy/Si structure is heated above T_ℓ, but below T_h, the low temperature metal silicide starts to form at the Si-alloy interface, the growth of the silicide being limited by the amount of this metal present. The thin silicide layer so formed is believed to determine the electrical behavior of the contact. Most studies have been on alloys comprising a near noble and refractory metal (e.g. PtCr, PtW, PtV, PdCr, PdW, PdV or NiCr (6-11). In general, the near noble metal silicide forms at the Si surface with a layer rich in the refractory metal silicide forming on top (7). Layered phase separation is also observed for IrV alloys (12) and is suggested for GdPt alloys on Si (13). In the particular case of TiW alloys, a ternary phase, $Ti_{0.3}W_{0.7}$ Si, has been observed to form (14).

The present study has been made to see whether layered phase separation can be achieved by melting of the alloy/Si interface using pulsed ion beam annealing. Here we compare the composition profiles of thermal and pulsed ion beam annealed alloy metal/Si structures:- PtCr/Si and CrTa/Si.

Mat. Res. Soc. Symp. Proc. Vol. 13 (1983) ©Elsevier Science Publishing Co., Inc.

EXPERIMENTAL

Codeposited alloy metal layers of Pt and Cr and Ta and Cr were e-gun
evaporated onto Si substrates. Thermal annealing was carried out in a
hydrocarbon free vacuum furnace at pressures $\sim 10^{-7}$ Torr. For pulsed ion
beam annealing, 300–380 keV protons at energy densities ~ 0.5–1.6 J/cm^2 were
used. Details of the pulsed ion beam equipment are given in references 2
and 15 and of the energy density calibrations in reference 5.
    Rutherford backscattering studies were performed using 2.7 MeV ^{4}He^{++} ions
scattered through an angle of 170°.

RESULTS

    $Pt_{30}Cr_{70}$/Si. Figure 1 shows the backscattering spectra of unannealed and
annealed $Pt_{30}Cr_{70}$ alloy films on Si. It is clear in this figure that layered
phase separation occurs with the segregation of a thin Pt-silicide layer at the
Si surface. This Pt segregation is in agreement with previous results (6,7).
However, in the case of pulsed ion beam annealed samples, shown in figure 2,
no clear evidence of layered phase separation exists. The steps in the Si,
Cr and Pt peaks (fig. 2) increase in width with increasing energy density,
without any accumulation of Pt occuring at the Si surface. This indicates
that the interface melt increases with increasing energy density without a
distinct layered phase separation.

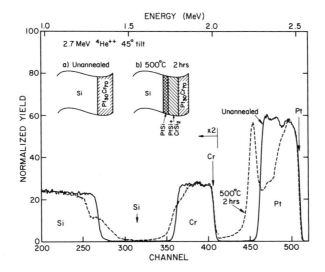

Fig. 1. Backscattering spectra of 1400 Å $Pt_{30}Cr_{70}$ alloy film on (100) Si
unannealed and after annealing at 500°C for 2 hrs.

    Furthermore, the backscattering spectra of both the unannealed and pulsed
ion beam annealed samples, shown in figure 2, show variations in the height
of Pt and Cr peaks. This arises from variations in alloy composition during
deposition. The fact that these composition variations remain after pulsed

ion beam annealing provides clear evidence that the alloy metal film itself
does not melt; the melt is limited solely to the interface region. This is
distinctly different from laser annealing, where the melt propagates from
the metal surface towards the interface.

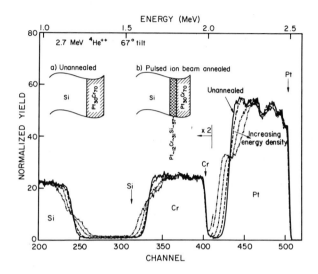

Fig. 2. Backscattering spectra of 1400 Å $Pt_{30}Cr_{70}$ alloy film on (100) Si
unannealed and after pulsed ion beam annealing using 300 keV protons at
increasing energy densities.

CrTa/Si. As in the case of PtCr alloys on Si, layered phase separation has
also been observed in CrTa alloys on Si (16). Figure 3 shows the backscattering
spectra for unannealed and annealed $Cr_{60}Ta_{40}/Cr_{77}Ta_{23}$/Si structures. Annealing
at 600°C resulted in Cr, but not Ta, starting to react with Si. In figure 3,
the Cr peak is broader than that of the Ta. The Ta peak can be seen to be
virtually identical to the unannealed case, indicating that the Cr has diffused
out of the alloy and into the Si. The position and shape of the high energy
edge of the Si shows that Si has diffused into the CrTa alloy. The spectrum
of the 700°C annealed sample also given in figure 3 shows that a layer of $CrSi_2$
has formed on the Si surface. Furthermore, the low energy edge of the Ta peak
has decreased in height relative to the unannealed one, indicating the possible
start of $TaSi_2$ formation on top of the $CrSi_2$ layer.

Pulsed ion beam annealing of the $Cr_{60}Ta_{40}/Cr_{77}Ta_{23}$/Si structure did not
result in layered phase separation; the backscattering spectra obtained from the
resulting structures are shown in figure 4. The spectrum of the 0.75 J/cm^2
annealed sample shows a small step in each of the Si, Cr and Ta peaks. With
increasing energy density, this step increases in width in all the peaks by an
equal amount, which is clear evidence for the lack of layered phase separation
observed in the thermal case.

718

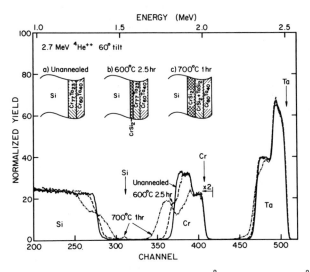

Fig. 3. Backscattering spectra of $Cr_{60}Ta_{40}$ (370 Å)/$Cr_{77}Ta_{23}$ (500 Å)/Si (100) structure unannealed and after annealing at 600°C for 2.5 hrs. and 700°C for 1 hr.

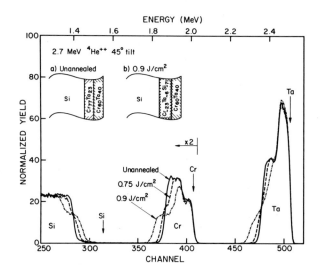

Fig. 4. Backscattering spectra of $Cr_{60}Ta_{40}$ (370 Å)/$Cr_{77}Ta_{23}$ (500 Å)/Si (100) unannealed and after pulsed ion beam annealing using 380 keV protons at energy densities of 0.75 J/cm^2 and 0.95 J/cm^2.

Figure 5 shows the backscattering spectra of unannealed and pulsed ion beam annealed $Cr_{93}Ta_7$/Si structures. Note that the steps in the Cr and Ta peaks remain relatively sharp for energy densities ~ 0.9 J/cm^2. Annealing with 1.2 J/cm^2 causes some tailing at the low energy edge of both peaks. The Ta peak shows a small rise at the high energy edge indicating accumulation of Ta at the surface. The amount of Ta accumulated at the surface increased dramatically for 1.5 J/cm^2 anneals. This surface accumulation probably arises from the melt extending throughout the alloy film and Ta being frozen out at the surface during resolidification of the melt. The large tails in both the low energy edges of Cr and Ta peaks arise from nonuniformity in the films.

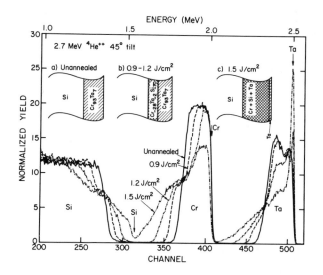

Fig. 5. Backscattering spectra of 1050 Å $Cr_{93}Ta_7$ alloy on (100) Si unannealed and after pulsed ion beam annealing using 380 keV protons at energy densities of 0.9, 1.2 and 1.5 J/cm^2.

SUMMARY AND CONCLUSIONS

Thermal annealing of $Pt_{30}Cr_{70}$/Si structures result in layered phase separation with the formation of Pt-silicide at the Si interface. Similarly, in the case of the $Cr_{60}Ta_{40}/Cr_{77}Ta_{23}$/Si structure, layered phase separation occured with $CrSi_2$ forming at the Si surface. However, in both cases pulsed ion beam annealing produced no observable layered phase separation.

Pulsed ion beam annealing involves an interface melt, the extent of which increases with increasing energy density. Once the melt reaches the surface, a nonuniformity tends to result.

Hence, pulsed ion beam annealing could be useful for initiating a controlled localized interface melt, which in turn may be of importance for shallow silicide contact technology, where a thin localized reaction is desired.

ACKNOWLEDGMENTS

Thanks are due to K. N. Tu and M. Eizenberg (IBM, Yorktown Heights) for supplying PtCr/Si prepared at the IBM Central Services Facility, D. Campbell

(IBM, East Fishkill) for supplying the CrTa/Si samples and D. A. Hammer (Cornell) for the use of the pulsed ion beam facility. The authors wish to acknowledge J. W. Mayer and J. Gyulai for encouragement and helpful discussions. This work was supported in part by the Defense Advanced Research Projects Agency through the U.S. Office of Naval Research (L. Cooper). The ion beam facility was supported in part by ONR (Contract N00173-79-C-0085).

REFERENCES

1. R. T. Hodgson, J. E. E. Baglin, R. Pal, J. M. Neri and D. A. Hammer, Appl. Phys. Lett. 37, 187 (1980).

2. J. E. E. Baglin, R. T. Hodgson, W. K. Chu, J. M. Neri, D. A. Hammer and L. J. Chen, Proc. of 5th Intern. Conf. on Ion Beam Analysis, Sydney, 1981.

3. L. J. Chen, L. S. Hung, J. W. Mayer and J. E. E. Baglin, Metastable Materials Formation by Ion Implantation, Materials Research Society, 319, (1982).

4. L. J. Chen, L. S. Hung, J. W. Mayer, J. E. E. Baglin, J. M. Neri and D. A. Hammer, Appl. Phys. Lett. 40, 595 (1982).

5. R. Fastow, J. Gyulai and J. W. Mayer. This conference.

6. K. N. Tu, W. N. Hammer and J. O. Olowolafe, J. Appl. Phys. 51, 1663 (1980).

7. G. Ottaviani, K. N. Tu, J. W. Mayer and B. Y. Tsaur, Appl. Phys. Lett. 36, 331 (1980).

8. J. W. Mayer, S. S. Lau and K. N. Tu, J. Appl. Phys. 50, 5855 (1979).

9. M. Eizenberg and K. N. Tu, J. Appl. Phys. 53, 1577 (1982).

10. J. O. Olowolafe, K. N. Tu and J. Angilello, J. Appl. Phys. 50, 6316 (1979).

11. M. Eizenberg, G. Ottaviani and K. N. Tu, Appl. Phys. Lett. 37, 87 (1980).

12. M. Eizenberg, Thin Solid Films 89, 355 (1982).

13. R. Thompson, M. Eizenberg and K. N. Tu, J. Appl. Phys. 52, 6763 (1981).

14. J. M. Harris, S. S. Lau, M.-A. Nicolet and R. S. Nowicki, J. Electrochem. Soc. 123, 120 (1976).

15. J. M. Neri, D. A. Hammer, G. Ginet and R. N. Sudan, Appl. Phys. Lett. 37, 101 (1980).

16. C. J. Palmstrom, J. Gyulai and J. W. Mayer. To be published.

IMPURITY REDISTRIBUTION STUDIES ON LASER-FORMED SILICIDES

A. S. WAKITA, T. W. SIGMON AND J. F. GIBBONS
Stanford Electronics Laboratories, Stanford, CA   94305

ABSTRACT

$^4He^+$ Backscattering and SIMS were used to study impurity redistribution during laser formation of refractory silicides. Thin films of Mo and W were evaporated on to <100> p-type silicon substrates, which were As or B implanted to doses of 1×10^{15} to 1×10^{16} cm^{-2}. These samples were laser reacted with multiple or single laser scans at various powers. Analysis of these films indicate impurity movement into the forming silicide layer. Impurity concentrations in the films were observed to be as high as 7.8 $\times 10^{20}$ cm^{-3} for As in WSi_2, however a reduction in this concentration occurred with subsequent thermal annealing.

INTRODUCTION

The current interest in metal silicides has instigated a wave of research on these materials ranging from their chemical properties to their applicability in high performance VLSI circuits. Many of these silicides are impervious to oxidizing ambients typically found in Si processing, and instead grow protective SiO_2 layers without detectable decomposition of the film. Thus, a layer of SiO_2 can be grown as insulation for multiple level device interconnects or for masking for further patterning. Snow plow effects of dopants such as arsenic from PtSi and Pd_2Si have been observed and are technologically useful for the formation of ohmic contacts with low contact resistance [1,2]. Overall, metal silicides have aroused interest because of the low sheet resistivity they have to offer in present interconnect technology.

Silicide films are commonly deposited (co-sputtered or co-evaporated) or formed by the interdiffusion of thin films of metal with silicon. Each technique has its advantages and disadvantages as well as providing films with different electrical and chemical properties. The refractory metal silicides are normally difficult to form by the latter method because of required heating of the entire substrate to high temperatures. This thermal cycle can result in significant dopant diffusion in the Si substrate. Sputtering offers the ability of low temperature deposition; however, the incorporation of Ar or other sputtering gases can interfere with the chemical and oxidation properties of the films. Reduced impurity incorporation can be achieved by metal-silicon interdiffusion, but again this is only feasible for silicides with low formation temperatures, such as Pd or Pt silicide. However, with laser processing, thin films of refractory metals can be interdiffused with silicon by heating the substrate to high temperatures for times on the order of milliseconds [2]. In our studies, we were concerned with the movement of dopants in the substrates as we formed $MoSi_2$ and WSi_2 from thin films of metal evaporated directly onto heavily doped substrates.

Mat. Res. Soc. Symp. Proc. Vol. 13 (1983) ©Elsevier Science Publishing Co., Inc.

EXPERIMENTAL

P-type Si <100> substrates of ~ 10 and 20 Ω-cm were implanted with arsenic at doses of 1x10^{16}, 6.9x10^{15} and 1.3x10^{15} and boron at doses of 3.9x10^{15} and 1.2x10^{15} cm^{-2}. The room temperature implants were performed at energies of 40 and 100 keV for arsenic and 20 keV for boron. These implanted wafers were preannealed at 550°C for 30 minutes to regrow the damaged silicon, then followed by a 900°C (30 mins for As, 15 min for B) anneal for electrical activation. The final measured doses and projected range, R_p, after preanneal are summarized in Table I.

TABLE I
Initial silicon substrate conditions prior to metallization

DOPANT	SUBSTRATE	ENERGY (keV)	MEASURED AFTER ANNEAL	
			DOSE cm^{-2}	R_p Å
As	p-type <100>	100 keV	1.3 x 10^{16}	760
			1 x 10^{16}	300
	20 Ω cm	40	6.9 x 10^{15}	300
			1.3 x 10^{15}	300
B	p-type <100>	20	3.9 x 10^{15}	750
	10 Ω cm		1.2 x 10^{15}	750

Following thermal anneal, the silicon wafers were cleaned and etched in dilute HF prior to evaporation in an oil free system. Approximately 400 Å of metal (molybdenum or tungsten) and 200 Å of silicon were e-beam evaporated in succession. The final silicon cap serves as a anti-reflection coating during laser annealing.

The samples were annealed with a cw argon laser, which was operated in the multiline, TEM$_{00}$ mode. Dwell times were typically 1 msec and the substrates were heated to a backside temperature of 350°C. Since the samples were usually scanned in air, the silicon cap would partially oxidize and retard excessive dopant outdiffusion. Various laser scans and subsequent thermal anneals were performed on these samples, which were then analyzed for dopant redistribution with backscattering spectrometry and SIMS.

The laser formed films show only the disilicide phase in the x-ray diffraction pattern. After laser formation, the grain size of these films was measured by TEM to be approximately 1000 Å .

RESULTS AND DISCUSSION

The redistribution of the arsenic profile during the formation of WSi$_2$ was followed by RBS by forming the silicide at low powers and with multiple scans. Figure 1 shows the backscattering spectra at high energy, taken with an incident ^{4}He$^+$ beam of 2.2 MeV. The initial As implant energy was 40 and 100 keV, which corresponds to arsenic peaks occurring at approximately 1.71 and 1.69 MeV, respectively in the as-deposited spectrum (———). The drop in height and broadening of the W signal following 1 laser scan at 0.8 the power to melt silicon indicates partial reaction of the silicon and tungsten. Glancing angle X-ray diffraction for this sample shows only the presence of polycrystalline W and WSi$_2$. It is apparent from the As signal (---) that the As has diffused into the metal film toward the surface. After three scans, the resulting spectra for a completely reacted WSi$_2$ film is also shown in Figure 1 (....).

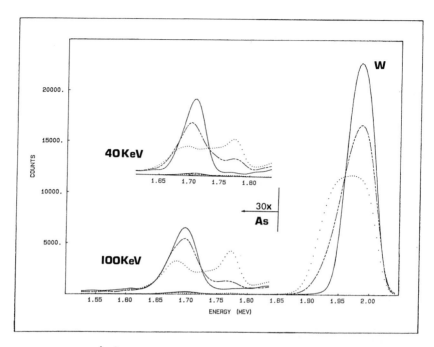

Fig. 1. 2.2 MeV ^{4}He$^+$ Backscattering spectra of laser formation of WSi$_2$ and corresponding Arsenic movement. (——) As-deposited. (——--——) Partially reacted W/WSi$_2$. (···) Complete formation of WSi$_2$.

For both 40 and 100 keV implants, the spectra for the completely reacted films show peaks at 1.77 MeV. These peaks correspond to the SiO$_2$/WSi$_2$ interface. The increase of arsenic at this interface may occur from the low diffusivity of arsenic in the SiO$_2$ or an interface segregation effect. A pronounced peak is observed at the Si/WSi$_2$ interface for the 100 keV implant but not for the 40 keV implant. If we take into account the shift of the 100 keV arsenic signal due to the formation in the oxide and the increase in the stopping power for WSi$_2$, the initial peak (no reaction) and the final peak (complete reaction) are at the same location. These results suggest that the second peak is primarily from the remaining implant profile. Nevertheless, we see from the final As profile that the As has remained only in the near surface of the Si and in the WSi$_2$ film. Therefore, as the WSi$_2$ film forms the impurity is incorporated into the film. The bulk concentrations of As in the WSi$_2$ were calculated to be 7.2×10^{20} and 7.8×10^{20} cm^{-3} for the 40 and 100 keV implants, respectively.

WSi$_2$ was also laser formed on Si substrates that were As-doped to levels of 6.9×10^{15} and 1×10^{15} cm^{-2}. These samples exhibit results similar to those discussed above.

A 1000 Å cap of SiO$_2$ was then deposited at 450°C on the laser formed WSi$_2$/100 keV initial implant sample and annealed for six hours at 1000°C. The subsequent thermal anneal served to show that: 1) the high concentration in the WSi$_2$ was only from the As incorporation during laser annealing

724

and 2) the low energy As peak was not a segregation effect. In Figure 2, the backscattering spectra of the sample before and after thermal annealing reveal that the As has diffused out of the WSi_2, decreasing the As concentration to 1×10^{20} cm^{-3}, and has diffused into the silicon substrate.

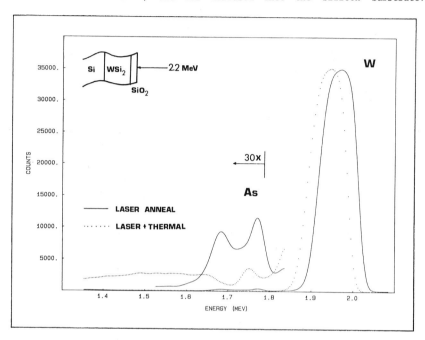

Fig. 2 $^{4}He^{+}$ Backscattering Spectra of laser annealed WSi_2 and the As profile before (——) and after (···) thermal anneal.

We also suspect As outdiffusion through the thin SiO_2 cap. However, the overlap of the As signal with the W and Si makes this difficult to quantify. Also, the appearance of the arsenic peak at the WSi_2/SiO_2 interface but not at the WSi_2/Si interface (~1.69 MeV) suggests a segregation effect at the WSi_2/SiO_2 interface. Therefore, the previous peak seen at 1.69 MeV was from the initial implant profile. From the data we also see that the amount of arsenic in the thermally annealed WSi_2 is less than that in the Si. This result has also been observed by Pan, et al. (4).

Molybdenum disilicide films were also formed by laser reaction of Mo films evaporated onto B or As doped silicon. These films were then scanned at laser powers of 6.0 and 7.2 W. The actual melt power of silicon for this case was 8.4 W. The backscattering spectra for these samples show uniform films with sharp interfaces and stoichiometric ratios (Si/Mo) of 2. In order to study the impurity redistribution during formation, the films were analyzed with SIMS.

Similar to the case for As in WSi_2, the impurities are incorporated into the film after laser reaction. For the heavily boron doped substrate, the B concentration in the silicide was estimated to be approximately 1×10^{20} cm^{-3}. Figure 3a shows the initial SIMS profile for the boron implant in the silicon

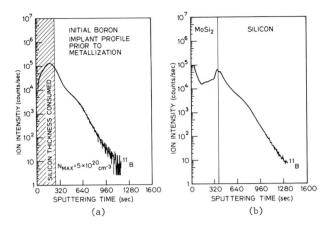

Fig. 3 a. SIMS Boron profile of initial substrate prior to metallization (left). b. SIMS Boron profile after laser formation of MoSi$_2$ (6.0 W).

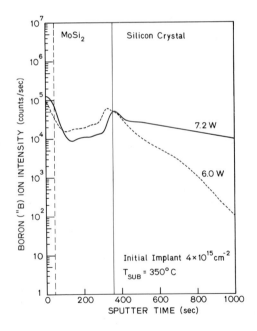

Fig. 4 SIMS Boron profile of MoSi$_2$ formed at laser powers of 7.2 W (——) and 6.0 W (···).

726

following pre-anneal and prior to metallization. The thickness of silicon consumed to form 1100 Å of $MoSi_2$ is marked on the initial profile. After a laser scan of 6.0 W (Fig. 3b), the concentration profile in the underlying, unconsumed silicon has not been changed. However, the SIMS profile for the $MoSi_2$ films formed at 7.2 W (Fig. 4) indicate substantial diffusion of the B into the silicon (to a depth of 0.6 microns). The lower boron intensity measured in the $MoSi_2$ films, formed at 7.2 W, indicates that formation of the films at higher laser powers incorporates less boron into the silicide due to increased boron outdiffusion through the thin oxide layer (<200 Å). In addition, since the boron diffuses into the metal film, it is expected to see a gradient of boron intensity with depth as exhibited in these profiles.

SIMS analysis of As in the $MoSi_2$ films gave results similar to those obtained for boron. That is, arsenic was incorporated into the film, and for higher laser powers, lower As concentration in the $MoSi_2$ and increased dopant diffusion in the underlying Si were observed (Fig. 4).

CONCLUSIONS

Formation of the refractory disilicides, WSi_2 and $MoSi_2$, from laser reaction of thin films of Mo or W with As or B doped silicon, incorporates the impurities into the silicide film during formation. At low laser powers, impurity incorporation is achieved without gross diffusion of the dopants into the underlying silicon. Therefore, silicide films can be laser formed without disturbing underlying impurity profiles essential to device fabrication. However, upon thermal annealing at high temperatures and long times, the concentration level in the silicide decreases as the dopant diffuses into the Si substrate and segregates to the SiO_2/WSi_2 interface.

ACKNOWLEDGEMENTS

We would like to thank the Defense Advanced Research Projects Agency, (MDA903-78-C-0290 and MDA903-78-C-0128), in particular S. Roosild and R. Reynolds for their support of this work.

REFERENCES

1. M. Wittmer and T. E. Seidel, J. Appl. Phys. 49 (12), 1978.

2. I. Ohdomari, K. N. Tu, K. Suguro, M. Akiyama, I. Kimura, and K. Yoneda, Appl. Phys. Lett. 38 (12), 1981.

3. T. Shibata, T. W. Sigmon, and J. F. Gibbons, Proc. of the Symposium on Laser and Electron Beam Processing of Electronic Materials, ECS, (520), 1980.

4. P. Pan, N. Hsieh, H. Geipel and G. Slusser, J. Appl. Phy. 53 (4), 1982.

SUBMICRON CONDUCTOR ARRAY FABRICATED FROM A RECRYSTALLIZED EUTECTIC THIN FILM

H. E. CLINE
General Electric Corporate Research and Development, P.O. Box 8, Schenectady,
NY 12301

ABSTRACT

A submicron conductor array was fabricated without using lithography by selectively etching a recrystallized $Al-Al_2Cu$ eutectic alloy thin film. The 2 micron thick eutectic films were deposited on glass substrates and directionally solidified with both a quartz-iodine lamp and a scanning laser at rates between .0016 and .14 cm/sec. The resulting structure consisted of alternate parallel stripes of the two eutectic phases with a spacing between .5 and 4 microns that was controlled by the solidification rate. An array of submicron Al- rich conductors was fabricated by selectively etching away the Al_2Cu phase. At solidification rates greater than .004 cm/sec using the lamp heater the solid-liquid interface became non-planar while with the laser the structure was well alligned at the highest rates used, .14 cm/sec. At the maximum theoretical solidification rate that produces a two phase aligned eutectic structure the width of the Al wires would be 100A.

INTRODUCTION

As the minimum dimension of electronic circuits is reduced below one micron, there has been increased interest in developing techniques for fabricating submicron structures. Photolithographic techniques are theoretically limited by diffraction to dimensions of about 0.5 micron. Electron beam techniques may approach 0.1 micron; however, at dimensions much below one micron the slow writing speed and the scattering of electrons becomes a problem.[1]

An alternative approach to fabricating submicron structures by selectively etching directionally solidified eutectic thin films does not require lithography to yield a periodic array of submicron size wires. Upon solidification the eutectic thin film liquid separates into alternate stripes of the two solid eutectic phases. The ratio of the widths of the strips of each phase are controlled by the alloy composition while the periodic spacing of the structure is controlled by the solidification rate. At higher solidification rates there is less time for diffusion in the liquid; consequently, the structure separates on a finer scale. An array of conductors of one of the two eutectic phases was fabricated by selectively etching away the other eutectic phase.

SUBMICRON ARRAY

Mat. Res. Soc. Symp. Proc. Vol. 13 (1983) ©Elsevier Science Publishing Co., Inc.

728

The Al–Al$_2$Cu eutectic alloy system, 33.2 w/o Cu, was chosen to demonstrate this process because the phase of aluminum saturated with copper is being used currently for the metallization of integrated circuits. Layers of Cu and Al were deposited on a Pyrex substrate with an electron beam evaporator in a vacuum of 10^{-6} torr at a rate of 20Å/sec. During evaporation the thicknesses of the layers were monitored to give a 2580Å thick copper layer and a 17420Å thick aluminum layer. These thicknesses were calculated to give a 2 micron thick alloyed film of the eutectic composition.

To directionally solidify the thin film a line of heat was moved at a constant rate along the film, the layers melted together in the thin molten zone and subsequently the liquid separated into the aligned lamellar geometry which was oriented normal to the solid-liquid interface. Two methods have been used to form the molten zone. In the first,[2] the heat from a quartz iodine lamp was focused to a 3 mm wide line of heat with an elliptical aluminum reflector. While this method is simple and easy to control, the thermal gradient is insufficient to yield aligned structures at solidification rates greater than about .004 cm/sec. A second method utilizes a laser to produce a much higher thermal gradient to yield aligned structures at the highest solidification rates tried thus far, .14 cm/sec, which corresponds to .16 micron wide strips of aluminum. The heat from a 90 watt CW Nd:YAG laser was focused to a .02 cm wide line with a 50 mm cylindrical lens, Figure 1.

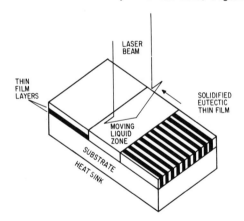

Fig. 1. Schematic of the solidification of a eutectic thin film by a moving line of heat produced by a laser. The thin film layers are alloyed together by the moving liquid zone which separates into a striped geometry during solidification.

The width of the Al phase was equal to the width of the Al$_2$Cu phase; consequently, the width of the Al-rich wires are one-half the periodic lamellar spacing. A plot of the lamellar spacing as a function of the growth velocity, Figure 2, shows the same dependence that was previously found for bulk Al–

Al$_2$Cu eutectic solidified at much slower rates,[3]

$$\lambda \;=\; A/\sqrt{V} \tag{1}$$

where the constant $A = 8.4 \times 10^{-6}$ cm$^{3/2}$-sec$^{-1/2}$ fits the previous data at low velocities, V.

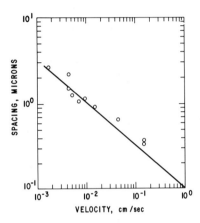

Fig. 2 The periodic spacing of the Al–Al$_2$Cu eutectic thin films solidified at different rates.

The scatter in the spacing data for these experiments was related to local fluctuations in the solidification velocity which may be improved in the future by better control of the heat source.

The Al$_2$Cu phase was selectively removed by electropolishing the film in a solution of 62 ml perchloric, 137 ml water, 700 ml ethanol, and 100 ml butylcellosolve at 35 volts. A scanning micrograph shows the resulting aluminum rich wires, Figure 3.

730

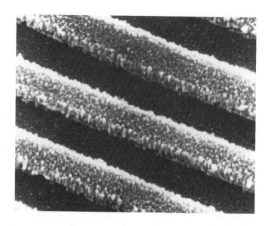

Fig. 3 A scanning micrograph of 1.2 micron wide aluminum-rich wires selectively etched from a Al-Al$_2$Cu film solidified at .0016 cm/sec.

In the case of a film solidified at .0016 cm/sec the wires are 1.2 microns wide. Some etching of the Al-rich phase occurred which yielded a rough surface.

Electron transmission microscopy of the unetched film shown in Figure 3 was used to demonstrate that the interphase boundary between the Al-rich phase and the Al$_2$Cu-phase was planar and oriented normal to the plane of the film, Figure 4.

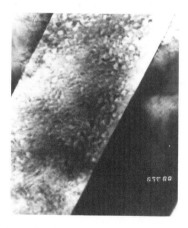

Fig. 4. Transmission electron micrograph of the unetched film shown in Figure 3, the interphase boundaries are planar and the oriented normal to the plane of the thin film, 70,000X.

Selective area electron diffraction was used to identify the phases and show that the fine wires were single crystals. Single crystals are advantageous for metallization at submicron dimensions to minimize electromigration and improve thermal stability.

ULTIMATE RESOLUTION

An array of submicron wide wires with a submicron wide spacing may be fabricated without lithography by selectively etching the natural periodic structure of directionally solidified eutectic thin films. Ultimately, the minimum feature size that may be obtained by this technique is 100A which previously has been demonstrated by splat cooling a thin foil of the $Al-Al_2Cu$ eutectic.[4] At more rapid solidification rates the eutectic alloy liquid transforms to a single non-equilibrium solid solution.[4]

The uniformity of a directionally solidified structure depends on the control of the crystallization process; consequently, it would not be expected to compete with the more precise optical and electron beam processes. However, as the need for submicron structures increases this aternative process of producing very fine geometries may find application in a limited number of cases that requires periodic arrays of fine wires.

Acknowledgements

The author thanks R. Marges for preparation of the thin films, Nathan Lewis for scanning electron microscopy and EF Koch for electron microscopy.

References

1. GR Brewer, "Electron-Beam Technology in Microelectronic Fabrication," Academic Press, New York (1980).

2. HE Cline, J. Appl. Phys., 52, 256 (1981).

3. JD Livingston, HE Cline, EF Koch and R Russel, Acta Met, 18, 399 (1970).

4. HA Davies and JB Hull, Mat. Sci. and Eng., 23, 193 (1976).

PROCESSING/MICROSTRUCTURE RELATIONSHIPS IN SURFACE MELTING

R. J. SCHAEFER AND R. MEHRABIAN
National Bureau of Standards, Washington, D.C. 20234

ABSTRACT

The development of predictive models for rapid surface
melting and resolidification requires coupling of realistic
heat flow models to emerging theories of rapid solidification
processing. Attainment of unique microstructures and phases,
for example through plane-front solidification and solute
trapping, can be correlated to solid/liquid interface velocity, temperature and temperature gradients, and to theories
of morphological stability. However, there are important
limitations on achievable solid/liquid interface velocity
depending upon the heating mode and heat flux distribution,
melt thickness and location of the interface within the
molten zone.
An overview is given of the emerging guidelines for prediction and control of rapid solidification conditions and
microstructures. Homogenization of the liquid by convection
and diffusion is also discussed. Electron beam surface
melting of alloy substrates is used as an example of these
processes.

INTRODUCTION

The generation of special microstructures through surface melting by directed
energy sources is strongly dependent upon the local interface velocities and
temperature gradients. In particular, the theory of morphological stability predicts in detail the velocity and gradient dependence of the transition from
microscopically planar to non-planar interfaces. The transition from cellular
to more complex dendritic interfaces is also influenced by velocities and
gradients, but the theory describing this transition is not so well developed,
particularly with respect to prediction of special behavior which may arise at
rapid rates of solidification.
When a laser or electron beam is used to create a transient melted layer on
the surface of a material, the solid-liquid interface experiences a cycle of
gradient and velocity conditions as it resolidifies, and a corresponding distribution of microstructures can be predicted. Because the solidification cycle
can be altered by selecting different modes of surface heating, such as pulsing
or scanning the energy beam, there exists the possibility of altering the distribution of microstructures by altering the heating mode.
We shall describe here these solidification cycles mainly as they apply to
the morphological stability theory of the transition from planar to non-planar
interfaces, because the quantitative description of this transition is so well
developed. The liquid is assumed to be initially homogeneous, and we consider
the processes which can eliminate in the liquid the microsegregation which was
present in the initial solid.

Mat. Res. Soc. Symp. Proc. Vol. 13 (1983) Published by Elsevier Science Publishing Co., Inc.

INTERFACE STABILITY

During solidification processes in which a positive temperature gradient is
present in the liquid and the latent heat of fusion is therefore removed
through the solid, plane-front solidification becomes decreasingly stable with
increasing growth velocity, so long as the velocity lies in the conventional
range of less than about 10^{-3} m/s. This trend and the approximate values of
the gradient, concentrations, and velocities at which planar interfaces become
unstable are predicted by the well-known principle of constitutional supercooling [1], which is based on the concept that the planar interface becomes
unstable if the temperature in a region ahead of the interface is lower than
the liquidus temperature as determined by the local solute concentration.
This theory predicts that planar interfaces will be stabilized by a temperature
gradient in the liquid greater than

$$G_L = mVC_\infty(k-1)/Dk \qquad (1)$$

where m is the liquidus slope, V is the solidification velocity, C_∞ is the
solute concentration in the liquid far from the interface, k is the equilibrium
partition coefficient and D is the solute diffusion constant in the liquid. In
most crystal growth processes the values of the parameters appearing in this
equation can be estimated with reasonable accuracy and the use of the constitutional supercooling concept is therefore widespread. However, morphological
stability analysis [2] of solidification processes has now been available for
many years and it not only gives a more detailed description of interfacial
behavior, it predicts effects, particularly at rapid solidification velocities,
which differ dramatically from the behavior expected on the basis of the constitutional supercooling concept. By considering the relatively simple expressions which morphological stability theory predicts under certain limiting
conditions, the influence of temperature gradients and interface velocity on
the stability is most clearly evident. Considering first a situation in which
all effects of interfacial surface energy and kinetics are neglected, the
morphological stability analysis predicts a relation very similar to Eq. (1):

$$\frac{G_L k_L + G_S k_S}{k_L + k_S} = \frac{mVC_\infty(k-1)}{Dk} \qquad (2)$$

where k_L and k_S are the thermal conductivity of the liquid and the solid,
respectively. Recognizing the expression for conservation of heat at the
solid-liquid interface,

$$VL = k_S G_S - k_L G_L \qquad (3)$$

where L is the latent heat of fusion per unit volume, Equation (2) may also
be written

$$\frac{2G_L k_L + VL}{k_L + k_S} = \frac{mVC_\infty(k-1)}{Dk} \qquad (4)$$

in which form it becomes clear that for $V << 2G_L k_L/L$, the critical concentration for instability at a given G_L is inversely proportional to V,

$$C_\infty = 2G_L k_L Dk/(k_L + k_S)m(k-1)V \qquad (5)$$

whereas for $V >> 2G_L k_L/L$ the critical concentration for instability approaches a minimum value

$$C_\infty = LDk/(k_L + k_S)(k-1)m \qquad (6)$$

independent of both G_L and V. The stabilization of the critical concentration at high velocity in this case is a direct result of the steep gradient in the solid which is required to extract the latent heat of fusion. The transition between these regimes occurs at velocities near $V = 2G_L k_L/L$. For a conventional (Czochralski or Bridgman) crystal growth process we might have $G_L = 5 \times 10^3$ K/m and using the properties from Table I we find $V = 8.4 \times 10^{-4}$ m/s for Al and $V = 1.5 \times 10^{-4}$ m/s for Si which is very rapid for such growth processes. When solid-liquid surface energy effects are included the morphological stability expressions become considerably more complex but equation (5) is still valid at small velocities whereas at high velocities a new limiting expression is obtained, which may be referred to as the absolute stability criterion,

$$C_\infty = k^2 T_M \Gamma V/m(k-1)D \qquad (7)$$

where T_M is the melting temperature and Γ is a surface energy parameter. Mullins and Sekerka [3] have shown that the critical concentration for morphological instability always lies above this value. Eq. (7) is a manifestation of the increasing tendency of surface energy to retard the growth of perturbations as the wavelength becomes shorter at high velocity. Note that now C_∞ is independent of G_L (although it is assumed that the quantity on the left side of Eq. (2) is positive) and is proportional to V. Eq. (7) therefore predicts, in comparison to the constitutional supercooling criterion or to morphological stability theory without surface energy, an extremely large stabilizing effect at high velocities. The limiting curves given by equations (5) and (7) intersect at the concentration

$$C_\infty = [2T_M \Gamma G_L k_L/(k_L + k_S)]^{1/2} k^{3/2}/m(k-1) \qquad (8)$$

Figures (1) and (2) show morphological stability curves for the solidification of silicon containing arsenic [4] and for aluminum containing silver [5], and the various limiting expressions of equations (1), (5), (6) and (7), computed using the material properties of Table I. A notable difference between these systems is that in the silicon based system the minimum concentration for morphological instability, Eq. (6), is considerably greater than the concentration at which the low velocity limit line, Eq. (5), and the absolute stability line, Eq. (7), intersect, whereas in aluminum this is not true. Thus in the silicon based system (and in other silicon-based systems [4]), the interface is stabilized by the temperature gradients due to latent heat emission before the absolute stability effects of surface energy become strong, whereas in aluminum the absolute stability effects are predicted to occur before the latent heat stabilization becomes significant. This difference is due to the large latent heat of fusion and the rapid solute diffusivity in liquid silicon compared to the values for metallic-based systems. The constitutional supercooling line lies above the low velocity limit of the morphological stability curve in aluminum-based systems and below it in silicon-based systems, because of the relative thermal conductivities of the solid and liquid phases.

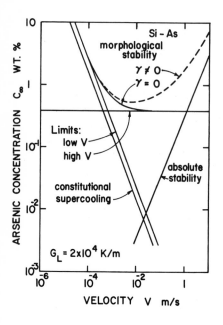

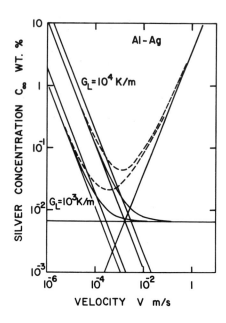

Fig. 1. Interfacial stability of Si containing As. The morphological stability curves with (dashed curve) and without (solid curve) surface energy are shown.

Fig. 2. Interfacial stability for two different gradients in Al containing Ag, showing the same curves as in Fig. 1.

Table I
Properties used in calculations

	k_L (J/s·m·k)	k_S (J/s·m·k)	L (J/m^3)	$T_M\Gamma$ (m·K)
Al	90.4	211	1.08 x 10^9	1.0 x 10^{-7}
Si	70	22	4.56 x 10^9	1.3 x 10^{-7}

	k	D (m^2/s)	m (K/wt.%)
Al–Ag	0.41	4.5 x 10^{-9}	-1.7
Al–Mn	0.79	2.1 x 10^{-9}	-1.0
Si–As	0.3	3 x 10^{-8}	-1.7

At solidification velocities of a few centimeters to a few meters per second, which are attainable by surface melting processes, non-equilibrium effects can be expected at the solid-liquid interface. Specifically it has been shown [6] that the solute distribution coefficient k can deviate from its equilibrium value toward unity at high solidification velocities. Narayan et al. [7] have analyzed the microstructure of laser melted ion implanted silicon and obtained, for isolated experimental points, a good fit to a morphological stability analysis in which a k value equal to that which prevailed at the solidification velocities occuring in their experiments was used. A recent more detailed analysis by Coriell and Sekerka [8], in which a velocity dependent k was used in the derivation of the morphological stability relations, showed that if the velocity dependence of k is sufficiently rapid the onset of interfacial instability may be different from that predicted using the procedure of Narayan et al. Coriell and Sekerka found that deviation of k from its equilibrium value toward unity is not always a stabilizing factor and if the variation of k with velocity is sufficiently rapid, oscillatory instabilities may occur.

The two systems considered in Figure 1 and 2 have similar rather high values of the distribution coefficient k. From Eq. (7) it is seen that at a given velocity the concentration at which absolute stability can be expected is strongly dependent on k. For example, when aluminum solidifies at a velocity of 10^{-2} m/s, it is predicted that 2 wt.% Mn (k = .79) is required to produce morphological instability, whereas 5.2×10^{-4} wt.% Fe (k = 2.8×10^{-2}), 8.3×10^{-5} wt.% Ni (k = 8×10^{-3}) or 4.4×10^{-5} wt.% Sn (k = 2×10^{-3}) will be sufficient to cause instability at the same velocity. If, however, the solidification velocity is sufficiently high to cause the very low partition coefficients to rise from their equilibrium values toward unity, the relative influence of the low k elements may be reduced.

The presentation of the morphological stability results in the format shown in Figures 1 and 2 is useful when considering steady state growth processes or when comparing the behavior of alloys with different compositions. During a surface melting process, however, the interfacial velocity and temperature gradients pass through a cycle of values and a presentation such as that shown in Figure 3, which applies to a specific concentration, is more useful. In the gradient vs. velocity presentation the cycle of conditions experienced during a surface melting process can be plotted as a trajectory. If this trajectory completely by-passes the region of predicted morphological instability, one can expect solidification to occur without microsegregation. Strictly, the theory of morphological stability applies to interfaces moving at steady state rather than under transient conditions, but in rapid solidification processes the solute diffusion fields are localized to an extremely thin region close to the interface and they can therefore change very rapidly as the interface accelerates. When a stationary interface starts to freeze at a velocity V, the characteristic distance over which the solute distribution field develops is s $\approx$ D/kV [1] and for V = 10^{-2} m/s we find in Al-Ag that s = 10^{-6} m, which is small compared to the melt depths which could produce this solidification velocity. In the next section we will consider some of the gradient and velocity cycles which occur during different types of surface melting operations, and their implications for morphological stability.

To estimate the effect which the velocity dependence of the partition coefficient k' would have on the curves shown in Fig. 3, we can calculate the absolute stability velocities for these curves with the assumption that the velocity dependence of the partition coefficient is given by

$$k' = \frac{k + \dfrac{V\lambda}{D}}{1 + \dfrac{V\lambda}{D}}$$

where λ is taken to be approximately 10 interatomic distances. With this formulation we find that the absolute stability limit for Al-1 wt% Ag is reduced from 2.7×10^{-1} m/s to 1.8×10^{-1} m/s but that the change of the absolute stability limits for the other curves in Fig. 3 is insignificant. Thus experiments at velocities of 10^{-2} m/s or less should not be affected by non-equilibrium partition coefficient effects. Where the solidification velocity is sufficiently high that $k' \to 1$ (solute trapping), partitionless solidification is expected at concentrations considerably greater than given by Eq. (7). This can occur only when the interface temperature is below T_o, the temperature at which solid and liquid have the same free energy.

ATTAINABLE FREEZING PARAMETERS

The combinations of melt depth, solidification velocity, and temperature gradient which can be attained by a surface melting operation are limited by heat flow, but are not identical for all modes of surface heating.

a) One-Dimensional Heat Flow

A one-dimensinal heat flow situation prevails when the surface is melted by a pulsed energy beam with radius, a, large compared to the distance over which heat can diffuse during the duration t of the pulse, i.e., $a > 4\sqrt{\alpha_s t}$ where α_s is the thermal diffusivity of the solid [9]. This condition commonly occurs during melting by Q-switched lasers or rapidly pulsed electron beams, but is difficult to attain during surface melting by CW lasers or by welding-type electron beams.

A useful characteristic of solidification with one-dimensional heat flow is that after the maximum melt depth has been attained the interface accelerates rapidly to a solidification velocity which remains almost constant until it reaches the surface. During the accelerating phase, the gradient in the liquid drops rapidly from a value which is comparable to the gradient in the solid when the interface velocity is zero at the maximum melt dept, to a value which is negligible compared to the gradient in the solid.

For the case of aluminum Hsu, Chakravarty and Mehrabian [10] found that if the surface is heated to the boiling temperature (2723 K) by a heat flux of intensity q (W/m^2) and the heat flux is then removed, the solidification velocity in meters per second is close to $V = 7 \times 10^{-11} q$ after the initial transient which accounts for about 20% of the melt depth. It was found that the maximum melt depth in meters for aluminum heated to its boiling point by a heat flux q is given by $d = 3.3 \times 10^5/q$. Combining these expressions gives $V = 2.3 \times 10^{-5}/d$ as the maximum velocity which may be attained for a surface melt of depth d with one-dimensional heat flow. Melting to the same depth with a lower power density and longer time will give lower velocities.

When the velocities and gradients which occur during solidification with one-dimensional heat flow are represented on the plot of Figure 3, they can completely by-pass the region of morphological instability of a dilute alloy. Near the maximum melt depth G_L/V is large and gradient stabilization is effective, whereas the subsequent precipitous decrease in G_L allows V to become large enough to stabilize the interface through absolute stability effects. Dashed line (a) in Figure 3 shows the velocity-gradient trajectory which occurs during solidification of aluminum following melting to the boiling point by an energy source of intensity 5×10^8 W/m^2.

b) Pulsed Spot

When a surface is melted by a pulsed energy beam with a diameter which is not large compared to the distance over which heat can diffuse during the length of the pulse, the relationships between solidification velocity, melt depth and

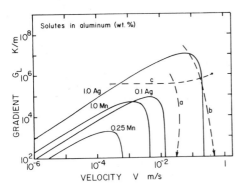

Fig. 3. Morphological stability of
Al containing Ag or Mn. The region
of instability lies below the curves.
Dashed lines show trajectories for
a) one-dimensional, b) pulsed spot
and c) moving spot melts.

absorbed power density are not as simple as in the one-dimensional case [9].

When a spot of radius a is heated by a uniform absorbed energy of intensity q the surface temperature at the center of the heated zone rises toward a steady state value which is approached very closely at a time $t = a^2/0.16\alpha_s$, which for the case of a 1 mm diameter spot on aluminum is 18 milliseconds. The temperature reached at the center of the spot at steady state is determined by the product qa, which must have a minimum value of 2.3×10^5 W/m to initiate surface melting in aluminum and a value of 4.2×10^5 W/m to bring the surface to the vaporization temperature at the center of the spot.

Melting by means of a pulsed spot allows solidification at a given velocity from a greater depth than is possible with one-dimensional heat flow. For example if $qa = 4.2 \times 10^5$ W/m and $a = 5 \times 10^{-4}$ m, the maximum melt depth is 3.25×10^{-4} m and the velocity of solidification rises from zero at the bottom of the pool to 10^{-1} m/s by the time the depth has decreased 5 percent. The interface then accelerates steadily to a velocity of almost 0.4 m/s as it approaches the surface. For the one-dimensional heat flow case a melt of the same depth would resolidify with an almost constant velocity of about 7×10^{-2} m/s (see Fig. 4).

Just as in the case of one-dimensional heat flow, the superheat in the liquid pool established by the pulsed spot diminishes rapidly after the heat source is turned off, and the gradient in the liquid becomes much smaller than that in the solid.

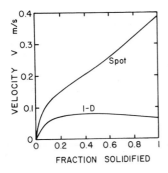

Fig. 4. Solidification velocities following melting to a depth of 3.25×10^{-4} m by a pulsed spot with $qa = 4.28 \times 10^5$ W/m and by one-dimensional heat flow.

For a pulsed spot melt, the gradients and velocities depend upon the location within the pool and the trajectory shown on a plot such as Figure 3 will be different for material growing straight up along the central axis of the pool when compared to that for material growing radially inward from the edges. Moreover, the maximum velocity of solidification in this type of melt develops only after the gradient in the liquid has fallen to a value much less than that which was present when solidification started. Typical trajectories may thus enter a region of instability and then pass into the absolute stability regime, as illustrated schematically by dashed curve (b) in Figure 3. For such cases it is important to make the distinction that morphological stability theory predicts the conditions under which a planar interface will become unstable but does not consider the conditions required for a cellular or dendritic interface to revert to a planar morphology. Thus there is no theoretical basis on which to predict that the maximum velocities experienced in the final stages of solidification after melting by a pulsed spot will result in microsegregation-free solid, if a region of instability has been encountered earlier in the process. To avoid morphological instability the critical region for the pulsed spot will be near the bottom of the melt pool where the interface has not reached its maximum velocity, but the gradient has decreased substantially. Even in this reigon, however, the velocity is higher than that obtainable with one-dimensional heat flow in a melt of the same depth.

c) Moving Spot

When a melt pool is established by a spot of directed energy moving across a surface of velocity U, the local solidification velocity is U cosθ, where θ is the angle between the normal to the interface and the direction of motion of the energy spot. Thus the interface velocities, which are always less than the velocity of motion of the spot, depend at steady state only on the location within the cross section of the melt and not on the position along the melt trail. Correlation of metallographically observed microstructures to local interface velocities is therfore relatively easy. In this situation the sequence of velocities and gradients experienced by the solid-liquid inteface are those lying along the line traced out by the normal to the interface as the melt pool moves. The shape of this line may often be deduced from observation of aligned microstructures in a longitudinal section of the melt.

Thermal analysis of the moving spot [11] is based on the parameters qa and $Ua/2\alpha$. When $Ua/2\alpha \ll 1$, the shape of the melt pool differs little from the steady state pool shape at U = 0, but as $Ua/2\alpha$ approaches unity the melt pool distortion becomes significant.

In this mode of melting the gradients in the liquid do not become very small as they do in the modes discussed above, because here the heat source is applied continuously. The value of G_L/V at the solidifying interface is shown for different values of qa and $Ua/2\alpha$ in [11] where it is seen that G_L/V decreases from infinity at the bottom of the pool (where V = 0) to a minimum value at the trailing edge of the pool (where V has its maximum value). The minimum value of G_L/V, at a given qa level, becomes larger as $Ua/2\alpha$ becomes smaller: when $Ua/2\alpha \ll 1$ and the shape of the melt pool is essentially independent of $Ua/2\alpha$, the minimum value of G_L/V is inversely proportional to U. Fig. 3 shows a trajectory (c) obtained with this type of melt for qa = 6.4 x 10^5 W/m, a = 5 x 10^{-4} m, and $Ua/2\alpha$ = 0.75.

HOMOGENIZATION OF THE MELT

Most surface melting processes will be carried out on materials which initially are microsegregated. The time in the liquid phase will in general be too short to allow elimination of this inhomogeneity by diffusion. An exception to this situation is possible with the pulsed spot, in which case the melt pool and thermal field reach a steady state configuration and if it is within the capability of the energy source the liquid phase may be held as long as desired to complete homogenization. For the case of a moving spot, if we estimate a diffusion distance $(Dt)^{1/2}$ for a melt pool 1 mm long we find that Ag in Al would move about 20 μm by diffusion in a melt moving at 10^{-2} m/s and only about 3μm in a melt moving at 0.5 m/s. In the latter case many microsegregation patterns would be diminished insignificantly by diffusional homogenization.

There is, however, abundant evidence that convection can be vigorous in the melt pools generated by directed energy sources. The surface topography features which are often present after melting have been attributed to the effect of this fluid flow [12]. A vigorous flow pattern can be expected to result from the influence of the temperature–dependent liquid surface energy (Marangoni convection) because gradients on the liquid surface can easily exceed 10^6 K/m. In addition, when a plasma is formed above the metal surface, the increased roughness of the surface suggests that a still more vigorous agitation of the fluid is present. In the next section we present some experimental evidence of the action of convective flow.

SURFACE MELTING OF ALUMINUM ALLOYS

We have used electron beams to melt the surface of aluminum alloys containing Ag and Mn, which have high partition coefficients and therefore should produce morphological instability only when present in relatively high concentrations. In pulsed spots and moving spots it is found that a microsegregation-free zone is present at the bottom of the melt pool but that cellular microstructure is present above this region even at velocities several times greater than those predicted to be required to produce absolute stability. The microsegregation-free zone is the region where G_L/R is highest but because there must be at least some growth of the solid between the place where the planar interface becomes unstable and the place where a metallographically visible cellular structure has developed, it is difficult to determine the exact G_L/R value at which the interface becomes unstable.

The most probable cause of the cellular substructure in aluminum alloys containing the high k solutes Ag and Mn at velocities where absolute stability was expected is the presence of much lower concentrations of elements with much lower k values. A concentration of 1-5 ppm of Fe or Ni would be sufficient to cause instability in the velocity range (approximately 10^{-2} to 0.3 m/s) which was investigated, but would be difficult to determine in a small zone especially in the presence of a large concentration of Mn.

It has been known for many years [13] that rapidly solidified Al–Mn alloys can contain Mn in solid solution far beyond the maximum equilibrium concentration of ~1.8 wt%. X-ray diffraction patterns of electron beam surface melted samples show at 7 wt.% Mn only a faint trace of second phase material present. This alloy appears to supersaturate so readily that the concentration differences resulting from cellular solidification are small and in many cases difficult to reveal. Electrolytic etching in a solution of 40% nitric acid and 60% methanol was found to be most effective at revealing cellular structures in these alloys.

While elements such as Fe or Ni might be responsible for causing the interface to become cellular, the microstructure becomes visible only because of the much higher concentration of Mn or Ag which also becomes segregated to the cell boundaries.

In some cases a swirled fluid flow pattern was frozen into the melt zone, which on close examination was revealed as differences in the cell spacing or pattern. These differences appeared to be remnants of the much larger scale cellular segregation pattern which was present in the arc melted substrate.

To obtain an indication of the rate at which mixing and homogenization occur in the melt zone, a thin Cu wire was pressed into the surface of a block of Al and reacted at slightly above the eutectic temperature to produce when it cooled a thin band containing primary Al in a eutectic matrix. The sample surface was then melted by electron beam spots 10^{-3} m in diameter moving perpendicular to the alloyed band at different power levels and velocities. Figures 5 through 8 show some of the results of this operation as seen on a polished section lying slightly below the original sample surface.

In Figure 5 the scan speed of the electron beam was 2×10^{-2} m/s. The power level was not sufficient to melt the pure Al at the plane of the section (although it did melt the surface) but the eutectic within the alloyed zone was melted and resolidified with a much finer lamellar structure. The lighter regions within the melt zone, streaming away from the large primary particles, contain a large volume fraction of finely divided primary phase. These streams can be seen to have carried material at least 200 µm during the approximately 5×10^{-2} s that the material was molten which implies that the flow velocity was at least 4×10^{-3} m/s, but possibly much larger because one cannot tell if the streams have merely dissipated beyond 200 µm in the melt.

With the same power level but ten times the scan velocity, there is little evidence of fluid flow (Fig. 6). The eutectic has been melted and refined in scale but the primary particles show only a shadowing around their perimeters which at higher magnification appears to be due to a partial melting in the form of numerous very small droplets. Thus the melting of the primary particles was insufficient to delineate any fluid flow which may have been present.

At higher power levels the fluid flow effects are much more conspicuous. In Figure 7, in which the electron beam was scanned at 2×10^{-2} m/s, all of the primary in the melt zone except that very close to the edges has been melted and thoroughly mixed. Copper has been transported upstream at least 10^{-3} m implying a fluid flow velocity of at least 2×10^{-2} m/s.

Figure 8 shows the result of scanning the beam at 0.2 m/s and here it is seen that although considerable fluid flow has taken place there has been insufficient time to thoroughly homogenize the molten zone. In this case the upstream motion of 10^{-3} m implies a fluid flow velocity of at least 0.2 m/s.

From these observations one can set the lower bounds on the fluid flow velocity, but one derives little information on upper bounds. It is clear that convective mixing is an important part of melt homogenization in microsegregated substrates. It will certainly play a dominant role in surface alloying [14].

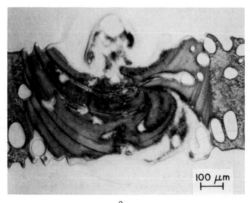

100 µm

Fig. 5.   Electron beam scan at 2×10^{-2} m/s across Al-Cu eutectic band in Al.

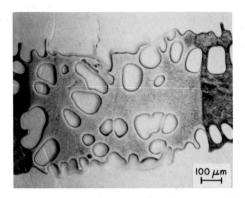

Fig. 6. Electron beam scan at 0.2 m/s across eutectic band.

Fig. 7. Electron beam scan at
2 x 10^{-2} m/s across eutectic band.

Fig. 8. Electron beam scan at
0.2 m/s across eutectic band.

SUMMARY

The temperature gradient and interface velocity cycles which occur during
different modes of surface melting are the most important controlling parameters
in the determination of solidification microstructures. These cycles can be
related to the conditions predicted to cause microstructural transitions, the
best developed of which is the theory of morphological stability of planar
interfaces. This theory predicts increasing stability at sufficiently high
rates of solidification, but clear evidence of this effect has been observed
only at velocities so high that the deviation of the solute partition coefficient from its equilibrium value plays an important role.

To obtain the ideal microstructures predicted by solidification theory the melt zone must be homogeneous. Convection plays a leading role in the elimination of the residual microsegregation patterns from the substrate. Complete homogenization may require a slow melt for homogenization followed by a faster melt for the final microstructural refinement.

ACKNOWLEDGMENTS

The authors thank S. R. Coriell for useful discussions and for calculation of morphological stability curves. We also thank the Defense Advanced Research Projects Agency and Technical Monitor Lt. Col. L. A. Jacobson for support of this work.

REFERENCES

1. W. A. Tiller, J. W. Rutler, K. A. Jackson and B. Chalmers, Acta Met. $\underline{1}$, 428 (1953).

2. W. W. Mullins and R. F. Sekerka, J. Appl. Phys. $\underline{34}$, 323 (1963)

3. W. W. Mullins and R. F. Sekerka, J. Appl. Phys. $\underline{35}$, 444 (1964).

4. J. W. Cahn, S. R. Coriell, and W. J. Boettinger in: Laser and Electron Beam Processing of Materials, C. W. White and P. S. Peercy eds. (Academic Press, New York 1980) pp. 89-103.

5. R. J. Schaefer, S. R. Coriell, R. Mehrabian, C. Fenimore and F. S. Biancaniello in: Rapidly Solidified Amorphous and Crystalline Alloys, B. H. Kear, B. C. Giessen and M. Cohen eds. (North-Holland, New York 1982) pp. 79-89.

6. C. W. White, S. R. Wilson, B. R. Appleton and J. Narayan in: Laser and Electron Beam Processing of Materials, C. W. White and P. S. Peercy eds. (Academic Press, New York 1980) pp. 124-129.

7. J. Narayan, J. Crystal Growth $\underline{59}$, 583 (1982).

8. S. R. Coriell and R. F. Sekerka, J. Crystal Growth (in press).

9. S. C. Hsu, S. Kou and R. Mehrabian, Met. Trans. $\underline{11B}$, 29 (1980).

10. S. C. Hsu, S. Chakravorty and R. Mehrabian, Met. Trans. $\underline{9B}$, 221 (1978).

11. S. Kou, S. C. Hsu and R. Mehrabian, Met. Trans. $\underline{12B}$, 33 (1981).

12. S. M. Copley, D. G. Beck, O. Esquivel and M. Bass in: Lasers in Metallurgy, K. Mukherjee and J. Mazumder eds. (The Metallurgical Society of AIME, Warrandale, Pa. 1981) pp. 11-19.

13. G. Falkenhagen and W. Hofmann, Z. Metallkunde $\underline{43}$, 69 (1952).

14. C. W. Draper, Journal of Metals $\underline{34}$, No. 6, 24 (1982).

METASTABLE Fe(Pd) ALLOYS FORMED BY PULSED ELECTRON BEAM MELTING[*]

D. M. FOLLSTAEDT AND J. A. KNAPP
Ion Solid Interactions, Division 1111
Sandia National Laboratories, Albuquerque, NM 87185

ABSTRACT

Fe(Pd) surface alloys formed by either ion implanting Pd into Fe or by vapor depositing Pd on Fe have been examined after pulsed ($\sim$70 nsec) electron
beam melting. TEM and ion backscattering/channeling showed that the implanted
alloys with up to 9.6 at.% Pd were bcc and epitaxial with the substrate, in
spite of predictions of initial solidification to the fcc phase by the Fe-Pd
phase diagram. The resolidified Pd films, however, were fcc as predicted by
the diagram. The defect structures observed in these alloys are discussed.

INTRODUCTION

The use of pulsed laser or electron beams is being increasingly studied as
a means of producing unique surface alloys on metal substrates. When a short
pulse (10-100 nsec) of electrons or laser light of the proper power level is
incident on the metal surface a thin layer ($\lesssim$ 1 μm) very rapidly melts and
resolidifies (within $\sim$0.01 to $\sim$1 μsec), and subsequent solid phase cooling
rates are quite high (10^8-10^{10} K/s). In many pure metals the resolidified
layer is found to be epitaxial with the substrate [1]. Highly metastable surface alloys can be formed by ion implanting other species or vapor depositing
a surface layer prior to pulsed melting [2].

Applying pulsed melting to Fe-based alloys may yield new surface alloys
with potential technological application. For instance, an engineering alloy
with desired bulk properties (eg.,strength, machineability) could be treated for
desired surface properties (eg., corrosion resistance, reduced friction and
wear). Using conventional alloy additions as a guideline, it is of interest to
examine additions of Cr and Ni in Fe. Determining the composition limits of
the bcc and fcc Fe-based phases after pulsed melting is of primary importance.

Pulse melted Fe is of interest for a second fundamental reason: the resolidified Fe is epitaxial with the bcc substrate [1,3] in spite of an allotropic transformation encountered during near-equilibrium cooling of Fe
(liquid$\rightarrow$bcc$\rightarrow$fcc$\rightarrow$bcc). If the just-resolidified bcc surface layer transforms to the fcc phase and then back to bcc, it might be expected to lose
epitaxy with the unaltered bcc substrate. One test which may determine whether
the transformation to the fcc phase occurs is to alloy an element which is
mutually soluble with fcc-Fe at concentrations for which the fcc phase is
stable at room temperature. If that composition can be solidified and retained
as bcc, this would indicate that the fcc phase had never formed.

Thus there are two important reasons for studying pulse melted Fe alloyed
with an fcc element. An obvious choice is Ni; however, the near surface concentrations and substitutionality are not easily determined with MeV ^{4}He$^+$ ion
backscattering/channeling. We therefore chose to study alloys with Pd, which
can be examined with ion beams. The Fe-Ni and Fe-Pd equilibrium phase diagrams
are very similar [4], and hence we expect the metallurgy of their melt-quenched

*This work performed at Sandia National Laboratories was supported by the U. S.
Department of Energy under contract number DE-AC04-76DP00789.

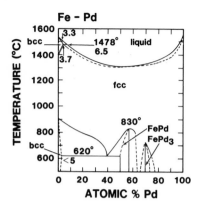

Fig. 1. Fe-Pd binary equilibrium phase diagram (Ref. 4).

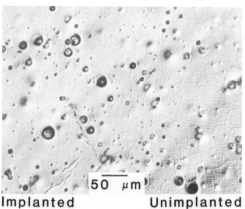

Implanted   Unimplanted

Fig. 2. Nomarski optical micrograph from <100> Fe surface implanted with 1×10^{16} Pd/cm^2 after pulsed melting. Both halves of the figure show surface flow but only the unimplanted half shows slip deformation (the criss-crossed pattern).

alloys to be similar. The Fe-Pd diagram is shown in Fig. 1. The fcc Fe phase forms continuous solid solutions with pure fcc Pd, which is stable at room temperature. At low Pd concentrations the diagram exhibits a peritectic reaction at 1478°C. Thus for concentrations <3.3 at.% Pd, initial solidification to the bcc phase is predicted, while for concentrations >6.5 at.% Pd, initial solidification to the fcc phase is predicted.

We have therefore examined three Fe(Pd) alloys. Implantation of 1×10^{16} Pd/cm^2 into pure Fe at 190 keV produced a peak concentration of 2.6 at.% Pd (at a range of 300 Å) for which initial solidification to bcc is predicted. Implantation of 5×10^{16} Pd/cm^2 at 190 keV gave a peak concentrations of 8–10 at.% for which initial solidification to fcc is predicted. Finally, a 400 Å layer of pure Pd was deposited on the Fe to give a surface layer concentration ∼100% Pd for which the fcc phase is stable at room temperature.

These three alloys were examined after pulsed melting with TEM at 120 kV and 2.0 MeV ^{4}He$^+$ ion backscattering/channeling. Samples for TEM used well annealed, electropolished, 125 μm thick, high purity Fe foil. Thin areas for TEM examination were prepared by jet electropolishing from the untreated side with a perchloric acid solution. Ion channeling analyses were done on polished <100> single crystals of high purity.

Pulsed electron beam melting was done on an instrument which we have previously described [3]. In that initial work on pure Fe, we found microstructural evidence of melting for deposited energies above 1.1 J/cm^2 with 50–100 nsec pulses. In Table I we list the pulse parameters for the samples used here. Finite element heat flow calculations predict melting in each case, and the melt parameters are also given in Table I. Nomarski optical microscopy showed surface changes (pock marks, surface flow, slip, etc.) which indicated that melting occurred for all samples.

Analysis of both ion implanted samples showed epitaxial resolidification to the bcc phase, while the layered sample showed fcc Pd on bcc Fe. Detailed observations on each type of sample are given below, and conclusions given in the final section.

TABLE I.
Pulse Melted Fe(Pd) Samples

	Deposited Energy (J/cm^2)	Pulse Width (nsec)	Melt Time (nsec)	Melt Depth (μm)	Solid Phase Cooling Rate (K/s)	Peak Pd Conc.(at.%) before/after
TEM Foil 1×10^{16}Pd/cm^2	1.51	70	160	0.6	5.9×10^9	2.6/2.0
<100> Crystal 1×10^{16}Pd/cm^2	2.18	80	350	1.0	3.2×10^9	2.6/2.5
TEM Foil 5×10^{16}Pd/cm^2	1.70	75	160	0.6	5.9×10^9	8.6/8.2
<100> Crystal 5×10^{16}Pd/cm^2	2.08	80	350	1.0	3.2×10^9	10.2/9.6
TEM Foil 400 Å Pd	1.58	70	300	0.9	6.2×10^9	~100
<100> Crystal 300 Å Pd	1.78	76				~100

ION IMPLANTATION RESULTS

A Nomarski optical micrograph from the surface of the <100> Fe single crystal implanted with 1×10^{16} Pd/cm^2 is shown in Fig. 2. The micrograph is from the interface between the implanted and unimplanted areas. Both sides show surface flow indicating that melting occurred. The unimplanted side shows slip traces which are also produced by melting; however, these are absent on the implanted side just as observed previously with Sn-implanted Fe [3]. Slip was also absent for the 5×10^{16} Pd/cm^2 implant, and was not seen at either fluence with TEM. The absence of slip in both Sn- and Pd-implanted Fe after pulsed melting demonstrates that its suppression with near surface alloying is not directly related to whether the alloy addition stabilizes the bcc (Sn) or fcc (Pd) phase. It is more likely that the solute strengthens the near surface so that the yield strength is not exceeded by thermal stress and slip deformation is thus avoided.

TEM examination of both implanted alloys showed bcc material with the same grain structure ($\sim$100 μm) as the initial Fe foil, thus indicating epitaxial resolidification. Electron diffraction patterns like those in Fig. 3 showed bcc Fe reflections with weaker superimposed reflections from surface oxides (γ-Fe$_2$O$_3$); no fcc reflections were observed. The grains showed dense dislocation networks like that seen in Fig. 4. A few small ($\sim$150 Å) dislocation loops can also be seen. Dislocation cells like those seen in pure Fe were not observed [3].

The 5×10^{16} Pd/cm^2 sample showed a new lattice defect not seen in the lower fluence sample nor in pure Fe: voids. Figure 5 demonstrates the presence of the voids which appear as small white areas ($\sim$50 Å) surrounded by a black ring in the underfocussed condition (Fig. 5a) but which change to a black ring just inside the void surrounded by a white ring outside when overfocussed (Fig. 5b) [5]. Void formation apparently requires a minimum concentration between 2.0 and 8.2 at.% Pd (see Table I).

Figure 6 shows the portion of the ^{4}He backscattering/channeling spectra due to scattering from Pd in the <100> Fe crystal implanted with 5×10^{16} Pd/cm^2. The e-beam treated sample shows a slightly broadened profile with a drop in the peak concentration which is consistent with liquid state diffusion. The small peak in front of the Pd surface is from Xe which was used in the ion implanter

Fig. 3. Electron diffraction patterns from Fe implanted with 5×10^{16} Pd/cm^2 at 190 keV and subsequently pulse melted. a) <100>, b) <110>, and c) <111>.

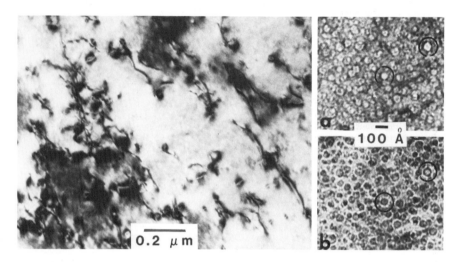

Fig. 4. Bright field TEM micrograph showing extended dislocations and dislocation loops in Fe implanted with 1×10^{16} Pd/cm^2 after pulsed melting.

Fig. 5. Bright field TEM images showing voids in Fe implanted with 5×10^{16} Pd/cm^2 and pulse melted. a) underfocussed; b) overfocussed.

to produce Pd$^+$ ions. The Pd substitutional fraction is essentially unchanged (93%) after pulsed melting, and the highest resulting substitutional Pd concentration is 8.9 at.%. Similar slight broadenings were observed with the other samples after pulsed melting. The Pd substitutional fractions were similar for 1×10^{16} Pd/cm^2, giving 2.3 at.% substitutional Pd.

Pd LAYER RESULTS

Optical microscopy of the samples with Pd layers also showed changes in surface features indicating that melting had occurred. The backscattering spectra from the TEM sample are shown in Fig. 7. The back edge of the Pd layer and the front edge of the Fe are broadened after e-beam treatment, consistent with interdiffusion at the Fe-Pd liquid interface. The single

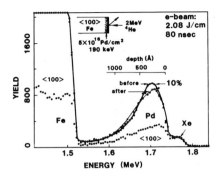

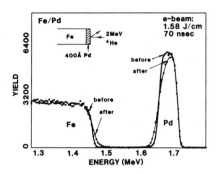

Fig. 6. ^{4}He backscattering spectra from Pd ion implanted into <100> Fe.

Fig. 7. ^{4}He backscattering spectra from an Fe foil with 400 Å of Pd.

crystal sample showed no channeling of the Pd layer after pulsed melting.

The microstructure of the Pd-rich layer after pulsed melting is shown in Fig.8a. The Pd-rich surface layer is the electron-optically thin (lighter) material, while the thicker (darker) area includes the bcc Fe substrate. This interpretation was confirmed with energy dispersive x-ray spectrometry. Figure 8b is an electron diffraction pattern taken from the center of the Pd-rich layer in Fig. 8a. The more intense spot pattern is a <110> zone from fcc Pd. Also seen is a partial (111) Pd ring and some weak arcs approximately in the position expected for twin reflections. Figure 8 shows linear features extending across the Pd layer on which pits are located. Dark field imaging with the reflections noted above from the misoriented fcc Pd shows that these correspond to small ($\sim 0.3 \mu$m) particles located along the linear features. The orientation of these features agrees with that of the traces of $\{111\}$ planes in the fcc material, and we thus interpret them to be traces of slip deformation and/or twinned material. Although a single orientation spot pattern is predominant in Fig. 8b, the Pd layer is not a single grain. Small, slightly misoriented subgrains are seen in the darkfield micrograph of Fig. 8c where the misorientations produce the contrast changes between the ~ 0.15μm subgrains.

Ignoring the misorientations noted above, one can find "grain boundaries" within the Pd layer at which major orientation changes occur. The "grains" within which most of the Pd has nearly the same orientation are similar in size to the Fe grains, and the grain boundaries in the underlying bcc continue into the fcc Pd. Thus there must be orientational relationships between the fcc Pd and bcc Fe lattices; however curvature of the foils and other experimental difficulties have prevented our determining the relationships.

CONCLUSIONS

We have demonstrated that bcc solutions with up to 8.9 at.% substitutional Pd can be formed on the surface of Fe by ion implantation and pulsed melting; it is probable that similar alloys with Ni can be formed. In another study, we have shown that pulse melted 304 stainless steel (with ~18 at.% Cr and ~ 9 at.% Ni) with the fcc lattice structure also resolidifies epitaxially on the unaltered substrate [6]. Thus epitaxial alloys of altered compositions may possibly be formed on both types of lattice substrate by pulsed melting. These may be useful for achieving desired surface properties in engineering alloys.

For 8-10 at.% Pd the Fe-Pd equilibrium phase diagram [4] (Fig. 1) clearly predicts initial solidification to the fcc phase. Our observations of only

750

Fig. 8 a) Bright field TEM image of ~ 400 Å Pd film on Fe substrate.
b) Electron diffraction pattern from the center of the Pd film in a).
c) Dark field image of subgrains in the Pd film.

epitaxial bcc material for both implanted samples suggest that the fcc phase
may never have been present, although the evidence is not conclusive. A similar
question arises for the observed epitaxy in 304 stainless steel where the
Fe-Cr-Ni ternary phase diagram predicts initial solidification to the bcc
phase [7]. For both alloys it is possible that the liquid resolidifies epitaxially on the substrate phase and the equilibrium phase is never nucleated.

The fcc Pd layer may mean that epitaxial resolidification on the substrate
phase is not possible for all concentrations, but we cannot rule out the possibility that the layer was initially bcc and then converted in the solid phase
to the equilibrium fcc structure. Additional work at concentrations between
10 and 100 at.% Pd is required to determine the Pd concentration at which the
change from bcc to fcc occurs.

REFERENCES

1. L. Buene, E. N. Kaufmann, C. M. Preece and C. W. Draper, in Laser and
    Electron-Beam Solid Interactions and Materials Processing, J. F. Gibbons,
    L. D. Hess and T. W. Sigmon, eds. (North Holland, New York 1981), p. 591.
2. S. T. Picraux and D. M. Follstaedt in Surface Modification Alloying, NATO
    Inst. Book, Eds. J. M. Poate and G. Foti (Plenum Press, New York, 1982).
3. J. A. Knapp and D. M. Follstaedt, in Laser and Electron-Beam Interactions
    With Solids, B. R. Appleton and G.K. Celler, eds. (North Holland, New York
    1982), p. 407.
4. M. Hansen, Constitution of Binary Alloys, (McGraw-Hill, New York 1958)
    p. 677 (Fe-Ni); R. P. Elliott, First Supplement (1965) p. 427 (Fe-Pd).
5. M. H. Loretto and R. E. Smallman, Defect Analysis in Electron Microscopy,
    (Chapman and Hall, London 1975) p. 57.
6. D. M. Follstaedt and J. A. Knapp, to be published.
7. J. C. Lippold and W. F. Savage, Welding Jour. 58 (12), Res. Supp. p.362s.

AMORPHOUS Al(Ni) ALLOY FORMATION BY PULSED ELECTRON BEAM QUENCHING[*]

S. T. PICRAUX AND D. M. FOLLSTAEDT
Sandia National Laboratories, Albuquerque, NM 87185

ABSTRACT

The microstructural transition from the epitaxial crystalline to amorphous state as a function of Ni concentration is examined in Al(Ni) alloys formed by ion implantation and e-beam melt quenching. The evolution from epitaxial crystalline Al at $\leq$ 3.5 at.% Ni to amorphous phase formation at 15-30 at.% Ni is characterized in detail for 8, 10 and 12% Ni uniform concentration samples by ion channeling and transmission electron microscopy. The transition occurs via complete loss of epitaxial regrowth at 9 at.% and the formation of a fine-grained polycrystalline Al, with grain size decreasing with increasing Ni from 20-100 Å at 10% Ni concentration to $\leq$ 10-50 Å at 12% Ni.

INTRODUCTION

In this paper we discuss the crystalline to amorphous transition in Al(Ni) formed by Ni implantation and electron beam (e-beam) pulsed melting. The microstructural evolution of this transition with increasing Ni concentration is examined by ion channeling/backscattering and transmission electron microscopy (TEM). We have previously shown that at low implanted Ni concentrations ($\leq$ 3.5 at.%) epitaxial regrowth of crystalline Al occurs [1], whereas at sufficiently high concentrations (~15-30 at.%) an amorphous alloy is formed [2]. This defines the composition regime in which to examine this crystalline to amorphous transformation in detail.

There are several special aspects to the use of ion implantation and e-beam melt quenching for studying rapid solidification. First, by multiple energy implantation of Ni we form surface layers (~0.1 μm thick) of uniform concentration with the Ni microscopically dispersed. Second, the heat is deposited uniformly into a layer a few microns deep on a thick substrate, so that the quench rates ($10^8 - 10^{10}$ K/sec) and interface velocities (1-10 m/sec) are among the highest of those studied in rapid solidification. Also, the sample geometry is better controlled than in splat quenching and the thermal history of the sample is readily calculated. Finally, by melting into the crystalline Al substrate, which always acts as a seed for planar epitaxial regrowth back to the surface, we exclude nucleation-limited transitions to the amorphous phase, and thus are able to focus just on the growth-limited case.

The rapid solidification of Al alloys by splat cooling and related techniques has been extensively studied, and greatly improved mechanical properties have been observed in many cases due to the microstructural refinement of the crystalline state [3]. In addition, evidence for amorphous phase formation has been seen in several binary aluminum alloys as a limiting case of high solidification rates [4]. In the case of the Al(Ni) system evidence has been presented for the enhanced solid solubility of Ni within crystalline Al to $\leq$ 3 at.% [5] and the formation of localized regions of an amorphous phase for rapid cooling at ~ 7 at.% [4]. Thus while these previous studies are consistent with the present results, the nonuniform splat cooling rates and the limited concentrations studied did not allow detailed examination of the crystalline to amorphous transition.

---

[*]This work performed at Sandia National Laboratories supported by the U. S. Department of Energy under contract number DE-ACO4-76DP00789.

Mat. Res. Soc. Symp. Proc. Vol. 13 (1983) Published by Elsevier Science Publishing Co., Inc.

EXPERIMENTAL TECHNIQUE

Single crystal Al samples of high purity (99.99%) and <110> orientation were implanted after electropolishing and vacuum annealing. Multiple energy Ni implantations from 15 to 160 keV were used to obtain two different conditions for the Ni concentration profiles: a) uniform Ni concentration to ~ 1000 Å depth, and b) monotonically decreasing Ni concentration with depth from the surface to ~ 1500 Å. The Ni concentrations were determined experimentally by 2 MeV ^{4}He backscattering. The e-beam pulsed melting was carried out under vacuum using a cold-cathode pulsed power system of our own design. All studies used single 65 nsec pulses, with peak accelerating voltage of 40 kV and a deposited energy of 1.7 or 2.4 J/cm^2. The microstructural analyses were carried out by 2 MeV ^{4}He ion channeling and 120 kV TEM using different areas of the same sample to allow direct comparison.

RESULTS

Previous e-beam melt quenching studies of Ni-implanted Al define the low and high concentration limits for the present study. At low Ni concentrations (~ 1 at.%) we observe metastable solutions of ≈ 0.4 at.% substitutional Ni (about 20 times the maximum equilibrium solid solubility), epitaxial recrystallization of Al and no segregation of the Ni to the surface [1]. At 3.5 at.% Ni some small clusters of a metastable phase related to the Al$_3$Ni phase are observed which show definite orientation with respect to the Al lattice. The high concentration limit has been defined as amorphous by the diffuse diffraction rings in TEM of a monotonically decreasing Ni profile of ≈ 30 at.% near the surface and ≈ 15 at.% at a depth of 1000 Å [2]. Also in this sample the crystalline phase AlNi was observed near the surface, consistent with the predicted nucleation of this phase above 27 at.% Ni.

The Al-rich part of the Al-Ni equilibrium phase diagram is shown in Fig. 1 by the solid and dotted lines. Under equilibrium conditions the intermetallic phase Al$_3$Ni would precipitate within liquid Al upon cooling for concentrations above 2.7 at.% and the Al$_3$Ni$_2$ phase would precipitate above 15.3 at.%. However, TEM analysis of the high concentration sample discussed above revealed that neither of these intermetallic phases are formed upon melt quenching. Using finite difference heat flow calculations for 2.38 J/cm^2, 65 nsec pulsed melting of pure Al and assuming no undercooling a total melt time ≈ 1.2 μsec is obtained. Once nucleated the precipitates would have grown to observable sizes in ~ 10 nsec. Hence we determine from the calculation that the nucleation times for precipitation within liquid Al are ≳ 0.75 μsec for Al$_3$Ni and ≳ 0.95 μsec for Al$_3$Ni$_2$. These lower limits to nucleation times contrast to the recently measured value of 15 ± 10 nsec for the precipitation of AlSb in liquid Al under similar experimental conditions [6]. The above results show that Al(Ni) melt quenching at 8-12 at.% Ni in the following studies (where AlNi was not observed) proceeds from a single phase liquid.

The influence of increasing Ni concentration on epitaxial growth during rapid solidification can be examined by ion channeling. In Fig. 2 results are shown for a shallow implanted Ni layer which decreases in concentration monotonically with depth. The analysis of Fig. 2 is for glancing angle detection at 23° from the surface to enhance the depth resolution. The depth at which the loss of channeling occurs (channeled and nonchanneled Al spectra coincide) corresponds to the loss of epitaxial growth. The drop and broadening of the Ni profile upon e-beam melting results from liquid phase diffusion. After melt quenching the loss of epitaxy occurs at point b on the Al spectrum in Fig. 2, which corresponds from the Ni spectrum to a concentration of 9 at.%. We also note that direct scattering in the Al spectrum indicating high local lattice distortions becomes appreciable at point c, which corresponds to less than 1 at.% Ni. From the large magnitude of this direct scattering we suggest

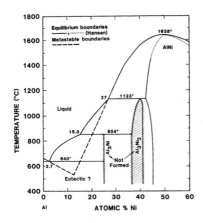

Fig. 1 Al-rich part of the Al-Ni
equilibrium phase diagram: the
dashed lines and the cross hatched
deletions show metastable modifications suggested by this work.

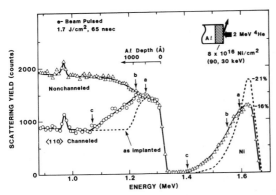

Fig. 2 Spectra for 2 MeV
^{4}He backscattering and
<110> axial channeling anal-
ysis at 23° glancing angle
to the surface and 113°
scattering angle for Al implanted to a total of 8x10^{16}
Ni/cm^2 at 90 and 30 keV and
e-beam melted. The non-
channeled and channeled
spectra coincide for the
Ni part of the spectrum.

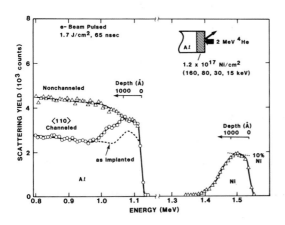

Fig. 3 The 2 MeV ^{4}He backscattering and <110>
channeling spectra at 160°
scattering angle from Al
implanted with 160, 80, 30
and 15 keV Ni to a total
fluence of 1.2 x 10^{17}/cm^2
to give ≈ 10 at.% Ni
after e-beam pulsed melting.

that this may, in part, be an indication that Ni produces local distortions on the Al lattice as a precursor to amorphous phase formation. In addition any metastable phase clusters, previously observed at ~ 3.5% Ni, will make some contribution to the direct scattering.

For the as-implanted sample in Fig. 2 the complete loss of channeling (point a) corresponds to Ni concentrations of 15 at.%. Since TEM of implanted alloys with > 15 at.% Ni shows an amorphous phase, the channeling result suggests that 15 at.% Ni is the critical concentration for amorphous phase formation during implantation. For the melt quenched samples the loss of epitaxy occurs at a significantly lower concentration of 9%. Previously we have suggested that this may indicate a lower critical Ni concentration for amorphous phase formation with melt quenching [2]. However as we show below, detailed examination by TEM indicates a more complex transition behavior.

To examine the transition in greater detail, we prepared thicker samples (~1000 Å) with nearly uniform Ni concentrations of 8, 10 and 12 ($\pm$ 1/2) at.%. In Fig. 3 we show 160° (non-high resolution) backscattering and <110> channeling analysis of the 10% sample. A loss of epitaxy is indicated over the ~ 1000Å depth corresponding to the region of 10% Ni. This contrasts with the as-implanted channeling spectrum (dashed line) which exhibits about a 15% channeling reduction in yield, indicating a heavily disordered region in which long range order with the substrate orientation is maintained. For the 8 and 12% samples (not shown), the as-implanted channeling reductions are 30% and 8%, respectively. The 8% sample exhibits only a very small, 4%, channeling reduction after melt quenching, consistent with the results of Fig. 2 where the complete loss of epitaxy occurs at 9 at.% Ni. As expected, complete loss of epitaxy is indicated for the 12% Ni after melt quenching.

Electron microscopy results for the 8, 10 and 12% uniform Ni implanted samples are shown in Fig. 4. We observe a transition from an epitaxial to a polycrystalline structure of decreasing grain size with increased Ni concentration. We now describe this evolution to microcrystallinity in detail.

For the 8% sample (a) crystalline Al with the substrate orientation is indicated by the <110> diffraction pattern. This oriented Al was found to extend to the sample surface. Also seen are partial rings of sharp (111) and (200) reflections from polycrystalline Al grains. These randomly-oriented crystallites are imaged in dark field in Fig. 4. Analysis indicates they range in size from 100 to 2000 Å. In addition, a uniform diffuse ring near the (200) Al position is seen in Fig. 4. Based on the earlier work at 3.5% Ni [1] where arcs are seen at a similar radius we suggest that this diffuse ring may be due to localized Ni clustering, perhaps indicating regions where short range Al-Ni order dominates sufficiently to induce a localized amorphous-like structure. Such regions may be responsible for the blocking of epitaxial growth which is just starting to occur at 8% Ni. At higher Ni concentrations the Al (200) ring is also diffuse, preventing the possibility of observing a Ni cluster ring.

For the 10% Ni sample (b) in Fig. 4 only polycrystalline Al rings are observed, indicating the complete loss of epitaxy in the Ni-alloyed layer. This is consistent with the channeling analysis which indicates complete loss of epitaxy at > 9% Ni. The diffraction rings are seen to be significantly broadened due to the small size of the crystallites, however the weaker (220) and (113) rings at larger radius are still observable. The dark field image indicates crystallite sizes from 20 to 100 Å. Upon increasing the Ni concentration to 12%, (c), the (111) and (200) Al polycrystalline diffraction rings nearly, but not quite, merge together, and the (200) and (113) rings are replaced by a single broad diffuse band, indicating even smaller crystallites. From the dark field image the crystallite density is higher and the observed sizes range from a few at ~ 50 Å to many at ~ 10 Å. The microstructural evolution is illustrated schematically in cross section at the bottom of Fig. 4. Future work at incrementally higher Ni concentrations will be carried out to determine more precisely the concentration at which Al(Ni) is quenched into the amorphous state.

DISCUSSION

The transition from crystalline to amorphous alloy formation with increasing Ni concentration proceeds via a loss of epitaxial growth and the formation of an extremely fine polycrystalline Al within the Al(Ni) alloy matrix. In the present case the interface velocity for the samples of Figs. 3 and 4 can be estimated from heat flow calculations to be ≈ 7.2 m/sec, with an initial cooling rate in a solid phase of 8×10^8 K/sec. Both these values are based on unalloyed crystalline Al and do not account for changes in the latent heat of fusion or for the presence of undercooling. The loss of epitaxy here can be contrasted to that in silicon, where a transition with increasing (111) interface velocity occurs from good epitaxial crystallization to amorphous quenched layers via a region of highly twinned and faulted crystalline Si upon laser melt quenching [7]. However, in the present case we are dealing with a binary metallic alloy and have characterized this transition with increasing Ni concentration rather than solidification velocity. The mechanisms producing our observed microstructures may be related to local

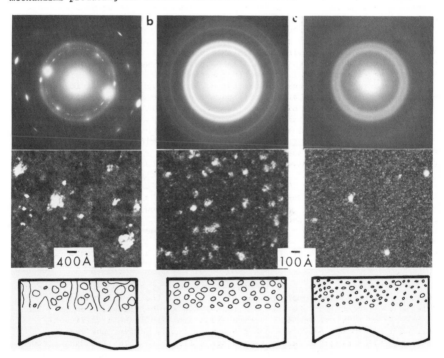

Fig. 4 Transmission electron microscopy results for uniform (left to right) 8, 10 and 12 at.% Ni-implanted Al samples after e-beam pulsed melting under identical conditions of a 1.7 J/cm², 65 nsec pulse. The upper panels show the electron diffraction patterns, the middle panels the dark field images of polycrystalline fcc Al grains and the lower panels the schematic cross section of the melt quenched layers.

variations in Ni concentration leading to short range ordering in the liquid and during solidification. Alternatively, initial solidification of the alloy into a fully amorphous phase may be followed by a localized solid state transformation into polycrystalline Al.

The extremely fine grain nature of this Al(Ni) alloy at 10 and 12% is noteworthy. One of the important results of the rapid solidification studies of Al alloys has been the grain refinement, which has led to significant improvement in mechanical properties of these alloys [3]. The majority of these results are adequately described by classical nucleation and growth theory with a grain size (in μm) given by $1.75 \times 10^7 \; R^{-0.9}$ for Al, where R is the cooling rate upon solidification [8]. This corresponds to a grain size $\approx 0.2 \, \mu$m at our cooling rates, about two orders of magnitude larger than our observed values at 12% Ni. There is some evidence for anomalously small grain sizes, down to $\approx 100 \, \overset{\circ}{A}$, for splat cooled Al(Ge) and Al(Fe) [8]. These and the present studies, which extend to nearly an order of magnitude smaller sizes and are in a growth-limited regime, clearly imply a different mechanism of polycrystallite formation. In the present case we show grain refinement down to ~10 $\overset{\circ}{A}$ and observe, for the first time, the variation with Ni concentration of crystallite size under uniform and identical pulsed heating conditions.

CONCLUSIONS

Preliminary results are reported for the microstructural evolution with concentration of rapidly solidified Al(Ni) through the transition region from the crystalline to the amorphous state. Under these growth-limited conditions at interface velocities ≈ 7.2 m/sec (neglecting undercooling and latent heat corrections) the transition occurs via the complete blockage of epitaxy at 9 at.% Ni and the formation of a very fine grained polycrystalline Al.

REFERENCES

1. S. T. Picraux, D. M. Follstaedt, J. A. Knapp, W. R. Wampler and E. Rimini, in Laser and Electron-Beam Solid Interactions, J. F. Gibbons, L. D. Hess and T. W. Sigmon eds. (North Holland, NY, 1981) p. 575.
2. D. M. Follstaedt, in "Laser and Electron-Beam Interactions with Solids", Ed. by B. R. Appleton and G. K. Celler (North Holland, NY, 1982) p. 377; S. T. Picraux and D. M. Follstaedt, in "Proceedings NATO Inst. on Surface Modification and Alloying" (Plenum Press, NY) in press.
3. "Amorphous and Metastable Microcrystalline Rapidly Solidified Alloys", National Materials Advisory Board, Publication NMAB-358. (National Academy of Sciences, Washington, DC, 1980) p. 119.
4. T. R. Anantharaman, P. Ramachandraras, C. Suryanarayana, S. Lele, K. Chattopadhyay, G. V. S. Sastry and H. A. Davies, in Proc. Rapidly Quenched Metals III Vol. I, University of Sussex, 1978, p. 126; K. Chattopadhyay et. al., in Rapidly Quenched Metals, N. J. Grant and B. C. Giessen eds. (Mass. Inst. of Tech., Cambridge, 1976) p. 157.
5. A. Fontaine, in Rapidly Quenched Metals, N. J. Grant and B. C. Giessen eds. (Mass. Inst. of Tech., Cambridge, 1976) p. 163.
6. P. S. Peercy, D. M. Follstaedt, S. T. Picraux and W. R. Wampler, in Laser and Electron-Beam Interactions in Solids, B. R. Appleton and G.K. Celler eds. (North Holland, NY, 1982) p. 401.
7. A. G. Cullis, H. C. Weber, N. G. Chew, J. M. Poate and P. Baeri, Phys. Rev. Lett. 49, 219 (1982).
8. P. G. Boswell and G. A. Chadwick, Scripta Metall. 11, 459 (1977); V. P. Furrer and H. Warlimont, Z. Metallkd. 64, 236 (1973).

Author Index

Adams, F. M., 393,633
Ahmed, H., 437,563,653
Aina, O., 117
Alestig, G., 517
Andrew, R., 665
Appleton, B. R., 281,297
Atwater, H., 491
Auvert, G. 165,177
Aydinli, A., 23

Baeri, P., 273
Baghdadi, M., 419
Baglin, J.E.E., 355,709
Barbier, D., 419,449
Barhorst, J. F., 43
Barraclough, K. G., 413
Battaglin, G., 703
Baufay, L., 665
Baumgart, H., 349
Bensahel, D., 165,177
Bensoussan, M., 329
Bentini, G. G., 241
Bertolotti, M., 217
Biegelsen, D. K., 211,537,605
Bloembergen, N., 3
Bok, J., 41
Bokor, J., 51
Bonnouvrier, F., 123
Borisenko, V. E., 375
Bosch, M. A., 581
Bouree, J. E., 221
Bourgoin, P. C., 647
Broutet, F., 311
Brueck, S.R.J., 191,593
Bucksbaum, P. H., 51
Bujatti, M., 641
Burgener, M. L., 613

Cachard, A., 419
Calder, I. D., 443
Campisano, S. U., 273
Cao, J. J., 117
Carnera, A., 703
Cass, T. R., 523
Celler, G. K., 349,575
Cetronio, A., 641
Chantre, A., 449
Chapman, R. L., 477,593
Chen, L. J., 709
Chi, Y. M., 337

Chiang, A., 605
Chiang, S. W., 549
Christie, W. H., 401,407
Ciceroni, S., 641
Clark, G. J., 303
Cleland, J. W., 407
Cline, H. E., 425,727
Combescot, M., 41
Compaan, A., 23
Cornish, B., 393
Correra, L. 241
Croset, M., 647
Cserhati, A., 185
Cullis, A. G., 75,303

D'Eustacchio, P., 641
Dass, S. C., 407
Davies, D. E., 683
Davis, J. R., 563
Delage, S., 647
Della Mea, G., 703
Desoyer, J. C., 311
Dieumegard, D., 647
Dimiduk, K. C., 671
Dorofeev, A. M., 375
Drowley, C. I., 511,523,529

Eby, R. E., 407
Eckhardt, G., 337
Edwards, C. B., 677
Ehrlich, D. J. 191

Fan, J.C.C., 477,593
Fastow, R., 69,455,715
Fauchet, P. M., 205
Fennell, L. E., 491,537,605
Ferrari, A., 217
Fogarassy, E., 311
Follstaedt, D. M., 745,751
Forchel, A., 105

Gallego, J. M., 677
Galvin, G. J., 57
Geis, M. W., 477,593
Gelpey, J. C. 355
Germain, P. J., 135
Gibbons, J. F., 463,671,721
Gilmer, G. H., 249
Goetz, G., 185
Gregory, R. B., 369
Greiner, M. E., 671
Guosheng, Z., 205
Gupta, A., 337
Gyulai, J., 69,455

Hamdi, A. H., 369
Harrison, H. B., 393
Hasselbeck, M. P., 97
Hawkins, G., 677
Hawkins, W. G., 211,537
Herbst, D., 581
Hess, L. D., 141,337
Hewett, C. A., 455
Hill, C., 381
Hirlimann, C., 13
Hodgson, R. T., 355
Holland, O. W., 155,281,287,297
Holmen, G., 517
Hu, C., 529
Hung, L. S., 709

Ito, C. R., 337

Jacobson, D. C., 303
Jain, A. K., 703
Jani, P., 217
Jellison, G. E., Jr., 35,407
Johnson, N. M., 491,537,605
Johnson, S. T., 393

Kabler, M. N., 89
Kamgar, A., 569
Kamins, T. I., 511,523
Katz, W., 117, 425
Kavanagh, K., 455
Kechouane, M., 449
Khondker, A. N., 431
Kim, D. M., 111,431
Knapp, J. A., 557,745
Knoell, R., 569
Kokorowski, S. A., 141,337
Kulkarni, V. N., 703
Kurz, H., 3
Kwok, H. S., 97
Kwong, D. L., 111

Larson, B. C., 43
Laude, L. D., 665
Laugier, A., 419,449
Laurich, B., 105
Lawson, E. M. 633
Lee, M. C., 23
Lemons, R. A., 581
Leray, C., 221
Lischner, D. J., 349,575
Liu, J. M., 3
Liu, Y. S., 425,549
Lo, H. W., 23
Long, J. P., 89
Lorenzo, J. P., 683
Lou, L. F., 337
Lowndes, D. H., 35,407

Maby, E. W., 491
Mader, S., 355
Mahler, G. 105
Malvezzi, G. M., 273
Maruani, A., 123
Mayer, J. W. , 69,455,709
Mazzoldi, P., 703
McDaniel, F. D., 369
McMahon, R. A., 563,653
McMillan, G. B., 437
McNally, 683
Mehrabian, R., 733
Michel, A. E., 355
Mills, D. M., 43
Miyao, M., 499
Moison, J. M., 329
Mountain, R. W., 477,593
Moyer, M. D., 491,537,605

Naem, A. A., 443
Naguib, H. H., 443
Nakaji, E. M., 337
Nandedkar, R. V., 703
Narayan, J., 155,177,281,287,297,361,401,407
Nemanich, R. J., 211
Nguyen, V. T., 177
Nilson, J. A., 407
Nissim, Y. I., 123,235
Noggle, T. S., 43
Norton, J., 117

Ohkura, M., 499
Olson, G. L., 141,155,337
Opyd, W. G., 671
Oudar, J. L., 235
Ozols, A., 17

Paesler, M. A., 135
Palmstrom, C. J., 455,715
Paquet, D., 123
Parks, H. G., 425,549
Paulson, W. M., 369
Peercy, P. S., 229
Pennycook, S. J., 281
Peterstrom, S., 517
Picraux, S. T., 557,751
Pigeolet, A., 665
Poate, J. M., 263,303
Pollard, C. J., 413
Pollock, J.T.A., 633
Possin, G. E., 425,549
Pribat, D., 647

Rai, A. K., 177
Reedy, R. E., 613
Reehal, H. S., 677
Rife, J. C., 89
Rimini, E., 273
Rivier, M., 185
Robinson, McD., 349,575
Rodot, M., 221
Rose, A., 633
Rose, K., 117
Roth, J. A., 141
Rothe, D. E., 401
Royt, T. R., 90
Rozgonyi, G. A., 177,569
Russo, G., 273
Ryan, T. G., 683

Sandstrom, R. L., 401
Sapriel, J., 235
Sayers, D. E., 135
Schaefer, R. J., 733
Scott, M. D., 633
Shah, R. R., 431
Shank, C. V., 13
Shannon, J. M., 437
Sheng, T. T. 349,575
Short, K. T., 393
Sibilia, C., 217
Siegman, A. E., 205
Siffert, P., 311
Sigmon, T. W., 671,721
Silversmith, D. J., 477,593
Sipe, J. E., 197
Smith, H. I., 477
Soda, K. J., 683
Sood, D. K., 703
Speight, J. D., 413
Srivastava, P. C., 647
Stein, H. J., 229
Stuck, R., 311
Stultz, T. J., 463
Sturm, J., 463

Thomas, C. B., 677
Thompson, M. J., 605
Thompson, M. O., 57,303,455
Tokuyama, T., 499
Tsao, J. Y., 191
Tsaur, B.-Y., 477,593
Tuan, H. C., 605
Turnbull, D., 131

Valdez, J. B., 337
van der Leeden, G. A., 401
van Driel, H. M., 197
von Allmen, M., 691
von der Linde, D., 17

Wakita, A. S., 721
Warabisako, T., 499
Wartmann, G., 17
Westbrook, R. D., 407
White, C. W., 43,287,297,317
Williams, J. S., 393,621,633
Williams, R. T., 89
Wilson, S. R., 287,369
Wood, R. F., 35,83,407

Yen, R.,13
Young, J. F., 197
Young, R. T., 361,401

Zehner, D. M., 287,317
Zellama, K., 135
Zorabedian, P., 523

Subject Index

absorption,
  coefficient, 111
  two-photon, 97
activation, incoherent focused radiation, 683
a-Ge, explosive crystallization, 165
Al, epitaxial crystalline, 751
ALAs, 665
aligned liquid and solid regions, 211
allotropic transformation, 745
AlSb, 665
amorphization,
  high velocity, 303
  Si, 57
amorphous,
  Al (Ni) alloy formation, 751
  crystalline (a-c) interface, 135
  layers, 361
  phase, 83
    of Ge, 297
    trapping, 83
  semiconducting (a-sc), 131
  silicon, 51
    crystallization, 263
    films, 141, 455
    formation, 235
    melting, 141, 263
  state, 751
annealing, 653
  arc lamp, 463
  CW laser, 211, 517
  electron beam, 155, 437
  flame, 361
  furnace, 349
  isothermal, 369
  laser, 43, 89, 191, 211, 229, 297,
    337, 381, 407, 529
    applications, 337
  liquid phase, 621
  picosecond laser, 235
  pulsed electron beam (PEBA), 419, 449
  pulsed laser (PLA), 23, 111
  pulsed ion beam (PIBA), 69, 455, 715
  rapid isothermal, 641
  rapid thermal (RTA), 349, 355
  scanning electron beam, 413
  selective laser, 443
  solid phase, 349, 621
  technique, 683
  thermal pulse, 671
  transient, 621
anti-reflection (AR) coating, 443
anti-Stokes/Stokes ratio, 17

applications of laser annealing, 337
arc lamp annealing, 463
argon lasers, 523
arsenic implant activation, 419
As^+ implant, 355
As implanted Si wafers, 369
As redistribution, 419
a-Si, explosive crystallization, 165
atomic
  steps, 329
  transport effects, 135
atomically clean surfaces, 317
Au surfaces, 329
AuGe/GaAs system, 117
autonucleation mechanism, 537
avalanche ionization, 97

B, 281
bcc, 745
beam homogenizer, 52
beam processing, 381
  scanning CW Hg lamp, 463
beam shaping, 523
bimaterial structures, 117
binary alloys, 691
bipolar transistor, 381
bulk amorphous substrates, 537

capitance-voltage (C-V), 647
  techniques, 491
carrier mobility, 499
CdSe, 665
CdTe, 665
cell foramtion, 297, 303
cellular structure, 311
chemical analysis, 569
circularly polarized light, 197
classical crystal growth theory, 83
CMOS, 593
CO_2-laser, 97
  crystallized, 605
coexisting liquid and solid regions, 211
coherent precipitate, 281
competing nucleation, 557
conduction, polycrystalline silicon, 431
convection, 733
  effects. 703
continuous wave incoherent light source
  (CWILS), 393
continuous wave laser-see CW laser
constitutional undercooling, 581

constitutional supercooling, 303
Co-Si thin film systems, 709
cross-section electron microscopy, 177
crystallizing interface, 177
crystalline semiconducting (c-sc), 131
crystallization,
  amorphous silicon films, 455
  charged dangling bonds, 135
  complete 3-inch wafers, 575
  front velocity, 165
  group iv semiconductors, 135
  growth velocities, 135
  pulsed ion beam annealing, 455
  silicon films on glass, 581
current-voltage (C-V), 647
  measurements, 511
CW
  annealing, 517
  laser,
    ar$^+$, 529
    Ar-ion, 523
    Si annealing, 211
  recrystallization, 511

deep level,
  defects, 449
  spectroscopic techniques, 491
  transient spectroscopy (DLTS), 449
"defect-free" annealing, 361
defect,
  generation processes, 75
  production, 303
  spectroscopy techniques, 647
deposition at back interface,
  nitrogen, 511
  SiNx, 511
depression in melting temperature, 263
device technology, 329
dielectric,
  encapsulation, 683
  isolation, 593
differential scanning calorimetrym 263
diffusion, 105
  effects, 437
dislocation loops, 361
dissociation, 633
distribution coefficient, 287
  kinetic, 249
DLTS, 449
dopant,
  activation, 425
  movement, 355, 721
  precipitation, 155
  profile spreading, 401
  redistribution, 297, 349, 621
  segregation, 155

dual electron beam processing, 563
dynamical diffraction theory, 43

e-beam, 751
EDAX, 569
efficiency, solar cells, 401
electrical,
  activation, 633
  properties, 401
    of GaAs crystals, 621, 647
electron beam, 653, 751
  annealing, 155, 437
  -induced-current-images, 491
  melting, 529
    surface, 733
electron channeling patterns, 557
electron-holeplasma, 13
  formation, 3
electronic,
  circuits, 727
  defects, 491
  surface properties, 317
encapsulation layer, 511
energy,
  -beam shaping, 523
  dispersion analysis by x-ray (EDAX), 569
  relaxation processes, 263
  transfer, 13
epitaxial, 745
  CoSi$_2$ layers, 709
  crystalline Al, 751
  regrowth, 751
  solar cells, 221
epitaxy, 463
eutectic,
  alloy, 727
  thin film, 727
explosive,
  crystallization, 165
    amorphous silicon, 185
  recrystallization, 177
extended,
  defects, 355
  radiative heat source, 575
  state mobility model, 431

Fe in Si, 221
femtosecond,
  optical pulse techniques, 13
  time resolved reflectivity, 13
fixed-charge density, 511
flame annealing, 361
flat panel displays, 581
free energy release, 165

furnace annealing, 349
fused silica, 605

GaAs surfaces, 329
gallium arsenide, implanted, 621
Ge, amorphous phase, 297
gettering, impurities, 375
Gibbs free energy, 263
glass, 581
  formation, 691
glassy phases, laser-generated, 691
glow discharge implantation, 401
grain,
  boundaries, 523
  structure, 523
graphite strip heaters, 593
graphoepitaxy, 593

Hall effect, 517
heat,
  diffusion equation, 111
  of crystallization (ΔH_{ac}), 263
heavy-metal impurities, 375
$Hg_{1-x}Cd_xTe$, 671
HgCdTe diodes, 337
Hg loss, 671
high regrowth velocities, 51
homogeneous materials, 117
hot electron device structures, 437
hydrochemical study, 41

impact ionization rate, 105
implant, 671
implanted,
  InP, 683
  silicon, 235
  Si wafers, 369
impurity,
  diffusion, 221
  distribution profiles, 311
  redistribution, 263, 721
  segregation, 263
  trapping, 303
-In and Bi, 297
incandescent sources, 683
incoherent,
  focused radiation, 683
  light,
    annealing, 349, 375
    source, 393
  tungsten radiation, 569
indium antimonide, 97
infrared absorption, 229
infrared detection, 671

inhomogeneities, 613
InP surfaces, 329
instability, electron-hole plasma, 41
interface,
  amorphous crystalline (a-c), 135
  instability, 303
ion,
  and electron spectroscopy, 89
  backscattering/channeling, 745
  beam structures, 709
  channeling, 281, 671, 751
  implantation, 297, 745, 751
    damage, 361, 621
  implanted, 297
    gallium arsenide, 393
    Ge, 297
    silicon, 407, 425
      wafers, 349
  mass spectroscopy, 89
integral transform method, 123
integrated circuit technology, 381
intergranular diffusion, 221
interface,
  melting, 709, 715
  properties (MOS), 499
  -state density, 491
  undercoolings, 83
interferometric measurements, 217
inverted metal-oxide silicon capacitors,
  491
Ising model simulators, 249
isothermal annealing, 369
isotherms, 613

kinetic rate model, 83
kinetics, laser-induced, 141
KrF* amplified excimer laser system, 52

laser, 523
  -annealed Si, 229
  annealing, 43, 89, 191, 211, 297, 381,
    407, 529, 549
    applications, 337
    pulsed, 43
    process, 191
    Si, 211
  beam shaping, 549
  -generated glassy phases, 691
  induced,
    diffusion, 311
    periodic surface structure, 197
    surface defects, 329
    temperature rise, 123
  irradiated media, 123
  melt quenching, 691

laser,
  melting, 703
  reacted, 721
  recrystallized silicon, 549
lateral epitaxial growth over oxide,
  (LEGO), 575
lateral epitaxy, 529
laterally seeded recrystallization, 523
lattice,
  heating, 3
  location, 155
  temperature, 111, 211
light-emitting devices, 677
line-source electron beam, 557
liquid,
  metal, 131
  -phase,
    crystallization, 177
    diffusion, 703
    epitaxial regrowth, 317
  silicon films, 51
  -solid interface, 51
  transition to amorphous phase, 83
low angle grain boundaries, 523, 581

material modifications, 111
maximum substitutional concentrations,
  155
measurements,
  dopant profile spreading, 401
  electrical properties, 401
  melt depth, 401
mechanical quenching, 691
melt,
  curves, 69
  depths, 69, 401
  durations, 69, 263
  threshold, 263
melting,
  model calculations, 407
  point, 43
  temperature depression, 263
metal alloy / Si reactions, 715
metal silicides, 709
90°metal wedge, 523
metallic,
  glasses, 703
  laser melting, 703
  liquid, 263
  substrates, 69
metastable
  Fe(Pd) alloys, 745
  surface structures, 317
microprobe laser analysis, 235
microstructural transition, 751
mobile dangling bonds (Dbs), 135

mode locked Nd-YAG laser, 235
morphological stability, 733
MOS (capacitors), 593
MOSFET, 463, 499, 549, 593
multilayer structures, 529

nanosecond laser pulses, 17
Nd-YAG, 647
Ni concentration, 751
nitrogen, 511
Nomarski optical microscopy, 557
nonlinear laser melting, 97
numerical model, 117

ohmic contacts (Al/Si), 337
optic phonon temperature, 23
optical,
  absorption coefficient, 111
  properties of silicon, 35
optically excited silicon, 13
ordered surface structures, 317
oriented nucleation
<100> oriented silicon crystals, 241
oxygen, 511
  donor complex, 413
  implanted, 229

patterning, 581
periodic crescent, 165
phase transitions, 3
phonon mode, 41
picosecond,
  laser,
    annealing, 235
    -induced surface transformations,205
    melting, 51
    pulses, 263
  Nd laser pulse irradiation, 263
  pulse duration, 3
plasma,
  annealing model, 135
  expansion, 42
  formation, 3
  self-confinement, 105
p-n junction formation, 407
Poisson ratio, 217
polycrystalline Al, 751
polysilicon, 511
  etching, 443
  laterally seeded recrystallization, 523
  /oxide capacitors, 337
  resistors, 337
precipitation, 311
processing/microstructure relationships,
  733

processing of shallow implanted layers, 437

radiation damage, 633
"radiation remnant", 197
Raman,
  scattering, 17, 235
  temperature measurements, 35
rapid,
  crystallization, 165
  isothermal annealing, 641
  solidification, 287
  thermal annealing (RTA), 349, 355
recombination centers, 375
recrystallization, 653
  pattern, 235
  threshold power, 529
recrystallized polysilicon, 563
redistribution, dopant, 721
reflectivity, piosecond, 3
regrowth velocities, 57, 69
retrograde solubility limits, 155
repetitive subthreshold illumination, 205
residual stress, 499
RHEED, 499
ripples,
  formation, 197
  spontaneous surface structures, 205
Rutherford,
  backscattering, 155
  channeling techniques, 155
ruby laser, 647

Sb, 281
scanning electron beam annealing, 413
scanning electron microscopic techniques, 491
scanning IR line source, 425
second harmonic, 75
seeded,
  crystallization, 575
  lateral epitaxial Si layers, 499
  lateral epitaxy, 499
segregation,
  coefficient, 303
  crystallization, 575
  effects, 311
selective laser,
  annealing, 443
  crystallization, 443
selectively etching, 727
SEM measurements, 703
semiconducting, amorphous (a-sc), 131
semiconductor, 653, 677

semiconductor,
  films, 477
  processing, 393
semi-insulating GaAs, 641
shaped laser beam, 523
sheet resistivity, 517
silicate glass substrates, 581
silicides, refractory, 721
silicon, 3, 23, 43, 281, 419, (see specific entries)
  amorphous, 51
  films, 581
    amorphous, 141, 455
  formation, 235
  implanted, 641
  laser annealing, 211, 229
  layers, 437
  nonlinear laser melting, 97
  solid phase epitaxial crystallization, 141
  surfaces, 329
  virgin (100) boron doped , 449
  wafers, 89, 369
    implanted with As, 369
    implanted with B, 369
    implanted with P, 369
SOI (silicon on insulator), 593, 613
  crystallization, 557
  films, 557
    substrates, 569
  structures, 499, 523
SOS (silicon on sapphire), 57, 517, 569
SIMS, 511
  techniques, 425
single crystal silicon films, 523
Sn, 297
softening of the T.A. phonon mode, 41
solar cells, 401
solar cells, efficiency, 401
  fabrication, 407
solidification, rapid, 249
solid,
  or liquid phase crystallization, 621
  -phase crystallization, 177
    epitaxial cryatallization, 177
    epitaxial growth, 155, 361
    epitaxy (SPE), 141
    laser annealing, 349
solubility and GaAs, 621
solute trapping, 287
  and metastable alloying, 297
Stokes/anti-Stokes ratios, 23
stress profile analysis, 241
strip-heater crystallized silicon films, 491
(1x1) structure, 329
subboundaries, 477

subgrain boundaries, 491
submicron conducotor, 727
substitutional solubilities, 287
substrate,
  latent heat effects, 117
  melting threshold power, 529
surface,
  disorder, 311
  dissociation, 621
  electromagnetic waves, 191
  melting, 111
  polarition, 191
  segregation, 297
  structures, metastable (lxl), 317
  studies, 317

temperature,
  -dependent material parameters, 111
  gradients, 43, 733
  profile(s), 185
    analysis, 241
    strip heater induced, 613
  rise(s),
    laser irradiated media, 123
    semiconductors, 123
temperatures, near surface, 43
thermal,
  melting model, 69
  profiles, 241
  pulse annealing, 671
thermalized electrons, 135
thermally,
  annealed, 281
  induced stresses, 241
  stable, 369
thermionic emission-diffusion, 431
thermodiffusion, 105
thermodynamics, crystallization, 131,
  263
thick Si films, 575
thin film(s), 677
  transistors (TFT), 605
three dimensional,
  circuits, 499
  devices, 499
threshold,
  damage introduction, 241
  influence, 329
time,
  -evolution, lattice temperature, 17
  -resolved,
    experiments, 205
    reflectivity measurements, 141
  reversal invariance, 23
Tl, 281
topography, 211

transformation, allotropic, 745
transient,
  annealing, 425, 633
    gallium arsenide, 621
  conductance, 57, 69
    technique, 263
  heating phenomena, 111
  ripple formation, 191
  stimulated Raman scattering, 52
transiently molten semiconductor layers,
  75
transistors, MOS/SOS, 337
transmission electron microscopy (TEM),
  281, 745, 751
TEM_{01}* mode, 523
transmission, piosecond time-resolved, 3
trapping, models, 249
tunable pulsed dye laser, 23
tungsten lamps, 349, 569, 575
two excitation pulses, 205
two photon absorption, 97

ultrafast energy transfer, 3
ultrahigh,
  power arc lamp. 355
  speed solidification, 75
UV irradiation, 57
uncapped processing, 633
undercooling, 177
unique damage structure, 297
universality characteristics, 197
unseeded crystallization, 177

vacancy trapping, 229
vacuum-cleaved silicon surfaces, 89
vapor depositing Pd, 745
variable bandgap semiconductor, 671
velocity, liquid/solid interface, 733
vertical,
  dilation, 217
  displacement, 217
visualization, temperature profiles, 185
VLSI, 499, 593
VLSI-related processes, 425
VLSIC applications, 443

wafer surface, 523
WSi_2, 721

ZnS films, 677
ZnS:Mn layers, 677
zone melting,
  poly-Si films, 593

zone melting,
  recrystallization (ZMR), 477
    semiconductor films, 477
  recrystallized Si films, 593
zone recrystallization, 463
zone refining, 655

$XeCl_2$ excimer laser, 303
x-ray diffraction, 43